QUANTUM MECHANICS

QUANTUM MECHANICS

SECOND EDITION

EUGEN MERZBACHER

University of North Carolina

JOHN WILEY & SONS

New York • Chichester • Brisbane • Toronto • Singapore

Library of Congress Catalogue Card Number: 74-88316

SBN 471   59670   1

Printed in the United States of America

To my Father

Preface

In a time of rapidly increasing specialization, physics appears to have suffered less damage from the fragmentation of knowledge than many other scientific disciplines. Such cohesion may be accounted for by the central role of quantum mechanics as a unifying principle in contemporary physics. This function of quantum mechanics is reflected in the training of physicists who take a common course, whether they expect to specialize in atomic, molecular, or nuclear physics, in solid state physics, or in the physics of elementary particles.

This book—a revised and expanded version of a book first published almost ten years ago—was written for such a course. It is intended to be a comprehensive introduction to the principles of quantum mechanics and to their application in a variety of fields to which physicists turn.

The material contained in this new edition is appropriate for three semesters (or four quarters). The first eighteen chapters make up the content of a two-semester (or three quarter) course on nonrelativistic quantum mechanics. Chapters 19–24 discuss the elements of formal scattering theory, second quantization, relativistic quantum theory, and other subjects, sometimes classified as "Advanced Quantum Mechanics," which provide the background for an understanding of quantum electrodynamics, theories of elementary particles, and many-body theories.

The book could also be divided by drawing a line between the first eleven (or thirteen) chapters, containing the elementary theory which is often taught in one semester to advanced undergraduates, and the remainder, starting with Chapter 12 or 14 and covering about two semesters (or three quarters) of graduate-level quantum mechanics, with standard wave mechanics (and perhaps simple spin theory) as a prerequisite.

Although the text has been revised in many places—even merging two chapters (old 10 and 11) into one (new 10) and dividing a long chapter (old 13) into two (new 12 and 13)—the first fourteen chapters of the two editions remain in close correspondence, with one exception. To Chapter 4 has been added a section on the variational principle which leads to the Schrödinger equation and to first-order perturbation theory of discrete stationary states. This device has made it possible to move many problems and illustrative applications to earlier chapters. A number of new problems have been added in the second edition.

As before, the book begins with a glance back into history and a preview of the ideas to be encountered and develops ordinary wave mechanics inductively with the aid of wave packets moving like classical particles. The Schrödinger equation is established, various operators are identified, and the probability doctrine emerges. Chapters 5, 6, and 7 provide practice in the use of fundamental concepts, including the construction of an elementary form of the scattering matrix from symmetry principles. Two-by-two matrices are employed in the calculations for transmission through a barrier by the WKB approximation.

Spin mechanics with matrices is developed from experimental facts in Chapter 12 as a companion to wave mechanics, and the general dynamics of two-level systems is presented in Chapter 13. The discussion of the spin, as a system with only two basis states, is more elaborate than usual. The inherent simplicity of the mathematical formalism permits attention to be focused on the essential features of quantum mechanics, the superposition principle and the physical meaning of a state. The connection between SU(2) and the rotation group is explicitly established at this elementary level, preparatory to a more general discussion of angular momentum in Chapter 16.

The general principles of quantum mechanics are the subject of Chapters 14 and 15. As a synthesis of wave mechanics and spin mechanics—both concrete examples of the abstract theory—these chapters contain the formulation of quantum dynamics in Hilbert space and the elements of quantal transformation theory. The essential mathematical background is provided, and the dual bra and ket spaces are used whenever convenient. The general formulation of quantum mechanics in terms of state vectors and linear operators is then used throughout the remainder of the book. As an example, the forced linear harmonic oscillator is treated in the Heisenberg picture, affording an introduction to the notion of "in" and "out" states and to the theory of coherent states.

From Chapter 15 on, the new edition of this book differs substantially from the first. With emphasis on rotations and angular momentum, Chapter 16 is devoted to symmetries, including reflection and time reversal, and to the rudiments or the theory of group representations. Bound state perturbation theory, with its applications to atoms, is covered in Chapter 17, and time-dependent perturbation theory is now consolidated in Chapter 18. Chapter 19 returns to the study of scattering, considered as a transition between free particle states, and is intended as a bridge from elementary quantum mechanics to modern collision theory. Those who do not intend to cross this bridge may omit Chapter 19.

Chapters 20–24 are entirely new. They are based on the belief that, for the quantum mechanics of identical particles, the second quantization formalism is conceptually simpler than either wave mechanics in configuration space or the canonical quantization of classical field equations. Chapter 20 is constructive in nature and attempts to spell out a set of simple assumptions from which the existence of only two kinds of statistics, with their characteristic commutation relations for creation and annihilation operators, can be deduced. In Chapter 21, the second quantization formalism is applied to a broad range of problems: angular momentum for systems of identical particles, including Schwinger's method of treating angular momentum by coupling spin 1/2 bosons in the stretched configuration; quantized spin waves in solids (magnons); the theory of atomic structure in outline, and a derivation of the Hartree-Fock equations; a basic account of the BCS theory of superconductivity, expressed in terms of the quasiparticles introduced by Bogoliubov; and a brief derivation of the distribution law for an ideal gas of bosons or fermions in thermal equilibrium.

Continuing in an ahistorial vein, Chapter 22 starts with photons as fundamental entities which compose the electromagnetic field with its local dynamical properties like energy and momentum. The most elementary consequences (emission and absorption) of the interaction between the quantized matter and radiation fields are worked out by perturbation theory in the nonrelativistic approximations, leaving all higher-order processes to more advanced textbooks on quantum electrodynamics. However, the emission and absorption of photons from a given current distribution, which can be described classically, is treated rigorously on the basis of the forced linear harmonic oscillator.

Similarly, in Chapter 23, positrons and electrons are taken as the elements of the relativistic theory of leptons and the Dirac equation is derived as the quantized field equation for charged spin 1/2 fermions moving in an externally prescribed electromagnetic field which may be approximated by a classical description. The relativistic covariance of the theory is shown, and the discrete symmetry operations, especially charge conjugation, are discussed. Finally, by a reversal of the usual arguments about hole theory, the one-particle Dirac theory of the electron (or positron) is obtained as an approximation to the complete many-electron-positron theory of Chapter 23. The magnetic moment of the Dirac electron and the fine structure of the hydrogen spectrum are derived in Chapter 24.

The reader of this book is assumed to know the basic facts of atomic and nuclear physics. No previous contact with quantum mechanics is explicitly required, but some earlier exposure to elementary quantum mechanics at the undergraduate level will be found advantageous. A good knowledge of classical mechanics and some familiarity with the elements of electromagnetic theory are also assumed.

There are almost four hundred exercises and problems, which form an integral part of the book. The exercises supplement the text and are woven into it. The problems, which appear at the end of each chapter, are usually more independent extensions and applications of the text and require more work.

One small departure from the first edition must still be mentioned. Since recent textbooks on electromagnetic theory give preference to Gaussian units, these are used exclusively in the new edition.

The quoted references are not intended to be exhaustive, but the footnotes should serve as a guide to further reading and bear witness that many sources have contributed to this book. Comments from many quarters were useful in the preparation of the new edition, and it is impossible to acknowledge the help of every student and colleague who has furnished constructive criticism. The author owes particular thanks to the following physicists: S. Borowitz, W. A. Bowers, B. Bransden, B. Chern, C. DeWitt, S. T. Epstein, N. and P. O. Fröman, B. Hoffmann, H. P. Kennedy, L. Parker, R. E. Peierls, C. P. Poole, and D. D. Sharma.

Most of the work on the new edition of this book was done at the University of North Carolina and, during a leave of absence, at the University of Washington. The material in the last four chapters was first developed in lectures given at the University of Colorado. Some sections were written with support from the Atomic Energy Commission. Chapel Hill, Boulder, and Seattle provided excellent environments for writing, and the encouragement of colleagues and friends in these places is gratefully acknowledged.

Eugen Merzbacher

Contents

QUANTUM MECHANICS

CHAPTER 1

Introduction to Quantum Mechanics

Quantum mechanics is the theoretical framework within which it has been found possible to describe, correlate, and predict the behavior of a vast range of physical systems, from elementary particles, through nuclei, atoms, and radiation, to molecules and solids. This introductory chapter aims at setting the stage with a brief review of the historical background and a preliminary discussion of some of the essential concepts which we shall encounter.

1. Quantum Theory and the Wave Nature of Matter. Matter at the atomic and nuclear level reveals the existence of a variety of particles which are identifiable by their distinct properties, such as mass, charge, spin, and magnetic moment. All of these seem to be of a quantum nature in the sense that they take on only certain discrete values. This discreteness of physical properties persists when elementary particles combine to form atoms and nuclei.

For instance, the notion that atoms and nuclei possess discrete energy levels is one of the basic facts of quantum physics. The experimental evidence for this fact is overwhelming and familiar. It comes most directly from observations on inelastic collisions (Franck-Hertz experiment) and selective absorption of radiation, and somewhat indirectly from the interpretation of spectral lines.

Consider as familiar an object as the hydrogen atom. The evidence that such an atom consists of a nucleus and an electron, bound to the former by forces of electrostatic attraction, is too well known to need recapitulation. The electron can be removed from the atom and identified by its charge, mass, and spin. It is equally well known that the hydrogen atom can be excited by absorbing certain discrete amounts of energy and that it can return the excitation energy by emitting light of discrete frequencies. These are empirical facts.

Niels Bohr discovered that any understanding of the observed discreteness requires, above all, the introduction of *Planck's constant*, $h = 6.6255 \times 10^{-27}$ erg sec. In old-fashioned language this constant is often called the *quantum of action.* By the simple relation

$$\Delta E = h\nu \tag{1.1}$$

it links the observed spectral frequency ν to the jump ΔE between discrete energy levels. And, divided by 4π, the same constant appears also as the unit of angular momentum, the discrete numbers $n(h/4\pi)(n = 0, 1, 2, \ldots)$ being the only values which a component of the angular momentum of a system can assume.[1]

More specifically, Bohr was able to calculate discrete energy levels of an atom by formulating a set of quantum conditions to which the canonical variables q_i and p_i of classical mechanics were to be subjected. For our purposes it is sufficient to remember that in this "old quantum theory" the classical phase (or action) integrals for a conditionally periodic motion were required to be quantized according to

$$\oint p_i \, dq_i = n_i h \tag{1.2}$$

where the *quantum numbers* n_i are integers, and each integral is taken over the full period of the generalized coordinate q_i. The quantum conditions (1.2) give good results in calculating the energy levels of simple systems but fail when applied to such systems as the helium atom.

Exercise 1.1. Calculate the quantized energy levels of a linear harmonic oscillator of angular frequency ω in the old quantum theory.

It is well known that (1.1) played an important role even in the earliest forms of quantum theory. Einstein used it to explain the photoelectric effect by inferring that light, which through the nineteenth century had been so well established as a wave phenomenon, can exhibit a particle-like nature and is emitted or absorbed only in quanta of energy. Thus, the concept of the photon as a particle with energy $E = h\nu$ emerged. The constant h connects the wave (ν) and particle (E) aspects of light.

De Broglie proposed that the wave-particle duality may not be a monopoly of light but a universal characteristic of nature which becomes evident when the magnitude of h cannot be neglected. He thus brought out a second fundamental fact, usually referred to as the *wave nature of matter*. This means that in certain experiments beams of massive particles give rise to interference and diffraction phenomena and exhibit a behavior very similar to that of light.[2]

---

[1] This is true for systems which are composed of several particles (atoms, nuclei) as well as for the elementary particles themselves. Of course, these particles may be no more "elementary" than atoms and nuclei. The ease with which they can be created or destroyed and converted into each other suggests that they may be composite structures.

[2] If we sometimes speak of *matter waves*, this term is not intended to convey the impression that the particles themselves are oscillating in space.

Although such effects were first produced with electron beams, they are now commonly observed with slow neutrons from a reactor. When incident on a crystal these behave very much like X rays.

From such experiments on the diffraction of free particles we infer the very simple law that the waves associated with the motion of a particle of momentum **p** propagate in the direction of motion and that their (de Broglie) wavelength is given by

$$\lambda = \frac{h}{p} \tag{1.3}$$

This relation establishes contact between the wave and the particle pictures. The finiteness of Planck's constant is the basic point here. For if h were zero, then no matter what momentum a particle had the associated wave would always correspond to $\lambda = 0$ and would follow the laws of *classical mechanics*, which can be regarded as the short wavelength limit of *wave mechanics* in the same way as *geometrical optics* is the short wavelength limit of *wave optics*. Indeed, a free particle would then not be diffracted but go on a straight rectilinear path, just as we expect classically.

Let us formulate this a bit more precisely. If x is a characteristic length involved in describing the motion of a body of momentum p, such as the linear dimension of an obstacle in its path, the wave aspect of matter will be hidden from our sight, if

$$\frac{\lambda}{x} = \frac{h}{xp} \ll 1 \tag{1.4}$$

i.e., if the quantum of action h is negligible compared with xp. Macroscopic bodies, to which classical mechanics is applicable, satisfy the condition $xp \gg h$ extremely well. To give a numerical example, we note that even as light a body as an atom moving with a kinetic energy corresponding to a temperature of $T = 10^{-6}\,°\mathrm{K}$ still has a wavelength no greater than about 10^{-3} cm! We thus expect that classical mechanics is contained in quantum mechanics as a limiting form ($h \to 0$).

Indeed, the gradual transition which we can make conceptually as well as practically from the atomic level with its quantum laws to the macroscopic level at which the classical laws of physics are valid suggests that quantum mechanics must not only be consistent with classical physics but should yield the classical laws in a suitable approximation. This requirement, which is a guide in discovering the correct quantum laws, is called the *correspondence principle*. We shall see later that the limiting process which establishes the connection between quantum and classical mechanics can be exploited to give a useful approximation of quantum mechanical problems (see WKB approximation, Chapter 7).

We may read (1.3) the other way around and infer that any wave phenomenon also has associated with it a particle, or quantum, of momentum

$p = h/\lambda$. Hence, if a macroscopic wave is to carry an appreciable amount of momentum, as a classical electromagnetic or an elastic wave may, there must be associated with the wave an enormous number of quanta, each contributing a very small momentum. A classically observable wave will result only if the elementary wavelets representing the individual quanta add coherently. For example, the waves of the electromagnetic field are accompanied by quanta (photons) for which the relation $E = h\nu$ holds. Since photons have no mass, their energy and momentum are related by $E = cp$. It follows that (1.3) is valid for photons as well as for material particles. At macroscopic wavelengths, corresponding to radio frequency, a very large number of photons is required to build up a field of macroscopically discernible intensity. Yet, such a field can be described in classical terms only if the photons can act coherently. This requirement, which will be discussed in detail in Chapter 22, leads to the peculiar conclusion that a state of *exactly* n photons cannot represent a classical field, even if n is arbitrarily large.[3]

Exercise 1.2. To what velocity would an electron (neutron) have to be slowed down, if its wavelength is to be 1 meter? Are matter waves of macroscopic dimensions a real possibility?

2. *Complementarity.* As we have seen, facing us at the outset is the fact that matter, say an electron, exhibits both particle and wave aspects.[4] This duplicity of character was described in deliberately vague language by saying that the de Broglie relation (1.3) "associates" a wavelength with a particle momentum. The vagueness reflects the fact that particle and wave aspects, when they show up in the same thing such as the electron, are incompatible with each other, unless we modify our traditional concepts to a certain extent. Particle means to us traditionally an object with a definite position in space. Wave means a periodically repeated pattern in space with no particular emphasis on any one crest or valley, and it is characteristic of a wave that it does not define a location or position sharply.

How a synthesis of these two concepts might be accomplished can, for a start, perhaps be understood if we recall that the quantum theory must give an account of the discreteness of certain physical properties, e.g., energy levels in an atom or nucleus. Yet discreteness is not new in physics. In classical macroscopic physics discrete, "quantized," physical quantities appear naturally as the frequencies of vibrating bodies of finite extension, such as

---

[3] The massless quanta corresponding to elastic (e.g., sound) waves are called *phonons* and behave similarly to photons, except that c is now the speed of sound.

[4] It will be convenient to use the term *electron* frequently when we wish to place equal emphasis on the particle and wave aspects of a constituent of matter. The electron has been chosen only for definiteness of expression. Quantum mechanics applies equally to protons, neutrons, mesons, etc.

strings, membranes, air columns. We speak typically of the natural modes of such systems. These phenomena have found their simple explanation in terms of interference between incident and reflected waves. Mathematically the discrete behavior is enforced by boundary conditions: the fixed ends of the string, the clamping of the membrane rim, etc. It is tempting likewise to see in the discrete properties of atoms the manifestations of bounded wave motion and to connect the discrete energy levels with standing waves. In such a picture the bounded wave must somehow be related to the confinement of the particle to its "orbit," but it is obvious that the conventional concept of an orbit as a trajectory covered with definite speed cannot be maintained.

A wave is generally described by its velocity of propagation, wavelength, and amplitude. Since in a standing wave it is the *wavelength* which assumes discrete values, it is evident that, if our analogy is meaningful at all, there must be a correspondence between the energy of the particle and the wavelength of the associated wave. For a free particle, one that is not bound in an atom, the de Broglie formula (1.3) has already given us a relationship connecting wavelength with energy (or momentum). The equation linking wavelength and momentum is likely to be much more complicated for an electron bound to a nucleus as in the hydrogen atom, and it is one of our tasks to uncover the appropriate connection.

For very loosely bound electrons, moving in circular orbits far from the nucleus, (1.3) should be a good approximation. Under these circumstances the wavelength is short compared with the distance between electron and nucleus. Hence, it should be possible to describe this type of almost-free motion either as a particle moving in a well-defined classical orbit around a center of force or as a wave which can be thought of as following this orbit. From Figure 1.1 it is clear that such a situation can be described by a wave only if the orbit accommodates exactly an integral number of wavelengths (Figure 1.1a). Otherwise (Figure 1.1b) the wave is out of phase with itself, and destructive interference will prevent a stable oscillation around the nucleus. Hence, the condition for having a standing wave is

$$\oint \frac{ds}{\lambda} = n \tag{1.5}$$

the integral to be taken over the path of the trajectory. In the classical limit n is a very large integer. Using (1.3), we obtain

$$\oint p \, ds = nh \tag{1.6}$$

which for the special case of a circular orbit ($ds = r \, d\varphi$) is easily seen to be identical with a quantum condition of the form (1.2), if the azimuth φ is used as the generalized coordinate, $p_\varphi = L_z = rp$ being the canonically conjugate momentum.

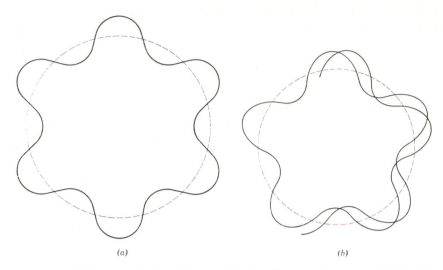

(a)   (b)

Figure 1.1. (*a*) Constructive interference of de Broglie waves in an atom distinguishes the allowed stable Bohr orbits. (*b*) Destructive interference of de Broglie waves in an atom disallows any orbit which fails to satisfy the quantum conditions.

Of course, conditions (1.2) and (1.6) are really useful and interesting only if the quantum numbers n are small integers, i.e., outside the domain of validity of the classical approximation. The bold extrapolation from large to small quantum numbers is in accordance with Bohr's demand, formulated in the correspondence principle, that a quantum equation which, like (1.6), has been established to be valid in the classical limit ($n \rightarrow \infty$) should be tentatively applied also in the unclassical quantum domain (small n). The fact that for $n = 1, 2, \ldots$ (1.6) yields the correct energy levels of the Bohr atom shows the usefulness of the correspondence principle and, more important for us, indicates that the wave-particle dualism contains the ingredients for a satisfactory description of atomic systems.

Exercise 1.3. Assuming that the electron moves in a circular orbit in a Coulomb field, derive the Balmer formula from the quantum condition (1.6).

It might be well to emphasize that the quantum conditions (1.2) or (1.6), although they give astonishingly good values for the energy levels of some quantum systems, are in general not accurate for small values of n. It should not be surprising that a naïve extrapolation from large to small quantum numbers can lead to incorrect results. The correspondence principle merely requires that quantum theory must be consistent with classical physics in the limit of large quantum numbers but does not specify uniquely the form of the new theory. In Chapter 7 it will be shown how the old quantum conditions are obtained from quantum mechanics in a semiclassical approximation.

We owe mainly to Bohr the discovery of the principles that allow us to understand how the two seemingly contradictory concepts of wave and particle, describing the same thing, may be reconciled. We shall see in later chapters that this reconciliation requires modification of some classical attributes with which we commonly endow waves and particles. When these concepts have been appropriately broadened, one of the two pictures alone will prove inadequate for the description of physical happenings. But their new broadened definitions will—owing to a departure from the strictly classical characterizations—carry within themselves enough freedom for the two aspects to coexist. Bohr, who was guided by the empirical fact of the dual nature of matter, has elevated this program to a *principle of complementarity*. According to this principle, wave and particle nature are considered complementary aspects of matter, both equally essential for a full description of the phenomena.

It may be argued that the reasoning which has led us so far utilizes classical and conventional notions, appropriate to macroscopic phenomena but patently inapplicable in the atomic domain. After all, Bohr's quantization rules, successful as they are, entail a whole set of assumptions which are in conflict with classical physics. Chief among these is the interdiction against the emission of radiation while the electron circulates about the nucleus. This insures the stability of the Bohr orbits, but it violates classical principles concerning the radiation from accelerated charges, and it fails to help us understand the occurrence, albeit infrequently, of transitions between energy levels which give rise to the observed spectral lines. Might we not be better off if we shed all pretext of making pictures of the quantum phenomena in terms of particles and waves and the like? Why not simply establish suitable mathematical laws for the description of the observations, as Newton urged for a branch of physics reaching maturity?

Such a point of view ignores the process by which new physical theories are arrived at; new ideas always depend on understanding the novel in terms of the familiar. Of course, as the recently "new" ideas become more familiar and we become accustomed to them, the need for the older points of reference diminishes. It is for this reason, for instance, that a complete review of the elaborate calculations of the "old" quantum theory is no longer regarded as particularly helpful in the study of quantum mechanics.

Nevertheless, when all this is recognized, there still remains the possibility—according to Bohr and his Copenhagen school the necessity—that all our experience of the quantum level of nature may have to be perceived and interpreted in classical, macroscopic terms. In support of this view it is pointed out that all physical quantities such as position, momentum, energy, angular momentum, etc., which are used in quantum mechanics as much as in classical mechanics, can be defined only by stating how they are measured. All such measurements, ultimately designed to convey information to our senses, must contain amplification mechanisms by which microscopic effects are translated into macroscopic effects accessible to our understanding. In the final analysis, atomic, nuclear, and particle experiments are all described in classical terms, since they invariably result in the reading of scales, the motion of large pointers, the activation of computer elements, etc., all of which are subject to the laws of classical physics.

According to this view, there is a subtle and profound relationship between quantum mechanics and classical physics. We have already encountered one facet of this relationship; it is the correspondence principle, which demands that quantum mechanics must be consistent with classical mechanics. Bohr went far beyond this obvious requirement and contended that quantum mechanics presupposes classical physics in a logical sense for its very formulation. Although it is not clear that this epistemological view of the relationship between classical and quantal physics and the associated doctrine of complementarity will remain central to the interpretation of quantum mechanics, an appreciation of Bohr's attitude is important because, through stimulation and provocation, it has greatly contributed to the development of the subject.

3. *The Wave Function and Its Interpretation.* So far we have considered only the length of matter waves, and we have found how the wavelength is connected with particle momentum. There are, of course, other important properties of these waves to be discussed. In particular, their amplitude may or may not have directional (i.e., polarization) properties. We shall see in Chapter 12 that the spin of the particles corresponds to the polarizability of the waves. However, for many purposes the dynamical effects of the spin are negligible in first approximation, especially if the particles move with nonrelativistic velocities.[5] We shall neglect the spin for the time being, much as in a simple theory of wave optical phenomena (interference, diffraction, geometrical limit, etc.) the transverse nature of light waves is neglected. Hence, we attempt to build up quantum mechanics by the use of *scalar* waves first. For particles with zero spin, e.g., pi mesons, this gives an appropriate description. For particles with nonzero spin, e.g., electrons, nucleons, mu mesons, suitable corrections must be made later (see Chapters 12 and 17).

Mathematically these waves are represented by a function $\psi(x, y, z, t)$ which in colorless terminology is called the *wave function*. Upon its introduction we immediately ask such questions as these: Is ψ a measurable quantity, and what precisely does it describe? In particular, what feature of the particle aspect is related to the wave function?

---

[5] This statement requires important corrections when we deal with systems of many identical particles (Chapter 20).

We cannot expect entirely satisfactory answers to these questions before we have become familiar with the properties of these waves and with the way in which ψ is used in calculations. As we progress through quantum mechanics, we shall become accustomed to ψ as an important addition to our arsenal of physical concepts. If its physical significance must as yet remain somewhat obscure to us, one thing seems certain: ψ must in some sense be a measure of the presence of a particle. Thus we do not expect to find a particle in those regions of space where $\psi = 0$. Conversely, in the regions of space where a particle may be found, ψ must be different from zero. But the function $\psi(x, y, z, t)$ itself cannot be a direct measure of the likelihood of finding a particle at position x, y, z at time t. For if it were that, it would have to be a positive number everywhere. Yet it is impossible that ψ be positive (or zero) everywhere, if destructive interference of the ψ waves is to account for the observed dark interference fringes and for the instability of any but the distinguished Bohr orbits in a hydrogen atom (Figure 1.1b).

In physical optics interference patterns are produced by the superposition of waves of $\mathbf{E}$ and $\mathbf{H}$, but the intensity of the fringes is measured by $\mathbf{E}^2$ and $\mathbf{H}^2$. In analogy to this situation we assume that the positive quantity $|\psi(x, y, z, t)|^2$ measures the probability of finding a particle at x, y, z. (The absolute value has been taken because it will turn out that ψ can have complex values.) The full meaning of this interpretation of ψ and its internal consistency will be discussed in detail in Chapters 4 and 8. Here we merely want to advance some general qualitative arguments for this so-called *probability interpretation* of the quantum wave function ψ. This interpretation was introduced into quantum mechanics by Born.

The discussion in the preceding section of the role that standing waves play in the hydrogen atom leads us to associate the wave ψ with a *single particle* as a representative of a statistical ensemble. The alternative would be to regard the wave as describing the behavior of a beam of particles, so that, for instance, the observed diffraction patterns would be the result of some collective phenomena. Owing to their characteristic properties, such as charge and mass, particles can, fortunately, be identified singly in the familiar detection devices of modern experimental physics: counters, bubble chambers, photographic plates, etc. With the aid of these tools it has been established abundantly that the interference fringes shown schematically in Figure 1.2 are the statistical result of the effect of a very large number of independent particles hitting the screen. The interference pattern appears only after many particles have passed through the slits, but each particle retains its discrete individuality and each goes its own way, being finally deposited somewhere on the screen. Note that the appearance of the interference effects does not require that a whole *beam* of particles go through the slits. In fact, particles can actually be accelerated and observed singly, and the interference pattern can be produced over a length of time, a particle hitting

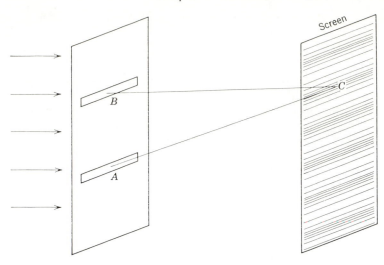

Figure 1.2. Double-slit interferences produced in analogy with Young's optical experiment. *C* is the location of a bright fringe on the screen if *AC* − *BC* equals an integral multiple of the wavelength.

the screen now here, now there in seemingly random fashion. When many particles have come through, a regular interference pattern will be seen to have formed. The conclusion is almost inevitable that ψ describes the behavior of single particles, but that it has an intrinsic *probabilistic* meaning. The quantity $|\psi|^2$ would appear to measure the chance of finding the particle at a certain place. In a sense, this conclusion was already implicit in our earlier discussion regarding a plane wave as representative of a particle with definite momentum (wavelength) but completely indefinite position. At least if ψ is so interpreted, the observations can be correlated effortlessly with the mathematical formalism.

The opposition to the probability interpretation of ψ originates in the discomfort felt by some (including very eminent) physicists with the lack of determinism which the probabilistic view is said to ascribe to nature. The idealized experiment shown in Figure 1.2 illustrates the argument. Single particles are subject to wave interference effects, and some are found deposited on the screen at locations which they could not reach if they moved along classical paths through either slit. The appearance of the interference fringes depends on the passage of the wave through both slits at once, just as a light wave goes through both slits in the analogous optical Young interference experiment. If the wave describes the behavior of a single particle, it follows that we cannot decide through which one of the two slits the particle has gone. If we try to avoid this consequence by determining experimentally with some clever monitoring device through which slit the particle has passed, we shall by the very action of the monitoring mechanism change the wave drastically and destroy the interference pattern. A single particle now goes definitely through one slit or the other, and the accumulation of a large number of particles on the screen will result in two well-separated traces. Exactly the same traces are obtained if we close one slit at a time, thereby predetermining the path of a particle. We conclude that the conditions under which the interference pattern is produced forbid a determination of the slit through which the particle passes.

More quantitative statements about this peculiar circumstance (Heisenberg uncertainty principle) will be made later, but the basic feature should be clear from the present example: The simultaneous appearance of wave and particle aspects compels us to be resigned to an inevitable indetermination in some of the conditions of an experiment. Wave aspect and particle aspect in one and the same thing are compatible only if we forgo asking certain questions of nature which no longer have any meaning (such as: "Let us see the interference fringes produced by particles whose paths through an arrangement of slits we have followed!"). Accepting this situation as inescapable, it is then not paradoxical to say that the interference fringes appear if the particle goes through both slits at once.

The probability doctrine of quantum mechanics asserts that the indetermination, of which we have just given an example, is a property inherent in nature and not merely a profession of our temporary ignorance from which we expect to be relieved by a future better and more complete theory. The conventional interpretation thus denies the possibility of an ideal theory which would encompass the present quantum mechanics but would be free of its supposed defects, the most notorious "imperfection" of quantum mechanics being the abandonment of strict classical determinism. Loose talk about these problems may even lead to the impression that quantum mechanics forsakes the classical goal of a wholly rational description of physical processes. Nothing could be further from the truth, as this book hopes to demonstrate.[6]

[6] Many references to the vast literature on quantum mechanics are found in the footnotes of this book. Other books that will contribute to an understanding of quantum mechanics are:

H. A. Bethe and R. W. Jackiw, *Intermediate Quantum Mechanics*, 2nd ed., W. A. Benjamin, New York, 1968.

D. Bohm, *Quantum Theory*, Prentice-Hall, New York, 1951.

P. A. M. Dirac, *The Principles of Quantum Mechanics*, 4th ed., Clarendon Press, Oxford, 1958; 1st ed.: 1930; 2nd ed.: 1935; 3rd ed.: 1947.

S. Gasiorowicz, *Elementary Particle Physics*, John Wiley and Sons, New York, 1967.

H. S. Green, *Matrix Mechanics*, P. Noordhoff Ltd., Groningen, 1965.

Footnote continued

J. M. Jauch, *Foundations of Quantum Mechanics*, Addison-Wesley Publishing Company, Reading, 1968.

L. D. Landau and E. M. Lifshitz, *Quantum Mechanics*, translated by J. B. Sykes and J. S. Bell, Addison-Wesley Publishing Company, Reading, 1958.

G. Ludwig, *Die Grundlagen der Quantenmechanik*, Springer, Berlin, 1954.

A. Messiah, *Quantum Mechanics*. Volume I translated by G. M. Temmer, Volume II translated by J. Potter, North-Holland Publishing Company, Amsterdam, 1961 and 1962.

W. Pauli, *Die allgemeinen Prinzipien der Wellenmechanik*, in S. Flügge, ed., *Encyclopedia of Physics*, vol 5/1, pp. 1–168, Springer Verlag, Berlin, 1958.

P. Roman, *Advanced Quantum Theory*, Addison-Wesley Publishing Company, Reading, 1965.

S. S. Schweber, *An Introduction to Relativistic Quantum Field Theory*, Row, Peterson and Company, Evanston, 1961.

A. J. W. Sommerfeld, *Atombau und Spektrallinien*, Volume II, F. Vieweg, Braunschweig, 1939.

S. Tomonaga, *Quantum Mechanics*, translated by Koshiba, Volume I, North-Holland Publishing Company, 1962; Volume II, John Wiley and Sons, New York, 1966.

B. L. van der Waerden, *Sources of Quantum Mechanics*, North-Holland Publishing Company, Amsterdam, 1967.

CHAPTER 2

Wave Packets and Free Particle Motion

1. The Principle of Superposition. We have learned that it is reasonable to suppose that a free particle of momentum **p** is associated with a harmonic plane wave. Defining a vector **k** which points in the direction of wave propagation and has the magnitude

$$k = \frac{2\pi}{\lambda} \tag{2.1}$$

we may write the fundamental de Broglie relation as

$$\mathbf{p} = \hbar\mathbf{k} \tag{2.2}$$

$\hbar$ is a common symbol in quantum physics and denotes the frequently recurring constant

$$\hbar = \frac{h}{2\pi} = 1.0545 \times 10^{-27} \text{ erg sec} = 6.5819 \times 10^{-22} \text{ MeV sec}$$

It is true that the diffraction experiment does not give us any direct information about the detailed dependence on space and time of the periodic disturbance which produces the alternatingly "bright" and "dark" fringes, but all the evidence points to the correctness of the simple inferences embodied in (2.1) and (2.2). The comparison with optical interference suggests that the fringes come about by linear superposition of two waves, a point of view which has already been stressed in Section 1.3 in the discussion of the simple double-slit interference experiment, Figure 1.2.

Mathematically these ideas are formulated in the following fundamental assumption about the wave function $\psi(x, y, z, t)$: if $\psi_1(x, y, z, t)$ and $\psi_2(x, y, z, t)$ describe two waves, their sum $\psi(x, y, z, t) = \psi_1 + \psi_2$ also describes a possible physical situation. This assumption is known as the *principle of superposition* and is illustrated by the interference experiment of Figure 1.2. The intensity produced on the screen by opening only one slit at a time is $|\psi_1|^2$ or $|\psi_2|^2$. When both slits are open, the intensity is determined by

$|\psi_1 + \psi_2|^2$. This differs from the sum of the two intensities, $|\psi_1|^2 + |\psi_2|^2$ by the interference terms $\psi_1\psi_2{}^* + \psi_1{}^*\psi_2$. (An asterisk will denote complex conjugation throughout this book.)

A careful analysis of the interference experiment would require detailed consideration of the boundary conditions at the slits,[1] but there is no need here for such a thorough treatment, because our purpose in describing the idealized double-slit experiment was merely to show how the principle of superposition accounts for some typical interference and diffraction phenomena. Such phenomena, when actually observed as in diffraction of particles through crystals, are impressive direct manifestations on a macroscopic scale of the wave nature of matter.

We therefore adopt the principle of superposition to guide us in developing quantum mechanics. The simplest type of wave motion to which it will be applied is an infinite harmonic plane wave propagating in the positive x-direction with wavelength $\lambda = 2\pi/k$ and frequency ω. Such a wave is associated with the motion of a free particle moving in the x-direction with momentum $p = \hbar k$. If we followed our first impulse, we would represent this plane wave by a function of the form

$$\psi(x, t) \propto \cos (kx - \omega t + \alpha) \qquad (2.3a)$$

and a plane wave moving in the negative x-direction would be written as

$$\psi(x, t) \propto \cos (kx + \omega t + \beta) \qquad (2.3b)$$

The coordinates y and z can obviously be omitted in describing these one-dimensional processes.

In order to decide if (2.3a) and (2.3b) offer satisfactory means of describing plane waves in quantum mechanics, it is necessary to give some thought to the time dependence of the wave function. Classical mechanics permits the prediction of what happens at time t, if the initial conditions (coordinates and velocities) of the system at $t = 0$ are known. The correspondence principle requires that quantum mechanics must be equally deterministic and that it should be possible to calculate the condition of a system at an arbitrary time t from knowledge about it at time $t = 0$. But we have no a priori knowledge about the degree of information needed at $t = 0$, if a complete prediction of the evolution of the system is to be made, and we do not know whether the initial wave function $\psi(x, y, z, 0)$ determines $\psi(x, y, z, t)$ at all times, or whether additional information about the state at $t = 0$ is necessary.

Judging from experience with other types of wave motion, it might at first sight seem questionable that knowledge of $\psi(x, y, z, 0)$ could suffice to

---

[1] This is similar to the situation found in wave optics, where the uncritical use of Huygens' principle must be justified by recourse to the wave equation and to Kirchhoff's approximation.

determine $\psi(x, y, z, t)$. For example, if ψ were an elastic displacement of a vibrating string, $\partial\psi/\partial t$ would have to be specified at $t = 0$ in addition to ψ. We need only to remember that a snapshot taken of a vibrating string at $t = 0$ tells us nothing about its subsequent behavior. We must also know what it looks like a little later in order to decide whether the wave is propagating to the right, or to the left, or whether it is a standing wave. This is clearly the reason why the differential equation governing the behavior of a vibrating string must be of the *second* order in the time derivative.

As an illustration we note that the initial wave form

$$\psi(x, 0) = \cos kx \tag{2.4}$$

is obtained from either (2.3*a*) or (2.3*b*) by setting $\alpha = \beta = 0$ and $t = 0$. The initial condition (2.4) does not tell us how much of the wave travels to the right, (2.3*a*), and how much travels to the left, (2.3*b*). In fact, the wave function

$$\psi(x, t) = a \cos (kx - \omega t) + (1 - a) \cos (kx + \omega t) \tag{2.5}$$

coincides with $\psi(x, 0) = \cos kx$ at $t = 0$ for any value of a. However, knowledge of the first time derivative of (2.5) at $t = 0$

$$\left(\frac{\partial\psi}{\partial t}\right)_{t=0} = (2a - 1)\omega \sin kx$$

determines a and hence the form of (2.5) unambiguously.

We conclude that we are confronted by the following choice of working hypotheses:

I. Equations (2.3*a*) and (2.3*b*) describe plane waves, as proposed, but $\psi(x, y, z, 0)$ does not determine the wave function $\psi(x, y, z, t)$ at other times.

II. $\psi(x, y, z, 0)$ determines $\psi(x, y, z, t)$, but the expressions (2.3*a*) and (2.3*b*) do not describe simple harmonic plane ψ waves.

Since, unlike elastic displacements or electric field vectors, the ψ waves of quantum mechanics cannot be observed directly, it is tempting to construct the theory by the use of working hypothesis II and to infer from the formalism the appropriate ψ for plane waves. The only way in which we can generalize the expression (2.3*a*) for a harmonic plane wave propagating in the positive x-direction is to adjoin to the cosine a sine term, thus forming

$$\psi_1(x, t) \propto \cos (kx - \omega t) + \delta_1 \sin (kx - \omega t) \tag{2.6}$$

Similarly, a wave traveling in the opposite direction is

$$\psi_2(x, t) \propto \cos (kx + \omega t) + \delta_2 \sin (kx + \omega t) \tag{2.7}$$

The constants δ_1 and δ_2 can be determined by the requirement that the two oppositely moving plane waves ψ_1 and ψ_2 should be linearly independent at all times,[2] and that an arbitrary displacement of x or t should not change the physical character of the wave. The linear independence insures that any harmonic plane wave can be written at all times as a linear combination of ψ_1 and ψ_2. It is then eminently reasonable to demand that the motion of a free particle moving in the positive x-direction should at all times be represented by a pure plane wave of the type ψ_1 without any admixture of ψ_2, which moves in the opposite direction. Hence, it must be required that ψ_1 at time t should be a simple multiple of ψ_1 evaluated at $t = 0$. Mathematically we must have

$$\cos (kx + \varepsilon) + \delta_1 \sin (kx + \varepsilon) = a_1(\varepsilon)(\cos kx + \delta_1 \sin kx)$$

for all values of x and ε. Comparing coefficients of $\cos kx$ and of $\sin kx$, this last equation leads to

$$\cos \varepsilon + \delta_1 \sin \varepsilon = a_1 \quad \text{and} \quad \delta_1 \cos \varepsilon - \sin \varepsilon = a_1 \delta_1$$

These equations are compatible for all ε only if

$$\delta_1{}^2 = -1 \quad \text{or} \quad \delta_1 = \pm i$$

Similarly, from (2.7) we obtain $\delta_2{}^2 = -1$. We choose $\delta_1 = i$ for the wave (2.6), and $\delta_2 = -i$ for (2.7), and we are thus led to the conclusion that ψ waves describing free particle motion must be complex quantities if working hypothesis II is to succeed.

Summarizing our conclusions, we see that in this scheme a harmonic plane wave propagating toward increasing x is

$$\psi_1(x, t) = A e^{i(kx - \omega t)} \qquad (2.8)$$

and a wave propagating toward decreasing x is

$$\psi_2(x, t) = B e^{-i(kx + \omega t)} \qquad (2.9)$$

The initial values of these waves are $\psi_1(x, 0) = A e^{ikx}$ and $\psi_2(x, 0) = B e^{-ikx}$ respectively. ω is the frequency of oscillation which so far we have not brought into connection with any physically observable phenomenon. Generally, it will be a function of k (see Section 2.3). We shall refer to wave functions like (2.8) and (2.9) briefly as *plane waves*.

A plane wave propagating in an arbitrary direction has the form

$$\psi(x, y, z, t) = A e^{i(\mathbf{k} \cdot \mathbf{r} - \omega t)} = A e^{i(k_x x + k_y y + k_z z - \omega t)} \qquad (2.10)$$

Equations (2.8) and (2.9) are special cases of this with $k_y = k_z = 0$ and $k_x = \pm k$.

Of course, it is always possible to replace a complex wave function by two equivalent real functions (e.g., the real and imaginary parts of ψ), but to do

---

[2] The waves (2.3a) and (2.3b) violate this requirement. For $t = (\alpha - \beta)/2\omega$ they are equal, hence linearly dependent.

so would violate the spirit of working hypothesis II. The essence of this hypothesis is that a single function ψ determines the future state of the system. We now recognize that a price must be paid for this assumption: Complex functions ψ must be admitted. There can be at this point hardly any objection to the use of such functions. In particular, since the diffraction pattern presumably measures only the amplitude of ψ, we have no physical reason for rejecting complex-valued wave functions.

However, it should be stressed that with the acceptance of complex values for ψ we are by no means excluding wave functions which are real. For example, wave function (2.4) can be written as the sum of two exponential functions:

$$\psi(x, 0) = \cos kx = \tfrac{1}{2}e^{ikx} + \tfrac{1}{2}e^{-ikx} \tag{2.11}$$

According to the principle of superposition, this must be an acceptable wave function. How does it develop in time? The principle of superposition, in conjunction with working hypothesis II, suggests (but does not logically require) a simple answer to this important question: Each of the two waves, into which ψ can be decomposed at $t = 0$, develops independently, as if the other component were not present. The wave function, which at $t = 0$ satisfies the initial condition $\psi(x, 0) = \cos kx$, thus becomes for arbitrary t:

$$\psi(x, t) = \tfrac{1}{2}e^{i(kx - \omega t)} + \tfrac{1}{2}e^{-i(kx + \omega t)}$$

It is seen that our rule, to which we shall adhere and which we shall generalize, assures that $\psi(x, 0)$ determines the future behavior of the wave uniquely. If this formulation is correct, we shall expect that the complex ψ waves may be described by a differential equation which is of the first order in time.[3]

2. Wave Packets and the Uncertainty Relation. The foregoing discussion points to the possibility that by allowing the wave function to be complex we might be able to describe the state of motion of a particle at time t completely by $\psi(x, t)$. The real test of this assumption is of course its success in accounting for experimental observations. However, strong support for it can be gained by demonstrating that the correspondence with classical mechanics can be established within the framework of this theory.

To this end we must find a way of making $\psi(x, t)$ describe, at least approximately, the classical motion of a particle which has both reasonably definite position and reasonably definite momentum. The plane waves (2.8) and (2.9) correspond to particle motions with momentum which is precisely defined by

---

[3] Maxwell's equations are of the first order even though the functions are real, but this is accomplished by using both **E** and **B** in coupled equations, instead of describing the field by just one function. The heat flow equation is of the first order in time and still describes the behavior of a real quantity, the temperature; but none of its solutions represent waves propagating without damping in one direction like (2.10).

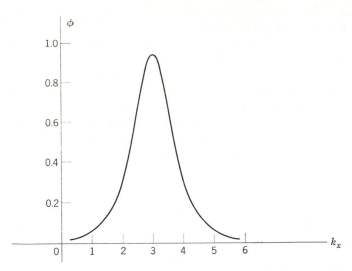

Figure 2.1. Example of a one-dimensional momentum distributicn. The function $\varphi(k_x) = \pi^{1/4}[2^{1/2} \cosh \pi^{1/2}(k_x - 3)]^{-1}$ represents a wave packet moving with mean wave number $\bar{k}_x = 3$. $\varphi(k_x)$ is normalized to unity.

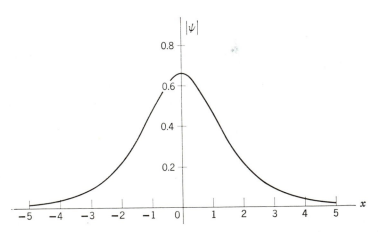

Figure 2.2. Normalized wave packet corresponding to the momentum distribution of Figure 2.1. The absolute value $|\psi(x, 0)| = \pi^{1/4}[2 \cosh (\pi^{1/2}x/2)]^{-1}$ is plotted.

(2.2); but, having amplitudes $|\psi| = $ const. for all x and t, these plane waves leave the position of the particle entirely unspecified. However, by super-position of several different plane waves a certain degree of localization can be achieved, as the fringes on the screen in the diffraction experiment attest.

The mathematical tools for such a synthesis of localized compact wave packets by the superposition of plane waves of different wave number are available in the form of *Fourier analysis*. Maintaining for simplicity the fiction that only one spatial coordinate, x, need be considered, we can write

$$\psi(x, 0) = \frac{1}{\sqrt{2\pi}} \int_{-\infty}^{+\infty} \varphi(k_x)e^{ik_x x}\, dk_x \tag{2.12}$$

and the inverse formula

$$\varphi(k_x) = \frac{1}{\sqrt{2\pi}} \int_{-\infty}^{+\infty} \psi(x, 0)e^{-ik_x x}\, dx \tag{2.13}$$

under very general conditions to be satisfied by $\psi(x, 0)$. Formulas (2.12) and (2.13) show that $\varphi(k_x)$ determines the initial wave function $\psi(x, 0)$, and vice versa.

We now apply the rule, established in the last section, that each component plane wave contained in (2.12) propagates independently of all the others according to the prescription of (2.10). The wave function at time t is thus:

$$\psi(x, t) = \frac{1}{\sqrt{2\pi}} \int_{-\infty}^{+\infty} \varphi(k_x)e^{i(k_x x - \omega t)}\, dk_x \tag{2.14}$$

Formula (2.14) is, of course, incomplete until we determine the dependence of ω on k. This determination will be made on physical grounds in Section 2.3.

For simplicity we now assume that φ is a real positive function and that it has a shape similar to Figure 2.1, i.e., an approximately symmetric distribution of k_x about a mean value $\bar{k}_x$. Making the simple change of variables

$$u = k_x - \bar{k}_x$$

we may write

$$\psi(x, 0) = \frac{1}{\sqrt{2\pi}} e^{i\bar{k}_x x} \int_{-\infty}^{+\infty} \varphi(u + \bar{k}_x)e^{iux}\, du \tag{2.15}$$

This is a wave packet whose absolute value is shown in Figure 2.2. The particle is most likely to be found at a position where ψ is appreciable. It is easy to see for any number of simple examples that the widths of the amplitude φ as a function of k_x and of the wave packet ψ stand in a reciprocal relationship:

$$\Delta x\, \Delta k_x \simeq 1 \tag{2.16}$$

A proof of this assertion will be supplied at the end of this chapter.

Exercise 2.1. Assume $\varphi(k_x) = \sqrt{2\pi}$ for $\bar{k}_x - \delta \leqslant k_x \leqslant \bar{k}_x + \delta$, and $\varphi = 0$ for all other values of k_x. Calculate $\psi(x, 0)$, plot $|\psi(x, 0)|^2$ for several values of δ and show that (2.16) holds if Δx is taken as the width at half-maximum.[4]

$\psi(x, 0)$ describes thus a particle which is localized within a distance Δx about the coordinate origin. This has been accomplished at the expense of combining waves of wave numbers in a range Δk_x about $\bar{k}_x$. The relation (2.16) shows that we can make the wave packet define a position more sharply only at the cost of broadening the spectrum of k_x-values which must be admitted. Any hope that these consequences of (2.16) might be averted by choosing $\varphi(k_x)$ more providentially has no basis in fact. On the contrary, we see easily (and we shall prove in Chapter 8 rigorously) that in general

$$\Delta x\,\Delta k_x \gtrsim 1 \tag{2.17}$$

where the $\simeq$ sign holds only if the absolute value of φ behaves as in Figure 2.1 and its phase is constant or a linear function of k_x.

Exercise 2.2. Assume $\varphi(k_x) = \sqrt{2\pi}$ for $\bar{k}_x - (\delta/2) \leqslant k_x \leqslant \bar{k}_x + (\delta/2)$, $\varphi(k_x) = -\sqrt{2\pi}$ for $\bar{k}_x - \delta \leqslant k_x < \bar{k}_x - (\delta/2)$ and for $\bar{k}_x + (\delta/2) < k_x \leqslant \bar{k}_x + \delta$, and $\varphi = 0$ for all other values of k_x. Calculate $\psi(x, 0)$ and show that, although the width of the k_x-distribution is the same as in Exercise 2.1, the width of ψ is greater in accordance with the inequality (2.17).

The fact that in quantum physics waves and particles both appear in the description of the same thing has already forced us to abandon the classical notion that position and momentum can be defined with perfect precision simultaneously. Equation (2.17) together with the equation $p_x = \hbar k_x$ expresses this characteristic property of quantum mechanics quantitatively:

$$\Delta x\,\Delta p_x \gtrsim \hbar \tag{2.18}$$

This relation suggests that when ψ is the sum of plane waves with different **k**'s the wave function cannot represent a particle of definite momentum (for if it did ψ would have to be a harmonic plane wave) and that its smeared-out momentum distribution is roughly pictured by the behavior of $|\varphi|$. The particle is most likely to have the momentum $\hbar \bar{k}_x$, since φ has its greatest magnitude at $\bar{k}_x$. Obviously, we shall have to make all these statements precise and quantitative later on.

Exercise 2.3. Convince yourself that (2.18) does not in practice impose any limitations on the precision with which position and momentum of a macroscopic body can be determined.

---

[4] Note that if $(\Delta x)^2$ is taken to be the variance of x, as in Section 8.6, its value is infinite for the wave packets defined in Exercises 2.1 and 2.2.

For the present it suffices to view a peak in ψ as a crudely localized particle, a peak in φ as a particle moving with an approximately defined velocity. The *uncertainty relation* (2.18) limits the accuracy with which position and momentum can be simultaneously ascribed to the particle. Both quantities are fuzzy and indeterminate (*Heisenberg uncertainty principle*).

We have already discussed in Chapter 1 what this means in terms of experiments. ψ, we recall, measures roughly the *probability* of finding the particle at position x. Yet upon measurement the particle will always be found to have a definite position. Likewise, quantum mechanics does not deny that precision measurements of momentum are feasible even when the particle is not represented by a plane wave of sharp momentum. Rather, φ is a rough measure of the probability of finding the particle to have momentum $\hbar k_x$. Again we emphasize that it is not permissible to consider φ as a measure of the relative frequency of finding various values of momentum in a large assembly of particles. Instead, φ is the attribute of a single particle. Naturally, whenever a theory predicts probabilities, the experimental verification of these probabilistic predictions requires the use of large numbers of identical systems.

Under the circumstances just described, must we admit that the particle has no definite momentum (or position) when we merely are unable to determine a single value consistently in making the same measurement on identically prepared systems? Could it be that we simply have not taken sufficient precautions to ensure that the systems were in truly identical conditions before the measurement was carried out? Or is there room in the theory for supplementing the quantal description by the specification of further ("hidden") variables, so that systems which are superficially in identical states may be found to be distinct in a more refined characterization?

Quantum mechanics asserts that for any given state the measurement of a particular physical quantity results with calculable probability in a numerical value belonging to an entire range of possible measured values. It presupposes that no technical or mathematical ingenuity can devise means of giving a sharper and more accurate account of a physical state than that permitted by the wave function and the uncertainty relation. These claims constitute a principle which by its very nature cannot be proved, but which is supported extraordinarily firmly by the enormous number of verified consequences derived from it.

Bohr has shown in detail in a number of interesting thought experiments[5]

[5] See the article by Niels Bohr, "Discussion with Einstein on Epistemological Problems in Atomic Physics," in P. A. Schilpp, editor, *Albert Einstein: Philosopher-Scientist*, p. 199, Tudor Publishing Co., New York, 1957; also reprinted in N. Bohr, *Atomic Physics and Human Knowledge*, John Wiley and Sons, New York, 1958, p. 32. Other thought experiments are discussed in W. Heisenberg, *The Physical Principles of the Quantum Theory*, translated by C. Eckart and C. Hoyt, University of Chicago Press, Chicago, 1930 (Dover Publications, New York, 1949).

how the finite value of $\hbar$ in the uncertainty relation makes the coexistence of wave and particle both possible and necessary. These idealized experiments demonstrate explicitly how any effort to design a measurement of the momentum component p_x with a precision set by a number Δp_x inescapably limits the precision of a simultaneous measurement of the coordinate x to $\Delta x > \hbar/\Delta p_x$, and vice versa.

Illuminating as the thought experiments are, they merely illustrate the important discovery that underlies quantum mechanics, namely, that the behavior of a material particle is described completely by its wave function ψ (suitably modified to include the spin and other degrees of freedom, if necessary), and generally not by its precise momentum or position. Quantum mechanics contends that the wave function contains the maximum amount of information that nature allows us concerning the behavior of electrons, photons, protons, and the like.

The mathematical formalism to be developed in this book, especially in Chapters 4 and 8, and in Chapters 12 to 15, faithfully mirrors the structure of the theory that has been outlined. Energy levels, multipole moments, transition rates, and cross sections can be calculated with high precision from the the formalism, although the theory itself contains no explicit instructions for linking the mathematical elements of the theory with observable quantities. To a certain extent, the physical significance of the theory is therefore subject to *interpretation*. Various interpretations of quantum mechanics must, of course, imply the same physically verifiable deductions from the theory, but they may and do differ in the treatment of the measurement process, where tests by actual (as distinguished from thought) experiments are difficult if not impossible. All competent interpretations of quantum mechanics must thus be judged at present on the basis of internal consistency and methodological appeal. A short explicit account of the commonly accepted "*Copenhagen interpretation*" is found in Chapter 13, but the general doctrine, in a variant mostly due to von Neumann, permeates all sections of this book. It must be conceded that even this interpretation leaves some important questions unanswered and implies an unconvincing, and certainly unsubstantiated, view of the process of cognition. Fortunately, these shortcomings do not in any way inhibit the successful elaboration and application of the quantum mechanical formalism.

3. *Motion of a Wave Packet*. Having placed the wave function in the center of our considerations, we must attempt to infer as much as possible about its properties. The wave packet (2.15) which we have constructed, using the real positive function $\varphi(k_x)$ of Figure 2.1, represents a function ψ which is a plane wave of wave number $\bar{k}_x$ with an x-dependent amplitude as shown in Figure 2.2. Since in this one-dimensional case ($|k_x| = k$, $k_y = k_z = 0$) $\varphi \approx 0$ for negative values of k_x, the mean value of the magnitude k is $\bar{k} = \bar{k}_x$. According to the correspondence principle, such a wave packet, centered around the coordinate origin, must describe classically the motion of a free particle which at $t = 0$ is at $x = 0$ and which has momentum $\hbar \bar{k}_x$ along the x-axis. Hence, the peak is expected to move according to the equation of motion

$$x_{\text{peak}} = \frac{\hbar \bar{k}_x}{\mu} t \tag{2.19}$$

where μ is the mass of the particle. The motion of the wave packet can be approximated if we expand $\omega(k)$ about $\bar{k}$,

$$\omega(k) = \omega(\bar{k}) + \left(\frac{d\omega}{dk}\right)_{k=\bar{k}} (k - \bar{k}) + \cdots = \bar{\omega} + \bar{\omega}'(k - \bar{k}) + \cdots \tag{2.20}$$

with obvious abbreviations.

For our special wave packet, which moves in the x-direction, we may set $k - \bar{k} = k_x - \bar{k}_x = u$. Assuming that we may break off the expansion with the linear term, we obtain from (2.14) and (2.15)

$$\psi(x, t) \simeq e^{i(\bar{k}_x x - \bar{\omega} t)} \frac{1}{\sqrt{2\pi}} \int_{-\infty}^{+\infty} \varphi(u + \bar{k}_x) e^{iu(x - \bar{\omega}' t)} \, du \tag{2.21}$$

$$= \exp\left(-i\bar{\omega} t + i\bar{\omega}' \bar{k}_x t\right) \psi(x - \bar{\omega}' t, 0)$$

Ignoring the phase factor in front, we see that this describes a wave packet which moves without any change of shape. In particular, the peak moves with the *group velocity* $\bar{\omega}'$:

$$x = \bar{\omega}' t \tag{2.22}$$

Equations (2.19) and (2.22) are expected to be identical equations if our interpretation of the wave packet and its motion is meaningful. This equialence must hold for an arbitrary choice of $\bar{k}$. Hence, we may omit the averaging bars and require

$$\frac{d\omega}{dk} = \frac{\hbar k}{\mu}$$

Integrating this, we obtain

$$\omega = \frac{\hbar k^2}{2\mu} + \text{const.}$$

or

$$\hbar \omega = \frac{p^2}{2\mu} = \text{const.} \tag{2.23}$$

Hence, $\hbar \omega$ must be the energy of the particle, apart from an arbitrary constant which corresponds to the usual failure of nonrelativistic mechanics to give meaning to the absolute value of the energy. We may therefore assume that the particle energy E is a multiple of the frequency:

$$E = \hbar \omega \tag{2.24}$$

A similar equation has already been found to hold for photons. This is not as surprising as it may seem, for (2.2) and (2.24), although obtained in this chapter for the special case of nonrelativistic particle mechanics, are Lorentz-invariant and thus remain valid in the relativistic case. Indeed, they merely express the proportionality of two four-vectors:

$$\left(\mathbf{p}, \frac{E}{c} \right) = \hbar \left(\mathbf{k}, \frac{\omega}{c} \right) \tag{2.25}$$

Hence their great universality.

It should be noted that for the nonrelativistic case considered here the expansion (2.20) actually terminates because

$$\omega(k) = \frac{\hbar k^2}{2\mu} = \frac{\hbar \bar{k}^2}{2\mu} + \frac{\hbar \bar{k}}{\mu}(k - \bar{k}) + \frac{\hbar}{2\mu}(k - \bar{k})^2$$

Thus, if quadratic terms are retained the behavior of the wave packet is given by an exact formula.

4. The Uncertainty Relations and the Spreading of Wave Packets. Now we must fill some of the gaps in the mathematical detail. First, let us show that the uncertainty relation (2.16) follows from the assumed properties of φ. In order to simplify the discussion let us suppose in addition that $\varphi(k_x)$ is symmetric with respect to $\bar{k}_x$. (There is no essential loss of generality in this assumption.) The amplitude of the wave packet,

$$\psi(x, 0) = \frac{1}{\sqrt{2\pi}} e^{i\bar{k}_x x} \int \varphi(u + \bar{k}_x) e^{iux} \, du$$

is then symmetric about the origin. Let us denote by $(-x_0, x_0)$ the range of x for which ψ is appreciably different from zero (see Figure 2.2). Since φ is appreciably different from zero only in a range Δk_x centered at $u = 0$, the phase of e^{iux} in the integrand varies from $-x \, \Delta k_x/2$ to $+x \, \Delta k_x/2$, i.e., by an amount $x \, \Delta k_x$ for any fixed value of x. Hence the largest variations which the phase ever undergoes are $\pm x_0 \, \Delta k_x$; this happens at the ends of the wave packet. If $x_0 \, \Delta k_x$ is less than $\simeq 1$, no appreciable cancellations in the integrand occur (since φ is positive definite). When $x_0 \, \Delta k_x \gg 1$, on the other hand, the phase goes through many periods as u ranges from $-\Delta k_x/2$ to $+\Delta k_x/2$. Hence, violent oscillations of the e^{iux} term occur, leading to destructive interference. Denoting by Δx the range $(-x_0, x_0)$, it follows that the widths of $|\psi|$ and φ are effectively related by

$$\Delta x \, \Delta k_x \simeq 1 \tag{2.16}$$

In order to prove the more general relation (2.17) we must define more precisely what is meant by the width of a function which does not have the simple behavior assumed here for φ. This will be done later (Chapter 8).

Next, we must show under what conditions it is legitimate to write the approximate equation (2.21), i.e., to neglect the quadratic term in (2.20). This quadratic term would contribute

$$\tfrac{1}{2}itu^2\left(\frac{d^2\omega}{dk^2}\right)_{k=\bar{k}}$$

to the exponent in the integrand of (2.21). An exponent can be neglected only if it is much less than unity in absolute value throughout the effective range of integration. This is assured if:

$$(\Delta k_x)^2\bar{\omega}''t \ll 1 \tag{2.26}$$

If $\omega = \hbar k^2/2\mu$, this condition becomes for nonrelativistic particles

$$t \ll \frac{\mu\hbar}{(\Delta p_x)^2} \tag{2.27}$$

or

$$t\,\Delta v_x = \frac{\Delta p_x}{\mu}\,t \ll \frac{\hbar}{\Delta p_x} \simeq \Delta x \tag{2.28}$$

$t\,\Delta v_x$ represents an uncertainty in the position of the particle at time t, over and beyond the initial uncertainty Δx. Hence, the second-order contributions to the phase can be neglected as long as the wave packet is not substantially broadened beyond its initial width. If condition (2.27) is violated, i.e., for fairly long times, our assumption about the rigid translation of the wave packet amplitude is no longer correct. It can then be shown that the quadratic term in the phase causes (2.16) to be replaced by the inequality

$$\Delta x(t)\,\Delta k_x > 1 \tag{2.29}$$

This means that the wave packet inevitably spreads, as time progresses.

Exercise 2.4. Consider a wave packet satisfying the relation $\Delta x\,\Delta p_x \simeq \hbar$. Show that if the packet is not to spread, while it passes through a fixed position, the condition $\Delta p_x \ll p_x$ must hold.

Exercise 2.5. Can the atoms in liquid helium at 4°K (interatomic distance about 1 Ångstrom) be adequately represented by nonspreading wave packets, so that their motion can be described classically?

The uncertainty relation (2.18) has a companion which relates uncertainties in time and energy. The kinetic energy is $E = p_x^2/2\mu$, and it is uncertain by an amount

$$\Delta E = \frac{p_x}{\mu}\,\Delta p_x = v_x\,\Delta p_x \tag{2.30}$$

If an observer is located at a fixed position, he "sees" the wave packet sweeping through his location. His determination of the time at which the particle passes through must then be uncertain by an amount:

$$\Delta t = \frac{\Delta x}{v_x} \simeq \frac{\hbar}{v_x \Delta p_x} = \frac{\hbar}{\Delta E}$$

Hence,

$$\Delta E \, \Delta t \simeq \hbar \tag{2.31}$$

In this derivation it was assumed that the wave packet does not spread appreciably in time Δt while it passes through the observer's position. According to (2.27), this is assured if

$$\frac{\hbar}{\Delta E} \simeq \Delta t \ll \frac{\mu \hbar}{(\Delta p_x)^2}$$

With (2.30) this condition is equivalent to $\Delta E \ll E$. But the latter condition must be satisfied if we are to be allowed to speak of the energy of the particles in a beam at all.

For purposes of completeness we finally note what changes have to be made in the equations of this chapter when the restriction to motion along the x-axis is removed. The initial wave packet in three dimensions can again be Fourier-analyzed:[6]

$$\psi(x, y, z, 0) \equiv \psi(\mathbf{r}, 0) = \frac{1}{(2\pi)^{3/2}} \int \varphi(\mathbf{k}) e^{i\mathbf{k}\cdot\mathbf{r}} \, d^3k \tag{2.32}$$

and the inverse formula is

$$\varphi(\mathbf{k}) = \frac{1}{(2\pi)^{3/2}} \int \psi(\mathbf{r}, 0) e^{-i\mathbf{k}\cdot\mathbf{r}} \, d^3r \tag{2.33}$$

If the particle which the initial wave function (2.32) describes is free, the wave function at time t becomes

$$\psi(\mathbf{r}, t) = \frac{1}{(2\pi)^{3/2}} \int \varphi(\mathbf{k}) e^{i(\mathbf{k}\cdot\mathbf{r} - \omega t)} \, d^3k \tag{2.34}$$

a generalization of (2.14). ω is the same function of k as before.

If ψ is of appreciable magnitude only in the coordinate ranges Δx, Δy, Δz, then ψ is sensibly different from zero only if k_x lies in a range Δk_x, k_y in a range Δk_y, and k_z in a range Δk_z, such that

$$\Delta x \, \Delta k_x \gtrsim 1, \qquad \Delta y \, \Delta k_y \gtrsim 1, \qquad \Delta z \, \Delta k_z \gtrsim 1 \tag{2.35}$$

---

[6] $d^3k = dk_x \, dk_y \, dk_z$ denotes the volume element in $\mathbf{k}$-space. $d^3r = dx \, dy \, dz$ is the volume element in (position) coordinate space. For brevity the symbol of $d\tau$ will also be used for the volume element in coordinate space.

By (2.2), $\Delta\mathbf{p} = \hbar\,\Delta\mathbf{k}$; hence in three dimensions the uncertainty relations are

$$\Delta x\,\Delta p_x \gtrsim \hbar, \qquad \Delta y\,\Delta p_y \gtrsim \hbar, \qquad \Delta z\,\Delta p_z \gtrsim \hbar \qquad (2.36)$$

The present chapter shows that our ideas about wave packets can be made quantitative and consistent with the laws of classical mechanics when the motion of *free* particles is considered. This is gratifying but it is not a sufficient test. Until we can show that the reasonable agreement between quantum and classical mechanics persists when forces are present, and until we demonstrate that the theory here developed yields results in accord with experiments in the atomic and nuclear domain, we cannot be entirely sure that we are on the right track.

Hence, we must now turn to an examination of the influence of forces on particle motion and wave propagation.

Problem

1. A one-dimensional initial wave packet with a mean wave number k_x and a Gaussian amplitude is given:

$$\psi(x, 0) = C \exp\left[-\frac{x^2}{4(\Delta x)^2} + ik_x x \right]$$

Calculate the corresponding k_x-distribution and $|\psi(x, t)|^2$, and show that the wave packet advances according to the classical laws but spreads in time. Apply the results to calculate the effect of spreading in some typical microscopic and macroscopic experiments. (See also Exercises 8.13 and 15.9, and Problem 8 in Chapter 8.)

CHAPTER 3

The Wave Equation

1. *General Remarks.* In the last chapter we discussed what might be considered the quantum analog of Newton's first law of motion, which is a statement about the motion of free particles. Now we must find the analog of Newton's second law, describing the wave behavior of a particle which moves under the influence of forces instead of moving freely.

Two obvious requirements must be met by the dynamical law which we are seeking: (a) As the force $\mathbf{F} \to 0$, it must go over into the quantum description of the motion of a free particle, and (b) according to the correspondence principle it must embody the classical law $\mathbf{F} = \mu\mathbf{a}$, when the behavior of wave packets is described—at least when the wavelength changes relatively slowly. However, these conditions do not lead us to an unambiguous formulation of the quantum mechanics of a particle, just as in classical mechanics knowing Newton's first law, i.e., $\mathbf{a} = 0$ if $\mathbf{F} = 0$, does not allow us to infer that $\mathbf{F} \propto \mathbf{a}$ although it makes that law a possibility. Rather, experience with actual known forces and their effect on the motion of a body is needed to arrive at Newton's second law.

Unfortunately, in the case of quantum mechanics the clues which experience with actual forces gives us are not easy to interpret. The reason for this is simple: The forces which are effective in the atomic and nuclear domain, where quantum effects are measurable, are either complicated or imperfectly known, or both. The simplest such force which we know accurately is the Coulomb force by which the electrons are held to the nucleus in an atom. Here the experimental evidence consists mainly of a knowledge of the atomic spectra to very high precision and of information obtained from scattering when a charged particle collides with an atom. It is not possible to infer the general dynamical law from such data without a daring guess at some point in the sequence of reasoning. In classical mechanics it took the genius of Newton to make the leap from a mass of empirical data to a compact dynamical law. But it must not be forgotten that the immediate experience with everyday types of forces—misleading as it can be—aided Newton considerably in formulating the second law. No comparable large-scale observations were available as guides when the quantum laws were discovered.

Under these circumstances it is astounding that a single generation of physicists was able to arrive with the requisite number of bold leaps at the dynamical equations which govern the motion of particles in the quantum domain. How this was done constitutes one of the most interesting chapters in the history of physics.[1]

Equipped with all the hindsight which the reading of this history provides, we can try to justify the dynamical law of quantum mechanics without following the historical development. Of course, a certain amount of guessing is inevitable in obtaining as general a law as we are hoping for, and it would be misleading to obscure this fact by pretending that the basic equations can be *derived*. The best that can be done here is to make the final dynamical law, the wave equation of quantum mechanics, appear reasonable. A more systematic and less intuitive procedure will be used in Chapter 15.

2. *Justification of the Wave Equation.* Let us consider again a broad wave packet, representing a rather sharply defined momentum and moving from a region of space in which the particle is free toward a region in which a force is present. For simplicity we shall continue to assume a one-dimensional description.[2] We also suppose that the forces are conservative, so that classically the particle has, and maintains throughout its motion, a constant energy:

$$E = \frac{p^2}{2\mu} + V(x) \tag{3.1}$$

Accordingly, the momentum is a function of position. It is again convenient to make a Fourier decomposition of the incident wave packet:

$$\psi(x, 0) = \frac{1}{\sqrt{2\pi}} \int_0^\infty \varphi(k)e^{ikx}\, dk \tag{3.2}$$

By omitting the range of integration from $-\infty$ to 0, we have included the assumption that the wave packet moves toward positive x, and contains no negative k-components. Using the substitution

$$k = \sqrt{\frac{2\mu E}{\hbar^2}} \tag{3.3}$$

we can transform expression (3.2) into an integral over the energy of the free particle:

$$\psi(x, 0) = \int_0^\infty f(E)e^{ik(E)x}\, dE \tag{3.4}$$

---

[1] For an informative and readable history of quantum mechanics, see M. Jammer, *The Conceptual Development of Quantum Mechanics*, McGraw-Hill Book Company, New York, 1966.

[2] Since in this chapter we confine ourselves to motion in one dimension, we drop subscripts and write p for p_x, k for k_x, etc.

$f(E)$ will be assumed to be a smoothly varying positive function of E with a single peak. Hence, (3.4) is a rather specialized wave packet, similar to that described by Figures 2.1 and 2.2. From the last chapter we know that, if this wave packet travels in force-free space, it develops in time as follows:

$$\psi(x, t) = \int f(E) \exp \left[\frac{i}{\hbar} (px - Et) \right] dE \tag{3.5}$$

We now ask: How will it move when it enters the region where a force is present?

The basic assumption which we shall make is that *the frequency is unaltered* by the approach to the region of changing potential energy. This assumption recommends itself by analogy with other types of wave motion. When a wave enters a different medium its wavelength (and its velocity of propagation) changes but the frequency of the oscillations is not affected. This behavior can usually be understood on the basis of a microscopic investigation of the structure of the medium and the simple nature of forced oscillations. Take, for instance, the case of elastic waves: If an atom of the medium vibrates with a certain frequency, the forces coupling it to its neighbors cause the neighbors to vibrate in the steady state with the same impressed frequency. (The amplitude of these vibrations depends, of course, on the proper frequencies of the atoms.)

Very similar language can be used to describe the propagation of light through a transparent refractive medium, if classical theory is used. The charges in the atoms are set into forced oscillation, *sympathetic oscillation*, as it used to be said nicely. The oscillating charges radiate with the impressed frequency, and the superposition of incident and reradiated waves results in an altered velocity of propagation.

Actually, this process requires a quantum description in which photons of the given frequency ω are absorbed and re-emitted by the atoms. Classically this means, as we have just remarked, that the atom is capable of vibrating with frequency ω, but from the particle point of view it means that the atom, owing to conservation of energy, acquires or loses energy $E = \hbar\omega$. This reasoning suggests that the relation

$$E = \hbar\omega \tag{2.24}$$

is much more universally valid than we have hitherto supposed. In a certain sense it now appears to hold also for electrons bound in atoms. Although the preceding arguments are of necessity vague and not entirely satisfactory, they indicate the general character of the theory which we are about to develop.

Of course, nothing as complicated as a theory about the interaction between light and matter is contemplated here. Rather, we are concerned with the behavior of a particle, say an electron, in a prescribed external force

field. Yet, ultimately an account must be given of the origin of any force
which might act on the particle. The forces are merely shorthand expressions
for the complex interactions between various wave-particle systems, which
in modern physics are usually referred to as *fields*. The only example of this
view of nature which has so far been almost completely elucidated is that of
the electromagnetic field in interaction with charged particles. The electromagnetic forces acting between electron and proton, for instance, can in a
sophisticated treatment be shown to arise as a result of the dynamical
behavior of photons, the quanta of the electromagnetic field. From this
point of view the constancy of the frequency of the matter waves is a consequence of the *principle of conservation of energy*, i.e., constant energy implies
constant frequency according to the equation $E = \hbar \omega$.

(There is no conflict between the assumption of strict conservation of
energy in the elementary interaction processes and the uncertainty relation
$\Delta E \, \Delta t \gtrsim \hbar$. This relation denies the possibility of both assigning to a system
a sharp value of the energy and using it as a more or less accurate time marker.
The ideal condition of a system with a precisely specified energy value could
only be achieved if the system were perfectly isolated and never interacted
with its surroundings. The energy-time uncertainty relation is entirely consistent with the assumption that if two systems, A and B, interact, the exact
amount of energy which is found to have been gained by B was lost by A, thus
insuring a strict energy balance.)

In the spirit of these remarks we may compare the motion of a particle
through a force field with the propagation of a wave through a strange
medium whose detailed microscopic properties we do not propose to
analyze here, but which can somehow be set into "sympathetic" vibrations
of the same frequency and which thus interacts with the incident wave to
modify its properties of propagation, i.e., its velocity of propagation and
its wavelength. Always admitting that we are at best giving an approximate and partial account of physical phenomena, we conjecture that the
wave packet which moves through a region of force is likely to be of the
form

$$\psi(x, t) = \int f(E) \exp\left\{\frac{i}{\hbar} \left[\varphi(x, E) - Et\right]\right\} dE \qquad (3.6)$$

This is a generalization of (3.5) in which $\varphi(x, E)$ is an as yet undetermined
function. This function must be chosen so that (3.6) represents a wave
packet which is moving classically. Since $f(E)$ is assumed to be a smooth
positive function of E, the peak of the amplitude of (3.6) occurs if x and t
are chosen so as to make the phase $\varphi(x, E) - Et$ stationary when $f(E)$ is
near its peak. This insures that the Fourier components near the peak of
$f(E)$ are as much in phase as possible and are therefore added constructively.

Mathematically, the phase is stationary if

$$\frac{\partial}{\partial E} [\varphi(x, E) - Et] = 0 \qquad (3.7)$$

After differentiation, E in this equation is to be evaluated at the energy which corresponds to the peak of $f(E)$. Equation (3.7) leads to the condition

$$t = \partial \varphi / \partial E \qquad (3.8)$$

On the other hand, classically for a conservative motion

$$t = \int^x \frac{dx'}{v(x')} = \int^x \frac{\partial p(x', E)}{\partial E} dx' \qquad (3.9)$$

where $p(x, E)$ is the momentum as defined by (3.1). Comparison of (3.8) with (3.9) gives the possible choice

$$\varphi(x, E) = \int^x p(x', E) \, dx' \qquad (3.10)$$

Thus we obtain

$$\psi(x, t) \simeq \int f(E) \exp\left[\frac{i}{\hbar}\left(\int^x p(x', E) \, dx' - Et \right) \right] dE \qquad (3.11)$$

According to Chapter 2, (3.11) is an exact equation, instead of an approximation, if the particle is free and moves with constant momentum p corresponding to a wavelength $\lambda = h/p$. It is reasonable to assume that (3.11) is the better an approximation the more nearly $\lambda(x) = h/p(x)$ is independent of x:

$$\left| \frac{d\lambda}{dx} \right| \ll 1 \qquad (3.12)$$

If this is true, the wave is quasi-harmonic and $\lambda(x)$ can be regarded as its slowly varying wavelength. Condition (3.12) can also be written as

$$h \left| \frac{dp}{dx} \right| \ll p^2 \qquad (3.12a)$$

or

$$\lambda(x) \left| \frac{dV}{dx} \right| \ll \frac{[p(x)]^2}{\mu} \qquad (3.12b)$$

This condition implies that the potential energy changes very little over the distance of a wavelength. The particle can then be considered as nearly free over a distance of many wavelengths, and it becomes legitimate to apply the criteria of Chapter 2 for the construction of wave packets corresponding to classical particle motion. For this reason, it will be assumed that $\hbar(dp/dx)$ can be neglected compared to p^2 in expressions derived from (3.11).

Equation (3.11) could be used to study the behavior of particles in quantum mechanics, when we are concerned with quantum corrections to the classical motion. We shall come back to this sort of approximate wave function in Chapter 7 when we discuss the WKB approximation. It suffices to point out here that some very typical quantum features are inherent in the form of ψ as given by (3.11). The most striking of these is the *penetration of a potential barrier.* Since

$$p(x, E) = \pm\sqrt{2\mu[E - V(x)]} \tag{3.13}$$

we see that an imaginary momentum results at points where $V(x) > E$. Classically, the particle cannot reach these points. Yet, if (3.11) is somewhat arbitrarily assumed to hold in this nonclassical region, the wave function does not vanish there. Hence, quantum mechanics allows for a tunneling of particles through classically inaccessible regions. Alpha decay of radioactive nuclei is the best-known example of penetration through a barrier.

One obstacle in the way of applying (3.11) extensively is that it may be quite inaccurate in regions where the kinetic energy varies rapidly with x, particularly near the classical turning points of the motion, where $p(x, E) = 0$, and where (3.12a) can never be satisfied. Hence, (3.11) does not provide us with an adequate formula for the description of the particle motion, where it is least classical and most interesting.

It is at this juncture that we ask ourselves whether a more generally satisfactory wave function could not be found if we abandoned the explicit construction of wave packets, which represent quite special semiclassical wave forms. A wave packet must always be given in terms of a particular initial wave $\psi(x, 0)$. Let us ask instead for some general relation which $\psi(x, t)$ must satisfy, no matter what the particular initial condition may be. Our experience with classical physics suggests that such a relation is most likely going to be a *differential equation* for $\psi(x, t)$. In fact, we are familiar with all sorts of wave equations, and the desirability of obtaining proper frequencies to account for the discrete energy spectra of atoms and nuclei suggests that one of the common types of partial differential equations, which are well known to yield discrete *eigenvalues* for appropriate boundary conditions, might be a suitable candidate. In searching for the correct equation, we are guided by the results of Chapter 2 for free particle motion. In general, we thus expect that ψ waves are complex quantities and that $\psi(x, 0)$ alone, without knowledge of the time derivative, determines the future of the wave. Hence, we shall attempt to obtain a first-order differential equation in t. It is well to admit that at this stage we are led more by analogy and guesswork than by compelling reasoning.

Following this idea, let us differentiate (3.11) with respect to time:

$$\frac{\partial \psi}{\partial t} \simeq -\int f(E) \frac{i}{\hbar} E \exp\left[\frac{i}{\hbar}\left(\int^x p\, dx' - Et\right)\right] dE \tag{3.14}$$

We would like to eliminate the initial condition $f(E)$ from the right-hand side of this equation. This can almost be accomplished if we note that

$$\frac{\partial \psi}{\partial x} \simeq \frac{i}{\hbar} \int f(E) p(x, E) \exp \left[\frac{i}{\hbar} \left(\int^x p \, dx' - Et \right) \right] dE$$

and

$$\frac{\partial^2 \psi}{\partial x^2} \simeq -\frac{1}{\hbar^2} \int f(E) p^2(x, E) \exp \left[\frac{i}{\hbar} \left(\int^x p \, dx' - Et \right) \right] dE$$

$$+ \frac{i}{\hbar} \int f(E) \frac{\partial p}{\partial x} \exp \left[\frac{i}{\hbar} \left(\int^x p \, dx' - Et \right) \right] dE \qquad (3.15)$$

But, according to (3.12*a*) the last term is negligible compared to the first. Using (3.13), (3.14), and (3.15), we obtain, at least as an approximation,

$$i\hbar \frac{\partial \psi}{\partial t} = -\frac{\hbar^2}{2\mu} \frac{\partial^2 \psi}{\partial x^2} + V(x)\psi \qquad (3.16)$$

The "bold leap" of quantization consists in asserting that this is the suitable place for going from the particular to the general and that (3.16) is valid even when (3.11) is not. In fact, we will make (3.16), the quantum mechanical *wave equation*, the basis of our theory.

CHAPTER 4

The Wave Function and the Schrödinger Equation

1. The Interpretation of ψ and the Conservation of Probability. Most of our work has so far been confined to the one-dimensional case. To insist that motion takes place in a three-dimensional space and to develop our theory accordingly would have been possible, but doing so would have required that we pay attention to details which are in no way essential to our purpose. It is far more practical to make the generalization to the case of three dimensions now that we are in possession of a partial differential equation which we believe governs the behavior of the wave function.

In Cartesian coordinates, the obvious generalization of the one-dimensional wave equation to three dimensions is

$$i\hbar \frac{\partial \psi}{\partial t} = -\frac{\hbar^2}{2\mu}\nabla^2\psi + V(x, y, z)\psi \tag{4.1}$$

V is assumed to be a real function. In simple cases it represents the potential energy.[1] Since the three-dimensional space in which the particle moves is isotropic[2] and since in the absence of external forces ($V = 0$) no direction in space is preferred, (4.1) is the only possible generalization of (3.16)—short of increasing the complexity of the differential equation.

We shall adopt (4.1) as the fundamental equation of nonrelativistic quantum mechanics for particles without spin. In subsequent chapters this theory will be generalized to encompass many different and more complicated systems, but *wave mechanics*, which is the branch of quantum mechanics with a dynamical law in the explicit form (4.1), remains one of the two simplest paradigms of the general theory, the other being the dynamics of a particle with spin as its only relevant degree of freedom (Chapter 13).

The ultimate justification for choosing (4.1) must, of course, come from agreement between predictions and experiment. Hence, we must examine the

---

[1] See Problem 1 of this chapter for the generalization to complex V.

[2] For a precise formulation and explanation of the isotropy of space see Section 16.1.

properties of this equation. Before going into mathematical details it would, however, seem wise to attempt to say precisely what ψ is. We are in the paradoxical situation of having obtained an equation which we believe is satisfied by this quantity, but of having so far given only a deliberately vague interpretation of its physical significance. We have regarded the wave function as "a measure of the probability" of finding the particle at time t at the position $\mathbf{r}$. How can this statement be made precise?

ψ itself obviously cannot be a probability. All hopes we might have entertained in that direction vanished when ψ became a complex function, since probabilities are real and *positive*.[3] In the face of this dilemma the next best guess is that the probability is proportional to $|\psi|^2$, the square of the amplitude of the wave function. There is an analogy here to the case of optical interference where the waves whose superposition produces the interference pattern are waves of $\mathbf{E}$ and $\mathbf{H}$, but the intensity of the fringes is measured by $\mathbf{E}^2$ and $\mathbf{H}^2$, as was mentioned in Chapter 1.

Of course, we were careless when we used the phrase "probability of finding the particle *at* position $\mathbf{r}$." Actually, all we can speak of is the probability that the particle is in a volume element d^3r which contains the point $\mathbf{r}$. Hence, we now try the interpretation that $|\psi(\mathbf{r}, t)|^2 \, d^3r$ is proportional to the probability that upon a measurement of its position the particle will be found in the given volume element. The probability of finding the particle in some finite region of space is proportional to the integral of $\psi^*\psi$ over this region.

Immediately doubts arise over the consistency of this probabilistic interpretation of the wave function. If the probability of finding the particle in some bounded region of space decreases as time goes on, then the probability of finding it outside of this region must increase by the same amount. The probability interpretation of the ψ waves can be made consistently only if this *conservation of probability* is guaranteed. This requirement is fulfilled, owing to Gauss' integral theorem, if it is possible to define a *probability current density* $\mathbf{j}$ which together with the *probability density* $\rho = \psi^*\psi$ satisfies a continuity equation

$$\frac{\partial \rho}{\partial t} + \nabla \cdot \mathbf{j} = 0 \tag{4.2}$$

exactly as in the case of the conservation of matter in hydrodynamics, or conservation of charge in electrodynamics.

A relation of the form (4.2) can easily be deduced from the wave equation.

---

[3] The solutions of (4.1) must in general be complex, because i appears explicitly in the differential equation. For $V = 0$ the quantum mechanical wave equation and the diffusion (or heat flow) equation become formally identical, but the presence of i assures that the former has solutions with wave character.

Multiply (4.1) on the left by ψ^*, and the complex conjugate of (4.1) by ψ on the right; subtract the two equations; and make a simple transformation. The resulting equation

$$\frac{\partial}{\partial t}(\psi^*\psi) + \frac{\hbar}{2\mu i}\nabla \cdot [\psi^* \nabla\psi - (\nabla\psi^*)\psi] = 0 \tag{4.3}$$

has the form of (4.2) if the identification

$$\rho = C\psi^*\psi \tag{4.4}$$

$$\mathbf{j} = C\frac{\hbar}{2\mu i}[\psi^* \nabla\psi - (\nabla\psi^*)\psi] \tag{4.5}$$

is made. C is a constant.

Exercise 4.1. Derive (4.3). Note that it depends on V being real.

Exercise 4.2. Generalizing (4.3), prove that

$$\frac{\partial}{\partial t}(\psi_1^*\psi_2) + \frac{\hbar}{2\mu i}\nabla \cdot [\psi_1^* \nabla\psi_2 - (\nabla\psi_1^*)\psi_2] = 0 \tag{4.6}$$

if ψ_1 and ψ_2 are two solutions of (4.1).

Exercise 4.3. Calculate ρ and $\mathbf{j}$ for a one-dimensional wave packet of the form (2.21) and show that in first approximation the relation

$$\mathbf{j} \simeq \frac{\bar{\mathbf{p}}}{\mu}\rho = \bar{\mathbf{v}}\rho$$

is obtained. Is this a reasonable result?

So far, we have only assumed that ρ and $\mathbf{j}$ are *proportional* to ψ and ψ^*. Since (4.1) determines ψ only to within a multiplicative constant, we may set $C = 1$,

$$\rho = \psi^*\psi \tag{4.4a}$$

$$\mathbf{j} = \frac{\hbar}{2\mu i}[\psi^* \nabla\psi - (\nabla\psi^*)\psi] \tag{4.5a}$$

and normalize the wave function by requiring

$$\int_{\text{all space}} \rho\, d\tau = \int_{\text{all space}} \psi^*\psi\, d\tau = 1 \tag{4.7}$$

provided that the integral of $\psi^*\psi$ over all space exists.[4] Equation (4.7)

---

[4] From here on all spatial integrations will be understood to extend over all space, unless otherwise stated, and the limits of integration will usually be omitted.

expresses the simple fact that the probability of finding the particle any-where in space at all is unity. Whenever this integral exists we shall assume that this normalization has been imposed. Functions ψ for which $\int \psi^*\psi \, d\tau$ exists are often called *quadratically integrable*.

If the probability current density falls to zero faster than $1/r^2$ as r becomes very large and if the integral (4.7) exists, then Gauss' theorem applied to (4.2) gives

$$\frac{\partial}{\partial t} \int \rho \, d\tau = 0$$

which simply tells us that the wave equation guarantees the conservation of normalization: If ψ was normalized at $t = 0$, it will remain normalized at all times.

In the light of the foregoing remarks it is well to emphasize that through-out this book all integrals which are written down will be assumed to exist. In other words, whatever conditions ψ must satisfy for a certain integral to exist will be assumed to be met by the particular wave function used. This understanding will save us laborious repetition. Examples will occur presently.

When the integral (4.7) does *not* exist, we may use only (4.4) and (4.5), and we must then speak of *relative* rather than absolute probabilities. Wave functions which are not normalizable to unity in the sense of (4.7) are as important as normalizable ones. The former appear when the particle is unconfined, and in principle they can always be avoided by the use of finite wave packets. However, such a policy would prevent us from using such simple wave functions as the infinite plane wave e^{ikx}, which describes a particle that is equally likely to be found anywhere in space at all, hence, of course, with zero absolute probability in any finite volume. Owing to the simple properties of Fourier integrals, this patently unphysical object is mathematically extremely convenient. It is an idealization of a finite, quadratically integrable, but very broad wave packet which contains very many wavelengths. If these limitations are recognized, the infinite plane matter wave is no more objectionable than an infinite electromagnetic wave which is also unphysical, because it represents an infinite amount of energy. We shall depend on the use of such wave functions.[5]

One further comment on terminology: Since $\psi(\mathbf{r}, t)$ is assumed to contain all information about the state of the physical system at time t, the terms *wave function*, ψ, and *state* may be used interchangeably. Thus we may speak without danger of confusion of a *normalized state* if (4.7) holds.

2. Review of Probability Concepts. Since the interpretation of the wave function is in terms of probabilities, a brief review of probability concepts may be helpful.

[5] For the normalization of such wave functions see Sections 6.3 and 8.4.

Let us suppose that each event E_k in a set of events E_1, E_2, ... is assigned a probability of occurrence P_k, with $\sum_k P_k = 1$. For example, in tossing a coin the two possible results, heads and tails, may be identified as event E_1 and event E_2, their probabilities being $P_1 = P_2 = 0.5$, if the coin is perfect. A variable X which takes on the value X_1 if E_1 occurs, X_2 if E_2 occurs, etc., is called a *random variable*. If in tossing the coin you are promised 1 penny for heads and 3 pennies for tails, your winnings constitute a random variable with values $X_1 = 1$, $X_2 = 3$.

The *expectation value* of X for the given probability distribution is defined to be the weighted sum

$$\langle X \rangle = \sum_k X_k P_k \tag{4.8}$$

In the example, $\langle X \rangle = 1 \times 0.5 + 3 \times 0.5 = 2$ pennies. Two pennies is the expected gain per tossing, whence the term *expectation value* for $\langle X \rangle$. (The game will be a fair one, if the ante is 2 pennies.)

If out of a large number of N trials conducted, N_1 lead to event E_1, N_2 to event E_2, etc., the relative frequencies N_k/N are expected to be approximated by the probabilities P_k. Hence, the *average value* of X, $\sum_k X_k N_k/N$, is approximately equal to the expectation value $\langle X \rangle$, and the two terms are often used synonymously. In quantum mechanics the term "expectation value" is preferred when it is desirable to emphasize the predictive nature of the theory and the fact that the behavior of a single particle is involved rather than of an ensemble of particles.

A related concept is the *variance*, $(\Delta X)^2$, of the random variable X. This is defined by

$$(\Delta X)^2 = \langle (X - \langle X \rangle)^2 \rangle = \sum_k (X_k - \langle X \rangle)^2 P_k = \langle X^2 \rangle - \langle X \rangle^2 \tag{4.9}$$

and measures the deviation from the mean. Variances will later be used for the rigorous formulation of the Heisenberg uncertainty relations (Chapter 8). In this connection we shall also encounter the *covariance* of two random variables X and Y:

$$\langle (X - \langle X \rangle)(Y - \langle Y \rangle) \rangle = \langle XY \rangle - \langle X \rangle \langle Y \rangle \tag{4.10}$$

If X and Y are independent (uncorrelated), the average of the product XY equals the product of the averages, $\langle X \rangle \langle Y \rangle$, and the correlation coefficient vanishes.

In this section it has been assumed that the events are discrete, whereas in the application to quantum mechanics continuous probability distributions like $|\psi|^2$ are common. The summations in (4.8) and (4.9) must then be replaced by integrations. For a normalized state the average or expectation value of the coordinate x, which is a random variable, is

$$\langle x \rangle = \int x \, |\psi|^2 \, d\tau \qquad (4.11)$$

The expectation value of the position vector, or, simply, the *center of the wave packet*, is defined by

$$\langle \mathbf{r} \rangle = \int \mathbf{r} \, |\psi|^2 \, d\tau \qquad (4.12)$$

An arbitrary function of $\mathbf{r}$ has the expectation value

$$\langle f(\mathbf{r}) \rangle = \int f(\mathbf{r}) \, |\psi|^2 \, d\tau = \int \psi^* f(\mathbf{r}) \psi \, d\tau \qquad (4.13)$$

The reason for writing $\langle f \rangle$ in the clumsy form $\int \psi^* f \psi \, d\tau$ will become clear when less elementary expectation values make their appearance.

3. Expectation Values of Dynamical Variables and Operators. Equations (4.12) and (4.13) define the expectation values of the position vector and any function thereof. The expectation values of such dynamical variables as velocity, momentum, and energy can be given equally satisfactory definitions by applying the correspondence principle. We require that the classical motion of a particle be approximated by the average behavior of a wave packet with a fairly sharp peak and as precise a momentum as the uncertainty principle permits and that the expectation values of the dynamical variables, calculated for such a wave packet, satisfy the laws of classical mechanics.

For example, we expect that the time derivative of $\langle \mathbf{r} \rangle$ will correspond to the classical velocity. This particular quantity can be calculated by the use of the continuity equation. For the x-component,

$$\frac{d}{dt} \langle x \rangle = \int x \frac{\partial \rho}{\partial t} \, d\tau = - \int x \nabla \cdot \mathbf{j} \, d\tau$$

$$= - \int \nabla \cdot (\mathbf{j}x) \, d\tau + \int j_x \, d\tau = \int j_x \, d\tau$$

or

$$\frac{d}{dt} \langle \mathbf{r} \rangle = \int \mathbf{j} \, d\tau \qquad (4.14)$$

where the divergence term has been removed under the assumption that ψ vanishes *sufficiently fast* at infinity. Using (4.5*a*) and integration by parts, we obtain finally

$$\mu \frac{d}{dt} \langle \mathbf{r} \rangle = \int \psi^* \frac{\hbar}{i} \nabla \psi \, d\tau \qquad (4.15)$$

The left-hand side is simply the mass times the classical velocity. According to the assumption that expectation values are to satisfy the laws of classical mechanics, the right-hand side of (4.15) must be equal to the expectation value of the particle momentum **p**. Hence we are led to the definition

$$\langle \mathbf{p} \rangle = \int \psi^* \frac{\hbar}{i} \nabla \psi \, d\tau \tag{4.16}$$

This identification receives further support if the time rate of change of $\langle \mathbf{p} \rangle$ is considered. Using the wave equation (4.1) we find for the x-component:

$$\frac{d}{dt} \langle p_x \rangle = -i\hbar \int \frac{\partial \psi^*}{\partial t} \frac{\partial \psi}{\partial x} \, d\tau - i\hbar \int \psi^* \frac{\partial}{\partial x} \left(\frac{\partial \psi}{\partial t} \right) d\tau$$

$$= -\frac{\hbar^2}{2\mu} \int \left[(\nabla^2 \psi^*) \frac{\partial \psi}{\partial x} - \psi^* \nabla^2 \frac{\partial \psi}{\partial x} \right] d\tau + \int \left[V \psi^* \frac{\partial \psi}{\partial x} - \psi^* \frac{\partial (V \psi)}{\partial x} \right] d\tau$$

The integral containing Laplacians can be transformed into a surface integral by Green's theorem.[6] It is assumed that this integral vanishes when taken over a very large surface. What remains is

$$\frac{d}{dt} \langle p_x \rangle = -\int \psi^* \frac{\partial V}{\partial x} \psi \, d\tau$$

or, more generally,

$$\frac{d}{dt} \langle \mathbf{p} \rangle = -\int \psi^* (\nabla V) \psi \, d\tau = -\langle \nabla V \rangle = \langle \mathbf{F} \rangle \tag{4.17}$$

This is simply Newton's second law, valid for expectation values, in accordance with the formulation of the correspondence principle at the beginning of this section. Equation (4.17) is known as *Ehrenfest's theorem*.

Since we have assumed conservative forces, we expect that the law of conservation of energy can also be written in terms of expectation values. It must have the form

$$\langle H \rangle = \langle T \rangle + \langle V \rangle = \langle T \rangle + \int \psi^* V \psi \, d\tau = \text{const.}$$

but we do not yet know how to write $\langle T \rangle$ or $\langle H \rangle$ in terms of the wave function.

---

[6] Green's theorem for two continuously differentiable functions u and v:

$$\int_S (u \nabla v - v \nabla u) \cdot d\mathbf{S} = \int_V (u \nabla^2 v - v \nabla^2 u) \, d\tau$$

where the surface S encloses the volume V.

A glance at (4.1) shows that this problem has a simple answer: Merely multiplying (4.1) on the left by ψ^* and integrating over all space, we obtain:

$$\int \psi^* i\hbar \frac{\partial \psi}{\partial t} \, d\tau = \int \psi^* \left(-\frac{\hbar^2}{2\mu} \nabla^2 \psi \right) d\tau + \int \psi^* V \psi \, d\tau \qquad (4.18)$$

By differentiating this equation with respect to t and applying the wave equation again, we find readily that the right-hand side of (4.18) is a constant of the motion. Hence, we identify the expectation value of the kinetic energy T with the expression:

$$\langle T \rangle = \int \psi^* \left(-\frac{\hbar^2}{2\mu} \nabla^2 \psi \right) d\tau \qquad (4.19)$$

The expectation values (4.12), (4.13), (4.16), and (4.19) all have the same form,

$$\int \psi^* A \psi \, d\tau$$

where A is a function or differential operator which is inserted between ψ^* and ψ. The integrand is the product of ψ^* and the function $A\psi$. Generally A is termed an *operator*.

From the equations mentioned we infer the operators to be used in the calculation of expectation values given in Table 4.1. The student of quantum mechanics must get used to the confusing practice of frequently denoting the operators with which we calculate expectation values of physical quantities by the same symbols as the physical quantities themselves. This practice is unfortunate because we shall shortly encounter other formulations of quantum mechanics in which the physical quantities are represented by entirely different operators. Yet, as in most matters of notation, it is not difficult to get used to these conventions, and the context will always establish the meaning of the symbols unambiguously.

Table 4.1

Physical quantity		Operator
Position	$\mathbf{r}$	$\mathbf{r}$
Momentum	$\mathbf{p}$	$\dfrac{\hbar}{i} \nabla$
Kinetic energy	T	$-\dfrac{\hbar^2}{2\mu} \nabla^2$
Potential energy	V	$V(\mathbf{r})$
Total energy	H	$-\dfrac{\hbar^2}{2\mu} \nabla^2 + V(\mathbf{r})$

Hence, we shall not hesitate to use the symbol **p** often when we really should use $(\hbar/i)\nabla$; similarly, we shall use T for $-\hbar^2\nabla^2/2\mu$. This peculiar and seemingly irresponsible identification of operators with physical quantities is encouraged by the observation that, for instance, the following operator identity holds:

$$T \equiv -\frac{\hbar^2}{2\mu}\nabla^2 = \frac{1}{2\mu}\left(\frac{\hbar}{i}\nabla\right)\cdot\left(\frac{\hbar}{i}\nabla\right) = \frac{\mathbf{p}\cdot\mathbf{p}}{2\mu} = \frac{p^2}{2\mu}$$

Thus the same relation holds between the operators T and **p** as between the classical entities known as kinetic energy and momentum. We have here a first suggestion of an alternative form of quantum mechanics in which the relations between the operators are similar to those between the corresponding classical physical quantities.[7]

All the operators listed in Table 4.1 share the property of being *linear*. An operator is said to be linear if its action on any two functions ψ_1 and ψ_2 is such that

$$A(\lambda\psi_1 + \mu\psi_2) = \lambda(A\psi_1) + \mu(A\psi_2) \tag{4.20}$$

where λ and μ are arbitrary complex numbers. Derivatives are obviously linear operators, as are mere multipliers. Most of the operators which are relevant to quantum mechanics are linear. Unless it is specifically stated that a given operator is not linear, the term *operator* will henceforth be reserved for linear operators.

The only other category of operators important in quantum mechanics are *antilinear* operators, characterized by the property

$$A(\lambda\psi_1 + \mu\psi_2) = \lambda^*(A\psi_1) + \mu^*(A\psi_2) \tag{4.20a}$$

Complex conjugation itself is an example of an antilinear operator.

Exercise 4.4. Construct some examples of linear and antilinear operators and of some operators which are neither linear nor antilinear.

4. *Stationary State Solutions.* We must now turn our attention to the wave equation and its solutions. The fundamental mathematical problem is to obtain a solution to (4.1) which for $t = 0$ agrees with a given *initial state* $\psi(\mathbf{r}, 0)$.

Since t appears nowhere explicitly in the differential equation, use of the method of separation of variables is clearly indicated. That is, we look for particular solutions of the form:

$$\psi(\mathbf{r}, t) = f(t)\psi(\mathbf{r}) \tag{4.21}$$

---

[7] However, it is clear that the formalism of this section is not adapted to such an approach, because an operator like $\mathbf{p} = (\hbar/i)\nabla$ does not depend on the time and could not be used in writing Newton's second law where $d\mathbf{p}/dt$ appears.

Again the use of the symbol $\psi(\mathbf{r})$ on the right-hand side of this equation for a function which is in general entirely different from $\psi(\mathbf{r}, t)$ appearing on the left-hand side is sanctioned by usage. Since this notation is very common, using a different symbol might spare the student a little confusion now but would make the reading of the literature more difficult later. Whenever necessary, care will be taken to distinguish the two functions by including the variables on which they depend.

Inserting (4.21) into (4.1), we easily obtain

$$\frac{i\hbar}{f}\frac{df}{dt} = \frac{1}{\psi(\mathbf{r})}\left[-\frac{\hbar^2}{2\mu}\nabla^2\psi(\mathbf{r}) + V\psi(\mathbf{r})\right]$$

The left-hand side of this equation depends only on t, the right-hand side only on x, y, z, hence, both must equal a constant. This constant must have the dimensions of an energy; hence, we denote it tentatively by E and obtain by integration:

$$f(t) = \exp\left(-iEt/\hbar\right) \tag{4.22}$$

On the other hand, the function $\psi(\mathbf{r})$ satisfies the equation

$$-\frac{\hbar^2}{2\mu}\nabla^2\psi(\mathbf{r}) + V\psi(\mathbf{r}) = E\psi(\mathbf{r}) \tag{4.23}$$

known as the *Schrödinger equation*. This equation is also called the *time-independent Schrödinger equation* in contrast to the *time-dependent Schrödinger equation* (4.1). The great importance of Equation (4.23) derives from the fact that the separation of variables yields not just some particular solution of (4.1), but generally yields *all* solutions of physical interest.

The crucial point is that under very mild boundary conditions imposed on the wave function we can prove that the separation constant E is a *real* number. The proof can be based on the continuity equation (4.2). Applying this to the state (4.21), we obtain from

$$\rho = \psi^*(\mathbf{r}, t)\psi(\mathbf{r}, t) = \exp\left[-\frac{i}{\hbar}(E - E^*)t\right]\psi^*(\mathbf{r})\psi(\mathbf{r})$$

upon substitution in (4.2) and integration over space:

$$(E - E^*)\int\psi^*(\mathbf{r}, t)\psi(\mathbf{r}, t)\,d\tau = \frac{\hbar}{i}\int\nabla\cdot\mathbf{j}\,d\tau = \frac{\hbar}{i}\int\mathbf{j}\cdot d\mathbf{S}$$

If the total probability flux over the boundary surface vanishes, we obtain $E = E^*$, and E must be real.

If E is real, the time-dependent factor (4.22) is purely oscillatory. Hence, in a separable state (4.21) the functions $\psi(\mathbf{r}, t)$ and $\psi(\mathbf{r})$ differ only by a time-dependent phase factor of constant amplitude. This justifies the occasional designation of the spatial factor $\psi(\mathbf{r})$ as a *time-independent wave function*.

For such a state the expectation value of any physical quantity which does not depend on time explicitly is constant and may be calculated from the time-independent wave function alone:

$$\langle A \rangle = \int \psi^*(\mathbf{r}, t) A \psi(\mathbf{r}, t) \, d\tau = \int \psi^*(\mathbf{r}) A \psi(\mathbf{r}) \, d\tau \tag{4.24}$$

If the wave function (4.21) with f given by (4.22) is substituted into (4.18), the relation

$$E = \langle H \rangle = \langle T \rangle + \langle V \rangle$$

is found, showing that E is the expectation value of the total energy in this state.

We thus see that the particular solutions

$$\psi(\mathbf{r}, t) = \exp\left(-iEt/\hbar\right)\psi_E(\mathbf{r}) \tag{4.25}$$

have intriguing properties.[8] $|\psi(\mathbf{r}, t)|^2$ and all probabilities are constant in time. For this reason wave functions of the form (4.25) are said to represent *stationary states*. Note that although the state is stationary, the particle so described is in motion and not stationary at all. Also, $\nabla \cdot \mathbf{j} = 0$ for such a state.

Exercise 4.5. Further explore and discuss the properties of stationary states. Using Equation (4.6), prove the *orthogonality* relation

$$\int \psi_1^*(\mathbf{r})\psi_2(\mathbf{r}) \, d\tau = 0 \tag{4.26}$$

for any two stationary states with energies $E_1 \neq E_2$ and suitable boundary conditions.

We have already encountered one special example of a stationary state. This is the state of the *free particle* of momentum $\mathbf{p}$ and consequently of energy $E = p^2/2\mu$, i.e., the plane wave

$$\psi(\mathbf{r}, t) = \exp\left(-iEt/\hbar\right) \exp\left(i\mathbf{p}\cdot\mathbf{r}/\hbar\right) \tag{4.27}$$

This wave function represents a state which has the definite value E for its energy. There is no uncertainty in the momentum and in the energy of this plane wave state; but there is, correspondingly, a complete ignorance of the position of the particle and of the transit time of the particle at a chosen position.

When conservative forces are introduced, the properties of momentum are, of course, drastically altered. But conservation of energy remains valid, as does—so we saw in Section 3.2—the relation according to which the

---

[8] The subscript E has been added in (4.25) to emphasize the time-independent factor.

energy equals $\hbar$ times the frequency. Since a stationary state (4.25) has the definite frequency $\omega = E/\hbar$, it follows that a stationary state is a state of well-defined energy, E being the definite value of its energy and not only its expectation value. This is to be understood in the sense that any determination of the energy of a particle which is in a stationary state always yields a particular value E and only that value. Such an interpretation conforms with the uncertainty relation $\Delta E \, \Delta t > \hbar$, which implies that a quantum state with a precise energy ($\Delta E = 0$) is possible only if there is unlimited time available to determine that energy. Stationary states are of just such nature in view of the constancy of $|\psi|$ in time.

The observed discrete energy levels of atoms and nuclei are approximated to a high degree of accuracy by stationary states. As will be seen in detail in Chapters 5 through 10, the simple regularity conditions which physical considerations lead us to impose on the wave function give rise in many cases to solutions of (4.23) which, instead of depending continuously on the parameter E, exist only for certain discrete values of E. This is entirely similar to the problem of a string with fixed end points, where the boundary conditions select the natural frequencies. In the quantum case, the relation $E = \hbar\omega$ associates a definite energy with every natural frequency, confirming our conclusion that the parameter E on which the stationary state wave function depends is actually *the* energy of the state. The values of E for which (4.23) has solutions consistent with the boundary conditions are said to constitute the *energy spectrum* of the Schrödinger equation. This spectrum may consist of isolated points (discrete energy levels) and continuous portions. If there is a lowest energy level, the corresponding stationary state is called the *ground state* of the system. All discrete higher energy levels are classified as *excited* states.

The reader will have noted the careful phrasing concerning the approximate nature of the connection between energy levels and stationary states. The point at issue here is that no excited state of an atom is truly discrete, stationary, and permanent. If it were, we would not see the spectral lines which result from its decay to some other level by photon emission. But the excited states are nevertheless stable to a high degree of approximation, for the period associated with an atomic state is of the order of the period of an electron moving in a Bohr orbit, i.e., about 10^{-15} sec, whereas the lifetime of the state is roughly 10^{-8} sec. Hence, during the lifetime of the state some 10^7 oscillations take place.[9] The situation is even more pronounced in the case of the so-called *unstable particles*, such as the lambda particle. Their natural periods are $\simeq 10^{-21}$ sec, their lifetimes $\simeq 10^{-10}$ sec. Thus they live for about 10^{11} periods. Comparing this with the fact that the earth has in its history completed some 10^9 revolutions around the sun—a motion which has always

[9] For a calculation of decay rates, see Chapter 22.

been considered one of the stablest—it can hardly be denied that the striking feature of the unstable particles is their stability.

5. General Solution of the Wave Equation. The stationary state wave functions are particular solutions of the wave equation. More general solutions of (4.1) can be constructed by superposition of such particular solutions. Summing over various allowed values of E, we get the solution

$$\psi(\mathbf{r}, t) = \sum_E c_E \exp\left(-\frac{i}{\hbar} Et\right) \psi_E(\mathbf{r}) \qquad (4.28)$$

Here as elsewhere in this book the summation may imply an integration if a part or all of the energy spectrum of E is continuous. Furthermore, the summation in (4.28) may include the same energy value more than once. This may occur if the Schrödinger equation has more than one linearly independent solution for the same E. E is then said to be *degenerate*.

The example of the free particle in one dimension illustrates these features. In this case (4.28) becomes

$$\psi(x, t) = \sum_E c_E \exp\left(-\frac{i}{\hbar} Et\right) \exp\left(\frac{i}{\hbar} px\right) \qquad (4.29)$$

There is degeneracy here, because, for any positive value of E, p can assume two values, $+\sqrt{2\mu E}$ and $-\sqrt{2\mu E}$. Since the energy spectrum is continuous and E can be any positive number, (4.29) must really be regarded as an integral. Negative values of E are excluded, because ψ would become singular for very large values of $|x|$. The exclusion of negative values of E on this ground is an example of the operation of the physically required boundary conditions.

Exercise 4.6. Show that (4.29) is equivalent to the Fourier integral (2.14).

Fourier analysis tells us that for any given initial wave packet, $\psi(x, 0)$, the expansion coefficients in (4.29) are uniquely defined. Hence, $\psi(x, t)$ in the form (4.29) is not only *a* solution of the wave equation

$$i\hbar \frac{\partial \psi}{\partial t} = -\frac{\hbar^2}{2\mu} \frac{\partial^2 \psi}{\partial x^2} \qquad (4.30)$$

but it is the *most general* solution which can be of physical utility when describing the motion of a free particle.

In generalization of this work we see that if the Schrödinger equation yields a set of solutions $\psi_E(\mathbf{r})$ which are *complete* in the sense that any initial state $\psi(\mathbf{r}, 0)$ of physical interest may be expanded in terms of them, i.e.,

$$\psi(\mathbf{r}, 0) = \sum_E c_E \psi_E(\mathbf{r}) \tag{4.31}$$

then we know automatically the solution of (4.1) for all t:

$$\psi(\mathbf{r}, t) = \sum_E c_E \exp\left(-\frac{i}{\hbar} Et\right) \psi_E(\mathbf{r}) \tag{4.32}$$

Thus, $\psi(\mathbf{r}, t)$ is a superposition of stationary states, if the initial wave function can be expanded in terms of solutions of the Schrödinger equation. This fact is responsible for the fundamental importance of stationary states. A discussion of the conditions under which an expansion such as (4.31) can be made will be given in Chapter 8. Here it suffices to state that the potential $V(x, y, z)$ appearing in the Schrödinger equation is usually such that the expandability of sufficiently general types of functions $\psi(\mathbf{r}, 0)$ can either be proved mathematically, as it can for the case $V = 0$, or it can be assumed to hold on physical grounds.

One more simple property of the wave equation (4.1), which can be discussed without specifying V in further detail, deserves to be mentioned. It is based on the observation that, if the complex conjugate of (4.1) is taken, and t replaced by $-t$, we get

$$i\hbar \frac{\partial \psi^*(\mathbf{r}, -t)}{\partial t} = -\frac{\hbar^2}{2\mu} \nabla^2 \psi^*(\mathbf{r}, -t) + V \psi^*(\mathbf{r}, -t) \tag{4.33}$$

provided only that V is real. This equation has the same form as (4.1). Hence, if $\psi(\mathbf{r}, t)$ is a solution of (4.1), $\psi^*(\mathbf{r}, -t)$ is also a solution. The latter is often referred to as the *time-reversed* solution of (4.1) with respect to $\psi(\mathbf{r}, t)$. The behavior of the wave equation exhibited by (4.33) is called its *invariance under time reversal*. For a stationary state, invariance under time reversal implies that, if $\psi_E(\mathbf{r})$ is a stationary wave function, $\psi_E^*(\mathbf{r})$ is also one. Hence, it follows that, if $\psi_E(\mathbf{r})$ is a nondegenerate solution of (4.23), $\psi_E(\mathbf{r})$ must be real, except for an arbitrary constant complex factor.

Exercise 4.7. Prove the statements of the last paragraph, and show that if degeneracy is present the solutions of the Schrödinger equation with real V may be chosen to be real.

Exercise 4.8. Prove that the Wronskian of two degenerate eigenfunctions of the one-dimensional Schrödinger equation is constant [see (4.6)].

6. *Approximate Solution of the Schrödinger Equation.* The Schrödinger equation can be regarded as the Euler-Lagrange equation for a variational problem which can be formulated as follows:

Find the functions $\psi(\mathbf{r})$ and $\psi^*(\mathbf{r})$, subject to the constraint

$$\int \psi^*(\mathbf{r})\psi(\mathbf{r})\,d\tau = 1 \tag{4.34}$$

which cause the variation of the expression

$$\langle H \rangle = \int \left(-\frac{\hbar^2}{2\mu}\,\psi^*\,\nabla^2\psi + V\psi^*\psi \right) d\tau = \int \left(\frac{\hbar^2}{2\mu}\,\nabla\psi^* \cdot \nabla\psi + V\psi^*\psi \right) d\tau \tag{4.35}$$

to vanish. It has been assumed that ψ and ψ^* satisfy certain regular boundary conditions to insure the equality of the two forms of the integral.

For an understanding of the procedure, it is sufficient to work with the one-dimensional case. According to the method of Lagrangian multipliers, the problem is equivalent to an unconstrained variational problem for the integral

$$\int F\,d\tau = \int \left(\frac{\hbar^2}{2\mu}\frac{d\psi^*}{dx}\frac{d\psi}{dx} + V\psi^*\psi - E\psi^*\psi \right) d\tau \tag{4.36}$$

where E is an as yet undetermined constant. If ψ and ψ^* are varied independently, with their conjugate relationship temporarily ignored, two differential (Euler) equations are obtained:[10]

$$\frac{\partial F}{\partial \psi^*} - \frac{d}{dx}\frac{dF}{\partial \psi^{*\prime}} = -\frac{\hbar^2}{2\mu}\frac{d^2\psi}{dx^2} + V\psi - E\psi = 0 \tag{4.37a}$$

$$\frac{\partial F}{\partial \psi} - \frac{d}{dx}\frac{\partial F}{\partial \psi'} = -\frac{\hbar^2}{2\mu}\frac{d^2\psi^*}{dx^2} + V\psi^* - E\psi^* = 0 \tag{4.37b}$$

Both of these have the form of the Schrödinger equation. Since ψ and ψ^* satisfy the appropriate regular boundary conditions, E is assured to be a real number. Hence, all equations are consistent if ψ^* is chosen as the complex conjugate of ψ.

The simultaneous use of ψ and ψ^* may seem like an unnecessary complication if V is real. However, it leads to a variational expression with a simple physical significance, since (4.35) denotes the expectation value of the energy. As the energy, this quantity has a lower bound, and the energy of the *ground state* of the system can therefore be determined as the absolute minimum of $\langle H \rangle$. Finally, the procedure described in this section is easily generalized to the Schrödinger equation for a particle in a magnetic field, which has no simple reality properties.

---

[10] For a review of the calculus of variations, see J. Mathews and R. L. Walker, *Mathematical Methods of Physics*, W. A. Benjamin, Inc., New York, 1964, Chapter 12.

The generalization to three dimensions is also straightforward. The solutions of the Schrödinger equation are those normalized functions ψ which render the expectation value (4.35) stationary (in the sense of the variational calculus) and equal to E. If a trial function $\psi + \delta\psi$, differing from an eigenfunction by $\delta\psi$, is used to calculate the expectation value of the energy, $E + \delta\langle H \rangle$ is obtained and, for small $\delta\psi$, the change $\delta\langle H \rangle$ is of order $(\delta\psi)^2$.

In practice, then, we can obtain various overestimates of the ground state energy by calculating $\langle H \rangle$ for suitably chosen trial functions. Obviously, altogether erroneous estimates are obtained unless the trial function is similar to the correct ground state wave function. Usually it is necessary to strike a compromise between the desire to improve the estimate of the ground state energy by choosing a "good" wave function and the requirement of ease of calculation.

The variational method is frequently applied to the Schrödinger equation of a complicated system by using for the calculation of $\langle H \rangle$ a trial wave function which contains one or more variable parameters $\alpha, \beta, \gamma, \ldots$. If the expectation value $\langle H \rangle$ is a differentiable function of these parameters, $\langle H \rangle$ is minimized (or maximized) with the help of the equations

$$\frac{\partial\langle H \rangle}{\partial\alpha} = \frac{\partial\langle H \rangle}{\partial\beta} = \frac{\partial\langle H \rangle}{\partial\gamma} = \cdots = 0$$

Evidently, the absolute minimum of $\langle H \rangle$ gives an upper limit for the lowest eigenvalue of H and may even be a fair approximation to this eigenvalue if the trial wave function is flexible enough to approximate the ground state eigenfunction ψ_0. The other, relative, extrema of $\langle H \rangle$ may correspond to excited states of the system, provided that the trial wave function is sufficiently adaptable.

As an important application, we suppose that a Schrödinger equation

$$-\frac{\hbar^2}{2\mu}\nabla^2\psi + V\psi = E\psi \tag{4.38}$$

with an unmanageable potential energy V is given, but that we know the normalized solutions of a similar Schrödinger equation

$$-\frac{\hbar^2}{2\mu}\nabla^2\psi^{(0)} + V_0\psi^{(0)} = E^{(0)}\psi^{(0)} \tag{4.39}$$

with V_0 differing from V by a small amount $\Delta V = V - V_0$. Owing to the variational property, the simple expression

$$\langle H \rangle = \int \psi^{(0)*}\left(-\frac{\hbar^2}{2\mu}\nabla^2 + V\right)\psi^{(0)}\, d\tau = E^{(0)} + \int \psi^{(0)*}\Delta V\psi^{(0)}\, d\tau \tag{4.40}$$

represents a sensible, and often quite accurate, estimate of the energy eigen-value of the Schrödinger equation (4.38). The difference ΔV is called a *perturbation* of the unperturbed potential energy V_0, and the change in the energy eigenvalue

$$\Delta E = E - E^{(0)} \approx \int \psi^{(0)*} \Delta V \psi^{(0)} \, d\tau \tag{4.41}$$

is seen to be approximately equal to the expectation value of the perturbation in the unperturbed state. The formula (4.41) is the most elementary and most fundamental result of quantum mechanical *perturbation theory*.

In every physical theory two trends are evident. On the one hand, we strive to formulate exact laws and equations which govern the phenomena; on the other hand, we are confronted with the need for obtaining more or less approximate solutions to the equations, because rigorous solutions can usually be found only for oversimplified models of the physical situation. They are nevertheless useful, because they often serve as a starting point for approximate solutions to the complicated equations of the actual system.

In quantum mechanics the perturbation theories are examples of this approach. Given a complex physical system, we choose, if possible, a simpler, similar comparison system whose quantum equations can be solved rigorously. The complicated actual system may then often be described to good approximation in terms of the familiar neighboring system.

Examples of the uses of perturbation theory are legion. For instance, in atomic physics the problem of the motion of an electron in a Coulomb field can be solved rigorously (Chapter 10), and we may regard the motion of an electron in a real many-electron atom as approximated by this simpler motion, perturbed by the interaction with the other electrons. Also, the change of the energy levels of an atom in an applied electric field can be calculated by treating the field as an added perturbation and the influence of an anharmonic term in the potential energy of an oscillator on the energy spectrum can be assessed by perturbation theory (Problem 3 in Chapter 5). In fact, in a large class of practical problems in the quantum domain pertur-bation theory provides at least a first qualitative orientation, even where its quantitative results may be inaccurate. The formula (4.41) provides our first introduction to perturbation theory. A more systematic treatment, including the calculation of higher-order approximations, will be given in later chapters.

Problem

1. Show that the addition of an imaginary part to the potential in the quantal wave equation describes the presence of sources or sinks of probability. (Work out the appropriate continuity equation.)

Solve the wave equation for a potential of the form $V = -V_0(1 + i\zeta)$, where V_0 and ζ are positive constants. If $\zeta \ll 1$, show that there are stationary state solutions which represent plane waves with exponentially attenuated amplitude, describing absorption of the waves. Calculate the absorption coefficient.

CHAPTER 5

The Linear Harmonic Oscillator

1. Preliminary Remarks. In order to learn how to solve the Schrödinger equation and to acquire a feeling for the peculiarities of nature at the quantum level, we proceed to treat a few simple problems. As usual in theoretical physics, a great mathematical simplification results from the assumption that ψ and V depend only on the x-coordinate. The Schrödinger equation corresponding to this one-dimensional problem is an ordinary differential equation:

$$-\frac{\hbar^2}{2\mu}\frac{d^2\psi(x)}{dx^2} + V(x)\psi(x) = E\psi(x) \tag{5.1}$$

We shall solve this equation for several special forms of the potential $V(x)$.

In this chapter the Schrödinger equation will be solved for the linear harmonic oscillator. Its potential energy is

$$V(x) = \tfrac{1}{2}\mu\omega^2 x^2 \tag{5.2}$$

We call ω loosely the (classical) frequency of the harmonic oscillator. Such a parabolic potential is actually of great practical importance, because it approximates any arbitrary potential in the neighborhood of a stable equilibrium position. $V(x)$ may be expanded in a Taylor series near $x = a$:

$$V(x) = V(a) + V'(a)(x - a) + \tfrac{1}{2}V''(a)(x - a)^2 + \cdots$$

If $x = a$ is a stable equilibrium position, $V(x)$ has a minimum at $x = a$, i.e., $V'(a) = 0$, and $V''(a) > 0$. If a is chosen as the coordinate origin and $V(a)$ as the zero of the energy scale, then (5.2) is the first approximation to $V(x)$. A familiar example is provided by the oscillations of the atoms in a diatomic molecule.

The linear harmonic oscillator is important for another reason. The behavior of most continuous physical systems, such as the vibrations of an elastic medium, or the electromagnetic field in a cavity, can be described by the superposition of an infinite number of simple harmonic oscillators. In the quantization of any such physical system we are then confronted by the quantum mechanics of many linear harmonic oscillators of various frequencies. For this reason, all modern field theories make use of the results which we are about to obtain.

The Schrödinger equation

$$-\frac{\hbar^2}{2\mu}\frac{d^2\psi}{dx^2} + \tfrac{1}{2}\mu\omega^2 x^2\psi = E\psi \qquad (5.3)$$

can be transformed into a convenient form, if we substitute:

$$\xi = \sqrt{\frac{\mu\omega}{\hbar}}\, x$$

We obtain[1]

$$\frac{d^2\psi}{d\xi^2} + \left(\frac{2E}{\hbar\omega} - \xi^2\right)\psi = 0 \qquad (5.4)$$

If a power series solution of this equation is attempted, a three-term recursion formula is obtained.[2] To get a differential equation whose power series solution admits a two-term recursion relation, which is simpler to analyze, we make the substitution

$$\psi = e^{-(\xi^2/2)}v(\xi) \qquad (5.5)$$

This yields the equation

$$\frac{d^2v}{d\xi^2} - 2\xi\frac{dv}{d\xi} + 2nv = 0 \qquad (5.6)$$

where n is defined by the relation

$$E = \hbar\omega(n + \tfrac{1}{2}) \qquad (5.7)$$

Exercise 5.1. Substituting power series with undetermined coefficients for ψ and v into (5.4) and (5.6), obtain the recursion relations and compare these.

2. Eigenvalues and Eigenfunctions. One simple and important property of (5.3) derives from the fact that V is an even function of x.[3] If the potential energy satisfies the condition $V(-x) = V(x)$, and if $\psi(x)$ is a solution of the

---

[1] By using the same symbol, ψ, in both (5.3) and (5.4) for two different functions of the variables x and ξ, we indulge in an inconsistency which, although deplorable, is sanctioned by custom.

[2] For a thorough discussion of (5.4), *Weber's equation*, see P. M. Morse and H. Feshbach, *Methods of Theoretical Physics*, McGraw-Hill Book Company, New York, 1953; also see M. Abramowitz and I. A. Stegun, *Handbook of Mathematical Functions*, National Bureau of Standards, Applied Mathematics Series 55, 1964.

[3] A function $f(x)$ is even if $f(-x) = f(x)$, and it is odd if $f(-x) = -f(x)$.

Schrödinger equation (5.1), then it follows that $\psi(-x)$ is also a solution of this equation. The Schrödinger equation with even V is said to be *invariant under reflection*, for if x is changed into $-x$, the equation is unaltered except for the replacement of $\psi(x)$ by $\psi(-x)$. Any linear combination of solutions of (5.1) also solves (5.1). Hence, if $\psi(x)$ is a solution of the Schrödinger equation, the following two functions must also be solutions:

$$\tfrac{1}{2}[\psi(x) + \psi(-x)] \tag{5.8a}$$

$$\tfrac{1}{2}[\psi(x) - \psi(-x)] \tag{5.8b}$$

These are the even and odd parts of $\psi(x)$. It is thus clear that in constructing the solutions of (5.1) we may confine ourselves to all even and all odd solutions. A state which is represented by a wave function that is even is said to have *even parity*; similarly, we speak of *odd parity* if $\psi(-x) = -\psi(x)$.

Exercise 5.2. Extend the notion of invariance under reflection (of all three Cartesian coordinates) to the Schrödinger equation in three dimensions. Show that if V depends merely on the radial distance r, only solutions of definite parity need be considered.

Since we are dealing with a linear second-order differential equation, it is easy to construct two linearly independent solutions by distinguishing two separate cases according to the "initial" conditions to be imposed:

$$(1) \quad v_1(0) = 1, \qquad v_1{}'(0) = 0$$
$$(2) \quad v_2(0) = 0, \qquad v_2{}'(0) = 1$$

v_1 is even in ξ, v_2 is odd. By substituting

$$v_1 = 1 + a_2\xi^2 + a_4\xi^4 + \cdots$$

or

$$v_2 = \xi + a_3\xi^3 + a_5\xi^5 + \cdots$$

into (5.6) and equating the coefficient of each power of ξ to zero (see Exercise 5.1), we obtain the power series expansions

$$v_1 = 1 - \frac{2n}{2!}\xi^2 + \frac{2^2 n(n-2)}{4!}\xi^4 - \frac{2^3 n(n-2)(n-4)}{6!}\xi^6 + \cdots \tag{5.9a}$$

$$v_2 = \xi - \frac{2(n-1)}{3!}\xi^3 + \frac{2^2(n-1)(n-3)}{5!}\xi^5 - \cdots \tag{5.9b}$$

The rule which governs these expansions is evident.

It follows that the general solution of (5.4) is

$$\psi = e^{-(\xi^2/2)}[Av_1(\xi) + Bv_2(\xi)] \tag{5.10}$$

In order to determine if such a solution can describe a physical state or not, we must consider the asymptotic behavior of v_1 and v_2. How do these functions behave for large values of $|\xi|$? For the purpose at hand we need not become involved in complicated estimates. Unless n is an integer both series are infinite, and it suffices to establish that asymptotically $v_1 \propto e^{+\xi^2}$ and $v_2 \propto \xi e^{+\xi^2}$. Indeed, for a fixed nonintegral value of n, all coefficients from a certain one on up have the same sign, and the ratio of the coefficient of ξ^k to that of ξ^{k-2} is

$$\frac{2(k - n - 2)}{k(k - 1)} \to \frac{2}{k} \quad \text{as} \quad k \to \infty$$

But this is precisely the ratio of the coefficients of ξ^k and ξ^{k-2} in the expansion of e^{ξ^2}

Hence, unless n is an integer, v_1 behaves as e^{ξ^2} for $\xi \to \pm \infty$. v_2, on the other hand, behaves as $\pm e^{\xi^2}$ for $\xi \to +\infty$ and as $\mp e^{\xi^2}$ for $\xi \to -\infty$. Consequently, there is no conceivable choice of the coefficients A and B which would prevent $\psi(\xi)$ in (5.10) from diverging as $e^{\xi^2/2}$ either for positive or for negative large values of ξ, or both. Such wave functions are physically not useful, because they describe a situation in which it is overwhelmingly probable that the particle is not "here" but at infinity. This behavior can only be avoided if n is an integer. In this case one of the two series, (5.9a) or (5.9b), terminates and becomes a polynomial of degree n. By setting the coefficient of the other one equal to zero in (5.10), we obtain physically meaningful solutions.

If n is even we get

$$\psi = e^{-(\xi^2/2)}v_1(\xi) \tag{5.11a}$$

and the state has even parity. If n is odd we get

$$\psi = e^{-(\xi^2/2)}v_2(\xi) \tag{5.11b}$$

and the state has odd parity. Both (5.11a) and (5.11b) are now finite everywhere and quadratically integrable.

We have come to a very important conclusion: n must be a nonnegative integer $(n = 0, 1, 2, \ldots)$; hence, E can assume only the discrete values $E_n = \hbar\omega(n + \frac{1}{2})$ or $E = \frac{1}{2}\hbar\omega, \frac{3}{2}\hbar\omega, \frac{5}{2}\hbar\omega, \ldots$. Classically, all nonnegative numbers are allowed for the energy of a harmonic oscillator. In quantum mechanics a stationary state of the harmonic oscillator can have only one of a discrete set of allowed energies! The energies are thus indeed quantized, and we may speak of a spectrum of *discrete energy levels* for the harmonic oscillator. The values of E or n which render the differential equation (5.3), (5.4), or (5.6) with its boundary conditions soluble are termed *eigenvalues*, and the corresponding solutions are called *eigenfunctions*. This terminology is of course not restricted to the linear harmonic oscillator but applies to the

Schrödinger equation in general. The eigenfunctions (5.11*a*) and (5.11*b*) strongly vanish at large distances, so that the particle is confined to the neighborhood of the center of force. The states described by such eigenfunctions are said to be *bound*. Note that the eigenvalues of the linear harmonic oscillator are not degenerate, since for each eigenvalue there exists only one eigenfunction, apart from an arbitrary constant factor.[4] That the eigenvalues are equidistant is one of the important peculiar features of the x^2 dependence of the oscillator potential.

The number n, whose integral values characterize and label the eigenfunctions and eigenvalues, is traditionally called a *quantum number*—a practical but not particularly sharply defined term, which we shall often use. $n = 0$ is the minimum value the quantum number n can assume and corresponds to the ground state, but the energy of the oscillator is still $\frac{1}{2}\hbar\omega$ and does not vanish as the lowest possible classical energy would. $\frac{1}{2}\hbar\omega$ is called the *zero point energy* of the harmonic oscillator. Being proportional to $\hbar$ it is obviously a quantum phenomenon, and indeed it can be understood on the basis of the uncertainty principle. Since the energy is $(p^2/2\mu) + \frac{1}{2}\mu\omega^2 x^2$, zero energy would mean vanishing p *and* vanishing x, hence a precise statement of both the coordinate and the momentum of the particle.[5] This would be in conflict with the uncertainty relation.

3. Study of the Eigenfunctions. In this section a few important mathematical properties of the harmonic oscillator eigenfunctions will be derived.

The finite polynomial solutions of (5.6) which can be constructed if n is an integer are known as *Hermite polynomials* of degree n. The complete eigenfunctions are of the form

$$\psi_n(x) = C_n H_n\left(\sqrt{\frac{\mu\omega}{\hbar}}\,x\right) \exp\left(-\frac{\mu\omega}{2\hbar}\,x^2\right) \tag{5.12}$$

where H_n denotes a Hermite polynomial of degree n, and C_n is an as yet undetermined normalization constant. But first the Hermite polynomials themselves must be normalized. It is traditional to define them so that the highest power of ξ appears with the coefficient 2^n. Hence, for even n

$$H_n(\xi) = (-1)^{n/2} \frac{n!}{(n/2)!} v_1(\xi) \tag{5.13a}$$

---

[4] This property is common to bound states in one-dimensional problems, provided that the potential is finite at finite values of x. Even if the particle is confined only from one side, there is still no degeneracy in one dimension.

[5] At first sight the argument given here may seem unsatisfactory, because it uses x and p as if they were classical variables, when really they are operators. However, the argument can be made rigorous by using expectation values. See Section 8.6 and Problem 4 in Chapter 8.

and for odd n

$$H_n(\xi) = (-1)^{(n-1)/2} \frac{2(n!)}{[(n-1)/2]!} v_2(\xi) \tag{5.13b}$$

Here is a list of the first few Hermite polynomials:

$$\begin{aligned}
H_0(\xi) &= 1 & H_3(\xi) &= -12\xi + 8\xi^3 \\
H_1(\xi) &= 2\xi & H_4(\xi) &= 12 - 48\xi^2 + 16\xi^4 \\
H_2(\xi) &= -2 + 4\xi^2 & H_5(\xi) &= 120\xi - 160\xi^3 + 32\xi^5
\end{aligned} \tag{5.14}$$

The first few harmonic oscillator eigenfunctions are plotted in Figure 5.1. They satisfy the differential equation

$$\frac{d^2 H_n}{d\xi^2} - 2\xi \frac{dH_n}{d\xi} + 2nH_n = 0 \tag{5.15}$$

Exercise 5.3. By differentiating (5.15), prove that

$$\frac{dH_n(\xi)}{d\xi} = 2nH_{n-1}(\xi) \tag{5.16}$$

A particularly simple representation of the Hermite polynomials is obtained by constructing the function

$$F(s, \xi) = \sum_{n=0}^{\infty} \frac{H_n(\xi)}{n!} s^n \tag{5.17}$$

As a consequence of (5.16), we see that

$$\frac{\partial F}{\partial \xi} = 2sF$$

This differential equation can be integrated:

$$F(s, \xi) = F(s, 0)e^{2s\xi}$$

The coefficient $F(s, 0)$ can be evaluated from (5.17) and (5.13):

$$F(s, 0) = \sum_{k=1}^{\infty} \frac{(-1)^k}{k!} s^{2k} = e^{-s^2}$$

and therefore the *generating function* $F(s, \xi)$ has the form

$$F(s, \xi) = e^{-s^2 + 2s\xi} = e^{\xi^2 - (s-\xi)^2} \tag{5.18}$$

The generating function $F(\xi, s)$ is very useful because it allows us to deduce a number of simple properties of the harmonic oscillator wave functions with

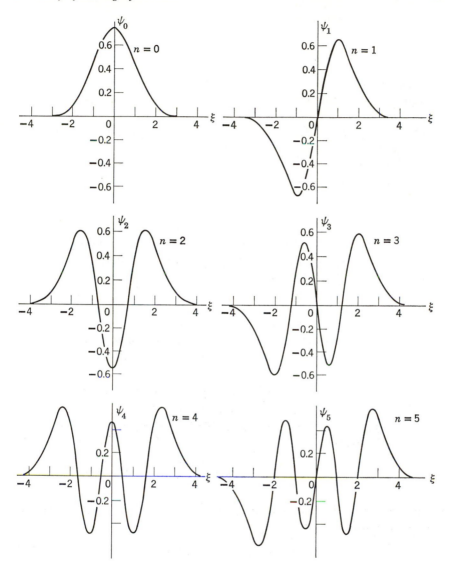

Figure 5.1. The eigenfunctions of the linear harmonic oscillator for the quantum numbers $n = 0$ to 5. ψ_n is plotted versus $\xi = \sqrt{\mu\omega/\hbar}\, x$, and all eigenfunctions are normalized as $\displaystyle\int_{-\infty}^{+\infty} |\psi_n(\xi)|^2 \, d\xi = 1$.

great ease. For example, by Taylor's expansion:

$$H_n(\xi) = \left[\frac{d^n}{ds^n} e^{\xi^2-(s-\xi)^2}\right]_{s=0} = (-1)^n e^{\xi^2} \frac{d^n}{d\xi^n} e^{-\xi^2} \tag{5.19}$$

a popular alternative form of definition of the Hermite polynomials.

From this definition it follows that all n roots of $H_n(\xi)$ must be real.

Proof. Assume that $d^{n-1}e^{-\xi^2}/d\xi^{n-1}$ has $n-1$ real roots. Since $e^{-\xi^2}$ and all its derivatives tend to zero as $\xi \to +\infty$ and $-\infty$, the derivative $d^n e^{-\xi^2}/d\xi^n$ must have at least n real roots. Being $e^{-\xi^2}$ times a polynomial, it can have no more than n such roots. The assumption holds for $n=1$, whence the assertion follows by induction.

The points in space at which a wave function goes through zero are called its *nodes*. Evidently, the oscillator eigenfunction (5.12) has n nodes.

A very important property of the Hermite polynomials is their orthogonality in the sense that for $n \neq k$

$$\int_{-\infty}^{+\infty} H_n(\xi)H_k(\xi)e^{-\xi^2} \, d\xi = 0 \tag{5.20}$$

It is convenient and economical to prove a rather more general formula by the use of the generating function. To this end we construct the expression

$$\int_{-\infty}^{+\infty} e^{\xi^2-(s-\xi)^2} e^{\xi^2-(t-\xi)^2} e^{2\lambda\xi-\xi^2} \, d\xi = \sum_{n=0}^{\infty} \sum_{k=0}^{\infty} \frac{s^n t^k}{n!\,k!} \int_{-\infty}^{+\infty} H_n(\xi)H_k(\xi)e^{2\lambda\xi-\xi^2} \, d\xi$$

where (5.17) and (5.18) have been used. The left-hand side can be integrated explicitly; it equals

$$e^{2st+\lambda^2+2\lambda(s+t)} \int_{-\infty}^{+\infty} e^{-(s+t+\lambda-\xi)^2} \, d\xi = \sqrt{\pi} \, e^{\lambda^2+2(st+\lambda s+\lambda t)}$$

Comparing the coefficients of equal powers of $s^n t^k \lambda^p$, we obtain the value of a useful integral:

$$\int_{-\infty}^{+\infty} H_n(\xi)H_k(\xi)e^{-\xi^2}\xi^p \, d\xi \tag{5.21}$$

In particular, for $p=0$ and $n \neq k$, the orthogonality relation (5.20) is derived. For $p=0$ and $n=k$ we obtain the integral

$$\int_{-\infty}^{+\infty} [H_n(\xi)]^2 e^{-\xi^2} \, d\xi = 2^n n! \sqrt{\pi} \tag{5.22}$$

If we recall that $\xi = \sqrt{\mu\omega/\hbar}\,x$, (5.20) and (5.22) can be written more simply in terms of the normalized eigenfunctions (5.12), and we have

$$\int_{-\infty}^{+\infty} \psi_n{}^*(x)\, \psi_k(x)\, dx = \delta_{nk} \tag{5.23}$$

if the normalization constant is chosen properly. A set of eigenfunctions satisfying relation (5.23) is said to be *orthonormal*. The orthonormal oscillator eigenfunctions are

$$\psi_n(x) = 2^{-n/2}(n!)^{-1/2}\left(\frac{\mu\omega}{\hbar\pi}\right)^{1/4} \exp\left(-\frac{\mu\omega}{2\hbar}\, x^2\right) H_n\left(\sqrt{\frac{\mu\omega}{\hbar}}\, x\right) \tag{5.24}$$

In (5.23) complex conjugation appears, although with the particular choice of the arbitrary phase factor embodied in (5.24) the eigenfunctions are real. However, in the form given, (5.23) is an example of the general orthogonality relation (4.26) for the eigenfunctions of the Schrödinger equation.

4. The Motion of Wave Packets. So far we have considered only the stationary states of the harmonic oscillator. We must now turn our attention to the behavior of a general wave $\psi(x, t)$ whose initial form $\psi(x, 0)$ is given. The wave equation

$$i\hbar\, \frac{\partial\psi(x, t)}{\partial t} = -\frac{\hbar^2}{2\mu}\, \frac{\partial^2\psi(x, t)}{\partial x^2} + \tfrac{1}{2}\mu\omega^2 x^2 \psi(x, t)$$

determines the time development of the wave. In Chapter 4 we saw that the solution of this equation can be obtained automatically, if the initial wave can be expanded in terms of the time-independent eigenfunctions of the corresponding Schrödinger equation. For the linear harmonic oscillator this is indeed possible to an approximation which can be made as close as we please.

To see this, we must ask whether an arbitrary initial wave function $\psi(x, 0)$ can be expanded as

$$\psi(x, 0) = \sum_{n=0}^{\infty} c_n C_n \exp\left(-\frac{\mu\omega}{2\hbar}\, x^2\right) H_n\left(\sqrt{\frac{\mu\omega}{\hbar}}\, x\right)$$

where C_n denotes the normalization constant given in (5.24).

It is convenient to transfer the exponential factor to the left side of the equation and to look at the function

$$\psi(x, 0) \exp\left(\mu\omega x^2/2\hbar\right) \tag{5.25}$$

Since x^n can always be expressed as a linear combination of the first $n + 1$ Hermite polynomials, it is seen that any function (5.25) which is analytic (i.e., can be expanded in a power series) can also be expanded in terms of Hermite polynomials.

Actually, a much larger class of functions can be expanded as

$$\psi(x, 0) = \sum_{n=0}^{\infty} c_n \psi_n(x) \tag{5.26}$$

if the notion of convergence is suitably relaxed. This will be discussed in Chapter 8. In any event, a finite number of terms in (5.26) approximates most physically pertinent initial wave packets very well.[6] Owing to the orthonormality, (5.26) gives us

$$c_n = \int_{-\infty}^{+\infty} \psi_n{}^*(x)\psi(x, 0)\, dx \tag{5.27}$$

If $\psi(x, 0)$ is normalized, the orthonormality of the eigenfunctions causes the expansion coefficients to satisfy the simple normalization relation

$$\sum_{n=0}^{\infty} |c_n|^2 = 1 \tag{5.28}$$

With these expansion coefficients and with $E_n = \hbar\omega(n + \tfrac{1}{2})$ we can now construct the wave packet at time t by the use of (4.28):

$$\psi(x, t) = \exp\left(-\frac{i\omega}{2}t\right) \sum_{n=0}^{\infty} c_n \psi_n(x) \exp\left(-in\omega t\right) \tag{5.29}$$

The expectation value of the energy in this state can be calculated from (4.18):

$$\langle H \rangle = \int_{-\infty}^{+\infty} \psi^*(x, t)\left(-\frac{\hbar^2}{2\mu}\frac{d^2}{dx^2} + \tfrac{1}{2}\mu\omega^2 x^2\right)\psi(x, t)\, dx \tag{5.30}$$

Substituting the expansion (5.29) into (5.30) and using the Schrödinger equation

$$\left(-\frac{\hbar^2}{2\mu}\frac{d^2}{dx^2} + \tfrac{1}{2}\mu\omega^2 x^2\right)\psi_n(x) = E_n\psi_n(x)$$

as well as the orthonormality relation (5.23), we obtain

$$\langle H \rangle = \sum_{n=0}^{\infty} E_n |c_n|^2 \tag{5.31}$$

Equations (5.28) and (5.31) taken together suggest an important and straightforward extension of the probability interpretation of quantum mechanics. These relations are in accord with the assumptions that

---

[6] The ambiguous phrase ". . . approximates . . . physically pertinent . . . very well" will be made precise in Chapter 8.

(a) the only possible results of measuring the energy of the system are the energy values $E_0, E_1, \ldots, E_n, \ldots$, and

(b) if the system is in state $\psi(x, t)$, the probability of obtaining the result E_n is equal to $|c_n|^2$.

These assumptions, which will find their proper place later in the general formulation of quantum mechanics, can be given further support by considering the motion of the wave packet in more detail.

The center of probability of the normalized wave packet, i.e., the expectation value of the position operator x, is according to (4.11)

$$\langle x \rangle = \int_{-\infty}^{+\infty} x \, |\psi(x, t)|^2 \, dx \tag{5.32}$$

Substituting (5.29) into (5.32) we find

$$\langle x \rangle = \sum_{n=0}^{\infty} \sum_{k=0}^{\infty} c_n^* c_k e^{i(n-k)\omega t} \int_{-\infty}^{+\infty} \psi_n^*(x) \, x \, \psi_k(x) \, dx \tag{5.33}$$

We see here that it is important to know the numbers

$$x_{nk} = \int_{-\infty}^{+\infty} \psi_n^*(x) \, x \, \psi_k(x) \, dx \tag{5.34}$$

As we shall see later in detail, the square array

$$\begin{pmatrix} x_{00} & x_{01} & x_{02} & \cdot & \cdot \\ x_{10} & x_{11} & x_{12} & \cdot & \cdot \\ x_{20} & x_{21} & x_{22} & \cdot & \cdot \\ x_{30} & \cdot & & \cdot & \\ \cdot & \cdot & \cdot & \cdot & \cdot \end{pmatrix} \tag{5.35}$$

formed by these numbers can be used conveniently as a matrix representing the operator x. In anticipation of these results we call the quantities x_{nk} the *matrix elements* of x.

Let us evaluate these matrix elements explicitly. Evidently

$$x_{nk} = 2^{-(n+k)/2}(n!)^{-1/2}(k!)^{-1/2}\left(\frac{\mu\omega}{\hbar\pi}\right)^{1/2}$$

$$\times \int_{-\infty}^{+\infty} \exp\left(-\frac{\mu\omega}{\hbar} x^2\right) H_n\left(\sqrt{\frac{\mu\omega}{\hbar}} x\right) x \, H_k\left(\sqrt{\frac{\mu\omega}{\hbar}} x\right) dx$$

$$= \sqrt{\frac{\hbar}{\mu\omega}} \, 2^{-(n+k)/2}(n!)^{-1/2}(k!)^{-1/2}\pi^{-1/2}\int_{-\infty}^{+\infty} H_n(\xi)H_k(\xi)\xi e^{-\xi^2} \, d\xi$$

By specializing (5.21) to $p = 1$, we obtain

$$\int_{-\infty}^{+\infty} H_n(\xi)H_k(\xi)\xi e^{-\xi^2}\, d\xi = \sqrt{\pi}\, 2^{n-1} n!\, [\delta_{k,n-1} + 2(n+1)\delta_{k,n+1}]$$

Hence,

$$x_{nk} = \sqrt{\frac{\hbar}{\mu\omega}} \left[\sqrt{\frac{n}{2}}\, \delta_{k,n-1} + \sqrt{\frac{n+1}{2}}\, \delta_{k,n+1} \right] \tag{5.36}$$

Using matrix notation and the rule that, if every element of a matrix is multiplied by the same number, this common factor can be taken out, the array (5.35) can be written:

$$(x) = \sqrt{\frac{\hbar}{2\mu\omega}} \begin{pmatrix} 0 & \sqrt{1} & 0 & 0 & \cdot \\ \sqrt{1} & 0 & \sqrt{2} & 0 & \cdot \\ 0 & \sqrt{2} & 0 & \sqrt{3} & \cdot \\ 0 & 0 & \sqrt{3} & 0 & \cdot \\ \cdot & \cdot & \cdot & \cdot & \cdot \end{pmatrix} \tag{5.37}$$

Two features are remarkable: Most matrix elements of (5.37) vanish, and (5.37) is a real symmetric matrix. Of these properties the first is peculiar to the harmonic oscillator, whereas the second is an example of a very general property (see Chapter 8).

Exercise 5.4. Using Exercise 5.3 and (5.23) and (5.36), calculate the matrix elements of the momentum operator,

$$p_{nk} = \int_{-\infty}^{+\infty} \psi_n{}^* \frac{\hbar}{i} \frac{d\psi_k}{dx}\, dx$$

and compare the result with the matrix for x.

Substituting the result (5.36) into (5.33), we obtain

$$\langle x \rangle = \sqrt{\frac{\hbar}{2\mu\omega}} \sum_{n=1}^{\infty} \sqrt{n}(c_n{}^* c_{n-1} e^{i\omega t} + c_{n-1}^* c_n e^{-i\omega t}) \tag{5.38}$$

If we set

$$c_n = |c_n| e^{i\varphi_n}$$

we can write

$$\langle x \rangle = \sqrt{\frac{2}{\mu\omega^2}} \sum_{n=0}^{\infty} \sqrt{n\hbar\omega}\, |c_n|\, |c_{n-1}| \cos(\omega t + \varphi_{n-1} - \varphi_n) \tag{5.39}$$

This expression is exact.

Exercise 5.5. By using the results of Section 4.3 show directly that for the linear harmonic oscillator $\langle x \rangle$ satisfies the classical equation of motion

$$\frac{d^2\langle x \rangle}{dt^2} + \omega^2 \langle x \rangle = 0$$

Solve this equation and show that it leads to the general form of (5.39).

If $\varphi_{n-1} - \varphi_n = \alpha$ is independent of n, if $|c_{n-1}| \simeq |c_n|$, and if c_n is appreciably different from zero only for large n, so that $E_n \simeq n\hbar\omega$, then we have approximately

$$\langle x \rangle \simeq \sqrt{\frac{2}{\mu\omega^2}} \left(\sum_{n=0}^{\infty} \sqrt{E_n} \, |c_n|^2 \right) \cos(\omega t + \alpha) \tag{5.40}$$

Equation (5.40) bears a close resemblance to the classical equation for a harmonic oscillator. If E is the energy of a classical oscillator, its instantaneous position $x(t)$ is

$$x(t) = \sqrt{\frac{2E}{\mu\omega^2}} \cos(\omega t + \alpha) \tag{5.41}$$

We see that the quantum equation (5.40) will relate expectation values and, in the correspondence limit of large quantum numbers, go over into the classical equations (5.41), if we write

$$\langle \sqrt{E} \rangle = \sum_{n=0}^{\infty} \sqrt{E_n} \, |c_n|^2 \tag{5.42}$$

But this is precisely what is to be expected, if the only allowed definite energy values of a harmonic oscillator are the eigenvalues E_n, and if $|c_n|^2$ is the probability for the occurrence of E_n in the given state [see assumptions (a) and (b) above]. The expectation value of any function of the energy would be

$$\langle f(E) \rangle = \sum_{n=0}^{\infty} f(E_n) |c_n|^2 \tag{5.43}$$

and (5.42) is a special case of this for $f(E) = \sqrt{E}$.

Exercise 5.6. Calculate the expectation value of momentum $\langle p \rangle$ for a wave packet in a linear harmonic oscillator. Make approximations similar to those leading to (5.40), and compare again with the appropriate classical equation.

5. The Double Oscillator. Preliminary Remarks. So far we have discussed in this chapter the simple linear harmonic oscillator. Now we shall supplement this discussion by study of a more complicated potential, pieced together from two harmonic oscillators. The discontinuities which arise in the shape of this potential will lead to a consideration of the question of boundary conditions whose general importance transcends the special example which we have chosen to illustrate them.

The example comes from molecular physics. There we frequently encounter motion in the neighborhood of a stable equilibrium configuration, approximated by a harmonic potential. To be sure, one-dimensional models are of limited utility. Even diatomic molecules rotate in space, besides vibrating along a straight line. Nevertheless, important qualitative features can be exhibited with a linear model, and some quantitative estimates can also be obtained.

Let us consider, as a model, two masses μ_1 and μ_2, constrained to move in a straight line and connected with each other by a spring whose force constant is k and whose length at equilibrium is a. If x_1 and x_2 are the coordinates of the two mass points, and p_1 and p_2 their momenta, it is shown in classical mechanics that the nonrelativistic two-body problem can be separated into the trivial motion of the center of mass and an equivalent one-body motion, executed by a particle of mass $\mu = \mu_1\mu_2/(\mu_1 + \mu_2)$ with a coordinate $x = x_1 - x_2$ about a fixed center under the action of the elastic force. Only the relative motion of the reduced mass μ will be considered in this chapter.[7]

The wave equation corresponding to the equivalent one-body problem is

$$ i\hbar \frac{\partial \psi(x,\,t)}{\partial t} = -\frac{\hbar^2}{2\mu} \frac{\partial^2 \psi(x,\,t)}{\partial x^2} + \tfrac{1}{2}k(|x| - a)^2 \psi(x,\,t) \tag{5.44}$$

If $a = 0$, (5.44) reduces to the wave equation for the simple linear harmonic oscillator. If $a \neq 0$, it is almost the equation of a harmonic oscillator whose equilibrium position is shifted by an amount a, but it is not quite that. For it is important to note that the *absolute value* of x appears in the potential energy $V = \tfrac{1}{2}k(|x| - a)^2$, if Hooke's law is assumed to hold for all values of the interparticle coordinate x. As shown in Figure 5.2, we have two parabolic potentials, one when particle 1 is to the right of particle 2 ($x > 0$), the other when the particles are in reverse order ($x < 0$). The parabolas are joined at $x = 0$ with the common value $V(0) = V_0 = \tfrac{1}{2}ka^2$. Classically, if $E < V_0$, we can assume that only one of these potential wells is present, for no penetration of barrier is possible. In quantum mechanics, even if $E < V_0$ the wave function may have a finite value at $x = 0$, which measures the

---

[7] The separation of the relative motion from the motion of the center of mass will be effected in Section 15.5. If the forces between two bodies are equal in magnitude and opposite in direction, classically the relative motion can be replaced by an equivalent one-body motion of the reduced mass. The correspondence principle suggests that these general dynamical features survive the transition to quantum mechanics.

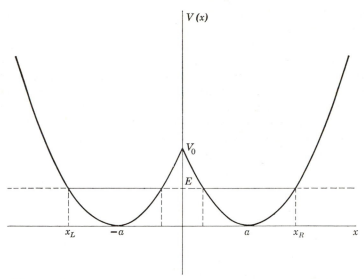

Figure 5.2. $V(x)$ for a double oscillator. x_L and x_R are the external classical turning points. If $E < V_0$, there are also two internal turning points, but the tunnel effect permits penetration of the barrier.

probability that the two particles are found in the same place; hence, the barrier can be penetrated.

The Schrödinger equation corresponding to (5.44) is

$$-\frac{\hbar^2}{2\mu}\frac{d^2\psi(x)}{dx^2} + \tfrac{1}{2}(k\,|x| - a)^2\psi(x) = E\psi(x) \qquad (5.45)$$

This equation must be supplemented by boundary conditions at $x \to \pm\infty$. For $|x| \gg a$, (5.45) approaches the Schrödinger equation (5.3) for the simple harmonic oscillator; hence, the considerations of Section 5.2 may be applied, and the physically acceptable eigenfunctions must be required to vanish as $|x| \to \infty$.

Before attempting to solve (5.45), we note that as the parameter a is varied from 0 to ∞, the potential changes from the limit (I) of a single harmonic oscillator well to the other limit (II) of two separate oscillator wells, divided by an infinitely high and broad potential wall. In case I we have nondegenerate energy eigenvalues

$$E_n = \hbar\omega(n + \tfrac{1}{2}) = \hbar\sqrt{\frac{k}{\mu}}(n + \tfrac{1}{2}) \qquad (n = 0, 1, 2 \cdots) \qquad (5.46)$$

In case II the energy values are the same as given by (5.46), but each is doubly degenerate, since the system may occupy an eigenstate of either the harmonic oscillator well on the left or the one on the right. As *a* is varied, energies and eigenfunctions must change continuously from limiting case I to limiting case II. It is customary to call this kind of continuous change of an external parameter an *adiabatic* change of the system. It is useful to note that as the potential is being distorted continuously certain features of the eigenfunctions remain unaltered. Such features are known as *adiabatic invariants*.

An example is provided by the number of zeros, or nodes, of the eigenfunctions. If an eigenfunction has *n* nodes, as the eigenfunction of potential I belonging to the eigenvalue E_n does, this number cannot change in the course of the transition to II. We prove this assertion in two steps:

(a) No two adjacent nodes disappear by coalescing as *a* changes, nor can new nodes ever be created by the inverse process. If two nodes did coalesce at $x = x_0$, the extremum of ψ between them would also have to coincide with the nodes. Hence, both ψ and its first derivative ψ' would vanish at this point. By the differential equation (5.45) this would imply that ψ'', ψ''', and all higher derivatives also vanish at this point. But a function all of whose derivatives vanish can only be $\psi \equiv 0$.[8]

(b) No node can wander off to or in from infinity as *a* changes. To show this we only need to prove that all nodes are confined to the region between the two extreme classical turning points of the motion, i.e., the region between x_L and x_R in Figure 5.2. Classically, the coordinate of a particle with energy E is restricted by $x_L \leqslant x \leqslant x_R$, where x_L and x_R are the smallest and largest roots respectively of $V(x) = E$. From Schrödinger's equation we infer

$$\int_{-\infty}^{x} \psi^*(x')[E - V(x')]\psi(x')\, dx'$$

$$= -\frac{\hbar^2}{2\mu}\left(\left.\psi^*(x')\psi'(x')\right|_{-\infty}^{x} - \int_{-\infty}^{x} |\psi'(x')|^2\, dx'\right) \quad (5.47)$$

Let $x \leqslant x_L$. Then the integral on the left is negative definite, hence

$$\psi^*(x)\psi'(x) > 0 \quad \text{if } x \leqslant x_L \quad (5.48)$$

Equation (5.48) holds similarly if $x \geqslant x_R$. It follows that outside the classical region there can be no node and no extremum. This can also be seen directly from the Schrödinger equation, since ψ''/ψ is positive outside the classical region (see Figure 6.4).

Being an adiabatic invariant, the number of nodes *n* characterizes the eigenfunctions of the double oscillator for any value of *a*. Figure 5.3 shows

---

[8] It will be shown in Section 5.6 that ψ and ψ' must be continuous. Hence, the presence of isolated singularities in V (as at $x = 0$ for the double oscillator) does not affect this conclusion.

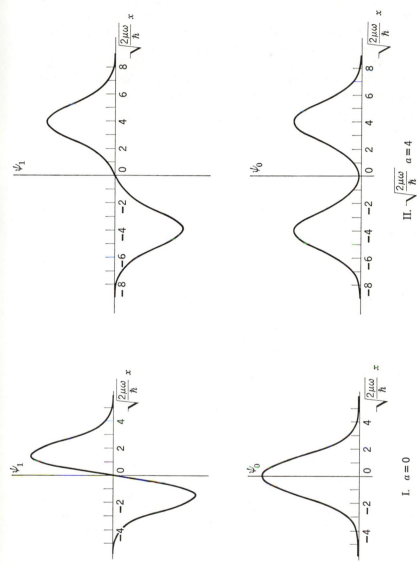

Figure 5.3. The two lowest energy eigenfunctions ψ_0 and ψ_1. In case I ($a = 0$) we have a simple harmonic oscillator, no barrier, and $E_1 - E_0 = \hbar\omega$. In case II ($\sqrt{2\mu\omega}/\hbar\, a = 4$) the double oscillator has a high barrier, and $E_1 - E_0 \approx 10^{-3}\hbar\omega$. The number of nodes characterizes the eigenfunctions.

this for the two lowest energy eigenvalues ($n = 0, 1$). Notice also that the eigenfunctions maintain their parity as a varies, since (5.45), owing to the presence of $|x|$ in the potential energy, is invariant under reflection.

The two eigenfunctions ψ_0 and ψ_1 correspond to $E = \frac{1}{2}\hbar\omega$ and $\frac{3}{2}\hbar\omega$ if $a = 0$ (case I). For $a \to \infty$ (case II) they become degenerate with the common energy $\frac{1}{2}\hbar\omega$. When a is very large, the linear combinations $\psi_0 + \psi_1$ and $\psi_0 - \psi_1$ are *approximate* eigenfunctions corresponding to the wave function being concentrated at $x = +a$ and at $x = -a$ respectively.

6. The Double Oscillator. Eigenvalues and Eigenfunctions. The rigorous solution of (5.45) is facilitated by introducing for positive x

$$z = \left(\frac{4\mu k}{\hbar^2}\right)^{1/4}(x - a) = \left(\frac{2\mu\omega}{\hbar}\right)^{1/2}(x - a) \qquad (x > 0)$$

and

$$E = \hbar\omega(\nu + \tfrac{1}{2}) \tag{5.49}$$

We obtain for positive x the differential equation

$$\frac{d^2\psi}{dz^2} + \left(\nu + \frac{1}{2} - \frac{z^2}{4}\right)\psi = 0 \tag{5.50}$$

For negative x the same equation is obtained except that now the substitution

$$z' = \left(\frac{2\mu\omega}{\hbar}\right)^{1/2}(x + a) \qquad (x < 0)$$

must be used. The differential equation valid for negative x is

$$\frac{d^2\psi}{dz'^2} + \left(\nu + \frac{1}{2} - \frac{z'^2}{4}\right)\psi = 0 \tag{5.51}$$

which has the same form as (5.50). For $a = 0$, the case of the simple linear harmonic oscillator, $z' = z$, and the two equations become identical. The boundary condition $\psi \to 0$ as $x \to \pm\infty$ implies that we must seek solutions of (5.50) and (5.51) which vanish as $z \to +\infty$ and $z' \to -\infty$.

Instead of proceeding to a detailed power series treatment of this differential equation, we refer to a standard treatise on mathematical analysis[2] for the solutions of (5.50). The particular solution of (5.50) which vanishes for very large positive values of z is called a *parabolic cylinder function*. It is denoted by $D_\nu(z)$ and defined as

$$D_\nu(z) = 2^{\nu/2}e^{-(z^2/4)}\left[\frac{\Gamma(\frac{1}{2})}{\Gamma[(1 - \nu)/2]}\,_1F_1\left(-\frac{\nu}{2}; \frac{1}{2}; \frac{z^2}{2}\right)\right.$$
$$\left. + \frac{z}{\sqrt{2}}\frac{\Gamma(-\frac{1}{2})}{\Gamma(-\nu/2)}\,_1F_1\left(\frac{1 - \nu}{2}; \frac{3}{2}; \frac{z^2}{2}\right)\right] \tag{5.52}$$

$_1F_1$ is the *confluent hypergeometric* (or Kummer) *function*. Its power series expansion is

$$_1F_1(a\,;c\,;z) = 1 + \frac{a}{c}\frac{z}{1!} + \frac{a(a+1)}{c(c+1)}\frac{z^2}{2!} + \cdots \tag{5.53}$$

If z is large and *positive* ($z \gg 1$ and $z \gg |\nu|$),

$$D_\nu(z) \sim e^{-(z^2/4)}z^\nu \left[1 - \frac{\nu(\nu-1)}{2z^2} + \frac{\nu(\nu-1)(\nu-2)(\nu-3)}{2\cdot 4\cdot z^4} \pm \cdots \right] \tag{5.54}$$

and if z is large and *negative* ($z \ll -1$ and $z \ll -|\nu|$),

$$D_\nu(z) \sim e^{-(z^2/4)}z^\nu \left[1 - \frac{\nu(\nu-1)}{2z^2} \pm \cdots \right]$$

$$- \frac{\sqrt{2\pi}}{\Gamma(-\nu)} e^{\nu\pi i}e^{z^2/4}z^{-\nu-1}\left[1 + \frac{(\nu+1)(\nu+2)}{2z^2} + \cdots\right] \tag{5.55}$$

Although the series in the brackets all diverge for any finite value of z, (5.54) and (5.55) are useful *asymptotic* expansions of $D_\nu(z)$.

Exercise 5.7. Using the identity

$$\Gamma(1+\nu)\Gamma(-\nu) = -\frac{\pi}{\sin \nu\pi} \tag{5.56}$$

obtain the eigenvalues of the simple harmonic oscillator from (5.55). Show the connection between the parabolic cylinder functions (5.52) and the eigenfunctions (5.12).

If $D_\nu(z)$ is a solution of (5.50), $D_\nu(-z)$ is also a solution of the same equation, and these two solutions are linearly independent unless ν is a nonnegative integer. Inspection of the asymptotic behavior shows that the particular solution of (5.51) which vanishes for very large negative values of z' must be $D_\nu(-z')$.

It follows that a double oscillator eigenfunction must be proportional to $D_\nu(z)$ for positive values of x, and proportional to $D_\nu(-z')$ for negative values of x. It remains to join these two solutions at $x = 0$, the point where the two parabolic potentials meet with discontinuous slope. This problem leads to a re-examination of the Schrödinger equation if the potential is nonanalytic at an isolated point x_0. We assume that such a singular potential function can be approximated by a regular one; the former can be regarded as the limiting case of the latter. Since the Schrödinger equation is a second-order differential equation, ψ and its first derivative must be continuous everywhere. Alternatively, if $x = x_0$ is the point of singularity, we get by integrating the Schrödinger equation from $x = x_0 - \varepsilon$ to $x = x_0 + \varepsilon$:

$$\psi'(x_0 + \varepsilon) - \psi'(x_0 - \varepsilon) = \int_{x_0-\varepsilon}^{x_0+\varepsilon} \frac{2\mu}{\hbar^2} [V(x) - E]\psi(x) \, dx$$

As long as $V(x)$ is finite, whether analytic or not, this equation implies that ψ' is continuous across the singularity. Hence, ψ itself must also be continuous.

Thus the joining condition to be assumed at a singular point of the potential, if V has a finite value, is that the wave function and its slope must be matched continuously. The probability current density, which is made up of ψ and ψ', must also be continuous—an evident physical requirement, if there are no sources and sinks of probability, i.e., of particles.

Armed with this knowledge we return to the double oscillator. Since it is invariant under reflection ($x \rightarrow -x$), the eigenfunctions can be assumed to have definite parity, even or odd. If an even function of x has a continuous slope at $x = 0$, as the joining condition requires, that slope must be zero. On the other hand, if an odd function of x is continuous at the origin, it must vanish there. Matching ψ and ψ' at $x = 0$ thus leads to

$$D_\nu{}'\left(-\sqrt{\frac{2\mu\omega}{\hbar}}\, a\right) = 0 \tag{5.57}$$

for even ψ, and

$$D_\nu\left(-\sqrt{\frac{2\mu\omega}{\hbar}}\, a\right) = 0 \tag{5.58}$$

for odd ψ. These are transcendental equations for ν.

Exercise 5.8. Show that if $a = 0$, the roots of (5.57) and (5.58) give the eigenvalues of the simple harmonic oscillator.

In general, it is difficult to calculate the roots ν of (5.57) and (5.58). Figure 5.4 shows how the lowest eigenvalues depend on the parameter $\sqrt{2\mu\omega/\hbar}\, a$. A few of the corresponding eigenfunctions are plotted in Figure 5.3. The unnormalized eigenfunctions are

$$\psi(x) = \begin{cases} D_\nu\left(\sqrt{\dfrac{2\mu\omega}{\hbar}}\,(x - a)\right) & (x \geqslant 0) \\[3mm] \pm D_\nu\left(-\sqrt{\dfrac{2\mu\omega}{\hbar}}\,(x + a)\right) & (x \leqslant 0) \end{cases} \tag{5.59}$$

where the upper (lower) sign is to be used if ν is a root of (5.57) [(5.58)].

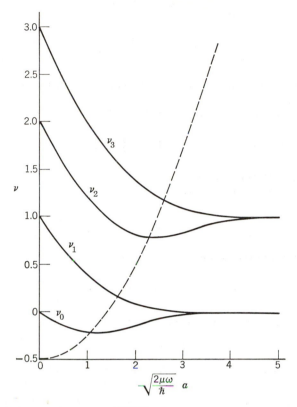

Figure 5.4. $v = (E/\hbar\omega) - \frac{1}{2}$ versus $\sqrt{2\mu\omega/\hbar}\, a$ for the four lowest eigenstates of the double oscillator. The dashed curve shows $(V_0/\hbar\omega) - \frac{1}{2}$ for a comparison of the energy levels with the barrier height.

Explicit approximate formulas for the eigenvalues can be given if $V_0 \gg E$ and

$$\sqrt{\frac{2\mu\omega}{\hbar}}\, a \gg 1 \qquad \text{or} \qquad V_0 = \tfrac{1}{2}\mu a^2\omega^2 \gg \tfrac{1}{4}\hbar\omega$$

These conditions represent the case of two almost completely separated oscillator wells, important in applications to molecules. For simplicity we shall only work out an approximation to the two roots which lie near $v = 0$, i.e., the lowest two energy eigenvalues. As $a \to \infty$, they become degenerate and equal to $\frac{1}{2}\hbar\omega$.

For the odd case we obtain from (5.55) with $z = -\sqrt{2\mu\omega/\hbar}\, a$ to first approximation ($\nu \approx 0$)

$$\nu_{\text{odd}} = \sqrt{\frac{\mu\omega}{\hbar\pi}}\, a \exp\left(-\frac{2V_0}{\hbar\omega}\right) = \sqrt{\frac{2V_0}{\hbar\omega\pi}} \exp\left(-\frac{2V_0}{\hbar\omega}\right) \qquad (5.60)$$

Exercise 5.9. Prove this result, noting that, if $\nu \approx 0$, $\Gamma(-\nu) \approx -1/\nu$.

For z negative, $|z| \gg 1$, and $\nu \approx 0$ we have

$$D_\nu'(z) \approx -\frac{z}{2}\exp\left(-\frac{z^2}{4}\right) + \sqrt{2\pi}\,\frac{\nu}{2}\exp\left(\frac{z^2}{4}\right)$$

from which it follows that for $\nu \approx 0$, $\nu_{\text{even}} \approx -\nu_{\text{odd}}$. The splitting of the two lowest energy levels is therefore

$$\Delta E = \hbar\omega(\nu_{\text{odd}} - \nu_{\text{even}}) = 2\hbar\omega\sqrt{\frac{2V_0}{\hbar\omega\pi}}\exp\left(-\frac{2V_0}{\hbar\omega}\right)$$

The frequency corresponding to this energy splitting is

$$\omega' = \frac{\Delta E}{\hbar} = 2\omega\sqrt{\frac{2V_0}{\hbar\omega\pi}}\exp\left(-\frac{2V_0}{\hbar\omega}\right) \qquad (5.61)$$

The physical significance of this frequency is best appreciated if we consider a system which is initially represented by a wave function proportional to $\psi_0 + \psi_1$, a wave function with a single peak around $x = +a$. If $x = x_1 - x_2$ is the coordinate difference for two elastically bound particles, as described in Section 5.5, this initial condition implies that at $t = 0$ particle 1 is definitely to the right of particle 2. What is the time development of this system? Neglecting questions of normalization, we have

$$\psi(x, t) = \exp\left(-\frac{i}{\hbar}E_0 t\right)\psi_0 + \exp\left(-\frac{i}{\hbar}E_1 t\right)\psi_1$$

$$= \tfrac{1}{2}\exp\left(-\frac{i}{\hbar}E_0 t\right)\{(\psi_0 + \psi_1)[1 + \exp(-i\omega' t)]$$

$$+ (\psi_0 - \psi_1)[1 - \exp(-i\omega' t)]\}$$

$$= \exp\left[-\frac{i(E_0 + E_1)}{2\hbar}t\right]\left[(\psi_0 + \psi_1)\cos\frac{\omega'}{2}t + i(\psi_0 - \psi_1)\sin\frac{\omega'}{2}t\right]$$

$$(5.62)$$

This last form shows clearly that the system shuttles back and forth between $\psi_0 + \psi_1$ and $\psi_0 - \psi_1$ with the frequency $\omega'/2$. It takes a time $\tau = \pi/\omega'$ for the system to change from its initial state $\psi_0 + \psi_1$ to the state $\psi_0 - \psi_1$, which means that the particles have exchanged their places in an entirely unclassical fashion, and the wave function is now peaked around $x = -a$.

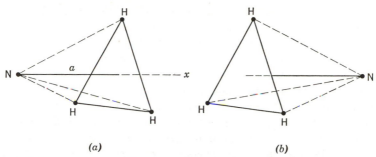

(a)   *(b)*

Figure 5.5. The ammonia (NH_3) molecule. The inversion spectrum arises from the motion of the nitrogen atom along the line perpendicular to the plane of the hydrogen atoms. A potential barrier prevents classical oscillations between equilibrium configurations *(a)* and *(b)*, but quantal barrier penetration can produce strong oscillations.

Since E was assumed to be much less than V_0, this exchange requires that the system tunnel with frequency $\omega'/2$ through the classically inaccessible barrier region indicated in Figure 5.2. The ratio of the time τ during which the exchange takes place to the period of the harmonic oscillator is given approximately by

$$\frac{\omega\tau}{2\pi} = \frac{\omega}{2\omega'} = \frac{1}{4}\sqrt{\frac{\hbar\omega\pi}{2V_0}}\,\exp\left(\frac{2V_0}{\hbar\omega}\right) \qquad (5.63)$$

If the barrier V_0 is high compared to $\hbar\omega$, the tunneling is strongly inhibited by the presence of the exponential factor.

A numerical example from molecular physics will serve as an illustration. Let the equilibrium distance of the particles in a molecule be $a = 10^{-8}$ cm, and the reduced mass $\mu = 10^{-23}$ gram. For $\omega = 2 \times 10^{15}$ sec^{-1} (corresponding to an infrared vibration spectrum common in many molecules) we have $V_0/\hbar\omega \simeq 10^3$. τ turns out to be superastronomical (about 10^{400} years); hence the exchange "hardly ever" takes place. On the other hand, if $\omega = 10^{12}$ sec^{-1}, corresponding to microwave vibrations of 0.2 cm wavelength, the value of ω' obtained from Figure 5.4 is approximately 8×10^{11} sec^{-1}, which is again in the microwave region.

The ammonia (NH_3) molecule is an example of a system in which transitions between the two states ψ_0 and ψ_1 can be induced by microwave radiation of frequency ω'.[9] There are two stable equilibrium positions of the nitrogen atom in the NH_3 molecule, symmetric with respect to the plane of the hydrogen atoms (Figure 5.5). One important vibrational mode consists of

[9] For this and other interesting applications of quantum mechanics to microwave spectroscopy see C. H. Townes and A. L. Schawlow, *Microwave Spectroscopy*, McGraw-Hill Book Co., New York, 1955.

oscillations of the nitrogen atom along the axis of the pyramid. If a is the distance from the equilibrium position to the symmetry plane, and x the instantaneous coordinate of the nitrogen atom on the axis, the potential energy will resemble Figure 5.2 closely, and our simple one-dimensional double oscillator model may be applied. The barrier which inhibits the inversion of the ammonia molecule from configuration a to b causes splitting of the vibration levels. Transitions between the split components give rise to a strong spectral line in the microwave region (*inversion spectrum*). Since the effect depends on quantum mechanical tunneling, we may well count the successful description of the inversion spectrum among the impressive accomplishments of quantum mechanics.

The double oscillator also provides a primitive model of a diatomic molecule. The force by which two similar atoms, such as two hydrogen atoms, are bound together in a diatomic molecule has its origin in the fact that an electron, which is classically confined to the vicinity of one or the other of the two nuclei, can be exchanged between them by quantal tunneling and is therefore shared by the two atoms.

We can get a qualitative understanding of this binding if we consider an electron in the field of two identical one-dimensional wells whose centers (the nuclei) are separated by a fixed distance X. For the example of the double oscillator potential of Figure 5.2 the dependence of the energy $E_e^{(i)}$ of each electronic eigenstate i on the interatomic distance $X = 2a$ can be inferred from Figure 5.4. If the nuclei were really fixed in their positions, the force between them would be $F^{(i)} = -[\partial E_e^{(i)}/\partial X]$. The quasi-static approximation which underlies the theory of molecular motion is based on the assumption that the nuclei move so slowly compared with the electrons that the force between the atoms can still be calculated by differentiation of the energies $E_e^{(i)}$. The function $E_e^{(i)}(X)$ thus serves as the potential energy for the motion of the nuclei if the electrons are in state i. Since X is regarded as an external parameter which changes slowly, allowing the electrons to remain in a quasi-stationary state throughout, the approximation is said to be *adiabatic*.

If μ_n is the reduced mass of the "homonuclear diatomic molecule," X the relative coordinate of the two "nuclei," μ the mass of the electron, and x its coordinate relative to the center of mass, the Schrödinger equation for our example may be written as

$$\left[-\frac{\hbar^2}{2\mu_n}\frac{\partial^2}{\partial X^2} - \frac{\hbar^2}{2\mu}\frac{\partial^2}{\partial x^2} + \tfrac{1}{2}\mu\omega^2\left(|x| - \frac{X}{2}\right)^2\right]\psi(x, X) = E\psi(x, X) \quad (5.64)$$

provided that the center of mass of the molecule is assumed to be at rest. We also assume that $X > 0$ and thus neglect the possibility of inversion of the molecule which was discussed above. For a real molecule, a repulsive inter-action $V(X)$ between the nuclei must be included in (5.64).

The adiabatic approximation[10] consists of assuming that the solution has the form

$$\psi(x, X) = \varphi_X^{(i)}(x)\eta(X) \tag{5.65}$$

so that the Schrödinger equation separates into a stationary state equation for the electron:

$$\left[-\frac{\hbar^2}{2\mu} \frac{\partial^2}{\partial x^2} + \tfrac{1}{2}\mu\omega^2\left(|x| - \frac{X}{2} \right)^2 \right] \varphi_X^{(i)}(x) = E_e^{(i)}(X)\varphi_X^{(i)}(x) \tag{5.66}$$

corresponding to a fixed value of X, and a second equation describing the motion of the nuclei:

$$\left[-\frac{\hbar^2}{2\mu_n} \frac{\partial^2}{\partial X^2} + E_e^{(i)}(X) \right]\eta(X) = E\eta(X) \tag{5.67}$$

By substituting the product wave function (5.65) in the Schrödinger equation (5.64), we see that (5.66) and (5.67) result, if we can neglect the terms

$$-\frac{\hbar^2}{\mu_n} \frac{\partial\varphi}{\partial X} \frac{\partial\eta}{\partial X} - \frac{\hbar^2}{2\mu_n} \frac{\partial^2\varphi}{\partial X^2}\eta \tag{5.68}$$

We shall estimate their magnitude presently.

By inspection of Figure 5.4 we see that the molecule can vibrate about a stable equilibrium configuration only if the electron is in one of the *symmetric* eigenstates. Confining ourselves to the electronic *ground* state ($i = 0$) we see that near the minimum of $E_e^{(0)}$ at $X = X_0$ we may write approximately

$$E_e^{(0)}(X) = E^0 + \tfrac{1}{2}\mu\omega^2 C(X - X_0)^2 \tag{5.69}$$

where C is a number of the order of unity. It follows that the nuclei perform harmonic oscillations with frequency $\simeq\sqrt{\mu/\mu_n}\,\omega$. Since actual molecules move in three dimensions and can rotate besides vibrating along the line joining the nuclei, their spectra exhibit more complex features, but the general nature of the approximations used is the same.

The accuracy of the adiabatic approximation is measured by the relative importance of the terms (5.68) compared with the kinetic energy $p_n^2/2\mu_n$ of the molecule. The first, and larger, of the two terms is of the order of magnitude $p_n p_e/2\mu_n$, since the wave function $\varphi_X^{(i)}(x)$ is equally sensitive to changes in X as in $2x$ so that

$$\left| \frac{\partial\varphi}{\partial X} \right| \simeq \frac{1}{2}\left| \frac{\partial\varphi}{\partial x} \right|$$

---

[10] The term *adiabatic approximation* applies to a general method. This particular form of it is referred to as the *Born-Oppenheimer approximation*.

As a necessary condition for the validity of the adiabatic approximation we thus obtain

$$\frac{p_n^2}{2\mu_n} \gg \frac{p_n p_e}{2\mu_n} \qquad \text{or} \qquad p_n \gg p_e$$

As estimates for the momenta we have (Problem 2)

$$p_e \simeq \sqrt{\mu\hbar\omega/2} \qquad \text{and} \qquad p_n \simeq \sqrt{\mu_n\hbar\sqrt{\mu/\mu_n}\omega/2}$$

Hence, the approximation may be expected to be good if

$$\left(\frac{\mu}{\mu_n}\right)^{1/4} \ll 1$$

which is reasonably well satisfied for molecules.

Problems

1. Calculate the matrix elements of x^2 with respect to the eigenfunctions of the harmonic oscillator and write down the first few rows and columns of the matrix. Can the same result be obtained directly by matrix algebra from a knowledge of the matrix elements of x?

2. Calculate the expectation values of the potential and kinetic energies in any stationary state of the harmonic oscillator. (See also Exercise 8.19.)

3. In first approximation, calculate the change in the energy levels of a linear harmonic oscillator that is perturbed by a potential gx^4.

4. Calculate the probability that the coordinate of a linear harmonic oscillator in its ground state has a value greater than the amplitude of a classical oscillator of the same energy.

5. Prove that

$$H_n(\xi) = \frac{2^n}{\sqrt{\pi}} \int_{-\infty}^{+\infty} (\xi + iu)^n e^{-u^2} du$$

is an integral representation for Hermite polynomials.

6. Show that if an ensemble of linear harmonic oscillators is in thermal equilibrium, governed by the Boltzmann distribution, the probability per unit length of finding a particle with displacement x is a Gaussian distribution. Plot the width of the distribution as a function of temperature. Check the results in the classical and the low temperature limits. (*Hint.* The formula in Problem 5 may be used.)

7. Show that as inadequate a variational trial function as

$$\psi(x) = \begin{cases} C\left(1 - \dfrac{|x|}{a}\right) & \text{for } |x| \leqslant a \\ 0 & \text{for } |x| > a \end{cases}$$

yields, for the optimum value of a, an upper limit to the ground state energy of the linear harmonic oscillator, which lies within less than 10 percent of the exact value.

8. Work out an approximation to the energy splitting between the second and third excited levels of the double oscillator, assuming the distance between the wells to be very large compared with the classical amplitude of the zero-point vibrations.

9. Apply the variational method to the double oscillator, using a real trial wave function of the form

$$\psi(x) = \alpha\psi_n(x - a) + \beta\psi_n(x + a)$$

where $\psi_n(x)$ denotes an eigenfunction of the simple linear harmonic oscillator. Show that the energy eigenvalues are approximated by[11]

$$\langle H \rangle = \frac{A_n \pm B_n}{1 \pm C_n}$$

where

$$A_n = \int \psi_n(x - a)H\psi_n(x - a)\, dx = \int \psi_n(x + a)H\psi_n(x + a)\, dx$$

$$B_n = \int \psi_n(x - a)H\psi_n(x + a)\, dx = \int \psi_n(x + a)H\psi_n(x - a)\, dx$$

and C_n is the *overlap integral*

$$C_n = \int \psi_n(x + a)\psi_n(x - a)\, dx$$

Calculate to a first approximation the splitting of the oscillator ground state for very large but finite values of a, and verify the result (5.61).

---

[11] The approximation method employed in this problem illustrates some of the techniques which are commonly used in the quantum theory of the chemical bond and parallels especially the theory of the positive H_2 ion. See C. A. Coulson, *Valence*, Clarendon Press, Oxford, 1952. Also, F. L. Pilar, *Elementary Quantum Chemistry*, McGraw-Hill Book Company, New York, 1968, and R. G. Parr, *Quantum Theory of Molecular Electronic Structure*, W. A. Benjamin, New York, 1963.

CHAPTER 6

Piecewise Constant Potentials in One Dimension

1. Introduction. Of all Schrödinger equations the one for a constant potential is mathematically the simplest. We know from Chapter 2 that the solutions are harmonic *plane waves*, with wave number

$$k = \sqrt{2\mu(E - V)}/\hbar$$

The reason for resuming the study of the Schrödinger equation with such a potential is that the qualitative features of a real physical potential can often be approximated reasonably well by a potential which is pieced together from a number of constant portions. For instance, although the forces acting between a proton and a neutron are not accurately known on theoretical grounds, it is known that, unlike the electrostatic forces which hold an atom together, the nuclear forces have a short range: they extend to some distance, and then drop to zero very fast. Figure 6.1 shows what these forces might look like and also how a rectangular potential well approximates them.

A potential with the shape shown in Figure 6.1 is frequently called a *square well*. The one-dimensional potential of this form simulates the more realistic three-dimensional potential square well which is a central potential with $V = 0$ for $r > a$ and $V = -V_0$ for $r < a$.

There are many other cases in which such a schematic potential approximates the real situation sufficiently to provide an orientation at the expense of comparatively little mathematical work. In solid state physics a periodic potential, as shown in Figure 6.2, exhibits some of the important features of any periodic potential seen by an electron in a crystal lattice.

Quite generally, the piecewise constant potential in one dimension offers an instructive example, and the behavior of a particle subject to such a potential is well worth studying in some detail.

$V(x) = $ const. for all x is the trivial case of the free particle. We recapitulate it here briefly. The Schrödinger equation is

$$\frac{d^2\psi}{dx^2} + \frac{2\mu(E - V)}{\hbar^2}\,\psi = 0 \tag{6.1}$$

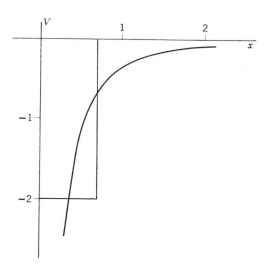

Figure 6.1. Possible shape of the potential representing the attractive part of the nuclear forces ($V = -e^{-x}/x$) and a square well approximating it.

If we write $p = \sqrt{2\mu(E - V)} = \hbar k$, the general solution of (6.1) is

$$\psi = Ae^{ikx} + Be^{-ikx} \tag{6.2}$$

For a solution to be physically useful k must not have any imaginary part. For if it did, ψ would grow beyond all bounds in the direction of either positive or negative x, or both. Hence, k must be real, but any real number will do. $E - V$ must be positive or zero, and the Schrödinger equation of the free particle has thus a *continuous* spectrum of eigenvalues, $E \geqslant V$. This means, of course, merely that the kinetic energy of a free particle must be positive definite. The totality of the solutions e^{ikx} and e^{-ikx}, where k can be any nonnegative number, is complete in the sense that any physically suitable wave packet can be expanded in terms of these eigenfunctions. This (Fourier) expansion theorem was the content of (2.12) and (2.13).

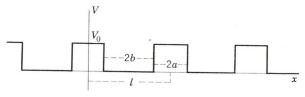

Figure 6.2. A periodic potential with rectangular sections (Kronig-Penney). The period has length $l = 2a + 2b$.

The eigenfunctions (6.2) are not quadratically integrable over all space. It is therefore impossible to speak of absolute probabilities and of expectation values for physical quantities in such a state. One way of avoiding this predicament would be to recognize the fact that physically no particle is ever absolutely free and that there is inevitably some confinement, the enclosure being for instance the wall of an accelerator tube or of the laboratory. *V* rises to infinity at the boundaries of the enclosure and does then not have the same value everywhere, the eigenfunctions are no longer infinite plane waves, and the eigenvalue spectrum is discrete rather than continuous (see Section 6.8). However, if the enclosure is very large compared with the de Broglie wavelength, the infinite plane wave is a useful idealization and an excellent approximation, and we shall gladly pay the price for using it.

Since normalization of $\int \psi^* \psi \, d\tau$ to unity is out of the question for infinite plane waves, we must decide on an alternative normalization for these wave functions.[1] A convenient tool in the discussion of such wave functions is the *delta function.*[2]

2. The Delta Function. The paradoxical feature of the delta function is that it is not a function at all. Rather, it is a symbol, $\delta(x)$, which for certain clearly defined purposes can be treated as if it were a function.

The delta function is a generalization of the Kronecker delta to the case of a continuous variable. The Kronecker delta δ_{ij} is defined for the discrete variables *i* and *j* as

$$\delta_{ij} = 0 \quad \text{if } i \neq j, \qquad \delta_{ii} = 1$$

Alternatively, it may be defined by

$$f(j) = \sum_{i=1}^{\infty} \delta_{ij} f(i) \tag{6.3}$$

if *i* and *j* take on the discrete values 1, 2, 3 $\cdots$.

In generalization of (6.3) the delta function is defined by the equation

$$f(x') = \int_{-\infty}^{+\infty} \delta(x, x') f(x) \, dx \tag{6.4}$$

where $f(x)$ is an arbitrary well-behaved function. Since $f(x)$ can be changed "almost anywhere" without affecting the value $f(x')$, it is clear that $\delta(x, x') = 0$ everywhere except when *x* is very close to *x'*. Owing to the definition (6.4),

[1] It is possible to normalize the eigenfunctions (6.2) by imposing periodic boundary conditions. This will be discussed in Section 6.3.

[2] Also known as the *Dirac delta function.* A useful compendium of the properties of the delta function and other related singular functions is M. J. Lighthill, *Introduction to Fourier Analysis and Generalised Functions*, University Press, Cambridge, 1958. A more comprehensive account is found in A. H. Zemanian, *Distribution Theory and Transform Analysis*, McGraw-Hill, New York, 1965.

$\delta(x, x')$ and $\delta(x + a, x' + a)$ are equivalent. a is arbitrary; if we set $a = -x'$, it becomes clear that $\delta(x, x')$ depends only on $x - x'$. From now on we may write $\delta(x, x') = \delta(x - x')$ and, instead of (6.4),

$$f(x') = \int_{-\infty}^{+\infty} \delta(x - x')f(x)\, dx \tag{6.5}$$

In particular, setting $x' = 0$, we get

$$f(0) = \int_{-\infty}^{+\infty} \delta(x')f(x')\, dx' \tag{6.6}$$

From this we see that effectively $\delta(x) = 0$ except at $x = 0$.
   If $f(x)$ is an odd function, $f(0) = 0$, hence $\delta(x)$ must be even:

$$\delta(-x) = \delta(x) \tag{6.7}$$

An equation like this, involving delta functions, is tacitly understood to imply that the expression on either side can be used equivalently in an integral like (6.6).

   It is in this sense that we may, for example, compare Fourier's integral formula

$$f(x') = \frac{1}{2\pi} \int_{-\infty}^{+\infty} du \int_{-\infty}^{+\infty} e^{iu(x-x')}f(x)\, dx \tag{6.8}$$

with (6.5) to infer the equivalence

$$\delta(x - x') = \frac{1}{2\pi} \int_{-\infty}^{+\infty} e^{iu(x-x')}\, du \tag{6.9}$$

The integral on the right-hand side of (6.9) does not exist by any of the conventional definitions of an integral. Yet it is convenient to admit this equality as meaningful with the important proviso that the entities which are being equated must be used only in conjunction with a well-behaved function $f(x)$ under an integral sign. Physically, (6.9) may be interpreted as the superposition with equal amplitudes and phases of simple harmonic oscillations of all frequencies. The contributions to the Fourier integral completely cancel by destructive interference unless the argument vanishes, i.e., $x = x'$.

   If condition (6.6) is applied to a simple function defined such that $f(x) = 1$ for $x_1 < x < x_2$, and $f(x) = 0$ outside the interval (x_1, x_2), we see that the delta function must satisfy the test

$$\int_{x_1}^{x_2} \delta(x)\, dx = \begin{cases} 0 & \text{if } x = 0 \text{ lies outside } (x_1, x_2) \\ 1 & \text{if } x_1 < 0 < x_2 \end{cases} \tag{6.10}$$

Exercise 6.1. By introducing suitable limiting processes in (6.9), show that the following expressions are all alternative representations of the delta function:

$$\delta(x) = \frac{1}{\sqrt{\pi}} \lim_{\varepsilon \to 0} \frac{1}{\sqrt{\varepsilon}} \exp\left(-\frac{x^2}{\varepsilon}\right) \qquad (6.11a)$$

$$\delta(x) = \frac{1}{\pi} \lim_{\varepsilon \to 0} \frac{\varepsilon}{x^2 + \varepsilon^2} \qquad (6.11b)$$

$$\delta(x) = \frac{1}{\pi} \lim_{N \to \infty} \frac{\sin Nx}{x} \qquad (6.11c)$$

(Do not worry about the legitimacy of freely interchanging integrations and limiting processes. Assume[3] that $\varepsilon > 0$.)

Show that the functions (6.11) as well as the representation

$$\delta(x) = \frac{1}{2} \frac{d^2}{dx^2} |x| \qquad (6.12)$$

meet the test (6.10). Invent a few other simple representations of the delta function.

Exercise 6.2. Prove that if a is a real constant, but $a \neq 0$, then

$$\delta(ax) = \frac{1}{|a|} \delta(x) \qquad (6.13)$$

Also show that

$$f(x)\delta(x - a) = f(a)\delta(x - a) \qquad (6.14)$$

A useful relation can be established for the delta function of a function $g(x)$. $\delta(g(x))$ vanishes except near the zeros of $g(x)$. If $g(x)$ is analytic near its zeros, x_i, the approximation $g(x) \approx g'(x_i)(x - x_i)$ may be used for $x \approx x_i$. From the definition (6.5) and from (6.13) we infer the equivalence

$$\delta(g(x)) = \sum_i \frac{1}{|g'(x_i)|} \delta(x - x_i) \qquad (6.15)$$

provided that $g'(x_i) \neq 0$.

Exercise 6.3. Prove that, if $a \neq b$,

$$\delta((x - a)(x - b)) = \frac{1}{|a - b|} [\delta(x - a) + \delta(x - b)] \qquad (6.16)$$

---

[3] Small ("infinitesimal") quantities like ε will be assumed to be *positive* everywhere in this book without special notice.

A very useful formula follows from the identity

$$\frac{1}{\omega \pm i\varepsilon} = \frac{\omega \mp i\varepsilon}{\omega^2 + \varepsilon^2} = \frac{\omega}{\omega^2 + \varepsilon^2} \mp \frac{i\varepsilon}{\omega^2 + \varepsilon^2} \tag{6.17}$$

The second term leads to a delta function [see (6.11b)] and the first term on the right-hand side becomes $1/\omega$ as $\varepsilon \to 0$, except if $\omega = 0$. If $f(\omega)$ is a well-behaved function, we have evidently

$$\lim_{\varepsilon \to 0} \int_{-\infty}^{+\infty} f(\omega) \frac{\omega}{\omega^2 + \varepsilon^2} \, d\omega$$

$$= \lim_{\varepsilon \to 0} \int_{-\infty}^{-\varepsilon} f(\omega) \frac{d\omega}{\omega} + \lim_{\varepsilon \to 0} \int_{\varepsilon}^{+\infty} f(\omega) \frac{d\omega}{\omega} + \lim_{\varepsilon \to 0} \int_{-\varepsilon}^{\varepsilon} f(\omega) \frac{\omega \, d\omega}{\omega^2 + \varepsilon^2}$$

$$= \mathsf{P} \int_{-\infty}^{+\infty} f(\omega) \frac{d\omega}{\omega} + f(0) \lim_{\varepsilon \to 0} \int_{-\varepsilon}^{\varepsilon} \frac{\omega \, d\omega}{\omega^2 + \varepsilon^2}$$

where P denotes the *Cauchy principal value*.[4] The last integral vanishes because the integrand is an odd function of ω. Hence, provided that the expression is used in an integrand in conjunction with a well-behaved function of ω, we have the effective equality

$$\lim_{\varepsilon \to 0} \frac{1}{\omega \pm i\varepsilon} = \mathsf{P} \frac{1}{\omega} \mp i\pi \, \delta(\omega) \tag{6.18}$$

In addition to the delta function, it is convenient to introduce the unit *step function* $\eta(x)$ by defining

$$\eta(x) = \int_{-\infty}^{x} \delta(x') \, dx' = \begin{cases} 0 & x < 0 \\ 1 & x > 0 \end{cases} \tag{6.19}$$

Exercise 6.4. Prove that ($\varepsilon > 0$)

$$\frac{1}{i\omega + \varepsilon} = \int_{-\infty}^{+\infty} e^{-(i\omega + \varepsilon)x} \eta(x) \, dx \tag{6.20}$$

and

$$\eta(x) = \frac{1}{2} + \frac{1}{2\pi i} \mathsf{P} \int_{-\infty}^{+\infty} \frac{e^{i\omega x}}{\omega} \, d\omega \tag{6.21}$$

3. *Normalization of Free Particle Wave Functions.* The delta function can now be used to fix the normalization of the free particle wave functions in a simple way. Let

$$\psi_k(x) = C(k)e^{ikx} \tag{6.22}$$

where k may be positive or negative.

---

[4] See E. T. Whittaker and G. N. Watson, *A Course of Modern Analysis*, p. 75, University Press, Cambridge, 1946.

These functions are said to be subject to *k-normalization* if

$$\int_{-\infty}^{+\infty} \psi_k{}^*(x)\psi_{k'}(x)\,dx = \delta(k - k')\tag{6.23}$$

Equation (6.23) gives not only the normalization; it also expresses the orthogonality of the eigenfunctions of (6.1) which belong to different values of k. Comparing (6.9) and (6.23), we find

$$|C(k)| = \frac{1}{\sqrt{2\pi}}$$

or, if the arbitrary phase of $C(k)$ is chosen to be zero,

$$C(k) = \frac{1}{\sqrt{2\pi}}\tag{6.24}$$

Similarly, we speak of *p-normalization* if

$$\psi_p(x) = \frac{1}{\sqrt{2\pi\hbar}}\exp\left(\frac{i}{\hbar}\,px\right) = \frac{1}{\sqrt{h}}\exp\left(\frac{i}{\hbar}\,px\right)\tag{6.25}$$

so that

$$\int_{-\infty}^{+\infty} \psi_p{}^*(x)\psi_{p'}(x)\,dx = \delta(p - p')$$

Still other normalizations are often used. A particularly common one is the *energy normalization* of the eigenfunctions

$$A(E)e^{ikx} \qquad\text{and}\qquad B(E)e^{-ikx} \qquad (k > 0)$$

the normalization conditions being

$$\int_{-\infty}^{+\infty} A^*(E)A(E')e^{i(k'-k)x}\,dx = \delta(E - E')\tag{6.26a}$$

and

$$\int_{-\infty}^{+\infty} B^*(E)B(E')e^{-i(k'-k)x}\,dx = \delta(E - E')\tag{6.26b}$$

Let us work out (6.26a), remembering that

$$\hbar k = \sqrt{2\mu E}, \qquad \hbar k' = \sqrt{2\mu E'}$$

With (6.9), (6.13), and (6.14) we can write (6.26a) as

$$2\pi\,|A(E)|^2\,\frac{\hbar}{\sqrt{2\mu}}\,\delta(\sqrt{E} - \sqrt{E'}) = \delta(E - E')$$

If $E \neq 0$, (6.15) gives

$$\delta(\sqrt{E} - \sqrt{E'}) = 2\sqrt{E}\,\delta(E - E')$$

Hence,

$$|A(E)|^2 = \sqrt{\frac{\mu}{2E}} \frac{1}{2\pi\hbar} = \frac{\mu}{hp} = \frac{1}{vh}$$

A similar calculation gives $|B(E)|^2$; hence, if the phases are chosen equal to zero, the energy normalized eigenfunctions for the free particle in one dimension are ($k > 0$)

$$\left(\frac{\mu}{8\pi^2\hbar^2 E}\right)^{1/4} e^{\pm ikx} \tag{6.27}$$

It is important to observe that any two eigenfunctions (6.27) with different signs in the exponent (representing waves propagating in opposite directions) are orthogonal, whether they belong to the same energy or not.

Again, as in the case of the harmonic oscillator, an arbitrary initial wave function can be expanded in terms of the free particle eigenfunctions. The *Fourier integral theorem* is the appropriate expansion formula, and the conditions under which it holds tell us how "arbitrary" the initial wave function can be.

Fourier analysis teaches us that there are alternative methods of constructing complete sets of free particle wave functions. They are defined by imposing certain boundary conditions on the solutions (6.2). A particularly useful set is obtained, if we require the eigenfunctions to satisfy *periodic boundary conditions* by demanding that the wave function assume the same value at any two points which are separated by a basic interval of length L. A suitable set of such eigenfunctions is again of the form (6.22), but the periodicity condition restricts the allowed values of k to a sequence of discrete numbers, expressible as

$$k = \frac{2\pi}{L} n \qquad (n = 0, \pm 1, \pm 2, \ldots) \tag{6.28}$$

Hence, if we subject the solutions of the Schrödinger equation to periodic boundary conditions, we obtain a discrete spectrum of energy eigenvalues

$$E = \frac{2\pi^2\hbar^2}{\mu L^2} n^2 \tag{6.29}$$

and each eigenvalue, except $E = 0$, is doubly degenerate. The eigenfunctions may now be normalized, and it is customary to do so by requiring that

$$\int_0^L |\psi|^2 \, dx = 1$$

Hence,

$$\psi = \frac{1}{\sqrt{L}} e^{ikx} \tag{6.30}$$

According to the theorems about *Fourier series*, any function defined between $x = 0$ and $x = L$ may be expanded in terms of these functions.

From (6.29) we obtain, for $n \gg \Delta n$,

$$\Delta E = E(n + \Delta n) - E(n) = \frac{2\pi^2 \hbar^2}{\mu L^2} [2n \Delta n + (\Delta n)^2] \sim \frac{2E}{n} \Delta n$$

We conclude that owing to the degeneracy the number of energy eigenstates per unit energy interval is approximately

$$\frac{2 |\Delta n|}{\Delta E} \sim \frac{n}{E} = \sqrt{\frac{2\mu}{E}} \frac{L}{h}$$

hence, is inversely proportional to the square root of the energy. The number of states per unit momentum interval is given by

$$\frac{\Delta n}{\Delta p} = \frac{1}{\hbar} \frac{\Delta n}{\Delta k} = \frac{L}{h}$$

In statistical mechanics this last equation is somewhat loosely interpreted by saying: "The number of states per unit momentum interval and per unit length is $1/h$." Conversely, a volume h in the phase space of x and p is ascribed to each state.

4. The Potential Step. Next in order of increasing complexity is the potential step $V(x) = V_0 \eta(x)$ as shown in Figure 6.3. There is no physically acceptable solution for $E < 0$ because of the general theorem that E can never be less than the absolute minimum of $V(x)$. Classically this is obvious. But, as the examples of the harmonic oscillator and the free particle have already shown us, it is also true in quantum mechanics despite the possibility of penetration

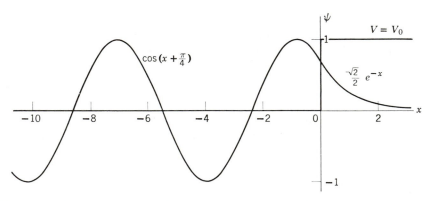

Figure 6.3. Eigenfunction for the potential step $V(x) = V_0 \eta(x)$ corresponding to an energy $E = V_0/2$.

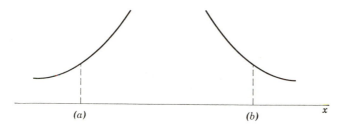

Figure 6.4. Shape of the wave function in the nonclassical region.

into classically inaccessible regions. We can prove the theorem by considering the real solutions of the Schrödinger equation (see Exercise 4.7)

$$-\frac{\hbar^2}{2\mu}\psi'' + [V(x) - E]\psi = 0$$

If $V(x) > E$ for *all* x, ψ'' has the same sign as ψ. Hence, if ψ is positive at some point x, the wave function has one of the two shapes shown in Figure 6.4, depending on whether the slope is positive or negative. In Figure 6.4*a* ψ can never bend down to be finite as $x \to \infty$. In Figure 6.4*b* ψ diverges as $x \to -\infty$. Hence, there must always be some region where $E > V(x)$ and where the particle could be found classically.

Now we consider the potential step with $0 < E < V_0$. Classically a particle of this energy, if it were incident from the left, would move freely until reflected at the potential step. Conservation of energy requires it to turn around, changing the sign of its momentum.

The Schrödinger equation has the solution

$$\psi(x) = \begin{cases} Ae^{ikx} + Be^{-ikx} & (x < 0) \\ Ce^{-\kappa x} & (x > 0) \end{cases} \qquad (6.31)$$

Here

$$\hbar k = \sqrt{2\mu E} \qquad \hbar\kappa = \sqrt{2\mu(V_0 - E)}$$

The second linearly independent solution for $x > 0$, $e^{\kappa x}$, has been omitted because it is in conflict with the boundary condition that ψ remain finite as $x \to \infty$. Since one of the two linearly independent solutions is excluded, the stationary states for $E < V_0$ are nondegenerate.[5]

By matching the wave function and its slope at the discontinuity of the potential, $x = 0$, we have

$$A + B = C$$

$$ik(A - B) = -\kappa C$$

[5] See Chapter 5, Footnote 4.

or

$$\frac{B}{A} = \frac{ik + \kappa}{ik - \kappa} = e^{i\alpha} \qquad (\alpha: \text{real})$$

$$\frac{C}{A} = \frac{2ik}{ik - \kappa} = 1 + e^{i\alpha} \tag{6.32}$$

Substituting these values into (6.31), we obtain

$$\psi = \begin{cases} 2Ae^{i\alpha/2} \cos\left(kx - \dfrac{\alpha}{2}\right) & (x < 0) \\ \\ 2Ae^{i\alpha/2} \cos\dfrac{\alpha}{2}\, e^{-\kappa x} & (x > 0) \end{cases} \tag{6.33}$$

in agreement with the remark made in Section 5 of Chapter 4 that the wave function in the case of no degeneracy is real except for an arbitrary constant factor. Hence, a graph may be drawn of such a wave function (Figure 6.3). The classical turning point ($x = 0$) is a point of inflection of the wave function. The oscillatory and the exponential portions can be joined smoothly at $x = 0$ for all values of E between 0 and V_0: the energy spectrum is continuous.

The solution (6.31) can be given a straightforward interpretation. It represents a plane wave incident from the left with an amplitude A and a reflected wave which propagates toward the left with an amplitude B. According to (6.32), $|A|^2 = |B|^2$; hence the reflection is *total*. Although ψ has a finite value in the region to the right of the potential step, there is no permanent penetration. A wave packet which is a superposition of eigenfunctions (6.31) could be constructed to represent a particle incident from the left. This packet would move classically, being reflected at the wall and giving again a vanishing probability of finding the particle in the region of positive x after the wave packet has receded.

These remarks can perhaps be better understood if we observe that for one-dimensional motion the conservation of probability leads to particularly transparent consequences. For a stationary state (4.2) reduces to $dj/dx = 0$. Hence, the current density

$$j = \frac{\hbar}{2\mu i}\left[\psi^* \frac{d\psi}{dx} - \frac{d\psi^*}{dx}\psi\right] \tag{6.34}$$

has the same value at all points x. j, when calculated with the wave function (6.33), is seen to vanish, as it does for any essentially real wave function. Hence, there is no net current anywhere at all. To the left of the potential step the relation $|A|^2 = |B|^2$ insures that incident and reflected probability currents cancel one another. If there is no current, there is no net momentum in the state (6.31). In order to observe the particle in the exponential tail, it must be localized within a distance of order $\Delta x \simeq 1/\kappa$. Hence, its momentum must be uncertain by

$$\Delta p > \hbar/\Delta x \simeq \hbar\kappa = \sqrt{2\mu(V_0 - E)}$$

The particle of energy E can thus be located in the nonclassical region only if it is given an energy $V_0 - E$, sufficient to raise it into the classically allowed region.

The case of an infinitely high potential barrier ($V_0 \to \infty$ or $\kappa \to \infty$) deserves special attention. From (6.31) it follows that in this limiting case $\psi(x) \to 0$ in the region under the barrier, no matter what value the coefficient C may have. According to (6.32), the joining conditions for the wave function at $x = 0$ now reduce formally to

$$\lim_{\kappa \to \infty} \frac{B}{A} = -1, \qquad \lim_{\kappa \to \infty} \frac{C}{A} = 0$$

or $A + B = 0$ and $C = 0$ as $V_0 \to \infty$. These equations show that at a point where the potential makes an infinite jump the wave function must vanish, whereas its slope may jump discontinuously from a finite value $2ikA$ to zero.[6]

We next examine the quantum mechanics of a particle which encounters the potential step in one dimension with an energy $E > V_0$. Classically this particle passes the potential step with altered velocity but no change of direction. The particle could be incident either from the right or from the left. The solutions of the Schrödinger equation are now oscillatory in both regions; hence, to each value of the energy correspond two linearly independent, degenerate eigenfunctions. For the physical interpretation their explicit construction is best accomplished by specializing the general solution:

$$\psi(x) = \begin{cases} Ae^{ikx} + Be^{-ikx} & (x < 0) \\ Ce^{ik_1x} + De^{-ik_1x} & (x > 0) \end{cases} \tag{6.35}$$

where $\hbar k = \sqrt{2\mu E}$ and $\hbar k_1 = \sqrt{2\mu(E - V_0)}$. Two useful particular solutions are obtained by setting $D = 0$, or $A = 0$. The first of these represents a wave incident from the left. Reflection occurs at the potential step, but there is also transmission to the right. The second particular solution represents incidence and transmission from right to left and reflection toward the right.

We consider here only the first case ($D = 0$). The remaining constants are related by the condition for smooth joining at $x = 0$,

$$A + B = C \qquad \text{and} \qquad k(A - B) = k_1C$$

---

[6] The discontinuity of the slope is not in conflict with the condition of smooth joining derived for *finite* jumps of the potential.

from which we solve

$$\frac{B}{A} = \frac{k - k_1}{k + k_1} \quad \text{and} \quad \frac{C}{A} = \frac{2k}{k + k_1} \tag{6.36}$$

The current density is again constant, but its value is no longer zero. Instead,

$$j = \frac{\hbar k}{\mu} (|A|^2 - |B|^2) \qquad (x < 0)$$

$$j = \frac{\hbar k_1}{\mu} |C|^2 \qquad (x > 0)$$

The equality of these values is assured by (6.36). We thus have

$$\frac{|B|^2}{|A|^2} + \frac{k_1}{k} \frac{|C|^2}{|A|^2} = 1 \tag{6.37}$$

In analogy to optics the first term in this sum is called the *reflection coefficient*, the second is the *transmission coefficient*. We have

$$R = \frac{|B|^2}{|A|^2} = \frac{(k - k_1)^2}{(k + k_1)^2} \tag{6.38}$$

$$T = \frac{k_1}{k} \frac{|C|^2}{|A|^2} = \frac{4kk_1}{(k + k_1)^2} \tag{6.39}$$

Equation (6.37) assures us that $R + T = 1$. R and T depend only on the ratio E/V_0.

For a wave packet incident from the left the presence of reflection means that the wave packet may, when it arrives at the potential step, split into two parts, provided that its average energy is close to V_0. This splitting up of the wave packet is a distinctly nonclassical effect which affords an argument against the early attempts to interpret the wave function as measuring the matter (or charge) density of the particle which it accompanies. For the splitting up of the wave packet would then imply a physical breakup of the particle, and this would be very difficult to reconcile with the facts of observation. After all, electrons and other particles are always found as complete entities with the same distinct properties. On the other hand, there is no contradiction between the splitting up of a wave packet and the probability interpretation of the wave function.

Exercise 6.5. Show that the coefficients for reflection and transmission at a potential step are the same for a wave incident from the right as for a wave incident from the left.

5. *The Rectangular Potential Barrier.* In our study of more and more complicated potential forms, we now reach a very important case, the rectangular *potential barrier* (Figure 6.5). There is a slight advantage in placing the coordinate origin at the center of the barrier so that $V(x)$ is an *even* function of x. Owing to the quantum mechanical penetration of a barrier, a case of great interest is that of $E < V_0$. The particle is free for $x < -a$ and $x > a$. For this reason the rectangular potential barrier simulates, albeit schematically, the scattering of a free particle from any potential.

We can immediately write down the general solution of the Schrödinger equation for $E < V_0$:

$$\psi(x) = \begin{cases} Ae^{ikx} + Be^{-ikx} & (x < -a) \\ Ce^{-\kappa x} + De^{\kappa x} & (-a < x < a) \\ Fe^{ikx} + Ge^{-ikx} & (a < x) \end{cases} \tag{6.40}$$

where again $\hbar k = \sqrt{2\mu E}$, $\hbar\kappa = \sqrt{2\mu(V_0 - E)}$. The boundary conditions at $x = -a$ require

$$Ae^{-ika} + Be^{ika} = Ce^{\kappa a} + De^{-\kappa a}$$

$$Ae^{-ika} - Be^{ika} = \frac{i\kappa}{k}(Ce^{\kappa a} - De^{-\kappa a}) \tag{6.41}$$

These linear homogeneous relations between the coefficients A, B, C, D are conveniently expressed in terms of matrices:

$$\begin{pmatrix} A \\ B \end{pmatrix} = \frac{1}{2}\begin{pmatrix} \left(1 + \dfrac{i\kappa}{k}\right)e^{\kappa a + ika} & \left(1 - \dfrac{i\kappa}{k}\right)e^{-\kappa a + ika} \\ \left(1 - \dfrac{i\kappa}{k}\right)e^{\kappa a - ika} & \left(1 + \dfrac{i\kappa}{k}\right)e^{-\kappa a - ika} \end{pmatrix}\begin{pmatrix} C \\ D \end{pmatrix}$$

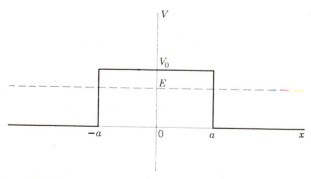

Figure 6.5. Rectangular potential barrier, height V_0, width $2a$.

The joining conditions at $x = a$ are similar. They yield

$$
\begin{pmatrix} C \\ D \end{pmatrix} = \frac{1}{2} \begin{pmatrix} \left(1 - \dfrac{ik}{\kappa}\right) e^{\kappa a + ika} & \left(1 + \dfrac{ik}{\kappa}\right) e^{\kappa a - ika} \\ \left(1 + \dfrac{ik}{\kappa}\right) e^{-\kappa a + ika} & \left(1 - \dfrac{ik}{\kappa}\right) e^{-\kappa a - ika} \end{pmatrix} \begin{pmatrix} F \\ G \end{pmatrix}
$$

Combining the last two equations, we obtain the relation between the wave function on both sides of the barrier:

$$
\begin{pmatrix} A \\ B \end{pmatrix} =
$$

$$
\begin{pmatrix} \left(\cosh 2\kappa a + \dfrac{i\varepsilon}{2} \sinh 2\kappa a\right) e^{2ika} & \dfrac{i\eta}{2} \sinh 2\kappa a \\ -\dfrac{i\eta}{2} \sinh 2\kappa a & \left(\cosh 2\kappa a - \dfrac{i\varepsilon}{2} \sinh 2\kappa a\right) e^{-2ika} \end{pmatrix} \begin{pmatrix} F \\ G \end{pmatrix}
$$

$$(6.42)$$

where the abbreviated notation

$$
\varepsilon = \frac{\kappa}{k} - \frac{k}{\kappa}, \qquad \eta = \frac{\kappa}{k} + \frac{k}{\kappa} \tag{6.43}
$$

has been used.

Exercise 6.6. Calculate the determinant of the 2 × 2 matrix in (6.42).

A particular solution of interest is obtained from (6.42) by letting $G = 0$. This represents a wave incident from the left, and transmitted through the barrier to the right. A reflected wave whose amplitude is B is also present. We calculate easily:

$$
\frac{F}{A} = \frac{e^{-2ika}}{\cosh 2\kappa a + i(\varepsilon/2) \sinh 2\kappa a} \tag{6.44}
$$

The square of the absolute value of this quantity is the transmission coefficient for the barrier. It assumes an especially simple form for a high and wide barrier which transmits poorly ($\kappa a \gg 1$). In first approximation,

$$
\cosh 2\kappa a \approx \sinh 2\kappa a \approx \tfrac{1}{2} e^{2\kappa a}
$$

hence,

$$
T = \left| \frac{F}{A} \right|^2 \approx 16 e^{-4\kappa a} \left(\frac{k\kappa}{k^2 + \kappa^2} \right)^2 \tag{6.45}
$$

The matrix which connects A and B with F and G in (6.42) has very simple properties. If we write the linear relations as

$$\begin{pmatrix} A \\ B \end{pmatrix} = \begin{pmatrix} \alpha_1 + i\beta_1 & \alpha_2 + i\beta_2 \\ \alpha_3 + i\beta_3 & \alpha_4 + i\beta_4 \end{pmatrix} \begin{pmatrix} F \\ G \end{pmatrix} \tag{6.46}$$

and compare this with (6.42), we observe that the eight real numbers α and β in the matrix satisfy the conditions

$$\alpha_1 = \alpha_4, \qquad \beta_1 = -\beta_4, \qquad \alpha_2 = \alpha_3 = 0, \qquad \beta_2 = -\beta_3 \tag{6.47}$$

These five equations reduce the number of independent variables on which the matrix depends from eight to three. As can be seen from (6.42) and Exercise 6.5, we must add to this an equation expressing the fact that the determinant of the matrix is equal to unity. Using (6.47), this condition reduces to

$$\alpha_1{}^2 + \beta_1{}^2 - \beta_2{}^2 = 1 \tag{6.48}$$

Hence, we are left with two parameters, as we must be, since the matrix depends explicitly only on the independent variables ka and κa.

Exercise 6.7. Show that (6.46) can be written as

$$\begin{pmatrix} A \\ B \end{pmatrix} = \begin{pmatrix} e^{i\mu} \cosh \lambda & i \sinh \lambda \\ -i \sinh \lambda & e^{-i\mu} \cosh \lambda \end{pmatrix} \begin{pmatrix} F \\ G \end{pmatrix} \tag{6.49}$$

where λ and μ are two real parameters.

In the next section it will be shown that the conditions (6.47) and (6.48) imposed on (6.46), rather than pertaining specifically to the rectangular-shaped potential, are consequences of very general symmetry properties of the physical system at hand.

6. Symmetries and Invariance Properties. Since the rectangular barrier of Figure 6.5 is a real potential and symmetric about the origin, the Schrödinger equation is invariant under time reversal and space reflection. We can exploit these properties to derive the general form of the matrix linking the incident with the transmitted wave.

We recapitulate the form of the general solution of the Schrödinger equation:

$$\psi(x) = \begin{cases} Ae^{ikx} + Be^{-ikx} & (x < -a) \\ Ce^{-\kappa x} + De^{\kappa x} & (-a < x < a) \\ Fe^{ikx} + Ge^{-ikx} & (a < x) \end{cases} \tag{6.40}$$

We need not reproduce here the joining conditions at $x = -a$ and $x = a$, because we want to see how far we can proceed without these conditions. Nevertheless, it is clear that the joining conditions lead to two linear homogeneous relations between the coefficients A, B, F, and G. If we regard the wave function on one side of the barrier, say for $x > a$, as given, then these relations are linear equations expressing the coefficients A and B in terms of F and G. Hence a matrix M exists such that

$$\begin{pmatrix} A \\ B \end{pmatrix} = \begin{pmatrix} M_{11} & M_{12} \\ M_{21} & M_{22} \end{pmatrix} \begin{pmatrix} F \\ G \end{pmatrix} \tag{6.50}$$

An equivalent representation expresses the coefficients B and F of the *outgoing* waves in terms of the coefficients A and G of the *incoming* waves by the matrix relation

$$\begin{pmatrix} B \\ F \end{pmatrix} = \begin{pmatrix} S_{11} & S_{12} \\ S_{21} & S_{22} \end{pmatrix} \begin{pmatrix} A \\ G \end{pmatrix} \tag{6.51}$$

Whereas the representation in terms of the S matrix is more readily generalized to three-dimensional situations, the M matrix is more appropriate in one-dimensional problems and is therefore used in Sections 5, 7, and 8 of this chapter. On the other hand, the symmetry properties are best formulated in terms of the S matrix.

The S and M matrices can be simply related if conservation of probability is invoked. As was shown in Section 6.4, in a one-dimensional stationary state the probability current density j must be independent of x. Applying expression (6.34) to the wave function (6.40), we obtain the condition

$$|A|^2 - |B|^2 = |F|^2 - |G|^2 \qquad \text{or} \qquad |B|^2 + |F|^2 = |A|^2 + |G|^2$$

as expected, since $|A|^2$ and $|F|^2$ measure the probability flow to the right, while $|B|^2$ and $|G|^2$ measure the flow in the opposite direction. Using matrix notation, we can write this as

$$(B^* \ F^*)\begin{pmatrix} B \\ F \end{pmatrix} = (A^* \ G^*)\tilde{S}^*S\begin{pmatrix} A \\ G \end{pmatrix} = (A^* \ G^*)\begin{pmatrix} A \\ G \end{pmatrix}$$

where $\tilde{S}$ denotes the transpose matrix of S, and S^* the complex conjugate. It follows that S must obey the condition

$$\tilde{S}^*S = 1 \tag{6.52}$$

with I denoting the unit matrix in two dimensions. If the *Hermitian conjugate* $S^\dagger$ of a matrix S is defined by

$$S^\dagger = \tilde{S}* = \begin{pmatrix} S_{11}* & S_{21}* \\ S_{12}* & S_{22}* \end{pmatrix} \tag{6.53}$$

equation (6.52) implies the statement that the inverse of S must be the same as its Hermitian conjugate. Such a matrix is said to be *unitary*.

The elements of the matrix S are therefore subject to the following constraints:

$$|S_{11}| = |S_{22}| \qquad \text{and} \qquad |S_{12}| = |S_{21}| \tag{6.54}$$

$$|S_{11}|^2 + |S_{12}|^2 = 1 \tag{6.55}$$

and

$$S_{11}S_{12}* + S_{21}S_{22}* = 0 \tag{6.56}$$

Since the potential is real, the Schrödinger equation has, according to Section 4.5, in addition to (6.40) the time-reversed solution

$$\psi_1(x) = \begin{cases} A*e^{-ikx} + B*e^{ikx} & (x < -a) \\ C*e^{-\kappa x} + D*e^{\kappa x} & (-a < x < a) \\ F*e^{-ikx} + G*e^{ikx} & (a < x) \end{cases} \tag{6.57}$$

Comparison of this solution with (6.40) shows that effectively the directions of motion have been reversed and the coefficient A has been interchanged with $B*$, and F with $G*$. Hence, in (6.51) we may make the replacements $A \leftrightarrow B*$ and $F \leftrightarrow G*$, and obtain an equally valid equation

$$\begin{pmatrix} A* \\ G* \end{pmatrix} = \begin{pmatrix} S_{11} & S_{12} \\ S_{21} & S_{22} \end{pmatrix} \begin{pmatrix} B* \\ F* \end{pmatrix} \tag{6.58}$$

Equations (6.58) and (6.51) can be combined to yield the condition

$$S*S = I \tag{6.59}$$

This condition in conjunction with the unitarity relation (6.52) implies that the S matrix must be *symmetric* as a consequence of time reversal symmetry. If S is unitary and symmetric, it is easy to verify by comparing equations (6.50) and (6.51) that the M matrix assumes the form:

$$M = \begin{pmatrix} \dfrac{1}{S_{12}} & \dfrac{S_{11}*}{S_{12}*} \\ \dfrac{S_{11}}{S_{12}} & \dfrac{1}{S_{12}*} \end{pmatrix} \tag{6.60}$$

$$\frac{1}{|S_{12}|^2} - \left|\frac{S_{11}}{S_{12}}\right|^2 = 1$$

$$\frac{|S_{12}|^2}{1 + |S_{11}|^2} = 1$$

with

$$\det M = (1 - |S_{11}|^2)/|S_{12}|^2 = 1$$

Since the potential is an even function of x, another solution is obtained by replacing x in (6.40) by $-x$. The substitution gives

$$\psi_2(x) = \begin{cases} Ae^{-ikx} + Be^{ikx} & (x > a) \\ Ce^{\kappa x} + De^{-\kappa x} & (a > x > -a) \\ Fe^{-ikx} + Ge^{ikx} & (-a > x) \end{cases} \quad (6.61)$$

If Ge^{ikx} is the wave incident on the barrier from the left, Be^{ikx} is the transmitted and Fe^{-ikx} the reflected wave in (6.61). Ae^{-ikx} is incident from the right. Hence, in (6.51) we may make the replacements $A \leftrightarrow G$ and $B \leftrightarrow F$, and obtain

$$\begin{pmatrix} F \\ B \end{pmatrix} = \begin{pmatrix} S_{11} & S_{12} \\ S_{21} & S_{22} \end{pmatrix} \begin{pmatrix} G \\ A \end{pmatrix}$$

This relation can also be written as

$$\begin{pmatrix} B \\ F \end{pmatrix} = \begin{pmatrix} S_{22} & S_{21} \\ S_{12} & S_{11} \end{pmatrix} \begin{pmatrix} A \\ G \end{pmatrix}$$

Hence, invariance under reflection implies the symmetry relations

$$S_{11} = S_{22} \quad \text{and} \quad S_{12} = S_{21} \quad (6.62)$$

If conservation of probability, time reversal invariance, and invariance under space reflection are simultaneously demanded, the matrix M has the structure

$$M_{11} = M_{22}{}^*, \qquad M_{12} = -M_{12}{}^* = -M_{21} = M_{21}{}^*, \qquad \det M = 1$$

The matrix in (6.42) and (6.49) satisfies precisely these conditions. We thus see that the conditions (6.47) and (6.48) are the result of very general properties, shared by all potentials that are symmetric with respect to the origin and vanish for large values of $|x|$. For all such potentials the solution of the Schrödinger equation must be asymptotically of the form

$$\psi(x) \sim Ae^{ikx} + Be^{-ikx} \quad \text{as } x \to -\infty$$

and

$$\psi(x) \sim Fe^{ikx} + Ge^{-ikx} \quad \text{as } x \to +\infty$$

By virtue of the general arguments just advanced, these two portions of the eigenfunction are related by the equation[7]

$$\begin{pmatrix} A \\ B \end{pmatrix} = \begin{pmatrix} \alpha_1 + i\beta_1 & i\beta_2 \\ -i\beta_2 & \alpha_1 - i\beta_1 \end{pmatrix} \begin{pmatrix} F \\ G \end{pmatrix}$$

with the real parameters α_1, β_1, and β_2 subject to the additional constraint

$$\alpha_1{}^2 + \beta_1{}^2 - \beta_2{}^2 = 1$$

Although the restrictions that various symmetries impose on the S or M matrix usually complement each other, they are occasionally redundant. For instance, in the simple one-dimensional problem treated in this section, invariance under reflection, if applicable, guarantees that the S matrix is symmetric [Equation (6.62)], thus yielding a condition that is equally prescribed by invariance under time reversal together with probability conservation.

A significant aspect of the matrix method of this section is that it allows a neat separation between the initial conditions, which can be adapted to the requirements of any particular problem, and the matrices M and S, which do not depend on the particular structure of the wave packet used. S and M depend only on the nature of the dynamical system, the forces, and the energy. Once either one of these matrices has been worked out as a function of energy, all problems relating to the potential barrier have essentially been solved. For example, the transmission coefficient T is given by $|F|^2/|A|^2$ if $G = 0$, and therefore

$$T = \frac{1}{|M_{11}|^2} = |S_{21}|^2 \tag{6.63}$$

We shall encounter other uses of the M and S matrices in subsequent sections. Eventually, in Chapter 19, we shall see that similar methods are pertinent in the general theory of collisions, where the S or *scattering matrix* plays a central role. The work of this section is S-matrix theory in its most elementary form.

---

[7] The same concepts can be generalized to include long-range forces. All that is needed to define a matrix M with the properties (6.47) and (6.48) is that the two linearly independent fundamental solutions of the Schrödinger equation have the asymptotic property

$$\psi_1(-x) = \psi_2(x) = \psi_1{}^*(x)$$

For a real even potential function $V(x)$ this can always be accomplished by choosing

$$\psi_{1,2}(x) = \psi_{\text{even}}(x) \pm i\psi_{\text{odd}}(x)$$

where ψ_{even} and ψ_{odd} are the real even and odd solutions.

Exercise 6.8. Noting that k appears in the Schrödinger equation only quadratically, prove that, as a function of k, the S matrix has the property

$$S(k)S(-k) = 1 \tag{6.64}$$

Derive the corresponding properties of the matrix M, and verify them in the example of Section 6.5.

Exercise 6.9. Using invariance under time reversal only, prove that at a fixed energy the value of the transmission coefficient is independent of the direction of incidence.

7. The Periodic Potential. With a little extra effort we can now proceed to solve the more complicated problem of a particle in the presence of a periodic potential composed of a succession of potential barriers.[8] Figure 6.2 shows a cut from this battlement-shaped potential which serves as a model of the potential to which an electron in a crystal lattice is exposed. The oversimplified shape treated here already exhibits the essential features of all such periodic potentials. As a useful idealization we assume that potential hills and valleys follow each other in periodic succession indefinitely in both directions, although in reality the number of atoms in a crystal is, of course, finite, if large. $l = 2a + 2b$ is the *period* of this potential.

The matrix method is especially well suited for treating this problem. The solution of the Schrödinger equation in the valleys, where $V = 0$ and $\hbar k = \sqrt{2\mu E}$, may be written in the form

$$\psi(x) = A_n e^{ik(x-nl)} + B_n e^{-ik(x-nl)} \tag{6.65}$$

for $a - l < x - nl < -a$. The coefficients belonging to successive values of n can be related by a matrix using the procedure and notation of the last section. Noting that the centers of the peaks have the coordinates $x = nl$, we obtain

$$\begin{pmatrix} A_n \\ B_n \end{pmatrix} = \begin{pmatrix} \alpha_1 + i\beta_1 & i\beta_2 \\ -i\beta_2 & \alpha_1 - i\beta_1 \end{pmatrix} \begin{pmatrix} A_{n+1} e^{-ikl} \\ B_{n+1} e^{ikl} \end{pmatrix}$$

This may also be written as

$$\begin{pmatrix} A_{n+1} \\ B_{n+1} \end{pmatrix} = P \begin{pmatrix} A_n \\ B_n \end{pmatrix} \tag{6.66}$$

---

[8] The periodic potential of Figure 6.2 is known as the *Kronig-Penney potential*. For a comprehensive treatment in the framework of solid state physics see C. Kittel, *Introduction to Solid State Physics*, third edition, John Wiley & Sons, New York, 1966.

where the matrix P is defined by

$$P = \begin{pmatrix} (\alpha_1 - i\beta_1)e^{ikl} & -i\beta_2 e^{ikl} \\ i\beta_2 e^{-ikl} & (\alpha_1 + i\beta_1)e^{-ikl} \end{pmatrix} \tag{6.67}$$

subject to the condition

$$\alpha_1{}^2 + \beta_1{}^2 - \beta_2{}^2 = 1 \tag{6.48}$$

By iteration we have

$$\begin{pmatrix} A_n \\ B_n \end{pmatrix} = P^n \begin{pmatrix} A_0 \\ B_0 \end{pmatrix} \tag{6.68}$$

Applying these considerations to an infinite periodic lattice, we must clearly demand that as $n \to \pm\infty$ the limit of P^n should exist. This is most conveniently discussed in terms of the eigenvalue problem of the matrix P.

The eigenvalues of P are roots of the characteristic equation

$$p^2 - p \text{ trace } P + \det P = 0$$

or

$$p^2 - 2(\alpha_1 \cos kl + \beta_1 \sin kl)\, p + 1 = 0$$

The roots are

$$p_\pm = \tfrac{1}{2}[\text{trace } P \pm \sqrt{(\text{trace } P)^2 - 4}]$$

If the roots are different, the two eigenvectors are linearly independent, and two linearly independent solutions of the Schrödinger equation are obtained by identifying the initial values $\begin{pmatrix} A_0 \\ B_0 \end{pmatrix}$ with these eigenvectors:

$$P \begin{pmatrix} A_0^{(\pm)} \\ B_0^{(\pm)} \end{pmatrix} = p_\pm \begin{pmatrix} A_0^{(\pm)} \\ B_0^{(\pm)} \end{pmatrix}$$

For these particular solutions (6.68) becomes

$$\begin{pmatrix} A_n^{(\pm)} \\ B_n^{(\pm)} \end{pmatrix} = p_\pm{}^n \begin{pmatrix} A_0^{(\pm)} \\ B_0^{(\pm)} \end{pmatrix} \tag{6.69}$$

If $|\text{trace } P| > 2$, p_+ and p_- are real and either $\lim_{n\to\infty} |p_\pm{}^n| \to \infty$ or $\lim_{n\to-\infty} |p_\pm{}^n| \to \infty$. Such solutions (6.69) are in conflict with the requirement that the wave function must remain finite. Hence, an acceptable solution is obtained and a particular energy value allowed only if

$$\tfrac{1}{2} |\text{trace } P| = |\alpha_1 \cos kl + \beta_1 \sin kl| \leqslant 1$$

If this condition holds, we may write

$$p_+ = e^{ikl}, \qquad p_- = e^{-ikl} \tag{6.70}$$

thus defining a real parameter k which is related to the energy by the condition

$$\cos kl = \alpha_1 \cos kl + \beta_1 \sin kl \qquad (6.71)$$

For the potential shape of Figure 6.2 this can, according to (6.42), be written as

$$\cos kl = \cosh 2\kappa a \cos 2kb + (\varepsilon/2) \sinh 2\kappa a \sin 2kb \qquad (6.72)$$

where $\hbar\kappa = \sqrt{2\mu(V_0 - E)}$ and $E < V_0$.

Since $|\cosh 2\kappa a| \geqslant 1$, it is readily seen that all energy values for which

$$2kb = N\pi \qquad (N = \text{integer}) \qquad (6.73)$$

are forbidden or are at edges of allowed bands. From the continuity of all the functions involved it follows that there must generally be forbidden ranges of energy values in the neighborhood of the discrete values determined by (6.73).

Exercise 6.10. Show that, if $E > V_0$, the eigenvalue condition for the periodic potential of Figure 6.2 becomes

$$\cos kl = \cos 2k'a \cos 2kb - \frac{k'^2 + k^2}{2kk'} \sin 2k'a \sin 2kb \qquad (6.74)$$

where $\hbar k' = \sqrt{2\mu(E - V_0)}$. Verify that the energies determined by the condition $(2k'a + 2kb) = N\pi$ are forbidden, and prove that $k \to k$ as $E \to \infty$.

From (6.72) and (6.74) the allowed values of the energy can be calculated in terms of the constants of the potential and the parameter k. Figure 6.6 is a plot of $k^2 \propto E$ versus k for a particular choice of constants, showing that the energy spectrum consists of continuous *bands* separated by forbidden gaps. The band structure of the energy levels of electrons in a periodic lattice is a direct consequence of the wave nature of matter. It is essential to an understanding of many basic properties of the solid state of matter and indispensable in the quantum theory of electric conduction in metals. Even in its crudest and most schematic form it accounts qualitatively for the distinction between conductors and insulators. In its refined form the band theory is able to correlate a large amount of experimental data quantitatively.[9] In applications to the solid state, quantum mechanics has scored some of its greatest triumphs.

We must now look briefly at the eigenfunctions of the Schrödinger equation. In the valleys they are of the form (6.65). The coefficients of the

[9] See reference in footnote 8.

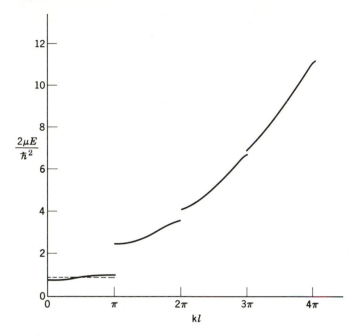

Figure 6.6. $k^2 = 2\mu E/\hbar^2$ versus k for the periodic potential of Figure 6.2 with $a = b = 1$ and $2\mu V_0/\hbar^2 = \pi^2/4$. The dashed line indicates the position of the bound energy level in a rectangular well of depth V_0 and width $2b$. The presence of infinitely many such wells in a periodic lattice produces a continuous narrow band of allowed energy levels.

plane waves for two fundamental solutions corresponding to the same energy are given by

$$\begin{pmatrix} A_n^{(+)} \\ B_n^{(+)} \end{pmatrix} = e^{inkl} \begin{pmatrix} A_0^{(+)} \\ B_0^{(+)} \end{pmatrix} \quad \text{and} \quad \begin{pmatrix} A_n^{(-)} \\ B_n^{(-)} \end{pmatrix} = e^{-inkl} \begin{pmatrix} A_0^{(-)} \\ B_0^{(-)} \end{pmatrix} \quad (6.75)$$

Because of invariance under time reversal, we may assume the relation

$$\frac{A_0^{(-)}}{B_0^{(-)}} = \left(\frac{B_0^{(+)}}{A_0^{(+)}} \right)^*$$

between the two solutions. From the eigenvalue equation for the matrix P, we obtain the ratio

$$\frac{A_0^{(+)}}{B_0^{(+)}} = \frac{\beta_2 e^{ikl}}{\alpha_1 \sin kl - \beta_1 \cos kl - \sin kl} \quad (6.76)$$

which can be used to construct the eigenfunctions:

$$\psi^{(+)} = [\psi^{(-)}]^*$$
$$= e^{inkl}\{\beta_2 e^{ik[x-(n-1/2)l]} + (\alpha_1 \sin kl - \beta_1 \cos kl - \sin kl)e^{-ik[x-(n-1/2)l]}\}$$
$$= e^{ikx}e^{-ik(x-nl)}\{\cdots\} = e^{ikx}u_k(x) \tag{6.77}$$

for $a - l < x - nl < -a$. Equation (6.77) gives the eigenfunctions in the valley portions of the potential in Figure 6.2. In the hill portions the harmonic waves are replaced by increasing and decreasing exponentials, the coefficients being determined by matching the eigenfunction at hill-valley boundaries. Since the function $u_k(x)$ defined by equation (6.77) has the periodicity property

$$u_k(x + l) = u_k(x)$$

the eigenfunctions (6.77) are products of harmonic plane waves with wave number k and functions $u_k(x)$ which have the same period as the potential (Figure 6.7).

The two solutions $\psi^{(+)}$ and $\psi^{(-)}$ are linearly independent unless $kl = N\pi$, N integer. In this important special case the two solutions become identical. Furthermore, since it can be shown that for these values of k

$$\alpha_1 \sin kl - \beta_1 \cos kl = \pm\beta_2 \tag{6.78}$$

the eigenfunctions (6.77) represent standing waves. These solutions correspond to the band edges.

Exercise 6.11. Verify (6.78) for $kl = N\pi$.

Although the results of this section were derived for a rectangular-shaped potential, any periodic potential yields eigenfunctions of the general form (6.77). This property is known as *Floquet's theorem* and functions of this

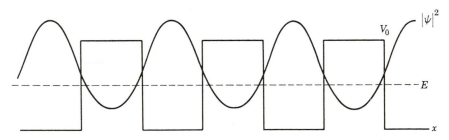

Figure 6.7. Sketch of $|\psi|^2$ for a periodic potential. $E = V_0/2$ was assumed. The curve consists of alternating sections of trigonometric and hyperbolic sine functions, joined smoothly at the discontinuities of V.

form, which play an important role in the electron theory of metals, are called *Bloch wave functions* (see Exercise 14.19).

Exercise 6.12. Prove that the periodic function $u_k(x)$ satisfies the equation

$$-\frac{\hbar^2}{2\mu}\frac{d^2 u_k}{dx^2} - \frac{i\hbar^2 k}{\mu}\frac{d u_k}{dx} + \frac{\hbar^2 k^2}{2\mu} u_k + V u_k = E u_k \tag{6.79}$$

8. The Square Well. Finally, we must discuss the so-called *square* (or rectangular) *well*. It is convenient to place the origin of the x-axis in the center of the potential well, so that $V(x)$ is again an even function of x: $V = -V_0$ for $-a < x < a$, and $V = 0$ for all other values of x, or

$$V(x) = V_0[\eta(x - a) - \eta(x + a)]$$

with $V_0 > 0$.

Depending on whether the energy is positive or negative, we distinguish two separate cases. If $E > 0$, the particle is unconfined and is *scattered* by the potential; if $E < 0$, it is confined and in a *bound* state. We treat this last case first.

$-V_0 \leqslant E < 0$. Reversing our earlier notation, we now set

$$\hbar k' = \sqrt{2\mu(E + V_0)}, \qquad \hbar\kappa = \sqrt{-2\mu E} \tag{6.80}$$

The Schrödinger equation can be written as

$$\frac{d^2\psi}{dx^2} + k'^2\psi = 0 \qquad \text{inside the well} \tag{6.81a}$$

and

$$\frac{d^2\psi}{dx^2} - \kappa^2\psi = 0 \qquad \text{outside the well} \tag{6.81b}$$

As for any even potential, we may restrict the search for eigenfunctions to those of definite parity.

Inside the well we have

$$\begin{aligned} \psi &= A' \cos k'x \qquad \text{for even parity} \\ \psi &= B' \sin k'x \qquad \text{for odd parity} \end{aligned} \tag{6.82}$$

Outside we have in either case only the decreasing exponential

$$\psi = C'e^{-\kappa|x|} \tag{6.83}$$

since the wave function must not become infinite at large distances.

It is necessary to match the wave function and its first derivative at $x = a$, i.e., to require that as $\varepsilon \to 0$

$$\psi(a - \varepsilon) = \psi(a + \varepsilon) \tag{6.84}$$

and

$$[\psi'(x)]_{a-\varepsilon} = [\psi'(x)]_{a+\varepsilon} \tag{6.85}$$

Since an overall constant factor remains arbitrary until determined by normalization, these two conditions are equivalent to demanding that the *logarithmic derivative* of ψ,

$$\frac{1}{\psi} \frac{d\psi}{dx} \tag{6.86}$$

shall be continuous at $x = a$. This is a very common way of phrasing the joining conditions.

The logarithmic derivative of the outside wave function, evaluated at $x = a$, is $-\kappa$; that of the inside wave function is $-k' \tan k'a$ for the even case, and $k' \cot an k'a$ for the odd case.

The transcendental equations

$$\begin{aligned} k' \tan k'a &= \kappa & \text{(even)} \\ k' \cot an k'a &= -\kappa & \text{(odd)} \end{aligned} \tag{6.87}$$

permit us to determine the allowed eigenvalues of the energy E.

The general symmetry considerations of Section 6.6 can also be extended to solutions of the Schrödinger equation with negative values of k^2 and the energy. In (6.40) we need only replace k by $i\kappa$ and κ by ik' in the regions $|x| > a$. The solution then takes the form

$$\psi(x) = \begin{cases} Ae^{-\kappa x} + Be^{\kappa x} & (x < -a) \\ Ce^{-ik'x} + De^{ik'x} & (-a < x < a) \\ Fe^{-\kappa x} + Ge^{\kappa x} & (a < x) \end{cases}$$

By applying invariance under time reversal and the principle of conservation of probability, we see that M must now be a real matrix with det $M = 1$. The boundary conditions at large distances require that $A = G = 0$. In terms of the matrix M, we must thus demand that

$$M_{11} = 0 \tag{6.88}$$

and this equation yields the energy eigenvalues. Reflection symmetry implies that $M_{12} = \pm 1$, giving us the even and odd solutions.

Exercise 6.13. Verify the statements in the last paragraph, and also show that the S matrix as a function of energy is singular at the bound states.

Exercise 6.14. By changing V_0 into $-V_0$ in (6.42), show that, for a square well, (6.88) is equivalent to the eigenvalue conditions (6.87).

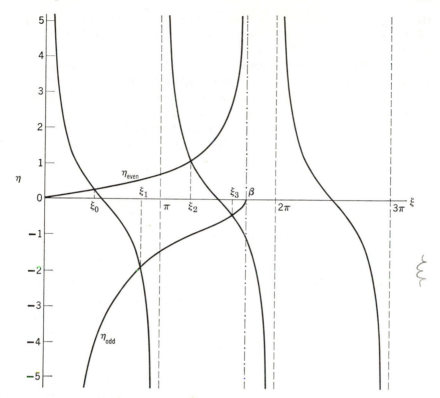

Figure 6.8. Graphical determination of the energy levels in a square well with $\beta = \sqrt{30}$.

A simple graphical method aids in visualizing the roots of (6.87). We set

$$\xi = k'a; \qquad \beta = \frac{\sqrt{2\mu V_0}}{\hbar}\, a; \qquad \sqrt{\beta^2 - \xi^2} = \kappa a$$

and

$$\eta = \cot an\ \xi$$

In Figure 6.8 η is plotted as a function of ξ. The required roots are found by determining the points of intersection of this curve with the curves

$$\eta_{\text{even}} = \xi(\beta^2 - \xi^2)^{-1/2} \qquad \text{(even)}$$

and

$$\eta_{\text{odd}} = \frac{-(\beta^2 - \xi^2)^{1/2}}{\xi} \equiv -\frac{1}{\eta_{\text{even}}} \qquad \text{(odd)} \qquad (6.89)$$

We see that the only pertinent parameter is the value of β. By inspection of Figure 6.8 we can immediately draw several conclusions: All bound states of the well are nondegenerate; even and odd solutions alternate as the energy increases; the number of bound states is finite and equal to $N + 1$, if $N\pi/2 < \beta \le (N + 1)\pi/2$; if the bound states are labeled in order of increasing energy by a quantum number $n = 0, 1, \ldots, N$, even values of n correspond to even parity, odd values of n correspond to odd parity, and n denotes the number of nodes; for any square well there is always at least one even state, but there can be no odd state unless $\beta > \pi/2$.

A case of special interest is the infinitely deep well, $V_0 \to \infty$, or $\beta \to \infty$. It is apparent that the roots of the equations expressing the boundary conditions are now

$$\xi = (n + 1)\frac{\pi}{2} \qquad (n = 0, 1, 2, \ldots)$$

or

$$E + V_0 = (n + 1)^2 \frac{\pi^2 \hbar^2}{8\mu a^2} \tag{6.90}$$

$E + V_0$, the distance in energy from the bottom of the well, is simply the kinetic energy of the particle in the well. Since $\kappa \to \infty$ as $V_0 \to \infty$ and $E \to -\infty$, the wave function itself must vanish at $x = \pm a$. There is in this limit no condition on the slope.[6] Taking into account a shift V_0 of the zero of energy and making the identification $2a = L$, the energy levels (6.90) for odd values of n coincide with the spectrum (6.29). Note that the number of states is essentially the same in either case, since there is double degeneracy in (6.29) for all but the lowest level, whereas (6.90) has no degeneracy, but between any two levels (6.29) there lies one given by (6.90) corresponding to even values of n. There is, however, no eigenstate of the infinitely deep well at $E + V_0 = 0[n = 0$ in (6.29)], because the corresponding eigenfunction vanishes.

To conclude this chapter we discuss briefly what happens to a particle incident from a great distance when it is scattered by a square well. Here $E > 0$. Actually this problem has already been solved. We may carry over the results for the potential barrier, replacing V_0 by $-V_0$ and κ by ik', where $\hbar k' = \sqrt{2\mu(E + V_0)}$. Equation (6.42) becomes

$$\begin{pmatrix} A \\ B \end{pmatrix} = \begin{pmatrix} \left(\cos 2k'a - \dfrac{i\varepsilon'}{2}\sin 2k'a\right)e^{2ika} & \dfrac{i\eta'}{2}\sin 2k'a \\[2ex] -\dfrac{i\eta'}{2}\sin 2k'a & \left(\cos 2k'a + \dfrac{i\varepsilon'}{2}\sin 2k'a\right)e^{-2ika} \end{pmatrix} \begin{pmatrix} F \\ G \end{pmatrix} \tag{6.91}$$

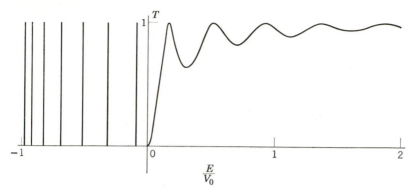

Figure 6.9. Transmission coefficient T versus E/V_0 for a square well with $\beta = 13\pi/4$. The spikes on the left are at the positions of the discrete energy levels.

where

$$\varepsilon' = \frac{k'}{k} + \frac{k}{k'}, \qquad \eta' = \frac{k}{k'} - \frac{k'}{k} \qquad (6.92)$$

Equation (6.91) defines the matrix M for the square well.
   According to (6.63), the transmission coefficient T is

$$T = \frac{1}{|M_{11}|^2} = \frac{1}{\cos^2 2k'a + \dfrac{\varepsilon'^2}{4} \sin^2 2k'a} \qquad (6.93)$$

As $E \to \infty$, $\varepsilon' \to 2$, and $T \to 1$, as expected.
   As a function of energy, the transmission coefficient rises from zero, fluctuates between maxima ($T = 1$) at $2k'a = n\pi$ and minima near $2k'a = (2n + 1)\pi/2$, and approaches the classical value $T = 1$ at the higher energies. Figures 6.9 and 6.10 show this behavior for two different values of $\beta = \sqrt{2\mu V_0 a^2}/\hbar$. The maxima occur when the distance $4a$ that a particle covers in traversing the well and back equals an integral number of de Broglie wavelengths, so that the incident wave and the wave which has been reflected inside the well are in phase reinforcing each other. If the well is deep and the energy E low (β and $\varepsilon' \gg 1$), the peaks stand out sharply between comparatively flat minima (see Figure 6.11). When the peaks in the transmission curve are pronounced, they are said to represent *resonances*. The energy values at which T reaches its maximum are determined by precisely the same condition as the truly discrete energy levels in an infinite square well of the same width [(6.90)].

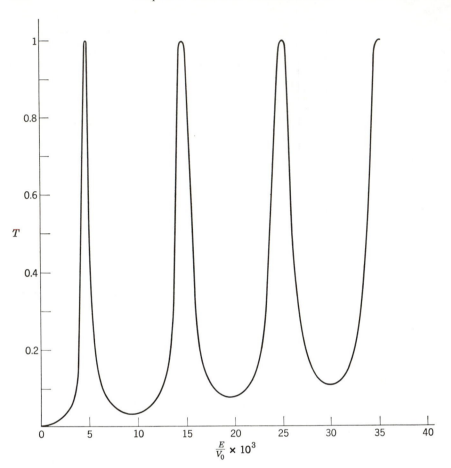

Figure 6.10. Transmission coefficient T versus E/V_0 for a square well with $\beta = 315$. As E increases, the resonances become broader.

It is instructive to consider the motion of a simple wave packet like (3.5) incident from the left upon the square well

$$\psi_{\text{inc}}(x, t) = \int_0^{\infty} f(k) e^{i(kx - \omega t)}\, dk \qquad (6.94)$$

for $x < -a$. Hence $\omega = \hbar k^2 / 2\mu$, and $f(k)$ is assumed to be a real positive function with a peak at $k = k_0$ (>0), corresponding to a velocity v_0 and energy E_0. As was discussed at length in Chapter 2, the location of the peak of the wave packet $\psi(x, t)$ is determined by the requirement of stationary phase evaluated at $k = k_0$. Thus the peak (6.94) moves uniformly toward the right according to the equation

$$x = \frac{\hbar k_0}{\mu} t = v_0 t \tag{6.95}$$

In the presence of a square well, (6.94) cannot be the complete wave function. For $x < -a$ there is also a reflected wave. If we set, with $G = 0$,

$$\frac{F}{A} = \frac{1}{M_{11}} = S_{12} = \sqrt{T} \, e^{-i\varphi} \tag{6.96}$$

$$\frac{B}{A} = \frac{M_{21}}{M_{11}} = S_{11} = \sqrt{1 - T} \, e^{-i(\varphi \pm \pi/2)} \tag{6.97}$$

thus defining a *phase shift* φ, the reflected wave may be written as

$$\psi_{\text{refl}}(x, t) = \int_0^\infty f(k)\sqrt{1 - T} \, e^{-i(kx + \omega t + \varphi \pm \pi/2)} \, dk \tag{6.98}$$

The peak of this wave packet is determined by

$$\frac{d}{dk_0} [k_0 x + \omega_0 t + \varphi(k_0)] = 0$$

which leads to the equation of motion

$$x = -v_0 \left[t + \hbar \left(\frac{\partial \varphi}{\partial E} \right)_{k=k_0} \right] \tag{6.99}$$

for $x < -a$. The time delay between arrival of the incident wave peak and the appearance of the reflected peak at the left edge of the well is $2a/v_0 - \hbar(\partial\varphi/\partial E)$, and Figure 6.11 displays this time interval (dotted curve minus solid curve).

Finally, there is to the right of the square well ($x > a$) the transmitted wave packet:

$$\psi_{\text{trans}}(x, t) = \int_0^\infty f(k)\sqrt{T} \, e^{i(kx - \omega t - \varphi)} \, dk \tag{6.100}$$

The equation of motion of its peak is

$$x = v_0 \left[t + \hbar \left(\frac{\partial \varphi}{\partial E} \right)_{k=k_0} \right] \tag{6.101}$$

for $x > a$.

The *phase shift* φ can be calculated from (6.91). We find

$$\varphi = 2ka - \arctan \left(\frac{\varepsilon'}{2} \tan 2k'a \right) \tag{6.102}$$

Exercise 6.15. Prove relation (6.102).

Figure 6.11 shows that the phase shifts also are subject to resonant behavior.

The interpretation of (6.101) is straightforward. If the potential well were absent, the coordinate of the center of the wave packet would be $x = v_0 t$ at all times, but since the particle moves faster inside the well than outside, it reaches the region $x > a$ earlier by an interval $\hbar(\partial\varphi/\partial E)$ than it would if V_0 were zero. If the well is deep and v_0 small, which is the condition for resonances to appear, this time interval would classically be approximately equal to $2a/v_0$ but $\hbar(\partial\varphi/\partial E)$ generally differs from the classical value. At the resonances it has the approximate value a/v_0, indicating that quantum mechanics introduces a time delay a/v_0 into the transmitted (and also the reflected) wave. This delay is a result of the wave nature of matter which causes repeated reflections back and forth in the well. Figure 6.12 gives us an idea of the number of times the wave packet bounces to and fro inside the well before it escapes. It should be noted that our estimate of the time delay introduced by the well is necessarily crude, since near resonance the shape of the wave

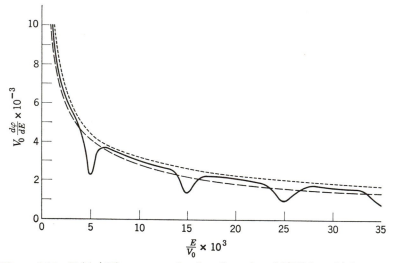

Figure 6.11. $V_0(d\varphi/dE)$ measures the time (in units of $\hbar/V_0$) by which a wave packet transmitted through a square well ($\beta = 315$) *precedes* a particle which, simultaneously released, moves with constant speed $v_0 = \sqrt{2E/\mu}$ throughout. The dashed curve corresponds to the advance of a classical particle moving through the well. The dotted curve is drawn for a classical particle assumed to move through the well with infinite speed. Since there can be no transmission before the incident particle has reached the well, the dotted curve is an upper limit for the time advance.

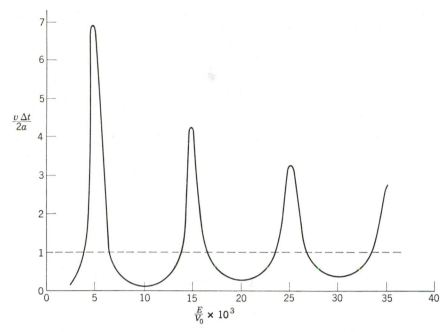

Figure 6.12. Average time Δt spent by a wave packet in the well ($\beta = 315$) in units of the time, $2a/v$, which it takes a particle of energy E to traverse the well. At higher energies the classical value, unity, is approached.

packet is drastically distorted and (6.101) represents an average position of the packet rather than the coordinate of its peak (see Section 7.4).

Exercise 6.16. Show that at a resonance of energy E_0,

$$\frac{\partial \varphi}{\partial E} = \frac{a}{\hbar v_0} \frac{1}{1 + (E_0/V_0)}$$

Resonance peaks in the transmission of particles are typical quantum features, and the classical picture is not capable of giving a simple account of such strong but smooth energy variations. Classically, depending on the available energy, T can only be zero or one, whereas in quantum mechanics T changes continuously between these limits.[10] Thus in a certain sense quantum mechanics attributes to matter more continuous and less abrupt characteristics than classical mechanics.

[10] The transmission coefficient for a potential barrier, Figure 6.5, is different from zero, although numerically small, for $E < V_0$ [(6.45)] and also varies continuously with energy. Classically, T jumps from 0 to 1 at $E = V_0$.

While these observations have general validity, it should be stressed that their verification by extending the solutions of the Schrödinger equation for discontinuous potentials to the classical limit meets with some obstacles.[11] For example, the reflection coefficient (6.38) is a function of the particle momentum only and, hence, it is apparently applicable to a particle moving under classical conditions. Yet classically, R is either 0 or 1. This paradox is resolved if we recognize that the correct classical limit of the quantum equations is obtained only if care is taken to keep the de Broglie wavelength short in comparison with the distance over which the fractional change of the potential is appreciable. The Schrödinger equation for the piecewise constant potential patently violates this requirement, but the next chapter will deal with potentials for which this condition is well satisfied.

Problems

1. Obtain the transmission coefficient for a rectangular potential barrier if the energy exceeds the height V_0 of the barrier. Plot the transmission coefficient as a function of E/V_0 (up to $E/V_0 = 3$) choosing $(2\mu V_0)^{1/2}a = 0.75h$.

2. Consider a potential $V = 0$ for $x > a$, $V = -V_0$ for $a \geqslant x \geqslant 0$, $V = +\infty$ for $x < 0$. Show that for $x > a$ the positive energy solutions of the Schrödinger equation have the form

$$e^{i(kx+2\delta)} - e^{-ikx}$$

Calculate the *scattering coefficient* $|1 - e^{2i\delta}|^2$ and show that it exhibits maxima (resonances) at certain discrete energies if the potential is sufficiently deep and broad.

3. Solve the Schrödinger equation in one dimension for an attractive delta function potential by regarding it as an infinitely deep and narrow square well such that V_0a remains finite. Show that there is one bound state and calculate its energy. Verify the result by integrating the Schrödinger equation directly. Also calculate the transmission coefficient for positive energies.

4. A particle of mass μ moves in the one-dimensional double well potential

$$V(x) = -g\,\delta(x - a) - g\,\delta(x + a)$$

If $g > 0$, obtain transcendental equations for the two energy eigenvalues of the system. Estimate the splitting between the energy levels in the limit of large a.

   If there is in addition a potential energy $V = \lambda g/(2a)$ corresponding to a mutual repulsion of the wells ("atoms"), show that, for a sufficiently small value of λ,

---

[11] For other interesting manifestations of abrupt changes in a potential, see the computer generated films on *Scattering in One Dimension*, Parts I to IV, produced by the MIT-Education Research Center and the Education Development Center, Newton, Mass. The fabrication of the films is described by their authors A. Goldberg, H. M. Schey, and J. L. Schwartz, *Am. J. Phys.*, **35**, 177 (1967). For information about these and other films on quantum mechanics, see W. R. Riley, Resource Letter BSPF-1, *Am. J. Phys.* **36**, 1 (1968).

the system ("molecule") is stable if the particle ("electron") is in the even parity state. Sketch the total potential energy of the system as a function of a.

If $g < 0$, calculate the transmission coefficient and show that it exhibits resonances.

5. In the periodic potential of Section 6.7 let the "hills" become delta functions by letting $a \to 0$ and $V_0 \to \infty$ such that $V_0 a$ remains finite. Show that the eigenvalue condition becomes

$$\cos kl = \cos 2kb + (C/2kb) \sin 2kb$$

and make a graph of the function on the right-hand side versus $2kb$ for $C = 3\pi/2$. Exhibit the occurrence of allowed and forbidden energy bands.

6. Prove that $kl = 2k'a + 2kb$ in the limit of high energies for the periodic potential of Section 6.7, and compare the numerical consequences of this relation with the exact curve in Figure 6.6.

CHAPTER 7

The WKB Approximation

1. The Method. If the potential energy does not have a very simple form, the solution of the Schrödinger equation even in one dimension,

$$\frac{d^2\psi}{dx^2} + \frac{2\mu}{\hbar^2}(E - V)\psi = 0 \tag{7.1}$$

is usually a complicated problem which requires the use of approximation methods. Some of these, for instance, perturbation and variational methods, briefly described in Section 4.6, are quite general and will be discussed in detail after we have freed ourselves from the limitation to one dimension. One particular method, however, is particularly suitable for obtaining approximate solutions to ordinary differential equations, and it is appropriate that we should take it up now. This is the so-called WKB method, named after its proponents in quantum mechanics, Wentzel, Kramers, and Brillouin. The WKB method can also be applied to three-dimensional problems, if the potential is spherically symmetric and a radial differential equation can be separated.

The basic idea is simple. If $V = \text{const.}$, (7.1) has the solutions $e^{\pm ikx}$. This suggests that if V, while no longer constant, varies only slowly with x, we might try a solution of the form

$$\psi(x) = e^{iu(x)} \tag{7.2}$$

except that the function $u(x)$ now is not simply proportional to x. Substitution of (7.2) into (7.1) gives us an equation for the x-dependent "phase," $u(x)$. This equation becomes particularly simple if we use the abbreviations

$$k(x) = \left\{ \frac{2\mu}{\hbar^2} [E - V(x)] \right\}^{1/2} \quad \text{if} \quad E > V(x) \tag{7.3a}$$

and

$$k(x) = -i \left\{ \frac{2\mu}{\hbar^2} [V(x) - E] \right\}^{1/2} = -i\kappa(x) \quad \text{if} \quad E < V(x) \tag{7.3b}$$

We then find that $u(x)$ satisfies the equation

$$i \frac{d^2u}{dx^2} - \left(\frac{du}{dx}\right)^2 + [k(x)]^2 = 0 \tag{7.4}$$

This differential equation is entirely equivalent to (7.1), but the boundary conditions are more easily expressed in terms of $\psi(x)$ than $u(x)$. The fact that (7.4) is a nonlinear equation, whereas the Schrödinger equation is linear, would usually be regarded as a drawback, but in this chapter we shall take advantage of the nonlinearity to develop a simple approximation method for solving (7.4). Indeed, an iteration procedure is suggested by the fact that u'' is zero for the free particle. We are led to suspect that this second derivative remains relatively small if the potential does not vary too violently. When we omit this term from the equation entirely, we obtain the first crude approximation, u_0, to u:

$$u_0'^2 = [k(x)]^2 \tag{7.5}$$

or, integrating this,

$$u_0 = \pm \int^x k(x)\, dx + C \tag{7.6}$$

This is the approximation to the wave function which in Chapter 3 was employed to establish the wave equation.

A successive approximation method can now be set up, if we cast (7.4) in the form[1]

$$\left(\frac{du}{dx}\right)^2 = k^2(x) + i \frac{d^2u}{dx^2} \tag{7.7}$$

We substitute the nth approximation on the right-hand side of this equation, and obtain from (7.7) the $(n+1)$th approximation by a mere quadrature:

$$u_{n+1}(x) = \pm \int^x \sqrt{k^2(x) + iu_n''(x)}\, dx + C_{n+1} \tag{7.8}$$

Thus, we have for $n = 0$

$$u_1(x) = \pm \int^x \sqrt{k^2(x) + iu_0''(x)}\, dx + C_1 = \pm \int^x \sqrt{k^2(x) \pm ik'(x)}\, dx + C_1 \tag{7.9}$$

Our hope that this procedure will yield a wave function which approximates

---

[1] If u_a and u_b are two particular solutions of (7.7) which differ not only by a constant, the general solution is

$$u(x) = u_a - i \log\left[1 + Ae^{i(u_b - u_a)}\right] + B$$

where A and B are arbitrary constants. The corresponding $\psi(x)$ is

$$\psi(x) = e^{iu(x)} = e^{iB}e^{iu_a} + Ae^{iB}e^{iu_b}$$

The two different signs in (7.6) give two particular u's, and therefore lead to the general solution of (7.1).

the correct $u(x)$ is baseless unless $u_1(x)$ is close to $u_0(x)$, i.e., unless

$$|k'(x)| \ll |k^2(x)| \tag{7.10}$$

In (7.9) both signs must be chosen the same as in the u_0 upon which u_1 is supposed to be an improvement. If condition (7.10) holds, we may expand the integrand and obtain

$$u_1(x) \simeq \int^x \left[\pm k(x) + \frac{i}{2}\frac{k'(x)}{k(x)} \right] dx + C_1 = \pm \int^x k(x)\, dx + \frac{i}{2} \log k(x) + C_1 \tag{7.11}$$

The constant of integration is of no moment, because it only affects the normalization of $\psi(x)$ which, if needed at all, is best accomplished after all the desired approximations have been made.

The approximation (7.11) to (7.4) is known as *WKB approximation*. It leads to the approximate wave function

$$\psi(x) \simeq \frac{1}{\sqrt{k(x)}} \exp \left[\pm i \int^x k(x)\, dx \right] \tag{7.12}$$

Condition (7.10) can be formulated in ways which are better suited to physical interpretation. If $k(x)$ is regarded as the effective wave number, we may for $E > V(x)$ define an effective wavelength

$$\lambda(x) = \frac{2\pi}{k(x)}$$

Condition (7.10) can then be written as

$$\lambda(x) \left| \frac{dp}{dx} \right| \ll |p(x)| \tag{7.13}$$

where $p(x) = \pm \hbar k(x)$ is the momentum which the particle would possess classically at point x. Condition (7.10) thus implies that the change of the momentum over a wavelength must be small compared to the momentum itself. This condition was already quoted in (3.12). It obviously breaks down if $k(x)$ vanishes or if $k(x)$ varies very rapidly. This certainly happens at the classical turning points for which

$$V(x) = E \tag{7.14}$$

or whenever $V(x)$ has a very steep behavior. In either case a more accurate solution must be used in the region where (7.10) breaks down. Yet the WKB method is not particularly useful unless we find ways to extend the wave function through these regions.

2. *The Connection Formulas.* Consider the most important problem arising from the breakdown of condition (7.10): Suppose that $x = a$ is a *classical turning point*[2] for the motion with the given energy E, as shown in Figure 7.1.

---

[2] In Figure 7.1 the barrier is to the right of the classical turning point. Analogous considerations hold if the barrier is to the left. For a summary of results see (7.25) and (7.26).

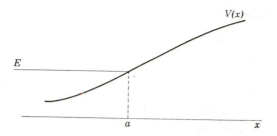

Figure 7.1. Classical turning point at $x = a$.

Assume that the WKB approximation is applicable except in the immediate neighborhood of the turning point. The discussion is considerably simplified if we change *dependent* as well as *independent* variables, introducing

$$v(x) = \sqrt{k(x)}\,\psi(x) \tag{7.15}$$

and

$$y = \int^{x} k(x)\,dx \tag{7.16}$$

By a little manipulation we obtain instead of the Schrödinger equation:

$$\frac{d^2v}{dy^2} + \left[\frac{1}{4k^2}\left(\frac{dk}{dy}\right)^2 - \frac{1}{2k}\frac{d^2k}{dy^2} + 1\right]v = 0 \tag{7.17}$$

From the earlier discussion it is clear that the WKB approximation is equivalent to replacing the bracket in (7.17) by unity. Indeed, if the first two terms in the bracket are neglected, we have the solutions $v(y) = e^{\pm iy}$, which by relation (7.15) give us the WKB wave functions (7.12). In the WKB region to the left of the classical turning point in Figure 7.1 y is real, and (7.17) has the approximate solution

$$v \simeq Ae^{iy} + Be^{-iy} \tag{7.18}$$

In the WKB region to the right of the turning point y is imaginary, and

$$v \simeq Ce^{|y|} + De^{-|y|} \tag{7.19}$$

We now ask the fundamental question: How are the coefficients C and D related to A and B if (7.18) and (7.19) are to represent the same state, albeit in different regions? The answer to this question can be found only if the unabridged equation (7.17) is integrated near the turning point. This requires that a somewhat special assumption be made about the behavior of the potential energy near the turning point.

We shall suppose that in the neighborhood of $x = a$ we may write

$$V(x) - E \approx \alpha(x - a) \tag{7.20}$$

where $\alpha > 0$. By (7.3a and b)

$$k(x) = \begin{cases} \left[\dfrac{2\mu\alpha}{\hbar^2}(a-x)\right]^{1/2} & \text{for } x < a \\[2ex] \exp\left(-\dfrac{i\pi}{2}\right)\left[\dfrac{2\mu\alpha}{\hbar^2}(x-a)\right]^{1/2} & \text{for } x > a \end{cases} \tag{7.21}$$

The multivaluedness of the fractional powers with which we have to deal here demands that attention must be paid to the phases. If this advice is not followed, inconsistencies arise which lead to wrong answers. All fractional powers of positive quantities are understood to be positive, and the phases have been chosen arbitrarily but definitely.

When the particular form of k given by (7.21) is substituted in (7.16), we can evaluate the integral. If the lower limit of integration is chosen to be $x = a$, i.e., such that $y(a) = 0$, y becomes a measure of the distance from the classical turning point. y is then small near the turning point because the two limits of integration are close to each other, and also because the integrand $k(x)$ is small near a. Conversely, at points far enough to the right or left of the turning point so that the WKB approximation becomes applicable, y is large in absolute value—again on both accounts. Explicitly,

$$y = \begin{cases} \dfrac{2}{3}\left(\dfrac{2\mu\alpha}{\hbar^2}\right)^{1/2}(a-x)^{3/2}e^{i\pi} & \text{for } x < a \\[2ex] \dfrac{2}{3}\left(\dfrac{2\mu\alpha}{\hbar^2}\right)^{1/2}(x-a)^{3/2}e^{-i\pi/2} & \text{for } x > a \end{cases} \tag{7.22}$$

If we express y in terms of k and then calculate the derivatives of y, (7.17) takes on a remarkably simple form:

$$\frac{d^2v}{dy^2} + \left(1 + \frac{5}{36y^2}\right)v = 0 \tag{7.23}$$

For large $|y|$ the second term in the parenthesis may be neglected. The asymptotic solutions of (7.23), $v = e^{\pm iy}$, yield again the WKB approximation. Equation (7.23) is accurate near the turning point, but the assumption is

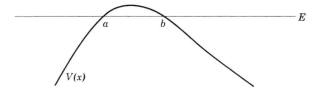

Figure 7.2. The near coincidence of two classical turning points requires special treatment in the WKB approximation.

made that it is also a good approximation to the Schrödinger equation in the intermediate region where y has moderate values.

Clearly, this entire approach breaks down if, for instance, the energy is close in value to an extremum of the potential (see Figure 7.2), because, proceeding from left to right, turning point b is reached before one gets sufficiently far away from a for the WKB approximation to hold. If, on the other hand, our procedure is valid, (7.23) can be used to connect the WKB wave functions across the classical turning point. To this end the asymptotic behavior of the solutions to (7.23) must be considered in detail. The mathematical work, using integral representations of the solutions of (7.17), is found in Section 7.5. Only the results will be quoted here.

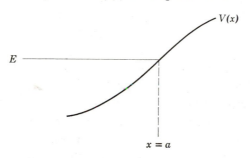

$$\frac{2}{\sqrt{k}} \cos\left(\int_x^a k\,dx - \tfrac{1}{4}\pi\right) \longleftrightarrow \frac{1}{\sqrt{\kappa}} \exp\left(-\int_a^x \kappa\,dx\right) \qquad (7.25a)$$

$$\frac{1}{\sqrt{k}} \sin\left(\int_x^a k\,dx - \tfrac{1}{4}\pi\right) \longleftrightarrow -\frac{1}{\sqrt{\kappa}} \exp\left(\int_a^x \kappa\,dx\right) \qquad (7.25b)$$

Figure 7.3. The turning point is to the right of the classical region.

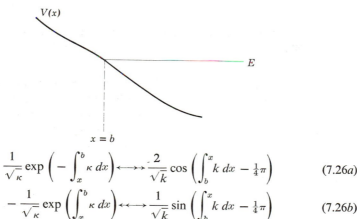

$$\frac{1}{\sqrt{\kappa}} \exp\left(-\int_x^b \kappa\,dx\right) \longleftrightarrow \frac{2}{\sqrt{k}} \cos\left(\int_b^x k\,dx - \tfrac{1}{4}\pi\right) \qquad (7.26a)$$

$$-\frac{1}{\sqrt{\kappa}} \exp\left(\int_x^b \kappa\,dx\right) \longleftrightarrow \frac{1}{\sqrt{k}} \sin\left(\int_b^x k\,dx - \tfrac{1}{4}\pi\right) \qquad (7.26b)$$

Figure 7.4. The turning point is to the left of the classical region.

The formulas connecting the wave functions to the left and right of the turning point in Figure 7.1 are

$$\frac{\cos\left(-y - \tfrac{1}{4}\pi\right)}{\sqrt{k}} \longleftarrow\!\!\!\longleftarrow \frac{1}{2}\frac{e^{-|y|}}{\sqrt{\kappa}} \qquad (7.24a)$$

$$\frac{\sin\left(-y - \tfrac{1}{4}\pi\right)}{\sqrt{k}} \longleftrightarrow\!\!\!\longrightarrow -\frac{e^{|y|}}{\sqrt{\kappa}} \qquad (7.24b)$$

We recognize these wave functions as the appropriate WKB solutions to the Schrödinger equation. Caution must be exercised in the use of the formulas. Suppose that we know the wave function is adequately represented far to the right of the turning point (Figure 7.3) by the increasing exponential in (7.24b). It is then in general not legitimate to infer that to the far left of the turning point the wave function is given by $\sin\left(-y - \tfrac{1}{4}\pi\right)/\sqrt{k}$. After all, an admixture of decreasing exponential would be considered negligible to the far right of the turning point although it might, according to (7.24a), contribute an appreciable amount of $\cos\left(-y - \tfrac{1}{4}\pi\right)/\sqrt{k}$ to the wave function on the left. Conversely, a minute admixture of $\sin\left(-y - \tfrac{1}{4}\pi\right)/\sqrt{k}$ to $\cos\left(-y - \tfrac{1}{4}\pi\right)/\sqrt{k}$ on the left (Figure 7.4) might be negligible there but might lead to a very appreciable exponentially increasing portion on the right, if the solutions are used for sufficiently large $|y|$. Thus we see that unless we have assured ourselves properly of the absence of the other linearly independent component in the wave function, the *connection formulas*, summarized for both kinds of classical turning points in equations (7.25a, b) and (7.26a, b), should be used only in the directions indicated by the double arrow if considerable error is to be avoided.[3]

Exercise 7.1. Show that the WKB approximation is consistent with conservation of probability, even across classical turning points.

3.   Application to Bound States. The WKB approximation can be applied to derive an equation for the energies of bound states. The basic idea emerges if we choose a simple well-shaped potential with two classical turning points as shown in Figure 7.5. The WKB approximation will be used in regions 1, 2, and 3 away from the turning points, and the connection formulas will serve near $x = a$ and $x = b$. The usual requirement that ψ must be finite dictates that the solutions which increase exponentially as one moves outward from the turning points must vanish rigorously. Thus, in region 1 the unnormalized

---

[3] For a detailed exposition, see N. Fröman and P. O. Fröman, *JWKB Approximation*, North-Holland Publishing Co., Amsterdam 1965.

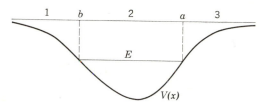

Figure 7.5. Simple potential well. Classically, a particle of energy E is confined to the region between a and b.

wave function is

$$\psi_1 \simeq \frac{1}{\sqrt{\kappa}} \exp\left(-\int_x^b \kappa\, dx\right) \qquad \text{for } x < b$$

Hence, by equation (7.26a),

$$\psi_2 \simeq \frac{2}{\sqrt{k}} \cos\left(\int_b^x k\, dx - \tfrac{1}{4}\pi\right) \qquad \text{for } b < x < a$$

This may also be written as

$$\psi_2 \simeq \frac{2}{\sqrt{k}} \cos\left(\int_b^a k\, dx - \int_x^a k\, dx - \tfrac{1}{4}\pi\right)$$

$$= -\frac{2}{\sqrt{k}} \cos\left(\int_b^a k\, dx\right) \sin\left(\int_x^a k\, dx - \tfrac{1}{4}\pi\right)$$

$$+ \frac{2}{\sqrt{k}} \sin\left(\int_b^a k\, dx\right) \cos\left(\int_x^a k\, dx - \tfrac{1}{4}\pi\right)$$

By (7.25) only the second of these two terms gives rise to a decreasing exponential satisfying the boundary conditions at infinity. Hence, the first term must vanish. We obtain the condition

$$\int_b^a k\, dx = (n + \tfrac{1}{2})\pi \tag{7.27}$$

where $n = 0, 1, 2, \ldots$. This equation determines the possible discrete values of E. E appears in the integrand as well as in the limits of integration, since the turning points a and b are determined such that $V(a) = V(b) = E$.

If we introduce the classical momentum $p(x) = \pm \hbar k(x)$ and plot $p(x)$ versus x in phase space, the bounded motion in a potential well can be pictured by a closed curve (Figure 7.6). It is then evident that condition (7.27) may be written as

$$J \equiv \oint p(x)\, dx = (n + \tfrac{1}{2})h \tag{7.28}$$

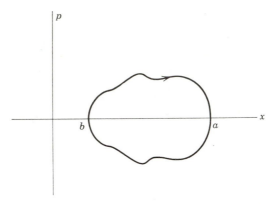

Figure 7.6. Phase space representation of the periodic motion of a particle confined between the classical turning points at $x = a$ and $x = b$.

This equation is very similar to the quantum condition (1.2) in the old quantum theory which occupied a position intermediate between classical and quantum mechanics.

The left-hand side of (7.28), which equals the area enclosed by the curve representing the motion in phase space, is called the *phase integral J* in classical terminology. It also measures the phase change which the oscillatory wave function ψ_2 undergoes between the turning points, for if the WKB approximation is used all the way from b to a,

$$\int_b^a k(x)\, dx = (n + \tfrac{1}{2})\pi$$

is the phase change across the well. Dividing this by 2π, we see that according to the WKB approximation $\tfrac{1}{2}n + \tfrac{1}{4}$ quasi wavelengths fit between b and a. Hence, n represents the number of nodes in the wave function, a fact which helps to visualize the elusive ψ.

According to (7.28), the area of phase space between one bound state and the next is equal to h. From this observation we infer the statement, often heard in statistical mechanics, that each quantum state occupies a volume h in phase space. This rule is useful in the domain where classical mechanics is actually applicable but some concession is to be made to the quantum structure of matter.

That the term *classical approximation* for the WKB method may not be amiss can be seen by noting that for high energies ψ_2 has a very short wavelength. It is a rapidly oscillating function of position but its amplitude is modulated slowly by a factor $1/\sqrt{k(x)}$. The probability, $|\psi_2|^2\, dx$, of finding the particle in an interval dx at x is proportional to the reciprocal of the classical velocity, $1/v(x)$. This is also classically the relative probability of finding the particle in the interval dx if a random determination of its position is made as the particle shuttles back and forth between the turning points. We see thus that the probability concepts used in quantum and classical mechanics, although basically different, are nevertheless related to each other in the limit in which the rapid phase fluctuations of quantum mechanics can be legitimately averaged to give the approximate classical motion.

Exercise 7.2. Show that the WKB approximation gives the energy levels of the linear harmonic oscillator correctly. Sketch the WKB approximation to the eigenfunctions for $n = 0$ and 1.

4. Transmission through a Barrier. The WKB method will now be applied to calculate the transmission coefficient for a barrier upon which particles are incident from the left with insufficient energy to pass to the other side classically. This problem is very similar to that of the rectangular potential barrier, Section 6.5, but no special assumption will be made here concerning the shape of the barrier.

If the WKB approximation is assumed to hold in the three regions indicated in Figure 7.7, the solution of the Schrödinger equation may be written as

$$\psi(x) = \begin{cases} \dfrac{A}{\sqrt{k(x)}} \exp\left(i\int_a^x k\,dx\right) + \dfrac{B}{\sqrt{k(x)}} \exp\left(-i\int_a^x k\,dx\right) & (x < a) \\[2ex] \dfrac{C}{\sqrt{\kappa(x)}} \exp\left(-\int_a^x \kappa\,dx\right) + \dfrac{D}{\sqrt{\kappa(x)}} \exp\left(\int_a^x \kappa\,dx\right) & (a < x < b) \\[2ex] \dfrac{F}{\sqrt{k(x)}} \exp\left(i\int_b^x k\,dx\right) + \dfrac{G}{\sqrt{k(x)}} \exp\left(-i\int_b^x k\,dx\right) & (b < x) \end{cases}$$

$$(7.29)$$

The connection formulas (7.25) and (7.26) can now be used to establish linear relations between the coefficients in (7.29) in much the same way as

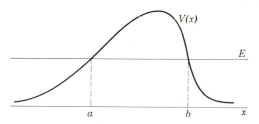

Figure 7.7. Potential barrier.

was done in Chapter 6 for the rectangular barrier. However, it must be observed that the connection formulas are applicable only if the ratios

$$\left| \frac{F - iG}{F + iG} \right| \quad \text{and} \quad \left| \frac{B - iA}{B + iA} \right|$$

are not too close to zero. The result of the calculation is remarkably simple and again best expressed in terms of a matrix M which connects F and G with A and B:

$$\begin{pmatrix} A \\ B \end{pmatrix} = \frac{1}{2} \begin{pmatrix} 2\theta + \dfrac{1}{2\theta} & i\left(2\theta - \dfrac{1}{2\theta}\right) \\ -i\left(2\theta - \dfrac{1}{2\theta}\right) & 2\theta + \dfrac{1}{2\theta} \end{pmatrix} \begin{pmatrix} F \\ G \end{pmatrix} \tag{7.30}$$

where the parameter

$$\theta = \exp\left(\int_a^b \kappa(x)\,dx\right) \tag{7.31}$$

measures the height and thickness of the barrier as a function of energy.

Exercise 7.3. Verify (7.30).

The *transmission coefficient* is defined as

$$T = \frac{|\psi_{\text{trans}}|^2\, v_{\text{trans}}}{|\psi_{\text{inc}}|^2\, v_{\text{inc}}} = \frac{|\psi_{\text{trans}}\sqrt{k_{\text{trans}}}|^2}{|\psi_{\text{inc}}\sqrt{k_{\text{inc}}}|^2} = \frac{|F|^2}{|A|^2}$$

assuming that there is no wave incident from the right, $G = 0$. From (7.30) we obtain

$$T = \frac{1}{|M_{11}|^2} = \frac{4}{\left(2\theta + \dfrac{1}{2\theta}\right)^2} \tag{7.32}$$

For a high and broad barrier $\theta \gg 1$, and

$$T \approx \frac{1}{\theta^2} = \exp\left(-2\int_a^b \kappa\,dx\right) \tag{7.33}$$

Hence, θ is a measure of the opacity of the barrier.

As an example we calculate θ for a one-dimensional model of a Coulomb repulsion barrier (Figure 7.8) such as a proton (charge Z_1e) has to penetrate to reach a nucleus (charge Z_2e). The essence of this calculation survives the

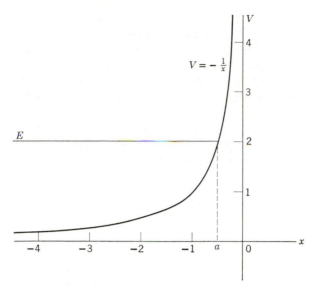

Figure 7.8. One-dimensional analog of a Coulomb barrier which repels particles incident from the left.

generalization to three dimensions (Section 11.8). Thus, let V be defined for $x < 0$ as

$$V = - \frac{Z_1 Z_2 e^2}{x} \tag{7.34}$$

The turning point a is determined by

$$E = - \frac{Z_1 Z_2 e^2}{a}$$

and we take $b = 0$. Evidently

$$\int_a^0 \kappa \, dx = \frac{\sqrt{2\mu E}}{\hbar} \int_a^0 \sqrt{\frac{a}{x} - 1} \, dx = \sqrt{\frac{2\mu}{E}} \frac{Z_1 Z_2 e^2}{\hbar} \int_0^1 \sqrt{\frac{1}{u} - 1} \, du = \frac{Z_1 Z_2 e^2}{\hbar v} \pi$$

Hence,

$$\frac{1}{\theta^2} = \exp \left(- \frac{2\pi Z_1 Z_2 e^2}{\hbar v} \right) \tag{7.35}$$

This transmission coefficient, which inhibits the approach of a positive charged particle to the nucleus, is called the *Gamow factor*. This quantity is also decisive in the description of nuclear alpha decay, since the alpha particle, once it is formed inside the nucleus, cannot escape unless it penetrates the surrounding Coulomb barrier.[4]

As a final application of the WKB method let us consider the passage of a particle through a potential well which is bounded by barriers as shown in Figure 7.9. It will be assumed that $V(x)$ is symmetric about the origin, which is located in the center of the well, and that $V = 0$ outside the interval between $-c$ and c.

In this section the effect of barrier penetration will be studied for a particle with an energy E below the peak of the barriers. We are particularly interested in the form of the free particle wave functions in regions 1 and 7:

$$\psi_1 = \frac{A_1}{\sqrt{k}} \exp(ikx) + \frac{B_1}{\sqrt{k}} \exp(-ikx)$$

$$\psi_7 = \frac{A_7}{\sqrt{k}} \exp(ikx) + \frac{B_7}{\sqrt{k}} \exp(-ikx)$$

$$(7.36)$$

When the WKB method is applied to connect the wave function in regions 1 and 7, the relation between the coefficients is again most advantageously

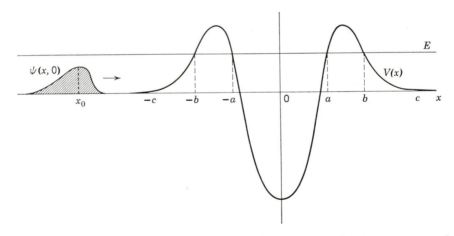

Figure 7.9. Potential barriers surrounding a well are favorable for the occurrence of narrow transmission resonances. We define as region 1: $x < -c$; region 2: $-c < x < -b$; region 3: $-b < x < -a$; region 4: $-a < x < a$; region 5: $a < x < b$; region 6: $b < x < c$; region 7: $c < x$. A wave packet is seen to be incident from the left.

[4] See J. M. Blatt and V. F. Weisskopf, *Theoretical Nuclear Physics*, John Wiley and Sons, New York, 1952. Alpha decay is treated in Chapter 11.

recorded in matrix notation:

$$\binom{A_1}{B_1} =$$

$$\frac{1}{2}\begin{pmatrix} e^{-2i\rho}\left[\left(4\theta^2 + \frac{1}{4\theta^2}\right)\cos L - 2i\sin L\right] & i\left(4\theta^2 - \frac{1}{4\theta^2}\right)\cos L \\ -i\left(4\theta^2 - \frac{1}{4\theta^2}\right)\cos L & e^{2i\rho}\left[\left(4\theta^2 + \frac{1}{4\theta^2}\right)\cos L + 2i\sin L\right] \end{pmatrix}$$

$$\times \binom{A_7}{B_7} \quad (7.37)$$

In writing these equations, the following abbreviations have been used:[5]

$$L = \frac{J}{2\hbar} = \int_{-a}^{a} k(x)\,dx, \qquad \rho = \int_{b}^{c} k(x)\,dx - kc \qquad (7.38)$$

It follows from the definition of L and from inspection of Figure 7.9 that

$$\frac{\partial L}{\partial E} > 0 \qquad (7.39)$$

The final matrix relation (7.37) has the form (6.46) subject to the conditions (6.47) and (6.48). This result is expected since, as was pointed out in Section 6.6, the matrix which links the asymptotic parts of a Schrödinger eigenfunction has the same general form for all potentials which are symmetric about the origin.

According to (6.63) the transmission coefficient is

$$T = \frac{1}{|M_{11}|^2} = \frac{4}{\left(4\theta^2 + \frac{1}{4\theta^2}\right)^2 \cos^2 L + 4\sin^2 L} \qquad (7.40)$$

This quantity reaches its maximum value, unity, whenever $\cos L = 0$, or

$$L = (2n + 1)\frac{\pi}{2}, \qquad J = (n + \tfrac{1}{2})h \qquad (7.41)$$

The condition determining the location of the transmission peaks is seen to be the same as the quantum condition (7.28) for bound states. If $\theta \gg 1$, so

[5] For a thorough discussion of barrier penetration in the WKB approximation, see D. Bohm, *Quantum Theory*, Prentice-Hall, New York, 1951. Our notation is adapted from Bohm's.

that penetration through the barriers is strongly inhibited, T has sharp, narrow *resonance* peaks at these energies. A graph of T in the resonance region will be similar to Figure 6.10.

Exercise 7.4. Show that the distance D between neighboring resonances (in terms of energy) is approximately

$$D \approx \frac{\pi}{\partial L/\partial E} \tag{7.42}$$

It is instructive to consider the motion of a simple broad wave packet incident from the left (see Figure 7.9). A wave packet, which at $t = 0$ is localized entirely in region 1 near the coordinate x_0 far to the left of the barrier and moving toward positive x, may be represented as

$$\psi(x, 0) = \int_0^\infty f(E)e^{ik(x-x_0)} \, dE \tag{7.43}$$

where $\hbar k = \sqrt{2\mu E}$. The amplitude $f(E)$ is a smoothly varying real function of energy with a fairly sharp peak and a width ΔE. If $\psi(x, 0)$ is normalized to unity, $f(E)$ satisfies the normalization condition

$$\int_0^\infty |f(E)|^2 \, v \, dE = \frac{1}{2\pi\hbar} \tag{7.44}$$

It can be shown that (7.43) may be replaced by an expansion in terms of eigenfunctions of the form (7.36). The first step in the proof is the observation that in region 1

$$\psi(x, 0) = \int_0^\infty f(E)e^{-ikx_0}\left(e^{ikx} + \frac{B_1}{A_1} e^{-ikx}\right) dE \qquad (x < -c) \tag{7.45}$$

the coefficients A_1 and B_1 being the same as those which appear in (7.36). Equation (7.45) holds because the integral

$$\int_0^\infty f(E) \frac{B_1}{A_1} e^{-ik(x+x_0)} \, dE \tag{7.46}$$

differs from zero only for values of x near $x = -x_0$. Hence, it vanishes in region 1.

If thus (7.45) represents the incident wave packet in region 1, it is necessary to check if the corresponding expansion in the other regions also gives the correct answer, $\psi(x, 0) = 0$, since the initial wave function is entirely confined in region 1. In particular, in region 7 we require

$$\int_0^\infty f(E)e^{-ikx_0}\left(\frac{A_7}{A_1} e^{ikx} + \frac{B_7}{A_1} e^{-ikx}\right) dE = 0 \qquad (x > c) \tag{7.47}$$

The first of these two integrals vanishes in region 7 for the same reason that (7.46) vanishes in region 1. However, the second integral in (7.47) leads to an unwanted contribution to $\psi(x, 0)$ in region 7, unless $B_7 = 0$ is chosen, eliminating any wave incident from the right.

From equation (7.37), for $B_7 = 0$,

$$\frac{A_7}{A_1} = \frac{1}{M_{11}} = \sqrt{T}\, e^{-i\varphi} = \frac{e^{2i\rho}}{\frac{1}{2}\left(4\theta^2 + \frac{1}{4\theta^2}\right)\cos L - i \sin L} \tag{7.48}$$

Hence, the initial wave function in region 7 may be expressed as

$$\psi(x, 0) = \int_0^\infty f(E)\sqrt{T}\, e^{-i\varphi} e^{ik(x-x_0)}\, dE \qquad (x > c) \tag{7.49}$$

By a somewhat more delicate argument it can also be shown that setting $B_7 = 0$ suffices to insure that the appropriate superposition of WKB wave functions gives no contribution to $\psi(x, 0)$ in the remaining internal regions.

Once the initial wave function has been written as a superposition of (approximate) eigenfunctions of the Schrödinger equation for the barrier and well, the wave function at arbitrary times is obtained simply by inserting the appropriate oscillatory phase. To the right of the barrier we obtain

$$\psi(x, t) = \int_0^\infty f(E)\sqrt{T}\exp(-i\varphi)\exp\left[ik(x - x_0) - \frac{i}{\hbar}Et\right] dE \qquad (x > c) \tag{7.50}$$

In order to study the behavior of the transmitted wave packet near a *resonance* we assume that the incident wave packet has a mean energy E_0 corresponding to a resonance and that the width ΔE of the packet considerably exceeds the width of the resonance (but is much smaller than the interval between neighboring resonances).

Under the conditions favorable for the occurrence of pronounced resonances $(\theta \gg 1)$ it may usually be assumed that in the vicinity of the resonance to a reasonable approximation

$$\cos L \simeq \mp \left(\frac{\partial L}{\partial E}\right)_{E=E_0} (E - E_0), \qquad \sin L \simeq \pm 1$$

Using these approximations and evaluating the slowly varying quantity θ at $E = E_0$, we get

$$\sqrt{T}\, e^{-i\varphi} \simeq \mp \frac{\Gamma/2}{E - E_0 + i(\Gamma/2)}\, e^{2i\rho} \tag{7.51}$$

where by definition

$$\Gamma = \frac{1}{\theta^2(\partial L/\partial E)_{E=E_0}} \tag{7.52}$$

Exercise 7.5. Apply the resonance approximation to the transmission coefficient T, and show that near E_0 it has the characteristic resonance shape, Γ being its width at half-maximum.

Except for uncommonly long-range potential barriers, the phase ρ may be assumed constant, and equal to ρ_0, over the width of the resonance. With all these approximations, the wave function in region 7 at any time t becomes

$$\psi(x, t) \simeq \mp f(E_0) \exp\left[i\frac{k_0}{z}(x - x_0) + 2i\rho_0\right]$$

$$\times \frac{\Gamma}{2} \int_{-\infty}^{\infty} \frac{\exp\left[\frac{i}{\hbar}E\left(\frac{x - x_0}{v_0} - t\right)\right]}{E - E_0 + i(\Gamma/2)}\, dE \quad (7.53)$$

In arriving at this form the approximation

$$k = k_0 + (k - k_0) = k_0 + \frac{k^2 - k_0^2}{k + k_0} \approx k_0 + \frac{k^2 - k_0^2}{2k_0} = \frac{k_0}{2} + \frac{E}{\hbar v_0}$$

has been used and the integration has been extended to $-\infty$ without appreciable error, assuming that t is not too large.

The integral in (7.53) is a well-known Fourier integral which can be evaluated by integration in the complex E-plane. The result is that in the region $x > c$

$$\psi(x, t) \simeq \begin{cases} \pm \pi i \Gamma f(E_0)e^{2i\rho_0} \exp\left[\frac{1}{\hbar}\left(iE_0 + \frac{\Gamma}{2}\right)\left(\frac{x - x_0}{v_0} - t\right)\right] & \text{if } t > \dfrac{x - x_0}{v_0} \\[2ex] 0 & \text{if } t < \dfrac{x - x_0}{v_0} \end{cases}$$

$$(7.54)$$

This wave function describes a wave packet with a discontinuous front edge at $x = x_0 + v_0 t$ and an exponentially decreasing tail to the left. After the pulse arrives at a point x the probability density decays according to the formula

$$|\psi(x, t)|^2 = \pi^2 \Gamma^2 |f(E_0)|^2 \exp\left[\frac{\Gamma}{\hbar}\left(\frac{x - x_0}{v_0} - t\right)\right]$$

We may calculate the probability that at time t the particle has been transmitted and is found in region 7. For a wave packet whose energy spread covers a single resonance such that

$$D \gg \Delta E \gg \Gamma \quad (7.55)$$

this probability is

$$\int_0^{x_0+v_0t} |\psi(x, t)|^2 \, dx = \pi^2 \hbar \Gamma v_0 \, |f(E_0)|^2 \left\{1 - \exp\left[-\frac{\Gamma}{\hbar}\left(t + \frac{x_0}{v_0}\right)\right]\right\}$$

$$\left(t > -\frac{x_0}{v_0}\right)$$

From (7.44) we obtain as a crude estimate

$$2\pi \hbar v_0 \, |f(E_0)|^2 \simeq \frac{1}{\Delta E}$$

Hence, an order of magnitude estimate for the probability that transmission has occurred is

$$\frac{\Gamma}{\Delta E}\left\{1 - \exp\left[-\frac{\Gamma}{\hbar}\left(t + \frac{x_0}{v_0}\right)\right]\right\} \tag{7.56}$$

The total transmission probability for the incident wave packet (7.43) is found by letting $t \to \infty$ and is thus approximately equal to $\Gamma/\Delta E$. Equation (7.56) leads to the following simple interpretation: The wave packet reaches the well at time $-x_0/v_0$. A fraction $\Gamma/\Delta E$ of the packet is transmitted according to an exponential time law with a mean lifetime

$$\tau = \frac{\hbar}{\Gamma} \tag{7.57}$$

and the remaining portion of the wave packet is reflected promptly. The study of resonance transmission has thus afforded us an example of the familiar *exponential decay law*, and the well with surrounding barriers can actually serve as a one-dimensional model of nuclear alpha decay. Decay processes will be encountered again in Chapters 11 and 18, but it is well to point out here that the exponential decay law can be derived only as an approximate, and not a rigorous, result of quantum mechanics and that it holds only if the decay process is essentially independent of the manner in which the decaying state was formed and of the particular details of the incident wave packet.

Exercise 7.6. Show that condition (7.55) implies that the time it takes the incident wave packet to enter the well must be long compared with the classical period of motion and short compared with the lifetime of the decaying state.

Exercise 7.7. Resonances in the double well may be also defined as quasi-bound states by requiring $A_1 = B_7 = 0$ (no incident wave), or $M_{11} = 0$

[(6.88)]. Show that this condition defines complex values of E whose real part is the resonance energy, whereas the imaginary part is one half the width. Construct and interpret the solution of the wave equation corresponding to one of these complex E values (decaying states).

5. Mathematical Detail. We now resume the study of the equation

$$\frac{d^2v}{dy^2} + \left(1 + \frac{5}{36y^2}\right)v = 0 \tag{7.23}$$

which we encountered in Section 7.2. We shall attempt to write the solution of this equation in the form

$$v(y) = y^\lambda \int e^{yt} f(t)\, dt \tag{7.58}$$

a modest generalization of a Laplace transform (which corresponds to the case $\lambda = 0$). Neither the limits of integration nor, since t can be regarded as a complex variable, the path of integration have yet been decided upon in (7.58). This choice, when made, will lead to various particular solutions of the differential equation.

Substituting (7.58) into (7.23), we obtain

$$\int [\lambda(\lambda - 1) + 2\lambda yt + y^2 t^2 + y^2 + \tfrac{5}{36}] e^{yt} f(t)\, dt = 0$$

We now choose λ such that the terms which are constant in y vanish. This requires that

$$\lambda(\lambda - 1) + \tfrac{5}{36} = 0$$

or

$$\lambda = \tfrac{1}{6}, \tfrac{5}{6}$$

The remaining expression can be written compactly as

$$\int f(t)\left[2\lambda t + (1 + t^2)\frac{d}{dt}\right] e^{yt}\, dt = 0 \tag{7.59}$$

By integration by parts this equation can be transformed into

$$\int \left\{2\lambda t f(t) - \frac{d}{dt}[(1 + t^2)f(t)]\right\} e^{yt}\, dt + \int \frac{d}{dt}[(1 + t^2)f(t)e^{yt}]\, dt = 0$$

If the integrand in the first integral can be made to vanish and the path of integration can be chosen so that the second integral disappears, we shall have succeeded in constructing a solution of the proposed form (7.58). Hence, we require

$$2\lambda t f = \frac{d}{dt}[(1 + t^2)f]$$

or

$$f(t) = f(0)(1 + t^2)^{\lambda - 1}$$

The limits of integration must then be chosen so as to make the term $(1 + t^2)^\lambda e^{yt}$ vanish. This will happen if $t = +i$ or $-i$, or, if, depending on the value of y, the path of integration in the complex plane of t leads in a suitable manner to the point at infinity. Thus, if y is negative real, t should go to infinity with a positive real part. If y is negative imaginary, t should go to infinity with a negative imaginary part. In this way various kinds of solutions are obtained, of which no more than two can be linearly independent, but every one of which may be suited for some particular purpose in solving the equation

$$\frac{d^2v}{dy^2} + \left[1 - \frac{\lambda(\lambda - 1)}{y^2}\right] v = 0 \tag{7.60}$$

Equation (7.23) is a special case of this. Note that if v_λ is a solution $v_{1-\lambda}$ is also a solution of this equation.

In deriving the WKB connection formulas for Figure 7.1 the variable y is negative imaginary in the classically inaccessible region $x > a$ [see (7.22)], and it is therefore convenient to consider the two particular solutions

$$v_\lambda^+(y) = y^\lambda \int_i^{-i\infty} \frac{e^{yt}}{(1 + t^2)^{1-\lambda}} \, dt \tag{7.61}$$

and

$$v_\lambda^-(y) = y^\lambda \int_{-i}^{-i\infty} \frac{e^{yt}}{(1 + t^2)^{1-\lambda}} \, dt \tag{7.62}$$

Since λ is not an integer ($\lambda = \frac{1}{6}$ or $\frac{5}{6}$) in this problem, $t = \pm i$ are *branch points* of the function $(1 + t^2)^{\lambda - 1}$. The cut, which is needed to enforce single-valuedness, may be placed on the imaginary axis outside the interval from $-i$ to $+i$. The phase of $(1 + t^2)^{\lambda - 1}$ is unambiguously determined by the choice arg $(1 + t^2) = 0$ when t is imaginary and lies between $-i$ and $+i$. We decide, again arbitrarily but definitely, that the path of integration in (7.61) and (7.62) shall lead along the cut in the right half of the complex plane, i.e., the phase of $1 + t^2$ shall be $-\pi$ as we go from $-i$ to $-i\infty$.

In the WKB approximation we need the asymptotic expansions of v^+ and v^- for large negative imaginary values of y. To this end we substitute in $v^\pm$

$$t = \pm i - \frac{z}{y}$$

and obtain

$$v_\lambda^\pm(y) = -y^{\lambda - 1} e^{\pm iy} \int_0^\infty \left(\frac{z^2}{y^2} \mp 2i \frac{z}{y}\right)^{\lambda - 1} e^{-z} \, dz \tag{7.63}$$

The asymptotic expansion of v is obtained by expanding the parenthesis in the integrand in powers of z/y and integrating term by term. If $|y|$ is large enough, a reasonable approximation to v is obtained by keeping only the first term in the expansion

$$v_\lambda^\pm(y) \sim \mp 2^{\lambda-1} i \Gamma(\lambda) \exp\left[\pm iy - i\lambda\left(\pi \mp \frac{\pi}{2}\right)\right] \tag{7.64}$$

or

$$v_\lambda^\pm(y) \sim \mp 2^{\lambda-1} i \Gamma(\lambda) \exp\left[-i\lambda\left(\pi \mp \frac{\pi}{2}\right)\right] \exp\left(\pm |y|\right) \tag{7.64a}$$

since y is negative imaginary. The form of (7.64a) is in accord with the WKB approximation in the region of negative kinetic energy where v is exponentially increasing or decreasing [(7.19)].

In the classically accessible region $x < a$ of Figure 7.1 the variable y is negative real [see (7.22)], and it is more convenient to use a different solution of (7.23), distinguished from the solution $v_\lambda^\pm(y)$ by the choice of different limits of integration on (7.58). A particularly useful solution is defined by

$$v_\lambda(y) = y^\lambda \int_{-i}^{+i} \frac{e^{yt}\,dt}{(1 + t^2)^{1-\lambda}} \tag{7.65}$$

The path of integration is best chosen as the dotted line in Figure 7.10, extending to infinity in the direction of positive real values of t. An asymptotic expansion of (7.65) for large negative real values of y can be easily made

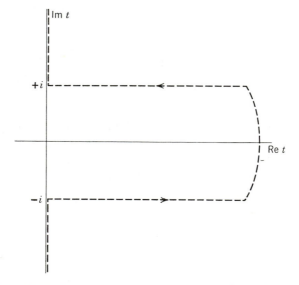

Figure 7.10. Paths of integration in the complex t-plane.

in the same way as was done for $v^{\pm}$. The result is

$$v_\lambda(y) \sim 2^\lambda i\Gamma(\lambda)e^{i\pi\lambda} \cos\left(y + \lambda\frac{\pi}{2}\right) \tag{7.66}$$

Again this result conforms with the WKB approximation in the region of positive kinetic energy, where the solutions are of an oscillatory nature [(7.18)]. Unless λ is half-integral, $v_\lambda(y)$ and $v_{1-\lambda}(y)$ are two linearly independent solutions of (7.60). Near $y = 0$ (the value corresponding to the turning point) these functions are easier to treat than $v^{\pm}$. However, the various kinds of solutions are related in a simple manner, as can be seen directly from the integral representations. Thus we have evidently from the definitions (7.61), (7.62), and (7.65)

$$v_\lambda(y) = v_\lambda^-(y) - v_\lambda^+(y) \tag{7.67}$$

and

$$v_{1-\lambda}(y) = v_{1-\lambda}^-(y) - v_{1-\lambda}^+(y) \tag{7.68}$$

Near the turning point $y = 0$ the integral (7.65) can be approximated readily, and we obtain an expansion in powers of y beginning with

$$v_\lambda(y) = i\frac{\Gamma(\tfrac{1}{2})\Gamma(\lambda)}{\Gamma(\lambda + \tfrac{1}{2})} \, y^\lambda[1 + 0(y^2)] \tag{7.69}$$

This formula shows clearly that the wave function

$$\psi = \frac{v}{\sqrt{k}} \propto \frac{v}{y^{1/6}} \tag{7.70}$$

is perfectly well behaved near the turning point despite the fact that $y = 0$ is a singularity of (7.23). Indeed, smooth joining of ψ is accomplished automatically if we let v in (7.70) be either v_λ or $v_{1-\lambda}$ (with $\lambda = \tfrac{1}{6}$ or $\tfrac{5}{6}$). Hence, we replace v^+ and v^- in (7.67) and (7.68) by their asymptotic expansions (7.64) for negative imaginary y and solve for $e^{|y|}$ and $e^{-|y|}$ in terms of v_λ and $v_{1-\lambda}$. We then divide the equations by $\sqrt{k}$, continue smoothly through the turning point, and expand v_λ and $v_{1-\lambda}$ asymptotically for negative real y. After a certain amount of manipulation the connection formulas (7.24) are obtained.

The functions v_λ, v_λ^+, and v_λ^- are closely related to *spherical Bessel and Hankel functions*. Formulas defining the connection will be found in Section 10.2.

Exercise 7.8. Determine the behavior of ψ as a function of $x - a$ near the turning point.

Problems

1. Apply the WKB method to a particle which falls with acceleration g in a uniform gravitational field directed along the z-axis and which is reflected from a perfectly elastic plane surface at $z = 0$. Indicate the nature of the rigorous solutions of this problem.

2. Apply the WKB approximation to the energy levels below the top of the barrier in a symmetric double well, and show that the energy eigenvalues are determined by a condition of the form

$$\tan \left(\int k \, dx + \alpha \right) = \pm 2\theta$$

where θ is the quantity defined in (7.31) for the barrier, α is a constant dependent on the boundary conditions, and the integral $\int k \, dx$ is to be extended between the classical turning points in one of the separate wells. Show that at low transmission the energy levels appear in close pairs with a level splitting approximately equal to $\hbar\omega/\pi\theta$ where ω is the classical frequency of oscillation in one of the single wells.

   As an example, apply the WKB approximation to the double harmonic oscillator, and contrast the energy level splitting with the results obtained in Section 5.6.

3. A particle of mass μ moves in one dimension between two infinitely high potential walls at $x = a$ and $x = -a$. In this interval the potential energy is $V = -C |x|$, C being a positive constant. In the WKB approximation, obtain an equation determining the energy eigenvalues $E \leqslant 0$. Estimate the minimum value of C required for an energy level with $E \leqslant 0$ to exist.

4. Apply the WKB approximation to a periodic potential in one dimension, and derive an implicit equation for E as a function of k. Estimate the width of the valence band.

5. Given the one-dimensional potential $V = C |x|$, find approximate values for the two lowest energy levels by (a) the WKB method and (b) the variational method, using a trial function with one variable parameter and minimizing $\langle H \rangle$ with respect to it ($C > 0$).

CHAPTER 8

The Principles of Wave Mechanics

1. Coordinate and Momentum Representations. The last three chapters were devoted to a discussion of various methods in use for solving the one-dimensional Schrödinger equation. The time has come to take up again the general formalism of quantum mechanics and develop it, guided by an awareness of the probability interpretation.

The wave nature of matter and the correspondence principle have led us to certain fundamental notions which we may now regard as *postulates of wave mechanics*. They are summarized here as follows:

The physical *state* of a particle at time t is described as fully as possible by the *normalized wave function* $\psi(\mathbf{r}, t)$. $|\psi(\mathbf{r}, t)|^2$ is the probability density for finding the particle at position $\mathbf{r}$. The expectation values of position and momentum[1] are calculated from the wave function by the formulas

$$\langle \mathbf{r} \rangle = \int \psi^*(\mathbf{r}, t)\, \mathbf{r}\, \psi(\mathbf{r}, t)\, d^3r \tag{8.1}$$

$$\langle \mathbf{p} \rangle = \int \psi^*(\mathbf{r}, t)\, \frac{\hbar}{i} \nabla \psi(\mathbf{r}, t)\, d^3r \tag{8.2}$$

The time development of the wave function is determined by the wave equation

$$i\hbar \frac{\partial \psi}{\partial t} = \left[-\frac{\hbar^2}{2\mu} \nabla^2 + V(\mathbf{r}) \right] \psi \tag{8.3}$$

if $V(\mathbf{r})$ is the classical potential energy of the particle. This completes the summary of fundamental results from the earlier chapters.

We can make substantial progress in our understanding of quantum mechanics if we make a Fourier analysis of the wave function. Generalizing

---

[1] $\langle \mathbf{p} \rangle$ can be calculated from $(\hbar/i)\nabla$ by (8.2) only if $\psi(\mathbf{r}, t)$ has a continuous derivative everywhere. If the wave function is not so well behaved, the momentum operator is more complicated.

(2.34), we may write

$$\psi(\mathbf{r}, t) = \frac{1}{(2\pi\hbar)^{3/2}} \int \varphi(\mathbf{p}, t) \exp\left(\frac{i}{\hbar}\mathbf{p}\cdot\mathbf{r}\right) d^3p \qquad (8.4)$$

$\varphi(\mathbf{p}, t)$ is the Fourier transform of $\psi(\mathbf{r}, t)$ at time t.

Some important properties of φ can be informally derived with the aid of the three-dimensional delta function. This is defined in analogy with (6.9) by

$$\delta(\mathbf{r}) = \frac{1}{(2\pi)^3} \int \exp(i\mathbf{k}\cdot\mathbf{r}) \, d^3k = \frac{1}{(2\pi\hbar)^3} \int \exp\left(\frac{i}{\hbar}\mathbf{p}\cdot\mathbf{r}\right) d^3p = \delta(x)\,\delta(y)\,\delta(z) \tag{8.5}$$

By inserting the Fourier representation of ψ and freely interchanging orders of integration, we can then prove the following assertions:

$$\int |\psi(\mathbf{r}, t)|^2 \, d^3r = \int |\varphi(\mathbf{p}, t)|^2 \, d^3p \qquad (8.6)$$

$$\langle \mathbf{p} \rangle = \int \mathbf{p}\, |\varphi(\mathbf{p}, t)|^2 \, d^3p \qquad (8.7)$$

$$\langle \mathbf{r} \rangle = -\int \varphi^*(\mathbf{p}, t)\, \frac{\hbar}{i}\, \nabla_{\mathbf{p}} \varphi(\mathbf{p}, t) \, d^3p \qquad (8.8)$$

$\nabla_{\mathbf{p}}$ denotes the gradient in $\mathbf{p}$-space.

An interpretation of the Fourier transform $\varphi(\mathbf{p}, t)$ is suggested by (8.4), (8.6), and (8.7). Equation (8.6) implies that, if ψ is normalized to unity for any t, φ is automatically also so normalized. If, in conformity with the discussion in Chapter 2, we assume that $\exp(i\mathbf{p}\cdot\mathbf{r}/\hbar)$ represents a state of definite momentum $\mathbf{p}$, φ according to (8.4) is the amplitude with which the momentum $\mathbf{p}$ appears represented in the wave function ψ. If φ is strongly peaked near a particular value of $\mathbf{p}$, the state tends to be one of rather definite momentum and similar to a plane wave. This observation, in conjunction with (8.7), leads us to the assumption that $|\varphi(\mathbf{p}, t)|^2 \, d^3p$ is the probability of finding the momentum of the particle in the volume element d^3p in the neighborhood of $\mathbf{p}$ at time t. $|\varphi(\mathbf{p}, t)|^2$ is the *probability density in momentum space.*[2]

There is an impressive similarity between (8.1) and (8.7) and between (8.2) and (8.8). Mathematically, this is simply a result of the reciprocity relations of Fourier integrals. Equation (8.4) has as its inverse the equation

$$\varphi(\mathbf{p}, t) = \frac{1}{(2\pi\hbar)^{3/2}} \int \psi(\mathbf{r}, t) \exp\left(-\frac{i}{\hbar}\mathbf{p}\cdot\mathbf{r}\right) d^3r \qquad (8.9)$$

[2] Since there are infinitely many different $|\varphi(\mathbf{p}, t)|^2$ giving the same $\langle\mathbf{p}\rangle$, the probability interpretation of φ is not a compelling result of (8.6) and (8.7). It is an assumption consistent with these equations.

The wave functions, ψ in (position) coordinate space and φ in momentum space, are both equally valid descriptions of the state of the system. Given either one of them, the other can be calculated from (8.4) or (8.9). Physically, this reciprocity is a symptom of complementarity and of the wave nature of matter.

Exercise 8.1. Starting from (8.4) and (8.5), derive (8.6) through (8.9).

From the postulates it follows that the expectation value of any (analytic) function of position is given by[3]

$$\langle f(\mathbf{r}) \rangle = \int f(\mathbf{r}) \, |\psi(\mathbf{r}, t)|^2 \, d^3 r \tag{8.10}$$

Substituting the Fourier integrals for ψ and ψ^* we can show that, if the wave function vanishes "sufficiently fast" at large distances (i.e., if the various integrations can be interchanged, and if the surface terms which appear can be neglected), this expression can be rewritten as

$$\langle f(\mathbf{r}) \rangle = \int \varphi^*(\mathbf{p}, t) f\left(-\frac{\hbar}{i} \nabla_{\mathbf{p}} \right) \varphi(\mathbf{p}, t) \, d^3 p \tag{8.11}$$

Similarly, (8.7) can be used to obtain for any (analytic) function of momentum

$$\langle g(\mathbf{p}) \rangle = \int g(\mathbf{p}) \, |\varphi(\mathbf{p}, t)|^2 \, d^3 p = \int \psi^*(\mathbf{r}, t) g\left(\frac{\hbar}{i} \nabla \right) \psi(\mathbf{r}, t) \, d^3 r \tag{8.12}$$

If the physical system under consideration is describable in the language of classical mechanics, all quantities associated with it, such as kinetic and potential energy, angular momentum, and the virial, can be expressed in terms of coordinates and momenta. The question then arises as how to compute the expectation value of an arbitrary function $F(\mathbf{r}, \mathbf{p})$. The expression for this expectation value can be derived to a certain extent from our previous work. For example, by carrying the Fourier analysis only part of the way we may define a mixed coordinate-momentum wave function by the integral

$$\chi(x, y, p_z, t) = \frac{1}{(2\pi\hbar)^{1/2}} \int \psi(x, y, z, t) \exp\left(-\frac{i}{\hbar} p_z z \right) dz \tag{8.13}$$

The square of the absolute value of this function, $|\chi(x, y, p_z, t)|^2$, can easily be seen to be the probability density of finding the particle to have coordinates x and y, but indeterminate z, whereas the z-component of its momentum is

---

[3] See Section 4.2 and (4.13).

equal to p_z. From this, it follows readily that the expectation value of any function of x, y, p_z is

$$\langle f(x, y, p_z) \rangle = \int f(x, y, p_z) \, |\chi(x, y, p_z, t)|^2 \, dx \, dy \, dp_z$$

$$= \int \psi^*(x, y, z, t) f\left(x, y, \frac{\hbar}{i} \frac{\partial}{\partial z}\right) \psi(x, y, z, t) \, d^3 r$$

$$= \int \varphi^*(p_x, p_y, p_z, t) f\left(-\frac{\hbar}{i} \frac{\partial}{\partial p_x}, -\frac{\hbar}{i} \frac{\partial}{\partial p_y}, p_z\right) \varphi(p_x, p_y, p_z, t) \, d^3 p$$

Similar considerations hold for other combinations of components of **r** and **p**, provided that the simultaneous specification of *conjugate* variables, such as x and p_x, is avoided. As a result we may conclude that

$$\langle F(\mathbf{r}, \mathbf{p}) \rangle = \begin{cases} \displaystyle \int \psi^*(\mathbf{r}, t) F\left(\mathbf{r}, \frac{\hbar}{i} \nabla\right) \psi(\mathbf{r}, t) \, d^3 r \\[12pt] \displaystyle \int \varphi^*(\mathbf{p}, t) F\left(-\frac{\hbar}{i} \nabla_{\mathbf{p}}, \mathbf{p}\right) \varphi(\mathbf{p}, t) \, d^3 p \end{cases} \tag{8.14}$$

However, we have not proved (8.14) generally, because no account has been taken of as simple a function of coordinates and momenta as $F = \mathbf{r} \cdot \mathbf{p} = x p_x + y p_y + z p_z$. The trouble here is that our analysis can never give us a wave function the square of whose absolute value would represent the probability density of finding the particle with coordinate x and momentum component p_x. Heisenberg's uncertainty principle is an articulation of this inability. One consoling fact is that few of the interesting physical quantities require us to deal with such products as $x p_x$ or the like.[4] If such products do appear we postulate that (8.14) holds, but it is then necessary to add a prescription about the proper ordering of factors. $x p_x$ is equal to $p_x x$ classically, but it is not clear at all what operator should represent this quantity quantum mechanically. We shall suppose, for the time being, that such ordering problems do not arise.[5]

Equation (8.14) informs us that with any physical quantity there is associated a linear operator which, when interposed between the wave function and its complex conjugate, gives upon integration the expectation value of that physical quantity. The explicit form of the operator depends evidently on whether the ψ or φ function is used to describe the state.

---

[4] See, however, (9.19) where $\mathbf{r} \cdot \mathbf{p}$ appears.

[5] For references to various prescriptions which have been proposed see J. R. Shewell, *Am. J. Phys.*, **27**, 16 (1959).

As an example of the equivalence of these two ways of doing quantum mechanics, we may derive the wave equation in momentum space from the wave equation in coordinate space:

$$ i\hbar \frac{\partial \psi(\mathbf{r}, t)}{\partial t} = -\frac{\hbar^2}{2\mu} \nabla^2 \psi(\mathbf{r}, t) + V(\mathbf{r})\psi(\mathbf{r}, t) $$

This is done most easily by differentiating (8.9) with respect to t:

$$ i\hbar \frac{\partial \varphi(\mathbf{p}, t)}{\partial t} = \frac{1}{(2\pi\hbar)^{3/2}} \int i\hbar \frac{\partial \psi(\mathbf{r}, t)}{\partial t} \exp\left(-\frac{i}{\hbar} \mathbf{p} \cdot \mathbf{r}\right) d^3r $$

or

$$ i\hbar \frac{\partial \varphi(\mathbf{p}, t)}{\partial t} = \frac{1}{(2\pi\hbar)^{3/2}} \int \exp\left(-\frac{i}{\hbar} \mathbf{p} \cdot \mathbf{r}\right) \left[-\frac{\hbar^2}{2\mu} \nabla^2 + V(\mathbf{r})\right] \psi(\mathbf{r}, t) d^3r $$

The first term of this last expression can be transformed by integrating by parts twice and assuming that ψ is subject to boundary conditions which permit the neglecting of the surface terms. If the Fourier formulas (8.4) and (8.9) are used again, we obtain after a little rearranging

$$ i\hbar \frac{\partial \varphi(\mathbf{p}, t)}{\partial t} = \frac{p^2}{2\mu} \varphi(\mathbf{p}, t) + \frac{1}{(2\pi\hbar)^3} \int\int \exp\left(-\frac{i}{\hbar} \mathbf{p} \cdot \mathbf{r}\right) V(\mathbf{r}) \exp\left(\frac{i}{\hbar} \mathbf{p}' \cdot \mathbf{r}\right) $$

$$ \times \varphi(\mathbf{p}', t) d^3r \, d^3p' \quad (8.15) $$

This is a mixed integral-differential equation for φ. It can be transformed into a differential equation if V is an analytic function, so that we may write

$$ \exp\left(-\frac{i}{\hbar} \mathbf{p} \cdot \mathbf{r}\right) V(\mathbf{r}) = V\left(-\frac{\hbar}{i} \nabla_\mathbf{p}\right) \exp\left(-\frac{i}{\hbar} \mathbf{p} \cdot \mathbf{r}\right) \quad (8.16) $$

Substituting this into (8.15) and removing the differential operator with respect to $\mathbf{p}$ from the integrals, the $\mathbf{r}$-integration can now be performed, and, according to (8.5), we are left with a delta function. This in turn makes the $\mathbf{p}'$-integration a trivial operation, and the result is a *wave equation in momentum space*:

$$ i\hbar \frac{\partial \varphi(\mathbf{p}, t)}{\partial t} = \frac{p^2}{2\mu} \varphi(\mathbf{p}, t) + V\left(-\frac{\hbar}{i} \nabla_\mathbf{p}\right) \varphi(\mathbf{p}, t) \quad (8.17) $$

Exercise 8.2. Work through the steps leading to (8.17).

This equation is, of course, equivalent to (8.3), and the choice between them is a matter of mathematical convenience.

For the case of the *harmonic oscillator*, (8.3) and (8.17) have identical structures, except that the roles of kinetic and potential energies are interchanged. The Schrödinger equations for the harmonic oscillator in coordinate and momentum space will then also be similar. Furthermore, the boundary conditions will be analogous: φ must vanish for large $|p|$, just as ψ was required to vanish for $|x| \to \infty$. It follows that the Fourier transform $\varphi_n(p)$ of an oscillator eigenfunction $\psi_n(x)$, corresponding to the quantum number n, must be simply related to $\psi_n(x)$.

Exercise 8.3. Show that

$$\varphi_n(p) = \frac{e^{i\alpha_n}}{\sqrt{\mu\omega}} \; \psi_n\!\left(\frac{p}{\mu\omega}\right) \tag{8.18}$$

(Closer investigation shows that the phase factor is $e^{i\alpha_n} = i^{-n}$. See Exercise 15.3.)

The symmetry of **r** and **p** which the present section exhibits is also evident in classical mechanics, for instance in the sophisticated Hamiltonian form, where (generalized) coordinates and momenta stand in a similar reciprocal relationship. However, in classical analytical dynamics the symmetry is primarily a formal one, for q and p together determine the state at time t, and they are independent quantities. Classical physics assumes tacitly that they can both be determined simultaneously with perfect precision. In quantum mechanics these two physical quantities are closely related to one another and can no longer be measured or chosen entirely independently. Position and momentum distributions, ψ and φ, are linked by Fourier integrals, and the quantities x and p_x, are subject to the Heisenberg uncertainty relation, a precise statement and mathematical proof of which will be given in Section 8.6.

Wave mechanics in coordinate space (ψ) and wave mechanics in momentum space (φ) are thus seen to be two equivalent descriptions of the same thing. In Chapters 14 and 15 it will become apparent that there is an advantage in regarding both of these descriptions as two special representations of a general and abstract formulation of quantum mechanics. The terms *coordinate representation* and *momentum representation* will then be seen to be apt for referring to these two forms of the theory. Other equivalent representations will be discussed.

In the remainder of this chapter the *coordinate representation* will be used, whenever an explicit representation is called for, but the equations can be transcribed into the momentum representation at any desired stage of the calculation.

2. Hermitian Operators. Their Eigenfunctions and Eigenvalues. We have learned that every physical quantity F can be represented by a linear operator, which for convenience is denoted by the same letter, F. The expectation value of F is given by the formula

$$\langle F \rangle = \int \psi^* F \psi \, d\tau \qquad (8.19)$$

Following the urge of the scientist to turn every question upside down, we now ask what general properties a *linear operator* must possess in order to be admitted as a candidate for representing a physical quantity.

Obviously it is necessary that the expectation value of such a linear operator, being the average of the measured values of a physical quantity, must be real. By (8.19) this implies that for all ψ which may represent possible states we must demand that

$$\int (F\psi)^* \psi \, d\tau = \int \psi^* F \psi \, d\tau \qquad (8.20)$$

Operators which have this property are called *Hermitian*.

Exercise 8.4. Prove that if F is a Hermitian operator,

$$\int (F\psi_1)^* \psi_2 \, d\tau = \int \psi_1^* F \psi_2 \, d\tau \qquad (8.20a)$$

for any two possible states ψ_1 and ψ_2.

The momentum $\mathbf{p}$ is an example of a Hermitian operator. It was shown in Section 4.3 that for the calculation of the expectation value $\langle \mathbf{p} \rangle$ the momentum may be represented by $(\hbar/i)\nabla$ if ψ vanishes sufficiently fast at large distances. But under the very same boundary conditions integration by parts gives

$$\int \left(\frac{\hbar}{i} \nabla \psi \right)^* \psi \, d\tau = - \frac{\hbar}{i} \int (\nabla \psi^*) \psi \, d\tau = \int \psi^* \frac{\hbar}{i} \nabla \psi \, d\tau$$

Hence, when $\mathbf{p}$ can be represented by $(\hbar/i)\nabla$, condition (8.20) is satisfied.[6]

Given an arbitrary, not necessarily Hermitian, operator F, it is useful to define its *(Hermitian) adjoint* $F^\dagger$ by the requirement

$$\int f^* F^\dagger g \, d\tau = \int (Ff)^* g \, d\tau \qquad (8.21)$$

where f and g are arbitrary functions (but subject to the condition that the integrals exist). The existence of $F^\dagger$ for operators of the type $F(\mathbf{r}, (\hbar/i)\nabla)$ can be seen by integration by parts. Comparing (8.21) with (8.20), we note

---

[6] $\mathbf{p}$ is represented by the Hermitian differential operator $(\hbar/i)\nabla$ also if ψ satisfies periodic boundary conditions (Sections 6.3 and 10.1).

that F is Hermitian if it is *self-adjoint*,

$$F^\dagger = F \tag{8.22}$$

Conversely, if an operator F is Hermitian,

$$\int (F\psi_1)^* \psi_2 \, d\tau = \int \psi_1^* F^\dagger \psi_2 \, d\tau = \int \psi_1^* F\psi_2 \, d\tau$$

The inference that a Hermitian operator F is self-adjoint can be drawn from the last equality only for a restricted class of operators. Since the physical interpretation requires that operators which represent measurable physical quantities must be self-adjoint, it has become customary in quantum mechanics to use the terms *Hermitian* and *self-adjoint* synonymously. We shall follow this mathematically improper usage.[7]

A number of simple theorems about operators can now be given. Their proofs, if not trivial, are indicated briefly.

1. The sum of two Hermitian operators is Hermitian.

2. The *identity operator 1*, which takes every function into itself, is Hermitian. If λ is a real number, $\lambda 1$ is Hermitian.

3. If F is non-Hermitian, $F + F^\dagger$ and $i(F - F^\dagger)$ are Hermitian. Hence, F can be written as a linear combination of two Hermitian operators:

$$F = \frac{F + F^\dagger}{2} + i \frac{F - F^\dagger}{2i} \tag{8.23}$$

4. If F and G are two arbitrary operators, the adjoint of their product is given by

$$(FG)^\dagger = G^\dagger F^\dagger \tag{8.24}$$

with an important reversal of the order of the factors.

Proof

$$\int \psi^* (FG)^\dagger \psi \, d\tau = \int (FG\psi)^* \psi \, d\tau = \int (G\psi)^* F^\dagger \psi \, d\tau = \int \psi^* G^\dagger F^\dagger \psi \, d\tau$$

But ψ is arbitrary, hence (8.24) follows. If F and G are Hermitian,

$$(FG)^\dagger = GF$$

Corollary: The product of two Hermitian operators is Hermitian if and only if they commute.

[7] Two recent books aim at providing a bridge between functional analysis and the theory of linear operators in Hilbert space on one hand and quantum mechanics on the other: T. F. Jordan, *Linear Operators for Quantum Mechanics*, John Wiley and Sons, New York, 1968; J. M. Jauch, *Foundations of Quantum Mechanics*, Addison-Wesley Publishing Co., Reading, Massachusetts, 1968.

5. The adjoint of a complex number λ is its complex conjugate λ^*.

6. The reader verifies without difficulty the Hermitian nature of such important physical operators as position, $\mathbf{r}$; momentum, $\mathbf{p} = (\hbar/i)\nabla$; energy, $H = (\mathbf{p}^2/2\mu) + V(\mathbf{r})$.

Since the weak requirement of being Hermitian is sufficient to establish the most important properties common to the operators which represent physical quantities, Dirac introduced the generic term *dynamical variable* for such operators. Thus, x and p_x are dynamical variables, but their product xp_x is not because the two operators fail to commute. Instead we have the fundamental operator equation

$$xp_x - p_x x = i\hbar 1 \tag{8.25}$$

The proof of such commutation relations for various dynamical variables is most safely accomplished by allowing the operators to act on an arbitrary function which is removed at the end of the calculation. Thus, (8.25) results because

$$(xp_x - p_x x)f = \left(x\frac{\hbar}{i}\frac{\partial}{\partial x} - \frac{\hbar}{i}\frac{\partial}{\partial x}x \right)f = -i\hbar\left[x\frac{\partial f}{\partial x} - \frac{\partial}{\partial x}(xf) \right] = i\hbar f$$

Similarly, we show that[8]

$$yp_y - p_y y = zp_z - p_z z = i\hbar \tag{8.25a}$$

All other products formed from Cartesian coordinates and their conjugate momenta are commutative, i.e., $xy = yx$, $xp_y = p_y x$, $xp_z = p_z x$, etc.

Exercise 8.5. Prove that the angular momentum operator $\mathbf{L} = \mathbf{r} \times \mathbf{p}$ is Hermitian.

It is not farfetched to suppose that the failure of x and p_x to commute is connected with the uncertainty relation for x and p_x, and with the incompatibility of precise simultaneous values for the x-coordinate and the x-component of momentum. As a first step toward an understanding of this connection we ask the following question: Given a dynamical variable A, how do we find and characterize those particular states of the system in which A has a *definite* value? If we make the convention that specific numerical *values* of dynamical variables (physical quantities) will be denoted by primed symbols (e.g., A') to distinguish them from the operators and physical quantities (e.g., A), the same question can be phrased in physical terms: If we measure A in a large number of separate systems, all replicas of each other

---

[8] $\lambda 1$ (λ: a number) is often written simply as λ, omitting the identity operator 1. Hence, $xp_x - p_x x = i\hbar$.

and each represented by the same wave function ψ, under what conditions will all these systems yield the same, hence accurately predictable, value A'? Phrased still differently, what kind of ψ corresponds to a probability distribution of the value of A that is peaked at A' and has no spread?

In the particular state ψ, in which the outcome of every measurement of A is the same, A', the expectation value, $\langle f(A) \rangle$, of any function of A must be equal to $f(A')$. Hence, we must demand that

$$\langle f(A) \rangle = f(A') \tag{8.26}$$

for any function of A. In particular, for $f(A) = A$ we demand

$$\langle A \rangle = A' \tag{8.27}$$

and for $f(A) = A^2$ we require that

$$\langle A^2 \rangle = \langle A \rangle^2 = A'^2 \tag{8.28}$$

According to (4.9), (8.28) expresses the vanishing of the *variance* $(\Delta A)^2$ of A:

$$(\Delta A)^2 = \langle (A - \langle A \rangle)^2 \rangle = \langle A^2 \rangle - \langle A \rangle^2 = 0$$

If ψ is assumed to be quadratically integrable, this condition implies

$$\int \psi^*(A - A')^2 \psi \, d\tau = 0$$

A, being a dynamical variable, is Hermitian, and A' is real. Hence, the last equation becomes

$$\int [(A - A')\psi]^*(A - A')\psi \, d\tau = \int |(A - A')\psi|^2 \, d\tau = 0$$

from which it follows that

$$A\psi = A'\psi \tag{8.29}$$

A quadratically integrable function ψ which satisfies (8.29) is called an *eigenfunction* of A, and the real number A' is the corresponding *eigenvalue*.[9] The totality of all eigenvalues of A is termed its *spectrum*. An eigenfunction of A is characterized by the fact that operating with A on it has no effect other than multiplication of the function by the eigenvalue A'. It may be noted that (8.27) and (8.28) are sufficient to satisfy condition (8.26) for any function f.

From this point of view the Schrödinger equation is but a special case of (8.29):

$$H\psi_n = E_n\psi_n \tag{8.30}$$

---

[9] The requirement of quadratic integrability for an eigenfunction is unnecessarily severe. It disallows, for instance, $\exp(ipx/\hbar)$ as eigenfunctions of the momentum operator $(\hbar/i)\partial/\partial x$. This defect will be remedied in Section 8.4.

The Schrödinger eigenfunctions ψ_n are the eigenfunctions of the energy operator

$$H = \frac{\mathbf{p}^2}{2\mu} + V(\mathbf{r}) = -\frac{\hbar^2}{2\mu}\nabla^2 + V(\mathbf{r})$$

If a system is in state ψ_n, measurement of its energy is certain to give the value E_n.

Our conclusion is that a system will reveal a definite value A', and no other value, when A is measured if and only if it is in a state represented by the eigenfunction $\psi_{A'}$. We often say that the system is in an *eigenstate* of A. The only definite values of A which a system can thus have are the eigenvalues A'.

The reality of the eigenvalues of Hermitian operators needs no further proof, but a most important property of the eigenfunctions, their orthogonality, does. Actually, the two properties follow from the same simple argument. Consider two eigenfunctions of A, ψ_1 and ψ_2, corresponding to the eigenvalues A_1' and A_2':

$$A\psi_1 = A_1'\psi_1 \tag{8.31}$$

$$A\psi_2 = A_2'\psi_2 \tag{8.32}$$

Multiply (8.31) on the left by ψ_2^*; take the complex conjugate of (8.32) and multiply it on the right by ψ_1. Then integrate in both equations over all space and subtract one from the other. Owing to the Hermitian property of A,

$$(A_1' - A_2'^*)\int \psi_2^*\psi_1 \, d\tau = 0 \tag{8.33}$$

If $\psi_2 = \psi_1$, then $A_2' = A_1'$, hence $A_1'^* = A_1'$, which demonstrates again that *all eigenvalues of A are real*. Using this result, it is then seen from (8.33) that if $A_2' \neq A_1'$ the two eigenfunctions are orthogonal in the sense that

$$\int \psi_2^*\psi_1 \, d\tau = 0 \tag{8.34}$$

Conclusion: *Eigenfunctions belonging to different eigenvalues are orthogonal.*

Equation (8.33) is trivially satisfied if two or more different eigenfunctions belong to a particular eigenvalue of A. An eigenfunction which is obtained from another one merely by multiplication with a constant is, of course, not considered "different." Rather, to be at all interesting the n eigenfunctions belonging to an eigenvalue A' must all be *linearly independent*. This means that none of them may be expressed as a linear combination of the others.

Although any two eigenfunctions belonging to the same eigenvalue may or may not be orthogonal, it is always possible to express all of them as linear combinations of n eigenfunctions which are orthogonal and belong to that same eigenvalue (*Schmidt orthogonalization* method).

Proof. Suppose that

$$A\psi_1 = A'\psi_1 \quad \text{and} \quad A\psi_2 = A'\psi_2$$

where ψ_2 is not a multiple of ψ_1, and

$$\int \psi_1{}^*\psi_2 \, d\tau = K \neq 0$$

Construct

$$\psi_2' = \alpha\psi_1 + \beta\psi_2$$

and demand that

$$\int \psi_1{}^*\psi_2' \, d\tau = 0$$

If ψ_1 and ψ_2 are assumed to be normalized to unity, ψ_2', similarly normalized is

$$\psi_2' = \frac{K\psi_1 - \psi_2}{\sqrt{1 - |K|^2}} \tag{8.35}$$

This new eigenfunction of A is orthogonal to ψ_1. If there are eigenfunctions of A which are linearly independent of ψ_1 and ψ_2, this process of orthogonalization can be continued systematically by demanding that ψ_3 be replaced by a function ψ_3', a linear combination of ψ_1, ψ_2', and ψ_3, so that ψ_3' is orthogonal to ψ_1 *and* ψ_2', etc. When the process is completed, any eigenfunction of A with eigenvalue A' will be expressible as a linear combination of $\psi_1, \psi_2', \psi_3', \ldots, \psi_n'$. This orthogonal set of functions is, of course, by no means unique. Its existence shows, however, that there is no loss of generality if we assume that *all* the eigenfunctions of the operator A are orthogonal. Any other eigenfunction of A can be written as a linear combination of orthogonal ones. Henceforth we shall therefore always assume that the eigenfunctions of A are all orthogonal and that the eigenfunctions, which are quadratically integrable, have been normalized to unity. Since

$$\int \psi_i{}^*\psi_j \, d\tau = \delta_{ij} \tag{8.36}$$

the solutions of the equation

$$A\psi_i = A_i'\psi_i \tag{8.37}$$

are said to form an *orthonormal set.*

3. *The Superposition of Eigenstates.* We have seen the physical significance of the eigenstates of A. What happens if the physical system is *not* in such an eigenstate? To find out, let us first look at a state which is a superposition of eigenstates of A, such that the wave function can be written as

$$\psi = \sum_i c_i \psi_i \tag{8.38}$$

Owing to the orthonormality of the eigenfunctions, (8.36), the expectation value of A in the state ψ is then given by

$$\langle A \rangle = \int \psi^* A \psi \, d\tau = \sum_i A_i' |c_i|^2 \tag{8.39}$$

and the normalization integral imposes the restriction

$$\int \psi^* \psi \, d\tau = \sum_i |c_i|^2 = 1 \tag{8.40}$$

Exercise 8.6. Prove (8.39) and (8.40).

More generally, for any function of A,

$$\langle f(A) \rangle = \sum_i f(A_i') |c_i|^2 \tag{8.41}$$

of which (8.39) and (8.40) are special cases.

From these equations we may infer that the eigenvalues A_i', which are characteristic of the operator A, are in fact the only values of A which can be found in the measurement of this physical quantity, even if ψ is not an eigenstate of A. If this interpretation is made, it then follows that $|c_i|^2$, which depends on the state, is the probability of finding the value A_i' when A is measured.[10] If ψ happens to be an eigenstate, ψ_i, of A, then $c_i = 1$, and all other $c_j = 0, j \neq i$. In this particular case (8.41) agrees with (8.26), showing the consistency of the present interpretation with our earlier conclusions.

Although the interpretation of (8.39), (8.40), and (8.41) is thus seen to be a perfectly natural one, it is not a rigorous deduction but remains an assumption. For it is asserted that the *possible* results of measuring a dynamical variable should depend only on the variable itself and not on the particular state of the system. We assume, in fact, that a line can be drawn between the nature of the system which determines the possible results of a measurement and the state of this system which determines the actual outcome of the measurement, at least in a probability sense.

Equation (8.38) shows how this division of the theory into physical quantities (operators) and particular states (wave functions), with its

[10] This conclusion comes about because if P_i and R_i are two probability distributions of the random variable A_i', such that the expectation values of *any* function of A_i' are the same for the two distributions, then $P_i \equiv R_i$.

attendant consequences for the interpretation of A_i' and $|c_i|^2$, is related to the possibility of superposing states by constructing linear combinations of eigenfunctions. The state ψ is in a certain sense intermediate between its component states ψ_i. It bears the mark of each of them, because a measurement of A in the state ψ may yield any of the eigenvalues A_i' which correspond to eigenfunctions ψ_i represented with nonvanishing contributions in the expansion (8.38) of ψ.

Now that the meaning of states obtained by the superposition of eigenstates of A has been established, a further question arises: Can the wave function ψ of an arbitrary state always be written as a linear combination of the eigenfunctions of A, or are there important states which cannot be so represented?

Cogitation alone cannot supply the answer to this question. It can only be answered by experience. And experience has shown that it is legitimate to assume that any state can be represented to a sufficient approximation by a superposition of eigenstates of those dynamical variables which are actually observed and measured in experiment. Mathematically, the superposition appears as a linear combination of the appropriate eigenfunctions, so that we may generalize (8.38) and write for any state ψ:

$$\psi \triangleq \sum_i c_i \psi_i \qquad (8.42)$$

The symbol $\triangleq$ indicates that for the calculation of any physical property $\sum_i c_i \psi_i$ may be used instead of ψ, even when the sum, which is usually an infinite series, does not converge to the state it represents.

However, if the series does converge to ψ we may write

$$\psi = \sum_{i=1}^{\infty} c_i \psi_i \qquad (8.42a)$$

If this series has only a finite number of terms, or is infinite but uniformly convergent, summations and integrations may be interchanged freely. Using the orthonormality condition (8.36), we can then determine the expansion coefficients from (8.42a):

$$c_i = \int \psi_i^* \psi \, d\tau \qquad (8.43)$$

and we have as in (8.40),

$$\int |\psi|^2 \, d\tau = \sum_{i=1}^{\infty} |c_i|^2 \qquad (8.44)$$

According to the discussion at the beginning of this section, $|c_i|^2$ is the probability of finding the value A_i' for the physical quantity A. The coefficient c_i is sometimes called a *probability amplitude*.

After these preliminaries the fundamental *expansion postulate* of quantum mechanics can be stated as follows: *Every physical quantity can be represented by a Hermitian operator with eigenfunctions* ψ_1, ψ_2, ..., $\psi_n \cdots$ *and every physical state* ψ *by a sum* $\Sigma_i\, c_i\psi_i$, *where the coefficients are defined by* (8.43). All quantum calculations can be made with the sum representing the state, and mathematical operations can be carried out as if this sum were finite, i.e., integrations, summations, and differentiations can be interchanged freely. $|c_i|^2$ *is the probability of finding the value* A_i' *for A in state* ψ.

According to this postulate, every physical operator has a sufficient number of eigenfunctions to represent an arbitrary state. Such a set of orthonormal eigenfunctions is therefore said to be *complete*. Briefly, the expansion postulate spells out the assumption that every physical operator has a complete set of orthogonal eigenfunctions. Following Dirac, we call *observable* any Hermitian operator which possesses a complete set of eigenfunctions.

From the mathematical point of view the preceding discussion is lacking in precision. This defect can be remedied by showing that $\Sigma\, c_i\psi_i$ is indeed a valid representation of ψ in the sense of the expansion postulate if[11]

$$\lim_{n \to \infty} \int |\psi - \sum_{i=1}^{n} c_i\psi_i|^2 \, d\tau = 0 \qquad (8.45)$$

The series which satisfies this requirement is said to *converge in the mean* to ψ. This convergence, which is much weaker than (8.42a), suffices to demonstrate that the expansion coefficients must be given by (8.43), that the so-called *completeness relation* (8.44) is satisfied, and that expectation values can be calculated as if the expansion were finite.

Equation (8.44) is called the *completeness relation* (or Parseval's formula), because generally *Bessel's inequality* holds,

$$\int |\psi|^2 \, d\tau \geqslant \sum_{i=1}^{n} |c_i|^2$$

Lest the reader question whether observables with complete sets of eigenfunctions really exist, the example of the harmonic oscillator eigenfunctions, Section 5.3, is recalled. It can be shown[12] that any quadratically integrable function $\psi(x)$ which is piecewise continuous can be expanded in a series of oscillator eigenfunctions for $-\infty \leqslant x \leqslant +\infty$ and that the series

---

[11] A careful mathematical treatment of these questions is found in N. I. Akhiezer and I. M. Glazman, *Theory of Linear Operators in Hilbert Space*, Volumes I and II, translated by M. Nestell, F. Ungar Publishing Co., New York, 1961 and 1963. P. M. Morse and H. Feshbach (in *Methods of Theoretical Physics*, McGraw-Hill Book Company, 1953, pp. 458, 709, 738) say that $\Sigma_i\, c_i\psi_i$ approximates ψ in the sense of a least square fit, if (8.45) holds.

[12] R. Courant and D. Hilbert, *Methods of Mathematical Physics*, Volume I, Interscience Publishers, New York, 1953, pp. 96–97.

converges in the mean to $\psi(x)$. In a *finite* interval the eigenfunctions of momentum, subject to periodic boundary conditions, provide a complete set for the well-known Fourier expansion. Other examples of complete sets will be encountered in the next few chapters.

4. The Continuous Spectrum. In the preceding work the assumption of quadratic integrability of the eigenfunctions was made. With this restriction only very few operators would have complete sets of eigenfunctions, and many common physical quantities would not qualify as observables, thus subjecting the theory to an intolerable limitation. The energy of a harmonic oscillator is an example of an operator with a complete set of quadratically integrable eigenfunctions. On the other hand, the momentum operator has no eigenfunctions which are quadratically integrable over all space. Yet a state ψ can be expanded in terms of these, as shown by the Fourier integral formulas, reviewed in Section 8.1. Many important operators, such as the energy of a square well with bound states, have some quadratically integrable eigenfunctions, but they do not make up a complete set, and it is in such cases easy to construct functions ψ, representative of real physical states, which cannot be expanded in terms of the quadratically integrable eigenfunctions alone. Strictly speaking, the expansion postulate demands the exclusion of such operators as observables.

If quadratic integrability is too narrow a restriction, what should be the criterion for selecting from the solutions of the equation

$$A\psi = A'\psi \tag{8.46}$$

the eigenfunctions that make up a complete set for all physical states? It is clear that the boundary conditions must be modified but cannot be completely relinquished. For example, for the momentum operator, $(\hbar/i)\,\partial/\partial x$, (8.46) has exponential solutions, $e^{\pm kx}$ (k real), but these would correspond to imaginary eigenvalues of the momentum and would not be orthogonal. They would therefore be useless for a probability interpretation of quantum mechanics and could never appear in the expansion of any ψ which represents an actual physical state. Such eigenfunctions must therefore be excluded.

Finiteness of the eigenfunctions everywhere is usually, but not always, a useful condition which draws the line between those solutions of (8.46) which must be admitted to the complete set of eigenfunctions and those which must be rejected.

Generally, it is best to assume that the expansion postulate holds and let nature tell us how wide we must make the set of eigenfunctions so that any physically occurring ψ can be expanded in terms of it. We thus tailor the mathematics to the requirements of the physics, the only assumption being that there is some cloth which will suit these requirements.

If the probability interpretation of quantum mechanics is to be maintained, all eigenvalues must be real, and eigenfunctions which are not quadratically integrable can appear in the expansion of a quadratically integrable wave function ψ only with infinitesimal amplitude. Hence, these functions are part of the complete set only if they belong to a *continuum* of real eigenvalues, depend on the eigenvalue continuously, and can be integrated over a finite range of eigenvalues. Thus, (8.42) must be generalized to read[13]

$$\psi \triangleq \sum c_i \psi_i + \int c(A')\psi_{A'}\, dA' \tag{8.47}$$

We now extend to the continuum the fundamental assumptions formulated in the expansion postulate and require that

$$\langle f(A)\rangle = \int \psi^* f(A)\psi\, d\tau = \sum_i |c_i|^2\, f(A_i') + \int |c(A')|^2 f(A')\, dA' \tag{8.48}$$

Substitution of (8.47) into $\int \psi^* f(A)\psi\, d\tau$ shows that (8.48) will result only if for the continuous spectrum we have[14]

$$\int \psi_{A'}{}^* \psi_{A''}\, d\tau = \delta(A' - A'') \tag{8.49}$$

We say that the eigenfunctions $\psi_{A'}$ are subject to A-normalization if (8.49) holds. This equation expresses orthogonality and delta function normalization of the continuum eigenfunctions. With the requirement that the eigenvalues must be real, it can be merged into the single condition

$$\int \psi_{A'}{}^* A\psi_{A''}\, d\tau - \int (A\psi_{A'})^* \psi_{A''}\, d\tau = (A'' - A')\, \delta(A' - A'') \tag{8.50}$$

But the right-hand side of this equation vanishes, since $x\, \delta(x) = 0$. Hence, we must demand that the Hermitian property of A,

$$\int \psi_{A'}{}^* A\psi_{A''}\, d\tau = \int (A\psi_{A'})^* \psi_{A''}\, d\tau \tag{8.51}$$

should hold in the usual sense of equations involving delta functions for the physically admissible eigenfunctions whether they are quadratically integrable or not. The orthonormality condition (8.49) permits evaluation of the

---

[13] In the following equations it is tacitly assumed that for each continuous eigenvalue A' there is only one eigenfunction. If there are several (possibly infinitely many) linearly independent eigenfunctions corresponding to A', more indices must be used and summations (or integrations) carried out over these.

[14] This result is only obtained if the order of integrations is changed. The license for such interchanges is implicit in the extended expansion postulate.

expansion coefficients in (8.47):

$$c(A') = \int \psi_{A'}{}^* \psi \, d\tau \tag{8.52}$$

in close analogy with (8.43).

We often write the equations of quantum mechanics as if all eigenvalues were discrete, the summations over the eigenvalues implying integrations over any continuous portions of the spectrum. This policy is suggested by the formal similarity of the expansion equations

$$\psi = \sum_i c_i \psi_i \qquad c_i = \int \psi_i{}^* \psi \, d\tau \tag{8.53a}$$

for the *discrete*, and

$$\psi = \int c(A') \psi_{A'} \, dA' \qquad c(A') = \int \psi_{A'}{}^* \psi \, d\tau \tag{8.53b}$$

for the *continuous* spectrum.

Moreover, it is always possible to make all eigenfunctions quadratically integrable, and the spectrum discrete, by imposing boundary conditions which confine the particle to a limited portion of space; for instance, impenetrable walls at great distances can be erected, or periodicity in a large unit cell required. As the size of the confining region increases, the discrete spectrum goes over into a continuous one. The transition is made by introducing a density, $\rho(A')$, of discrete states. This is the number of discrete states ψ_i per unit A' interval

$$\rho(A') = \frac{\Delta n}{\Delta A'} \tag{8.54}$$

In the limit of an infinitely large confining region the density ρ becomes infinite, since then the eigenvalues are continuous; but usually ρ is proportional to the volume of the enclosure, and it is then possible to speak of a finite *density of states per unit volume*. An example of this behavior has already appeared in Section 6.3.

It is easy to see that consistency between the expansion formulas (8.53a) and (8.53b) is achieved if the relations

$$\psi_{A'} = \psi_i \sqrt{\rho(A')} \tag{8.55}$$

$$c(A') = c_i \sqrt{\rho(A')} \tag{8.56}$$

are adopted. Hence, if the equations are written with the notation of the discrete spectrum, it is a simple matter to take proper cognizance of the continuous spectrum at any stage of the calculation.

The continuous spectrum is thus seen to be as important as the discrete spectrum, and both are needed to make a set of eigenfunctions complete. A useful condition which a given set of orthonormal functions must satisfy, if it is to be complete, can be derived from the identity

$$\psi(\mathbf{r}) = \sum_i c_i\psi_i(\mathbf{r}) = \sum_i \psi_i(\mathbf{r})\int \psi_i{}^*(\mathbf{r}')\psi(\mathbf{r})\,d\tau' = \int\left[\sum_i \psi_i{}^*(\mathbf{r}')\psi_i(\mathbf{r})\right]\psi(\mathbf{r}')\,d\tau'$$

Since this must be true for any function ψ, we infer that

$$\sum_i \psi_i{}^*(\mathbf{r}')\psi_i(\mathbf{r}) = \delta(\mathbf{r} - \mathbf{r}') \tag{8.57}$$

which is known as the *closure relation*.

Exercise 8.7. Show that the eigenfunctions of momentum satisfy the closure relation.

5. Simultaneous Measurements and Commuting Operators.

The expansion postulate answers all questions concerning the result of a measurement of an observable A. Since a definite value A' can be assigned to a system only if it is in an eigenstate of A, it is clear that we can assign values of two different observables, A and B, to the system only when it is in an eigenstate of *both* operators. If we want the observables A and B to be compatible, i.e., simultaneously assignable to the system for all eigenvalues, it is necessary that the two operators have a common complete set of eigenfunctions.

If this is so, each eigenfunction can be labeled by two eigenvalues, A_i' of A and B_j' of B. Thus,

$$A\psi_{A_i'B_j'} = A_i'\psi_{A_i'B_j'} \tag{8.58a}$$

$$B\psi_{A_i'B_j'} = B_j'\psi_{A_i'B_j'} \tag{8.58b}$$

If (8.58a) is multiplied by B and (8.58b) by A, we obtain

$$(AB - BA)\psi_{A_i'B_j'} = 0 \tag{8.59}$$

If this is to hold for all functions belonging to the complete set, $AB - BA$ must be equal to the null operator, or

$$AB = BA \tag{8.60}$$

Hence, a necessary condition for two observables to be *simultaneously measurable* is that they *commute*.

The converse of this theorem is also true, albeit in a slightly modified form: If A and B are two commuting observables, a common set of eigenfunctions can be selected for them.

Although this theorem has quite general validity, it is useful to give a simple constructive proof which is applicable only if one of the operators, say B, has a finite number, n, of linearly independent eigenfunctions corresponding to an eigenvalue B'. We denote these eigenfunctions by $\psi_{B',i}$ with i ranging from 1 to n. From the commutivity of A with B and

$$B\psi_{B',i} = B'\psi_{B',i} \tag{8.61}$$

it follows that

$$BA\psi_{B',i} = AB\psi_{B',i} = B'A\psi_{B',i}$$

This equation shows that $A\psi_{B',i}$ is also an eigenfunction of B with eigenvalue B'; hence,

$$A\psi_{B',i} = \sum_{k=1}^{n} c_{ik}\psi_{B',k} \tag{8.62}$$

In general, (8.62) does not allow us to conclude that the $\psi_{B',i}$ are eigenfunctions of A. However, it is possible to construct a new set of eigenfunctions $\bar{\psi}_{B',i}$, of B with eigenvalue B', which are linear combinations of the old ones, and which are at the same time eigenfunctions of A. To see this we form

$$\bar{\psi}_{B',j} = \sum_{l=1}^{n} d_{jl}\psi_{B',l} \tag{8.63}$$

and demand that

$$A\bar{\psi}_{B',j} = \sum_{l} d_{jl} \sum_{k} c_{lk}\psi_{B',k} = A_j'\bar{\psi}_{B',j} = A_j' \sum_{l} \sum_{k} d_{jl}\psi_{B',k} \, \delta_{lk}$$

We thus obtain n linear homogeneous equations for the n unknown d_{jl}:

$$\sum_{l=1}^{n} (c_{lk} - A_j' \, \delta_{lk}) d_{jl} = 0 \qquad (j \text{ fixed}, \ k = 1, 2, \ldots, n) \tag{8.64}$$

This system of equations possesses nontrivial solutions only if the determinant of the coefficients vanishes:

$$\det (c_{lk} - A_j'\delta_{lk}) = 0 \tag{8.65}$$

The n roots of this characteristic equation give us the eigenvalues $A_1', A_2', \ldots, A_n'$. Equation (8.64) can then be used to calculate the coefficients d_{jl}. The new functions $\bar{\psi}$ are obviously simultaneous eigenfunctions of both operators A and B.

6. The Heisenberg Uncertainty Relations. We have seen that *commuting observables* can be specified simultaneously. If A and B are two Hermitian operators which do not commute, the physical quantities A and B cannot both be sharply defined simultaneously. The degree to which an inevitable lack of precision in A and B has to be admitted is measured by their *commutator*

$$AB - BA = iC \tag{8.66}$$

Exercise 8.8. Prove that C is a Hermitian operator.

The variances of A and B are defined as[15]

$$(\Delta A)^2 = \int \psi^*(A - \langle A \rangle)^2 \psi \, d\tau \qquad (8.67a)$$

$$(\Delta B)^2 = \int \psi^*(B - \langle B \rangle)^2 \psi \, d\tau \qquad (8.67b)$$

Their positive square roots, ΔA and ΔB, are called the *uncertainties* in A and B. For these quantities we shall prove that

$$(\Delta A)^2 (\Delta B)^2 \geqslant (\tfrac{1}{2}\langle C \rangle)^2 \qquad (8.68)$$

or

$$\Delta A \, \Delta B \geqslant |\tfrac{1}{2}\langle C \rangle| \qquad (8.68a)$$

Proof. Since A is Hermitian we can write

$$(\Delta A)^2 = \int [(A - \langle A \rangle)\psi]^*(A - \langle A \rangle)\psi \, d\tau$$

$$= \int |(A - \langle A \rangle)\psi|^2 \, d\tau$$

and similarly for B. We now apply the Schwarz inequality

$$\left(\int |f|^2 \, d\tau \right) \left(\int |g|^2 \, d\tau \right) \geqslant \left| \int f^*g \, d\tau \right|^2 \qquad (8.69)$$

a proof of which will be found in Section 8.10. This inequality is valid for any two functions f and g. Letting $f = (A - \langle A \rangle)\psi$ and $g = (B - \langle B \rangle)\psi$, we get

$$(\Delta A)^2 (\Delta B)^2 \geqslant \left| \int \psi^*(A - \langle A \rangle)(B - \langle B \rangle)\psi \, d\tau \right|^2 \qquad (8.70)$$

where the equality sign holds if and only if $f \propto g$, or

$$(B - \langle B \rangle)\psi = \lambda(A - \langle A \rangle)\psi \qquad (8.71)$$

Now we use the simple identity, based on (8.23),

$$(A - \langle A \rangle)(B - \langle B \rangle) = \frac{(A - \langle A \rangle)(B - \langle B \rangle) + (B - \langle B \rangle)(A - \langle A \rangle)}{2} + i(\tfrac{1}{2}C)$$

$$= F + i(\tfrac{1}{2}C)$$

by which the operator on the left is written as a linear combination of two Hermitian operators, F and C. Since their expectation values are real, we can

---

[15] For definitions of probability concepts, see Section 4.2.

write (8.70) as

$$(\Delta A)^2(\Delta B)^2 \geqslant \langle F \rangle^2 + \tfrac{1}{4}\langle C \rangle^2 \geqslant \tfrac{1}{4}\langle C \rangle^2$$

which proves the theorem (8.68). The last equality holds if and only if[16]

$$\langle F \rangle = 0 \tag{8.72}$$

As a special case, let $A = x$, $B = p_x$. Then

$$xp_x - p_x x = i\hbar 1 \tag{8.25}$$

hence $C = \hbar 1$, and it follows from (8.68a) that

$$\Delta x \, \Delta p_x \geqslant \tfrac{1}{2}\hbar \tag{8.73}$$

making specific and quantitative the somewhat vague statements of Chapter 2. The Heisenberg uncertainty relation is thus seen to be a direct consequence of the noncommutivity of the position and momentum operators.

It is of interest to study the particular state ψ, for which (8.68) becomes an equality. This is the state in which the product of the uncertainties in A and B is as small as the noncommutivity allows:

$$\Delta A \, \Delta B = \tfrac{1}{2}|\langle C \rangle|$$

Such a *minimum uncertainty state* obeys the conditions (8.71) and (8.72). From (8.71) we can obtain two simple relations:

$$\int \psi^*(A - \langle A \rangle)(B - \langle B \rangle)\psi \, d\tau = \lambda(\Delta A)^2$$

and

$$\int \psi^*(B - \langle B \rangle)(A - \langle A \rangle)\psi \, d\tau = \frac{1}{\lambda}(\Delta B)^2$$

Adding the left-hand sides yields $2\langle F \rangle$; hence by (8.72)

$$\lambda(\Delta A)^2 + \frac{1}{\lambda}(\Delta B)^2 = 0 \tag{8.74}$$

Subtracting the left-hand sides gives $i\langle C \rangle$; hence

$$\lambda(\Delta A)^2 - \frac{1}{\lambda}(\Delta B)^2 = i\langle C \rangle \tag{8.75}$$

From (8.74) and (8.75) we obtain

$$\lambda = \frac{i\langle C \rangle}{2(\Delta A)^2} \tag{8.76}$$

---

[16] $\langle F \rangle = \langle \tfrac{1}{2}(AB + BA) - \langle A \rangle \langle B \rangle \rangle$ can be regarded as the quantum of the covariance of the random variables A and B as defined in (4.10).

The example of $A = x$, $B = p_x$ will show how we can use (8.71) and (8.76) to determine the form of the minimum uncertainty state completely. In one dimension, (8.71) becomes with (8.76)

$$\left(\frac{\hbar}{i}\frac{d}{dx} - \langle p_x \rangle\right)\psi = \frac{i\hbar}{2(\Delta x)^2}(x - \langle x \rangle)\psi \tag{8.77}$$

This differential equation is solved by the normalized wave function

$$\psi(x) = [2\pi(\Delta x)^2]^{-1/4}\exp\left[-\frac{(x - \langle x \rangle)^2}{4(\Delta x)^2} + \frac{i\langle p_x \rangle x}{\hbar}\right] \tag{8.78}$$

For this *minimum uncertainty wave packet*, which is a plane wave, modulated by a Gaussian amplitude function, the equation

$$\Delta x\,\Delta p_x = \tfrac{1}{2}\hbar$$

holds. Since λ is imaginary, the expression (8.78) is an eigenfunction of the non-Hermitian operator $p_x - \lambda x$. The interesting properties of these eigenfunctions will be considered repeatedly in later sections of this book.[17] The ground state of the harmonic oscillator coincides with the eigenfunction corresponding to eigenvalue zero.

7. The Equation of Motion. So far little has been said in this chapter about the time development of the physical system. The wave equation can be concisely written in the form

$$i\hbar\frac{\partial\psi}{\partial t} = H\psi \tag{8.79}$$

The Hamiltonian operator H is simply the energy operator, $H = (\mathbf{p}^2/2\mu) + V$, if the system is a particle in a conservative field, as was assumed in (8.3). For a particle (charge q) in a magnetic field, described by a vector potential $\mathbf{A}$, the Hamiltonian operator

$$H = \frac{1}{2\mu}\left(\mathbf{p} - \frac{q}{c}\mathbf{A}\right)^2 + V \tag{8.80}$$

must be used.[18] *Hamiltonians*, as these operators are often briefly called, will be constructed for more complicated systems when the occasion arises.

Exercise 8.9. Show that the Hamiltonian must be Hermitian if the equation of motion (8.79) is to guarantee conservation of probability.

[17] See Exercise 8.15, the problems at the end of this chapter, and the discussion of *coherent states* in Sections 15.9 and 15.10.

[18] See H. Goldstein, *Classical Mechanics*, Addison-Wesley Publishing Co., Reading, 1953, p. 222.

Exercise 8.10. Derive an expression for the probability current density which satisfies the continuity equation in the presence of a vector potential.

The stationary state wave function

$$\psi_k(\mathbf{r}) \exp\left(-\frac{i}{\hbar} E_k t\right)$$

is a solution of (8.79) if ψ_k is an eigenfunction of H, i.e., a solution of the Schrödinger equation

$$H\psi_k(\mathbf{r}) = E_k\psi_k(\mathbf{r}) \tag{8.81}$$

or

$$-\frac{\hbar^2}{2\mu}\left(\nabla - \frac{iq}{\hbar c}\mathbf{A}\right)^2 \psi_k + V\psi_k = E_k\psi_k \tag{8.81a}$$

with eigenvalue E_k. If the potentials V or $\mathbf{A}$ are spatially discontinuous but finite, we can infer by extending the argument presented in Section 5.6 that

$$\psi_k \qquad \text{and} \qquad \left(\nabla - \frac{iq}{\hbar c}\mathbf{A}\right)\psi_k$$

must be continuous across the discontinuities of the potential. If appropriate boundary conditions are imposed, the eigenfunctions form a complete set of orthonormal functions and the initial wave function $\psi(\mathbf{r}, 0)$ can be expanded as

$$\psi(\mathbf{r}, 0) = \sum_k c_k\psi_k(\mathbf{r}) \tag{8.82}$$

where

$$c_k = \int \psi_k{}^*(\mathbf{r})\psi(\mathbf{r}, 0)\, d\tau \tag{8.83}$$

The solution of (8.79) which is adapted to the initial condition (8.82) is

$$\psi(\mathbf{r}, t) = \sum_k c_k\psi_k(\mathbf{r}) \exp\left(-\frac{i}{\hbar} E_k t\right) \tag{8.84}$$

To exhibit the relation between the initial wave function and the wave function at time t, we substitute (8.83) into (8.84) to obtain

$$\psi(\mathbf{r}, t) = \sum_k \int \psi_k{}^*(\mathbf{r}')\psi(\mathbf{r}', 0)\, d\tau'\, \psi_k(\mathbf{r}) \exp\left(-\frac{i}{\hbar} E_k t\right)$$

$$= \int \sum_k \psi_k{}^*(\mathbf{r}')\psi_k(\mathbf{r}) \exp\left(-\frac{i}{\hbar} E_k t\right) d\tau'\, \psi(\mathbf{r}', 0) \tag{8.85}$$

$$= \int K(\mathbf{r}', \mathbf{r}; t)\psi(\mathbf{r}', 0)\, d\tau'$$

where the *Green's function* or *kernel*, K, is defined as

$$K(\mathbf{r}', \mathbf{r}; t) = \sum_k \psi_k^*(\mathbf{r}')\psi_k(\mathbf{r}) \exp\left(-\frac{i}{\hbar} E_k t\right) \tag{8.86}$$

Obviously, K is a solution of the equation of motion (8.79). At $t = 0$,

$$K(\mathbf{r}', \mathbf{r}; 0) = \delta(\mathbf{r} - \mathbf{r}') \tag{8.87}$$

by closure, or directly from (8.85). Hence, K can be characterized as the particular solution of the wave equation which equals the delta function at $t = 0$.

The related Green's function

$$G(\mathbf{r}', t'; \mathbf{r}, t) = -i\eta(t - t')K(\mathbf{r}', \mathbf{r}; t - t')$$

is often called the (nonrelativistic) *propagator*, and this integral kernel is the backbone of Feynman's formulation of quantum mechanics.[19]

Exercise 8.11. Show that the propagator satisfies the equation

$$i\hbar \frac{\partial G}{\partial t} = HG + \delta(\mathbf{r} - \mathbf{r}')\,\delta(t - t')$$

As an example, consider again the free particle in one dimension with a Hamiltonian

$$H = \frac{p^2}{2\mu} = -\frac{\hbar^2}{2\mu}\frac{\partial^2}{\partial x^2} \tag{8.88}$$

The Green's function K is the solution of

$$i\hbar \frac{\partial K(x, x', t)}{\partial t} = -\frac{\hbar^2}{2\mu}\frac{\partial^2 K(x, x', t)}{\partial x^2} \tag{8.89}$$

which satisfies the initial condition

$$K(x, x', 0) = \delta(x - x') \tag{8.90}$$

It is easy to verify that the required solution is

$$K(x, x', t) = \left(\frac{\mu}{2\pi i h t}\right)^{1/2} \exp\left[-\frac{\mu(x - x')^2}{2iht}\right] \tag{8.91}$$

Exercise 8.12. Verify that (8.91) solves (8.89) and agrees with (8.90). Show that the same result follows from the use of (8.86) directly.

---

[19] An authoritative source is the book by R. P. Feynman and A. R. Hibbs, *Quantum Mechanics and Path Integrals*, McGraw-Hill Book Company, New York, 1965.

If the system is initially represented by the minimum uncertainty wave packet [see (8.78) and Problem 1, Chapter 2]

$$\psi(x, 0) = [2\pi(\Delta x)_0^2]^{-1/4} \exp\left[-\frac{x^2}{4(\Delta x)_0^2} + ik_0 x\right] \tag{8.92}$$

we will, according to (8.85), have

$$\psi(x, t) = \left(\frac{\mu}{2\pi i\hbar t}\right)^{1/2} [2\pi(\Delta x)_0^2]^{-1/4}$$

$$\times \int_{-\infty}^{+\infty} \exp\left[-\frac{\mu(x - x')^2}{2i\hbar t} - \frac{[x' - 2ik_0(\Delta x)_0^2]^2}{4(\Delta x)_0^2} - k_0^2(\Delta x)_0^2\right] dx'$$

If the integration is carried out,[20] this expression reduces to

$$\psi(x, t) = [2\pi(\Delta x)_0^2]^{-1/4}\left[1 + \frac{i\hbar t}{2(\Delta x)_0^2\mu}\right]^{-1/2}$$

$$\times \exp\left[\frac{-\dfrac{x^2}{4(\Delta x)_0^2} + ik_0 x - ik_0^2\dfrac{\hbar t}{2\mu}}{1 + \dfrac{i\hbar t}{2(\Delta x)_0^2\mu}}\right] \tag{8.93}$$

Exercise 8.13. Calculate $|\psi(x, t)|^2$ from (8.93), and show that the wave packet moves uniformly and at the same time spreads so that

$$(\Delta x)_t^2 = (\Delta x)_0^2\left[1 + \frac{\hbar^2 t^2}{4(\Delta x)_0^4\mu^2}\right] \tag{8.94}$$

All these results are in agreement with the conclusions derived in Section 2.4. (See also Problem 1, Chapter 2, and Problem 8, Chapter 8.)

The linear harmonic oscillator is a second example. The Green's function solution of the equation

$$i\hbar\frac{\partial K(x, x', t)}{\partial t} = -\frac{\hbar^2}{2\mu}\frac{\partial^2 K(x, x', t)}{\partial x^2} + \tfrac{1}{2}\mu\omega^2 x^2 K(x, x', t) \tag{8.95}$$

is found to be

$$K(x, x', t) = \left(\frac{\mu\omega}{2\pi i\hbar \sin \omega t}\right)^{1/2}$$

$$\times \exp\left[-\frac{\mu\omega}{2i\hbar \sin \omega t}(x^2 \cos \omega t - 2xx' + x'^2 \cos \omega t)\right] \tag{8.96}$$

---

[20] $\displaystyle\int_{-\infty}^{+\infty} \exp(-ax^2)\, dx = \left(\frac{\pi}{a}\right)^{1/2}.$

Exercise 8.14. Verify that (8.96) is the Green's function for the harmonic oscillator.[21]

Exercise 8.15. Show that, if the initial state of a harmonic oscillator is

$$\psi(x, 0) = N \exp\left[-\frac{\mu\omega}{2\hbar}(x - a)^2\right]$$

the state at time t is

$$\psi(x, t) = N \exp\left[-\frac{\mu\omega}{2\hbar}(x - a\cos\omega t)^2\right.$$

$$\left. - i\left(\frac{\omega}{2}t + \frac{\mu\omega}{\hbar}ax\sin\omega t - \frac{\mu\omega}{4\hbar}a^2\sin 2\omega t\right)\right] \quad (8.97)$$

Show that $|\psi(x, t)|$ oscillates without any change of shape.

Finally, how does the expectation value of an observable A change in time? From (8.84),

$$\langle A \rangle = \int \psi^*(\mathbf{r}, t)A\psi(\mathbf{r}, t)\, d\tau$$

$$= \sum_l \sum_k c_l^* c_k \exp\left(-\frac{i}{\hbar}(E_k - E_l)t\right)\int \psi_l^*(\mathbf{r})A\psi_k(\mathbf{r})\, d\tau$$

If we denote the matrix elements of A with respect to the eigenstates of H by

$$A_{lk} = \int \psi_l^* A\psi_k\, d\tau \quad (8.98)$$

we have

$$\langle A \rangle = \sum_k |c_k|^2 A_{kk} + \sum_l \sum_{\substack{k \\ (l \neq k)}} c_l^* c_k \exp\left(-\frac{i}{\hbar}(E_k - E_l)t\right)A_{lk} \quad (8.99)$$

Equation (8.99) exhibits the time dependence of $\langle A \rangle$. A more implicit but very useful form is obtained by differentiating $\langle A \rangle$ directly:

$$\frac{d\langle A \rangle}{dt} = \int \frac{\partial\psi^*}{\partial t}A\psi\, d\tau + \int \psi^* A\frac{\partial\psi}{\partial t}\, d\tau + \int \psi^* \frac{\partial A}{\partial t}\psi\, d\tau$$

where we have included in the last term the possibility of an explicit time dependence of A. Using the wave equation (8.79), this can be transformed into

$$i\hbar\frac{d\langle A \rangle}{dt} = \int \psi^*(AH - HA)\psi\, d\tau + i\hbar\int \psi^*\frac{\partial A}{\partial t}\psi\, d\tau \quad (8.100)$$

---

[21] The Green's function (8.96) may, for instance, be obtained from the harmonic oscillator eigenfunctions by the use of Mehler's formula, P. M. Morse and H. Feshbach, *Methods of Theoretical Physics*, McGraw-Hill Book Company, New York, 1953, p. 781, Problem 6.21.

If A commutes with H and does not explicitly depend on t, $\langle A \rangle$ is constant in time for any state. The physical observable, A, is then termed a *constant of the motion* and is said to be *conserved*.

Thus the linear momentum is a constant of the motion if $\mathbf{p}$ commutes with H. Since $\mathbf{p}$ commutes with the kinetic energy operator $\mathbf{p}^2/2\mu$, it is only necessary to examine whether it also commutes with V. All three components of momentum commute with V only if V is constant. Hence, in quantum mechanics as in classical mechanics, momentum is conserved only if the system is not acted upon by external forces.

8. Commutator Algebra. The uncertainty principle and the equation of motion (8.100) show that the commutator $AB - BA$ of two operators plays a very prominent role in quantum mechanics. It is therefore appropriate to conclude this chapter with the introduction of a simple and useful notation for commutators and a review of some rules of operator algebra. All of these rules apply, in particular, to square matrices.

By definition, the *commutator* is written as a bracket:

$$[A, B] = AB - BA \tag{8.101}$$

With this notation the following elementary rules of calculation are almost self-evident. All are easy to verify.

$$[A, B] + [B, A] = 0 \tag{8.102a}$$

$$[A, A] = 0 \tag{8.102b}$$

$$[A, B + C] = [A, B] + [A, C] \tag{8.102c}$$

$$[A + B, C] = [A, C] + [B, C] \tag{8.102d}$$

$$[A, BC] = [A, B]C + B[A, C] \tag{8.102e}$$

$$[AB, C] = [A, C]B + A[B, C] \tag{8.102f}$$

$$[A, [B, C]] + [C, [A, B]] + [B, [C, A]] = 0 \tag{8.102g}$$

Exercise 8.16. If A and B are two operators which both commute with their commutator, $[A, B]$, prove that

$$[A, B^n] = nB^{n-1}[A, B] \tag{8.103a}$$

$$[A^n, B] = nA^{n-1}[A, B] \tag{8.103b}$$

Note the similarity of this process with differentiation. Apply to the special case $A = x$, $B = p_x$. Prove that

$$[p_x, f(x)] = \frac{\hbar}{i} \frac{df}{dx} I$$

if $f(x)$ can be expanded in a power series.

Exercise 8.17 If the Hamiltonian has the form (8.80), show from the commutators of **r** and **p** with H that the expectation values of **r** and **p** satisfy the classical equations of motion.

Another operator relation, which is frequently used, involves the function e^A, defined by the power series

$$e^A = 1 + \frac{A}{1!} + \frac{A^2}{2!} + \frac{A^3}{3!} + \cdots \tag{8.104}$$

The identity

$$e^A B e^{-A} = B + [A, B] + \frac{1}{2!}[A, [A, B]] + \frac{1}{3!}[A, [A, [A, B]]] + \cdots \tag{8.105}$$

can be proved as follows:

Consider

$$f(\lambda) = e^{\lambda A} B e^{-\lambda A}$$

Make a Taylor series expansion of $f(\lambda)$, observing that

$$\frac{df}{d\lambda} = Af(\lambda) - f(\lambda)A = [A, f(\lambda)]$$

$$\frac{d^2 f}{d\lambda^2} = \left[A, \frac{df(\lambda)}{d\lambda}\right] = [A, [A, f(\lambda)]]$$

etc. Since $f(0) = B$, we get

$$f(\lambda) = B + \frac{\lambda}{1!}[A, B] + \frac{\lambda^2}{2!}[A, [A, B]] + \cdots$$

from which (8.105) follows by setting $\lambda = 1$.

If the operators A and B satisfy the condition $[A, B] = \gamma B$, where γ is a number, (8.105) reduces to

$$e^A B e^{-A} = e^\gamma B \tag{8.106}$$

Exercise 8.18. If A and B are two operators which both commute with their commutator, $[A, B]$, prove the identity,

$$e^A e^B = e^{A+B+(1/2)[A,B]} \tag{8.107}$$

(Hint: Consider $f(\lambda) = e^{\lambda A} e^{\lambda B} e^{-\lambda(A+B)}$, establish the differential equation $df/d\lambda = \lambda[A, B]f$, and integrate this.)

Equations (8.105) and (8.107) caution us against the indiscriminate application of well-known mathematical rules, when the variables are noncommuting operators. Thus (8.107) shows that generally for operators $e^A e^B \neq e^{A+B}$.

9. The Virial Theorem. A simple example will illustrate the principles of the last two sections. Consider the equation of motion for the operator $\mathbf{r} \cdot \mathbf{p}$. According to (8.100),

$$i\hbar \frac{d}{dt} \langle \mathbf{r} \cdot \mathbf{p} \rangle = \langle [\mathbf{r} \cdot \mathbf{p}, H] \rangle$$

By applying several of the rules (8.102) and the fundamental commutation relation (8.25), $xp_x - p_x x = i\hbar$, we obtain

$$[xp_x, H] = \left[xp_x, \frac{p_x^2}{2\mu} + V \right] = \left[xp_x, \frac{p_x^2}{2\mu} \right] + [xp_x, V]$$

$$= [x, p_x^2] \frac{p_x}{2\mu} + x[p_x, V]$$

$$= p_x[x, p_x] \frac{p_x}{2\mu} + [x, p_x] \frac{p_x^2}{2\mu} + x[p_x, V] = i\hbar \frac{p_x^2}{\mu} + x[p_x, V]$$

Similar relations hold for the y- and z-components. Combining these results, we get

$$\frac{d}{dt} \langle \mathbf{r} \cdot \mathbf{p} \rangle = \left\langle \frac{\mathbf{p}^2}{\mu} \right\rangle - \int \psi^* \mathbf{r} \cdot [\nabla(V\psi) - V \nabla \psi] \, d\tau$$

$$= \left\langle \frac{\mathbf{p}^2}{\mu} \right\rangle - \int \psi^* \mathbf{r} \cdot (\nabla V)\psi \, d\tau \tag{8.108}$$

For a *stationary state* it follows that

$$2\langle T \rangle = \langle \mathbf{r} \cdot \nabla V \rangle \tag{8.109}$$

which is known as the *virial theorem*.

Exercise 8.19. Apply the virial theorem to the harmonic oscillator. Compare with Exercise 8.3 and with the classical form of the virial theorem.

10. The Schwarz Inequality. The proof of the Schwarz inequality (8.69) must yet be supplied. It is convenient to derive this inequality as a special case of a more general theorem which has occasional use in quantum mechanics.

Let us define a *positive definite* operator A by requiring that its expectation value in any state be real and nonnegative, i.e.,

$$\int \psi^* A \psi \, d\tau \geqslant 0 \tag{8.110}$$

for any function ψ. Since its expectation values are real, A must be Hermitian.

Now it follows from (8.110) that for a function

$$\psi = f + \lambda g \tag{8.111}$$

we have

$$\int (f + \lambda g)^* A(f + \lambda g) \, d\tau$$

$$= \int f^* Af \, d\tau + \lambda \int f^* Ag \, d\tau + \lambda^* \int g^* Af \, d\tau + |\lambda|^2 \int g^* Ag \, d\tau \geqslant 0 \tag{8.112}$$

where f and g are arbitrary functions and λ may be any complex number. In particular, the last inequality must be true for the choice

$$\lambda = -\frac{\int g^* Af \, d\tau}{\int g^* Ag \, d\tau} = -\frac{\left(\int f^* Ag \, d\tau \right)^*}{\int g^* Ag \, d\tau}$$

Substituting this value of λ into (8.112), we obtain the inequality

$$\left(\int f^* Af \, d\tau \right) \left(\int g^* Ag \, d\tau \right) \geqslant \left| \int f^* Ag \, d\tau \right|^2 \tag{8.113}$$

for any positive definite operator. The Hermitian nature of A has been used in going from (8.112) to (8.113).

For the trivial positive definite operator $A = 1$, we get from (8.113) the *Schwarz inequality*

$$\left(\int |f|^2 \, d\tau \right) \left(\int |g|^2 \, d\tau \right) \geqslant \left| \int f^* g \, d\tau \right|^2 \tag{8.114}$$

This much on the foundations of wave mechanics will suffice for the present. We shall return to the general theory in Chapters 14 and 15 to reformulate quantum mechanics at a higher level of abstraction. Meanwhile, the principles developed in this chapter will be applied to a number of concrete physical problems.

Problems

1. Carry out a numerical integration to test the uncertainty relation for the wave packet defined by Figures 2.1 and 2.2.

2. Assuming a particle to be in one of the stationary states of an infinitely high one-dimensional box, calculate the uncertainties in position and momentum, and show that they agree with the Heisenberg uncertainty relation. Also show that in the limit of very large quantum numbers the uncertainty in x equals the rms deviation of the position of a particle moving classically in the enclosure.

3. Calculate the value of $\Delta x \, \Delta p$ for a linear harmonic oscillator in one of its eigenstates.

4. Using the uncertainty relation, but not the explicit solutions of the eigenvalue problem, show that the expectation value of the energy of a harmonic oscillator can never be less than the zero-point energy.

5. Rederive the one-dimensional minimum uncertainty wave packet by using the variational calculus to minimize the expression $I = (\Delta x)^2 (\Delta p)^2$ subject to the condition

$$\int |\psi|^2 \, dx = 1$$

Show that the solution ψ of this problem satisfies a differential equation which is equivalent to the Schrödinger equation for the harmonic oscillator, and calculate the minimum value of $\Delta x \, \Delta p$.

6. If a particle of mass μ and charge q is constrained to move in the xy-plane on a circular orbit of radius ρ around the origin 0, but is otherwise free, determine the energy eigenvalues and the eigenfunctions.

   A magnetic field, represented by the vector potential $\mathbf{A} = \Phi \hat{\mathbf{k}} \times \mathbf{r}/[2\pi(\hat{\mathbf{k}} \times \mathbf{r})^2]$ is now introduced. (a) Show that the magnetic field approximates that of a long thin solenoid with flux Φ placed on the z-axis. (b) Determine the new energy spectrum and show that it coincides with the spectrum for $\Phi = 0$ if the flux assumes certain quantized values. Note that the energy levels depend on the strength of a magnetic field $\mathbf{B}$ to which the particle is not exposed (*Aharonov-Bohm effect*).

7. An electromagnetic field which is defined by the potentials ϕ and $\mathbf{A}$ can be equally well derived from the set of potentials

$$\mathbf{A}' = \mathbf{A} + \nabla f, \qquad \phi' = \phi - \frac{\partial f}{\partial t}$$

where f is an arbitrary function of $\mathbf{r}$ and t. The two sets of potentials are said to differ by a *gauge transformation*.

   For each set of potentials there is a wave function describing the state and a corresponding wave equation. Show how ψ and ψ', both describing the same state, are related, and compare the expectation values of $\mathbf{r}$ and $\mathbf{p}$ in the two gauges. Also compare the current densities.

8. For a free particle in one dimension calculate the time development of $(\Delta x)^2$ without explicit use of the wave function by applying (8.100) repeatedly and show that

$$(\Delta x)_t^2 = (\Delta x)_0^2 + \frac{2}{\mu}\left[\frac{1}{2}\langle xp + px \rangle_0 - \langle x \rangle_0 \langle p \rangle\right]t + \frac{(\Delta p)^2}{\mu^2}t^2$$

Compare this result with the value of $(\Delta x)^2$ as a function of time for a beam of classical particles whose initial positions and momenta have distributions with variances $(\Delta x)_0^2$ and $(\Delta p)^2$.

   As a special case consider $(\Delta x)_t^2$ for a minimum uncertainty wave packet.

9. Apply the virial theorem to a central potential of the form $V = Ar^n$, and express the expectation value of V for bound states in terms of the energy.

10. Using the Schrödinger equation in the momentum representation, calculate the bound state energy eigenvalue and the corresponding eigenfunction for the potential $-g\,\delta(x)$. Compare with the results of Problem 3 in Chapter 6.

CHAPTER 9

Central Forces and Angular Momentum

1. Orbital Angular Momentum. We now turn to the motion of a particle in ordinary three-dimensional space. Quantum mechanically, as classically, a very important class of dynamical problems arises with *central forces*. These are forces which are derivable from a potential that depends only on the distance r of the moving particle from a fixed point, usually the coordinate origin. The Hamiltonian operator is

$$H = \frac{\mathbf{p}^2}{2\mu} + V(r) = -\frac{\hbar^2}{2\mu} \nabla^2 + V(r) \tag{9.1}$$

Since central forces produce no torque about the origin, the (orbital) *angular momentum*

$$\mathbf{L} = \mathbf{r} \times \mathbf{p} \tag{9.2}$$

is conserved. This is the statement of Kepler's second law in classical mechanics.

According to the correspondence principle, we must expect that angular momentum will play an equally essential role in quantum mechanics. The operator which represents angular momentum in the coordinate representation is obtained from (9.2) by replacing $\mathbf{p}$ by $(\hbar/i)\nabla$:

$$\mathbf{L} = \mathbf{r} \times \frac{\hbar}{i} \nabla \tag{9.3}$$

No difficulty arises here with operators which fail to commute, because only products like xp_y, yp_z, etc., appear (see Exercise 8.5).

In view of the great importance of angular momentum as a physical quantity it is well to derive some of the properties of the operator $\mathbf{L}$, using the basic commutation relations between the components of $\mathbf{r}$ and $\mathbf{p}$ and the rules of Section 8.8. For example,

$$[L_x, y] = [yp_z - zp_y, y] = -z[p_y, y] = i\hbar z$$
$$[L_x, p_y] = [yp_z - zp_y, p_y] = [y, p_y]p_z = i\hbar p_z \tag{9.4}$$
$$[L_x, x] = 0, \qquad [L_x, p_x] = 0$$

Similar relations hold for all other commutators between **L** and **r** and between **L** and **p**. From these relations we can further deduce the commutation relations between the various components of **L**,

$$[L_x, L_y] = [L_x, zp_x - xp_z] = [L_x, z]p_x - x[L_x, p_z]$$
$$= -i\hbar yp_x + i\hbar xp_y = i\hbar L_z \qquad (9.5a)$$

and by cyclic permutation $(x \to y \to z \to x)$ of this result,

$$[L_y, L_z] = i\hbar L_x, \qquad [L_z, L_x] = i\hbar L_y \qquad (9.5b)$$

Since the components of **L** do not commute, the system cannot in general be assigned definite values for all angular momentum components simultaneously.

   If we apply the relations (9.5) to the general theorem (8.68), we obtain the inequalities

$$\Delta L_x \Delta L_y \geqslant \frac{\hbar}{2} |\langle L_z \rangle| \quad \text{et cycl.} \qquad (9.6)$$

An exact simultaneous determination of all three components of **L** is possible only if $\langle L_x \rangle = \langle L_y \rangle = \langle L_z \rangle = 0$, or,

$$\langle \mathbf{L} \rangle = 0 \qquad (9.7)$$

Since

$$(\Delta L_x)^2 = \langle L_x{}^2 \rangle - \langle L_x \rangle^2 \quad \text{etc.}$$

it follows that $(\Delta \mathbf{L}) = 0$ only if besides (9.7) we have also

$$\langle L_x{}^2 \rangle = \langle L_y{}^2 \rangle = \langle L_z{}^2 \rangle = 0 \qquad (9.8)$$

Hence, the special state, ψ, for which an exact value for all three components of **L** can be specified, must be an eigenstate of each component of **L** with eigenvalue zero:

$$\mathbf{L}\psi = 0 \qquad (9.9)$$

   Another remarkable quantum property of angular momentum is obtained if we define the square of the magnitude of **L** by

$$\mathbf{L}^2 = L_x{}^2 + L_y{}^2 + L_z{}^2 \qquad (9.10)$$

$L_x{}^2$ and $L_y{}^2$ are positive definite Hermitian operators; hence,

$$\langle \mathbf{L}^2 \rangle \geqslant \langle L_z{}^2 \rangle$$

The *equality* holds only if $\langle L_x{}^2 \rangle = \langle L_y{}^2 \rangle = 0$. In this exceptional case, $L_x\psi = L_y\psi = 0$; hence by (9.5a) also $L_z\psi = 0$, or $\mathbf{L}\psi = 0$. Since

$$(\Delta L_z)^2 = \langle L_z{}^2 \rangle - \langle L_z \rangle^2 \geqslant 0$$

we conclude that

$$\langle \mathbf{L}^2 \rangle > \langle L_z \rangle^2 \qquad (9.11)$$

for all states, except those satisfying condition (9.9). We thus see that, except in the trivial case in which all components of **L** vanish, the magnitude of angular momentum exceeds the maximum value of any one of its components. If $\langle L_z \rangle$ cannot generally be made equal to the magnitude of **L**, we may well ask if L^2 can at all be determined simultaneously with a component, such as L_z. That it can follows from the commutivity of L^2 with any component of **L**.

Proof

$$[\mathbf{L}^2, L_z] = [L_x{}^2 + L_y{}^2 + L_z{}^2, L_z] = [L_x{}^2 + L_y{}^2, L_z]$$
$$= L_x[L_x, L_z] + [L_x, L_z]L_x + L_y[L_y, L_z] + [L_y, L_z]L_y$$
$$= -i\hbar L_x L_y - i\hbar L_y L_x + i\hbar L_y L_x + i\hbar L_x L_y = 0$$

Hence, generally

$$[\mathbf{L}^2, \mathbf{L}] = 0 \tag{9.12}$$

Insight into the nature of the angular momentum operator is gained by noting its connection with (*rigid*) *rotations*. Suppose $f(\mathbf{r})$ is an arbitrary differentiable function in space. If the value $f(\mathbf{r})$ of the function is displaced by **a** to the new point $\mathbf{r} + \mathbf{a}$, a new function $F(\mathbf{r})$ is obtained such that

$$F(\mathbf{r} + \mathbf{a}) = f(\mathbf{r})$$

For an *infinitesimal* displacement,

$$F(\mathbf{r}) = f(\mathbf{r} - \mathbf{a}) = f(\mathbf{r}) - \mathbf{a} \cdot \nabla f(\mathbf{r}) \tag{9.13}$$

If the displacement is an *infinitesimal rotation* by an angle $\delta\varphi$ about an axis through the origin, we can express **a** as

$$\mathbf{a} = \delta\boldsymbol{\varphi} \times \mathbf{r}$$

where $\delta\boldsymbol{\varphi}$ is a vector of length $\delta\varphi$ pointing in the direction of the axis of rotation with an orientation defined by a right-handed screw. The change of the function f is then

$$\delta f(\mathbf{r}) = F(\mathbf{r}) - f(\mathbf{r}) = -(\delta\boldsymbol{\varphi} \times \mathbf{r}) \cdot \nabla f = -\delta\boldsymbol{\varphi} \cdot (\mathbf{r} \times \nabla f)$$

or

$$\delta f = -\frac{i}{\hbar} \delta\boldsymbol{\varphi} \cdot \mathbf{L} f \tag{9.14}$$

The operator $\mathbf{L}/\hbar$ is called the *generator of infinitesimal rotations*.

Exercise 9.1. Show that $\mathbf{p}/\hbar$ is the *generator of infinitesimal translations*.

From (9.14) and (9.9) it follows that the state ψ in which L_x, L_y, *and* L_z are definite does not change under an arbitrary small rotation, or

$$\delta\psi(\mathbf{r}) = 0 \tag{9.15}$$

Hence, ψ depends on r only, and not on the angles at all. Such spherically symmetric states are generally designated as *S-states*. According to (9.9), there is zero angular momentum associated with them.

The fundamental relation (9.14) can be used to derive the commutation relations once more. For instance, from (9.14)

$$\delta(xf) = x\,\delta f + f\,\delta x = -\frac{i}{\hbar}\,\delta\boldsymbol{\varphi}\cdot\mathbf{L}(xf)$$

Let us consider in particular a rotation about the z-axis. Since

$$\delta x = -a_x = y\,\delta\varphi$$

we have

$$-x\frac{i}{\hbar}\,\delta\varphi\,L_z f + fy\,\delta\varphi = -\frac{i}{\hbar}\,\delta\varphi\,L_z xf$$

or

$$L_z x - xL_z = i\hbar y \tag{9.16}$$

which is indeed one of the relations implied by (9.4).

If f in (9.14) is replaced by $V(r)g$, we obtain under a rotation

$$\delta[V(r)g(\mathbf{r})] = V(r)\,\delta g(\mathbf{r})$$

or

$$-\frac{i}{\hbar}\,\delta\boldsymbol{\varphi}\cdot\mathbf{L}(Vg) = -\frac{i}{\hbar}V\,\delta\boldsymbol{\varphi}\cdot\mathbf{L}g$$

Since $\delta\varphi$ is an arbitrary vector and $g(\mathbf{r})$ an arbitrary function, we conclude that

$$\mathbf{L}V(r) - V(r)\mathbf{L} = 0 \tag{9.17}$$

Hence, the angular momentum commutes with the potential energy for central forces. If we can show that $\mathbf{L}$ commutes with the kinetic energy, we shall have proved that it is a constant of the motion.

2. Kinetic Energy and Angular Momentum. The operator $\mathbf{L}^2$ is useful in showing that the kinetic energy commutes with $\mathbf{L}$. At the same time we shall acquire the tools necessary to reduce the central force problem to an equivalent one-dimensional one. This procedure is familiar from classical mechanics, where use is made of the identity

$$\mathbf{L}^2 = (\mathbf{r} \times \mathbf{p})^2 = r^2 p^2 - (\mathbf{r}\cdot\mathbf{p})^2 \tag{9.18}$$

or

$$p^2 = \frac{\mathbf{L}^2}{r^2} + \left(\frac{\mathbf{r}\cdot\mathbf{p}}{r}\right)^2 \tag{9.19}$$

expressing the kinetic energy in terms of a constant of the motion, $\mathbf{L}^2$, and the radial component of momentum. Unfortunately (9.19) cannot be taken over into quantum mechanics, because the operator $\mathbf{r} \cdot \mathbf{p}$ is not Hermitian. However, the correct relation can be worked out readily from the operator identity

$$L^2 = (\mathbf{r} \times \mathbf{p}) \cdot (\mathbf{r} \times \mathbf{p}) = r^2 p^2 - r(\mathbf{r} \cdot \mathbf{p}) \cdot \mathbf{p} + 2i\hbar \mathbf{r} \cdot \mathbf{p} \qquad (9.20)$$

where the middle term is to be understood to mean

$$\mathbf{r}(\mathbf{r} \cdot \mathbf{p}) \cdot \mathbf{p} = \sum_{i=1}^{3} \sum_{k=1}^{3} r_i r_k p_k p_i$$

(Occasionally subscripts 1, 2, 3 will be used to denote the Cartesian x, y, z components of vectors.)

Exercise 9.2. Prove the identity (9.20). Discuss the conditions under which the classical limit (9.18) is obtained from (9.20).

The component of the gradient ∇f in the direction of $\mathbf{r}$ is $\partial f / \partial r$; hence

$$\mathbf{r} \cdot \mathbf{p} = \frac{\hbar}{i} r \frac{\partial}{\partial r}$$

and consequently

$$L^2 = r^2 p^2 + \hbar^2 r^2 \frac{\partial^2}{\partial r^2} + 2\hbar^2 r \frac{\partial}{\partial r}$$

or more compactly

$$L^2 = r^2 p^2 + \hbar^2 \frac{\partial}{\partial r} \left(r^2 \frac{\partial}{\partial r} \right) \qquad (9.21)$$

Since $\mathbf{L}$ and therefore also L^2 commutes with any function of r, we may use (9.21) to write

$$T = \frac{p^2}{2\mu} = \frac{L^2}{2\mu r^2} - \frac{\hbar^2}{2\mu r^2} \frac{\partial}{\partial r} \left(r^2 \frac{\partial}{\partial r} \right) \qquad (9.22)$$

We shall make frequent use of this fundamental relation between T and L^2. $\mathbf{L}$ commutes with any radial derivative, since it is patently irrelevant whether differentiation with respect to r is performed before or after a rotation. Hence $\mathbf{L}$ and T commute, and thus the proof that for central forces

$$[H, \mathbf{L}] = 0 \qquad (9.23)$$

is complete. This commutation relation insures that angular momentum is a constant of the motion.

3. *Reduction of the Central Force Problem.* Since H and L^2 commute and since L^2 is assumed to have a complete set of eigenfunctions, it is possible to select the eigenfunctions of H to be eigenfunctions of L^2 also. These particular eigenfunctions must have the property

$$\mathbf{L}^2\psi_\lambda = \lambda\hbar^2\psi_\lambda \tag{9.24}$$

where $\lambda\hbar^2$ denotes an eigenvalue of $\mathbf{L}^2$. λ is dimensionless because $\hbar$ has the dimensions of angular momentum.

Using (9.22) and (9.24), the Schrödinger equation

$$H\psi \equiv [T + V(r)]\psi = E\psi \tag{9.25}$$

for central forces can thus be reduced to

$$\left[-\frac{\hbar^2}{2\mu r^2}\frac{\partial}{\partial r}\left(r^2\frac{\partial}{\partial r}\right) + \frac{\lambda\hbar^2}{2\mu r^2} + V(r)\right]\psi_\lambda = E\psi_\lambda \tag{9.26}$$

Since $\mathbf{L}^2$ commutes with any function of r, it follows from (9.24) and (9.26), using spherical polar coordinates, that *all* solutions of the Schrödinger equation are obtained as linear combinations of separable solutions of the form

$$\psi(\mathbf{r}) = R(r)Y(\theta, \varphi) \tag{9.27}$$

Here $Y(\theta, \varphi)$ is a solution of (9.24), and the radial eigenfunction $R(r)$ satisfies the ordinary differential equation

$$\left[-\frac{\hbar^2}{2\mu r^2}\frac{d}{dr}\left(r^2\frac{d}{dr}\right) + \frac{\lambda\hbar^2}{2\mu r^2} + V(r)\right]R(r) = ER(r) \tag{9.28}$$

which is easier to solve than the original partial differential equation (9.25). Our procedure has been entirely equivalent to the familiar *separation of variables* of the Laplacian operator in spherical polar coordinates with emphasis on the physical meaning of the method.

By making the substitution

$$u = rR(r) \tag{9.29}$$

we find that u obeys the *radial* equation

$$-\frac{\hbar^2}{2\mu}\frac{d^2u}{dr^2} + \left[\frac{\lambda\hbar^2}{2\mu r^2} + V(r)\right]u = Eu \tag{9.30}$$

This equation is identical in form with the one-dimensional Schrödinger equation except for the addition of the term $\lambda\hbar^2/2\mu r^2$ to the potential energy. This term is sometimes called the *centrifugal potential*, since it represents the potential whose negative gradient is the centrifugal force.

Although (9.30) is similar to the one-dimensional Schrödinger equation, the boundary conditions to be imposed on the solutions are quite different, since r is never negative. For instance, if ψ is to be finite everywhere, u must vanish at $r = 0$, according to (9.29). A detailed discussion of these questions requires specific assumptions about the shape of the potential energy, and in Chapter 10 the radial equation (9.28) will be solved for several particular cases. In this chapter we devote our attention to the general features common to all central force problems. Hence we return to an explicit consideration of the angular momentum operators and their eigenvalues.

4. The Eigenvalue Problem for L². It is convenient to express the angular momentum as a differential operator in terms of spherical polar coordinates defined by

$$x = r \sin \theta \cos \varphi, \qquad y = r \sin \theta \sin \varphi, \qquad z = r \cos \theta$$

The calculations become transparent if we note that the gradient operator can be written in terms of the unit vectors of spherical polar coordinates as[1]

$$\nabla = \hat{\mathbf{r}} \frac{\partial}{\partial r} + \hat{\boldsymbol{\varphi}} \frac{1}{r \sin \theta} \frac{\partial}{\partial \varphi} + \hat{\boldsymbol{\theta}} \frac{1}{r} \frac{\partial}{\partial \theta} \tag{9.31}$$

where

$$\hat{\mathbf{r}} = \sin \theta \cos \varphi \, \hat{\mathbf{i}} + \sin \theta \sin \varphi \, \hat{\mathbf{j}} + \cos \theta \, \hat{\mathbf{k}} \tag{9.32a}$$

$$\hat{\boldsymbol{\varphi}} = -\sin \varphi \, \hat{\mathbf{i}} + \cos \varphi \, \hat{\mathbf{j}} \tag{9.32b}$$

$$\hat{\boldsymbol{\theta}} = \cos \theta \cos \varphi \, \hat{\mathbf{i}} + \cos \theta \sin \varphi \, \hat{\mathbf{j}} - \sin \theta \, \hat{\mathbf{k}} \tag{9.32c}$$

From (9.31) it is evident that the three spherical polar components of the momentum operator $(\hbar/i)\nabla$, unlike its Cartesian components, do not commute. The angular momentum may now be expressed as

$$\mathbf{L} = \mathbf{r} \times \mathbf{p} = r\hat{\mathbf{r}} \times \frac{\hbar}{i} \nabla = \frac{\hbar}{i} \left(\hat{\boldsymbol{\varphi}} \frac{\partial}{\partial \theta} - \hat{\boldsymbol{\theta}} \frac{1}{\sin \theta} \frac{\partial}{\partial \varphi} \right) \tag{9.33}$$

From this simple form it is again apparent that **L** commutes with any function of r and with any derivative with respect to r.

From (9.33) and (9.32) we obtain

$$L_x = \frac{\hbar}{i} \left(-\sin \varphi \frac{\partial}{\partial \theta} - \cos \varphi \cotan \theta \frac{\partial}{\partial \varphi} \right) \tag{9.34a}$$

$$L_y = \frac{\hbar}{i} \left(\cos \varphi \frac{\partial}{\partial \theta} - \sin \varphi \cotan \theta \frac{\partial}{\partial \varphi} \right) \tag{9.34b}$$

$$L_z = \frac{\hbar}{i} \frac{\partial}{\partial \varphi} \tag{9.34c}$$

---

[1] We use carets to denote unit vectors.

and from here

$$\mathbf{L}^2 = L_x{}^2 + L_y{}^2 + L_z{}^2$$

$$= -\hbar^2 \left[\frac{1}{\sin^2 \theta} \frac{\partial^2}{\partial \varphi^2} + \frac{1}{\sin \theta} \frac{\partial}{\partial \theta} \left(\sin \theta \frac{\partial}{\partial \theta} \right) \right] \tag{9.35}$$

The eigenvalue problem (9.24) for **L²** now can be formulated explicitly as follows:

$$\left[\frac{1}{\sin^2 \theta} \frac{\partial^2}{\partial \varphi^2} + \frac{1}{\sin \theta} \frac{\partial}{\partial \theta} \left(\sin \theta \frac{\partial}{\partial \theta} \right) \right] Y(\theta, \varphi) = -\lambda Y(\theta, \varphi) \tag{9.36}$$

This partial differential equation can be solved by finding separated solutions of the form

$$Y(\theta, \varphi) = \Phi(\varphi)\Theta(\theta) \tag{9.37}$$

When we substitute (9.37) into (9.36), we obtain by a little algebra

$$-\frac{1}{\Phi} \frac{d^2\Phi}{d\varphi^2} = \frac{\sin^2 \theta}{\Theta} \left[\frac{1}{\sin \theta} \frac{d}{d\theta} \left(\sin \theta \frac{d\Theta}{d\theta} \right) + \lambda\Theta \right]$$

indicating a neat separation of the two variables. Since each side of this equation must be equal to a constant, we deduce the ordinary differential equations,

$$\frac{d^2\Phi}{d\varphi^2} + m^2\Phi = 0 \tag{9.38}$$

$$\frac{1}{\sin \theta} \frac{d}{d\theta} \left(\sin \theta \frac{d\Theta}{d\theta} \right) - \frac{m^2}{\sin^2 \theta} \Theta + \lambda\Theta = 0 \tag{9.39}$$

The separation constant has been taken to be m^2 with forethought, since in this way m will soon be seen to be a *real* number by virtue of the boundary conditions which must now be formulated.

What conditions must we impose on the solutions of (9.38) and (9.39) to give us physically acceptable wave functions? It is implicit in the fundamental postulates of quantum mechanics that the wave function for a particle without spin must have a definite value at every point in space.[2] Hence, we must demand that the wave function be a *single-valued* function of the particle's position. In particular, ψ must take on the same value whether the azimuth of a point is given by φ or $\varphi + 2\pi$.

Applied to (9.38), the condition $\Phi(\varphi + 2\pi) = \Phi(\varphi)$ restricts us to those solutions

$$\Phi(\varphi) = e^{im\varphi} \tag{9.40}$$

for which $m = 0, \pm 1, \pm 2, \ldots$, i.e., an *integer*.

[2] For an analysis of the requirement that wave functions must be single-valued, see E. Merzbacher, *Am. J. Phys.*, **30**, 237 (1962).

5. *The Eigenvalue Problem for L_z.* The eigenfunctions (9.40) could also have
been written in the form cos $m\varphi$ and sin $m\varphi$—and these, being real, have
something to recommend them—but the exponentials of (9.40) have the
weighty advantage of being eigenfunctions of the operator L_z, whereas
cos $m\varphi$ and sin $m\varphi$ are not eigenfunctions of L_z. Indeed, according to (9.34c):

$$L_z e^{im\varphi} \equiv \frac{\hbar}{i} \frac{\partial}{\partial \varphi} e^{im\varphi} = m\hbar e^{im\varphi} \qquad (9.41)$$

Since L_z commutes with $\mathbf{L}^2$, we know, of course, that these two operators
admit of simultaneous eigenfunctions. [So do $\mathbf{L}^2$ and L_x, or $\mathbf{L}^2$ and L_y, but the
corresponding eigenfunctions are more complicated to write down when the
z-axis is used as the axis of the spherical polar coordinates, because (9.34a)
and (9.34b) contain θ, whereas L_z is expressed in terms of φ alone.]

Equation (9.41) shows explicitly that the eigenvalues of L_z are $m\hbar$. Thus,
a measurement of L_z can yield as its result only the values $0, \pm\hbar, \pm2\hbar, \dots$.
Since the z-axis points in an arbitrarily chosen direction, it must be true that
the angular momentum about any axis is *quantized* and can upon measurement only reveal one of these discrete values.

The term *magnetic quantum number* is frequently used for the integer m
because of the part played by this number in describing the effect of a uniform
magnetic field $\mathbf{B}$ on a charged particle moving in a central field. The vector
potential $\mathbf{A} = \frac{1}{2}\mathbf{B} \times \mathbf{r}$, which satisfies $\nabla \cdot \mathbf{A} = 0$, represents this field simply.
The Schrödinger equation (8.81a) becomes

$$\left[\frac{p^2}{2\mu} - \frac{q}{2\mu c} \mathbf{B} \times \mathbf{r} \cdot \mathbf{p} + \frac{q^2}{8\mu c^2} (\mathbf{B} \times \mathbf{r})^2 + V(r)\right]\psi = E\psi \qquad (9.42)$$

and, if terms quadratic in the magnetic field are neglected, this can be written
as

$$\left[\frac{p^2}{2\mu} - \frac{q}{2\mu c} \mathbf{B} \cdot \mathbf{L} + V(r)\right]\psi = E\psi \qquad (9.43)$$

showing that such a field can be taken into account to a first approximation
by adding to the Hamiltonian operator a magnetic energy term $-\mathbf{m} \cdot \mathbf{B}$.
Here $\mathbf{m}$ is the effective magnetic moment of the system, linked to the angular
momentum by the relation

$$\mathbf{m} = \frac{q}{2\mu c} \mathbf{L}$$

where q is the charge of the particle and μ its mass.

If the z-axis is chosen parallel to the external field, the magnetic perturbation contains only the operator L_z, and the eigenfunctions of (9.25) will also
automatically be eigenfunctions of (9.43) if $\Phi = e^{im\varphi}$ is adopted rather than

cos $m\varphi$ and sin $m\varphi$. The energy values are, of course, changed in the presence of a magnetic field. It is easily seen that

$$E' = E - \frac{q}{2\mu c} Bm\hbar \qquad (9.44)$$

This naïve theory of the *Zeeman effect* of a magnetic field must be refined for most applications, because the spin is usually important and **B** fields cannot be made so huge that other corrections to the energy could be neglected compared with the magnetic term (see Section 17.6). Yet, (9.44) does indicate the origin of the term *magnetic quantum number*. For an electron, $q = -e$, and the energy change becomes proportional to the *Bohr magneton* $e\hbar/2\mu c$.

6. *Eigenvalues and Eigenfunctions of* L². Now we turn to (9.39). By a change of variables,

$$\xi = \cos\theta, \qquad F(\xi) = \Theta(\theta) \qquad (9.45)$$

(9.39) is transformed into

$$\frac{d}{d\xi}\left[(1 - \xi^2)\frac{dF}{d\xi}\right] - \frac{m^2}{1 - \xi^2}F + \lambda F = 0 \qquad (9.46)$$

For the particular case $m = 0$, (9.46) assumes an especially simple form, familiar in many problems of mathematical physics,

$$\frac{d}{d\xi}\left[(1 - \xi^2)\frac{dF}{d\xi}\right] + \lambda F = 0 \qquad (9.47)$$

and known as *Legendre's differential equation.* Its examination follows a conventional pattern.

Equation (9.47) does not change its form when $-\xi$ is substituted for ξ. Hence, we need only look for solutions of (9.47) which are even or odd functions of ξ. Since $\xi \to -\xi$ implies $\theta \to \pi - \theta$, these functions are symmetric or antisymmetric with respect to the xy-plane.

The solution of (9.47) which is regular can be expanded in a power series,

$$F(\xi) = \sum_{k=0}^{\infty} a_k \xi^k$$

and substitution into (9.47) yields the recursion relation

$$a_{k+2} = \frac{k(k + 1) - \lambda}{(k + 1)(k + 2)} a_k \qquad (9.48)$$

Equation (9.48) shows that $a_{-2} = 0$ for the even case and $a_{-1} = 0$ for the odd case, in agreement with the assumption that F is regular at $\xi = 0$. If the series does not terminate at some finite value of k, $a_{k+2}/a_k \to k/(k + 2)$ as

$k \rightarrow \infty$. The series thus behaves like $\Sigma\, n^{-1}$ for even or odd n, and diverges for $\xi = \pm 1$, i.e., for $\theta = 0$ and π. For the same reason we exclude the second linearly independent solution of (9.47). It has logarithmic singularities at $\xi = \pm 1$. Such singular functions, although solutions of the differential equation for almost all values of ξ, are not acceptable eigenfunctions of $\mathbf{L}^2$.

We conclude that the power series must terminate at some finite value $k = l$, where l is a nonnegative integer, and all higher powers vanish. According to (9.48), this will happen if λ has the value.

$$\lambda = l(l + 1) \tag{9.49}$$

We have thus arrived at the celebrated law that the eigenvalues of $\mathbf{L}^2$, hence its measured values, can only be 0, $2\hbar^2$, $6\hbar^2$, $12\hbar^2$, ..., as l, the *orbital angular momentum quantum number*, assumes the values 0, 1, 2, 3, ... It is customary to designate the corresponding angular momentum states by the symbols $S, P, D, F, \ldots$. If there are several particles in a central field, lower case letters $s, p, d, \ldots$ will be used to identify the angular momentum state of each particle, and capital letters $S, P, D, \ldots$ will be reserved for the total orbital angular momentum.

The conventional form of the polynomial solutions of (9.47) is

$$P_l(\xi) = \frac{1}{2^l \cdot l!} \frac{d^l}{d\xi^l} (\xi^2 - 1)^l \tag{9.50}$$

These are called *Legendre polynomials*. The coefficient of ξ^k in the expansion of (9.50) is easily seen to be

$$a_k = \frac{(-1)^{(l-k)/2}(l + k)(l + k - 1) \cdots (k + 2)(k + 1)}{2^l \cdot l!} \binom{l}{\frac{1}{2}(l + k)}$$

where the last factor is a binomial coefficient. We verify readily that (9.48) is satisfied by the coefficients a_k, hence that $P_l(\xi)$ indeed solves (9.47). The peculiar constant in (9.50) has been adopted, because it gives

$$P_l(\pm 1) = (\pm 1)^l \tag{9.51}$$

The first few Legendre polynomials are

$$P_0(\xi) = 1 \qquad\qquad P_3(\xi) = \tfrac{1}{2}(5\xi^3 - 3\xi)$$

$$P_1(\xi) = \xi \qquad\qquad P_4(\xi) = \tfrac{1}{8}(35\xi^4 - 30\xi^2 + 3) \tag{9.52}$$

$$P_2(\xi) = \tfrac{1}{2}(3\xi^2 - 1) \qquad P_5(\xi) = \tfrac{1}{8}(63\xi^5 - 70\xi^3 + 15\xi)$$

Since $P_l(\cos\theta)$ is an eigenfunction of the Hermitian operator $\mathbf{L}^2$, it is clear from the general theorems of Chapter 8 that the Legendre polynomials must be orthogonal. Only the integration over the polar angle θ concerns us here—not the entire volume integral—and we expect that

$$\int_0^\pi P_l(\cos\theta)P_{l'}(\cos\theta)\sin\theta\,d\theta = 0 \qquad \text{if} \qquad l' \neq l \qquad (9.53)$$

No complex conjugation is needed, as the Legendre polynomials are real functions. The orthogonality relation

$$\int_{-1}^{+1} P_l(\xi)P_{l'}(\xi)\,d\xi = 0 \qquad \text{if} \qquad l' \neq l \qquad (9.53a)$$

can also be proved directly, using the definition (9.50) and successive integrations by parts. The normalization of these orthogonal polynomials can likewise be obtained easily by *l*-fold integration by parts:

$$\int_{-1}^{+1} [P_l(\xi)]^2\,d\xi = \left(\frac{1}{2^l \cdot l!}\right)^2 \int_{-1}^{+1} \left[\frac{d^l}{d\xi^l}(\xi^2-1)^l\right]\left[\frac{d^l}{d\xi^l}(\xi^2-1)^l\right]d\xi$$

$$= (-1)^l \frac{1}{(2^l \cdot l!)^2} \int_{-1}^{+1} \left[\frac{d^{2l}}{d\xi^{2l}}(\xi^2-1)^l\right](\xi^2-1)^l\,d\xi$$

$$= (-1)^l \frac{(2l)!}{(2^l \cdot l!)^2} \int_{-1}^{+1} (\xi^2-1)^l\,d\xi = \frac{2}{2l+1} \qquad (9.54)$$

Exercise 9.3. Prove the orthogonality relation (9.53a) directly, using the definition (9.50).

As usual in the study of special functions, it is helpful to introduce a generating function for Legendre polynomials. The claim is made that such a generating function is

$$(1 - 2\xi s + s^2)^{-1/2} = \sum_{n=0}^\infty a_n(\xi)s^n \qquad (|s| < 1) \qquad (9.55)$$

To prove the identity of the coefficients $a_n(\xi)$ in (9.55) with the Legendre polynomials $P_n(\xi)$, let us derive a simple recursion relation by differentiating (9.55) with respect to s:

$$(\xi - s)(1 - 2\xi s + s^2)^{-3/2} = \sum_{n=0}^\infty na_n(\xi)s^{n-1} \qquad (9.56)$$

or, by the use of (9.55),

$$(\xi - s)\sum_{n=0}^\infty a_n(\xi)s^n = (1 - 2\xi s + s^2)\sum_{n=0}^\infty na_n(\xi)s^{n-1}$$

Equating the coefficients of each power of s, we obtain

$$(n + 1)a_{n+1}(\xi) = (2n + 1)\xi a_n(\xi) - na_{n-1}(\xi) \qquad (9.57)$$

By substituting $s = 0$ in (9.55) and (9.56) we see that

$$a_0(\xi) = 1, \qquad a_1(\xi) = \xi$$

$a_0(\xi)$ and $a_1(\xi)$ are thus identical with $P_0(\xi)$ and $P_1(\xi)$ respectively. The reader may verify from the definition (9.50) the recurrence formula

$$(n + 1)P_{n+1}(\xi) = (2n + 1)\xi P_n(\xi) - nP_{n-1}(\xi) \qquad (9.57a)$$

which is to be compared with (9.57). Hence, $a_n(\xi) = P_n(\xi)$ for all n, proving that indeed

$$(1 - 2\xi s + s^2)^{-1/2} = \sum_{n=0}^{\infty} P_n(\xi)s^n \qquad (9.58)$$

Having solved (9.47), it is not difficult to obtain also the physically acceptable solutions of (9.46) with $m \neq 0$. If Legendre's equation (9.47) is differentiated m times and if the *associated Legendre functions*

$$P_l{}^m(\xi) = (1 - \xi^2)^{m/2} \frac{d^m P_l(\xi)}{d\xi^m} = \frac{1}{2^l \cdot l!} (1 - \xi^2)^{m/2} \frac{d^{l+m}}{d\xi^{l+m}} (\xi^2 - 1)^l \quad (9.59)$$

are defined for positive integers $m \leqslant l$, we deduce readily that

$$\frac{d}{d\xi} \left[(1 - \xi^2) \frac{dP_l{}^m(\xi)}{d\xi} \right] - \frac{m^2}{1 - \xi^2} P_l{}^m(\xi) + l(l + 1)P_l{}^m(\xi) = 0 \quad (9.60)$$

which is identical with (9.46) for $\lambda = l(l + 1)$. The associated Legendre functions with $m \leqslant l$ are the only nonsingular and physically acceptable solutions of (9.46).[3]

Exercise 9.4. Use the inequality (9.11) to show that the magnetic quantum number cannot exceed the orbital angular momentum quantum number.

The associated Legendre functions are also orthogonal in the sense that

$$\int_{-1}^{+1} P_l{}^m(\xi)P_{l'}{}^m(\xi) \, d\xi = 0 \qquad \text{if} \qquad l \neq l' \qquad (9.61)$$

Note that in this relation the two superscripts are the same. Legendre functions with different values of m are generally not orthogonal. For purposes of normalization we note that

$$\int_{-1}^{+1} [P_l{}^m(\xi)]^2 \, d\xi = \frac{2}{2l + 1} \frac{(l + m)!}{(l - m)!} \qquad (9.62)$$

---

[3] These functions are also called *associated Legendre functions of the first kind* to distinguish them from the second kind, $Q_l{}^m(\xi)$, which is the singular variety.

We leave the proof to the interested reader. When ξ is changed to $-\xi$, $P_l{}^m(\xi)$ merely retains or changes its sign, depending on whether $l + m$ is an even or odd integer.

It is natural to supplement the definition (9.59) by defining the associated Legendre functions for $m = 0$ as

$$P_l{}^0(\xi) = P_l(\xi) \tag{9.63}$$

Returning now to (9.36) and (9.37), we see that the solutions of (9.36) which are separable in spherical polar coordinates are products of $e^{im\varphi}$ and $P_l{}^m(\cos\theta)$. Since (9.39) is unchanged if m is replaced by $-m$, and since $P_l{}^m$ is the only admissible solution of this equation, it follows that the same associated Legendre function must be used for a given absolute value of m.

Exercise 9.5. Justify the following alternative definition of associated Legendre functions:

$$P_l{}^m(\xi) = \frac{(-1)^m}{2^l \cdot l!} \frac{(l + m)!}{(l - m)!} (1 - \xi^2)^{-m/2} \frac{d^{l-m}}{d\xi^{l-m}} (\xi^2 - 1)^l \tag{9.59a}$$

The first few associated Legendre functions are

$$P_1{}^1(\xi) = \sqrt{1 - \xi^2}, \qquad P_2{}^1(\xi) = 3\xi\sqrt{1 - \xi^2}, \qquad P_2{}^2(\xi) = 3(1 - \xi^2) \tag{9.64}$$

7. Spherical Harmonics. It is convenient to define the separable solutions of (9.36) as functions which are normalized with respect to an integration over the entire solid angle. They are called *spherical harmonics* and are given for $m \geqslant 0$ as

$$Y_l{}^m(\theta, \varphi) = \sqrt{\frac{2l + 1}{4\pi} \frac{(l - m)!}{(l + m)!}} (-1)^m e^{im\varphi} P_l{}^m(\cos\theta) \tag{9.65}$$

Spherical harmonics with negative superscripts (subject to the restriction $-l \leqslant m \leqslant l$) will be defined by

$$Y_l{}^m = (-1)^m Y_l^{-m*} \tag{9.65a}$$

The spherical harmonics are normalized simultaneous eigenfunctions of L_z and $\mathbf{L}^2$ such that

$$L_z Y_l{}^m = \frac{\hbar}{i} \frac{\partial Y_l{}^m}{\partial\varphi} = m\hbar Y_l{}^m \tag{9.66}$$

$$\mathbf{L}^2 Y_l{}^m = l(l + 1)\hbar^2 Y_l{}^m \tag{9.67}$$

The first few spherical harmonics are listed below:

$$Y_0{}^0 = \frac{1}{\sqrt{4\pi}}$$

$$Y_1{}^0 = \sqrt{\frac{3}{4\pi}} \cos \theta = \sqrt{\frac{3}{4\pi}} \frac{z}{r}$$

$$Y_1{}^{\pm 1} = \mp \sqrt{\frac{3}{8\pi}} e^{\pm i\varphi} \sin \theta = \mp \sqrt{\frac{3}{8\pi}} \frac{x \pm iy}{r}$$

$$Y_2{}^0 = \sqrt{\frac{5}{16\pi}} (3 \cos^2 \theta - 1) = \sqrt{\frac{5}{16\pi}} \frac{2z^2 - x^2 - y^2}{r^2} \qquad (9.68)$$

$$Y_2{}^{\pm 1} = \mp \sqrt{\frac{15}{8\pi}} e^{\pm i\varphi} \cos \theta \sin \theta = \mp \sqrt{\frac{15}{8\pi}} \frac{(x \pm iy)z}{r^2}$$

$$Y_2{}^{\pm 2} = \sqrt{\frac{15}{32\pi}} e^{\pm 2i\varphi} \sin^2 \theta = \sqrt{\frac{15}{32\pi}} \frac{(x \pm iy)^2}{r^2}$$

It is worth noting that under a coordinate reflection, or *inversion*, through the origin ($\varphi \to \varphi + \pi$, $\theta \to \pi - \theta$) $e^{im\varphi}$ is multiplied by $(-1)^m$, and $P_l{}^m(\cos \theta)$ by $(-1)^{l+m}$. Hence, $Y_l{}^m(\theta, \varphi)$ is multiplied by $(-1)^l$, when $\mathbf{r}$ is changed to $-\mathbf{r}$. Consequently, $Y_l{}^m$ has the parity of the angular momentum quantum number l. (See Exercise 5.2.)

The spherical harmonics form an orthonormal set, since

$$\int_0^{2\pi} \int_0^{\pi} Y_l{}^{m*}(\theta, \varphi) Y_{l'}{}^{m'}(\theta, \varphi) \sin \theta \, d\theta \, d\varphi = \delta_{l,l'} \delta_{m,m'} \qquad (9.69)$$

where m is an integer, $-l \leqslant m \leqslant l$.

Although no detailed proof will be given here, it is important to note that the spherical harmonics do form a complete set for the expansion of wave functions.[4] Roughly this can be seen from the following facts:

(a) The eigenfunctions $e^{im\varphi}$ of L_z are complete in the sense of Fourier series in the range $0 \leqslant \varphi \leqslant 2\pi$. Hence, a very large class of functions of φ can be expanded in terms of them.

(b) The Legendre polynomials $P_0(\xi), P_1(\xi), P_2(\xi), \ldots$, are the orthogonal polynomials which are obtained by applying the orthogonalization procedure described in Section 8.2 to the sequence of monomials $1, \xi, \xi^2, \ldots$, requiring that there be a polynomial of every degree and that they be orthogonal in the interval $-1 \leqslant \xi \leqslant +1$. Hence, ξ^k can be expressed in terms of Legendre polynomials, and any function which can be expanded in a uniformly converging power series of ξ can also be expanded in terms of Legendre

[4] R. Courant and D. Hilbert, *Methods of Mathematical Physics*, Volume I, Interscience Publishers, New York, 1953, pp. 512–513.

polynomials. The same is true, although less obviously, for the associated Legendre functions of order m, which also form a complete set as l varies from 0 to ∞.

Hence, any wave function which depends on the angles θ and φ can be expanded in the mean in terms of spherical harmonics (see Section 8.3).

Exercise 9.6. Construct $P_2(\xi)$ by the orthogonalization described above.

Some of the most frequently used expansions of angular functions in terms of spherical harmonics may be quoted without proof.[5]

$$\cos\theta\, Y_l{}^m(\theta,\varphi) = \sqrt{\frac{(l+m+1)(l-m+1)}{(2l+1)(2l+3)}}\, Y_{l+1}^m$$

$$+\sqrt{\frac{(l+m)(l-m)}{(2l+1)(2l-1)}}\, Y_{l-1}^m \quad (9.70a)$$

$$\sin\theta\, e^{i\varphi}\, Y_l{}^m(\theta,\varphi) = -\sqrt{\frac{(l+m+1)(l+m+2)}{(2l+1)(2l+3)}}\, Y_{l+1}^{m+1}$$

$$+\sqrt{\frac{(l-m)(l-m-1)}{(2l+1)(2l-1)}}\, Y_{l-1}^{m+1} \quad (9.70b)$$

$$\sin\theta\, e^{-i\varphi}\, Y_l{}^m(\theta,\varphi) = \sqrt{\frac{(l-m+1)(l-m+2)}{(2l+1)(2l+3)}}\, Y_{l+1}^{m-1}$$

$$\sqrt{\frac{(l+m)(l+m-1)}{(2l+1)(2l-1)}}\, Y_{l-1}^{m-1} \quad (9.70c)$$

The effect of the operators L_x and L_y on $Y_l{}^m$ is conveniently studied by introducing the operators

$$L_+ = L_x + iL_y, \qquad L_- = L_x - iL_y \quad (9.71)$$

which, according to (9.34), may be written as

$$L_+ = \hbar e^{i\varphi}\left(\frac{\partial}{\partial\theta} + i\cot\theta\,\frac{\partial}{\partial\varphi}\right) \quad (9.72a)$$

$$L_- = -\hbar e^{-i\varphi}\left(\frac{\partial}{\partial\theta} - i\cot\theta\,\frac{\partial}{\partial\varphi}\right) \quad (9.72b)$$

The effect of $\partial/\partial\varphi$ on $Y_l{}^m$ is known from (9.66). To determine $\partial Y/\partial\theta$ we note

[5] Equations (9.70) are special cases of (16.89).

that from the definitions (9.59) and (9.59a),

$$\frac{dP_l{}^m}{d\xi} = \frac{1}{\sqrt{1-\xi^2}} P_l^{m+1} - \frac{m\xi}{1-\xi^2} P_l{}^m$$

$$= -\frac{(l+m)(l-m+1)}{\sqrt{1-\xi^2}} P_l^{m-1} + \frac{m\xi}{1-\xi^2} P_l{}^m$$

With $\xi = \cos\theta$ and the definition (9.65) it is then easy to derive the relations

$$L_+ Y_l{}^m(\theta, \varphi) = \hbar\sqrt{(l-m)(l+m+1)}\, Y_l^{m+1}(\theta, \varphi) \qquad (9.73a)$$

$$L_- Y_l{}^m(\theta, \varphi) = \hbar\sqrt{(l+m)(l-m+1)}\, Y_l^{m-1}(\theta, \varphi) \qquad (9.73b)$$

Finally, the addition theorem for spherical harmonics will be derived. Consider two coordinate systems, xyz and $x'y'z'$. The addition theorem is the formula expressing the eigenfunction $P_l(\cos\theta')$ of angular momentum about a z'-axis in terms of the eigenfunctions $Y_l{}^m(\theta, \varphi)$ of L_z. Figure 9.1

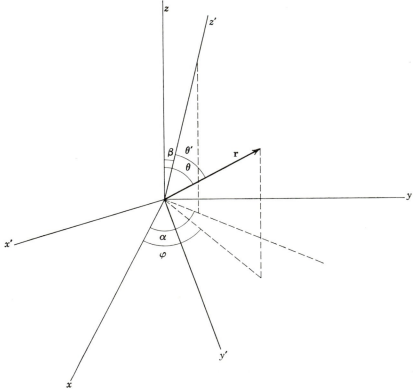

Figure 9.1. Angles used in the addition theorem of spherical harmonics.

indicates the various angles: The position vector **r** has angular coordinates θ, φ, and θ', φ' in the two coordinate systems. The direction of the z'-axis in space is specified by its polar angle β and its azimuth α with respect to the unprimed system. Since P_l is an eigenfunction of $\mathbf{L}^2$, only spherical harmonics with the same subscript l can appear in the expansion. An interchange of θ, φ, and β, α must leave the expansion unchanged. Hence, $P_l(\cos\theta')$ can also be expanded in terms of $Y_l{}^m(\beta, \alpha)$. In a rigid rotation of the figure about the z-axis, α and φ change by equal amounts, and θ' remains constant. Hence, $P_l(\cos\theta')$ must be a function of $\varphi - \alpha$. All these requirements can be satisfied only if

$$P_l(\cos\theta') = \sum_{l=-m}^{m} c_m Y_l^{-m}(\beta, \alpha) Y_l{}^m(\theta, \varphi) \tag{9.74}$$

The coefficients c_m can be determined by using the condition

$$L_{z'} P_l(\cos\theta') = 0 \tag{9.75}$$

Since

$$L_{z'} = \sin\beta\cos\alpha L_x + \sin\beta\sin\alpha L_y + \cos\beta L_z$$
$$= \tfrac{1}{2}\sin\beta e^{-i\alpha}L_+ + \tfrac{1}{2}\sin\beta e^{i\alpha}L_- + \cos\beta L_z \tag{9.76}$$

(9.70) and (9.73) may be used to evaluate $L_z P_l(\cos\theta')$. If the linear independence of the spherical harmonics is invoked, we obtain, after some calculation, the simple result

$$c_{m\pm1} = -c_m$$

Thus, $c_m = (-1)^m c_0$, and only c_0 need be determined. For this purpose we specialize to $\beta = 0$, or $\theta = \theta'$. Since, from the definitions of $Y_l{}^m$ and $P_l{}^m$,

$$Y_l{}^m(0, \varphi) = \sqrt{\frac{2l+1}{4\pi}}\,\delta_{m0} \tag{9.77}$$

and

$$Y_l{}^0(\theta, \varphi) = \sqrt{\frac{2l+1}{4\pi}}\,P_l(\cos\theta) \tag{9.78}$$

it follows easily that

$$c_0 = \frac{4\pi}{2l+1}$$

Using (9.65a), this proves the *addition theorem*,

$$P_l(\cos\theta') = \frac{4\pi}{2l+1}\sum_{m=-l}^{l} Y_l{}^{m*}(\beta, \alpha) Y_l{}^m(\theta, \varphi) \tag{9.79}$$

The discussion of orbital angular momentum given in this chapter is adequate for an analysis of the central force problem. But the notion of angular momentum, being a very fundamental one because of its connection with rigid rotations, will engage our attention again and again, and we shall see, especially in Chapters 12, 15, and 16, that the concept of angular momentum can and must be generalized.

Problems

1. Classically we have for central forces

$$H = \frac{p_r^2}{2\mu} + \frac{L^2}{2\mu r^2} + V(r)$$

where $p_r = (1/r)(\mathbf{r} \cdot \mathbf{p})$. Show that for translation into quantum mechanics we must write

$$p_r = \frac{1}{2}\left[\frac{1}{r}(\mathbf{r} \cdot \mathbf{p}) + (\mathbf{p} \cdot \mathbf{r})\frac{1}{r}\right]$$

and that this gives the correct Schrödinger equation with the Hermitian operator

$$p_r = \frac{\hbar}{i}\left(\frac{\partial}{\partial r} + \frac{1}{r}\right)$$

whereas $(\hbar/i)(\partial/\partial r)$ is not Hermitian.

2. Apply the WKB method to the radial Schrödinger equation for a central potential. Show that the quantum condition for bound S-states in an attractive well which increases monotonically from $r = 0$ to $r = \infty [V(\infty) = 0]$ is

$$2\int_0^a \sqrt{2\mu(E - V)}\,dr = (N + \tfrac{3}{4})h$$

where $r = a$ is the radial coordinate of the classical turning point.

Show that this quantum condition has infinitely many solutions for an attractive central potential which behaves as r^{-n} for large r, with $n \leqslant 2$.

CHAPTER 10

Piecewise Constant Potentials in Three Dimensions and The Hydrogen Atom

1. *The Free Particle in Three Dimensions*. The simplest dynamical problem in three dimensions is the motion of the *free particle*, corresponding to $V = 0$ everywhere in space. The Schrödinger equation for $E > 0$,

$$-\frac{\hbar^2}{2\mu} \nabla^2 \psi = E\psi \tag{10.1}$$

has, of course, plane wave solutions. These are obtained by separation of variables when ∇^2 is expressed in Cartesian coordinates. Thus we set

$$\psi(x, y, z) = X(x)Y(y)Z(z) \tag{10.2}$$

and replace (10.1) by three separate equations

$$\frac{d^2X}{dx^2} + k_x{}^2 X = 0, \qquad \frac{d^2Y}{dy^2} + k_y{}^2 Y = 0, \qquad \frac{d^2Z}{dz^2} + k_z{}^2 Z = 0 \tag{10.3}$$

so that

$$k_x{}^2 + k_y{}^2 + k_z{}^2 = \frac{2\mu}{\hbar^2} E \tag{10.4}$$

Hence, the solutions are plane waves

$$\psi_{\mathbf{k}}(\mathbf{r}) = C(\mathbf{k})e^{i\mathbf{k}\cdot\mathbf{r}} \tag{10.5}$$

where the three components of the propagation vector $\mathbf{k}$ must be real if the eigenfunctions are to remain finite at large distances. The plane wave states of energy E are eigenstates of the momentum operator $(\hbar/i)\nabla$, and $\hbar\mathbf{k}$ is the momentum eigenvalue.

Since every positive value of E is allowed, the spectrum is evidently continuous, and there is an infinite degree of degeneracy associated with each energy eigenvalue, since the condition (10.4) merely restricts the magnitude of the vector $\mathbf{k}$, but not its direction.

Generalizing (6.23) to three dimensions, we commonly introduce the **k**-normalization for the free particle eigenfunctions,

$$\int \psi_{\mathbf{k}}^*(\mathbf{r})\psi_{\mathbf{k}'}(\mathbf{r})\, d^3r = \delta(\mathbf{k} - \mathbf{k}') \tag{10.6}$$

By use of the delta function (8.5), the absolute value of the normalization constant $C(\mathbf{k})$ can now be determined. Choosing the arbitrary phase to be real, we have the **k**-normalized wave functions

$$\psi_{\mathbf{k}}(\mathbf{r}) = (2\pi)^{-3/2}e^{i\mathbf{k}\cdot\mathbf{r}} \tag{10.7}$$

The (Fourier) expansion of an arbitrary state in terms of the complete set of eigenfunctions (10.7) has already been discussed thoroughly in Chapter 8. The closure relation (8.57) applied to these eigenfunctions is

$$\int \psi_{\mathbf{k}}^*(\mathbf{r})\psi_{\mathbf{k}}(\mathbf{r}')\, d^3k = \delta(\mathbf{r} - \mathbf{r}')$$

or

$$\frac{1}{(2\pi)^3}\int e^{i\mathbf{k}\cdot(\mathbf{r}'-\mathbf{r})}\, d^3k = \delta(\mathbf{r} - \mathbf{r}')$$

in agreement with the definition (8.5) of the delta function.

If we prefer to deal with discrete eigenvalues, we can avail ourselves, as in the one-dimensional case (Section 6.3), of the mathematically convenient periodic boundary conditions, requiring that upon translation by L in the direction of any one of the three coordinate axes all eigenfunctions of the Schrödinger equation shall resume their previous values. The eigenfunctions are still of the form (10.5) but the eigenvalues are now, in addition to (10.4), subject to the restrictions

$$k_x = \frac{2\pi}{L}\, n_x, \qquad k_y = \frac{2\pi}{L}\, n_y, \qquad k_z = \frac{2\pi}{L}\, n_z \tag{10.8}$$

where n_x, n_y, n_z are integers (positive, negative, or zero). The wave functions may be normalized in the basic cube of side length L, thus:

$$\int_0^L \int_0^L \int_0^L |\psi_{\mathbf{k}}(\mathbf{r})|^2\, dx\, dy\, dz = 1$$

Hence,

$$\psi_{\mathbf{k}}(r) = L^{-3/2}e^{i\mathbf{k}\cdot\mathbf{r}} \tag{10.9}$$

The number, $\Delta n/\Delta E$, of such eigenstates per unit energy interval can be calculated by the method familiar in counting the normal modes of any three-dimensional continuum.[1] For any but the lowest energies we obtain to

[1] C. Kittel, *Introduction to Solid State Physics*, 3rd ed., John Wiley and Sons, New York, 1966, p. 174.

good approximation

$$\frac{\Delta n}{\Delta E} = \frac{\mu^{3/2}\sqrt{E}\,L^3}{\sqrt{2}\,\pi^2\hbar^3} \tag{10.10}$$

This expression is easily seen to be equivalent to the famous rule that each state occupies a volume h^3 in phase space, or

$$\frac{L^3\,\Delta p_x\,\Delta p_y\,\Delta p_z}{\Delta n} = h^3 \tag{10.11}$$

The main difference between this density of states and its one-dimensional analog (Section 6.3) is that the latter is proportional to $1/\sqrt{E}$ whereas the former increases with energy as $\sqrt{E}$. This different behavior is due to the fact that the degree of degeneracy of each energy value is independent of E and equal to 2 for one dimension, whereas in three-dimensional space it increases linearly with E. As the energy becomes greater, the number of ways in which three perfect squares can be added to yield

$$n_x^{\,2} + n_y^{\,2} + n_z^{\,2} = \frac{\mu EL^2}{2\pi^2\hbar^2} \tag{10.12}$$

rises rapidly.

Closely related to the foregoing formulation of the free particle problem is the problem of the "*particle in a box.*" Here we assume zero potential energy inside a finite volume at whose boundaries V jumps to an infinitely high value. As shown in Section 6.4, this situation is properly characterized by requiring that the wave function must vanish on the walls of the box. Choosing the box, as a convenient example, to be a cube of length L, situated in the first coordinate octant with one corner at the origin, we see by separation of variables as in (10.2) and (10.3) that the eigenfunctions are

$$\psi = C \sin\frac{l_x\pi}{L}\,x \sin\frac{l_y\pi}{L}\,y \sin\frac{l_z\pi}{L}\,z \tag{10.13}$$

with l_x, l_y, l_z satisfying the condition

$$l_x^{\,2} + l_y^{\,2} + l_z^{\,2} = \frac{2\mu EL^2}{\pi^2\hbar^2} \tag{10.14}$$

l_x, l_y, l_z are restricted to be positive integers, since a change of sign in (10.13) produces no new linearly independent eigenfunction. Hence, the degree of degeneracy of any given energy value is less here by a factor of 8 than in the case of periodic boundary conditions. However, by virtue of the different right-hand sides of (10.12) and (10.14), for a given energy E there are now more ways of finding suitable values to satisfy (10.14). These two effects compensate each other, and as a result the *density of states* is the same for the case of periodic boundary conditions and the box. In either case the eigenfunctions constitute a complete set, in terms of which an arbitrary wave packet can be expanded by Fourier series within the basic cube.

2. $V = 0$ *as a Central Force Problem.* The free particle problem can also be regarded as a *central force* problem, since $V = 0$ is a special, if trivial, case of a potential that depends only on the radial coordinate r. From the point of view of Chapter 9, the eigenfunctions must be separable in spherical polar coordinates, and the angular momentum must be a constant of the motion. We therefore assume the eigenfunctions in the form (9.27) or

$$\psi(r, \theta, \varphi) = R(r) Y_l^m(\theta, \varphi) \tag{10.15}$$

and for $V = 0$ these must satisfy the radial equation

$$\left[-\frac{\hbar^2}{2\mu r^2} \frac{d}{dr} \left(r^2 \frac{d}{dr} \right) + \frac{l(l + 1)\hbar^2}{2\mu r^2} \right] R(r) = ER(r) \tag{10.16}$$

The differential equation (10.16) can be solved by standard power series techniques, but it is more economical to observe that a closely related equation has already been encountered earlier in discussing the WKB method. If we substitute

$$\rho = \sqrt{\frac{2\mu E}{\hbar^2}} r = kr, \qquad v(\rho) = rR(r) \tag{10.17}$$

and

$$\lambda = l + 1 \qquad \text{or} \qquad -l \tag{10.18}$$

(10.16) becomes

$$\frac{d^2 v}{d\rho^2} + \left[1 - \frac{\lambda(\lambda - 1)}{\rho^2} \right] v = 0 \tag{10.19}$$

This equation was discussed in detail in Section 7.5. The solution of (10.19) which vanishes at $\rho = 0$ is given by (7.60) and (7.65) as

$$v(\rho) = \rho^{l+1} \int_{-i}^{+i} e^{\rho t} (1 + t^2)^l \, dt \tag{10.20}$$

This solution is closely related to the *spherical Bessel function* j_l, defined by[2]

$$j_l(z) = \frac{z^l}{2^{l+1} \cdot l!} \int_{-1}^{+1} e^{izs} (1 - s^2)^l \, ds \tag{10.21}$$

for a complex variable z. The first term in the series expansion of (10.21) in

[2] Since ρ is positive by definition, the variable z will be used in this chapter in all formulas which are also valid outside of the real positive axis of the complex plane.

powers of z is

$$j_l(z) = \frac{2^l \cdot l!}{(2l+1)!} z^l + O(z^{l+2}) \tag{10.22}$$

We thus see that the spherical Bessel function $j_l(kr)$ is the solution of the radial equation

$$-\frac{1}{r^2}\frac{d}{dr}\left(r^2 \frac{dR}{dr}\right) + \frac{l(l+1)}{r^2} R = k^2 R \tag{10.23}$$

which is finite at the origin and is said to be *regular*. The radial eigenfunction of the Schrödinger equation for the free particle is thus

$$R(r) = C j_l(kr) \tag{10.24}$$

A useful formula is obtained by integrating (10.21) by parts l times and using the definition (9.50) of the Legendre polynomials:

$$j_l(z) = \frac{1}{2i^l} \int_{-1}^{+1} e^{izs} P_l(s)\,ds \tag{10.25}$$

The asymptotic form of the spherical Bessel functions can be derived from this expression by further integration by parts, and the leading term is

$$j_l(\rho) \sim \frac{\cos\left[\rho - (l+1)\pi/2\right]}{\rho} \qquad (|\rho| \gg 1 \quad \text{and} \quad |\rho| \gg l) \tag{10.26}$$

We also note the relation between the *spherical* and the *ordinary Bessel functions*:

$$j_l(z) = \sqrt{\frac{\pi}{2z}} J_{l+1/2}(z) \tag{10.27}$$

All other solutions of (10.23) are singular at the origin and are not admissible as energy eigenfunctions for the free particle. Nevertheless they have important uses and will be introduced here.[3]

Two particularly useful singular solutions of the equation

$$-\frac{1}{z^2}\frac{d}{dz}\left(z^2 \frac{dR}{dz}\right) + \frac{l(l+1)}{z^2} R = R$$

are the *spherical Hankel functions* of the *first* and *second kind*, defined for $0 \leqslant \arg z \leqslant \pi/2$ by the relations

$$h_l^{(1)}(z) = \frac{z^l}{i 2^l \cdot l!} \int_{\infty e^{i\alpha}}^{i} e^{zt}(1+t^2)^l\,dt \tag{10.28a}$$

---

[3] Some of the related mathematical work is found in Section 7.5. For details see P. M. Morse and H. Feshbach, *Methods of Theoretical Physics*, McGraw-Hill Book Company, 1953, Chapter 5.

and

$$h_l^{(2)}(z) = \frac{z^l}{i2^l \cdot l!} \int_{-i}^{\infty e^{i\alpha}} e^{zt}(1 + t^2)^l \, dt \tag{10.28b}$$

where $\pi/2 < \alpha < \pi$. The Hankel functions diverge as z^{-l-1} near the origin:

$$h_l^{(1,2)}(z) = \pm \frac{(2l)!}{i2^l \cdot l!} z^{-l-1} + O^{(1,2)}(z^{-l}) \tag{10.29}$$

Exercise 10.1. Prove (10.29) by substituting $x = zt$ in (10.28a) and (10.28b).

The asymptotic behavior of the Hankel functions for large positive ρ can be inferred from (7.64) by an appropriate change of symbols:

$$h_l^{(1)}(\rho) \sim \frac{1}{\rho} \exp\left\{i\left[\rho - \frac{\pi}{2}(l + 1)\right]\right\} \tag{10.30}$$

$$h_l^{(2)}(\rho) \sim \frac{1}{\rho} \exp\left\{-i\left[\rho - \frac{\pi}{2}(l + 1)\right]\right\} \tag{10.31}$$

Comparing (10.28a) and (10.28b) with (10.21), we can express the regular solution in terms of the Hankel functions:

$$j_l(z) = \tfrac{1}{2}[h_l^{(1)}(z) + h_l^{(2)}(z)] \tag{10.32}$$

It is useful to define also the *spherical Neumann function* $n_l(z)$ as

$$n_l(z) = \frac{1}{2i} [h_l^{(1)}(z) - h_l^{(2)}(z)] \tag{10.33}$$

Hence,

$$h_l^{(1)}(z) = j_l(z) + in_l(z) \qquad \text{and} \qquad h_l^{(2)}(z) = j_l(z) - in_l(z)$$

For $z \approx 0$,

$$n_l(z) = -\frac{(2l)!}{2^l \cdot l!} z^{-l-1} + O(z^{-l+1}) \tag{10.34}$$

and for large positive ρ,

$$n_l(\rho) \sim \frac{\sin [\rho - (l + 1)\pi/2]}{\rho} \tag{10.35}$$

This particular singular solution of the radial equation, being asymptotically out of phase by $\pi/2$ compared to the regular solution (10.25), is sometimes distinguished as "the" irregular solution, although any linear combination of j_l and n_l that contains a nonvanishing portion of n_l is singular at the origin.

The explicit forms of the spherical Bessel, Hankel, and Neumann functions for $l = 0$, 1, and 2, are given below

$$j_0(z) = \frac{\sin z}{z}$$

$$j_1(z) = \frac{\sin z}{z^2} - \frac{\cos z}{z} \tag{10.36}$$

$$j_2(z) = \left(\frac{3}{z^3} - \frac{1}{z}\right) \sin z - \frac{3}{z^2} \cos z$$

$$n_0(z) = - \frac{\cos z}{z}$$

$$n_1(z) = - \frac{\cos z}{z^2} - \frac{\sin z}{z} \tag{10.37}$$

$$n_2(z) = - \left(\frac{3}{z^3} - \frac{1}{z}\right) \cos z - \frac{3}{z^2} \sin z$$

$$h_0^{(1)}(z) = -i \frac{e^{iz}}{z}$$

$$h_1^{(1)}(z) = \left(-\frac{i}{z^2} - \frac{1}{z}\right) e^{iz} \tag{10.38}$$

$$h_2^{(1)}(z) = \left(-\frac{3i}{z^3} - \frac{3}{z^2} + \frac{i}{z}\right) e^{iz}$$

$$h_0^{(2)}(z) = i \frac{e^{-iz}}{z}$$

$$h_1^{(2)}(z) = \left(\frac{i}{z^2} - \frac{1}{z}\right) e^{-iz} \tag{10.39}$$

$$h_2^{(2)}(z) = \left(\frac{3i}{z^3} - \frac{3}{z^2} - \frac{i}{z}\right) e^{-iz}$$

3. Spherical Harmonic Expansion of Plane Waves. The regular radial eigenfunctions of the Schrödinger equation for $V = 0$ constitute a complete set, as a consequence of a fundamental theorem concerning Sturm-Liouville differential equations[4] of which (10.19) is an example. Hence, we have before us two alternative complete sets of eigenfunctions of the free particle

---

[4] P. M. Morse and H. Feshbach, *Methods of Theoretical Physics*, McGraw-Hill Book Company, New York, 1953, p. 738.

Hamiltonian. They are the plane waves $e^{i\mathbf{k}\cdot\mathbf{r}}$ and the spherical waves $j_l(kr)Y_l^m(\theta, \varphi)$, where $k = \sqrt{2\mu E}/\hbar$. Both sets exhibit an infinite degree of degeneracy, but it is noteworthy that for a given value of the energy the number of plane waves is indenumerable, yet the number of spherical waves is denumerable, corresponding to the countability of the integer quantum numbers l and m. Nevertheless, these two sets of eigenfunctions are equivalent, and one kind must be capable of expansion in terms of the other. Thus it is an interesting, and for future applications an important, problem to determine the coefficients in the expansion

$$e^{i\mathbf{k}\cdot\mathbf{r}} = \sum_{l=0}^{\infty} \sum_{m=-l}^{+l} c_{l,m}j_l(kr)Y_l^m(\theta, \varphi) \tag{10.40}$$

Actually, it is sufficient to specialize (10.40) to the case where $\mathbf{k}$ points along the z-axis and attempt to determine the expansion

$$e^{ikz} = e^{ikr \cos\theta} = \sum_{l=0}^{\infty} a_l j_l(kr) P_l(\cos\theta) \tag{10.41}$$

The more general expansion (10.40) is obtained from (10.41) by use of the addition theorem for spherical harmonics, (9.79). From the orthogonality and normalization properties of Legendre polynomials we obtain (with $s = \cos\theta$)

$$a_l j_l(kr) = \frac{2l + 1}{2} \int_{-1}^{+1} e^{ikrs} P_l(s) \, ds \tag{10.42}$$

which we may compare with (10.25) to establish the important identity

$$e^{ikz} = \sum_{l=0}^{\infty} (2l + 1)i^l j_l(kr) P_l(\cos\theta) \tag{10.43}$$

This formula is especially useful in scattering theory (see Chapter 11).

4. The Square Well Potential. The spherically symmetric *square well* in three dimensions is of interest because it is mathematically straightforward and approximates a number of real physical situations. In contradistinction to the Coulomb potential, which gives rise to infinitely many discrete energy levels for bound states, the square well, owing to its finite range and finite depth, possesses only a finite number of such levels.

A square well is a central potential composed of two constant pieces: $V = -V_0$ for $r < a$, and $V = 0$ for $r > a$. The particle is free inside and outside the well, and subject to a force only at the discontinuity at $r = a$. In this section the emphasis will be on the *bound states* of a particle in such a potential. The radial wave equation for a state of angular momentum l is

$$-\frac{\hbar^2}{2\mu r^2}\frac{d}{dr}\left(r^2\frac{dR}{dr}\right)+\frac{\hbar^2 l(l+1)}{2\mu r^2}R=(E+V_0)R \qquad \text{for } r<a \quad (10.44a)$$

$$-\frac{\hbar^2}{2\mu r^2}\frac{d}{dr}\left(r^2\frac{dR}{dr}\right)+\frac{\hbar^2 l(l+1)}{2\mu r^2}R=ER \qquad \text{for } r>a \quad (10.44b)$$

For bound states $-V_0 \leqslant E \leqslant 0$.

The condition of regularity at the origin restricts us again to the spherical Bessel function for the solution inside the well. All the results of Section 10.2 apply provided that we take into account the fact that E must be replaced by the kinetic energy, $E - V = E + V_0$. Thus

$$R(r)=Aj_l\left(\sqrt{\frac{2\mu(E+V_0)}{\hbar^2}}\,r\right)\qquad \text{for } r<a \qquad (10.45)$$

Outside the well we must exclude any solution of (10.44b) which would increase exponentially at large distances. Since $E<0$ for bound states, (10.44b) has the same solutions as (10.23), but k is now an imaginary number. If we define

$$k=i\sqrt{\frac{-2\mu E}{\hbar^2}}$$

it is easily verified from the integral representations in Section 10.2 that only the Hankel function of the first kind decreases exponentially. The eigenfunction outside the well must thus be of the form

$$R(r)=Bh_l^{(1)}\left(i\sqrt{\frac{-2\mu E}{\hbar^2}}\,r\right)\qquad \text{for } r>a \qquad (10.46)$$

The interior and exterior solutions must be matched at $r = a$. In conformity with the analogous one-dimensional problem (see Sections 5.6 and 8.7), the radial wave function and its derivative are required to be continuous at the discontinuity of the potential. Hence, the logarithmic derivative, $(1/R)\,dR/dr$ or $[(1/u)\,du/dr]$, must be continuous. This condition, applied to (10.45) and (10.46), yields an equation for the allowed discrete energy eigenvalues.

The solutions for positive E are asymptotically oscillatory and correspond to scattering states in which the particle can go to infinity with a finite kinetic energy. They will be studied in Chapter 11.

Exercise 10.2. Compare the energy eigenvalues for S-states in the three-dimensional square well with the energy eigenvalues of a one-dimensional square well of the same depth and width.

5. The Radial Equation and the Boundary Conditions. We now return to a general discussion of the radial equation for central forces. From Section 9.3 we know that the solutions of the Schrödinger equation can be constructed as

$$\psi(r, \theta, \varphi) = \frac{u(r)}{r} Y_l^m(\theta, \varphi) \tag{10.47}$$

Since r does not change under reflection, these wave functions have the same parity as Y_l^m. Hence, for even l we have states of even parity, and for odd l we have states of odd parity.

The radial wave function $u(r)$ must satisfy the equation

$$-\frac{\hbar^2}{2\mu} \frac{d^2u}{dr^2} + \left[\frac{\hbar^2 l(l+1)}{2\mu r^2} + V(r)\right]u = Eu \tag{10.48}$$

The general principles of quantum mechanics require that the eigenfunctions (10.47) be normalizable. Since the spherical harmonics are normalized to unity, the eigenfunctions corresponding to discrete eigenvalues must satisfy the condition

$$\iint \psi^*\psi \, r^2 \, dr \, d\Omega = \int_0^\infty u^*u \, dr = 1 \tag{10.49}$$

If E lies in the continuous part of the spectrum, the eigenfunctions must be normalized in the sense of (8.49), or

$$\int_0^\infty u_E^* u_{E'} \, dr = \delta(E - E') \tag{10.50}$$

Most situations of practical interest are covered if we assume that $V(r)$ is finite everywhere except possibly at the origin, and that near $r = 0$ it can be represented by[5]

$$V(r) \approx cr^\alpha \tag{10.51}$$

with α an integer and $\alpha \geqslant -1$. Furthermore, we assume that $V \to 0$ as $r \to \infty$.

We must not forget that, since division by r is invoked, (9.22) is not a representation of the kinetic energy at the coordinate origin. For the same reason (10.48) is valid only for $r \neq 0$ and must be supplemented by a boundary condition at $r = 0$. Without going into detail, we note that the appropriate boundary condition is obtained by demanding that the Hamiltonian, or energy, operator must be self-adjoint in the sense of (8.51). This

---

[5] For potentials which are more singular at the origin see K. M. Case, *Phys. Rev.*, **80**, 797 (1950). See also P. M. Morse and H. Feshbach, *Methods of Theoretical Physics*, McGraw-Hill Book Company, New York, 1953, pp. 1665–1667.

is the condition which consistency of the probability interpretation of quantum mechanics imposes on the eigenfunctions of H.[6]

Applying this requirement to the operator

$$H = -\frac{\hbar^2}{2\mu r^2}\frac{\partial}{\partial r}\left(r^2\frac{\partial}{\partial r}\right) + \frac{\mathbf{L}^2}{2\mu r^2} + V(r)$$

we find by integrating by parts that any two physically admissible eigen-solutions of (10.48) must satisfy the condition

$$\lim_{r\to 0}\left(u_{k_1}{}^*\frac{du_{k_2}}{dr} - \frac{du_{k_1}{}^*}{dr}u_{k_2}\right) = 0 \tag{10.52}$$

In applications this condition usually may be replaced by the much simpler one requiring that $u(r)$ vanish at the origin,[7]

$$u(0) = 0 \tag{10.53}$$

In most cases this boundary condition singles out correctly the set of eigenfunctions which pass the test (10.52).

If in the immediate vicinity of the origin V can be neglected in comparison with the centrifugal term ($\sim 1/r^2$), (10.48) reduces near $r = 0$ to

$$\frac{d^2u}{dr^2} - \frac{l(l+1)}{r^2}u = 0 \tag{10.54}$$

for states with $l \neq 0$. Potentials of the form (10.51), including the square well and the Coulomb potential, are examples of this. The general solution of (10.54) is

$$u = Ar^{l+1} + Br^{-l} \tag{10.55}$$

and, since $l \geqslant 1$, the boundary condition (10.52) or (10.53) eliminates the second solution; hence, $B = 0$. Thus for any but S-states, u must be proportional to r^{l+1} at the origin, and ψ must behave as r^l. Hence, a power series solution of (10.48) must have the form

$$u(r) = r^{l+1}(a_0 + a_1 r + a_2 r^2 + \cdots) \tag{10.56}$$

For S-states the potential cannot be neglected, and a separate investigation is required to obtain the behavior of the wave function near the origin.

[6] For a precise formulation see W. Pauli, *Die allgemeinen Prinzipien der Wellenmechanik*, in S. Flügge, ed., *Encyclopedia of Physics*, vol. 5/1, pp. 1–168, Springer, Berlin, 1958.

[7] Condition (10.53) is usually "derived" by requiring that ψ be finite everywhere. However, finiteness of ψ is not really a necessary requirement, and mildly singular wave functions are occasionally encountered (e.g., in the relativistic theory of the hydrogen atom, Section 24.4).

Assuming that the potential energy vanishes at great distances, the radial equation (10.48) reduces to

$$\frac{d^2u}{dr^2} + \frac{2\mu E}{\hbar^2} u = 0 \tag{10.57}$$

as $r \to \infty$. Equation (10.57) possesses oscillatory solutions for positive E and exponential solutions for negative E, with the increasing exponential excluded by the condition that ψ must be normalizable in the sense of (10.49) or (10.50).

If $E < 0$, the eigenfunctions have the asymptotic behavior

$$u(r) \sim \exp \left(-\frac{\sqrt{-2\mu E}}{\hbar} r \right)$$

representing spatially confined, or *bound*, states. The boundary conditions will in general allow only certain discrete energy eigenvalues.

For bound states, the radial equation is conveniently transformed by the introduction of the dimensionless variable

$$\rho = \sqrt{\frac{-2\mu E}{\hbar^2}} r = \kappa r \tag{10.58}$$

Sometimes it is also convenient to remove from the unknown dependent variable the portions which describe its behavior at $r = 0$ and $r = \infty$. Thus we introduce a new function $w(\rho)$ by setting

$$u(r) = \rho^{l+1} e^{-\rho} w(\rho) \tag{10.59}$$

Substituting this expression into (10.48), we obtain

$$\frac{d^2w}{d\rho^2} + 2\left(\frac{l+1}{\rho} - 1\right) \frac{dw}{d\rho} + \left[\frac{V}{E} - \frac{2(l+1)}{\rho}\right] w = 0 \tag{10.60}$$

Of the solutions of this equation we seek those which satisfy the boundary conditions at infinity and at the origin.

6. The Coulomb Potential. We must now specify the form of V. Let us suppose that V is the potential energy of the Coulomb attraction between a fixed charge Ze and a moving particle of charge $-e$:

$$V(r) = -\frac{Ze^2}{r} \tag{10.61}$$

For the hydrogen atom $Z = 1$. According to the discussion of the last section, and especially (10.55), the radial wave function $u(r)$ must behave as r^{l+1} near the origin. This is also true for S-states, as shown in Exercise 10.3.

Exercise 10.3. Show that for a potential $V = -k/r$ and angular momentum $l = 0$ the general solution of (10.48) is of the form

$$u = A\left(r - \frac{k}{2}r^2 + \cdots\right) + B(1 - kr \log r + \cdots)$$

for small values of r, and infer that for S-states, $B = 0$.

The energy levels and eigenfunctions of *bound hydrogenic* states will be discussed in this section and the next. The energy continuum $(E > 0)$ of a particle in a Coulomb potential is the subject of Section 11.8.

For convenience we introduce a parameter

$$\rho_0 = \frac{Ze^2\kappa}{|E|} = \sqrt{\frac{2\mu}{|E|}}\frac{Ze^2}{\hbar} \tag{10.62}$$

such that

$$\frac{E}{V} = \frac{\rho}{\rho_0} \tag{10.63}$$

The differential equation (10.60) can then be written as

$$\rho\frac{d^2w}{d\rho^2} + 2(l + 1 - \rho)\frac{dw}{d\rho} + [\rho_0 - 2(l + 1)]w = 0 \tag{10.64}$$

A simple two-term recursion relation is found if we assume $w(\rho)$ expanded in a power series:

$$w(\rho) = a_0 + a_1\rho + a_2\rho^2 + \cdots \tag{10.65}$$

Let us substitute (10.65) into (10.64) and equate to zero the coefficient of ρ^k The result is

$$(k + 1)ka_{k+1} + 2(l + 1)(k + 1)a_{k+1} - 2ka_k + [\rho_0 - 2(l + 1)]a_k = 0$$

or

$$\frac{a_{k+1}}{a_k} = \frac{2(k + l + 1) - \rho_0}{(k + 1)(k + 2l + 2)} \tag{10.66}$$

This recursion relation shows that $a_{-1} = 0$; hence the power series (10.65) begins with a constant term, $a_0 \neq 0$. We shall now see that this power series must terminate at some finite maximum power. If it failed to do so, the coefficients of very high powers would satisfy the abbreviated recursion relation

$$a_{k+1}/a_k \to 2/k \qquad (k \text{ very large})$$

This ratio permits the conclusion that for large ρ the function $w(\rho)$ behaves asymptotically as $e^{2\rho}$. Such a divergence at large distances is incompatible with the proper behavior of u at infinity. Indeed, a factor of $w = e^{2\rho}$ in

(10.59) undoes the asymptotic behavior which was already built into that equation. Hence, the series (10.65) must terminate, and $w(\rho)$ must be a polynomial. Let us suppose its degree to be N, i.e., $a_{N+1} = 0$, but $a_N \neq 0$. Equation (10.66) leads to the condition

$$\rho_0 = 2(N + l + 1) \tag{10.67}$$

where $l = 0, 1, 2 \cdots$ and $N = 0, 1, 2 \cdots$.

Exercise 10.4. Assume, contrary to the conventional procedure, that the radial eigenfunctions for a bound state of the hydrogen atom can be written as

$$u = \rho^k e^{-\rho}\left(b_0 + \frac{b_1}{\rho} + \frac{b_2}{\rho^2} + \cdots\right) \tag{10.68}$$

Obtain the recursion relations for the coefficients, and show that the boundary conditions give the same eigenvalues and eigenfunctions as usual.[8]

It is amusing to contemplate that as innocuous an equation as (10.67) is equivalent to the Balmer formula for the energy levels in hydrogenic atoms. To see this, we merely substitute (10.62) into (10.67) and define the *principal quantum number*

$$n \equiv N + l + 1 = \frac{\rho_0}{2} \tag{10.69}$$

The result is

$$E_n = -\frac{Z^2 \mu e^4}{2\hbar^2 n^2} \tag{10.70}$$

As is well known from the elementary Bohr theory, the length

$$a = \hbar^2/\mu e^2 \tag{10.71}$$

plays an important role in the quantum description of the hydrogen atom. a is termed the first *Bohr radius* of hydrogen if μ is the mass and $-e$ the charge of the electron. Its numerical value is

$$a = 0.529167 \times 10^{-8} \text{ cm}$$

Using this quantity, the energy can be written simply as

$$E_n = -\frac{Z^2 e^2}{2an^2} \tag{10.72}$$

Also we see that

$$\kappa = Z/na$$

[8] A series like (10.68) in descending powers of r can be useful for obtaining approximate wave functions if V behaves as $1/r$ at large, but not at small, distances.

Since N is a nonnegative integer by its definition, it is obvious from (10.69) that n must be a positive integer subject to the restriction

$$n > 1 \qquad\qquad (10.73)$$

The *ground state* of the hydrogen atom corresponds to n = 1, $l = 0$, with an energy of approximately -13.6 ev. There are infinitely many discrete energy levels; they have a point of accumulation at $E_n = 0$ (n $\to \infty$).

The fact that the energy depends only on the quantum number n implies that there are in general several linearly independent eigenfunctions of the Schrödinger equation for hydrogen corresponding to the same value of the energy. In the first place, $2l + 1$ different eigenfunctions of the same energy are obtained by varying the magnetic quantum number m in integral steps from $-l$ to l. Secondly, there are n values of l ($l = 0, 1, 2, \ldots, n - 1$) consistent with a given value of n. Hence, the number of linearly independent stationary states of hydrogen having the energy (10.70) is

$$\sum_{l=0}^{n-1} (2l + 1) = n^2$$

Hence, all energy levels with n > 1 are degenerate, in the terminology of Section 4.5. In the present case the degree of degeneracy is said to be n^2. For example, in standard spectroscopic notation (n followed by the symbol for l) the first excited level of hydrogen (n = 2) is fourfold degenerate and consists of the $2S$ state and three $2P$ states. (The degeneracy is doubled if the spin is taken into account.)

The occurrence of degeneracy can often be ascribed to some transparent symmetry property of the physical system. For instance, the degeneracy with respect to magnetic quantum numbers will clearly be present for any central potential. It has its origin in the absence of a preferred spatial direction in such systems and reflects their invariance with regard to rotations about the origin. (See also Chapter 16). The degeneracy of states with the same n but different values of l is peculiar to the Coulomb potential. Any departure of the potential from a strict $1/r$ dependence will remove this degeneracy.[9]

Exercise 10.5. Show that the addition of a small $1/r^2$ term to the Coulomb potential removes the degeneracy of states with different l. The energy levels are still given by a Balmer-like formula (10.70), but n differs from an

---

[9] The degeneracy of states with the same n but different l is related to the separability of the Schrödinger equation for $V = c/r$ in parabolic coordinates (Section 11.8) in addition to its separability in spherical polar coordinates which is common to all central potentials. The term *accidental degeneracy*, sometimes used to describe this degeneracy, is misleading. A subtle symmetry in momentum space is responsible for the degeneracy peculiar to the Coulomb potential. See V. Fock, *Z. Physik*, **98**, 145 (1935); also H. V. McIntosh, *Am. J. Phys.* **27**, 620 (1959), and M. Bander and C. Itzykson, *Rev. Mod. Phys.*, **38**, 330 (1966).

integer by an *l*-dependent quantity, the *quantum defect* in the terminology of one-electron (alkali) spectra.

7. Study of the Eigenfunctions. The most important properties of the radial wave functions for the Coulomb potential will be summarized in this section. These functions can be expressed in terms of confluent hypergeometric functions, as can be seen readily when the value for ρ_0 is put into (10.66):

$$\frac{a_{k+1}}{a_k} = \frac{2(k - N)}{(k + 1)(k + 2l + 2)} \tag{10.74}$$

The confluent hypergeometric function has already been defined in (5.53) as

$$_1F_1(a; c; z) = 1 + \frac{a}{c} \frac{z}{1!} + \frac{a(a + 1)}{c(c + 1)} \frac{z^2}{2!} + \cdots \tag{10.75}$$

Comparing its coefficients with (10.74), we see that

$$w(\rho) \propto {}_1F_1(-N; 2l + 2; 2\rho) \tag{10.76}$$

This can also be seen by comparing the differential equation

$$z \frac{d^2w}{dz^2} + (c - z) \frac{dw}{dz} - aw = 0 \tag{10.77}$$

which $w = {}_1F_1(a; c; z)$ satisfies with the radial equation (10.64) if the latter by the use of (10.69) is cast in the form

$$\frac{\rho}{2} \frac{d^2w}{d\rho^2} + (l + 1 - \rho) \frac{dw}{d\rho} + Nw = 0 \tag{10.78}$$

In quantum mechanics it has become customary to relate the functions (10.76) to a certain class of orthogonal polynomials, known as the *associated Laguerre polynomials*, $L_q^p(z)$, which will now be discussed briefly.

We first note that, if we obtain a solution of the simple differential equation

$$z \frac{d^2w}{dz^2} + (1 - z) \frac{dw}{dz} + qw = 0, \tag{10.79}$$

the *p*th derivative of this solution satisfies the equation

$$z \frac{d^2w}{dz^2} + (p + 1 - z) \frac{dw}{dz} + (q - p)w = 0 \tag{10.80}$$

as can be seen by differentiating (10.79) *p* times. Equation (10.80) is the same as (10.77) and (10.78) if we make the identifications

$$z = 2\rho, \quad p = c - 1 = 2l + 1, \quad q = c - a - 1 = N + 2l + 1 = n + l \tag{10.81}$$

Obviously, we shall be interested only in the values $p = 1, 3, 5, \ldots$ and $q = 1, 2, 3, \ldots$ and $q - p = N = 0, 1, 2, \ldots$.

A compact solution of (10.80) can be obtained by using Laplace transforms. We attempt to solve (10.80) by choosing the integrand and path of integration in an integral of the form

$$w(z) = \int_C e^{-zt} f(t) \, dt \tag{10.82}$$

Substituting this into (10.80), we get in several easy steps

$$\int f(t)[zt^2 - (p + 1 - z)t + q - p]e^{-zt} \, dt = 0$$

or

$$\int f(t)\left[-t(1 + t)\frac{d}{dt} + q - p - (p + 1)t\right]e^{-zt} \, dt = 0$$

Integrating by parts,

$$\int \frac{d}{dt}[-e^{-zt}t(1 + t)f(t)] \, dt$$

$$+ \int e^{-zt}\left\{\frac{df}{dt}t(1 + t) + f[q - p + (1 - p)t + 1]\right\} \, dt = 0$$

If the first term can be made to vanish to a judicious choice of the path of integration in the complex t-plane, we are led to ask that f should satisfy the equation

$$\frac{1}{f}\frac{df}{dt} = -\frac{q - p + 1 + (1 - p)t}{t(t + 1)}$$

This can be integrated immediately:

$$f(t) = A\frac{(1 + t)^q}{t^{q+1-p}} \qquad (p = 1, 3, 5, \ldots ; q = 1, 2, 3, \ldots) \tag{10.83}$$

The expression (10.82) will be a solution of (10.80) with $f(t)$ given by (10.83), provided that the integrated part above vanishes. This is accomplished if we choose a closed path of integration around the pole of the integrand at $t = 0$. Hence,

$$w(z) = A \oint \frac{e^{-zt}}{t^{q+1-p}}(1 + t)^q \, dt \tag{10.84}$$

where the path of integration C encircles the origin counterclockwise. Since

$$\oint t^n \, dt = 0$$

for all integers n except $n = -1$ and

$$\oint \frac{dt}{t} = 2\pi i$$

it is clear that (10.84) represents the polynomial (and not the singular) solution of (10.80).

The solution of (10.79) obtained by setting $p = 0$ in (10.84) is said to represent a *Laguerre polynomial*, $w(z) = L_q(z)$ if $L_q(0) = q!$. This normalization determines A in (10.84) and gives

$$L_q(z) = \frac{q!}{2\pi i} \oint \frac{e^{-zt}}{t^{q+1}} (1 + t)^q \, dt \qquad (10.85)$$

From this we can infer a simple formula by noting the steps

$$L_q(z) = \frac{q!}{2\pi i} e^z \frac{d^q}{dz^q} \oint \frac{e^{-z(t+1)}}{t^{q+1}} (-1)^q \, dt = (-1)^q \frac{q!}{2\pi i} e^z \frac{d^q}{dz^q} \left[\oint \frac{e^{-zt}}{t^{q+1}} \, dt \, e^{-z} \right]$$

from which it follows, by expanding the exponential under the integral, that

$$L_q(z) = e^z \frac{d^q}{dz^q} (e^{-z} z^q) \qquad (10.86)$$

Despite the appearance of the exponential function, $L_q(z)$ is, of course, a polynomial of degree q.

Exercise 10.6. Show that $L_q(z)$ has q distinct real positive zeros.

Equations (10.77) and (10.80) are solved by the *associated Laguerre polynomials*

$$L_{-a}^{c-1}(z) = L_{q-p}^p(z) = (-1)^p \frac{d^p}{dz^p} L_q(z) \qquad (10.87)$$

Since L_q is a polynomial with q distinct zeros on the positive real axis, L_{q-p}^p has $q - p$ such zeros. By (10.85)

$$L_{q-p}^p(z) = \frac{q!}{2\pi i} \oint \frac{e^{-zt}}{t^{q-p+1}} (1 + t)^q \, dt \qquad (10.88)$$

in conformity with (10.84). Hence,

$$L_{q-p}^p(0) = \frac{(q!)^2}{(q - p)! \, p!}$$

With (10.75), (10.77), and (10.80), this result permits the identification

$$L_{q-p}^{p}(z) = \frac{(q!)^2}{(q-p)!\, p!}\, {}_1F_1(p-q; p+1; z) \tag{10.89}$$

or

$$L_{q-p}^{p}(z) = \frac{[\Gamma(q+1)]^2}{\Gamma(q-p+1)\Gamma(p+1)}\, {}_1F_1(p-q; p+1; z) \tag{10.89a}$$

The last equation defines the generalized Laguerre functions for arbitrary values of p and q.

Exercise 10.7. Show that the generating function for the associated Laguerre polynomials is (for $|s| \leqslant 1$)

$$\frac{\exp\left(-\dfrac{sz}{1-s}\right)}{(1-s)^{p+1}} = \sum_{n=0}^{\infty} \frac{L_n^{\,p}(z)}{(n+p)!}\, s^n \tag{10.90}$$

The unnormalized eigensolution of the radial equation (10.78) is thus

$$w(\rho) = L_N^{2l+1}(2\rho) = L_{n-l-1}^{2l+1}(2\rho) \tag{10.91}$$

The number of its zeros is $q - p = N$. The complete normalized wave function can only be obtained if we know the value of the normalization integral for the associated Laguerre polynomials. It can be shown that

$$\int_0^\infty e^{-x} x^{2l+2} [L_{n-l-1}^{2l+1}(x)]^2 \, dx = \frac{2n[(n+l)!]^3}{(n-l-1)!} \tag{10.92}$$

Since the spherical harmonics are normalized to unity, the complete eigenfunction $\psi(r, \theta, \varphi)$ of (10.47) is normalized to unity, if

$$\int_0^\infty |u(r)|^2 \, dr = 1$$

Collecting all the conclusions of this chapter together we can write the hydrogen eigenfunctions, normalized to unity, as

$$\psi_{n,l,m}(r, \theta, \varphi) = \left\{ (2\kappa)^3 \frac{(n-l-1)!}{2n[(n+l)!]^3} \right\}^{1/2} e^{-\kappa r}(2\kappa r)^l L_{n-l-1}^{2l+1}(2\kappa r) Y_l^{\,m}(\theta, \varphi)$$

$$= \frac{e^{-\kappa r}(2\kappa r)^l}{(2l+1)!} \left[(2\kappa)^3 \frac{(n+l)!}{2n(n-l-1)!} \right]^{1/2}$$

$$\times {}_1F_1(-n+l+1; 2l+2; 2\kappa r) Y_l^{\,m}(\theta, \varphi) \tag{10.93}$$

Exercise 10.8. Obtain explicitly the radial hydrogen wave functions for n = 1, 2, and 3, and sketch them graphically.[10]

A few remarks may be appropriate concerning the properties of the hydrogenic eigenfunctions. The wave function possesses a number of nodal surfaces. These are the surfaces on which $\psi = 0$. For these considerations it is customary to refer instead of (10.93) to the real eigenfunctions

$$\psi_{n,l,m} \propto r^l e^{-\kappa r} L_{n-l-1}^{2l+1}(2\kappa r) P_l^m(\cos\theta) \begin{cases} \cos m\varphi \\ \sin m\varphi \end{cases}$$

There are $l - m$ values of θ for which $P_l^m(\cos\theta)$ vanishes. $\cos m\varphi$ and $\sin m\varphi$ vanish at m values of the azimuth, and $L_{n-l-1}^{2l+1}(2\kappa r)$ vanishes at $n - l - 1$ values of r. For $l \neq 0$, r^l has a node at $r = 0$. Hence, except in S-states, the total number of nodal surfaces is n, if $r = 0$ is counted as a surface. The fact that the wave function vanishes at the origin except when $l = 0$ (S-states) has very important consequences. For instance, the capture of an atomic electron by a nucleus can occur with appreciable probability only from a level with $l = 0$, because these are the only states for which ψ has a finite value at the position of the nucleus. Similarly, in the phenomenon of internal conversion an atomic s-electron interacts with the nucleus and is imparted enough energy to be ejected from the atom.

The quantum mechanical significance of the Bohr radius, $a = \hbar^2/(\mu e^2)$, can be appreciated by observing that the wave function for the ground state is

$$\psi_{1,0,0}(r, \theta, \varphi) = \left(\frac{Z^3}{\pi a^3}\right)^{1/2} \exp\left(-\frac{Zr}{a}\right) \tag{10.94}$$

The expectation value of r in this state is

$$\langle r \rangle = 4\left(\frac{Z}{a}\right)^3 \int_0^\infty \exp\left(-\frac{2Z}{a}r\right) r^3 \, dr = \frac{3}{2}\frac{a}{Z} \tag{10.95}$$

The maximum of the probability density for finding the particle in the ground state with a radial separation r from the nucleus, i.e., the maximum of the function

$$\left|\exp\left(-\frac{Z}{a}r\right)\right|^2 r^2$$

is located at a/Z. a is inversely proportional to the mass of the particle which moves around the nucleus. Hence, a mu or pi meson in a mesic atom is much closer to the nucleus than an electron is in an ordinary atom. The finite size

[10] The reader will enjoy the graphic representations by H. E. White, *Phys. Rev.*, **37**, 1416 (1931). See also L. C. Pauling and B. E. Wilson, Jr., *Introduction to Quantum Mechanics with Applications to Chemistry*, McGraw-Hill Book Company, New York, 1935, Section 5.21.

of the nucleus will thus be expected to affect the discrete energy levels of mesic atoms, whereas the nucleus can usually be assumed to have no extension when we deal with electrons in bound states.

There are, of course, many other corrections which must be taken into account when comparing the simple Balmer formula (10.70) with the fantastically accurate results of modern atomic spectroscopy. Most obviously, we must correct the error made in assuming that the nucleus is infinitely massive and therefore fixed. Since for central forces the actual two-body problem can be replaced by an effective one-body problem, if we substitute the reduced mass $\mu_1\mu_2/(\mu_1 + \mu_2)$ for μ, this correction can be applied accurately and without difficulty. It gives rise to small but significant differences between the spectra of hydrogen and deuterium. For a positronium atom, composed of an electron and a positron which have equal masses, all energy levels are changed by a factor $\frac{1}{2}$ compared to hydrogen. Further, and often more important, corrections are due to the presence of the electron spin and the high speed of the electron, which necessitate a relativistic calculation (Section 24.4); hyperfine structure effects arise from the magnetic properties of the nucleus; and, finally, there are small but measurable effects owing to the interaction between the electron and the electromagnetic field (Lamb shift). The theory of some of these effects will be discussed later in this book; others lie outside its scope. But all are overshadowed in magnitude by the basic gross structure of the spectrum as obtained in this chapter by the application of nonrelativistic quantum mechanics to the Coulomb potential.

Problems

1. Plot the ten lowest energy eigenvalues of a particle in an infinitely deep spherically symmetric square well, and label the states by their appropriate quantum numbers. (Use mathematical tables or program for a computer.)

2. If H is the sum of a Hermitian operator H_0 and a positive definite perturbation, prove by a variational argument that the ground state energy of H_0 lies below the ground state energy of H.

   Apply this theorem to prove that in a central potential the ground state of a bound particle is an S-state.

3. If the ground state of a particle in a square well is just barely bound, show that the well depth and radius parameters V_0 and a are related to the binding energy by the expansion

$$\frac{2\mu V_0 a^2}{\hbar^2} = \frac{\pi^2}{4} + 2\eta + \left(1 - \frac{4}{\pi^2}\right)\eta^2 + \cdots$$

where

$$\eta = \frac{\sqrt{-2\mu E}\, a}{\hbar}$$

The deuteron is bound with an energy of 2.226 MeV and has no discrete excited states. If the deuteron is represented by a nucleon, with reduced mass, moving in a square well with $a = 1.5$ fermi, estimate the depth of the potential.

4. Given an attractive central potential of the form

$$V = -V_0 e^{-r/a}$$

solve the Schrödinger equation for the S-states by making the substitution

$$\xi = e^{-r/2a}$$

Obtain an equation for the eigenvalues. Estimate the value of V_0, if the state of the deuteron is to be described with an exponential potential (see Problem 3 for data).

5. Show that, if a square well just binds an energy level of angular momentum $l(\neq 0)$, its parameters satisfy the condition

$$j_{l-1}\left(\sqrt{\frac{2\mu V_0}{\hbar^2}}\, a\right) = 0$$

(Use recurrence formulas for Bessel functions from standard texts.)

6. Assuming the eigenfunctions for the hydrogen atom to be of the form $r^\beta e^{-\alpha r} Y_l{}^m$, with undetermined parameters α and β, solve the Schrödinger equation. Are all eigenfunctions and eigenvalues obtained in this way?

7. Apply the WKB method to an attractive Coulomb potential, and derive an approximate formula for the $l = 0$ energy levels of hydrogen.

8. Find the probability that the electron in a hydrogen atom will be found at a distance from the nucleus greater than its energy would permit on the classical theory. Assume the atom to be in its ground state.

9. Calculate the probability distribution for the momentum of the electron in the ground state of a hydrogen atom. Obtain the expectation value of $p_x{}^2$ from this or from the virial theorem. Also calculate $\langle x^2 \rangle$ from the ground state wave function, and verify the uncertainty principle for this state.

10. Solve the Schrödinger equation for a three-dimensional isotropic harmonic oscillator, $V = \frac{1}{2}\mu\omega^2 r^2$, by separation of variables in Cartesian and in spherical polar coordinates. In the latter case assume the eigenfunctions to be of the form

$$\psi = r^l \exp\left(-\frac{\mu\omega}{2\hbar} r^2\right) f(r) Y_l{}^m(\theta, \varphi)$$

and show that $f(r)$ can be expressed as an associated Laguerre polynomial with half-integral indices of the variable $\mu\omega r^2/\hbar$. Obtain the eigenvalues and establish the correspondence between the two sets of quantum numbers. For the lowest two eigenvalues show the relation between the eigenfunctions obtained by the two methods. Obtain a formula for the degree of degeneracy in terms of the energy and compare the density of energy eigenstates in the oscillator and in a cubic box.

11. The initial ($t = 0$) state of an isotropic harmonic oscillator is known to be an eigenstate of L_z with eigenvalue zero and a superposition of the ground and first excited states. Assuming that the expectation value of the coordinate z has at this time its largest possible value, determine the wave function for all times as completely as possible.

12. Discuss the two-dimensional isotropic harmonic oscillator. Assume that the eigenfunctions are of the form

$$\psi = \rho^m \exp\left(-\frac{\mu\omega}{2\hbar}\,\rho^2\right) f(\rho) \exp\left(\pm im\varphi\right)$$

where ρ and φ are plane polar coordinates. Show that $f(\rho)$ can be expressed as an associated Laguerre polynominal of the variable $\mu\omega\rho^2/\hbar$ and determine the eigenvalues. Solve the same problem in Cartesian coordinates, and establish the correspondence between the two methods. Discuss a few simple eigenfunctions.

13. Apply the variational method to the ground state ($l = 0$) of a particle moving in an attractive central potential $V(r) = Ar^n$ (integer $n \geqslant -1$), using

$$R(r) = e^{-\beta r}$$

as a trial wave function with a variational parameter β. For $n = -1$ and $+2$ compare the results with the exact ground state energies.

14. Apply the variational method to the ground state ($l = 0$) of a particle moving in an attractive (Yukawa or screened Coulomb) potential

$$V(r) = -V_0 \frac{e^{-r/a}}{r/a}$$

with $V_0 > 0$. Use as a trial wave function

$$R(r) = e^{-\beta r/a}$$

with a variable parameter β. Obtain the "best" trial wave function of this form and deduce a relation between β and $2\mu V_0 a^2/\hbar^2$. Evaluate β and calculate an upper limit to the energy for the value $2\mu V_0 a^2/\hbar^2 = 2.7$.

   Are there any bound excited states?

   Show that in the limit of the Coulomb potential ($V_0 \to 0$, $a \to \infty$, $V_0 a$ finite) the correct energy and wave function for the hydrogenic atom are obtained.

15. Using first-order perturbation theory, estimate the correction due to the finite size of the nucleus to the ground state energy of a hydrogenic atom. Under the assumption that the nucleus is much smaller than the atomic Bohr radius, show that the energy change is approximately proportional to the nuclear mean square radius. Evaluate the correction for a uniformly charged spherical nucleus

of radius R. Is the level shift due to the finite nuclear size observable? Consider also mu mesic atoms.

16. An electron is moving in the Coulomb field of a point charge Ze, modified by the presence of a uniformly charged spherical shell of charge $-Z'e$ and radius R, centered at the point charge. Perform a first-order perturbation calculation of the hydrogenic $1S$, $2S$, and $2P$ energy levels. For some representative values of $Z = Z'$, estimate the limit that must be placed on R so that none of the lowest three levels shift by more than five percent of the distance between the unperturbed first excited and ground state energy levels.

CHAPTER 11

Scattering

1. *Introduction.* Much of what we know about the forces and interactions in atoms and nuclei has been learned from scattering experiments, in which atoms in a target are bombarded with beams of particles. Usually we know the nature of the particles used as projectiles, their momentum, and perhaps their polarization (see Chapter 12). These particles are scattered by the target atoms and are subsequently detected by devices that may give us the intensity as a function of the scattering angle (and possibly of the energy of the scattered particles, if inelastic processes are involved).

Thus, nucleons scattered from nuclei at various energies reveal information about the nuclear forces as well as about the structure of nuclei. Electrons of high energy, hence short wavelength, are particularly well suited to probe the charge distribution in nuclei, and indeed within nucleons. Electrons and heavier projectiles of low energy are scattered from atoms to obtain data which can serve as input information for calculations of kinetic processes in gases where low energy collisions predominate. And mesons and other elementary particles are scattered from protons to tell us about interactions of which we have no other direct information. These are just a few examples of the utility of scattering in studying the internal structure of atoms and nuclei and the interactions which govern systems of elementary particles.

From a theoretical point of view the most significant aspect of scattering processes is that we are now concerned with the *continuous* part of the energy spectrum. We are free to choose the value of the incident particle energy arbitrarily. If the zero of energy is chosen in the usual manner, this corresponds to positive eigenvalues of the Schrödinger equation and to eigenfunctions of unbound states. But it is not enough merely to solve the differential equation for various potentials, as we did for the bound states. There the interest was focused on the discrete energy *eigenvalues* which allow a direct comparison of theory and experiment. In the continuous part of the energy spectrum, as it comes into play in scattering, the energy is given by the incident beam, and *intensities* are the object of measurement and prediction. These, being measures of the likelihood of finding a particle at certain places, are of course related to the *eigenfunctions* rather than the eigenvalues. But the connection between eigenfunctions and measured intensities is recondite and indirect. Unlike the effortless comparison of calculated discrete eigenvalues with measured spectral frequencies, which requires only that we assume conservation of energy, i.e., $E_1 - E_2 = h\nu$, scattering data can be compared with theoretical predictions only if we elucidate carefully the various stages of a scattering process. Relating observed intensities to calculated wave functions is the first problem of scattering theory.

In an idealized scattering experiment a single fixed scattering center is bombarded by particles incident along the z-axis.[1] It is assumed that the effect of the scattering center on the particles can be represented by a potential energy $V(\mathbf{r})$ which is appreciably different from zero only within a finite region. Although this assumption excludes as common a force as the Coulomb field, represented by an inverse square law, this limitation is not really severe. In actual fact the Coulomb field is screened at large distances, for instance, by the presence of electrons in atoms, and for large r the potential falls off faster than $1/r$.

By limiting ourselves to scattering from a potential we specialize to the case of *elastic scattering*, which is scattering without energy loss or gain by the projectile. However, many of the concepts developed here will be found useful in the discussion of inelastic collision processes and more general reactions.

The experiment illustrated in Figure 11.1 involves a collimated homogeneous beam of monoenergetic particles moving toward the scatterer from a great distance. The width of the beam is determined by slits, which, although quite narrow from an experimental point of view, are nevertheless very wide compared with the spatial extension of the scattering region. Experimentally, in the interest of securing "good statistics," i.e., a maximum number of counts in a given period of operation, it is desirable to employ intense beams. Yet the beam density must be low enough so that it can safely be assumed that the incident particles do not interact with one another.

For simplicity we shall suppose that the particles in the beam are all represented by very broad and very long wave packets and that, before they reach the neighborhood of the scatterer, these packets can be described approximately by plane waves $e^{i(kz - \omega t)}$, although strictly speaking the waves do not extend to infinity either in width or in length.

After scattering, the particles are detected at a great distance from the

---

[1] We need not be concerned with complications which arise from the fact that experimental targets are macroscopic samples containing many atoms to be bombarded simultaneously. Except for such phenomena as neutron diffraction in crystals, these are instrumental problems (effects of target thickness, thermal motion of the target atoms, etc.) and not germane to the quantum theoretical discussion.

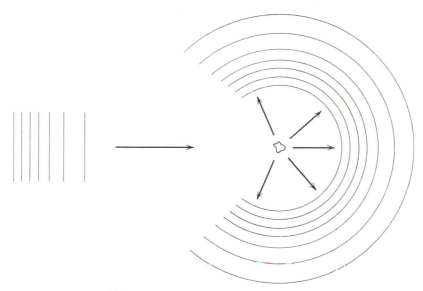

Figure 11.1. Scattering of a wave packet.

scatterer. The detector subtends a cone of solid angle $d\Omega$ at the origin, and the particles scattered into this cone are counted. If I_0 is the number of particles incident from the left per unit area and $I\, d\Omega$ the number of these scattered into the cone, then we define the *differential scattering cross section* as

$$\frac{d\sigma}{d\Omega} = \frac{I(\theta, \varphi)}{I_0} \tag{11.1}$$

This is the quantity which the experimentalist delivers to the theoretician. The latter interprets the cross section in terms of probabilities calculated from wave functions.

The concept of a cross section, based on the assumption that an ensemble of noninteracting particles is incident upon the scatterer, is a statistical one, whether the description is quantal or classical. A brief review of classical scattering theory may help to clarify the fundamental notions involved.

In the classical limit each incident particle can be assigned an *impact parameter*, i.e., a distance, ρ, from the z-axis, parallel to which it approaches from infinity, and an azimuth angle, φ, which together with z define its position in cylindrical coordinates. It is the essence of scattering cross section measurements that no effort is made to determine the path of an individual particle. Rather, particles are projected from the left (Figure 11.2) with random lateral positions (i.e., with random values of impact parameter and azimuth angle) in a broad homogeneous beam, and the number of hits

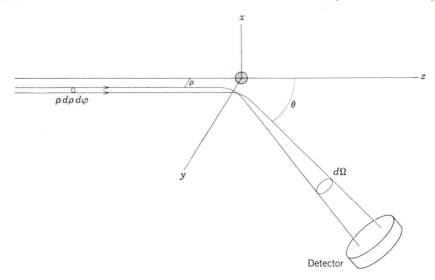

Figure 11.2. Scattering through an angle θ of a classical particle beam incident at impact parameter ρ and azimuth φ.

is recorded at the detector. The differential cross section $d\sigma/d\Omega$, as defined in (11.1), is simply equal to the size of the area which, when placed at right angles to the incident beam, would be traversed by as many particles as are scattered into the unit solid angle around a direction characterized by the angles θ and φ. Figure 11.2 illustrates the point. If the scattering potential is spherically symmetric, $V = V(r)$, the orbits lie in planes through the center of force, and the scattering becomes independent of the azimuth angle φ. Therefore, classically,

$$d\sigma = \rho\, d\rho\, d\varphi = \rho \left| \frac{d\rho}{d(\cos\theta)} \right| \left| d(\cos\theta) \right| d\varphi = \rho \left| \frac{d\rho}{d(\cos\theta)} \right| d\Omega$$

or

$$\frac{d\sigma}{d\Omega} = \rho \left| \frac{d\rho}{d(\cos\theta)} \right| = \frac{\rho}{\sin\theta} \left| \frac{d\rho}{d\theta} \right| \tag{11.2}$$

Hence, the differential cross section is calculable if, for the given energy, we know ρ as a function of the scattering angle θ. To determine this function from Newton's second law is the problem of classical scattering theory.

Equation (11.2) is a good approximation only if the de Broglie wavelength of the incident particles is much smaller than the dimensions of the scattering region. It ceases to be useful when the wave description can no longer be adequately approximated by the geometrical optics limit. As the wavelength increases, quantum features make their appearance, and the quantum uncertainty begins to limit the simultaneous knowledge of, say, p_x which for a well-collimated beam should be close to zero, and x, which is proportional to the impact parameter. The wave nature of matter must be taken into account, and we must learn how to calculate cross sections from wave functions.

Quantum mechanics represents the particles in the beam by wave packets of various shapes and sizes. Only comparison between theory and experiment can tell us whether our assumption that all the particles in the beam can be represented by very broad and long wave packets reflects truly the properties of a real beam, consisting, as it does, of particles which were emitted from a source, or which have undergone some acceleration, and which may have been selected as to energy and momentum by analyzers and slits. The problem here is to construct an ensemble of wave packets such that its average properties coincide with the observed average properties of a real beam of whose complex constitution we are partially ignorant. Fortunately, the mathematical analysis of the next section will show the cross section to be in general independent of the particularities of the incident wave packet, provided only that the latter is large compared with the scattering region toward which it is directed.

2. The Scattering of a Wave Packet.[2] Let us assume that the motion of the particles is described by a Hamiltonian

$$H = \frac{\mathbf{p}^2}{2\mu} + V = H_0 + V \tag{11.3}$$

where V is appreciably different from zero only within a sphere of radius a surrounding the origin. At $t = 0$ a particle is represented by a wave packet of the general form

$$\psi(\mathbf{r}, 0) = \frac{1}{(2\pi)^{3/2}} \int \varphi(\mathbf{k}) \exp\left[i\mathbf{k} \cdot (\mathbf{r} - \mathbf{r}_0)\right] d^3k \tag{11.4}$$

where φ is a smooth function of narrow width, $\Delta\mathbf{k}$, centered around a mean momentum $\mathbf{k}_0$. $\psi(\mathbf{r}, 0)$, as expressed by (11.4), is thus an extended wave packet located in the vicinity of $\mathbf{r}_0$, but with a fairly sharply defined momentum. We assume that $\mathbf{k}_0$ is parallel to $\mathbf{r}_0$, but in an opposite direction, so that the wave packet, if unhindered in its motion, would move freely toward the origin; and we assume r_0 to be so large that ψ at $t = 0$ lies in its entirety well to the left of the scatterer (Figure 11.1).

The dynamical problem to be solved is this: What is the shape of the wave packet (11.4) at a much later time, when the packet has hit the scatterer and

---

[2] The content of this section and the next leans heavily on unpublished lecture notes by F. E. Low.

been eventually dispersed by it? In principle the answer can be given easily, if we succeed in expanding $\psi(\mathbf{r}, 0)$ in terms of eigenfunctions, $\psi_n(\mathbf{r})$, of H. Indeed, if we can establish an expansion

$$\psi(\mathbf{r}, 0) = \sum_n c_n \psi_n(\mathbf{r})$$

then the wave packet at time t is

$$\psi(\mathbf{r}, t) = \sum_n c_n \psi_n(\mathbf{r}) \exp\left(-\frac{i}{\hbar} E_n t\right)$$

How can this program be applied to a wave packet of the form (11.4)? True, it is an expansion in terms of a complete orthonormal set, the plane waves $e^{i\mathbf{k}\cdot\mathbf{r}}$, but these are eigenfunctions of the operator H_0, and not of H. However, it will be proved that it is possible to replace the plane wave functions in the expansion by particular eigenfunctions of H which we shall designate as $\psi_{\mathbf{k}}^{(+)}(\mathbf{r})$, if $\psi(\mathbf{r}, 0)$ is a wave packet of the kind described. Asymptotically, the eigenfunctions $\psi_{\mathbf{k}}^{(+)}(\mathbf{r})$ bear a considerable resemblance to plane waves, since they are of the form

$$\psi_{\mathbf{k}}^{(+)}(\mathbf{r}) \sim \frac{1}{(2\pi)^{3/2}}\left(e^{i\mathbf{k}\cdot\mathbf{r}} + f_{\mathbf{k}}(\hat{\mathbf{r}})\frac{e^{ikr}}{r}\right) \qquad (r \text{ large}) \qquad (11.5)$$

This differs from a plane wave at large r only by an *outgoing spherical wave*. We shall see in Section 11.3 that such solutions of the Schrödinger equation

$$H\psi_{\mathbf{k}}^{(+)}(\mathbf{r}) = E\psi_{\mathbf{k}}^{(+)}(\mathbf{r}) \qquad (11.6)$$

do indeed exist. In order not to interrupt the argument we assume their existence in this section. It will also be shown in Section 11.3 that the special wave packet $\psi(\mathbf{r}, 0)$ has the same expansion coefficients, whether it is expanded in terms of plane waves, as in (11.4), or in terms of the set $\psi_{\mathbf{k}}^{(+)}(\mathbf{r})$:

$$\psi(\mathbf{r}, 0) = \int \varphi(\mathbf{k}) \exp\left(-i\mathbf{k}\cdot\mathbf{r}_0\right)\psi_{\mathbf{k}}^{(+)}(\mathbf{r})\, d^3k \qquad (11.7)$$

In other words, the outgoing wave in (11.5) makes no contribution to the initial wave packet.

With these observations an important part of our program has been accomplished: The initial wave function $\psi(\mathbf{r}, 0)$ has been expanded in terms of eigenfunctions of H. Hence, at any time t,

$$\psi(\mathbf{r}, t) = \int \varphi(\mathbf{k}) \exp\left(-i\mathbf{k}\cdot\mathbf{r}_0 - i\omega t\right)\psi_{\mathbf{k}}^{(+)}(\mathbf{r})\, d^3k \qquad (11.8)$$

where

$$\hbar\omega = \frac{\hbar^2 k^2}{2\mu}$$

Assuming that the scattering detectors are at a macroscopic distance of the order of r_0 from the scatterer, after a time

$$T \simeq \frac{2\mu r_0}{\hbar k_0} \qquad (11.9)$$

the broad pulse will be traveling through the position of the detectors (see Figure 11.1). When we examine the pulse at the position of the detectors, $\psi_{\mathbf{k}}^{(+)}$ can again be represented by its asymptotic expansion, but since the phases have changed with time the outgoing spherical wave may no longer be neglected. However, we can make an approximation based on the identity

$$\omega = \frac{\hbar}{2\mu} k^2 = \frac{\hbar}{2\mu} [\mathbf{k}_0 + (\mathbf{k} - \mathbf{k}_0)]^2 = \frac{\hbar}{2\mu} [2\mathbf{k}_0 \cdot \mathbf{k} - k_0{}^2 + (\mathbf{k} - \mathbf{k}_0)^2] \quad (11.10)$$

In order to be able to neglect the last term in the bracket, when ω is substituted into (11.8), we require that T, although large, should still satisfy the inequality

$$\frac{\hbar}{2\mu} (\mathbf{k} - \mathbf{k}_0)^2 T \ll 1$$

or, using (11.9)

$$\frac{(\Delta k)^2 r_0}{k_0} \ll 1$$

This condition, familiar from Section 2.4, implies that the wave packet does not spread appreciably when it is displaced by the macroscopic distance r_0.[3]

Exercise 11.1. Show, by numerical example, that in scattering of elementary particles from atoms and nuclei the condition of no spreading can be easily attained by a minimum uncertainty wave packet ($\Delta k \, \Delta r \sim 1$) which is large compared with the scattering region but small compared with r_0.

Approximately we can thus write (11.8) as

$$\psi(\mathbf{r}, t) = \int \varphi(\mathbf{k}) \exp \left[-i\mathbf{k} \cdot (\mathbf{r}_0 + \mathbf{v}_0 t) + i\omega_0 t \right]$$
$$\times \frac{1}{(2\pi)^{3/2}} \left(e^{i\mathbf{k} \cdot \mathbf{r}} + f_{\mathbf{k}}(\hat{\mathbf{r}}) \frac{e^{ikr}}{r} \right) d^3 k \quad (11.11)$$

where

$$\mathbf{v}_0 = \frac{\hbar \mathbf{k}_0}{\mu}, \qquad \hbar \omega_0 = \tfrac{1}{2} \mu v_0{}^2$$

Comparing (11.11) with (11.4), and assuming that f, unlike φ, is a slowly

---

[3] The wave packet inevitably spreads, and it is therefore strictly speaking impossible to make it vanish in the scattering region for all times before $t = 0$. However, the spreading can be made negligible for the purpose at hand.

varying function of $\mathbf{k}$, we can write

$$\psi(\mathbf{r}, t) = \psi(\mathbf{r} - \mathbf{v}_0 t, 0)e^{i\omega_0 t}$$

$$+ \frac{1}{(2\pi)^{3/2}} \frac{f_{\mathbf{k}_0}(\hat{\mathbf{r}})e^{i\omega_0 t}}{r} \int \varphi(\mathbf{k}) \exp \left\{i[kr - \mathbf{k} \cdot (\mathbf{r}_0 + \mathbf{v}_0 t)]\right\} d^3k$$

with carets denoting unit vectors. Since φ is appreciably different from zero only for $\hat{\mathbf{k}} \simeq \hat{\mathbf{k}}_0$, we can write in the exponent effectively

$$kr \simeq \mathbf{k} \cdot \hat{\mathbf{k}}_0 r$$

and consequently, again by comparison with (11.4),

$$\psi(\mathbf{r}, t) = \psi(\mathbf{r} - \mathbf{v}_0 t, 0)e^{i\omega_0 t} + \frac{f_{\mathbf{k}_0}(\hat{\mathbf{r}})}{r} \psi(r\hat{\mathbf{k}}_0 - \mathbf{v}_0 t, 0)e^{i\omega_0 t} \qquad (11.12)$$

Except for the phase factor $e^{i\omega_0 t}$, the first term on the right-hand side represents the initial wave packet displaced without change of shape, as if no scattering had occurred, i.e., as if V were absent from the Hamiltonian. The second term is a *scattered spherical wave packet*, indicating that, if the detector lies outside of the incident wave packet, it will intercept a radially expanding replica of the initial wave packet as "seen" by the scatterer at the origin, but retarded by the interval r/v_0, reduced in amplitude by the factor $1/r$, and modulated by the angular amplitude $f_{\mathbf{k}_0}(\hat{\mathbf{r}})$. Sensibly, the latter is called the *scattering amplitude*.

The assumption, made in deriving (11.12), of a slow variation of the scattering amplitude with $\mathbf{k}$ excludes from this treatment some of the so-called scattering *resonances* which are characterized by a very rapid change of the scattering amplitude with the energy of the incident particle. When a resonance is so sharp that the scattering amplitude changes appreciably even over the narrowest tolerable momentum range Δk, the scattered wave packet may have a shape which is considerably distorted from the incident wave packet, because near a resonance different momentum components scatter very differently. (See Section 11.6 for some discussion of resonances.)

The probability of observing the particle at the detector in the time interval between t and $t + dt$ is

$$v_0 \frac{|f_{\mathbf{k}_0}(\hat{\mathbf{r}})|^2}{r^2} |\psi(r\hat{\mathbf{k}}_0 - \mathbf{v}_0 t, 0)|^2 r^2 \, d\Omega \, dt$$

Hence, the total probability for detecting it is

$$v_0 |f_{\mathbf{k}_0}(\hat{\mathbf{r}})|^2 \, d\Omega \int_{-\infty}^{+\infty} |\psi([r - v_0 t]\hat{\mathbf{k}}_0, 0)|^2 \, dt$$

$$= |f_{\mathbf{k}_0}(\hat{\mathbf{r}})|^2 \, d\Omega \int_{-\infty}^{+\infty} |\psi(\xi\hat{\mathbf{k}}_0, 0)|^2 \, d\xi \qquad (11.13)$$

where the limits of integration may be taken to be $-\infty$ and $+\infty$ with impunity, since ψ describes a wave packet of finite length. On the other hand, the probability that the incident particle will pass through a unit area located perpendicular to the beam in front of the scatterer is

$$\int_{-\infty}^{+\infty} |\psi(\xi\hat{\mathbf{k}}_0, 0)|^2 \, d\xi \tag{11.14}$$

There is no harm in extending this integration from $-\infty$ to $+\infty$, since at $t = 0$ the wave packet is entirely in front of the scatterer.

If the ensemble contains N particles, all represented by the same general type of wave packet, the number $I \, d\Omega$ of particles scattered into the solid angle $d\Omega$ is

$$I \, d\Omega = |f_{\mathbf{k}_0}(\hat{\mathbf{r}})|^2 \, d\Omega \sum_{i=1}^{N} \int_{-\infty}^{+\infty} |\psi_i(\xi\hat{\mathbf{k}}_0, 0)|^2 \, d\xi$$

where

$$I_0 = \sum_{i=1}^{N} \int_{-\infty}^{+\infty} |\psi_i(\xi\hat{\mathbf{k}}_0, 0)|^2 \, d\xi$$

gives the number of particles incident per unit area. Hence, by definition (11.1), we arrive at the fundamental result:

$$\frac{d\sigma}{d\Omega} = |f_{\mathbf{k}_0}(\hat{\mathbf{r}})|^2 \tag{11.15}$$

3. Green's Functions in Scattering Theory. To complete the discussion it is necessary to show that

(a) eigenfunctions of the asymptotic form (11.5) exist, and
(b) the expansion (11.7) is correct.

These two problems are not unconnected, and (b) will find its answer after we have constructed the solutions $\psi_{\mathbf{k}}^{(+)}(\mathbf{r})$. The method of the *Green's function* by which this may be accomplished is much more general than the immediate problem would suggest.

The Schrödinger equation to be solved is

$$\left(-\frac{\hbar^2}{2\mu} \nabla^2 + V \right) \psi = E\psi$$

or

$$(\nabla^2 + k^2)\psi = U\psi \tag{11.16}$$

where $k^2 = 2\mu E/\hbar^2$ and $U = 2\mu V/\hbar^2$. It is useful to replace the differential equation (11.16) by an integral equation. The transformation to an integral equation is performed most efficiently by regarding $U\psi$ on the right-hand side of (11.16) temporarily as a given inhomogeneity, even though it contains the unknown function ψ. Formally, then, a particular "solution" of (11.16) is conveniently constructed in terms of the Green's function $G(\mathbf{r}, \mathbf{r}')$ which is a solution of the equation

$$(\nabla^2 + k^2)G(\mathbf{r}, \mathbf{r}') = -4\pi\delta(\mathbf{r} - \mathbf{r}') \qquad (11.17)$$

Indeed, the expression

$$-\frac{1}{4\pi} \int G(\mathbf{r}, \mathbf{r}')U(\mathbf{r}')\psi(\mathbf{r}') \, d^3r' \qquad (11.18)$$

solves (11.16) by virtue of the properties of the delta function. To this we may add an arbitrary solution of the homogeneous equation

$$(\nabla^2 + k^2)\psi = 0 \qquad (11.19)$$

which is the Schrödinger equation for a free particle (no scattering). Choosing a suitable normalization factor, we thus establish the integral equation

$$\psi_{\mathbf{k}}(\mathbf{r}) = \frac{1}{(2\pi)^{3/2}} e^{i\mathbf{k}\cdot\mathbf{r}} - \frac{1}{4\pi} \int G(\mathbf{r}, \mathbf{r}')U(\mathbf{r}')\psi_{\mathbf{k}}(\mathbf{r}') \, d^3r' \qquad (11.20)$$

for a particular set of solutions of the Schrödinger equation (11.16). $\mathbf{k}$ has a definite magnitude, fixed by the energy eigenvalue, but its direction is undetermined, thus exhibiting an infinite degree of degeneracy, which corresponds physically to the possibility of choosing an arbitrary direction of incidence.

Even if a particular vector $\mathbf{k}$ is selected, (11.20) is by no means completely defined yet: The Green's function could be any solution of (11.17), and there are infinitely many different ones. The choice of a particular $G(\mathbf{r}, \mathbf{r}')$ imposes definite boundary conditions on the eigenfunctions $\psi_{\mathbf{k}}(\mathbf{r})$. Two particularly useful Green's functions are

$$G_{\pm}(\mathbf{r}, \mathbf{r}') = \frac{\exp(\pm ik |\mathbf{r} - \mathbf{r}'|)}{|\mathbf{r} - \mathbf{r}'|} \qquad (11.21)$$

A host of Green's functions of the form

$$G(\mathbf{r}, \mathbf{r}') = G(\mathbf{r} - \mathbf{r}') \qquad (11.22)$$

may be obtained by applying a Fourier transformation to the equation

$$(\nabla^2 + k^2)G(\mathbf{r}) = -4\pi\delta(\mathbf{r}) \qquad (11.23)$$

which is a simplified version of (11.17). If we introduce

$$G(\mathbf{r}) = \int g(\mathbf{k}')e^{i\mathbf{k}'\cdot\mathbf{r}} \, d^3k'$$

and

$$\delta(\mathbf{r}) = \frac{1}{(2\pi)^3} \int e^{i\mathbf{k}'\cdot\mathbf{r}} \, d^3k'$$

we obtain by substitution into (11.23)

$$g(\mathbf{k}') = \frac{1}{2\pi^2} \frac{1}{k'^2 - k^2}$$

hence, the Fourier representation

$$G(\mathbf{r}) = \frac{1}{2\pi^2} \int \frac{e^{i\mathbf{k}'\cdot\mathbf{r}}}{k'^2 - k^2} \, d^3k' \qquad (11.24)$$

Integrating over the angles, we obtain after a little algebra the convenient form

$$G(\mathbf{r}) = -\frac{1}{\pi r} \frac{d}{dr} \int_{-\infty}^{+\infty} \frac{e^{ik'r}}{k'^2 - k^2} \, dk' \qquad (11.25)$$

Since the integrand has simple poles on the real axis in the complex k'-plane at $k' = \pm k$, the integral in (11.25) does not exist—suggesting that our attempt to represent the solutions of (11.17) as Fourier integrals has failed. This approach is nevertheless potent, because the integral in (11.25) can be replaced by another one which does exist, thus

$$G_{+\eta}(r) = -\frac{1}{\pi r} \frac{d}{dr} \int_{-\infty}^{+\infty} \frac{e^{ik'r}}{k'^2 - (k^2 + i\eta)} \, dk' \qquad (11.26)$$

where η is a small positive number. $G_{+\eta}(r)$ exists but is, of course, no longer a solution to (11.23). The trick is to evaluate the expression (11.26) for $\eta \neq 0$ and to let $\eta \to 0$, i.e., $G_{+\eta}(r) \to G_+(r)$, after the integration has been performed.[4]

The integral

$$\int_{-\infty}^{+\infty} \frac{e^{ik'r}}{k'^2 - (k^2 + i\eta)} \, dk'$$

is most easily performed by using the complex k'-plane as an auxiliary device (see Figure 11.3). The poles of the integrand are at

$$k' = \pm\sqrt{k^2 + i\eta} \approx \pm\left(k + \frac{i\eta}{2k}\right)$$

---

[4] Alternatively, we might say that a unique solution of (11.23) does not exist, for if it did G would be the inverse of the operator $-(1/4\pi)(\nabla^2 + k^2)$. But this operator has no inverse, because the equation $(\nabla^2 + k^2)\psi = 0$ does have solutions. However, the inverse of $-(1/4\pi)(\nabla^2 + k^2 + i\eta)$ exists and is the Green's function $G_{+\eta}(r)$. See also Sections 17.2 and 19.3.

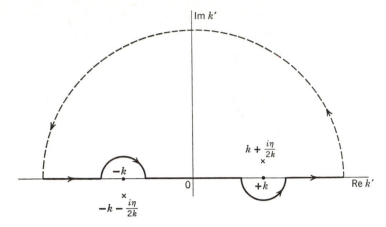

Figure 11.3. Path of integration in the complex k'-plane.

for small η. The path of integration leads along the real axis from $-\infty$ to $+\infty$. A closed contour may be used if we complete the path by a semicircle of very large radius through the upper half plane (because $r > 0$). It encloses the pole in the right half plane. The result of the integration is not altered by introducing detours avoiding the two points $k' = +k$ and $k' = -k$. In fact, if this is done, the limit $\eta \to 0$ can be taken prior to the integration[5] and we may write

$$G_+(r) = \lim_{\eta \to 0} G_{+\eta}(r) = -\frac{1}{\pi r}\frac{d}{dr}\oint \frac{e^{ik'r}}{k'^2 - k^2}\,dk' \qquad (11.27)$$

which by the use of the residue at $k' = +k$ becomes

$$G_+(r) = \frac{e^{ikr}}{r} \qquad (11.28)$$

If we replace η in (11.26) by $-\eta(\eta > 0)$, we obtain a second Green's function

$$G_-(r) = \lim_{\eta \to 0} G_{-\eta}(r) = \frac{e^{-ikr}}{r} \qquad (11.29)$$

in agreement with (11.21). Still other Green's functions can be found by treating differently the singularities which appear in (11.25).

---

[5] If the reader feels uncomfortable about the promiscuous interchanges of integration, differentiation, and limit processes, he may reassure himself by verifying that the result (11.28) is indeed a solution of (11.23).

Exercise 11.2. Show that the Green's function

$$G_1(r) = \frac{\cos kr}{r} \tag{11.30}$$

is obtained if the integral in (11.25) is replaced by its (Cauchy) principal value.[6]

When the special form (11.28) or (11.29) is substituted in (11.20), two distinct eigensolutions, denoted by $\psi^{(+)}$ and $\psi^{(-)}$, result. They satisfy the integral equation

$$\psi_{\mathbf{k}}^{(\pm)}(\mathbf{r}) = \frac{1}{(2\pi)^{3/2}} e^{i\mathbf{k}\cdot\mathbf{r}} - \frac{1}{4\pi} \int \frac{\exp\left(\pm ik\,|\mathbf{r} - \mathbf{r}'|\right)}{|\mathbf{r} - \mathbf{r}'|} U(\mathbf{r}')\psi_{\mathbf{k}}^{(\pm)}(\mathbf{r}')\,d^3r' \tag{11.31}$$

For large r the integrand can be closely approximated in view of the fact that $U \neq 0$ only for values of $r' < a$. In the exponent we expand in powers of $\mathbf{r}'$:

$$k\,|\mathbf{r} - \mathbf{r}'| = k\sqrt{r^2 - 2\mathbf{r}\cdot\mathbf{r}' + r'^2} = kr - k\hat{\mathbf{r}}\cdot\mathbf{r}' + \frac{k(\hat{\mathbf{r}} \times \mathbf{r}')^2}{2r} + \cdots$$

If r is chosen so large that

$$ka^2/r \ll 1$$

then the quadratic term in the exponent can be neglected. If, further, $\mathbf{r}'$ in the denominator of the integrand is neglected, we obtain for large r:

$$\psi_{\mathbf{k}}^{(\pm)}(\mathbf{r}) \sim \frac{1}{(2\pi)^{3/2}} e^{i\mathbf{k}\cdot\mathbf{r}} - \frac{e^{\pm ikr}}{4\pi r} \int e^{\mp i\mathbf{k}'\cdot\mathbf{r}'} U(\mathbf{r}')\psi_{\mathbf{k}}^{(\pm)}(\mathbf{r}')\,d^3r' \tag{11.32}$$

where we have set

$$\mathbf{k}' = k\hat{\mathbf{r}} \tag{11.33}$$

The asymptotic expression (11.32) can be written as

$$\psi_{\mathbf{k}}^{(\pm)}(\mathbf{r}) \sim \frac{1}{(2\pi)^{3/2}} \left(e^{i\mathbf{k}\cdot\mathbf{r}} + \frac{e^{\pm ikr}}{r} f_{\mathbf{k}}^{(\pm)}(\hat{\mathbf{r}}) \right) \qquad (r\text{ large}) \tag{11.34}$$

where

$$f_{\mathbf{k}}^{(\pm)}(\hat{\mathbf{r}}) = -\frac{(2\pi)^{3/2}}{4\pi} \int e^{\mp i\mathbf{k}'\cdot\mathbf{r}'} U(\mathbf{r}')\psi_{\mathbf{k}}^{(\pm)}(\mathbf{r}')\,d^3r' \tag{11.35}$$

Equation (11.34) shows why $\psi^{(+)}$ and $\psi^{(-)}$, when supplemented by

---

[6] The Green's functions (11.28), (11.29), and (11.30) may be identified as *outgoing*, *incoming*, and *standing* waves. To appreciate this terminology we need only multiply them by the time factor $\exp\left(-iEt/\hbar\right)$. The description of scattering in terms of wave packets suggests the designation *retarded* and *advanced* for G_+ and G_-.

exp $(-iEt/\hbar)$, are called the *outgoing* and *incoming* solutions of the Schrödinger equation: they satisfy the appropriate boundary conditions at infinity. The outgoing solution is asymptotically of the form (11.5), thus verifying our previous assumption. It is customary to omit the symbol $(+)$ qualifying the outgoing scattered amplitude and to write $f_{\mathbf{k}}(\hat{\mathbf{r}})$ for $f_{\mathbf{k}}^{(+)}(\hat{\mathbf{r}})$.

To prove assertion (b) made at the beginning of this section we must employ the exact form (11.31) and demonstrate that for the initial wave packet

$$\int \varphi(\mathbf{k}) \exp\left(-i\mathbf{k} \cdot \mathbf{r}_0\right) d^3k \int \frac{\exp\left(ik\,|\mathbf{r} - \mathbf{r}'|\right)}{|\mathbf{r} - \mathbf{r}'|} U(\mathbf{r}')\psi_{\mathbf{k}}^{(+)}(\mathbf{r}')\, d^3r' = 0$$

Since $U = 0$ for $r' > a$, it is sufficient to show that

$$\int \varphi(\mathbf{k}) \exp\left(-i\mathbf{k} \cdot \mathbf{r}_0 + ik\,|\mathbf{r} - \mathbf{r}'|\right)\psi_{\mathbf{k}}^{(+)}(\mathbf{r}')\, d^3k = 0$$

for $r' < a$. It may usually be assumed that in this integral the variation of $\psi_{\mathbf{k}}^{(+)}(\mathbf{r}')$ with $\mathbf{k}$ can be neglected.[7] $\varphi(\mathbf{k})$ is appreciably different from zero only for vectors $\mathbf{k}$ near the direction of $\mathbf{k}_0$. Hence, we may approximate

$$k\,|\mathbf{r} - \mathbf{r}'| \approx \hat{\mathbf{k}}_0 \cdot \mathbf{k}\,|\mathbf{r} - \mathbf{r}'|$$

hence, the left-hand side of the above equation is nearly equal to

$$\psi_{\mathbf{k}_0}^{(+)}(\mathbf{r}')\int \varphi(\mathbf{k}) \exp\left[i\mathbf{k} \cdot (\hat{\mathbf{k}}_0\,|\mathbf{r} - \mathbf{r}'| - \mathbf{r}_0)\right] d^3k = (2\pi)^{3/2}\psi(\hat{\mathbf{k}}_0\,|\mathbf{r} - \mathbf{r}'|, 0)\psi_{\mathbf{k}_0}^{(+)}(\mathbf{r}')$$

The right-hand side of this equation vanishes, because $\hat{\mathbf{k}}_0|\mathbf{r} - \mathbf{r}'|$ points to a position behind the scatterer where the wave packet was assumed to vanish at $t = 0$. Hence, assertion (b) is proved.

4. *The Born Approximation.* Before we proceed let us summarize the results of the preceding sections. If particles with an average momentum $\hbar\mathbf{k}$ are incident upon a scatterer represented by the potential $V(\mathbf{r})$, the differential cross section is given by

$$\frac{d\sigma}{d\Omega} = |f_{\mathbf{k}}(\hat{\mathbf{r}})|^2$$

where the scattering amplitude $f_{\mathbf{k}}(\hat{\mathbf{r}})$ is defined as the coefficient of the outgoing wave in the asymptotic solution

$$\psi_{\mathbf{k}}^{(+)}(\mathbf{r}) \sim \frac{1}{(2\pi)^{3/2}} \left(e^{i\mathbf{k}\cdot\mathbf{r}} + f_{\mathbf{k}}(\hat{\mathbf{r}})\frac{e^{ikr}}{r}\right)$$

---

[7] It can be seen from the integral equation for $\psi_{\mathbf{k}}^{(+)}$ that this is true if $a\,\Delta k \gg 1$ and if we are not at an inordinately narrow resonance for which the scattering amplitude varies extremely rapidly with $\mathbf{k}$. Most physical situations meet these conditions.

of the Schrödinger equation

$$(\nabla^2 + k^2)\psi = \frac{2\mu}{\hbar^2} V \psi$$

The scattering amplitude is given by the formula

$$f_k(\hat{\mathbf{k}}') = - \frac{\sqrt{2\pi\mu}}{\hbar^2} \int e^{-i\mathbf{k}'\cdot\mathbf{r}'} V(\mathbf{r}')\psi_k^{(+)}(\mathbf{r}')\, d\tau' \qquad (11.35a)$$

but, since $\psi^{(+)}$ appears under the integral sign, this is, of course, not an explicit expression. However, it can be used to obtain an approximate value of the scattering amplitude if we replace the exact eigenfunction $\psi^{(+)}$ in the integrand by a plane wave, neglecting the scattered wave. In this approximation we obtain

$$f_k(\hat{\mathbf{k}}') \approx - \frac{\mu}{2\pi\hbar^2} \int e^{-i\mathbf{k}'\cdot\mathbf{r}'} V(\mathbf{r}')\, e^{i\mathbf{k}\cdot\mathbf{r}'}\, d\tau' \qquad (11.36)$$

which is known as the scattering amplitude in the (first) *Born approximation*. Here the scattering amplitude appears proportional to the matrix element of the scattering potential between the two plane waves, $e^{i\mathbf{k}\cdot\mathbf{r}'}$ and $e^{i\mathbf{k}'\cdot\mathbf{r}'}$, representing the free particle before and after the scattering. It should be evident that the Born approximation corresponds to a first iteration of (11.31), where the plane wave is substituted under the integral. The iteration may be continued to obtain higher order Born approximations. (See Section 19.5.)

A reliable estimate of the accuracy of the Born approximation is not easy to obtain, but for a smooth central force potential, $V(r)$, of short range a crude estimate may be given. Since $\psi^{(+)}$ appears in the integrand of (11.35a) evaluated only at those values of the coordinates where $V \neq 0$, i.e., near the origin, we can estimate the scattered wave in (11.31) by setting $r = 0$:

$$- \frac{1}{4\pi} \int \frac{e^{ikr'}}{r'} U(r') \frac{e^{i\mathbf{k}\cdot\mathbf{r}'}}{(2\pi)^{3/2}}\, d\tau' = - \frac{1}{(2\pi)^{3/2}} \int_0^\infty \frac{e^{ikr'}}{r'} U(r') \frac{\sin kr'}{kr'} r'^2\, dr'$$

This term can be neglected in comparison with the plane wave term, if

$$\frac{2\mu}{k\hbar^2} \left| \int_0^\infty e^{ikr'} \sin kr'\, V(r')\, dr' \right| \ll 1 \qquad (11.37)$$

It is seen that the Born approximation is likely to be valid for weak potentials and high incident energies.[8]

---

[8] For a discussion of the convergence of the Born approximation and many other details of scattering theory, see M. L. Goldberger and K. M. Watson, *Collision Theory*, John Wiley and Sons, New York, 1964, and R. G. Newton, *Scattering of Waves and Particles*, McGraw-Hill Book Company, New York, 1966.

As an application of the Born approximation consider the potential

$$V(r) = V_0 \frac{e^{-\alpha r}}{\alpha r} \tag{11.38}$$

where the length $1/\alpha$ may be considered as the range of the potential. In the Born approximation, after a simple integration over angles we find

$$
\begin{aligned}
f_{\mathbf{k}}(\hat{\mathbf{k}}') &\simeq -\frac{2\mu}{\hbar^2} V_0 \int_0^\infty \frac{\sin|\mathbf{k} - \mathbf{k}'| r'}{|\mathbf{k} - \mathbf{k}'| r'} \frac{e^{-\alpha r'}}{\alpha r'} r'^2 \, dr' \\
&= -\frac{2\mu V_0}{\hbar^2 \alpha} \frac{1}{(\mathbf{k} - \mathbf{k}')^2 + \alpha^2} \\
&= -\frac{2\mu V_0}{\hbar^2 \alpha} \frac{1}{4k^2 \sin^2(\theta/2) + \alpha^2} \tag{11.39}
\end{aligned}
$$

if we denote the scattering angle between $\hat{\mathbf{k}}$ and $\hat{\mathbf{k}}'$ by θ, and note that $k' = k$ for elastic scattering. The differential scattering cross section is obtained by taking the square of the amplitude (11.39).

The Coulomb potential between two charges q_1 and q_2 is a limiting case of the potential (11.38) for $\alpha \to 0$ and $V_0 \to 0$ with $V_0/\alpha = q_1 q_2$. Hence, in the Born approximation,

$$\frac{d\sigma}{d\Omega} = \frac{\mu^2 q_1^2 q_2^2}{4p^4 \sin^4(\theta/2)} = \frac{q_1^2 q_2^2}{16E^2 \sin^4(\theta/2)} \tag{11.40}$$

This result, even though obtained by an approximation method, is in exact agreement with both the classical Rutherford cross section and the exact quantum mechanical evaluation of the Coulomb scattering cross section—one of the many coincidences peculiar to the pure Coulomb potential. (See also Section 11.8.)

The validity condition (11.37) applied to the potential (11.38) gives

$$\frac{\mu |V_0|}{k\hbar^2 \alpha} \left| \int_0^\infty (e^{2ikr'} - 1) \frac{e^{-\alpha r'}}{r'} \, dr' \right| \ll 1$$

Evaluating the integral, we obtain

$$\frac{\mu |V_0|}{k\hbar^2 \alpha} \sqrt{\left(\log\sqrt{1 + \frac{4k^2}{\alpha^2}}\right)^2 + \left(\arctan\frac{2k}{\alpha}\right)^2} \ll 1 \tag{11.41}$$

If $k/\alpha \ll 1$, this simplifies to

$$\frac{2\mu |V_0|}{\hbar^2 \alpha^2} \ll 1 \tag{11.42}$$

and thus becomes independent of the velocity. Using the range of the forces

$a = 1/\alpha$, we may write condition (11.42) as

$$|V_0| \, a^2 \ll \frac{\hbar^2}{2\mu} \tag{11.43}$$

The existence of a bound state requires that the wavelength of a particle with kinetic energy V_0 be less than the range of the potential. Broadly speaking, we may therefore say that in the limit of long wavelengths the Born approximation is valid if the potential is too weak to produce a bound state.

If $k/\alpha \gg 1$, condition (11.41) simplifies to

$$\frac{|V_0| \, a}{\hbar v} \log 2ka \ll 1 \tag{11.44}$$

In this limit the velocity enters the condition explicitly.

Generally, the Born approximation affords a rapid estimate of scattering cross sections and is accurate for reasonably high energies in comparison with the interaction energy. Because of its simplicity, it has enjoyed enormous popularity in atomic and nuclear physics. Its usefulness does not, however, vitiate the need for an exact method of calculating scattering cross sections. To this task we must now devote our attention.[9]

5. Partial Waves and Phase Shifts. Let us assume that V is a central potential. It is to be expected that for a spherically symmetric potential the solutions of (11.31), representing an incident and a scattered wave, should exhibit cylindrical symmetry about the direction of incidence. Indeed, we can also see formally from (11.31) that $\psi_{\mathbf{k}}^{(+)}(\mathbf{r})$ depends on k, r, and on the angle between $\mathbf{k}$ and $\mathbf{r}$ only, if V is a function of r alone. Hence, we may without loss of generality assume that $\mathbf{k}$ points in the positive z-direction, and that, for a given value of k, $\psi^{(+)}$ is a function of r and the scattering angle θ only.

Exercise 11.3. Show that for a central potential $\psi_{\mathbf{k}}^{(+)}$ is an eigenfunction of the component of $\mathbf{L}$ in the direction of $\mathbf{k}$, with eigenvalue zero, and discuss the significance of this fact for the scattering of a wave packet.

We must thus look for solutions of the Schrödinger equation

$$-\frac{\hbar^2}{2\mu} \nabla^2 \psi + V(r)\psi = \frac{\hbar^2 k^2}{2\mu} \, \psi$$

[9] Interesting and useful new methods for calculating scattering amplitudes are described in F. Calogero, *Variable Phase Approach to Potential Scattering*, Academic Press, New York, 1967, and R. G. Glauber, *High Energy Collisions* in *Lectures in Theoretical Physics* edited by W. E. Brittin and L. G. Dunham, John Wiley, New York, 1959.

which have the asymptotic form

$$\psi_k^{(+)} \sim \frac{1}{(2\pi)^{3/2}} \left(e^{ikr \cos \theta} + f_k(\theta) \frac{e^{ikr}}{r} \right) \tag{11.45}$$

It is clearly desirable to establish the connection between these solutions and the separable solutions (9.27) of the central force problem,

$$R_{l,k}(r) Y_l^m(\theta, \varphi) = \frac{u_{l,k}(r)}{r} Y_l^m(\theta, \varphi) \tag{11.46}$$

which are common eigenfunctions of H, L^2, and L_z. The radial functions $R_{l,k}(r)$ and $u_{l,k}(r)$ satisfy the differential equations

$$\left[-\frac{1}{r^2} \frac{d}{dr} \left(r^2 \frac{d}{dr} \right) + \frac{l(l+1)}{r^2} + \frac{2\mu}{\hbar^2} V(r) - k^2 \right] R_{l,k}(r) = 0 \tag{11.47}$$

and

$$\left[-\frac{d^2}{dr^2} + \frac{2\mu}{\hbar^2} V(r) + \frac{l(l+1)}{r^2} - k^2 \right] u_{l,k} = 0 \tag{11.48}$$

respectively, as well as a boundary condition at the origin. In general, this boundary condition depends on the shape of V but, as we saw in Section 10.5, in most practical cases it reduces to the requirement that the wave function $R_{l,k}(r)$ be finite at the origin, from which it follows that

$$u_{l,k}(0) = 0 \tag{11.49}$$

We shall restrict ourselves to potentials which are in accord with this boundary condition. The slope $u'_{l,k}(0)$ at the origin is proportional to the normalization constant and may be chosen to be a real number. Then $u_{l,k}(r)$ is a real function.

The radial equation for the external region $r > a$, where the scattering potential vanishes, is identical with equation (10.23) which was solved in the last chapter. The general solution of this equation is a linear combination of the regular and irregular solutions and has the form

$$R_{l,k}(r) = \frac{u_{l,k}(r)}{r} = A_l j_l(kr) + B_l n_l(kr) \qquad r > a \tag{11.50}$$

Using the asymptotic approximations (10.26) and (10.35), we get for large kr

$$R_{l,k}(r) = \frac{u_{l,k}(r)}{r} \sim A_l \frac{\sin (kr - l\pi/2)}{kr} - B_l \frac{\cos (kr - l\pi/2)}{kr} \qquad (kr \text{ large}) \tag{11.51}$$

In the complete absence of a scattering potential ($V = 0$ everywhere) the boundary condition at the origin would exclude the irregular solution, and we would have $B_l = 0$. Hence, the magnitude of B_l compared with A_l is a measure of the intensity of scattering. The value of B_l/A_l must be determined by solving the Schrödinger equation inside the scattering region ($r < a$), subject to the boundary condition (11.49), and by joining the interior solution smoothly onto the exterior solution (11.50) at $r = a$. To do this we must know V explicitly and solve (11.48), by numerical methods, if necessary.

A very useful expression for the cross section can be derived by introducing the ratios B_l/A_l as parameters. Since for a real potential $u_{l,k}(r)$ may be assumed to be real, these parameters are real numbers, and we may set

$$B_l/A_l = -\tan \delta_l \tag{11.52}$$

where δ_l is a real angle which vanishes for all l if $V = 0$ everywhere. The name *scattering phase shift* is thus appropriate for δ_l, particularly if we note that (11.51) can now be written as

$$\frac{u_{l,k}(r)}{r} \sim C_l \frac{\sin (kr - l\pi/2 + \delta_l)}{kr}$$

It is convenient to fix the normalization of the radial functions by setting $C_l = 1$, so that

$$\frac{u_{l,k}(r)}{r} \sim \frac{\sin (kr - l\pi/2 + \delta_l)}{kr} \tag{11.53}$$

δ_l measures the amount by which the phase of the radial function for angular momentum l differs from the no-scattering case ($\delta_l = 0$). Each phase shift is, of course, a function of the energy, or of k.

Exercise 11.4. Show that (11.53) implies the normalization

$$\int_0^\infty u_{l,k}(r)u_{l,k'}(r)\, dr = \frac{\pi}{2k^2} \delta(k - k') \tag{11.54}$$

for the radial eigenfunctions. (Hint: Use the radial Schrödinger equation and integrations by parts.)

In order to express the differential scattering cross section, or the scattering amplitude, through the phase shifts we must expand $\psi_k^{(+)}$ in terms of the separable solutions of the form (11.46), which are assumed to constitute a complete set of orthonormal eigenfunctions. Thus we set

$$\psi_k^{(+)}(r, \theta) = \sum_{l=0}^\infty a_l(k)P_l(\cos \theta) \frac{u_{l,k}(r)}{r} \tag{11.55}$$

where use has been made of the fact that $\psi^{(+)}$ depends only on the angle between **k** and **r** and not on the directions of each of these vectors separately (see Exercise 11.3).

The expansion coefficients $a_l(k)$ can be determined by comparing the two sides of (11.55) for large distances r. We make use of the asymptotic expansion of the plane wave, as derived from (10.26) and (10.43):

$$e^{ikz} = e^{ikr \cos \theta} \sim \sum_{l=0}^{\infty} (2l + 1)i^l \frac{\sin (kr - l\pi/2)}{kr} P_l(\cos \theta) \qquad (11.56)$$

Substituting (11.56) into (11.45), and (11.53) into (11.55), we may compare the coefficients of the terms of form

$$P_l(\cos \theta) \frac{\exp (-ikr + il\pi/2)}{kr}$$

The result is

$$a_l = \frac{(2l + 1)i^l}{(2\pi)^{3/2}} e^{i\delta_l} \qquad (11.57)$$

Hence, asymptotically,

$$\psi_k^{(+)}(r, \theta) \sim \sum_{l=0}^{\infty} (2l + 1)i^l e^{i\delta_l} \frac{\sin (kr - l\pi/2 + \delta_l)}{(2\pi)^{3/2}kr} P_l(\cos \theta) \qquad (11.58)$$

which differs from a plane wave by the presence of δ_l and is called a *distorted plane wave*. Comparing now the coefficients of e^{ikr}/r in (11.58) and (11.45), with expression (11.56) representing the plane wave in the latter equation, we obtain

$$f_k(\theta) = \sum_{l=1}^{\infty} (2l + 1) \frac{e^{2i\delta_l} - 1}{2ik} P_l(\cos \theta)$$

or

$$f_k(\theta) = \frac{1}{k} \sum_{l=0}^{\infty} (2l + 1)e^{i\delta_l(k)} \sin \delta_l(k) P_l(\cos \theta) \qquad (11.59)$$

This important formula gives the scattering amplitude in terms of the phase shifts by making what is known as a *partial wave analysis* of the scattering amplitude. If we remember that each term (partial wave) in the sum (11.59) corresponds to a definite value of angular momentum l, the formula may be seen in a more physical light. If the scattering potential is strongest at the origin and decreases in strength as r increases, then we may expect the low angular momentum components, which classically correspond to small impact parameters and therefore close collisions, to scatter more intensely than the high angular momentum components. More quantitatively, this semiclassical argument suggests that, if the impact parameter

$$\rho = \frac{l\hbar}{p} = \frac{l}{k}$$

exceeds the range a of the potential, or when $l > ka$, no appreciable scattering occurs. Thus, if $ka \gg 1$ to make the classical argument applicable, we expect the phase shifts δ_l for $l > ka$ to be vanishingly small. But this argument is of a more general nature. For, suppose that $ka \ll 1$, as is the case for nuclear scattering at low energies. Then the incident waves are long compared with the scattering region, and at a given instant the phase of the wave changes by only very little across the scattering region. Hence, all spatial sense of direction is lost, and the scattering amplitude must become independent of the angle θ. By (11.59) this implies that all phase shifts vanish, except for $\delta_0(k)$, corresponding to $l = 0$. When this occurs we say that there is only S-wave scattering, and this is isotropic, since $P_0(\cos \theta) = 1$. (See Section 11.7 for a better estimate of δ_l.)

We conclude that the partial wave sum (11.59) is a particularly useful representation of the scattering amplitude, if on physical grounds only a few angular momenta are expected to contribute significantly. In fact if V is not known beforehand, we may attempt to determine the phase shifts as functions of k empirically by comparing scattering data at various energies with the formula

$$\frac{d\sigma}{d\Omega} = |f_k(\theta)|^2 = \frac{1}{k^2} \left| \sum_{l=0}^{\infty} (2l + 1)e^{i\delta_l} \sin \delta_l \, P_l(\cos \theta) \right|^2 \qquad (11.60)$$

By integrating, we obtain the *total scattering cross section:*

$$\sigma = \int \frac{d\sigma}{d\Omega} d\Omega = \frac{4\pi}{k^2} \sum_{l=0}^{\infty} (2l + 1) \sin^2 \delta_l \qquad (11.61)$$

To this sum each angular momentum contributes at most a partial cross section,

$$(\sigma_l)_{\text{max}} = \frac{4\pi}{k^2} (2l + 1) \qquad (11.62)$$

This value is of the same order of magnitude as the maximum classical scattering cross section per unit $\hbar$ of angular momentum. Indeed, if we use the estimate $\rho = l/k$ for the impact parameter, we obtain for the contribution to the total cross section from a range $\Delta l = 1$, or $\Delta \rho = 1/k$,

$$\sigma_l = 2\pi\rho \, \Delta\rho = 2\pi \frac{l}{k^2}$$

For large l this agrees with (11.62) except for a factor 4. The difference is due to the inevitable presence of diffraction effects for which the wave nature of matter is responsible.

6. Determination of the Phase Shifts and Scattering Resonances. The theoretical determination of phase shifts requires that we solve the

Schrödinger equation for the given potential and obtain the asymptotic form of the solutions. If the interior wave function is calculated and joined smoothly onto the exterior solutions (11.50) at $r = a$, the phase shifts can be expressed in terms of the logarithmic derivatives at the boundary $r = a$:

$$\beta_l = \left(\frac{a}{R_l}\frac{dR_l}{dr}\right)_{r=a} \tag{11.63}$$

It is useful to inquire about the energy dependence of β_l. If equation (4.6), which is an immediate consequence of the Schrödinger equation, is applied to two separable solutions of the form $R_{l,k_1}(r)Y_l^{m_1}(\theta, \varphi)$ and $R_{l,k_2}(r)Y_l^{m_2}(\theta, \varphi)$ with energies E_1 and E_2, but both corresponding to the same angular momentum, a simple relation between the radial wave functions can be deduced:[10]

$$(E_2 - E_1)R_{l,k_1}(r)R_{l,k_2}(r)r^2 +$$

$$\frac{\hbar^2}{2\mu}\frac{d}{dr}\left\{r^2\left[R_{l,k_1}(r)\frac{dR_{l,k_2}(r)}{dr} - \frac{dR_{l,k_1}(r)}{dr}R_{l,k_2}(r)\right]\right\} = 0$$

If this equation is integrated from the origin to a and then divided by the product $R_{l,k_1}(r)R_{l,k_2}(r)$, we find that two nonsingular values of the logarithmic derivative are related by the formula

$$\beta_l(E_2) - \beta_l(E_1) = -\frac{2\mu(E_2 - E_1)\int_0^a R_{l,k_1}R_{l,k_2}r^2\,dr}{\hbar^2 a R_{l,k_1}(a)R_{l,k_2}(a)} \tag{11.64}$$

Hence, the logarithmic derivative $\beta_l(E)$ is a monotonically decreasing function of the energy.

Using (11.50) and (11.52), we find easily the desired relation between the phase shift and the logarithmic derivative at $r = a$:

$$\beta_l(k) = ka\frac{j_l'(ka)\cos\delta_l - n_l'(ka)\sin\delta_l}{j_l(ka)\cos\delta_l - n_l(ka)\sin\delta_l} \tag{11.65}$$

We may solve this relation and obtain

$$e^{2i\delta_l} = -\frac{j_l - in_l}{j_l + in_l}\left(1 + ka\frac{\dfrac{j_l' + in_l'}{j_l + in_l} - \dfrac{j_l' - in_l'}{j_l - in_l}}{\beta_l - ka\dfrac{j_l' + in_l'}{j_l + in_l}}\right) \tag{11.66}$$

[10] It is convenient to use the representation (9.31) of the ∇ operator in reducing (4.6) to the radial equation given here. Alternatively, this equation can be derived directly from equation (11.47).

where all the spherical cylinder functions and their derivatives are to be evaluated at the argument ka.

From (11.53) it can be seen that

$$S_l(k) = e^{2i\delta_l} \tag{11.67}$$

is the ratio of the coefficients multiplying the outgoing wave, $\sim e^{i(kr - l\pi/2)}$, and the incoming wave, $\sim e^{-i(kr - l\pi/2)}$. In Section 19.7 the quantities S_l will be identified as the eigenvalues of the S matrix.

We define a real phase angle ξ_l by

$$e^{2i\xi_l} = - \frac{j_l - in_l}{j_l + in_l} \tag{11.68}$$

and introduce the real parameters Δ_l and s_l:

$$ka \frac{j_l' + in_l'}{j_l + in_l} = \Delta_l + is_l \tag{11.69}$$

From Section 10.2 it is easy to verify that

$$j_l(z)n_l'(z) - j_l'(z)n_l(z) = 1/z^2$$

Hence, the quantity s_l is positive definite and equal to

$$s_l = \frac{1}{ka[(j_l)^2 + (n_l)^2]} \tag{11.70}$$

With these definitions we obtain the simple form

$$e^{2i(\delta_l - \xi_l)} = \frac{\beta_l - \Delta_l + is_l}{\beta_l - \Delta_l - is_l} \tag{11.71}$$

Exercise 11.5. Show that

$$e^{i\delta_l} \sin \delta_l = e^{2i\xi_l} \left(\frac{s_l}{\beta_l - \Delta_l - is_l} + e^{-i\xi_l} \sin \xi_l \right) \tag{11.72}$$

The quantities ξ_l have an interesting physical significance, since they are the phase shifts if $\beta_l \to \infty$. An infinite logarithmic derivative at $r = a$ occurs for all values l if the potential represents a hard spherical core of radius a, so that the wave function cannot penetrate into the interior at all. The angles ξ_l are therefore often referred to as the *hard sphere phase shifts*.

Let us apply these results now to the special case of the square well of range a and depth V_0. From (10.45)

$$R_l(r) = \frac{u_l(r)}{r} = j_l(k'r)$$

where $\hbar k' = \sqrt{2\mu(E + V_0)}$. Hence,

$$\beta_l = k'a \frac{j_l'(k'a)}{j_l(k'a)}$$

For $l = 0$ we find from (10.36) and (10.37) the simple expressions $\xi_0 = -ka$, $\Delta_0 = -1$, $s_0 = ka$, and $\beta_0 = k'a \cot k'a - 1$ which is indeed a monotonically decreasing function of energy. Using (11.71) and (11.72), we may calculate the S-wave phase shift and the corresponding scattering amplitude. The latter turns out to be

$$f_0 = \frac{1}{k} e^{i\delta_0} \sin \delta_0 = \frac{e^{-2ika}}{k} \left(\frac{k}{k' \cot k'a - ik} - e^{ika} \sin ka \right)$$

In the limit $E \to 0$, $k \to 0$, this gives a nonvanishing isotropic S-wave cross section,

$$\sigma_0 \to 4\pi a^2 \left(\frac{\tan k_0'a}{k_0'a} - 1 \right)^2, \qquad k_0' = \sqrt{\frac{2\mu V_0}{\hbar^2}}$$

Exercise 11.6. Work out ξ_1, Δ_1, s_1, β_1, and f_1 for the square well and examine their dependence on the energy.

Figures 11.4 and 11.5, calculated for a particular square well, illustrate some common important features of scattering cross sections. The phase shifts δ_0, δ_1, and δ_2, which are determined only to within a multiple of π, were normalized so as to go to 0 as $E \to \infty$, when the particle is effectively free. At low energies P-waves (and waves of higher angular momentum) are scattered less than S-waves, because the presence of the centrifugal potential makes it improbable for a particle to be found near the center of force. Generally, the partial cross sections tend to decrease with increasing angular momentum and increasing energy, but the figures show also that the smooth variation of the phase shifts and cross sections is interrupted by a violent change in one of the phase shifts and a corresponding pronounced fluctuation in the partial cross section. Thus, for the particular values of the parameters on which Figures 11.4 and 11.5 are based, the P-wave phase shift rises sharply near $ka = 0.7$, and since it passes through the value $3\pi/2$ in this energy range, $\sin^2 \delta_1$ becomes equal to unity, and the partial cross section σ_1 goes through its maximum value, $12\pi/k^2$.

If the quantities ξ_l, Δ_l, and s_l, which characterize the external wave function, vary slowly and smoothly with energy, such rapid changes of a phase shift are usually to be attributed to a strong energy dependence of the logarithmic derivative β_l. It may happen that the rapid change of β_l in a

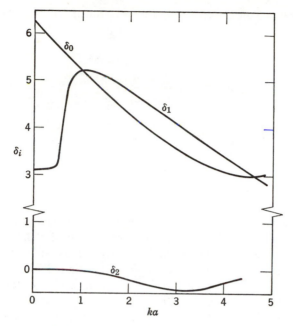

Figure 11.4. *S*-, *P*-, and *D*-wave phase shifts (δ_0, δ_1, δ_2) for scattering from a square well of radius a with $k_0'a = \sqrt{2\mu V_0/\hbar^2}\, a = 6.2$.

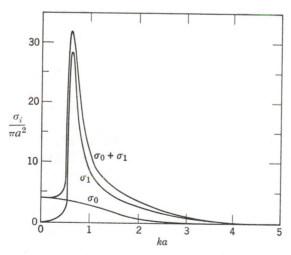

Figure 11.5. Momentum dependence of the partial cross sections (σ_0 and σ_1) for *S*- and *P*-waves corresponding to the phase shifts of Figure 11.4. The cross sections are given in units of πa^2.

small energy range can be represented by a linear approximation,

$$\beta_l(E) = c + bE \tag{11.73}$$

If this approximation is substituted in (11.71), we get

$$e^{2i(\delta_l - \xi_l)} = \frac{E - E_0 - i(\Gamma/2)}{E - E_0 + i(\Gamma/2)} \tag{11.74}$$

where

$$E_0 = \frac{\Delta_l - c}{b} \quad \text{and} \quad \frac{\Gamma}{2} = -\frac{s_l}{b} \tag{11.75}$$

Hence,

$$\tan(\delta_l - \xi_l) = \frac{\Gamma}{2(E_0 - E)} \tag{11.76}$$

Since β_l is a decreasing function of energy, b must be negative. By its definition, s_l is positive, and it thus follows that the quantity Γ defined in (11.75) is positive. The expressions (11.74) and (11.76) are useful if E_0 and Γ are reasonably constant and if the linear approximation (11.73) is accurate over an energy range large compared with Γ. Under these circumstances it can be seen from (11.74) that the phase, $2(\delta_l - \xi_l)$, changes by 2π over this energy range. Hence, if ξ_l is also nearly constant in this interval, the phase shift δ_l changes by π, and the partial cross section σ_l, which is proportional to $\sin^2 \delta_l$, changes abruptly. Such sudden variations in the phase shifts are called *resonances* with E_0 being the *resonant energy* and Γ the *width* of the resonance.[11]

If the phase shift δ_l is near resonance, the contribution of the corresponding partial wave to the scattering amplitude can be written according to (11.72) as

$$f_l = \frac{2l + 1}{k} e^{2i\xi_l} \left(\frac{\Gamma/2}{E_0 - E - i(\Gamma/2)} + e^{-i\xi_l} \sin \xi_l \right) P_l(\cos \theta) \tag{11.77}$$

giving a neat separation of the resonant part of the amplitude from the nonresonant part, which depends only on ξ_l. If, as is the case for low energies or high angular momenta, the hard sphere phase shifts are negligible, the resonant term predominates, and the resonant partial wave contributes an amount

$$\sigma_l = \frac{4\pi}{k^2} (2l + 1) \sin^2 \delta_l = \frac{4\pi(2l + 1)}{k^2} \frac{\Gamma^2}{4(E - E_0)^2 + \Gamma^2} \tag{11.78}$$

to the total cross section. For small width Γ this represents a sharp maximum

[11] If the resonance, although narrow, is wide enough so that wave packets with $\Delta E \ll \Gamma$ can be constructed, the cross-section formulas (11.15) and (11.60) are still valid (see Footnote 7). For wave packets with $\Delta E \gg \Gamma$ more care is required. The fate of such a wave packet is discussed at the end of this section.

centered at E_0 with a shape similar to that of the transmission resonance peaks in Figure 6.10.[12]

Since we have discussed phase shifts and resonances in this section entirely in terms of the logarithmic derivatives at the boundary of an interior region within which the scattering forces are concentrated, it should be clear that our conclusions are independent of the particular mechanism which operates inside this region. From this point of view the interior region is a "black box," characterized only by the values of the logarithmic derivatives at the surface. In particular, there is no need for the phase shifts to result from an interaction with a simple, fixed potential V. Rather, the scattering may arise from much more complicated interactions within the black box, and resonances may occur as the result of constructive interference at certain discrete frequences, if some mechanism impedes the escape of the wave from the box. In Sections 6.8 and 7.4, where transmission resonances were considered, we saw how an abrupt change in the potential or a high potential barrier can produce narrow resonances, but in nuclear reactions resonances come about because when, for example, a slow neutron enters a complex nucleus it interacts in a complicated way with all the nucleons and its reemission is thereby considerably delayed. Owing to the wave properties of matter such metastable states can occur only at certain discrete energies, and these are the resonant energies. At low energies the neutron scattering cross section thus exhibits sharp maxima which may be regarded as the remnants of the discrete level structure which exists for $E < 0$.

The width Γ of a resonance is a direct measure of the stability of the resonant state or of the delay between the time of absorption of the particle by the black box and its re-emission. To see this we note from (11.53), (11.55), and (11.57) that the contribution from angular momentum l to the eigenfunction $\psi^{(+)}$ can be written as

$$\frac{(2l+1)i^l}{(2\pi)^{3/2}} e^{i\delta_l}[\cos\delta_l \, j_l(kr) + \sin\delta_l \, n_l(kr)]$$

for $r \geqslant a$. Assuming that we have an l-wave resonance at low energy so that $\xi_l \ll 1$, $\cos\delta_l j_l(kr)$ is negligible near the resonance for $r = a$, but the second term is large and strongly dependent on the energy. Neglecting all non-resonant partial waves, we see from the condition of matching the internal and external wave functions at $r = a$ that the eigenfunction for $r \leqslant a$ must have the form

$$\psi^{(+)} \simeq e^{i\delta_l}\sin\delta_l \, g_k(r) \simeq \frac{\Gamma/2}{E_0 - E - i(\Gamma/2)} \, g_k(r)$$

[12] The *P*-wave resonance in Figure 11.5 is not accurately described by the simple formula (11.78), because the conditions under which this formula was derived ($\xi_1 \ll 1$, Δ_1 and s_1 reasonably constant) are not well satisfied.

where $g_k(r)$ is a slowly varying function of the energy. As in Section 11.2, we now construct by superposition of eigenfunctions $\psi^{(+)}$ a wave packet incident upon the black box. In order to define with good precision the time at which the particle enters the internal region it is necessary to construct a sharply pulsed packet involving a broad range, ΔE, of energies, and we therefore assume that $\Delta E \gg \Gamma$. Choosing $t = 0$ as the time at which the incident wave packet passes the scattering region, we may write the internal wave function as

$$\psi(0) = \int_{\Delta E} C(E) \frac{\Gamma/2}{E_0 - E - i(\Gamma/2)} g_k(r) \, dE \qquad (r < a)$$

where $C(E)$ is a slowly varying function of the energy. At time t this internal wave function becomes

$$\psi(t) = C(E_0)g_{k_0}(r) \int_{\Delta E} \frac{\Gamma/2}{E_0 - E - i(\Gamma/2)} \exp\left(-\frac{i}{\hbar} Et\right) dE \qquad (r < a)$$

the assumption being that ΔE is broad compared to the resonance at E_0 but small compared with the distance between successive resonances. Hence, we may approximate the integral to get

$$\psi(t) \simeq C(E_0)g_{k_0}(r) \frac{\hbar\Gamma}{2} \int_{-\infty}^{+\infty} \frac{e^{-i\omega t}}{[E_0 - (i\Gamma/2)] - \hbar\omega} \, d\omega$$

and, as in (7.54),

$$\psi(t) \simeq \begin{cases} C(E_0)g_{k_0}(r)\Gamma\pi i \exp\left(-\frac{\Gamma}{2\hbar} t\right) \exp\left(-\frac{i}{\hbar} E_0 t\right) & t > 0 \\ 0 & t < 0 \end{cases} \qquad (11.79)$$

Thus, the probability of finding the particle inside the black box decreases exponentially with time, and the mean lifetime, $\tau = \hbar/\Gamma$, is inversely proportional to the width of the resonance. Hence, a narrow resonance corresponds to a situation in which the incident particle spends a long time in the region of interaction.

Exercise 11.7. Using Figure 11.5, estimate the mean lifetime of the metastable state responsible for the *P*-wave resonance in units of the period associated with the motion of the particle in the square well.

For $t > 0$, the *decaying* state (11.79) may be regarded as a solution of the Schrödinger equation corresponding to the complex eigenvalue $E_0 - i\Gamma/2$, but the wave function must of course violate the asymptotic boundary conditions for a *stationary* state. If the values of E and k are continued into the complex plane, equation (11.74) shows that $E_0 - i\Gamma/2$ is a pole of

$S_l(k) = e^{2i\delta_l}$ and that the state with this complex E-value has no incoming wave. Since

$$\hbar k = \sqrt{2\mu(E_0 - i\Gamma/2)} \simeq \sqrt{2\mu E_0} - i\sqrt{\mu/8E_0}\,\Gamma$$

the outgoing wave increases as $\exp(\sqrt{\mu/8E_0\hbar^2}\,\Gamma r)$ for large r but $r < \sqrt{2E_0/\mu}\,t$.

7. Phase Shifts and Green's Functions. Although the relation between the phase shifts and the logarithmic derivatives of the radial wave functions is a very general and simple one, it does not shed any direct light on the dependence of δ_l on the scattering potential. This connection can be elucidated if a partial wave analysis is applied to the integral equation (11.31). To this end the outgoing Green's function must first be expanded in terms of Legendre polynomials. The result is the formula

$$G_+(\mathbf{r}, \mathbf{r}') = \frac{\exp(ik\,|\mathbf{r} - \mathbf{r}'|)}{|\mathbf{r} - \mathbf{r}'|} = ik\sum_{l=0}^{\infty}(2l+1)P_l(\hat{\mathbf{r}}\cdot\hat{\mathbf{r}}')j_l(kr')h_l^{(1)}(kr) \quad (11.80)$$

which is valid if $r > r'$.

Proof. Since $G_+(\mathbf{r}, \mathbf{r}')$ is a solution of the equations

$$(\nabla^2 + k^2)G_+ = 0 \quad \text{and} \quad (\nabla'^2 + k^2)G_+ = 0$$

if $r \neq r'$, it is seen from the separation of these equations in spherical coordinates that the partial wave expansion for $r > r'$ must be of the form

$$G_+(\mathbf{r}, \mathbf{r}') = \sum_{l=0}^{\infty} q_l(k)P_l(\hat{\mathbf{r}}\cdot\hat{\mathbf{r}}')j_l(kr')h_l^{(1)}(kr) \quad (11.81)$$

where the particular choice of the spherical cylinder functions is dictated by the regular behavior of G_+ at $r' = 0$ and its asymptotic behavior, $G_+ \to e^{ikr}/r$, as $r \to \infty$. The remaining unknown coefficients, $q_l(k)$, in the series (11.81) can be determined by letting $r \approx r' \approx 0$. By using the first approximations (10.22) and (10.29) for the Bessel and Hankel functions, (11.81) simplifies to

$$\frac{1}{|\mathbf{r} - \mathbf{r}'|} = \sum_{l=0}^{\infty}\frac{q_l(k)}{(2l+1)ik}\frac{r'^l}{r^{l+1}}P_l(\hat{\mathbf{r}}\cdot\hat{\mathbf{r}}')$$

This has to be compared with the expansion

$$\frac{1}{|\mathbf{r} - \mathbf{r}'|} = \sum_{l=0}^{\infty}\frac{r'^l}{r^{l+1}}P_l(\hat{\mathbf{r}}\cdot\hat{\mathbf{r}}')$$

which comes from the generating function (9.58) for Legendre polynomials. Comparison of the last two equations yields the values of $q_l(k)$ and (11.80).

We now substitute (11.80) and the partial wave expansion of $\psi_k^{(+)}$ from Section 11.5 into the integral equation (11.31) and carry out the integration over the direction of $\mathbf{r}'$. For $r > a$ we obtain the radial integral equations,

$$e^{i\delta_l} \frac{u_{l,k}(r)}{r} = j_l(kr) - ike^{i\delta_l} \int_0^r j_l(kr')h_l^{(1)}(kr)u_{l,k}(r')U(r')r' \, dr' \quad (11.82)$$

Letting $r \to \infty$ and replacing $u_{l,k}(r)$ and the cylinder functions of argument kr by their asymptotic expansions, we finally arrive at the simple formula

$$\sin \delta_l = -k \int_0^\infty j_l(kr')U(r')u_{l,k}(r')r' \, dr' \quad (11.83)$$

This is an explicit expression for the phase shifts in terms of the potential and the radial eigenfunctions.

Exercise 11.8. Show that for *all* values of r the radial wave function $u_{l,k}(r)$ satisfies the integral equation

$$u_{l,k}(r) = r \cos \delta_l \, j_l(kr) + kr \int_0^r j_l(kr')n_l(kr)U(r')u_{l,k}(r')r' \, dr'$$

$$+ kr \int_r^\infty j_l(kr)n_l(kr')U(r')u_{l,k}(r')r' \, dr' \quad (11.84)$$

Exercise 11.9. Verify (11.83) by applying a partial wave analysis directly to the scattering amplitude (11.35a).

Some useful estimates of phase shifts may be based on (11.83) and (11.84). For instance, these coupled equations may be solved by successive approximations in an iteration procedure. The zeroth approximation to the wave function is

$$u_{l,k}^{(0)}(r) = r \cos \delta_l \, j_l(kr)$$

When this wave function is substituted in (11.83) we get the approximate phase shift

$$\tan \delta_l \simeq -k \int_0^\infty [j_l(kr')]^2 U(r')r'^2 \, dr' \quad (11.85)$$

Higher approximations may be obtained by iteration, but this is usually cumbersome.

Exercise 11.10. Show that a partial wave analysis of the scattering amplitude in the first Born approximation, (11.36), gives the same estimate as (11.85) if $\delta_l \ll 1$.[13]

For values of $l > ka$ the spherical Bessel functions in the integral (11.85) may be approximated by the first term in their series expansion [see (10.22)]. We thus obtain

$$\tan \delta_l \simeq - \frac{2^{2l}(l!)^2}{[(2l+1)!]^2} k^{2l+1} \int_0^\infty U(r')r'^{2l+2}\, dr' \qquad (11.86)$$

as an estimate for those phase shifts for which $\tan \delta_l$ is small. From this formula we deduce the rule of thumb that for low energies and high angular momenta $\tan \delta_l \propto k^{2l+1}$.

8. Scattering in a Coulomb Field. Since a Coulomb field, $V = c/r$, has an infinite range, many of the results which we have derived in this chapter under the assumption of a potential with a finite range are not immediately applicable without a separate investigation. For example, the concept of a total scattering cross section is surely meaningless for such a potential, because every incident particle, no matter how large its impact parameter, is scattered, hence, the total scattering cross section is infinite.

Exercise 11.11. Show that the total cross section obtained from the differential cross section (11.40) diverges.

Owing to its great importance in atomic and nuclear physics, a vast amount of work has been done on the continuum ($E > 0$) eigenstates of the Coulomb potential.[14] In this section a particular method of solving this problem will be presented because of its intrinsic interest. This method depends on the observation that the Schrödinger equation with a Coulomb potential is separable in *parabolic coordinates* which are defined by the equations

$$\xi = r + z, \qquad \eta = r - z, \qquad \varphi = \arctan \frac{y}{x} \qquad (11.87)$$

---

[13] Since the Born approximation agrees with (11.85) only if *all* phase shifts are $\ll 1$, it is possible to improve substantially upon the Born approximation by using (11.85) for those values of l (usually the higher angular momenta) for which $\delta_l \ll 1$ and determine the remaining few δ_l exactly by solving the radial equations.

[14] See H. A. Bethe and E. E. Salpeter, *Quantum Mechanics of One- and Two-Electron Systems* in *Encyclopedia of Physics*, Vol. 35, pp. 88–436, Springer-Verlag, Berlin, 1957.

The Schrödinger equation in these coordinates becomes

$$-\frac{\hbar^2}{2\mu}\left[\frac{4}{\xi+\eta}\frac{\partial}{\partial\xi}\left(\xi\frac{\partial}{\partial\xi}\right)+\frac{4}{\xi+\eta}\frac{\partial}{\partial\eta}\left(\eta\frac{\partial}{\partial\eta}\right)+\frac{1}{\xi\eta}\frac{\partial^2}{\partial\varphi^2}\right]\psi$$
$$+\frac{2q_1q_2}{(\xi+\eta)}\psi = E\psi \quad (11.88)$$

where the Coulomb potential energy has been assumed in the form

$$V(r) = \frac{q_1q_2}{r} \quad (11.89)$$

describing the interaction between two charges q_1 and q_2.

For scattering we are interested in a solution with axial symmetry about the direction of incidence which we choose to be the z-axis. Hence, we must look for solutions which are independent of φ and have the form

$$\psi = f_1(\xi)f_2(\eta) \quad (11.90)$$

Separation of the variables ξ and η is achieved easily and the Schrödinger equation replaced by the two ordinary differential equations,

$$\frac{d}{d\xi}\left(\xi\frac{df_1}{d\xi}\right)+\frac{k^2}{4}\xi f_1-c_1 f_1 = 0 \quad (11.91a)$$

$$\frac{d}{d\eta}\left(\eta\frac{df_2}{d\eta}\right)+\frac{k^2}{4}\eta f_2-c_2 f_2 = 0 \quad (11.91b)$$

where $\hbar^2 k^2 = 2\mu E$ and the constants of separation, c_1 and c_2, are related by the condition

$$c_1 + c_2 = \frac{q_1q_2\mu}{\hbar^2} \quad (11.92)$$

Scattering is described by a wave function which asymptotically has an incident plane wave part,

$$\exp(ikz) = \exp\left[i\frac{k}{2}(\xi-\eta)\right] \quad (11.93)$$

and a radially outgoing scattered part,

$$\exp(ikr) = \exp\left[i\frac{k}{2}(\xi+\eta)\right] \quad (11.94)$$

The behavior of these two portions of the wave function suggests that we look for a particular solution with

$$f_1(\xi) = \exp(ik\xi/2) \quad (11.95)$$

Substituting this into (11.91a), we see that $\exp(ik\xi/2)$ is indeed a solution if we choose

$$c_1 = i\frac{k}{2}$$

The remaining equation for f_2,

$$\frac{d}{d\eta}\left(\eta\frac{df_2}{d\eta}\right) - \left(\frac{q_1q_2\mu}{\hbar^2} - i\frac{k}{2}\right)f_2 + \frac{k^2}{4}\eta f_2 = 0$$

is conveniently transformed if we set

$$f_2(\eta) = \exp(-ik\eta/2)g(\eta) \tag{11.96}$$

Using also the abbreviation[15]

$$n' = -\frac{q_1q_2\mu}{\hbar^2 k} = -\frac{q_1q_2}{\hbar v} \tag{11.97}$$

we obtain the equation

$$\eta\frac{d^2g}{d\eta^2} + (1 - ik\eta)\frac{dg}{d\eta} + n'kg = 0 \tag{11.98}$$

which should be compared with (10.77):

$$z\frac{d^2w}{dz^2} + (c - z)\frac{dw}{dz} - aw = 0$$

We thus see that the solution of (11.98), which is regular at the origin, is the confluent hypergeometric function

$$g(\eta) = {}_1F_1(in'; 1; ik\eta) \tag{11.99}$$

For purely positive imaginary argument z, the asymptotic expansion of this function has the leading terms[16]

$${}_1F_1(a; c; z) \sim \frac{\Gamma(c)}{\Gamma(c - a)}|z|^{-a}\exp\left(ia\frac{\pi}{2}\right)$$

$$+ \frac{\Gamma(c)}{\Gamma(a)}|z|^{a-c}\exp\left[i(a - c)\frac{\pi}{2}\right]e^z \qquad (|z| \text{ large}) \tag{11.100}$$

---

[15] n' defined for $E > 0$ is the analog of the principal quantum number n in terms of which the discrete energy eigenvalues ($E < 0$) were expressed.

[16] See P. M. Morse and H. Feshbach, *Methods of Theoretical Physics*, McGraw-Hill Book Company, New York, 1953, equation (5.3.51).

Applying this approximation to $g(\eta)$, we obtain for large values of η

$$g(\eta) \sim \frac{\exp\left(-n'\frac{\pi}{2}\right)}{|\Gamma(1 + in')|} \left\{ \exp[-i(n'\log k\eta - \sigma_{n'})] \right.$$

$$\left. + \frac{n'}{k\eta} \exp[i(k\eta + n'\log k\eta - \sigma_{n'})] \right\} \quad (11.101)$$

where $\Gamma(1 + in') = |\Gamma(1 + in')|e^{i\sigma_{n'}}$.

Collecting all these results together, we conclude that the particular solution

$$\psi(\xi, \eta, \varphi) = \exp\left[i\frac{k}{2}(\xi - \eta)\right]{}_1F_1(in'; 1; ik\eta) \quad (11.102)$$

of the Schrödinger equation has the asymptotic form

$$\psi \sim \frac{\exp\left(-n'\frac{\pi}{2}\right)}{\Gamma|(1 + in')|} \left\{ \exp\{i[kz - n'\log k(r - z) + \sigma_{n'}]\} \right.$$

$$\left. + \frac{n'}{k(r - z)} \exp\{i[kr + n'\log k(r - z) - \sigma_{n'}]\} \right\} \quad (11.103)$$

for large values of $r - z$, i.e., at large distances from the scattering center, except in the forward direction. Equation (11.103) shows clearly why it is impossible to obtain for a Coulomb potential an asymptotic eigenfunction which has the simple form

$$\psi \sim e^{ikz} + f_k(\theta)\frac{e^{ikr}}{r}$$

deduced for a potential of finite range. The Coulomb force is effective even at very great distances and prevents the incident wave from ever approaching a plane wave; similarly, the scattered wave fails to approach the simple free particle form, e^{ikr}/r, as $r \to \infty$. However, it is important to note that the modifications, for which the Coulomb potential is responsible at large distances from the scatterer, affect only the *phases* of the incident and scattered waves. Since the asymptotic contributions from the Coulomb potential to the phases are logarithmic functions of r, hence vary but slowly, the analysis of the fate of an incident wave packet in a scattering process can still be carried through, and the result is precisely the same as before. The differential cross section is [see (11.15)]

$$\frac{d\sigma}{d\Omega} = |f_k(\theta)|^2$$

where the scattering amplitude $f_k(\theta)$ is the factor multiplying the radial part of the asymptotically outgoing wave, provided the incident (modified) plane wave is normalized so that the probability density is unity. Hence, for all finite values of the scattering angle θ,

$$f_k(\theta) = \frac{n'}{k(1 - \cos\theta)} \exp\left[in' \log(1 - \cos\theta)\right] \tag{11.104}$$

and the differential scattering cross section is

$$\frac{d\sigma}{d\Omega} = \frac{n'^2}{4k^2 \sin^4(\theta/2)} = \frac{q_1^2 q_2^2}{16E^2} \frac{1}{\sin^4(\theta/2)} \tag{11.105}$$

in exact agreement with the classical Rutherford scattering cross section and the Born approximation (11.40). An angle-independent addition to the phase has been omitted in (11.104).

Since $_1F_1(a; c; 0) = 1$, the normalization of the eigenfunction (11.102) is such that $\psi = 1$ at the scattering center. On the other hand, the incident wave in the large brace of (11.103) is normalized to probability density unity. Hence, the relative probability of finding the particle at the origin to the probability of finding it in the incident beam is determined by the coefficient in front of (11.103), and we have

$$\frac{|\psi(0)|^2}{|\psi(\infty)|^2} = |\Gamma(1 + in')|^2 e^{n'\pi} = \frac{2\pi n'}{1 - e^{-2\pi n'}} \tag{11.106}$$

where in obtaining the last expression use was made of (5.56) and the definition of the Γ function: $\Gamma(1 + z) = z\Gamma(z)$.

If the Coulomb potential is repulsive ($q_1 q_2 > 0$ and $n' < 0$) and strong compared with the kinetic energy of the particle, we have

$$\frac{|\psi(0)|^2}{|\psi(\infty)|^2} = -2\pi n' e^{2\pi n'} \qquad (n' \ll 0)$$

The significance of the Gamow factor, $\exp(2\pi n') = \exp(-2\pi q_1 q_2/\hbar v)$, was already discussed in connection with (7.35) for a one-dimensional model of Coulomb scattering.

Exercise 11.12. Discuss other limiting cases (fast particles, attractive potentials) for (11.106).

Exercise 11.13. Calculate the wave function (11.102) in the forward direction. What physical conclusions can you draw from its form?

Since the Schrödinger equation is separable in parabolic coordinates only if the potential behaves strictly as $1/r$ at all distances, the method of this section is not appropriate if the potential is Coulombic only at large distances but, as in the case of nuclear interactions, has a different radial dependence near the origin. It is then preferable to use spherical polar coordinates and attempt a phase shift analysis. By expanding the eigenfunction (11.102) in terms of Legendre polynomials, phase shifts can be calculated for the Coulomb potential, and the theory of Section 11.5 can be extended to include the presence of a long-range $1/r$ potential.

Problems

1. Using the first three partial waves, compute and display on a polar graph the differential cross section for an impenetrable hard sphere when the de Broglie wavelength of the incident particle equals the circumference of the sphere. Evaluate the total cross section and estimate the accuracy of the result. Also discuss what happens if the wavelength becomes very large compared with the size of the sphere.

2. Calculate the total scattering cross section for the *screened Coulomb potential* (11.38) in the Born approximation and discuss the accuracy of this result.

3. Apply the Born approximation to the scattering from a square well. Evaluate and plot the differential and total scattering cross sections.

4. Obtain the differential scattering cross section in the Born approximation for the potential of Problem 4 in Chapter 10.

5. If $V = C/r^n$, obtain the functional dependence of the Born scattering amplitude on the scattering angle. Discuss the reasonableness of the result qualitatively. What values of n give a meaningful answer?

6. If the scattering potential has the translation invariance property $V(\mathbf{r} + \mathbf{R}) = V(\mathbf{r})$, where $\mathbf{R}$ is a constant vector, show that in Born approximation scattering occurs only in the directions defined by the condition

$$(\mathbf{k} - \mathbf{k}') \cdot \mathbf{R} = 2\pi n$$

where n is an integer.

CHAPTER 12

The Spin

1. *The Intrinsic Angular Momentum.* Up to this point we have been concerned with the quantum description of a particle as a mass point without internal structure. It has been assumed that the state of the particle can be completely specified by giving the wave function ψ as a function of the spatial coordinates x, y, z. In the language of Chapter 8, these three dynamical variables were postulated to constitute a complete set. Alternatively, and equivalently, the momentum components p_x, p_y, p_z also form a complete set of dynamical variables, since $\varphi(p_x, p_y, p_z)$ contains just as much information about the state as $\psi(x, y, z)$. The Fourier integral links the two equivalent descriptions and allows us to calculate φ from ψ, and vice versa.

It is important to stress here that mathematical completeness of a set of dynamical variables is to be understood with reference to a model of the physical situation, but it would be presumptuous and quite unsafe to attribute completeness in any other sense to the mathematical description of a physical system. For, no matter how complete the description of a state may seem today, history teaches us that sooner or later new experimental facts will come to light which will require us to improve the model to give a more detailed and, usually, more complex description.

Thus, the wave mechanical description of the preceding chapters is complete with reference to the simple model of a point particle in a given external field, and it is remarkable how many fundamental problems of atomic and nuclear physics can be solved with such a gross picture. Yet this achievement must not blind us to the fact that the simple model which we have used so far is incapable of accounting for many of the finer details.

A whole host of these can be understood on the basis of the discovery that many elementary particles, including electrons, protons, and neutrons, are not adequately described by the model of a massive point whose position in space exhausts its dynamical properties. Rather, all the empirical evidence points to the need for attributing to these particles an *intrinsic angular momentum*, or *spin*, and, associated with this, a *magnetic moment*.

What is the experimental evidence for the spin and intrinsic magnetic moment? Although it was not realized at the time, the latter was first measured by Stern and Gerlach in experiments[1] the basic features of which are interesting here because they illustrate a number of concepts important in the interpretation of quantum mechanics. The particles (entire atoms or molecules) whose magnetic moment is to be measured are sent through a nonuniform magnetic field **B**. They are deflected by a force which according to classical physics is given by

$$\mathbf{F} = \nabla(\mathbf{m} \cdot \mathbf{B}) \qquad (12.1)$$

The arrangement is such that in the region through which the beam passes the direction of **B** varies only slowly, but its magnitude B is strongly dependent on position. Hence, if the projection on **m** in the direction of **B** is denoted by m_B, we have approximately

$$\mathbf{F} = m_B \nabla B \qquad (12.2)$$

By measuring the deflection, through inspection of the trace which the beam leaves on a screen, we can determine this force, hence the component of the magnetic moment in the direction of **B**. Figure 12.1 shows the outline of such an experiment.

The results of these experiments were striking. Classically, we would have expected a single continuous trace, corresponding to values of m_B ranging from $-m$ to $+m$. Observations showed instead a number of *equidistant distinct traces*, giving clear proof of the discrete quantum nature of the

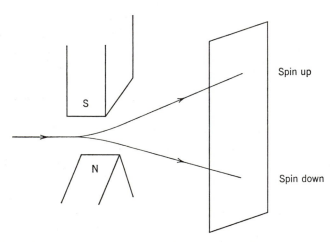

Spin up

Spin down

Figure 12.1. Measurement of the vertical component of the magnetic moment of atoms in a Stern-Gerlach experiment. Silver atoms produce two traces corresponding to "spin up" and "spin down."

[1] See N. Ramsey, *Molecular Beams*, Clarendon Press, Oxford, 1956, p. 100.

magnetic moment. Since the vector **m** seemed to be capable of assuming only certain discrete directions in space, it became customary to speak of *space quantization*.

Stern and Gerlach also obtained quantitative results. They measured the allowed values of m_B to moderate accuracy and found that the values of m_B appeared to range in equal steps from a minimum, $-m$, to a maximum, m. The value m of the maximum projection of **m** is conventionally regarded as *the* magnetic moment of a particle.

In order to interpret these results we recall Ampère's hypothesis that the magnetic properties of matter are entirely attributable to electric currents of one form or another. Thus, the circulating currents due to electrons in atoms produce an orbital angular momentum **L** and a magnetic moment **m** connected by the classical relation,

$$\mathbf{m} = -\frac{e}{2\mu c}\,\mathbf{L} \tag{12.3}$$

which, being a simple proportionality, is expected to survive in quantum mechanics also. Since any component of **L** has $2l + 1$ eigenvalues, we expect the projection of **m** in a fixed direction, such as on **B**, to possess also $2l + 1$ distinct eigenvalues and to be expressible as

$$m_B = -\frac{e\hbar}{2\mu c}\,m = -\beta_0 m \tag{12.4}$$

where the magnetic quantum number m can assume the values $-l$, $-l + 1$, $\ldots$, $l - 1$, l. For an electron, the Bohr magneton

$$\beta_0 = \frac{e\hbar}{2\mu c}$$

has the value 9.2732×10^{-21} erg/gauss.

Since l is an integer, we expect an *odd* number $(2l + 1)$ of traces in a Stern-Gerlach experiment. It is well known that the classical experiment with a beam of silver atoms, passing through an inhomogeneous magnetic field, yielded instead *two* traces, i.e., an even number, and a value of m equal to

$$\mathbf{m} = \frac{e\hbar}{2\mu c} = \beta_0 \tag{12.5}$$

The extraordinary implications of this experiment were not immediately understood. Efforts were made to interpret the results in terms of the relations (12.3) and (12.4) by supposing that the single valence electron, which an Ag atom was known to have outside of spherically symmetric closed shells (and by virtue of which it is placed in the first column of the periodic table), was in a *P*-state ($l = 1$), but that the state $m = 0$ was somehow suppressed so that only two instead of the expected $2l + 1 = 3$ traces were observed. This assumption, unnatural and *ad hoc* though it was, accounted for the duplicity of traces and for the measured value of the magnetic moment, but it remained an implausible assumption, particularly since the lowest energy state in an attractive potential is usually an *S*-state ($l = 0$) and not a *P*-state (see Problem 2, Chapter 10).

We may ask if the semiclassical arguments used above are valid when we have quantum phenomena before us. Equation (12.2) is purely classical, and we may wonder if its application to quantized magnetic moments has not led us astray. The answer to these questions is that every experiment has components which are properly and correctly described by the laws of classical physics. For these are the laws which govern the experiences of our senses by which we ultimately, if indirectly, make contact with what happens inside atoms and nuclei. If the particles which the inhomogeneous field in a Stern-Gerlach experiment deflects are sufficiently massive, their motion can be described by wave packets that spread very slowly; hence, this motion can be approximated by a classical description. This is the reason why the electron, which is the actual object of our curiosity in this experiment, must travel attached to an atom. In a Stern-Gerlach experiment with free electrons, quantal interference effects would wash out the entire pattern of distinct traces.[2] We would obtain a pattern which does not allow as simple and direct an interpretation based on the use of classical concepts.

The correct interpretation was given to the Stern–Gerlach observations only after Goudsmit and Uhlenbeck were led by a wealth of spectroscopic evidence to hypothesize the existence of an electron spin and intrinsic magnetic moment. If one assumes that the electron is in an *S*-state in the Ag atom, then (12.5) measures the maximum value of a component of the *intrinsic* magnetic moment, and, unlike a magnetic moment arising from charged particles moving in spatial orbits, this magnetic moment may be assumed to have only two projections,

$$m_B = \pm \frac{e\hbar}{2\mu c}$$

According to the Goudsmit-Uhlenbeck hypothesis, we must envisage the electron to be a point charge with a finite magnetic dipole moment,[3] the projection of which can only take on two discrete values.

[2] See N. F. Mott and H. S. W. Massey, *The Theory of Atomic Collisions*, 3rd ed., Clarendon Press, Oxford, 1965, p. 214.

[3] It is now known that for electrons (and also for mu mesons) m is not exactly equal to the Bohr magneton. The measured value of the magnetic moment of the electron is 9.2839×10^{-21} erg/gauss.

Goudsmit and Uhlenbeck also postulated that the electron has an intrinsic angular momentum (*spin*), but this quantity is not nearly as easy to measure directly as the magnetic moment. Without appealing to the original arguments for the electron spin, which were based on experience with the theory of atomic spectra, we can marshal two reasons for the assumption that an electron has intrinsic angular momentum:

(a) It does have a magnetic moment, and, if this magnetic moment is due to some (admittedly unanalyzable or, at least, unanalyzed) internal circulating currents of charged matter, then the appearance of an angular momentum is expected together with a magnetic moment.

(b) Unless the electron, moving in the electric field of the nucleus, possesses intrinsic angular momentum, conservation of angular momentum cannot be maintained for an isolated system such as an atom.

To elaborate argument (b), we note that, just as a moving point charge is subject to a force in a magnetic field, so a moving magnetic moment, such as the intrinsic electron moment is envisaged to be, is also acted on by forces in an electric field. The potential energy associated with these forces is

$$\mathbf{m} \cdot \frac{\mathbf{v}}{c} \times \mathbf{E}$$

which, for a central field [$\mathbf{E} = f(r)\mathbf{r}$], is proportional to $\mathbf{m} \cdot \mathbf{v} \times \mathbf{r}$, or to

$$\mathbf{m} \cdot \mathbf{r} \times \mathbf{p} = \mathbf{m} \cdot \mathbf{L} \tag{12.6}$$

The factor of proportionality depends only on the radial coordinate r. The energy of the electron thus depends on the relative orientation of the magnetic moment and the orbital angular momentum; the Hamiltonian operator contains an interaction term proportional to $\mathbf{m} \cdot \mathbf{L}$ in addition to the central potential. It is apparent that $\mathbf{L}$, whose components do not commute, can no longer be a constant of the motion. Conservation of angular momentum can only be restored if the electron can participate in the transfer of angular momentum by virtue of an intrinsic spin associated with $\mathbf{m}$. (See Section 12.7.)

It is legitimate to ask whether we can understand the occurrence of an intrinsic angular momentum and magnetic moment in terms of the structure of the elementary particles. The Dirac theory of the electron, in Chapters 23 and 24, will provide us with a deeper understanding of these properties for some particles. However, at a comparatively unsophisticated level and in describing interactions which are too weak to disturb the internal structure of the particles appreciably, we may treat mass, charge, intrinsic angular momentum, and magnetic moment as given properties, and we may characterize the particles, whether composite or elementary, by their moments which are overall features and only indirectly measures of internal detail.

2. *The Polarization of ψ Waves*. As the magnitude β_0 of the magnetic moment shows, the presence of an intrinsic spin and magnetic moment is directly related to the finite value of $\hbar$; hence, this is a quantum effect, and classical physics cannot be expected to be of much help in guiding us to the proper description of the spin. So far we have only found that an electron cannot be completely described by its coordinates x, y, z. It has in addition an orientation in space, and we must find an appropriate way of including this in the quantum description of the electron. The fact that wave mechanics was developed in Chapters 2 and 3 with relative ease on the basis of the correspondence between the momentum of a particle and its wavelength suggests that in our effort to construct a mathematical theory we shall be aided by first determining what *wave* feature corresponds to the electron spin.

A scattering experiment can be designed to bring out the directional properties of waves. If a homogeneous beam of particles, described by a *scalar* wave function $\psi(x, y, z, t)$, is incident on a scatterer, and if the target is composed of spherically symmetric or randomly oriented scattering centers (atoms or nuclei), we expect the scattered intensity to depend on the scattering angle θ but *not* on the azimuthal angle φ which defines the orientation of the scattering plane with respect to some fixed reference plane. In actual fact, if the beam in such experiments with electrons, protons, or neutrons is suitably prepared, a marked azimuthal dependence is observed, including a *right-left asymmetry* between particles scattered at the same angle θ but on opposite sides of the target. It is empirically found that the scattered intensity can be represented by the simple formula

$$I = a(\theta) + b(\theta) \cos \varphi \tag{12.7}$$

provided that a suitable direction is chosen as the origin of the angle measure φ. The simplest explanation of this observation is that ψ is not a scalar field, i.e., that *electron waves can be polarized*.[4]

Figure 12.2 shows the essential features of one particular polarization experiment. A beam I_1 of unpolarized particles is incident on an unpolarized scatterer A. The particles, scattered at an angle θ_1 from the direction of incidence, are scattered again through the angle θ_2 by a second unpolarized scatterer B, and the intensity of the so-called second scattered particles is measured as a function of the azimuthal angle φ, which is the angle between the first and second planes of scattering. Owing to the axial symmetry with respect to the z-axis, the intensities I_2 and I_2' are equal, but $I_R \neq I_L$, and the

---

[4] We speak only nominally of electrons here. Polarization experiments are frequently conducted with protons or neutrons.

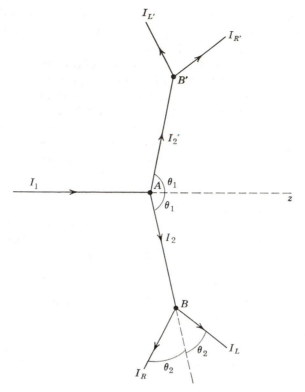

Figure 12.2. Arrangement in a double scattering experiment. The first scattering at A polarizes the beam, and the second scattering at B and B' analyzes the polarization. The second plane of scattering, formed by I_2, I_R, and I_L, need not coincide with the first plane of scattering formed by I_1 and I_2.

azimuthal dependence of the second scattered particle beam can be fitted by an expression of the form (12.7).

It is instructive to compare these conclusions with the results of the analogous double scattering experiment for initially unpolarized X rays. With the same basic arrangement as in Figure 12.2, no right-left asymmetry of X rays is observed, but the polarization manifests itself in a $\cos^2 \varphi$ dependence of the second scattered intensity. Since intensities are calculated as squares of amplitudes, such a behavior suggests that electromagnetic waves may be represented by a *vector* field which is transverse and whose projection on the scattering plane, when squared, determines the intensity.

The presence of a $\cos \varphi$, instead of a $\cos^2 \varphi$, term precludes a similar conclusion for the electron waves and shows that, if their polarization can be represented by a vector, the intensity must depend on this vector *linearly* and not quadratically. Hence, the wave function, whose *square* is related to the intensity, is not itself a vectorial quantity, and the polarization vector (**P**) will have to be calculated from it indirectly.

In summary, the polarization experiments suggest that the wave must be represented by a wave function which under spatial rotations transforms neither as a scalar nor as a vector, but in a more complicated way. On the other hand, the interpretation of the Stern-Gerlach experiment (Section 12.1) requires that, in addition to x, y, z, the wave function must depend on at least one other dynamical variable to permit the description of a magnetic moment and intrinsic angular momentum which the electron possesses. Since both the polarization of the waves and the lining up of the particle spins are aspects of an *orientation* in space of the electron, whether it be wave or particle, it is not far-fetched to suppose that the same extension of the formalism of wave mechanics may account for both observations.

Similarly, we shall see in Chapter 22 that the vector properties of electromagnetic waves are closely related with the intrinsic angular momentum (spin 1) of photons.

3. The Spin as a Dynamical Variable. To the dynamical variables x, y, z, describing the position of the electron, we now add a fourth variable which we denote by σ. This *spin variable* is given a physical meaning by associating the two possible projections of the magnetic moment **m**, as measured in the Stern-Gerlach experiment, with two distinct values of σ.[5] Thus we associate arbitrarily:

$$\sigma = +1, \qquad \mathsf{m}_B = -\frac{e\hbar}{2\mu c} \tag{12.8a}$$

$$\sigma = -1, \qquad \mathsf{m}_B = +\frac{e\hbar}{2\mu c} \tag{12.8b}$$

Often, $\sigma = +1$ is referred to as "spin up" and $\sigma = -1$ as "spin down" (see Figure 12.1). The wave function depends on σ in addition to x, y, z, and we shall assume that the basic postulates of quantum mechanics (Chapter 8) apply to the new independent variable in very much the same way as to the old ones. For instance, $|\psi(x, y, z, \pm 1)|^2 \, dx \, dy \, dz$ is assumed to measure not only the probability of finding the particle near x, y, z, but also the probability of its having the values $\mathsf{m}_B = \mp \beta_0$ respectively for the projection of its magnetic moment in the direction of **B**.

---

[5] We first relate the spin variable to the intrinsic *magnetic moment* rather than to the intrinsic *angular momentum* of the electron, because the former is measured more directly than the latter. The choice of the values of σ is, however, made in anticipation of the ultimate identification of the spin with the intrinsic angular momentum. The two vectors will be seen to be proportional but pointing in opposite directions.

There is no a priori reason to expect that such a modest generalization of the previous formulation will be adequate, but the insistent appearance of merely two traces in the Stern-Gerlach experiment and, as we shall see later, the splitting of the spectral lines of one-electron atoms into narrow doublets make it reasonable to assume that a variable which can take on only two different values[6] may be a sufficiently inclusive addition to the theory.

The formulas and theorems of Chapter 8 can almost all be extended without difficulty from wave mechanics *without* spin to wave mechanics *with* spin. Wherever we had previously an integration over the continuously infinitely many values of the position variables, such as in the normalization integral,

$$\int_{-\infty}^{+\infty} \int_{-\infty}^{+\infty} \int_{-\infty}^{+\infty} |\psi(x, y, z)|^2 \, dx \, dy \, dz$$

we must now instead introduce an additional summation over the pair of values which σ assumes:

$$\int |\psi(x, y, z, +1)|^2 \, d\tau + \int |\psi(x, y, z, -1)|^2 \, d\tau$$

Again we postulate that to every physical quantity there corresponds a linear operator F, i.e., a prescription by which to any function $\psi(x, y, z, \sigma)$ we assign another function $\varphi(x, y, z, \sigma)$,

$$\varphi(x, y, z, \sigma) = F\psi(x, y, z, \sigma)$$

such that if

$$\varphi_1 = F\psi_1 \quad \text{and} \quad \varphi_2 = F\psi_2$$

we have

$$\lambda \varphi_1 + \mu \varphi_2 = F(\lambda \psi_1 + \mu \psi_2)$$

where λ and μ are arbitrary complex constants.

Among all the wave functions $\psi(x, y, z, \sigma)$ there is a special class which is particularly easy to treat. These are the *separable* wave functions of the form

$$\psi(x, y, z, \sigma) = \vartheta(x, y, z)\chi(\sigma) \tag{12.9}$$

We shall be especially interested in two remarkably simple functions $\chi(\sigma)$, which we designate by $\alpha(\sigma)$ and $\beta(\sigma)$ and define as follows:

$$\chi(\sigma) = \alpha(\sigma) \quad \text{if } \chi(+1) = 1, \quad \chi(-1) = 0 \tag{12.10a}$$

$$\chi(\sigma) = \beta(\sigma) \quad \text{if } \chi(+1) = 0, \quad \chi(-1) = 1 \tag{12.10b}$$

---

[6] A variable which can take on only two distinct values is sometimes called a *dichotomic* variable.

Thus, α corresponds to "spin up," and a wave function (12.9) whose spin factor is α describes a particle with a definite value of $\sigma = +1$, and zero probability of finding $\sigma = -1$. Similarly, β corresponds to "spin down."

The most general state can be written as a linear combination of separable ones, and this fact is responsible for enormous simplifications. Thus, for an arbitrary $\psi(x, y, z, \sigma)$,

$$\psi(x, y, z, \sigma) = \psi(x, y, z, +1)\alpha(\sigma) + \psi(x, y, z, -1)\beta(\sigma) \qquad (12.11)$$

We can draw an important conclusion from the preceding discussion: The variables x, y, z on the one hand, and σ on the other, can be studied separately and merged together easily at any stage. We may therefore temporarily disregard the position variables and develop a *spin quantum mechanics* for a particle whose state is determined by a wave function χ dependent only on a single variable σ, which can take on but two values, $\sigma = +1$, and $\sigma = -1$. There are actually situations in which the bodily motion of a particle can be ignored and only its spin degree of freedom considered. The study of nuclear magnetism is an example, since we can discuss many experiments by assuming that the nuclei are at fixed positions and only their spins are subject to change owing to the interaction with the magnetic field.

Study of the spin formalism in isolation from all other degrees of freedom serves as a paradigm for the behavior of any quantum system whose states can be described as *linear superpositions* of only a finite number of independent states—two, α, and β, in the case of spin $\frac{1}{2}$. There are innumerable problems in quantum mechanics where such a two-state formalism is applicable to good approximation, but which have nothing to do with intrinsic angular momentum. Examples are the coupling of the $2S$ and $2P$ states of the hydrogen atom through the Stark effect, the magnetic quenching of the triplet state of positronium, the isospin description of a nucleon, and the life and death of a neutral K meson.

It is important to stress that much of the theory developed here is applicable to the quantum mechanics of any dynamical variable with only a finite number of linearly independent basis states. The two-state formalism is, of course, particularly transparent and adaptable to a thorough discussion of the interpretation of quantum mechanics and the theory of measurement. Untroubled by mathematical complexities, the student can concentrate here on the novel and strange features of the theory.

4. The Spin Theory in Matrix Form. It was discovered early, and we saw in Chapters 6 and 7, that the use of matrix algebra is extraordinarily convenient for some purposes in quantum mechanics, especially when a dynamical variable can take on only a finite number of values. The spin offers an excellent illustration.

To begin with, we now represent the special states α and β by one-column matrices:

$$\alpha = \begin{pmatrix} 1 \\ 0 \end{pmatrix} \qquad \beta = \begin{pmatrix} 0 \\ 1 \end{pmatrix} \tag{12.12}$$

and the general spin state χ by

$$\chi = \begin{pmatrix} \chi(+1) \\ \chi(-1) \end{pmatrix} = \chi(+1)\alpha + \chi(-1)\beta = \begin{pmatrix} c_1 \\ c_2 \end{pmatrix} = c_1\alpha + c_2\beta \tag{12.13}$$

where we have set

$$c_1 = \chi(+1), \qquad c_2 = \chi(-1) \tag{12.14}$$

c_1 and c_2 are, of course, complex numbers. As an object with well-defined transformation properties under spatial rotations, which will be explained in the next section, the matrix χ with one column and two rows is called a *spinor*. It can also be regarded as a *vector* with complex components, c_1 and c_2, in a two-dimensional linear vector space. However, the latter terminology must not mislead us to think of χ as having anything to do with a vector in the ordinary three-dimensional space. $|c_1|^2$ is the probability of finding the particle with spin up, $|c_2|^2$ the probability of finding it with spin down. Hence, we must require the normalization

$$|c_1|^2 + |c_2|^2 = (c_1^* \quad c_2^*)\begin{pmatrix} c_1 \\ c_2 \end{pmatrix} = 1 \tag{12.15}$$

This can be written as

$$\chi^\dagger\chi = 1 \tag{12.16}$$

if we define

$$\chi^\dagger = (c_1^* \quad c_2^*) \tag{12.17}$$

Given two spinors, χ and χ', frequent use is made of the (complex scalar) product

$$\chi^\dagger\chi' = (c_1^* \quad c_2^*)\begin{pmatrix} c_1' \\ c_2' \end{pmatrix} = c_1^*c_1' + c_2^*c_2' \tag{12.18}$$

Two spinors are said to be *orthogonal* if this product is zero. Thus, α and β are two orthogonal spinors, and they are normalized, as $\alpha^\dagger\alpha = \beta^\dagger\beta = 1$. Such pairs of orthogonal spinors are said to form a *basis*, in terms of which arbitrary spinors may be conveniently expanded [(12.13)].

We have postulated that physical quantities must be represented by *linear operators*. If F is a linear operator, its effect can be most readily defined by what it does to the special states represented by the basis spinors, α and β:

$$F\alpha = F_{11}\alpha + F_{21}\beta, \qquad F\beta = F_{12}\alpha + F_{22}\beta \tag{12.19}$$

The coefficients on the right characterize the operator F completely, as seen from the equations

$$\xi = F\chi = c_1 F\alpha + c_2 F\beta = (c_1 F_{11} + c_2 F_{12})\alpha + (c_1 F_{21} + c_2 F_{22})\beta$$
$$= d_1\alpha + d_2\beta \tag{12.20}$$

where the components of ξ are denoted by d_1, d_2. Equation (12.20) can equally well be written in matrix notation:

$$\begin{pmatrix} d_1 \\ d_2 \end{pmatrix} = \begin{pmatrix} F_{11} & F_{12} \\ F_{21} & F_{22} \end{pmatrix} \begin{pmatrix} c_1 \\ c_2 \end{pmatrix} \tag{12.21}$$

Hence, $\xi = F\chi$ can be interpreted directly as a matrix equation, the same symbol being used for a state and the spinor which represents it; similarly, the same letter is used for a physical quantity and the matrix (operator) which represents it.

As in (6.53), the (Hermitian) conjugate $F^\dagger$ of a matrix F is defined by the equation

$$F^\dagger = \begin{pmatrix} F_{11}{}^* & F_{21}{}^* \\ F_{12}{}^* & F_{22}{}^* \end{pmatrix} \tag{12.22}$$

and the *expectation value* of F is defined in analogy with (8.19) as

$$\langle F \rangle = (c_1{}^* \quad c_2{}^*) \begin{pmatrix} F_{11} & F_{12} \\ F_{21} & F_{22} \end{pmatrix} \begin{pmatrix} c_1 \\ c_2 \end{pmatrix} = \chi^\dagger F\chi \tag{12.23}$$

This is, of course, a basic postulate of quantum mechanics.

Since, for a physical quantity, $\langle F \rangle$ must be real we require that

$$\langle F \rangle = \langle F \rangle^* = (c_1 \quad c_2) \begin{pmatrix} F_{11}{}^* & F_{12}{}^* \\ F_{21}{}^* & F_{22}{}^* \end{pmatrix} \begin{pmatrix} c_1{}^* \\ c_2{}^* \end{pmatrix}$$

If we take the transpose of this equation, we get for any state χ:

$$\langle F \rangle = \langle F \rangle^* = (c_1{}^* \quad c_2{}^*) \begin{pmatrix} F_{11}{}^* & F_{21}{}^* \\ F_{12}{}^* & F_{22}{}^* \end{pmatrix} \begin{pmatrix} c_1 \\ c_2 \end{pmatrix}$$

Comparing with (12.23), we obtain the conditions

$$F_{11}{}^* = F_{11}, \qquad F_{22}{}^* = F_{22}, \qquad F_{12}{}^* = F_{21}, \qquad F_{21}{}^* = F_{12} \tag{12.24}$$

which show that the matrix F must be *Hermitian*, or self-conjugate, i.e., $F^\dagger = F$. We can now follow Chapter 8 and restate most of the theorems listed there. For instance, theorems 1 through 5 in Section 8.2 can be translated simply by substituting the term *matrix* for *operator*.

By arguments paralleling those of Chapter 8, it can be shown that the only possible result of the measurement of a physical quantity represented by a Hermitian 2×2 matrix A is one of the two eigenvalues, A_1' or A_2'. The eigenvalue problem is explicitly formulated by the equations

$$\begin{pmatrix} A_{11} & A_{12} \\ A_{21} & A_{22} \end{pmatrix} \begin{pmatrix} u_1 \\ u_2 \end{pmatrix} = A_1' \begin{pmatrix} u_1 \\ u_2 \end{pmatrix} \tag{12.25a}$$

$$\begin{pmatrix} A_{11} & A_{12} \\ A_{21} & A_{22} \end{pmatrix} \begin{pmatrix} v_1 \\ v_2 \end{pmatrix} = A_2' \begin{pmatrix} v_1 \\ v_2 \end{pmatrix} \tag{12.25b}$$

Exercise 12.1. Prove that the eigenvalues of any Hermitian 2×2 matrix are real and its eigenvectors orthogonal if $A_1' \neq A_2'$. What happens if $A_1' = A_2'$?

An arbitrary state χ can be expanded in terms of the orthonormal eigen-spinors, u and v, of any Hermitian matrix:

$$\chi = d_1 u + d_2 v \tag{12.26}$$

The complex expansion coefficients d_1 and d_2 determine the state χ completely, $|d_1|^2$ and $|d_2|^2$ being the probabilities of finding A_1' and A_2', respectively, as a result of measuring A. Using the orthonormality, we find in analogy with (8.43):

$$d_1 = u^\dagger \chi = (u_1^* \quad u_2^*) \begin{pmatrix} c_1 \\ c_2 \end{pmatrix}$$

$$d_2 = v^\dagger \chi = (v_1^* \quad v_2^*) \begin{pmatrix} c_1 \\ c_2 \end{pmatrix} \tag{12.27}$$

In many ways the formalism developed here is much simpler than that of Chapter 8. Because the maximum number of linearly independent spinors is finite (two), such questions as completeness are automatically answered in the affirmative, and others, such as the appearance of continuous eigenvalues, do not arise at all.

So much for the purely mathematical framework. Our next task must be to fill it with physical content. In particular, we have spoken of Hermitian matrices as being representative of physical quantities, but we have not yet identified these physical quantities. That must now be done.

5. *Spin and Rotations.* One physical quantity of interest is evident. It is the two-valued variable which started us on our way: the component of the electron magnetic moment in the direction of the magnetic field. Since **B** can be chosen to point in any direction whatever, we first select this to be the z-axis of the spatial Cartesian coordinate system. Then the z-component of the magnetic moment is evidently represented by the Hermitian matrix

$$m_z = -\beta_0 \begin{pmatrix} 1 & 0 \\ 0 & -1 \end{pmatrix} \tag{12.28}$$

since the eigenvalues of m_z, i.e., its measured values, are to be $\mp\beta_0 = \mp e\hbar/2\mu c$ and the corresponding states are represented by the basis spinors α and β.

How are the other components of $\mathbf{m}$ represented? $\mathbf{m}$ has three spatial components, m_x, m_y, m_z, and by choosing a different direction for $\mathbf{B}$ we can measure any projection, m_B, of $\mathbf{m}$. If our two-dimensional formalism is adequate to describe the physical situation, any such projection m_B must be represented by a Hermitian matrix with the eigenvalues $-\beta_0$ and $+\beta_0$. In order to determine the matrices m_x and m_y we stipulate that the three components of $\langle\mathbf{m}\rangle$ must under a rotation transform as the components of a *vector*. Since an expectation value, such as $\langle m_x\rangle = \chi^\dagger m_x\chi$, is calculated from matrices and spinors, we cannot say how the components of $\langle\mathbf{m}\rangle$ transform unless we establish the transformation properties of a spinor χ under rotation. To this task we shall give our attention next.

Figure 12.3 indicates a rotation about the z-axis.[7] Let a physical system be represented by χ before the rotation, and by χ' after the system has been rotated by an angle θ:

$$\chi = \begin{pmatrix} c_1 \\ c_2 \end{pmatrix}, \qquad \chi' = \begin{pmatrix} c_1' \\ c_2' \end{pmatrix} \tag{12.29}$$

The relation between c_1', c_2' and c_1, c_2 may be assumed to be linear.[8] This will preserve the linear superposition of two states; i.e., if in a rotation χ_1' corresponds to χ_1, and χ_2' to χ_2, then $a\chi_1' + b\chi_2'$ corresponds to $a\chi_1 + b\chi_2$. This postulate leads to the condition that

$$\begin{pmatrix} c_1' \\ c_2' \end{pmatrix} = U\begin{pmatrix} c_1 \\ c_2 \end{pmatrix} \tag{12.30}$$

where U is a matrix whose elements depend on the parameters of rotation only, e.g., on the angles of rotation.

We require that χ' should be normalized if χ is normalized; hence

$$\chi'^\dagger\chi' = (c_1'^* \quad c_2'^*)\begin{pmatrix} c_1' \\ c_2' \end{pmatrix} = (c_1^* \quad c_2^*)U^\dagger U\begin{pmatrix} c_1 \\ c_2 \end{pmatrix}$$

$$= (c_1^* \quad c_2^*)\begin{pmatrix} c_1 \\ c_2 \end{pmatrix} = \chi^\dagger\chi$$

---

[7] The physical system is given a right-handed rotation. The coordinate axes remain fixed.
[8] It will be shown in Section 16.1 that this assumption involves no loss of generality.

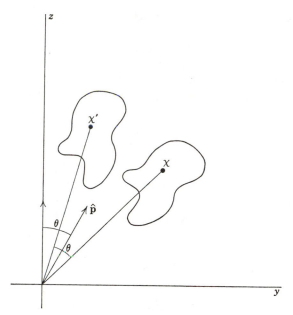

Figure 12.3. Symbolic representation of the rotation of a state χ. The state χ is related to the vector $\hat{\mathbf{p}}$ in the same way as χ' is related to the unit vector in the z-direction (see Section 13.5).

Since χ is arbitrary, this gives the condition on U that

$$U^\dagger U = I \tag{12.31}$$

i.e., U must be a *unitary* matrix.

Note that, according to (12.31),

$$\det U^\dagger \det U = |\det U|^2 = 1 \tag{12.32}$$

Hence, a unitary matrix has a unique inverse, $U^{-1} = U^\dagger$, and

$$UU^\dagger = I \tag{12.33}$$

The unitary matrix U, which corresponds to the rotation that takes χ into χ', is said to *represent* this rotation. If U_1 represents a rotation R_1, and U_2 represents a rotation R_2, then $U_2 U_1$ represents a rotation R_3, obtained by performing first R_1 and then R_2. In this way χ is first transformed to $\chi' = U_1\chi$, which subsequently is transformed to $\chi'' = U_2\chi' = U_2 U_1\chi$. Alternatively, we could have obtained the same physical state directly from χ by performing a single rotation R_3, represented by U_3. Hence, it is to be required that the unitary rotation matrices have the property

$$U_3 = e^{i\varphi} U_2 U_1 \tag{12.34}$$

The phase factor has been put in, because all spinors $e^{i\varphi}\chi$ represent the same state.

Our goal, the construction of U corresponding to a given rotation R, will be considerably facilitated if we consider *infinitesimal rotations* first. A small rotation must correspond to a matrix very near the unit matrix, and thus for a small rotation we write to a first approximation the first two terms in a Taylor series

$$U = 1 - \frac{i}{2}\varepsilon\hat{\mathbf{n}} \cdot \boldsymbol{\sigma} \tag{12.35}$$

where $\hat{\mathbf{n}}$ is the axis of rotation, ε the angle of rotation about this axis, and where $\boldsymbol{\sigma}$ represents three constant matrices σ_x, σ_y, σ_z, whose detailed structure is yet to be determined. The factor $i/2$ has been introduced so that $\boldsymbol{\sigma}$ will have certain simple and desirable properties. In particular, the imaginary coefficient insures that, if U is to be unitary to first order in ε, then $\boldsymbol{\sigma}$ must be Hermitian, i.e.,

$$\sigma_x^{\dagger} = \sigma_x, \qquad \sigma_y^{\dagger} = \sigma_y, \qquad \sigma_z^{\dagger} = \sigma_z \tag{12.36}$$

If the σ matrices were known, U for any finite rotation could be constructed from (12.35) by successive application of many infinitesimal rotations, i.e., by "integration" of (12.35).

This integration is easily accomplished if we restrict ourselves to rotations about a fixed axis (constant $\hat{\mathbf{n}}$), with a finite angle $\theta = N\varepsilon$. This is legitimate because Euler's famous theorem states that for any rotation in three dimensions one can find a fixed axis, so that the rotation may be reduced to a rotation about that axis. Carrying out $N = \theta/\varepsilon$ infinitesimal rotations (12.35) in succession, we get for the final product

$$U_R \approx \left(1 - \frac{i\theta}{2N}\hat{\mathbf{n}} \cdot \boldsymbol{\sigma}\right)^N \tag{12.37}$$

or, in the limit $N \to \infty$,

$$U_R = \lim_{N \to \infty}\left(1 - \frac{i\theta}{2N}\hat{\mathbf{n}} \cdot \boldsymbol{\sigma}\right)^N = \exp\left(-\frac{i}{2}\theta\hat{\mathbf{n}} \cdot \boldsymbol{\sigma}\right) \tag{12.38}$$

as in elementary calculus, even though U_R and $\hat{\mathbf{n}} \cdot \boldsymbol{\sigma}$ are matrices. The exponential function with a matrix in the exponent is defined by (12.38), or by the usual power series expansion.

We still have to derive the conditions under which a matrix of the form (12.38) is actually the solution to our problem, i.e., represents the rotation R and satisfies the basic requirement (12.34). The application of the condition (12.34) will lead to severe restrictions on the possible form of the matrices σ_x, σ_y, σ_z, which so far have not been specified at all. However, it is convenient not to attack this problem directly, but instead to discuss first the rotational transformation properties of a vector $\langle \mathbf{A} \rangle$, where A_x, A_y, A_z are three matrices (operators) such that the expectation values $\langle A_x \rangle$, $\langle A_y \rangle$, $\langle A_z \rangle$ transform as the components of a vector. As stated at the beginning of this section, the components of the magnetic moment $\langle \mathbf{m} \rangle$ are an example of matrices which must satisfy this condition. Generally, a set of three matrices A_x, A_y, A_z is called a *vector operator* $\mathbf{A}$ if the expectation values of A_x, A_y, A_z transform under rotation like the components of a vector.

It is of decisive importance to note that $\boldsymbol{\sigma}$ itself is a vector operator. This follows from the equation valid for infinitesimal rotations:

$$\chi' = \left(1 - \frac{i}{2}\varepsilon \hat{\mathbf{n}} \cdot \boldsymbol{\sigma} \right) \chi$$

Multiplying on the left by $\chi^{\dagger}$, we obtain

$$\chi^{\dagger}\chi' = \chi^{\dagger}\chi - \frac{i}{2}\varepsilon \hat{\mathbf{n}} \cdot \langle \boldsymbol{\sigma} \rangle$$

where the expectation value $\langle \boldsymbol{\sigma} \rangle$ is taken with respect to the state χ. Now let us apply an arbitrary *finite* rotation simultaneously to the states χ and χ'. The products $\chi^{\dagger}\chi'$ and $\chi^{\dagger}\chi$ are invariant under the unitary transformation which represents this finite rotation. Hence, the scalar product $\hat{\mathbf{n}} \cdot \langle \boldsymbol{\sigma} \rangle$ is a *rotational invariant*. Since $\hat{\mathbf{n}}$ is a vector, $\langle \boldsymbol{\sigma} \rangle$ must also be a vector, and thus $\boldsymbol{\sigma}$ is a vector operator.

Mathematically, a vector operator is characterized as follows. If in ordinary three-dimensional space the rotation is represented by an orthogonal matrix

$$R = \begin{pmatrix} R_{11} & R_{12} & R_{13} \\ R_{21} & R_{22} & R_{23} \\ R_{31} & R_{32} & R_{33} \end{pmatrix} \tag{12.39}$$

we should expect:

$$\begin{pmatrix} \chi'^{\dagger}A_x\chi' \\ \chi'^{\dagger}A_y\chi' \\ \chi'^{\dagger}A_z\chi' \end{pmatrix} = R \begin{pmatrix} \chi^{\dagger}A_x\chi \\ \chi^{\dagger}A_y\chi \\ \chi^{\dagger}A_z\chi \end{pmatrix} \tag{12.40}$$

Hence, as $\chi' = U\chi$,

$$U^{\dagger}A_x U = R_{11}A_x + R_{12}A_y + R_{13}A_z \tag{12.41}$$

Similar equations hold for $U^{\dagger}A_y U$ and $U^{\dagger}A_z U$.

Consider the special case of an infinitesimal rotation about the z-axis.

This implies

$$R = \begin{pmatrix} 1 & -\varepsilon & 0 \\ \varepsilon & 1 & 0 \\ 0 & 0 & 1 \end{pmatrix}$$

and

$$U = 1 - \frac{i}{2}\varepsilon\sigma_z$$

Hence, by substituting these quantities into (12.41) and its sister equations, we obtain the *necessary conditions*

$$\sigma_z A_x - A_x \sigma_z = 2iA_y$$
$$\sigma_z A_y - A_y \sigma_z = -2iA_x \qquad (12.42a)$$
$$\sigma_z A_z - A_z \sigma_z = 0$$

Similarly, rotations about the x- and y-axes give the relations

$$\sigma_x A_y - A_y \sigma_x = 2iA_z$$
$$\sigma_x A_z - A_z \sigma_x = -2iA_y \qquad (12.42b)$$
$$\sigma_x A_x - A_x \sigma_x = 0$$

and

$$\sigma_y A_z - A_z \sigma_y = 2iA_x$$
$$\sigma_y A_x - A_x \sigma_y = -2iA_z \qquad (12.42c)$$
$$\sigma_y A_y - A_y \sigma_y = 0$$

Since $\boldsymbol{\sigma}$ is a vector operator, we may make the identification $\mathbf{A} = \boldsymbol{\sigma}$ in (12.42). In this way we obtain a set of commutation relations which the hitherto unrestricted Hermitian matrices σ_x, σ_y, σ_z must obey:

$$\sigma_x \sigma_y - \sigma_y \sigma_x = 2i\sigma_z$$
$$\sigma_y \sigma_z - \sigma_z \sigma_y = 2i\sigma_x \qquad (12.43)$$
$$\sigma_z \sigma_x - \sigma_x \sigma_z = 2i\sigma_y$$

These fundamental commutation relations are sometimes combined symbolically in the single equation

$$\boldsymbol{\sigma} \times \boldsymbol{\sigma} = 2i\boldsymbol{\sigma} \qquad (12.43a)$$

Equations (12.43) encompass the long-awaited restrictions which the σ matrices must obey if matrices of the form

$$U_R = \exp\left(-\frac{i}{2}\theta\hat{\mathbf{n}} \cdot \boldsymbol{\sigma}\right)$$

are to represent rotations. It can be shown that these conditions are both necessary and sufficient.[9]

None of the arguments presented in this section have depended on the dimensionality of the matrices involved. If χ and χ' in (12.29) had been matrices of one column but n rows, we would have obtained exactly the same results. In particular, the commutation relations (12.43) would then have to be satisfied by three $n \times n$ matrices. These observations will become pertinent in Chapters 15 and 16 where higher spin values and general angular momentum operators are discussed. In this chapter, however, we restrict ourselves to the case $n = 2$, and we must now find Hermitian 2×2 matrices $\sigma_x, \sigma_y, \sigma_z$ which satisfy the commutation relations (12.43).

6. *Determination of* $\sigma_x, \sigma_y, \sigma_z$. The matrix σ_z is already largely determined by the choice of the basis spinors, for we specified in Section 12.4 that $|c_1|^2$ was to be the probability of finding $m_z = -\beta_0$ (spin up), and $|c_2|^2$ the probability that $m_z = \beta_0$ (spin down). A rotation about the z-axis can surely have no effect on these probabilities, since the orientation of the x- and y-axes is irrelevant when the z-component of $\mathbf{m}$ is measured. Hence, if a rotation about the z-axis by an angle ε is performed, $|c_1|^2$ and $|c_2|^2$ must not change, and

$$U = 1 - \frac{i\varepsilon}{2} \sigma_z$$

must be a matrix in diagonal form. It follows that σ_z must itself be a diagonal matrix. Furthermore, by taking the trace of both sides of the first equation (12.43), we see that

$$\text{trace } \sigma_z = 0 \tag{12.44}$$

since the trace of AB equals the trace of BA. Hence, σ_z must have the form

$$\sigma_z = \begin{pmatrix} \lambda & 0 \\ 0 & -\lambda \end{pmatrix} \tag{12.45}$$

It remains to determine the matrices σ_x and σ_y, and the value of λ in σ_z. To this end it is advantageous to construct the auxiliary matrices

$$\sigma_+ = \sigma_x + i\sigma_y, \qquad \sigma_- = \sigma_x - i\sigma_y \tag{12.46}$$

Using the commutator notation $[A, B] = AB - BA$, we establish from (12.43) readily that

$$[\sigma_z, \sigma_+] = 2\sigma_+$$
$$[\sigma_z, \sigma_-] = -2\sigma_- \tag{12.47}$$
$$[\sigma_+, \sigma_-] = 4\sigma_z$$

---

[9] For an indication of how to prove the sufficiency see Exercise 12.3.

If we substitute the matrix

$$\sigma_+ = \begin{pmatrix} a & b \\ c & d \end{pmatrix}$$

and (12.45) in the first equation (12.47), we get

$$\begin{pmatrix} 0 & 2\lambda b \\ -2\lambda c & 0 \end{pmatrix} = 2\begin{pmatrix} a & b \\ c & d \end{pmatrix}$$

and hence we conclude that $a = d = 0$ and that $(\lambda - 1)b = 0$ and $(\lambda + 1)c = 0$. From these equations we infer that λ must be chosen equal to $+1$ or -1, the choice being arbitrary, since the other value is automatically present in (12.45). Let us choose $\lambda = +1$, $c = 0$. From (12.46) we see that σ_- is the Hermitian adjoint of σ_+. Hence, we have as a result

$$\sigma_+ = \begin{pmatrix} 0 & b \\ 0 & 0 \end{pmatrix}, \qquad \sigma_- = \begin{pmatrix} 0 & 0 \\ b^* & 0 \end{pmatrix}$$

Substituting these matrices into the last equation (12.47), we find that $|b|^2 = 4$. The phase of b can be chosen arbitrarily, for it cannot be inferred from (12.43). Choosing simply $b = 2$, we are finally led to the result

$$\sigma_x = \begin{pmatrix} 0 & 1 \\ 1 & 0 \end{pmatrix}, \qquad \sigma_y = \begin{pmatrix} 0 & -i \\ i & 0 \end{pmatrix}, \qquad \sigma_z = \begin{pmatrix} 1 & 0 \\ 0 & -1 \end{pmatrix} \qquad (12.48)$$

These are the celebrated *Pauli spin matrices*. We see that, apart from the order of rows and columns ($\lambda = +1$ or -1) and the phase of b, they are completely determined by the commutation relations and by our selection of the eigenspinors of σ_z as the basis α, β.

A few simple properties of the Pauli matrices are easily derived. The proofs are left to the reader. Thus we have

$$\sigma_x{}^2 = \sigma_y{}^2 = \sigma_z{}^2 = 1 \qquad (12.49)$$

and

$$\sigma_x\sigma_y = i\sigma_z, \qquad \sigma_y\sigma_z = i\sigma_x, \qquad \sigma_z\sigma_x = i\sigma_y \qquad (12.50)$$

Hence, these matrices, besides being Hermitian, are also *unitary*.

Note further that any two different Pauli matrices *anticommute*,

$$\sigma_x\sigma_y + \sigma_y\sigma_x = 0 \qquad (12.51)$$

The traces of all Pauli matrices vanish:

$$\text{trace } \sigma_x = \text{trace } \sigma_y = \text{trace } \sigma_z = 0 \qquad (12.52)$$

The four matrices 1, σ_x, σ_y, σ_z are linearly independent, and any 2×2 matrix A can be represented as

$$A = \lambda_0 1 + \lambda_1 \sigma_x + \lambda_2 \sigma_y + \lambda_3 \sigma_z \qquad (12.53)$$

Exercise 12.2. Work out the eigenvalues and eigenspinors of $A = \lambda_0 1 + \boldsymbol{\lambda} \cdot \boldsymbol{\sigma}$ in terms of λ_0 and $\boldsymbol{\lambda}$. Specialize to the case $\lambda_0 = 0$, $\boldsymbol{\lambda} = \hat{\mathbf{n}}$, where $\hat{\mathbf{n}}$ is an arbitrary unit vector.

If A is Hermitian, all coefficients must be real. If U is a unitary matrix, we can always write

$$U = e^{i\gamma}(1 \cos \omega + i\hat{\mathbf{n}} \cdot \boldsymbol{\sigma} \sin \omega) \qquad (12.54)$$

where γ and ω are real angles, and $\hat{\mathbf{n}}$ is a unit vector.

If $\mathbf{A}$ and $\mathbf{B}$ are two vectors which commute with $\boldsymbol{\sigma}$, we have the useful identity

$$(\boldsymbol{\sigma} \cdot \mathbf{A})(\boldsymbol{\sigma} \cdot \mathbf{B}) = \mathbf{A} \cdot \mathbf{B} + i\boldsymbol{\sigma} \cdot (\mathbf{A} \times \mathbf{B}) \qquad (12.55)$$

Applying this to the power series expansion of an exponential, we obtain a generalized de Moivre formula for (12.54):

$$U = \exp(i\gamma + i\omega\hat{\mathbf{n}} \cdot \boldsymbol{\sigma}) \qquad (12.56)$$

Any unitary 2×2 matrix can be written in this form. Comparing (12.56) and (12.38), we see that every unitary matrix with $\gamma = 0$ represents a rotation. The angle of rotation is $\theta = -2\omega$, and $\hat{\mathbf{n}}$ is the axis of rotation. For $\gamma = 0$ we have det $U = 1$, and the matrix U is said to be *unimodular*. The set of all unitary unimodular 2×2 matrices constitutes the *group* SU(2). The connection between this group and three-dimensional rotations will be made precise in Chapter 16.

We may now write (12.38) in the form

$$U_R = \exp\left(-\frac{i}{2}\theta\hat{\mathbf{n}} \cdot \boldsymbol{\sigma}\right) = 1 \cos\frac{\theta}{2} - i\hat{\mathbf{n}} \cdot \boldsymbol{\sigma} \sin\frac{\theta}{2} \qquad (12.57)$$

One simple consequence of this equation is that for $\theta = 2\pi$ we get $U = -1$. A full rotation by $360°$ about a fixed axis thus changes the sign of every spinor component. Vectors (and tensors in general) behave differently: they return to their original values upon such a rotation. However, this sign change of spinors under rotation is no obstacle to their usefulness, since all expectation values and matrix elements depend quadratically on spinors.

Exercise 12.3. Using the properties of the Pauli matrices, prove that

$$U_R{}^\dagger \boldsymbol{\sigma} U_R = \hat{\mathbf{n}}(\hat{\mathbf{n}} \cdot \boldsymbol{\sigma}) - \hat{\mathbf{n}} \times (\hat{\mathbf{n}} \times \boldsymbol{\sigma}) \cos\theta + \hat{\mathbf{n}} \times \boldsymbol{\sigma} \sin\theta$$

if U_R is given by (12.57). This result may be used to show that the commutation relations (12.43) are *sufficient* to assure that U_R represents a rotation.

To contrast the different behavior of spinors and vectors under rotation, we consider the special case of a rotation about the x-axis by an angle θ. From (12.57), (12.48), and (12.30) we obtain

$$c_1' = c_1 \cos \frac{\theta}{2} - i c_2 \sin \frac{\theta}{2}$$

$$c_2' = -i c_1 \sin \frac{\theta}{2} + c_2 \cos \frac{\theta}{2}$$
(12.58)

The components of a vector $\mathbf{A}$, on the other hand, transform according to

$$A_x' = A_x$$
$$A_y' = A_y \cos \theta - A_z \sin \theta$$
$$A_z' = A_y \sin \theta + A_z \cos \theta$$

We saw in Section 12.5 that $\boldsymbol{\sigma}$ is a vector operator. It is a remarkable property of spinor space that, apart from multiplication by a constant, this is the only possible vector operator.

Proof. Take the trace of the equations (12.42) which characterize a vector operator $\mathbf{A}$. Since the trace of any commutator vanishes, we find that

$$\text{trace } A_x = \text{trace } A_y = \text{trace } A_z = 0$$

Now, the most general 2×2 matrix with vanishing trace can be written as a linear combination of σ_x, σ_y, σ_z. Thus, for instance,

$$A_x = a_x \sigma_x + b_x \sigma_y + c_x \sigma_z, \text{ etc.}$$

Substituting these linear combinations into the commutation relations (12.42), we determine easily that $a_x = b_y = c_z$, but all other coefficients vanish. Hence,

$$\mathbf{A} = k\boldsymbol{\sigma}$$
(12.59)

where k is an arbitrary constant. Equation (12.59) is the simplest illustration of the Wigner-Eckart theorem which will be derived in Chapter 16.

Exercise 12.4. Distinguish those results of this section which are valid only for $n = 2$ from the properties which hold for $\boldsymbol{\sigma}$ matrices of arbitrary dimensions, if the equations (12.43) are used as defining relations.

7. The Magnetic Moment and Spin Operators. Since the vector operator $\boldsymbol{\sigma}$ is essentially unique, we conclude from (12.28) that the magnetic moment is given by

$$\mathbf{m} = -\beta_0\boldsymbol{\sigma} \tag{12.60}$$

thus completing the program of determining the components of $\mathbf{m}$ upon which we embarked in Section 12.5.

What about the *intrinsic angular momentum* of the electron, its *spin*? It was shown in Section 12.1 that conservation of angular momentum is destroyed unless the electron is endowed with an intrinsic angular momentum. Guided by the heuristic notion that the law of conservation of angular momentum has a universal validity, we inferred qualitatively the existence of an electron spin.

The interaction energy responsible for disturbing the spherical symmetry of the central forces is proportional to $\mathbf{m} \cdot \mathbf{L}$. If we assume that two-component spinors are the appropriate tools for describing the intrinsic degrees of freedom of an electron and that the electron wave function ψ can be written in the form

$$\psi = \begin{pmatrix} \psi_1(x, y, z) \\ \psi_2(x, y, z) \end{pmatrix}$$

this interaction energy is, according to (12.60), proportional to $\boldsymbol{\sigma} \cdot \mathbf{L}$. The operator $\mathbf{L}$ acts only on the coordinates x, y, z, but $\boldsymbol{\sigma}$ couples the two spinor components.

A term of the form $\boldsymbol{\sigma} \cdot \mathbf{L}$ is often referred to as the *spin-orbit interaction*. As was explained in Section 12.1, an interaction of this form arises in atoms as a magnetic and relativistic correction to the electrostatic central potential. In nuclei the spin-orbit energy has its origin in strong interactions and has, therefore, very conspicuous effects.

In the presence of a spin-orbit interaction, $\mathbf{L}$ is no longer a constant of the motion. It is our hope that an *intrinsic angular momentum* $\mathbf{S}$ can be defined in such a manner that, when it is added to the orbital angular momentum $\mathbf{L}$, the total angular momentum $\mathbf{J} = \mathbf{L} + \mathbf{S}$ will again be a constant of motion. Since $\mathbf{S}$, like $\mathbf{L}$, must be a vector operator, we have, according to Section 12.6, no choice but to assume

$$\mathbf{S} = a\boldsymbol{\sigma} \tag{12.61}$$

The constant a can be determined by requiring that

$$\mathbf{J} = \mathbf{L} + \mathbf{S} \tag{12.62}$$

shall commute with $\boldsymbol{\sigma} \cdot \mathbf{L}$, so that $\mathbf{J}$ will be a constant of the motion. For example, we demand for the z-component:

$$[\boldsymbol{\sigma} \cdot \mathbf{L}, J_z] = \sigma_x[L_x, L_z] + \sigma_y[L_y, L_z] + a[\sigma_x, \sigma_z]L_x + a[\sigma_y, \sigma_z]L_y = 0$$

Hence,

$$-\sigma_x i\hbar L_y + \sigma_y i\hbar L_x - 2ia\sigma_y L_x + 2ia\sigma_x L_y = 0$$

from which it follows that $a = \hbar/2$ and

$$\mathbf{S} = \hbar\boldsymbol{\sigma}/2 \tag{12.63}$$

Evidently, any component of this intrinsic angular momentum vector has the two eigenvalues $+\hbar/2$ and $-\hbar/2$. The maximum value of a component of $\mathbf{S}$ in units of $\hbar$ is $\frac{1}{2}$, and we say that *the electron has spin* $\frac{1}{2}$. Further, we note that

$$\mathbf{S}^2 = \frac{\hbar^2}{4}\,\boldsymbol{\sigma}\cdot\boldsymbol{\sigma} = \frac{3\hbar^2}{4}\,1 \tag{12.64}$$

Hence, any spinor is an eigenspinor of $\mathbf{S}^2$, with eigenvalue $\frac{3}{4}\hbar^2$.

Exercise 12.5. Obtain the commutation relations for the components of $\mathbf{S}$, and compare these with (9.5). Since $\mathbf{L}$ and $\mathbf{S}$ commute, what conclusions can be drawn for the commutators $[J_x, J_y]$, etc.?

It is interesting to note that the unitary operator which transforms the ψ of an electron with spin under an infinitesimal three-dimensional rotation must be given by [see (9.14) and (12.35)]

$$U = 1 - \frac{i}{2}\varepsilon\hat{\mathbf{n}}\cdot\boldsymbol{\sigma} - \frac{i}{\hbar}\varepsilon\hat{\mathbf{n}}\cdot\mathbf{L}$$

or

$$U = 1 - \frac{i}{\hbar}\varepsilon\hat{\mathbf{n}}\cdot(\mathbf{S} + \mathbf{L}) = 1 - \frac{i}{\hbar}\varepsilon\hat{\mathbf{n}}\cdot\mathbf{J} \tag{12.65}$$

We see that when the spin of the electron is taken into account, $\mathbf{J}$ is the generator of infinitesimal rotations (in units of $\hbar$), and conservation of angular momentum is merely a consequence of the invariance of the Hamiltonian under rotations. This broad viewpoint allows further generalizations of the concept of angular momentum (see Chapter 16).

8. Spin and Reflections. We have learned how spinors behave under rotation. A few words must be added concerning their behavior under *reflection* of the space coordinates. We attempt to relate by a unitary transformation the spinors χ and χ' which represent states that are mirror images of each other. Consider, for example, the reflection

$$x' = -x, \qquad y' = y, \qquad z' = z \tag{12.66}$$

and suppose that it induces a transformation U_x such that

$$\chi' = U_x\chi$$

At first sight there appears to be an embarrassing freedom in the choice of U_x, since the transformation properties of the spin under reflection are largely a matter of definition; but the choice is narrowed if we consider reflections in conjunction with rotations.

A reflection (12.66) commutes with a rotation about the x-axis. On the other hand, the result of a reflection (12.66) followed by a rotation through an angle θ about the y-axis is equal to a rotation about this axis through the angle $-\theta$ followed by the x-reflection; and similarly for rotations about the z-axis. Hence, we are led to ask for a matrix U_x satisfying the conditions

$$\exp\left(-\frac{i}{2}\theta\sigma_x\right)U_x = U_x \exp\left(-\frac{i}{2}\theta\sigma_x\right)$$

$$\exp\left(-\frac{i}{2}\theta\sigma_y\right)U_x = U_x \exp\left(\frac{i}{2}\theta\sigma_y\right)$$

$$\exp\left(-\frac{i}{2}\theta\sigma_z\right)U_x = U_x \exp\left(\frac{i}{2}\theta\sigma_z\right)$$

or, equivalently,

$$U_x^\dagger \sigma_x U_x = \sigma_x, \qquad U_x^\dagger \sigma_y U_x = -\sigma_y, \qquad U_x^\dagger \sigma_z U_x = -\sigma_z \qquad (12.67)$$

Condition (12.67) can be satisfied by

$$U_x = b\sigma_x, \qquad |b| = 1 \qquad (12.68)$$

By taking the expectation values of the operator equations (12.67), we see that under reflection $\boldsymbol{\sigma}$ behaves as an *axial* vector operator. (This follows also from the fact that $\hat{n} \cdot \langle \boldsymbol{\sigma} \rangle$ is a scalar and the rotation vector $\hat{n}$ an axial vector.)

The simultaneous reflection of all three coordinates (inversion), or *parity* operation,

$$x' = -x, \qquad y' = -y, \qquad z' = -z$$

induces a transformation in spinor space which is the product of three matrices proportional to σ_x, σ_y, σ_z. But, $\sigma_x \sigma_y \sigma_z = i\sigma_z^2 = i1$. Hence apart from an uninteresting phase factor, the parity operator for spinors is the unit matrix.[10]

Exercise 12.6. Show that the only matrix which commutes with σ_x, σ_y, and σ_z is a multiple of the unit matrix. Also show that no matrix exists which anticommutes with all three Pauli matrices.

[10] See Section 16.10 for more on reflections and parity.

CHAPTER 13

Dynamics of Two-Level Systems

1. The Equation of Motion. The fundamental dynamical assumption in the quantum mechanics of systems with two linearly independent states is made in analogy with the quantum mechanical wave equation. We postulate that the time development of a two-component state $\chi(t)$ is governed by the equation of motion,

$$i\hbar \frac{\partial \chi(t)}{\partial t} = H\chi(t) \tag{13.1}$$

where the Hamiltonian H is in this instance a 2×2 matrix characteristic of the physical system under consideration.

Obviously, if there are no time-dependent external influences acting and the system is invariant under translation in time, H must be a constant matrix, independent of t. Under these conditions, equation (13.1) can be integrated, giving

$$\chi(t) = e^{-(i/\hbar)Ht}\chi(0) \tag{13.2}$$

in terms of the initial state $\chi(0)$.

It is convenient to introduce the eigenvalues of H, which are defined as the roots of the characteristic equation

$$\det (H - \lambda I) = 0 \tag{13.3}$$

If there are two distinct roots $\lambda = E_1, E_2$, with $E_1 \neq E_2$, we have

$$H\chi_1 = E_1\chi_1 \quad \text{and} \quad H\chi_2 = E_2\chi_2 \tag{13.4}$$

and an arbitrary two-component spinor may be expanded as

$$\chi = c_1\chi_1 + c_2\chi_2 \tag{13.5}$$

If $f(z)$ is a function of a complex variable, the function $f(H)$ of the matrix H is defined by the relation

$$f(H)\chi = c_1 f(H)\chi_1 + c_2 f(H)\chi_2 = c_1 f(E_1)\chi_1 + c_2 f(E_2)\chi_2 \tag{13.6}$$

The following important equality is easily seen to hold:

$$f(H) = f(E_1) \frac{E_2 1 - H}{E_2 - E_1} + f(E_2) \frac{E_1 1 - H}{E_1 - E_2} \tag{13.7}$$

If the characteristic equation has only one distinct root, so that $E_2 = E_1$, the preceding equation degenerates into

$$f(H) = f(E_1) 1 + f'(E_1)(H - E_1 1) \tag{13.8}$$

Exercise 13.1. In the limit $E_2 \to E_1$, derive (13.8) from (13.7).

Equation (13.7) may be applied to expand the time development operator $f(H) = e^{-(i/\hbar)Ht}$ in the form

$$e^{-(i/\hbar)Ht} = \frac{1}{E_2 - E_1} (E_2 e^{-(i/\hbar)E_1 t} - E_1 e^{-(i/\hbar)E_2 t})$$

$$+ \frac{H}{E_2 - E_1} (e^{-(i/\hbar)E_2 t} - e^{-(i/\hbar)E_1 t}) \tag{13.9}$$

if $E_1 \neq E_2$. A system whose Hamiltonian has two distinct eigenvalues may be called a *two-level system*. The formula (13.9) answers all questions about its motion.

From (13.1) it follows readily that

$$\frac{d}{dt} [\chi^\dagger(t)\chi(t)] = -\frac{i}{\hbar} \chi^\dagger(t)(H - H^\dagger)\chi(t) \tag{13.10}$$

and if H is constant in time this may be integrated to give

$$\chi^\dagger(t)\chi(t) = \chi^\dagger(0)e^{(i/\hbar)H^\dagger t} e^{-(i/\hbar)Ht}\chi(0) \tag{13.11}$$

If the matrix H is Hermitian, $\chi^\dagger \chi$ is constant and probability is conserved. This must certainly happen if H represents the energy. If H is Hermitian, E_1 and E_2 are real.

If the Hermitian matrix

$$i(H - H^\dagger) = \Gamma \tag{13.12}$$

is positive definite, the system undergoes decay, since according to (13.10) the time derivative of $\chi^\dagger(t)\chi(t)$ is then negative. The eigenstates χ_1 and χ_2 give rise to the simple fundamental solutions of (13.1),

$$e^{-(i/\hbar)E_1 t}\chi_1 \quad \text{and} \quad e^{-(i/\hbar)E_2 t}\chi_2$$

The general solution of (13.1) is a linear combination of these. Unless the imaginary parts of E_1 and E_2 are equal, the state does not generally follow a pure exponential decay law.

The utility of equation (13.9) is evident, for instance, if we are required to calculate the probability amplitude for a transition from an initial state α to a state β. One gets immediately

$$\beta^\dagger e^{-(i/\hbar)Ht}\alpha = (0 \quad 1)e^{-(i/\hbar)Ht}\begin{pmatrix}1\\0\end{pmatrix} = \frac{H_{21}}{E_2 - E_1}(e^{-(i/\hbar)E_2 t} - e^{-(i/\hbar)E_1 t}) \quad (13.13)$$

The probability obtained from this expression clearly exhibits an interference term in addition to the sum of two exponentials. Such nonexponential decay has been observed in atomic spectroscopy and in the behavior of neutral K mesons.

2. Density Matrix and Polarization. In discussing two-level systems, we have so far characterized the states in terms of two-component spinors. In this section we shall consider some other methods of specifying a state. The spinor (γ_1, γ_2 real)

$$\chi = \begin{pmatrix}c_1\\c_2\end{pmatrix} = \begin{pmatrix}|c_1|e^{i\gamma_1}\\|c_2|e^{i\gamma_1}\end{pmatrix} \quad (13.14)$$

characterizes a particular state. However, the same physical state can be described by different spinors, since χ depends on four real parameters, but the probability interpretation of quantum mechanics claims that measurements can give us only two parameters; these are the relative probabilities $|c_1|^2 : |c_2|^2$ and the relative phase, $\gamma_1 - \gamma_2$, of c_1 and c_2. If χ is normalized to unity

$$|c_1|^2 + |c_2|^2 = 1 \quad (13.15)$$

the only remaining redundancy is the common phase factor of the components of χ, and this is acknowledged by postulating that χ and $e^{i\alpha}\chi$ (α:arbitrary, real) represent the same state.

An elegant and very useful way of representing the state without the phase arbitrariness is to characterize it by a Hermitian matrix ρ, defined by

$$\rho = \begin{pmatrix}|c_1|^2 & c_1 c_2{}^*\\c_2 c_1{}^* & |c_2|^2\end{pmatrix} \quad (13.16)$$

subject to condition (13.15) which requires that

$$\text{trace } \rho = 1 \quad (13.17)$$

This matrix is called the *density matrix*. According to the probability doctrine of quantum mechanics its knowledge exhausts all that we can find out about the state.

To demonstrate this assertion we show how, using ρ, we obtain the expectation value $\langle A \rangle$ of an operator A:

$$\langle A \rangle = \chi^\dagger A \chi = (c_1{}^* \quad c_2{}^*) \begin{pmatrix} A_{11} & A_{12} \\ A_{21} & A_{22} \end{pmatrix} \begin{pmatrix} c_1 \\ c_2 \end{pmatrix}$$

$$= |c_1|^2 A_{11} + c_1 c_2{}^* A_{21} + c_2 c_1{}^* A_{12} + |c_2|^2 A_{22}$$

$$= \mathrm{trace} \begin{pmatrix} |c_1|^2 & c_1 c_2{}^* \\ c_2 c_1{}^* & |c_2|^2 \end{pmatrix} \begin{pmatrix} A_{11} & A_{12} \\ A_{21} & A_{22} \end{pmatrix}$$

Hence, we have the fundamental formula

$$\langle A \rangle = \mathrm{trace}\,(\rho A) = \mathrm{trace}\,(A\rho) \tag{13.18}$$

Exercise 13.2. If A is a Hermitian matrix with eigenspinors u and v, corresponding to the distinct eigenvalues A_1' and A_2', show that the probability of finding A_1' in a measurement of A on the state χ is given by

$$|(u^\dagger \chi)|^2 = \mathrm{trace}\left(\rho \frac{A_2' - A}{A_2' - A_1'}\right) = \mathrm{trace}\,(\rho u u^\dagger)$$

Like any 2×2 matrix, ρ can be expanded in terms of σ_x, σ_y, σ_z, and 1. Since ρ is Hermitian and its trace equals unity, it can most generally be represented as

$$\rho = \tfrac{1}{2}(1 + \mathbf{P} \cdot \boldsymbol{\sigma}) \tag{13.19}$$

where P_x, P_y, P_z are three real numbers given by

$$\begin{aligned} P_x &= 2\,\mathrm{Re}\,(c_1{}^* c_2) \\ P_y &= 2\,\mathrm{Im}\,(c_1{}^* c_2) \\ P_z &= |c_1|^2 - |c_2|^2 \end{aligned} \tag{13.20}$$

It is immediately verified that ρ has eigenvalues 0 and 1. The eigenspinor which corresponds to the latter eigenvalue is χ itself, i.e.,

$$\rho\chi = \chi \tag{13.21}$$

The other eigenvector must be orthogonal to χ. ρ applied to it gives zero. Hence, when ρ is applied to an arbitrary state φ, we have

$$\rho\varphi = \chi(\chi^\dagger \varphi) \tag{13.22}$$

if χ is assumed to be normalized to unity. As φ in (13.22) is arbitrary, we infer the identity

$$\rho = \chi\chi^\dagger = \begin{pmatrix} c_1 \\ c_2 \end{pmatrix} (c_1{}^* \quad c_2{}^*) \tag{13.23}$$

in agreement with the definition (13.16). We see that ρ projects φ in the "direction" of χ. It follows that

$$\rho^2 = \rho \tag{13.24}$$

which can also be verified directly from (13.16). Equation (13.24) expresses the fact that ρ is *idempotent*. Applied to (13.19) and using (12.55), this gives us the condition

$$\mathbf{P} \cdot \mathbf{P} = P_x{}^2 + P_y{}^2 + P_z{}^2 = \mathbf{P}^2 = 1 \tag{13.25}$$

Hence, the state is characterized by two independent parameters, as it should be.

The expectation value of σ_x in the state χ is

$$\langle \sigma_x \rangle = \text{trace} \, (\rho \sigma_x) = \tfrac{1}{2} \, \text{trace} \, \sigma_x + \tfrac{1}{2} \mathbf{P} \cdot \text{trace} \, (\boldsymbol{\sigma} \sigma_x) = \tfrac{1}{2} P_x \, \text{trace} \, (\sigma_x{}^2)$$

where use is made of the fact that trace $\boldsymbol{\sigma} = 0$. Since $\sigma_x{}^2 = 1$, we get from this and analogous equations for σ_y and σ_z the simple result

$$\mathbf{P} = \langle \boldsymbol{\sigma} \rangle = \text{trace} \, (\rho \boldsymbol{\sigma}) = \text{trace} \, (\boldsymbol{\sigma} \rho) \tag{13.26}$$

proving that $\mathbf{P}$ indeed transforms like a vector under rotations.

Owing to (13.19) and (13.21), the spinor χ is an eigenspinor of the matrix $\mathbf{P} \cdot \boldsymbol{\sigma}$:

$$\mathbf{P} \cdot \boldsymbol{\sigma} \chi = \chi \tag{13.27}$$

Hence, the unit vector $\mathbf{P}$ may legitimately be said to point in the direction of the particle's spin. The vector $\mathbf{P}$ is also known as the *polarization vector* of the state.

Exercise 13.3. Given a spinor

$$\chi = \begin{pmatrix} e^{i\alpha} \cos \delta \\ e^{i\beta} \sin \delta \end{pmatrix}$$

calculate the polarization vector and construct the matrix U_R which rotates this state into $\begin{pmatrix} 1 \\ 0 \end{pmatrix}$.

The language we have used in describing the properties of $\mathbf{P}$ refers to spin and rotations in ordinary space. However, the concepts have more general applicability, and the formalism allows us to define a polarization "vector" corresponding to the state of any two-level system. $\mathbf{P}$ is then a "vector" in an abstract three-dimensional Euclidean space, and the operator

$$\exp \left(-\frac{i}{2} \theta \hat{\mathbf{n}} \cdot \boldsymbol{\sigma} \right)$$

induces "rotations" in this space. (*Example.* The isospin space in nuclear physics.)

The development in time of the density matrix ρ can be inferred from the equation of motion for χ,

$$i\hbar \frac{d\chi}{dt} = H\chi \qquad (13.28)$$

where H is assumed to be a Hermitian 2×2 matrix. Using the definition (13.23) of the density matrix, we obtain

$$\frac{d\rho}{dt} = \frac{\partial \chi}{\partial t}\chi^{\dagger} + \chi \frac{\partial \chi^{\dagger}}{\partial t} = \frac{1}{i\hbar} H\chi\chi^{\dagger} - \frac{1}{i\hbar}\chi\chi^{\dagger}H$$

or

$$i\hbar \frac{d\rho}{dt} = H\rho - \rho H \qquad (13.29)$$

Exercise 13.4. Derive the properties of the density matrix which represents a stationary state.

From (13.18) and (13.29) the equation of motion for any expectation value $\langle A \rangle$ can be derived:

$$i\hbar \frac{d\langle A \rangle}{dt} = i\hbar \,\text{trace}\left(\frac{d\rho}{dt} A\right) + i\hbar \,\text{trace}\left(\rho \frac{\partial A}{\partial t}\right)$$

$$= \text{trace}\,[(H\rho - \rho H)A] + i\hbar \left\langle \frac{\partial A}{\partial t}\right\rangle$$

$$= \text{trace}\,[\rho(AH - HA)] + i\hbar \left\langle \frac{\partial A}{\partial t}\right\rangle = \langle AH - HA \rangle + i\hbar \left\langle \frac{\partial A}{\partial t}\right\rangle$$

$$(13.30)$$

in agreement with the similar equation (8.100).

It is instructive to derive the equation of motion for the vector $\mathbf{P} = \langle \boldsymbol{\sigma} \rangle$. To obtain a simple formula it is convenient to represent the Hamiltonian operator H as

$$H = \tfrac{1}{2}(Q_0 I + \mathbf{Q}\cdot\boldsymbol{\sigma}) \qquad (13.31)$$

where Q_0 and the three components of $\mathbf{Q}$ are real numbers. (All these equations hold, even if H is a function of time.) By (13.30), (13.31), and (12.43a),

$$\frac{d\mathbf{P}}{dt} = \frac{d\langle \boldsymbol{\sigma} \rangle}{dt} = \frac{1}{i\hbar}\langle \boldsymbol{\sigma}H - H\boldsymbol{\sigma}\rangle = \frac{1}{2i\hbar}\langle \boldsymbol{\sigma}\mathbf{Q}\cdot\boldsymbol{\sigma} - \mathbf{Q}\cdot\boldsymbol{\sigma}\boldsymbol{\sigma}\rangle$$

$$= \frac{1}{2i\hbar}\langle \mathbf{Q} \times (\boldsymbol{\sigma} \times \boldsymbol{\sigma})\rangle = \frac{1}{\hbar}\mathbf{Q} \times \langle \boldsymbol{\sigma} \rangle$$

or

$$\hbar \frac{d\mathbf{P}}{dt} = \mathbf{Q} \times \mathbf{P} \qquad (13.32)$$

Such a simple equation of motion is familiar from the classical mechanics of the spinning top. Since

$$\frac{d\mathbf{P}^2}{dt} = \frac{d(\mathbf{P} \cdot \mathbf{P})}{dt} = 2\mathbf{P} \cdot \frac{d\mathbf{P}}{dt} = \frac{2}{\hbar} \mathbf{P} \cdot (\mathbf{Q} \times \mathbf{P}) = 0 \qquad (13.33)$$

the vector $\mathbf{P}$ maintains a constant length. This is merely an alternative way of saying that the normalization of χ is conserved during the motion.

If $\mathbf{Q}$ is a *constant* vector, (13.32) implies that $\mathbf{P}$ precesses about $\mathbf{Q}$ with a constant angular velocity

$$\boldsymbol{\omega}_Q = \frac{\mathbf{Q}}{\hbar} \qquad (13.34)$$

If

$$\mathbf{P}(0) = \mathbf{P}_0 \qquad \text{and} \qquad \mathbf{Q}/Q = \hat{\mathbf{Q}} \qquad (13.35)$$

the solution of (13.32) is

$$\mathbf{P}(t) = \hat{\mathbf{Q}}(\mathbf{P}_0 \cdot \hat{\mathbf{Q}}) + [\mathbf{P}_0 - \hat{\mathbf{Q}}(\mathbf{P}_0 \cdot \hat{\mathbf{Q}})] \cos \omega_Q t + \hat{\mathbf{Q}} \times \mathbf{P}_0 \sin \omega_Q t \qquad (13.36)$$

Exercise 13.5. Show that, if $\mathbf{Q}$ is constant, $\mathbf{Q} \cdot \mathbf{P}$ and $(d\mathbf{P}/dt)^2$ are constants of the motion. Verify that (13.36) is the solution of (13.32) (See also Exercise 12.3.)

If $\mathbf{Q}$ is a constant vector and $\mathbf{P}_0$ parallel to $\mathbf{Q}$, it is seen from (13.36) and Figure 13.1 that $\mathbf{P}$ is constant and equal to $\hat{\mathbf{Q}}$ or $-\hat{\mathbf{Q}}$. These two vectors represent the two stationary states of the system. The energies of these levels are given by the eigenvalues of H, but only their difference, ΔE, is of physical interest. Since $\hat{\mathbf{Q}} \cdot \boldsymbol{\sigma}$ has the eigenvalues $+1$ and -1, the eigenvalues of H are

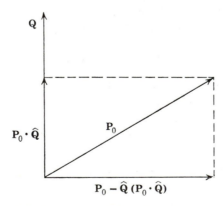

Figure 13.1. Precession of the spin polarization vector.

$\frac{1}{2}(Q_0 \pm Q)$ and

$$\Delta E = Q = \hbar \omega_Q \qquad (13.37)$$

The similarity between (13.37) and the familiar Bohr frequency condition suggests that transitions can be induced between the two stationary states if the spin system is exposed to an oscillating field which has the same frequency as the spin precession. For example, if a spin $\frac{1}{2}$ particle, whose degrees of freedom other than the spin can be neglected, is in a magnetic field **B**, the Hamiltonian can be written as

$$H = -\mathbf{m} \cdot \mathbf{B} = -\gamma \frac{\hbar}{2} \boldsymbol{\sigma} \cdot \mathbf{B} \qquad (13.38)$$

The quantity γ is the *gyromagnetic ratio*, and the vector **Q** is given by

$$\mathbf{Q} = -\hbar \gamma \mathbf{B}$$

A constant field $\mathbf{B}_0$ causes a precession of **P** with angular velocity $\omega_0 = -\gamma B_0$.[1] If in addition an oscillating magnetic field with the same (or nearly the same) frequency is applied, the system will absorb or deliver energy, and the precession motion of **P** will be changed. These general principles are at the basis of all the magnetic resonance techniques so widely used in atomic and nuclear studies.

A special case of an *oscillating* field, for which a solution of the equation of motion can be obtained easily, is that in which the vector **Q**, instead of being constant in time, rotates uniformly about a fixed axis. Suppose that $\boldsymbol{\omega}$ is its angular velocity. It is then advantageous to change over to a frame of reference which is rotating with this same angular velocity. Viewed from the rotating frame of reference, **Q** is a constant vector. If we denote the time rate of change of **P** with respect to the fixed system by $d\mathbf{P}/dt$, and with respect to the rotating system by $\partial \mathbf{P}/\partial t$, we have

$$\frac{d\mathbf{P}}{dt} = \frac{\partial \mathbf{P}}{\partial t} + \boldsymbol{\omega} \times \mathbf{P}$$

as is well known from the kinematics of rigid bodies; hence,

$$\hbar \frac{\partial \mathbf{P}}{\partial t} = (\mathbf{Q} - \hbar \boldsymbol{\omega}) \times \mathbf{P} \qquad (13.39)$$

Since in the rotating frame of reference $\mathbf{Q} - \hbar \boldsymbol{\omega}$ is a constant vector, the problem has effectively been reduced to the previous one. Equation (13.39) can therefore be solved by transcribing the solution of (13.32) appropriately.

---

[1] This is a quantum analog of the classical Larmor precession. H. Goldstein, *Classical Mechanics*, Addison-Wesley Publishing Company, Cambridge, 1953, Section 5-8.

Exercise 13.6. If **Q** rotates uniformly about a fixed axis, the equation of motion (13.28) may conveniently be transformed to a frame of reference which rotates similarly. Derive the new Hamiltonian and show that it corresponds effectively to precession about the constant vector **Q** $-$ $\hbar\boldsymbol{\omega}$.

Exercise 13.7. If a constant magnetic field $\mathbf{B}_0$, pointing along the z-axis, and a field $\mathbf{B}_1$, rotating with angular velocity ω in the xy-plane, act in concert on a spin system (gyromagnetic ratio γ), calculate the polarization vector **P** as a function of time. Assume **P** to point in the z-direction at $t = 0$. Calculate the average probability that the particle has "spin down," and plot this probability as a function of ω/ω_0 for a value of $B_1/B_0 = 0.1$. (This arrangement is a model for all magnetic resonance experiments.)

3. Polarization and Scattering. The theory of scattering was developed in Chapter 11, neglecting the spin entirely. However, the forces that cause a beam of particles to be scattered may be spin-dependent, and it is then necessary to supplement the theory accordingly. The incident particles, which are assumed to have spin $\frac{1}{2}$, are represented by a wave function of the form

$$e^{ikz}\chi_{\text{inc}} = e^{ikz}\begin{pmatrix} c_1 \\ c_2 \end{pmatrix}, \qquad \chi_{\text{inc}}^{\dagger}\chi_{\text{inc}} = 1 \tag{13.40}$$

Following the procedure of Chapter 11, we must look for asymptotic solutions of the Schrödinger equation which have the form

$$e^{ikz}\chi_{\text{inc}} + f(\theta, \varphi)\frac{e^{ikr}}{r} \tag{13.41}$$

but the scattering amplitude $f(\theta, \varphi)$ is now a two-component spinor. Spin-dependent scattering occurs, for instance, if the Hamiltonian has the form

$$H = \frac{\mathbf{p}^2}{2\mu} + V(r) + W(r)\mathbf{L} \cdot \boldsymbol{\sigma} \tag{13.42}$$

representing a spin-orbit interaction term in addition to a central force. Since **L** and $\boldsymbol{\sigma}$ are axial vectors, H as given by (13.42) is a scalar invariant under rotations and reflections.

The superposition principle—and more specifically, the linearity of the Schrödinger equation—allows us to construct the solution (13.41) from the two particular solutions which correspond to $\chi_{\text{inc}} = \alpha$ and $\chi_{\text{inc}} = \beta$. These two special cases describe incident beams which are polarized along the direction of the initial momentum and opposite to it. The polarization is said to be *longitudinal*. We are thus led to look for two solutions of the asymptotic form

$$\psi_1 \sim e^{ikz}\alpha + (S_{11}\alpha + S_{21}\beta)\,\frac{e^{ikr}}{r} \tag{13.43a}$$

$$\psi_2 \sim e^{ikz}\beta + (S_{12}\alpha + S_{22}\beta)\,\frac{e^{ikr}}{r} \tag{13.43b}$$

The quantities in parentheses are the appropriate scattering amplitudes.

Exercise 13.8. Show that the incident waves $e^{ikz}\alpha$ and $e^{ikz}\beta$ are eigenstates of J_z. What are the eigenvalues?

Multiplying (13.43a) by c_1, and (13.43b) by c_2, and adding the two equations, we obtain the more general solution

$$e^{ikz}\begin{pmatrix} c_1 \\ c_2 \end{pmatrix} + \frac{e^{ikr}}{r}\begin{pmatrix} S_{11} & S_{12} \\ S_{21} & S_{22} \end{pmatrix}\begin{pmatrix} c_1 \\ c_2 \end{pmatrix} = \left(e^{ikz} + \frac{e^{ikr}}{r}\,S \right)\chi_{\mathrm{inc}} \tag{13.44}$$

Here S stands for the 2×2 *scattering matrix*

$$S = \begin{pmatrix} S_{11} & S_{12} \\ S_{21} & S_{22} \end{pmatrix} \tag{13.45}$$

S depends on the angles θ and φ, and on the momentum k. The scattering problem is solved if S can be determined as a function of these variables.

The form of S can be largely predicted by very general invariance arguments, although its dependence on the scattering angle θ can be worked out only by a detailed calculation, such as a phase shift analysis. Here we shall be satisfied to deduce the general form of the scattering matrix. The basic idea is to utilize the obvious constants of the motion which the symmetries of the problem generate. If A commutes with the Hamiltonian, then if ψ is an eigenfunction of H, $A\psi$ is likewise an eigenfunction of H, and both belong to the same energy. $A\psi$ may represent the same scattering state as ψ, or a different one of the same energy, depending on the asymptotic form of ψ.

Let us assume that, owing to spherical symmetry of the scattering potential, H is invariant under rotations and, according to Section 12.7, commutes with the components of **J**. Expression (13.42) shows an example of a spin-dependent Hamiltonian with rotational symmetry. The incident waves in (13.43a) and (13.43b) are eigenstates of J_z with eigenvalues $+\hbar/2$ and $-\hbar/2$, respectively (Exercise 13.8). Since the operator J_z leaves the radial dependence of the scattered wave unchanged, the solutions (13.43) must both be eigenfunctions of J_z. By requiring that

$$J_z\psi_1 = \hbar\left(\frac{1}{i}\frac{\partial}{\partial\varphi} + \frac{1}{2}\sigma_z \right)\psi_1 = \frac{\hbar}{2}\psi_1$$

and

$$J_z \psi_2 = \hbar \left(\frac{1}{i} \frac{\partial}{\partial \varphi} + \frac{1}{2} \sigma_z \right) \psi_2 = -\frac{\hbar}{2} \psi_2$$

it is easily seen that S_{11} and S_{22} can be functions of θ only and that the off-diagonal elements of the scattering matrix have the form

$$S_{12} = e^{-i\varphi} \cdot \text{function of } \theta, \qquad S_{21} = e^{i\varphi} \cdot \text{function of } \theta \qquad (13.46)$$

Furthermore, H is invariant under a reflection with respect to any coordinate plane. It was shown in Section 12.8 that the operator for reflection in the yz-plane is $P_x \sigma_x$, where P_x simply changes x into $-x$, and σ_x has the effect

$$\sigma_x \alpha = \beta, \qquad \sigma_x \beta = \alpha$$

Under this operation the incident wave $e^{ikz}\alpha$ changes into $e^{ikz}\beta$ and e^{ikr}/r is invariant. Hence, (13.43a) must go over into (13.43b). In terms of spherical polar coordinates P_x has the effect of changing φ into $\pi - \varphi$. It follows from this and (13.46) that

$$S_{11} = S_{22} = g(\theta), \qquad S_{21}(-\varphi, \theta) = -S_{12}(\varphi, \theta) = -e^{-i\varphi}h(\theta)$$

Consequently, we may write

$$S = \begin{pmatrix} g(\theta) & h(\theta)e^{-i\varphi} \\ -h(\theta)e^{i\varphi} & g(\theta) \end{pmatrix} = g(\theta)1 + ih(\theta)(\sigma_y \cos \varphi - \sigma_x \sin \varphi) \quad (13.47)$$

The vector $\hat{n}(-\sin \varphi, \cos \varphi, 0)$ is the normal to the plane of scattering. Hence, we conclude that the scattering matrix has the form

$$S = g(\theta)1 + ih(\theta)\hat{n} \cdot \boldsymbol{\sigma} \qquad (13.48)$$

This is as much as we can infer about the scattering matrix from the invariance properties of the Hamiltonian. The functions $g(\theta)$ and $h(\theta)$ can be determined only by solving the Schrödinger equation.

Knowing S, it is easy to calculate the intensity of the beam for a given direction. If (13.44) is the asymptotic form of the wave function, then by a straightforward generalization of the results of Chapter 11 the differential scattering cross section is found to be

$$\frac{d\sigma}{d\Omega} = (S\chi_{\text{inc}})^\dagger S\chi_{\text{inc}} = \chi_{\text{inc}}^\dagger S^\dagger S\chi_{\text{inc}} \qquad (13.49)$$

which is merely the analog of $|f(\theta)|^2$ for a particle with spin. If the density matrix ρ_{inc} describes the state of polarization of the incident beam, this expression may also be written as

$$\frac{d\sigma}{d\Omega} = \text{trace}(\rho_{\text{inc}} S^\dagger S) \qquad (13.50)$$

Using the form (13.48) for the scattering matrix and

$$\rho_{inc} = \tfrac{1}{2}(1 + \mathbf{P}_0 \cdot \boldsymbol{\sigma})$$

for the incident density matrix, we obtain the cross section in terms of the polarization $\mathbf{P}_0$ of the incident beam:

$$\frac{d\sigma}{d\Omega} = (|g|^2 + |h|^2)\left(1 + i\,\frac{g^*h - gh^*}{|g|^2 + |h^2|}\,\mathbf{P}_0 \cdot \hat{\mathbf{n}}\right) \tag{13.51}$$

If the y-axis is chosen to be along the direction of the transverse component of the polarization, $\mathbf{P}_0 - \mathbf{P}_0 \cdot \hat{\mathbf{k}}\hat{\mathbf{k}}$, we may write $\mathbf{P}_0 \cdot \hat{\mathbf{n}} = |\mathbf{P}_0 - \mathbf{P}_0 \cdot \hat{\mathbf{k}}\hat{\mathbf{k}}| \cos \varphi$. With these conventions, formula (13.51) shows that the scattered intensity depends on the azimuthal angle as $I = a(\theta) + b(\theta) \cos \varphi$, in agreement with the empirical statement (12.7) in Section 12.2. In this way we find substantiated our original supposition that the right-left asymmetry in the scattering of polarized beams is a consequence of the particle spin.

4. Polarization in an Ensemble. Although we have found the density matrix to be an attractive tool for describing a definite state χ, its full advantages become apparent only when it is applied to a *statistical ensemble* composed of N systems, which all have identical structure but which may be in different quantum states. Each system τ is therefore represented by its own spinor χ_τ and its own density matrix ρ_τ. If such an ensemble is used to measure a physical quantity A, the maximum information that can be obtained, without identifying individual members of the ensemble, consists of a kind of census, telling us that the fraction N_1/N has the value A_1', N_2/N the value A_2', etc. If A_i' is the eigenvalue of A corresponding to an eigenstate ξ_i, the statistical data are related to the quantum states by the equation

$$\frac{N_i}{N} = \frac{1}{N}\sum_{\tau=1}^{N}|\xi_i{}^\dagger\chi_\tau|^2 = \frac{1}{N}\sum_{\tau=1}^{N}\xi_i{}^\dagger\chi_\tau\chi_\tau{}^\dagger\xi_i = \frac{1}{N}\xi_i{}^\dagger\left(\sum_{\tau=1}^{N}\rho_\tau\right)\xi_i$$

If we define an *average density matrix* for the ensemble

$$\bar{\rho} = \frac{1}{N}\sum_{\tau=1}^{N}\rho_\tau \tag{13.52}$$

the relative populations of the various eigenstates of A are simply expressed as

$$\frac{N_i}{N} = \xi_i{}^*\bar{\rho}\xi_i = \langle\bar{\rho}\rangle_i \tag{13.53}$$

where $\langle \ \rangle_i$ denotes the expectation value in the state ξ_i.

The expectation value of A, averaged over the ensemble, can also be expressed in terms of the matrix $\bar{\rho}$. Indeed, we have

$$\overline{\langle A \rangle} = \frac{1}{N} \sum_{\tau=1}^{N} \langle A \rangle_\tau = \frac{1}{N} \sum_{\tau=1}^{N} \text{trace }(\rho_\tau A) = \text{trace }(\bar{\rho}A) \qquad (13.54)$$

Since quantal probabilities and expectation values depend on the density matrix linearly, whereas they depend on the state χ quadratically, it is clear that statistical ensemble averages can be entirely expressed in terms of the matrix $\bar{\rho}$.

A polarization vector for the ensemble, such as a beam of spin $\frac{1}{2}$ particles, can be defined by

$$\bar{\mathbf{P}} = \frac{1}{N} \sum_{\tau=1}^{N} \langle \boldsymbol{\sigma} \rangle_\tau \qquad (13.55)$$

This is related to the matrix $\bar{\rho}$ by the equation

$$\bar{\rho} = \tfrac{1}{2}(I + \bar{\mathbf{P}} \cdot \boldsymbol{\sigma}) \qquad (13.56)$$

The proof parallels that given for the density matrix in Section 13.2. However, it is important to note that $\bar{\rho}$ is usually not idempotent, nor is $\bar{\mathbf{P}}$ generally a unit vector as for a single system.

From (13.57) we obtain

$$(\bar{\rho})^2 = \bar{\rho} + \tfrac{1}{4}(\bar{P}^2 - 1) \qquad (13.57)$$

where $\bar{P}$ denotes the magnitude of $\bar{\mathbf{P}}$.

By its definition $\bar{\rho}$ is a positive definite Hermitian matrix with

$$\text{trace }\bar{\rho} = 1 \qquad (13.58)$$

For any such matrix, owing to the positive definiteness of the eigenvalues,

$$0 \leqslant \text{trace }(\bar{\rho})^2 \leqslant (\text{trace }\bar{\rho})^2 = 1 \qquad (13.59)$$

If this inequality is applied to the trace of equation (13.58), we conclude that $0 \leqslant \bar{P} \leqslant 1$.

If $\bar{P} = 0$, $\bar{\rho} = \tfrac{1}{2}I$, and we say that the ensemble is *unpolarized*. If $\bar{P} = 1$, $(\bar{\rho})^2 = \bar{\rho}$, and the eigenvalues of $\bar{\rho}$ must be 0 and 1. In this case, the ensemble can be uniquely regarded as being composed of N systems, each of which is in the same two-component state, the eigenstate of $\bar{\rho}$ which corresponds to the eigenvalue 1. Such an ensemble is said to be *homogeneous* or in a statistically *pure* state. If $\bar{P} < 1$, the ensemble is sometimes referred to as a *mixture*.

Much of this chapter has been concerned with the description of the state of a system in terms of a vector $\mathbf{P}$. The possibility of such a simple description is confined to dynamical variables which have only two states. Thus, whereas the state of a particle of spin $\frac{1}{2}$ is completely defined by a polarization vector, tensors of higher rank are required to characterize fully the states of particles with higher spin. However, the density matrix techniques are not limited to dichotomic variables. They can be generalized to any number of states and are particularly useful for the description of ensembles of particles with higher intrinsic angular momentum.

Exercise 13.9. If an unpolarized beam of spin $\frac{1}{2}$ particles is incident on a spherically symmetric scatterer, prove that the polarization of the scattered beam is given by

$$\mathbf{P} = i \frac{g^*h - gh^*}{|g|^2 + |h|^2} \,\hat{\mathbf{n}} \tag{13.60}$$

5. *Measurements and Probabilities.* The spin formalism is so easy to survey that it lends itself particularly well to a demonstration of how quantum mechanics is to be interpreted. We have already seen that in the expansion of an arbitrary spinor,

$$\chi = c_1\alpha + c_2\beta \tag{13.61}$$

the quantities $|c_1|^2$ and $|c_2|^2$ give the probability of finding the system with spin "up" or spin "down" respectively. The only possible results of a measurement of S_z (or of any other component of $\mathbf{S}$) are $+\hbar/2$ or $-\hbar/2$. By measuring S_z for a large number of replicas of the system, all in the same state, i.e., by using a homogeneous ensemble, we can determine the absolute values of c_1 and c_2 experimentally.

This discussion shows again that the predictions of quantum mechanics are statistical. Although the wave function must be regarded as describing a single system, this wave function can be compared with experimental measurements only by the use of a large number of identical systems. The situation is analogous to the one encountered customarily in the use of probability concepts. For instance, the probability of tossing heads with a slightly unsymmetrical coin might be 0.48, that of tossing tails 0.52. These probabilities refer to the particular coin and to the individual tossings, but the experimental verification of the statement "the probability for tossing heads is 48 %" requires that we toss many identical coins, or the same coin many times in succession.

A measurement of S_z yields the absolute values of c_1 and c_2, but their phases cannot be determined by such a measurement. Other components of $\mathbf{S}$ must be measured in addition to S_z if further information is to be gained about the wave function. If we wish to determine the probabilities of finding the values $+\hbar/2$ or $-\hbar/2$ as a result of measuring $\hat{\mathbf{p}} \cdot \mathbf{S}$, we must expand χ in terms of the eigenspinors of $\hat{\mathbf{p}} \cdot \mathbf{S}$. This problem has already been solved in Exercise 12.2. Consider the special case where $\hat{\mathbf{p}}$ is a unit vector in the yz-plane making an angle θ with the z-axis and $90° - \theta$ with the y-axis (Figure 12.3). $\hat{\mathbf{p}}$ is obtained from the unit vector in the z-direction by a rotation through an angle $-\theta$ about the x-axis. The state χ is related to $\hat{\mathbf{p}}$

in exactly the same way as a state $\chi' = U\chi$ to the z-axis, provided that U represents a right-handed rotation through the angle θ about the x-axis. Hence, by (12.58):

$$\begin{pmatrix} c_1' \\ \\ c_2' \end{pmatrix} = \begin{pmatrix} \cos\dfrac{\theta}{2} & -i\sin\dfrac{\theta}{2} \\ \\ -i\sin\dfrac{\theta}{2} & \cos\dfrac{\theta}{2} \end{pmatrix} \begin{pmatrix} c_1 \\ \\ c_2 \end{pmatrix} \tag{13.62}$$

The probability that $\hat{\mathbf{p}} \cdot \mathbf{S}$ is $\hbar/2$ is given by

$$|c_1'|^2 = \left| c_1 \cos\frac{\theta}{2} - ic_2\sin\frac{\theta}{2} \right|^2$$

$$= |c_1|^2 \cos^2\frac{\theta}{2} + |c_2|^2 \sin^2\frac{\theta}{2} - |c_1||c_2|\sin(\gamma_1 - \gamma_2)\sin\theta \tag{13.63}$$

if

$$c_1 = |c_1|e^{i\gamma_1}, \qquad c_2 = |c_2|e^{i\gamma_2} \tag{13.64}$$

Similarly, the probability that $\hat{\mathbf{p}} \cdot \mathbf{S}$ is $-\hbar/2$ is given by

$$|c_2'|^2 = |c_1|^2 \sin^2\frac{\theta}{2} + |c_2|^2 \cos^2\frac{\theta}{2} + |c_1||c_2|\sin(\gamma_1 - \gamma_2)\sin\theta \tag{13.65}$$

These equations exhibit the fact that a measurement of $\hat{\mathbf{p}} \cdot \mathbf{S}$ on an ensemble of systems, all represented by χ, can be used to determine the relative phase $\gamma_1 - \gamma_2$. It is entirely satisfactory that we cannot determine γ_1 and γ_2 separately but only their difference, for c_1 and c_2 can be multiplied by a common arbitrary phase factor without affecting the physical state or the result of any measurement.

The complete determination of the spin wave function thus requires that we measure two different components of the spin vector $\mathbf{S}$, for example, S_z and S_y. A molecular beam experiment of the Stern-Gerlach type has traditionally been regarded as the prototype of a measurement, fundamental to a proper understanding of quantum mechanics. When, as depicted in Figure 12.1, the z-component of the spin is measured, there is a bodily separation of the particles that are subjected by the experimenter to the question "Is the spin up or down?" The beam splits into two components made up respectively of those particles which respond with "up" or with "down" to this experimental question. A careful analysis of the Stern-Gerlach experiment shows that near the magnet the state of the particles can be represented as

$$c_1\psi_{\mathrm{upper}}(x, y, z)\alpha + c_2\psi_{\mathrm{lower}}(x, y, z)\beta \tag{13.66}$$

where $\psi_{\mathrm{upper}}(\psi_{\mathrm{lower}})$ is a normalized function which differs from zero only in the region traversed by the upper (lower) beam. The upper component of the wave function is said to be *correlated* with α and the lower component with β. In the measurement a particle reveals a spin "up" or "down" with probabilities equal to $|c_1|^2$ and $|c_2|^2$.

Before the particle interacts with the measuring apparatus, its state may be assumed to have the simple product form

$$\psi(x, y, z)(c_1\alpha + c_2\beta) \tag{13.67}$$

The interaction causes this state to change into the more complicated correlated state of the form (13.66). We thus see that the act of measurement has a profound effect on the state itself.

In this connection it is interesting to give some thought to a multiple Stern-Gerlach experiment in which two or more spin measurements are carried out in series. Let us assume again that S_z is measured in the first experiment. If in the second experiment S_z is remeasured, we will find that every particle in the upper beam has spin up, and every particle in the lower beam has spin down. Neither beam is split any further, confirming merely that the Stern-Gerlach experiment is conceptually a particularly simple kind of measurement, in which the measurement itself does not alter the value of the measured quantity (S_z), and that this quantity does not change between measurements. If in the second measurement the inhomogeneous magnetic field has a different direction, and thus a different component of the spin, say S_y, is measured, we will find that each beam is split into two components of equal intensity, corresponding to the values $+\hbar/2$ and $-\hbar/2$ for S_y (Figure 13.2).

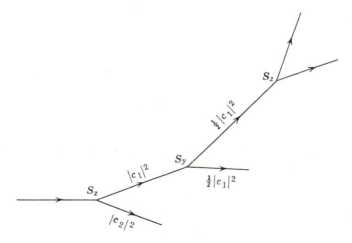

Figure 13.2. Double and triple Stern-Gerlach experiment. A measurement of S_z is followed by a measurement of S_y and another measurement of S_z.

This example shows the unavoidable effect which in the quantum domain a measurement has on the system upon which the measurement is carried out. If $\chi = c_1\alpha + c_2\beta$ is the spin state before the measurement, and S_y, rather than S_z, is measured in a first experiment, then according to (13.63) the probability of finding $S_y = +\hbar/2$ is ($\theta = 90°$)

$$|c_1'|^2 = \tfrac{1}{2}|c_1|^2 + \tfrac{1}{2}|c_2|^2 - |c_1||c_2| \sin(\gamma_1 - \gamma_2) = \tfrac{1}{2} - |c_1||c_2| \sin(\gamma_1 - \gamma_2)$$

whereas, if we precede this S_y measurement by a measurement of S_z, the probability of finding S_y to be $+\hbar/2$ is simply $\tfrac{1}{2}|c_1|^2 + \tfrac{1}{2}|c_2|^2 = \tfrac{1}{2}$, in accordance with the common rule of compounding conditional probabilities. The probability $|c_1'|^2$ differs from $\tfrac{1}{2}|c_1|^2 + \tfrac{1}{2}|c_2|^2$ by an interference term which the intervening S_z measurement must wipe out if the probability interpretation of quantum mechanics is to be consistent. If in a third successive Stern-Gerlach measurement S_z is measured again (Figure 13.2), we find anew a splitting of the beam, showing that the intervening measurement of S_y has undone what the first S_z measurement had accomplished.

In an arrangement of this kind two observables A and B are termed *compatible* if for any state of the system the results of a measurement of A are the same, whether a measurement of B precedes that of A or not. In other words, A and B are compatible if measuring B does not destroy the result of the determination of A. Clearly, this can happen only if the eigenfunctions of A are simultaneously also eigenfunctions of B. In the spin quantum mechanics we can show in analogy with Section 8.5 that the necessary and sufficient condition for this is that the matrices representing A and B commute:

$$AB - BA = 0$$

Two observables are compatible if and only if the Hermitian matrices representing them commute.

For example, S_z and S_y are incompatible, for they do not commute; a state cannot simultaneously have a definite value of S_z and S_y. If we wish to measure S_z *and* S_y for a state χ, two separate ensembles must be used. The two components of the spin cannot be measured simultaneously on the same system.

A measurement of the simple kind just described (and sometimes called a measurement of the first kind), such as the spatial separation of the two spin components in a Stern-Gerlach experiment, results in a homogeneous ensemble corresponding to a correlated state like (13.66). Yet, for many purposes this ensemble may eventually be replaced by the mixture in which a fraction $|c_1|^2$ of the particles is definitely in state α, and a fraction $|c_2|^2$ is in state β. It is sometimes said that the act of measurement puts the system into an eigenstate of the dynamical variable which is being measured, and the various eigenstate projections are prevented from interfering after the measurement, either through spatial separation, as in a Stern-Gerlach experiment, or in some other way.[2] The replacement of the original correlated state by one or the other of its components with definite probabilities is conventionally referred to as the *reduction of the state* (or wave packet). After this reduction has taken place, the system has a definite value of the observable, namely, the eigenvalue determined by the measurement. A repetition of the measurement of the same quantity will now yield with certainty this very eigenvalue.

Although much has been written about the quantum theory of measurement,[3] it is still not clear how in detail and precisely at what stage of the measurement process the reduction of the state takes place. It is therefore satisfactory that, wherever and whenever the reduction of the state may actually occur, in practical terms the experimental arrangement usually tells us unambiguously when the homogeneous ensemble of a correlated state may to very high approximation be replaced by a suitably weighted mixture of its components.

The brief discussion of this section was intended to illustrate the fundamental connection between the principles of quantum mechanics and the theory of measurement, emphasizing particularly the operation of the principle of superposition of states. Thus, a homogeneous ensemble represented by the spinor χ with $c_1 = \sqrt{\frac{1}{3}}$ and $c_2 = \sqrt{\frac{2}{3}}$ is not a collection of particles, one-third of which have definitely "spin up," the other two-thirds "spin down." Rather, each of the particles is represented by this wave function and is potentially able to go, upon measurement, into the "spin up" state (probability 1/3) or the "spin down" state (probability 2/3).

Exercise 13.10. Determine and compare the density matrices for the two ensembles, pure and mixed, just described.

---

[2] The great variety in the design of actual experiments defies any effort to classify all measurements systematically. Most measurements are more difficult to analyze than the Stern-Gerlach experiment, but for an understanding of the physical significance of quantum states it is sufficient to consider the simplest prototypes.

[3] For a recent review of the quantum theory of measurement see E. P. Wigner, *Am. J. Phys.*, **31**, 6 (1963).

CHAPTER 14

Linear Vector Spaces in Quantum Mechanics

1. *Introduction.* Wave mechanics in either the coordinate or the momentum representation and the matrix description of the spin have provided us with insight into two different, and yet in many respects similar, ways of treating a system by quantum methods. With these two approaches guiding us, we can now construct a much more general and abstract form of quantum mechanics which includes both wave mechanics and the matrix formulation as special cases. From a physicist's point of view mathematical abstractness would hardly seem a virtue, were it not that with the increased elegance a great deal of understanding can be gained. Besides allowing us to merge wave mechanics and the spin theory into a unified description of a particle with spin, the general formalism will prepare us for the application of quantum mechanics to different types of systems. Examples are systems composed of several particles, or systems with new degrees of freedom, such as the isospin. But above all there are physical systems, like the electromagnetic field, which although they are quite different from simple mechanical systems can nevertheless be treated in the same general framework.

The fundamental assumption which is common to the wave mechanics of a particle and the matrix theory of a system with spin $\frac{1}{2}$ is that the maximum information about the outcome of physical measurements on the system is contained in a function ψ.[1] The nature of the physical system determines on how many variables, such as x, y, z, σ_z, . . . , the function ψ depends. As we saw, these variables may take on a continuous range of values from $-\infty$ to $+\infty$, as x does, or merely a discrete and possibly even finite set of values, as for instance σ_z does. In the first case ψ is an ordinary function as studied in calculus, subject to differentiation and integration; in the second case it is convenient to use matrix notation to represent ψ and the linear operators which operate on it.

---

[1] In Chapters 12 and 13 on spin the state of the system was denoted by χ rather than ψ. We now drop the distinct notation in order to stress the similarity of wave mechanics and spin mechanics.

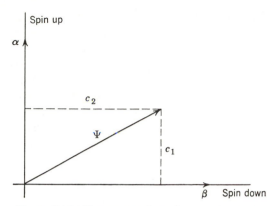

Figure 14.1. Representation of two spin states.

In both cases we saw that it was important that ψ could be regarded as a linear combination, or *superposition*, of certain basic functions, such as for instance plane waves of definite momentum, or "spin up" (α) and "spin down" (β) functions (spinors). The mathematical theory was brought in contact with physical reality by the assumption that the squares of the absolute value of the expansion coefficients in the superposition give the probabilities of finding the system, upon observation and measurement, in one of these basic states.

In the present chapter a mathematical structure will be introduced which can be made to contain all these concepts in a unified way. We shall consider the values of the function ψ as the components of a *vector* Ψ in a space whose "coordinate axes" are labeled by the values which the independent variables can take on. Specifically, we may call Ψ a *state vector*, or sometimes briefly a *state*. Such a vector is most easily visualized in the case of a system with spin $\frac{1}{2}$ and no other pertinent degrees of freedom. Figure 14.1 illustrates how Ψ can be considered as a linear combination of the basic vectors, α and β. A general vector has two components, c_1 and c_2, $|c_1|^2$ and $|c_2|^2$ being respectively the probabilities of finding the spin up and down. It is important not to confuse Figure 14.1 with a drawing in ordinary space, such as Figure 12.3. In fact, a figure like 14.1 is really inadequate, because the components c_1 and c_2 must in general be complex numbers, and no simple graphic representation is possible. Nevertheless, it is convenient to use the language of geometry, to introduce a product analogous to the scalar product of two vectors, and to say in the example given above that the two states α and β are orthogonal.

If ψ is a function of a variable, such as x, which takes on any real value between $-\infty$ and $+\infty$, the set of axes in the space in which Ψ is a vector is *indenumerably* infinite. Yet the analogy with ordinary geometry is still close, and the use of vector terminology remains appropriate.

Most of the discussion of this chapter will be confined to the *complex linear vector space* of n dimensions, even though n may in many cases eventually be allowed to become infinitely great, for in the application to quantum mechanics we shall be interested only in those properties and theorems for $n \to \infty$ which are straightforward generalizations of the finite dimensional theory. If, for these generalizations to hold, the vectors and operators of the space must be subjected to certain additional restrictive conditions, we shall suppose that these conditions are enforced. By confining ourselves to the complex linear vector space in n dimensions, we shall avoid questions which concern the convergence of sums over infinitely many terms, the interchangeability of several such summations, or the legitimacy of certain limiting processes.

Of course, it is necessary to show that the mathematical conclusions which we shall draw for the infinitely dimensional space by analogy with n-dimensional space can be rigorously justified. The following sections do not pretend to satisfy this quest for rigor. They are intended only to provide a working knowledge of the mathematical structure which underlies quantum mechanics.[2]

2. Vectors and Operators. Our abstract vector space is defined as a collection of vectors, denoted by Ψ, any two of which, say Ψ_a and Ψ_b, can be combined to define a new vector, denoted as the sum $\Psi_a + \Psi_b$, with the properties

$$\Psi_a + \Psi_b = \Psi_b + \Psi_a \tag{14.1}$$

$$\Psi_a + (\Psi_b + \Psi_c) = (\Psi_a + \Psi_b) + \Psi_c \tag{14.2}$$

These rules define the addition of vectors.

We also define the multiplication of a vector by an arbitrary complex number λ. This is done by associating with any vector Ψ a vector $\lambda\Psi$, subject to the rules

$$\mu(\lambda\Psi_a) = (\mu\lambda)\Psi_a \qquad (\lambda, \mu : \text{complex numbers}) \tag{14.3}$$

$$\lambda(\Psi_a + \Psi_b) = \lambda\Psi_a + \lambda\Psi_b \tag{14.4}$$

The vector space contains the *null vector* $\Psi_0 = 0$ such that

$$\Psi_a + \Psi_0 = \Psi_a \tag{14.5}$$

for any vector Ψ_a.

The k vectors $\Psi_1, \Psi_2, \ldots, \Psi_k$ are said to be *linearly independent* if no relation

$$\lambda_1\Psi_1 + \lambda_2\Psi_2 + \cdots + \lambda_k\Psi_k = 0$$

exists between them, except the trivial one with $\lambda_1 = \lambda_2 = \cdots = \lambda_k = 0$.

---

[2] For guidance toward more careful treatments, see the references cited in footnote 7 of Chapter 8.

The vector space is said to be *n-dimensional* if there exist n linearly independent vectors, but if no $n + 1$ vectors are linearly independent.

In an *n*-dimensional space we may choose a set of n linearly independent vectors $\Psi_1, \Psi_2, \ldots, \Psi_n$. We shall refer to these vectors as the members of a *basis*, or as *basis vectors*. They are said to *span* the space, or to form a *complete set of vectors*, since an arbitrary vector Ψ_a can be expanded in terms of these:

$$\Psi_a = \sum_{i=1}^{n} a_i \Psi_i \tag{14.6}$$

The coefficients a_i are complex numbers. They are called the *components* of the vector Ψ_a. The components determine the vector completely. The components of the sum of two vectors are equal to the sums of the components: If $\Psi_a = \sum a_i \Psi_i$ and $\Psi_b = \sum b_i \Psi_i$,

$$\Psi_a + \Psi_b = \sum_i (a_i + b_i)\Psi_i \tag{14.7}$$

and similarly

$$\lambda\Psi_a = \sum_i (\lambda a_i)\Psi_i \tag{14.8}$$

by the above rules for addition and multiplication.

Next we introduce a *scalar*[3] *product* between two vectors, denoted by the symbol (Ψ_a, Ψ_b). This is a complex number with the following properties:

$$(\Psi_b, \Psi_a) = (\Psi_a, \Psi_b)^* \tag{14.9}$$

where the asterisk denotes complex conjugation.

We further require that

$$(\Psi_a, \lambda\Psi_b) = \lambda(\Psi_a, \Psi_b) \tag{14.10}$$

From (14.9) and (14.10) it follows that

$$(\lambda\Psi_a, \Psi_b) = \lambda^*(\Psi_a, \Psi_b) \tag{14.11}$$

We also postulate that

$$(\Psi_a, \Psi_b + \Psi_c) = (\Psi_a, \Psi_b) + (\Psi_a, \Psi_c) \tag{14.12}$$

and that

$$(\Psi_a, \Psi_a) \geqslant 0 \tag{14.13}$$

with the equality sign holding if and only if Ψ_a is the null vector. (Ψ_a, Ψ_a) is called the *norm*, and $\sqrt{(\Psi_a, \Psi_a)}$ the "length" of the vector Ψ_a. A vector for which $(\Psi, \Psi) = 1$ is called a *unit* vector, and such vectors will henceforth be distinguished by carets over them.

---

[3] This is also variously known as a Hermitian or complex scalar or inner product. Since no confusion is likely to arise, we may plainly call it *scalar product*.

Two vectors, neither of which is a null vector, are said to be *orthogonal* if their scalar product vanishes.

It is possible to construct sets of n vectors which satisfy the orthogonality and normalization conditions (often briefly referred to as the *orthonormality* property)

$$(\hat{\Psi}_i, \hat{\Psi}_j) = \delta_{ij} \qquad (i, j = 1, \ldots, n) \tag{14.14}$$

Since orthogonal vectors are automatically linearly independent, an orthonormal set can serve as a suitable basis. We shall assume throughout that the basis vectors form an orthonormal set.

With (14.14) we obtain for the scalar product of two arbitrary vectors

$$(\Psi_a, \Psi_b) = \left(\sum_i a_i\Psi_i, \sum_j b_j\Psi_j\right) = \sum_i \sum_j a_i{}^*b_j(\Psi_i, \Psi_j) = \sum_i a_i{}^*b_i \tag{14.15}$$

and in particular

$$(\Psi_a, \Psi_a) = \sum_i |a_i|^2 \tag{14.16}$$

These relations emphasize the similarity between our (complex) scalar product and the ordinary scalar product of two (real) vectors in space:

$$\mathbf{A} \cdot \mathbf{B} = |\mathbf{A}| |\mathbf{B}| \cos (\mathbf{A}, \mathbf{B}) = \sum_i A_i B_i$$

Since the cosine of any angle lies between $+1$ and -1, we have

$$|\mathbf{A} \cdot \mathbf{B}| \leqslant |\mathbf{A}| |\mathbf{B}|$$

The analog of this theorem is

$$|(\Psi_a, \Psi_b)| \leqslant \sqrt{(\Psi_a, \Psi_a)(\Psi_b, \Psi_b)} \tag{14.17}$$

or, squaring this,

$$|(\Psi_a, \Psi_b)|^2 \leqslant (\Psi_a, \Psi_a)(\Psi_b, \Psi_b) \tag{14.17a}$$

known as the *Schwarz inequality*. This inequality may be used to define an "angle" α between two vectors Ψ_a and Ψ_b, thus

$$\cos \alpha = \frac{|(\Psi_a, \Psi_b)|}{\sqrt{(\Psi_a, \Psi_a)(\Psi_b, \Psi_b)}}$$

Proof of Schwarz inequality: Construct

$$\Psi = \Psi_a + \lambda\Psi_b$$

where λ is an undetermined parameter. Now by (14.13), $(\Psi, \Psi) \geqslant 0$; hence,

$$(\Psi, \Psi) = (\Psi_a, \Psi_a) + \lambda(\Psi_a, \Psi_b) + \lambda^*(\Psi_b, \Psi_a) + \lambda\lambda^*(\Psi_b, \Psi_b) \geqslant 0$$

The "best" inequality is obtained if λ is chosen so as to minimize the left-hand side. By differentiation, the value of λ which accomplishes this is found to be

$$\lambda = -\frac{(\Psi'_b, \Psi'_a)}{(\Psi'_b, \Psi'_b)}$$

Substitution of this value in the above inequality yields the Schwarz inequality.[4]

We note also that the equality sign holds if and only if $(\Psi, \Psi) = 0$, i.e., $\Psi'_a + \lambda \Psi'_b = 0$. Hence, the equality holds if and only if Ψ'_a and Ψ'_b are multiples of each other, or are "parallel."

It follows from the Schwarz inequality that (Ψ'_a, Ψ'_b) is finite if the norms of Ψ'_a and Ψ'_b are finite.

Exercise 14.1. If $f(\Psi'_a)$ is a complex scalar function of the vector Ψ'_a with the linearity property

$$f(\lambda \Psi'_a + \mu \Psi'_b) = \lambda f(\Psi'_a) + \mu f(\Psi'_b)$$

show that f defines a unique vector Ψ such that $f(\Psi'_a) = (\Psi, \Psi'_a)$ for every Ψ'_a in the space.

We are now in a position to define *operators* in the vector space. An operator A is a prescription by which every vector Ψ in the space is associated with a vector Ψ'' in the space. Thus, Ψ'' is a function of Ψ, and the notation

$$\Psi'' = A(\Psi) \qquad (14.18)$$

is employed. The special class of operators which satisfies the conditions

$$A(\Psi'_a + \Psi'_b) = A(\Psi'_a) + A(\Psi'_b) \qquad (14.19)$$

$$A(\lambda \Psi) = \lambda A(\Psi) \qquad (\lambda: \text{arbitrary complex number}) \qquad (14.20)$$

is most important to us. Such operators are called *linear*. For linear operators the parenthesis in (14.18) can be dropped, and we may simply write

$$\Psi'' = A\Psi \qquad (14.21)$$

thus stressing that the application of a linear operator is in many ways similar to ordinary multiplication of a vector by a number.

On occasion we shall also deal with *antilinear* operators. These share the property (14.19) with linear operators, but (14.20) is replaced by

$$A(\lambda \Psi) = \lambda^* A\Psi \qquad (14.20a)$$

Unless it is specifically stated that a particular operator is not linear, we shall assume every operator to be linear and usually omit the adjective "linear."

Two operators, A and B, are equal if $A\Psi = B\Psi$ for every Ψ. Just as

---

[4] See also Section 8.10.

numbers can be added and multiplied, it is also sensible to define sums and products of operators by the relations:

$$(A + B)\Psi = A\Psi + B\Psi \tag{14.22}$$

$$(AB)\Psi = A(B\Psi) \tag{14.23}$$

The last equation says that the operator AB acting on Ψ produces the same vector which would be obtained if we first let B act on Ψ and then A on the result of the previous operation. But, whereas with numbers, $ab = ba$, there is no need for operators to yield the same result if they are applied in the reverse order. Hence, in general $AB \neq BA$, although in exceptional cases two operators may, of course, commute.

A trivial example of a linear operator is the *identity operator*, denoted by *1*, with the property that

$$\Psi = 1\Psi$$

for every Ψ. The operator $\lambda 1$, where λ is a number, merely multiplies each vector by the constant factor λ. Hence, this operator may be simply written as λ.

A less trivial example of a linear operator is provided by the equation

$$\Psi'' = \hat{\Psi}_a(\hat{\Psi}_a, \Psi) \tag{14.24}$$

where $\hat{\Psi}_a$ is a given unit vector. This equation associates a vector Ψ'' with every Ψ. The association is a linear one, and we write (14.24) as

$$\Psi'' = P_a\Psi \tag{14.25}$$

Exercise 14.2. Prove that P_a is a linear operator.

P_a is termed a *projection operator*; reasonably so, since all Ψ'' are in the direction of $\hat{\Psi}_a$, and the length of Ψ'' equals the component $(\hat{\Psi}_a, \Psi)$ of Ψ in that direction.

A fundamental property of projection operators is derived as follows:

$$P_a\hat{\Psi}_a = \hat{\Psi}_a$$

Hence, for any vector Ψ,

$$P_a^2\Psi = P_a(P_a\Psi) = P_a\hat{\Psi}_a(\hat{\Psi}_a, \Psi) = \hat{\Psi}_a(\hat{\Psi}_a, \Psi) = P_a\Psi$$

Thus, projection operators are *idempotent*, i.e.,

$$P_a^2 = P_a \tag{14.26}$$

In particular, for the projections on the basis vectors we have

$$P_i\Psi = \hat{\Psi}_i(\hat{\Psi}_i, \Psi) = \hat{\Psi}_i a_i \tag{14.27}$$

Hence,

$$P_i P_j \Psi = a_j P_i \hat{\Psi}_j = 0 \qquad \text{if } i \neq j$$

Consequently, the projection operators for the basis vectors have the property

$$P_i P_j = P_j P_i = 0 \qquad \text{if } i \neq j \tag{14.28}$$

Note also that for every Ψ

$$\sum_{i=1}^{n} P_i \Psi = \sum_{i=1}^{n} a_i \hat{\Psi}_i = \Psi$$

hence,

$$\sum_{i=1}^{n} P_i = 1 \tag{14.29}$$

When a basis is given in the space, an operator A can be characterized by its effect on the basis vectors. Indeed, being again a vector in the space, $A\hat{\Psi}_j$ can obviously be expanded as

$$A\hat{\Psi}_j = \sum_{i=1}^{n} \hat{\Psi}_i A_{ij} \qquad (j = 1, 2, \ldots, n) \tag{14.30}$$

where the A_{ij} are n^2 numbers which, owing to the linearity of A, completely specify the effect of A on any vector. To see this we note

$$\Psi_b = A\Psi_a = A \sum_j a_j \hat{\Psi}_j = \sum_j a_j A\hat{\Psi}_j = \sum_j \sum_i a_j \hat{\Psi}_i A_{ij}$$

$$= \sum_i \hat{\Psi}_i \left(\sum_j A_{ij} a_j \right)$$

Hence,

$$b_i = \sum_j A_{ij} a_j \tag{14.31}$$

proving the contention that the effect of A on any vector is known if all A_{ij} are known. Equation (14.31) can be written conveniently in *matrix form* as

$$\begin{pmatrix} b_1 \\ b_2 \\ \cdot \\ \cdot \\ \cdot \\ b_n \end{pmatrix} = \begin{pmatrix} A_{11} & A_{12} & \cdots & A_{1n} \\ A_{21} & A_{22} & \cdots & A_{2n} \\ \cdot & \cdot & & \cdot \\ \cdot & \cdot & & \cdot \\ \cdot & \cdot & & \cdot \\ A_{n1} & A_{n2} & \cdots & A_{nn} \end{pmatrix} \begin{pmatrix} a_1 \\ a_2 \\ \cdot \\ \cdot \\ \cdot \\ a_n \end{pmatrix} \tag{14.32}$$

The possibility of using matrix notation here is not the result of any strange coincidence. Rather, the peculiar rule by which matrices are multiplied was invented for the theory of linear transformations, and matrices are therefore naturally adapted to any calculation in which linear quantities play a fundamental role.

The scalar product of two vectors can also be written in matrix notation. According to (14.15), we have

$$(\Psi_a, \Psi_b) = \sum_i a_i^* b_i = (a_1^* a_2^* \cdots a_n^*) \begin{pmatrix} b_1 \\ b_2 \\ \cdot \\ \cdot \\ \cdot \\ b_n \end{pmatrix} \tag{14.33}$$

The choice of a particular basis determines the matrices in (14.32) and (14.33). The column matrices

$$\begin{pmatrix} a_1 \\ a_2 \\ \cdot \\ \cdot \\ \cdot \\ a_n \end{pmatrix} \quad \text{and} \quad \begin{pmatrix} b_1 \\ b_2 \\ \cdot \\ \cdot \\ \cdot \\ b_n \end{pmatrix}$$

represent the vectors Ψ_a and Ψ_b, and the square matrix

$$\begin{pmatrix} A_{11} & A_{12} \cdots A_{1n} \\ A_{21} & A_{22} \cdots A_{2n} \\ \cdot & \cdot & \cdot \\ \cdot & \cdot & \cdot \\ \cdot & \cdot & \cdot \\ A_{n1} & A_{n2} \cdots A_{nn} \end{pmatrix}$$

represents the operator A. For this reason we say that all these matrices constitute a *representation* in the space. The *matrix elements* of A in a given representation can be calculated by the formula

$$A_{ij} = (\hat{\Psi}_i, A\hat{\Psi}_j) \tag{14.34}$$

which follows immediately from (14.30) and (14.14).

As an example, if $\hat{\Psi}_a$ is an arbitrary unit vector with components a_i, the matrix elements of the projection operator P_a are

$$(P_a)_{ij} = (\hat{\Psi}_i, \hat{\Psi}_a(\hat{\Psi}_a, \hat{\Psi}_j)) = (\hat{\Psi}_i, \hat{\Psi}_a)(\hat{\Psi}_a, \hat{\Psi}_j) = a_i a_j^*$$

Exercise 14.3. Show that the density matrix ρ of Section 13.2 represents a projection operator.

Exercise 14.4. Show that for a linear operator

$$(\Psi_b, A\Psi_a) = \sum_i \sum_j b_i^* A_{ij} a_j \tag{14.35}$$

and write this equation in matrix form. If A were antilinear, how would the corresponding expansion look?

The matrix representing the sum of two operators is obtained by adding the matrices representing the two operators:

$$(A + B)_{ij} = A_{ij} + B_{ij} \tag{14.36}$$

For the matrix of the product of two operators we have

$$(AB)_{ij} = (\hat{\Psi}_i, AB\hat{\Psi}_j) = (\hat{\Psi}_i, A \sum_k \hat{\Psi}_k B_{kj})$$

$$= \sum_k (\hat{\Psi}_i, A\hat{\Psi}_k)B_{kj} = \sum_k A_{ik}B_{kj} \tag{14.37}$$

This result shows that the matrix of an operator product equals the product of the matrices representing the operators, taken in the same order. We shall denote the matrix representing an operator A by the symbol A. Hence, if

$$C = AB$$

then (14.37) shows that likewise

$$C = AB$$

again emphasizing the parallelism between linear operators and matrices. However, the reader should avoid a complete identification of the two concepts, tempting as it may be, because the matrix A depends on the particular choice of basis vectors, whereas the operator A is a geometric entity, represented by a different matrix in every representation. We shall return to this point when we consider the connection between different bases.

Exercise 14.5. If $f(\Psi_a, \Psi_b)$ is a complex scalar function of the vectors Ψ_a and Ψ_b with the linearity properties

$$f(\lambda\Psi_a + \mu\Psi_c, \Psi_b) = \lambda^* f(\Psi_a, \Psi_b) + \mu^* f(\Psi_c, \Psi_b)$$

$$f(\Psi_a, \lambda\Psi_b + \mu\Psi_d) = \lambda f(\Psi_a, \Psi_b) + \mu f(\Psi_a, \Psi_d)$$

show that f defines a unique linear operator A such that $f(\Psi_a, \Psi_b) = (\Psi_a, A\Psi_b)$ for Ψ_a and Ψ_b.

It follows from Exercise 14.5 that corresponding to any given linear operator A we may define another linear operator, denoted by $A^\dagger$ and called the (Hermitian) *adjoint* of A, which has the property that for any two vectors,

Ψ'_a and Ψ'_b,

$$(A\Psi'_a, \Psi'_b) = (\Psi'_a, A^\dagger \Psi'_b) \tag{14.38}$$

Specializing to $\Psi'_a = \hat{\Psi}_i$, $\Psi'_b = \hat{\Psi}_j$, we see that

$$(A^\dagger)_{ij} = (A)_{ji}{}^*$$

Thus the matrix representing $A^\dagger$ is obtained from the matrix representing A by complex conjugation and transposition

$$\mathsf{A}^\dagger = \tilde{\mathsf{A}}^* \tag{14.39}$$

where the symbol $\tilde{\mathsf{A}}$ is used for the transpose of A, and $\tilde{\mathsf{A}}^*$ is called the *Hermitian conjugate* of A. Note also that

$$(\Psi_a, A\Psi_b) = (A\Psi_b, \Psi_a)^* = (\Psi_b, A^\dagger \Psi_a)^* = (A^\dagger \Psi_a, \Psi_b) \tag{14.40}$$

From this and (14.38) we see that an operator can be moved at will from its position as multiplier of the postfactor in a scalar product to a new position as multiplier of the prefactor, and vice versa, provided that the operator's adjoint is taken.

An important theorem concerns the adjoint of a product:

$$(AB)^\dagger = B^\dagger A^\dagger \tag{14.41}$$

The proof is left to the reader.

A linear operator which is identical with its adjoint is said to be *Hermitian* (or self-adjoint). For a Hermitian operator,

$$A^\dagger = A \tag{14.42}$$

Hermitian operators thus are generalizations of real numbers (which are identical with their complex conjugates).

If A is a Hermitian operator, the corresponding matrix A satisfies the condition

$$\tilde{\mathsf{A}}^* = \mathsf{A} \tag{14.43}$$

i.e., the matrix elements which are located symmetrically with respect to the main diagonal are complex conjugates of each other. In particular, the diagonal matrix elements of a Hermitian operator are real. Matrices which satisfy condition (14.43) are called Hermitian.

From (14.40) it follows that for a Hermitian operator

$$(\Psi_a, A\Psi_a) = (\Psi_a, A\Psi_a)^* = \text{real} \tag{14.44}$$

The physical interpretation makes important use of the reality of this scalar product which is brought into correspondence with the expectation value of a physical quantity represented by the Hermitian operator A.

An example of a Hermitian operator is afforded by the projection operator P_a. Indeed,

$$(\Psi_b, P_a\Psi_c) = (\Psi_b, \hat{\Psi}_a(\hat{\Psi}_a, \Psi_c)) = (\hat{\Psi}_a, \Psi_c)(\Psi_b, \hat{\Psi}_a)$$

$$= (\hat{\Psi}_a, \Psi_b)^*(\hat{\Psi}_a, \Psi_c) = (\hat{\Psi}_a(\hat{\Psi}_a, \Psi_b), \Psi_c) = (P_a\Psi_b, \Psi_c)$$

Exercise 14.6. Show that if A is an antilinear operator, the equation

$$(\Psi_b, A\Psi_a) = (\Psi_a, \tilde{A}\Psi_b)$$

defines an antilinear operator $\tilde{A}$ which may be called the transpose of A.

A linear operator A, which is defined by

$$\Psi'' = A\Psi' \tag{14.45}$$

may or may not have an *inverse*. An operator B which reverses the action of A, such that

$$\Psi' = B\Psi'', \tag{14.46}$$

exists only if (14.45) associates different vectors, Ψ_a' and Ψ_b', with any two different vectors, Ψ_a and Ψ_b, or, in other words, if the operator A preserves linear independence. Hence, as Ψ' ranges through the entire n-dimensional vector space, Ψ'' does the same.[5] We may substitute (14.45) in (14.46), or vice versa, and conclude that

$$AB = 1 \quad \text{and} \quad BA = 1 \tag{14.47}$$

The operator B is unique, for if there were another operator B' with the property $AB' = 1$, we would have $A(B' - B) = 0$, or $BA(B' - B) = 0$; according to the second of the equations (14.47) this implies $B' - B = 0$. It is therefore legitimate to speak of the *inverse* of A and use the notation $B = A^{-1}$. Evidently,

$$(A_1A_2)^{-1} = A_2^{-1}A_1^{-1} \tag{14.48}$$

It is worth noting that, if an operator A has an inverse, there can be no vector Ψ' (other than the null vector) such that

$$A\Psi' = 0 \tag{14.49}$$

Conversely, it can be shown that, if there is no nontrivial vector which satisfies (14.49), then A has an inverse. A projection operator is an example of an operator which has no inverse.

---

[5] If $n \to \infty$ this conclusion does not follow from the preceding discussion. In infinitely dimensional space it is quite possible for the domain of Ψ'' to exceed the space in which Ψ' ranges. For some of the resulting qualifications in the case of $n \to \infty$ see Section 14.6.

The matrix which represents A^{-1} is the inverse of the matrix A. Hence, $AA^{-1} = A^{-1}A = I$. A necessary and sufficient condition for the existence of the inverse matrix is that det $A \neq 0$.

A linear operator whose inverse and adjoint are identical is called *unitary*. Such operators are generalizations of complex numbers of absolute value 1, i.e., e^{ia}. For a unitary operator U:

$$U^{-1} = U^\dagger \qquad \text{or} \qquad UU^\dagger = U^\dagger U = I \tag{14.50}$$

Exercise 14.7. Prove that products of unitary operators are also unitary.

Evidently,

$$(\Psi_a, \Psi_b) = (U\Psi_a, U\Psi_b) \tag{14.51}$$

Hence, a unitary operator, applied to all the vectors of the space, preserves the "lengths" of the vectors and the "angles" between any two of them. In this sense U can be regarded as defining a "rotation" in the abstract vector space. In fact, the matrix representing U satisfies the condition of unitarity

$$U\tilde{U}^* = I \tag{14.52}$$

which, but for the complex conjugation, is analogous to the ordinary orthogonality relation. If U is a real matrix, condition (14.52) becomes identical with the orthogonality relation in Euclidean space, emphasizing the formal analogy between unitary operators in the complex vector space and rotations in ordinary space.

The matrix U of Section 12.5, which serves to rotate spinors, is an example of a unitary operator. In this example there was an intimate connection between the formal "rotations" in the complex vector space (of two dimensions) and the physical rotations in ordinary space,[6] but in general the "rotations" defined by a unitary operator need not have anything to do with rotations in ordinary space.

The reader will discover that many of the concepts and propositions established in this chapter are straightforward generalizations of similar notions encountered in earlier chapters. In particular, for the special case of $n = 2$ and a fixed basis (spanned by the states "spin up" and "spin·down") the essence of this section was already contained in Chapter 12. On the other hand, the precise connection with wave mechanics remains yet to be displayed, but it should be apparent from the similarity of such equations as

$$(A\Psi_a, \Psi_b) = (\Psi_a, A^\dagger \Psi_b) \tag{14.38}$$

---

[6] In the language of group theory this connection may be attributed to an isomorphism (one-to-one correspondence) between the real orthogonal group in three dimensions and the group of unitary unimodular (i.e., det $U = 1$) transformations in two dimensions.

and

$$\int (Ff)^* g \, d\tau = \int f^* F^\dagger g \, d\tau \tag{8.21}$$

that Hermitian operators in abstract vector space are intended to resemble the Hermitian operators introduced in Chapter 8.

3. *Change of Basis.* In the last section the similarity between the geometry of the abstract complex vector space and geometry in ordinary Euclidean space was emphasized. A representation in the former space corresponds to the introduction of a coordinate system in the latter. Just as we study rotations of coordinate systems in analytic geometry, we must now consider the transformation from one representation to another in the general space.

Along with the old unprimed basis we consider a new primed basis. The new basis vectors may be expressed in terms of the old ones:

$$\hat{\Psi}_k{}' = \sum_i \hat{\Psi}_i S_{ik} \tag{14.53}$$

The matrix of the transformation coefficients

$$S = \begin{pmatrix} S_{11} & S_{12} \cdots S_{1n} \\ S_{21} & S_{22} \cdots S_{2n} \\ \cdot & \cdot \cdots \\ \cdot & \cdot \cdots \\ \cdot & \cdot \cdots \\ S_{n1} & S_{n2} \cdots S_{nn} \end{pmatrix}$$

defines the change of basis.

A succession of two such basis changes, S and R, performed in this order, is equivalent to a single one whose matrix is simply the product matrix RS.

To obtain the new components of an arbitrary vector we write

$$\Psi_a = \sum_i a_i \hat{\Psi}_i = \sum_k a_k{}' \hat{\Psi}_k{}'$$

Substituting $\hat{\Psi}_k{}'$ from (14.53), we get

$$a_i = \sum_k S_{ik} a_k{}' \tag{14.54}$$

or

$$\begin{pmatrix} a_1 \\ a_2 \\ \cdot \\ \cdot \\ \cdot \\ a_n \end{pmatrix} = S \begin{pmatrix} a_1{}' \\ a_2{}' \\ \cdot \\ \cdot \\ \cdot \\ a_n{}' \end{pmatrix} \tag{14.55}$$

We must also determine the connection between the matrices A and A′ representing the operator A in the old and new representations. Evidently the new matrix elements are defined by

$$A\hat{\Psi}_j{'} = \sum_i \hat{\Psi}_i{'} A_{ij}{'} = \sum_i \sum_k \hat{\Psi}_k S_{ki} A_{ij}{'}$$

But on the other hand,

$$A\hat{\Psi}_j{'} = A \sum_l \hat{\Psi}_l S_{lj} = \sum_l \sum_k \hat{\Psi}_k A_{kl} S_{lj}$$

Comparing the two right-hand sides of these equations, we obtain in matrix notation

$$SA' = AS$$

or

$$A' = S^{-1}AS \tag{14.56}$$

We say that A′ is obtained from A by a *similarity transformation*.

Exercise 14.8. If $f(A, B, C, \ldots)$ is any function which is obtained from the matrices A, B, C, . . . by algebraic processes involving numbers (but no other, constant, matrices), show that

$$f(S^{-1}AS, S^{-1}BS, S^{-1}CS, \ldots) = S^{-1}f(A, B, C, \ldots)S$$

Give three examples.

If, as is usually the case, the old and new bases are both orthonormal, we have the additional conditions

$$(\hat{\Psi}_i, \hat{\Psi}_j) = \delta_{ij} \qquad \text{and} \qquad (\hat{\Psi}_k{'}, \hat{\Psi}_l{'}) = \delta_{kl}$$

If (14.53) is used with these conditions, it is found that the coefficients S_{ik} must satisfy the relations

$$\sum_i S_{ik}{}^* S_{il} = \delta_{kl}$$

and that

$$S_{ik} = (\hat{\Psi}_i, \hat{\Psi}_k{'}) \tag{14.57}$$

Hence, S must be a unitary matrix

$$S^\dagger S = I \tag{14.58}$$

and we often refer to such changes of representation as *unitary transformations*. (Again the analogy with orthogonal transformations in ordinary space is evident.)

Using the unitarity condition, we may rewrite the similarity transformation equation (14.56) for a matrix representing the operator A in the form

$$A' = S^\dagger AS \tag{14.59}$$

An alternative interpretation of a unitary transformation consists of keeping the basis fixed and regarding $S^{-1} = S^\dagger$ as the matrix of a unitary operator U which changes every vector Ψ into a vector $\Psi'' = U\Psi$. The operator A' which takes the unitary transform $U\Psi$ of Ψ into the transform $UA\Psi$ of $A\Psi$ is defined by the equation

$$A'(U\Psi) = U(A\Psi)$$

Hence, we have the operator equation

$$A' = UAU^\dagger \tag{14.59a}$$

which agrees with the matrix equation (14.59), since $U = S^\dagger$.

It is not surprising that the roles of the transformation matrix (operator) and its Hermitian conjugate have been interchanged in going from (14.59) to (14.59a). The two "rotations," one affecting only the basis, and the other keeping the basis fixed while rotating all vectors and operators, are equivalent only if they are performed in opposite "directions," i.e., if one is the inverse of the other.

Exercise 14.9. Show that under a unitary transformation a Hermitian operator remains Hermitian, and a unitary operator remains unitary. Also show that a symmetric matrix does not in general remain symmetric under such a transformation.

4. Dirac's Bra and Ket Notation. A different notation, which we owe to Dirac, has the advantage of great convenience when we consider eigenvalue problems for Hermitian and unitary operators. This elegant notation is based on the observation that the order of the two factors in a (complex) scalar product is important, since in general

$$(\Psi_a, \Psi_b) \neq (\Psi_b, \Psi_a)$$

although the absolute values of the two products are the same. Rules (14.10) and (14.12) show that the scalar product is linear with respect to the post-factor, but because of (14.9) and (14.11) it is not linear with respect to the prefactor. In fact, we have from (14.9), (14.11), and (14.12) the two relations

$$(\Psi_a + \Psi_b, \Psi_c) = (\Psi_a, \Psi_c) + (\Psi_b, \Psi_c)$$

and

$$(\lambda\Psi_a, \Psi_b) = \lambda^*(\Psi_a, \Psi_b)$$

The scalar product is said to depend on the prefactor in an *antilinear* fashion. This apparent asymmetry can be avoided if we think of the two factors as belonging to two different spaces. Each space is linear in itself, but they are related to each other in an antilinear manner. We thus have a space of post-factor vectors, and another space of prefactor vectors, but they are not independent of one another.

The two spaces are said to be *dual* to each other. Clearly, we must invent a new notation, because by merely writing Ψ'_a we would not know whether this is to be the prefactor or the postfactor in a scalar product. To make the distinction we might consider a notation like $\Psi'_a)$ for a postfactor and $(\Psi'_a$ for a prefactor. Dirac has stylized this notation and introduced the two kinds of vectors

$$|a\rangle \quad \text{and} \quad \langle a|$$

for post- and prefactors respectively. We assume that to every $|a\rangle$ in the postfactor space there corresponds a $\langle a|$ in the prefactor space, and vice versa, subject to the conditions

$$|a\rangle + |b\rangle \leftrightarrow \langle a| + \langle b| \tag{14.60}$$

and

$$\lambda |a\rangle \leftrightarrow \lambda^* \langle a| \tag{14.61}$$

where the arrow indicates the correspondence between the two spaces. Taken by itself each one of the two spaces is a linear vector space satisfying postulates (14.1)–(14.5). The connection between the dual spaces is given by defining the scalar product of a prefactor vector with a postfactor vector such that

$$\langle a || b \rangle = (\Psi'_a, \Psi_b)$$

expressing the new notation in terms of the old. It is customary to omit the double bar and to write

$$\langle a \,|\, b \rangle = (\Psi'_a, \Psi_b) \tag{14.62}$$

This notation has led to the colorful designation of the $\langle a|$ vector as a *bra* and the $|a\rangle$ vector as a *ket*.

Evidently from our previous rules

$$\langle b \,|\, a \rangle = \langle a \,|\, b \rangle^*$$

The vector equations of Section 14.2 can be transcribed in terms of kets without any major change. Thus, a linear operator associates with every ket $|a\rangle$ another ket,

$$|b\rangle = A \,|a\rangle$$

such that

$$A(|a_1\rangle + |a_2\rangle) = A \,|a_1\rangle + A \,|a_2\rangle$$

and

$$A(\lambda \,|a\rangle) = \lambda(A \,|a\rangle)$$

Some unnecessary bars on the left of certain kets have been omitted for reasons of economy in notation.

By letting the equation

$$\langle c | \{A | a\rangle\} = \{\langle c | A\} | a\rangle \tag{14.63}$$

define a bra $\langle c | A$, we may allow a linear operator also to act from right to left, or "backwards." Since,

$$(\langle c_1 | + \langle c_2 |)A = \langle c_1 | A + \langle c_2 | A$$

and

$$(\langle c | \lambda)A = (\langle c | A)\lambda$$

A is a linear operator in bra space as well as in ket space, and we may write the expression (14.63) unambiguously as $\langle c | A | a\rangle$. We can consider the operator as acting either to the right or to the left, whichever is convenient, thus emphasizing the symmetry of the dual spaces.

The definition (14.38) of the Hermitian adjoint operator becomes, in bra-ket notation,

$$\langle a | A^\dagger | c\rangle = \langle c | A | a\rangle^* \tag{14.64}$$

Hence, the general correspondence

$$|b\rangle = A | a\rangle \leftrightarrow \langle b | = \langle a | A^\dagger$$

is established.

A *Hermitian* operator A is characterized in this notation by

$$\langle a | A | b\rangle = \langle b | A | a\rangle^* \tag{14.65}$$

With these rules and conventions our old notation and Dirac's bra-ket notation become equivalent. A student of quantum mechanics should be familiar with both notations and should be able to use them interchangeably as convenience dictates, even to the point of mixing the two notations and using both in the same equation.

5. *The Eigenvalue Problem for Operators.* A ket $|A'\rangle$ is called an *eigenvector*, or *eigenket*, of the operator A if

$$A | A'\rangle = A' | A'\rangle \tag{14.66}$$

The number A' which characterizes the eigenvector is called an *eigenvalue*. The effect of A on $|A'\rangle$ is merely multiplication by a number. Since we are not interested in the most general case, only eigenvalue problems for Hermitian and unitary operators will be taken up here. In the remainder of this chapter all primed (or multiply primed) *italic* letters will denote *numbers*. An eigenvalue enclosed in a ket $| \ \rangle$, as in $|A'\rangle$, denotes the eigenket belonging to that eigenvalue. Here we assume that there is only one linearly independent eigenket for each eigenvalue. The possibility of several linearly independent eigenkets corresponding to the same eigenvalue will be considered later.

To show that the eigenvectors of any linear operator A (whether Hermitian, unitary, or otherwise) are linearly independent if they belong to *different* eigenvalues, assume that $k - 1$ of the eigenvectors are linearly independent but that the kth eigenvector depends linearly on these. Then, if $A\Psi_j = A_j'\Psi_j$,

$$\Psi_k = \sum_{i=1}^{k-1} \lambda_i \Psi_i$$

and

$$A\Psi_k = A_k'\Psi_k = A_k' \sum \lambda_i \Psi_i = \sum \lambda_i A_i'\Psi_i$$

Hence $A_k' = A_i'$, contrary to the assumption.

Let A be a *Hermitian* operator and

$$A\,|A'\rangle = A'\,|A'\rangle \quad \text{and} \quad A\,|A''\rangle = A''\,|A''\rangle$$

Then by multiplication of the first equation on the left with a bra,

$$\langle A''|\,A\,|A'\rangle = A'\langle A''\,|\,A'\rangle \tag{14.67}$$

From the second equation, by the Hermitian property of A,

$$\langle A''|\,A = A''{}^*\langle A''|$$

hence,

$$\langle A''|\,A\,|A'\rangle = A''{}^*\langle A''\,|\,A'\rangle \tag{14.67a}$$

Combining (14.67) and (14.67a),

$$(A' - A''{}^*)\langle A''\,|\,A'\rangle = 0 \tag{14.68}$$

If we let $A'' = A'$, and recall that $\langle A'\,|\,A'\rangle > 0$, it follows that

$$A' = A'{}^* = \text{real}$$

i.e., all eigenvalues of a Hermitian operator are real. Furthermore, (14.68) may now be written as

$$(A' - A'')\langle A''\,|\,A'\rangle = 0$$

which shows that eigenvectors belonging to two different eigenvalues ($A'' \neq A'$) are orthogonal:

$$\langle A''\,|\,A'\rangle = 0$$

Owing to the linearity of A we can normalize the eigenvectors arbitrarily. We shall usually assume that

$$\langle A''\,|\,A'\rangle = \delta_{A',A''} \tag{14.69}$$

Although we have thus demonstrated the reality of the eigenvalues and the orthogonality of the eigenvectors if they exist, there is so far no evidence to show that in fact any solutions of the eigenvalue problem (14.66) do exist. Our hope is, of course, not only that solutions exist, but that there are n linearly independent solutions so that these eigenvectors might be used to span the space, and can serve as basis vectors if they are orthogonal. Two proofs will be given of the existence of solutions of the eigenvalue problem for a Hermitian operator A.

Proof I. The idea here is to construct the *real* quantity

$$\lambda(\Psi') = \frac{(\Psi', A\Psi')}{(\Psi', \Psi')} \tag{14.70}$$

which is a function of the vector Ψ', and to look for the particular Ψ' which minimizes (or maximizes) λ. By dividing $(\Psi', A\Psi')$ by (Ψ', Ψ') we have made λ independent of the length of Ψ' and dependent only on its "direction." We note immediately that, if Ψ' is an eigenvector of A, such that

$$A\Psi' = A'\Psi'$$

then $\lambda = A'$. Let us assume that λ has a greatest lower bound λ_0, which it assumes for a particular vector Ψ'_0:

$$\lambda(\Psi'_0) = \lambda_0 = \frac{(\Psi'_0, A\Psi'_0)}{(\Psi'_0, \Psi'_0)} \qquad (\lambda \geqslant \lambda_0)$$

Let us calculate λ for $\Psi' = \Psi'_0 \pm \varepsilon\Phi$, where ε is a small positive number and Φ is an arbitrary vector. Since λ_0 is the greatest lower bound, we have

$$\lambda(\Psi'_0 \pm \varepsilon\Phi) \geqslant \lambda_0$$

Upon substitution we obtain the result

$$\pm \varepsilon[(\Phi, (A - \lambda_0)\Psi'_0) + (\Psi'_0, (A - \lambda_0)\Phi) \pm \varepsilon(\Phi, (A - \lambda_0)\Phi)] \geqslant 0$$

Since $Q = (\Phi, (A - \lambda_0)\Phi) \geqslant 0$, we find by applying the above inequality that

$$-\varepsilon Q \leqslant (\Phi, (A - \lambda_0)\Psi'_0) + (\Psi'_0, (A - \lambda_0)\Phi) \leqslant \varepsilon Q$$

Now let $\varepsilon \to 0$. Then

$$(\Phi, (A - \lambda_0)\Psi'_0) + (\Phi, (A - \lambda_0)\Psi'_0)^* = 0$$

owing to the Hermitian nature of A. Since Φ is arbitrary, we may choose it equal to $(A - \lambda_0)\Psi'_0$ and thus conclude

$$((A - \lambda_0)\Psi'_0, (A - \lambda_0)\Psi'_0) = 0$$

which implies necessarily that

$$A\Psi'_0 = \lambda_0\Psi'_0 \tag{14.71}$$

Thus a vector Ψ'_0 which makes λ of (14.70) a minimum is an eigenvector of A, and λ_0 is the corresponding eigenvalue. Evidently it must be the least of all eigenvalues, allowing the identification $\lambda_0 = A_1'$ if $A_1' \leqslant A_2' \leqslant \cdots$.

Leaving still aside for a moment the possibility that more than one linearly independent vector might minimize λ, we now consider a new variational problem. We construct

$$\mu(\Psi'') = \frac{(\Psi'', A\Psi'')}{(\Psi'', \Psi'')}$$

where Ψ'' is the totality of all the vectors which are orthogonal to Ψ'_0:

$$\Psi'' = \Psi - \hat{\Psi}_0(\hat{\Psi}_0, \Psi) \tag{14.72}$$

As Ψ ranges through the entire n-dimensional space, Ψ'' ranges through a subspace of $n - 1$ dimensions, orthogonal to Ψ'_0. The same argument as before gives for the minimum μ_0 of μ:

$$A\Psi''_0 = \mu_0 \Psi''_0$$

Since eigenvectors belonging to two different eigenvalues are orthogonal, and $\mu_0 \neq \lambda_0$, Ψ''_0 must be the eigenvector belonging to the second smallest eigenvalue, $\mu_0 = A_2'$. In this manner we may continue and will eventually exhaust the entire n-dimensional space after n steps. Hence, there are n orthogonal eigenvectors if all eigenvalues are different.

Proof II. The second proof makes use of a representation in the space. The eigenvalue problem then appears as a matrix equation [See (14.32)]:

$$\begin{pmatrix} A_{11} & A_{12} \cdots A_{1n} \\ A_{21} & A_{22} \cdots A_{2n} \\ \cdot & \cdots \\ \cdot & \cdots \\ \cdot & \cdots \\ A_{n1} & A_{n2} \cdots A_{nn} \end{pmatrix} \begin{pmatrix} x_1 \\ x_2 \\ \cdot \\ \cdot \\ x_n \end{pmatrix} = \lambda \begin{pmatrix} x_1 \\ x_2 \\ \cdot \\ \cdot \\ x_n \end{pmatrix} \tag{14.73}$$

$x_1, x_2, \ldots, x_n$ are the components of the eigenvector which belongs to the eigenvalue λ. Equation (14.73) is a set of n linear homogeneous equations which possess nontrivial solutions only if

$$\begin{vmatrix} A_{11} - \lambda & A_{12} & \cdots & A_{1n} \\ A_{21} & A_{22} - \lambda & \cdots & \cdot \\ \cdot & \cdot & & \cdot \\ \cdot & \cdot & & \cdot \\ \cdot & \cdot & & \cdot \\ A_{n1} & \cdot & \cdots & A_{nn} - \lambda \end{vmatrix} = 0$$

or

$$\det (A_{ij} - \lambda \delta_{ij}) = 0 \tag{14.74}$$

This equation of nth degree in the unknown λ is called the *secular* or *characteristic equation*. If its n complex roots are all distinct, then the system (14.73) yields n linearly independent eigenvectors. The roots of (14.74), $\lambda = A_i'$, are the eigenvalues of A.

Although proof II introduces a particular representation, it is easy to verify that the eigenvalues, which are the roots of (14.74), do not depend on the choice of the basis. Indeed, if we choose some other basis, linked to the previous one by the similarity transformation (14.56),

$$A' = S^{-1}AS$$

the new secular equation is

$$\det (A' - \lambda I) = \det [S^{-1}(A - \lambda I)S] = \det (A - \lambda I) = 0$$

In this proof the property of determinants

$$\det AB = \det A \det B \tag{14.75}$$

has been used. Hence, the eigenvalues of A as defined by (14.74) are independent of the representation. It follows that, if we expand the secular equation explicitly,

$$(-\lambda)^n + (\text{trace } A)(-\lambda)^{n-1} + \cdots + \det A = 0 \tag{14.76}$$

the coefficient of each power of λ must be independent of the choice of representation.

It is easy to prove these properties for the trace and the determinant of A directly. Since for finite dimensional matrices

$$\text{trace } AB = \text{trace } BA \tag{14.77}$$

it follows that

$$\text{trace } (ABC) = \text{trace } (CAB) = \text{trace } (BCA)$$

hence,

$$\text{trace } A' = \text{trace } (S^{-1}AS) = \text{trace } A \tag{14.78}$$

Similarly, using (14.75),

$$\det A' = \det (S^{-1}AS) = \det A \tag{14.79}$$

It is therefore legitimate to consider trace A and det A to be properties of the operator A itself and to attach a representation-independent meaning to the quantities trace A and det A.

Furthermore, the theory of equations allows us to conclude from (14.76) that

$$\text{trace } A = A_1' + A_2' + \cdots + A_n' = \text{sum of the eigenvalues of } A \tag{14.80}$$

and

$$\det A = A_1' \times A_2' \cdots \times A_n' = \text{product of the eigenvalues of } A \quad (14.81)$$

As an application, let a matrix B be represented as

$$B = e^A \quad (14.82)$$

where the exponential function is defined as

$$e^A = I + \frac{1}{1!} A + \frac{1}{2!} A^2 + \frac{1}{3!} A^3 + \cdots = \lim_{N \to \infty} \left(I + \frac{A}{N} \right)^N$$

Equation (14.82) holds for the eigenvalues as well as the matrices: $B_i' = e^{A_i'}$. Hence, we have the useful relation

$$\det B = \prod_i B_i' = \exp\left(\sum_i A_i' \right) = \exp(\text{trace } A) \quad (14.83)$$

Exercise 14.10. Give an independent proof of (14.83) without using the eigenvalues.

If $f(z)$ is an analytic function, the function $B = f(A)$ of the operator A may be expanded very simply if all n eigenvalues are distinct. In this case any vector may be expanded in terms of the n eigenvectors:

$$\Psi = \sum_{k=1}^{n} c_k \,|A_k'\rangle$$

Since, for fixed j and k,

$$\prod_{i \neq j} \frac{A_i' I - A}{A_i' - A_j'} \,|A_k'\rangle = \delta_{jk} \,|A_k'\rangle \quad (14.84)$$

and

$$f(A) \,|A_k'\rangle = f(A_k') \,|A_k'\rangle \quad (14.85)$$

it is evident that for any function $f(z)$ whose singularities do not coincide with any eigenvalue of A, we may write[7]

$$f(A) = \sum_{j=1}^{n} f(A_j') \prod_{i \neq j} \frac{A_i' I - A}{A_i' - A_j'} \quad (14.86)$$

Equation (13.7) is a special case of this formula, applied to matrices.

---

[7] If all eigenvalues are not distinct, the expansion of $f(A)$ in powers of A is more complicated (*Sylvester's formula*), but $f(A)$ can still be expressed as a polynomial in A of degree less than n. See F. R. Gantmacher, *The Theory of Matrices*, 2 volumes, translated by K. A. Hirsch, Chelsea Publishing Co., New York, 1959, and E. Merzbacher, *Am. J. Phys.*, **36**, 814 (1968).

Let us compare briefly the two rather different proofs for the existence of eigenvalues which we have given. Proof I exploits the connection between the eigenvalue problem of a *Hermitian* operator and a variational (i.e., an extremum) problem. It is elegant because it avoids the use of a representation, and, since it makes no essential use of the assumption that n is finite, it can be generalized to the case $n \to \infty$. The generalization requires only that the operator A be bounded at least from one side. A is said to be *bounded* if λ as defined in (14.70) is bounded $(-\infty < \lambda < +\infty)$. From the Schwarz inequality it follows that boundedness of A is assured if, for a given A, there exists a positive number C, independent of Ψ, such that

$$(A\Psi, A\Psi) \leqslant C(\Psi, \Psi)$$

for every Ψ. Many operators common in quantum mechanics, such as the energy of a harmonic oscillator and L^2, have only a lower bound. The variational property of eigenvalues, on which proof I is based, can be exploited in practical calculations (see Section 4.6). However, note that this proof makes essential use of the Hermitian nature of the operator A.

Proof II, on the other hand, makes no reference at all to the nature of the linear operator A. The secular equation has always n roots, and these are the eigenvalues of the operator. The secular equation provides a means of calculating the eigenvalues to any desired approximation, but this task can be prohibitively complicated for high values of n.

Finally, in this section we must discuss the possibility of *repeated* eigenvalues of a Hermitian operator. By this we refer to the occurrence of more than one linearly independent eigenvector belonging to the same eigenvalue. Suppose, for example, that an eigenvalue A' corresponds to two linearly independent eigenvectors which we take to be normalized:

$$A\hat{\Psi}_1 = A'\hat{\Psi}_1, \qquad A\hat{\Psi}_2 = A'\hat{\Psi}_2$$

Evidently any linear combination $c_1\hat{\Psi}_1 + c_2\hat{\Psi}_2$ is also an eigenvector of A with eigenvalue A'. In this example the totality of all eigenvectors belonging to the eigenvalue A' spans a two-dimensional subspace of the original vector space. We may always assume that the two eigenvectors spanning this space are orthogonal. For if $\hat{\Psi}_1$ and $\hat{\Psi}_2$ do not happen to be orthogonal to begin with, we may construct the eigenvector

$$\Psi_2' = \hat{\Psi}_2 - \hat{\Psi}_1(\hat{\Psi}_1, \hat{\Psi}_2) = \hat{\Psi}_2 - P_{(\Psi_1)}\hat{\Psi}_2$$

which also belongs to the eigenvalue A'. Since Ψ_2' is obtained by subtracting from $\hat{\Psi}_2$ its component along $\hat{\Psi}_1$, it is easy to see that $\hat{\Psi}_1$ and Ψ_2' must be orthogonal:

$$(\hat{\Psi}_1, \Psi_2') = 0$$

Ψ_2' may be normalized if desired. If there are more than two linearly independent eigenvectors, this procedure can be continued. Such a successive orthogonalization algorithm, known as the *Schmidt* orthogonalization method, was already described in Section 8.2.

It is clear that the presence of repeated eigenvalues does not affect proof I in any essential way. If, for instance, the lowest eigenvalue λ_0 admits two linearly independent and, as we now know, orthonormal eigenfunctions $\hat{\Psi}_{01}$ and $\hat{\Psi}_{02}$, only (14.72) requires modification. From the original space we subtract the two-dimensional subspace spanned by $\hat{\Psi}_{01}$ and $\hat{\Psi}_{02}$ and instead of (14.72) we obtain the $n - 2$ dimensional subspace of the vectors

$$\Psi' = \Psi - \hat{\Psi}_{01}(\hat{\Psi}_{01}, \Psi) - \Psi_{02}(\hat{\Psi}_{02}, \Psi)$$

The conclusions are obviously unaltered, and we can now formulate the general theorem that:

The eigenvectors of a Hermitian operator can all be assumed to be orthonormal. They are n in number and span the space, i.e., every vector in the space can be expressed as a linear combination of the eigenvectors. To express this, we say that the eigenvectors of a Hermitian operator are *complete*.

From the point of view of proof II and the secular equation, the multiplicity of eigenvalues appears as a coincidence of several of the roots of (14.74). Whether such multiplicity occurs or not, it can be shown that an n-dimensional matrix A which commutes with its Hermitian adjoint has n orthogonal, and thus linearly independent, eigenvectors. A matrix with the property $[A, A^\dagger] = 0$ is said to be *normal*, and this class includes Hermitian as well as unitary matrices.[8]

Exercise 14.11. If A' and A'' are two different simple, nondegenerate eigenvalues of a normal matrix A corresponding to the eigenvectors $|A'\rangle$ and $|A''\rangle$, show that $\langle A'' | A'\rangle = 0$. Also prove that $|A'\rangle$ is an eigenvector of $A^\dagger$ with eigenvalue A'^*.

Actually, the problem of multiple eigenvalues can be circumvented if, when it occurs, we add to A a normal operator εB (where ε is a small parameter) chosen suitably such that the eigenvalues of $A + \varepsilon B$ are all distinct. This operator has n orthogonal eigenvectors. We now let $\varepsilon \to 0$. If certain continuity requirements are fulfilled, as is the case in all physical problems, the eigenvalues will go over into those of A, and the eigenvectors will remain orthogonal as $\varepsilon \to 0$. Hence, whether there is multiplicity of eigenvalues or not, all eigenvectors may be assumed to be orthogonal.

[8] An example of a matrix for which it is not true is

$$\begin{pmatrix} 1 & 1 \\ 0 & 1 \end{pmatrix}$$

Since the eigenvectors of a Hermitian operator form a complete ortho-normal set, the possibility of using such sets of eigenvectors as basis vectors of representations arises. These basis vectors are conveniently characterized by the eigenvalues to which they belong. Thus, if all eigenvalues of A are distinct, we may label the eigenvectors by $|A'\rangle$. But if more than one linearly independent eigenvector belongs to a particular eigenvalue, then the symbol $|A'\rangle$ is not sufficient to characterize the ket. A further mark of distinction must be added for such repeated eigenvalues, say, $|A'1\rangle$, $|A'2\rangle$. An alternative notation is based on the fact that two Hermitian operators A and B which commute may be assumed to have the same set of eigenvectors.[9] Hence, if we can find a second Hermitian operator B, commuting with A, such that

$$B\,|A'1\rangle = B'\,|A'1\rangle, \qquad B\,|A'2\rangle = B''\,|A'2\rangle$$

but with $B' \neq B''$, then the eigenvalues of B may serve to distinguish the eigenvectors, and we write $|A'1\rangle \equiv |A'B'\rangle$, $|A'2\rangle \equiv |A'B''\rangle$.

If we can find a set of *commuting Hermitian operators* A, B, $C \cdots$, whose n common eigenvectors can be characterized completely by the eigenvalues A', B', $C' \cdots$, such that no two eigenvectors have exactly identical sets of eigenvalues, this set of operators is said to be *complete*. We assume for the eigenvectors:

$$\langle A'B' \cdots | A''B'' \cdots\rangle = \delta_{A'A''}\delta_{B'B''} \cdots \tag{14.87}$$

but often we shall write for this simply the orthonormality condition

$$\langle K' | K''\rangle = \delta_{K'K''} \tag{14.88}$$

K here symbolizes the complete set A, B, $C \cdots$, and K' is a symbol for the set of eigenvalues $A'B' \cdots$. In particular, $K' = K''$ means that $A' = A''$, $B' = B'', \ldots$.

The completeness of the set of eigenvectors implies that for every ket $|a\rangle$,

$$|a\rangle = \sum_{K'} |K'\rangle\langle K' | a\rangle \tag{14.89}$$

Each term in the sum is a projection, and

$$P_{K'} = |K'\rangle\langle K'| \tag{14.90}$$

is the projection operator for $|K'\rangle$. Evidently from (14.89) we have

$$\sum_{K'} P_{K'} = \sum_{K'} |K'\rangle\langle K'| = 1 \tag{14.91}$$

This equation is called the *closure relation*.

With these simple equations the eigenvalue problem can be reformulated. If A is one of the Hermitian operators whose common set of eigenvectors is

---

[9] The proof of Section 8.5 may be transcribed in bra-ket notation.

denoted by $|A'B'C' \cdots \rangle$, then

$$A \, |A'B'C' \cdots \rangle = A' \, |A'B'C' \cdots \rangle$$

hence, using the closure relation,

$$A = \sum_{A'B'C'\cdots} A \, |A'B'C'\cdots\rangle\langle A'B'C' \cdots| = \sum_{A'B'C'\cdots} A' \, |A'B'C'\cdots\rangle\langle A'B'C'\cdots|$$

or, noting that A' is a function of K', more simply

$$A = \sum_{K'} A' \, |K'\rangle\langle K'| \tag{14.92}$$

The eigenvalue problem for A can thus be expressed as follows: Given a Hermitian operator A, decompose the space into a complete set of orthonormal vectors $|K'\rangle$ such that A is a linear combination of the projection operators $|K'\rangle\langle K'|$. The coefficients in this expansion are the eigenvalues of A.

Denoting by

$$P_{A'} = \sum_{K' \in A'} |K'\rangle\langle K'|$$

the partial sum of all those projection operators which correspond to the same eigenvalue A', we may write

$$A = \sum_{A'} A' P_{A'}, \qquad \sum_{A'} P_{A'} = I \tag{14.93}$$

Equations (14.93) define what is called the *spectral decomposition* of the Hermitian operator A. This form of the problem is sometimes convenient because the operators $P_{A'}$ are unique, whereas the eigenvectors belonging to a repeated eigenvalue A' contain an element of arbitrariness.

Henceforth the basis of a matrix representation will always be characterized by a complete set of commuting operators, of which the basis vectors are eigenvectors, and by the corresponding eigenvalues. An arbitrary ket appears in such a representation as

$$|a\rangle = \sum_{K'} |K'\rangle\langle K' \, | \, a\rangle \tag{14.94}$$

the numbers $\langle K' \, | \, a\rangle$ being the components of $|a\rangle$ in the representation defined by the complete set K. In another representation, defined by the complete set L,

$$|a\rangle = \sum_{L'} |L'\rangle\langle L' \, | \, a\rangle$$

For brevity we may speak of the K and L representations.

For both representations we have orthonormality,

$$\langle K' \, | \, K''\rangle = \delta_{K'K''}, \qquad \langle L' \, | \, L''\rangle = \delta_{L'L''} \tag{14.95a}$$

and closure,

$$\sum_{K'} |K'\rangle\langle K'| = \sum_{L'} |L'\rangle\langle L'| = 1 \qquad (14.95b)$$

With the aid of these relations the connection between the two representations is easily established. We have

$$\langle L' \mid a \rangle = \langle L' | \left(\sum_{K'} |K'\rangle\langle K'| \right) |a\rangle = \sum_{K'} \langle L' \mid K'\rangle\langle K' \mid a \rangle \qquad (14.96)$$

which is the Dirac version of (14.54). The complex numbers $\langle L' \mid K' \rangle$ are the *transformation coefficients* for the basis change from K to L. The ortho-normality and completeness of both bases assures that this is a unitary transformation, i.e., that

$$\sum_{K'} \langle L' \mid K'\rangle\langle K' \mid L'' \rangle = \langle L' \mid L'' \rangle = \delta_{L'L''} \qquad (14.97a)$$

and

$$\sum_{L'} \langle K' \mid L'\rangle\langle L' \mid K'' \rangle = \langle K' \mid K'' \rangle = \delta_{K'K''} \qquad (14.97b)$$

This is the transcription into the Dirac notation of

$$SS^\dagger = S^\dagger S = I$$

where S denotes the transformation matrix with rows labeled by L' and columns by K'. The Dirac form, $\langle L' \mid K' \rangle$, of the matrix elements of S shows explicitly how the transformation connects the old basis with the new.

In the Dirac bra-ket notation all such formulas as (14.96) and (14.97) are generated almost trivially and automatically by the use of the relations (14.95). As an example we work out the matrix elements of an operator M in the two representations. In the K representation these matrix elements are defined by

$$M |K'\rangle = \sum_{K''} |K''\rangle M_{K''K'} \qquad (14.98)$$

By taking the scalar product of the ket $M |K'\rangle$ with $\langle K''|$, we obtain

$$M_{K''K'} = \langle K''| M |K'\rangle \qquad (14.99)$$

for the matrix elements of M with respect to basis vectors labeled by K' and K''. The matrices of the operators A, B, $C \cdots$ which make up the set K are, of course, "*diagonal*" in the K representation, i.e., all off-diagonal elements are zero.

The matrix elements in two different (K and L) representations are linked by the equation

$$\langle K'| M |K''\rangle = \sum_{L',L''} \langle K' \mid L'\rangle\langle L'| M |L''\rangle\langle L'' \mid K'' \rangle \qquad (14.100)$$

This equation is the Dirac form of (14.59).

Starting with an arbitrary representation K, the eigenvalue problem for a Hermitian operator M,

$$M \,|M'\rangle = M' \,|M'\rangle$$

appears in the K representation as

$$\sum_{K'} \langle K''| \, M \, |K'\rangle\langle K' \,|\, M'\rangle = M'\langle K'' \,|\, M'\rangle \qquad (14.101)$$

This is just another way of writing the matrix equation (14.73).

Padding out the trivial equation

$$\langle M''| \, M \, |M'\rangle = M'\delta_{M'M''}$$

we get

$$\sum_{K',K''} \langle M'' \,|\, K''\rangle\langle K''| \, M \, |K'\rangle\langle K' \,|\, M'\rangle = M'\langle M'' \,|\, M'\rangle = M'\,\delta_{M'M''} \quad (14.102)$$

Equations (14.101) and (14.102) show two ways by which, if the matrix $\langle K''| \, M \, |K'\rangle$ is given, we may obtain the transformation coefficients $\langle K' \,|\, M'\rangle$ which take us from an arbitrary K representation to the representation in which M is diagonal. The diagonal elements of the transformed matrix are the eigenvalues of M.

Exercise 14.12. Calculate the sum over a complete orthonormal set $|K'\rangle$ of the quantities $|\langle a| \, A \, |K'\rangle|^2$. What value is obtained if A is unitary?

Exercise 14.13. Show by use of the bra-ket notation that

$$\text{trace } A = \sum_{K'} \langle K'| \, A \, |K'\rangle$$

is independent of the choice of the basis $|K'\rangle$ and that trace $AB = $ trace BA.

Exercise 14.14. Show that

$$\sum_{K'} \sum_{N'} |\langle K'| \, A \, |N'\rangle|^2$$

is independent of the bases $|K'\rangle$ and $|N'\rangle$ (which need not be the same).

We finally note that the $|A'\rangle$ are also eigenkets of any function of A which can be written as a power series, including negative powers if the inverse of A exists. The eigenvalues of $f(A)$ are the numbers $f(A')$. It follows from (14.93) that $f(A)$ can be written as

$$f(A) = \sum_{A'} f(A')P_{A'} \qquad (14.103)$$

Conversely, this equation may be regarded as the definition of an operator function $f(A)$ even if no series expansion is possible.

Equation (14.103) has the same structure as (14.86), but the latter is valid for any operator with distinct eigenvalues whereas (14.103) has been written under the assumption that the operators $P_{A'}$ effect orthogonal projections as is appropriate if A is Hermitian (or, more generally, normal).

Exercise 14.15. Show that an operator $A^{1/2}$ can be defined such that its square is A. How much arbitrariness is there in the definition of $A^{1/2}$?

As an example of a function of a Hermitian operator, consider an arbitrary *unitary* operator U. It can be written in the form

$$U = \frac{U + U^\dagger}{2} + i \frac{U - U^\dagger}{2i} = A + iB \qquad (14.104)$$

where A and B are both Hermitian. The property, $UU^\dagger = U^\dagger U$, which a unitary operator shares with all normal operators, guarantees that A and B commute. Hence, they can be simultaneously diagonalized, and their orthogonal eigenvectors $|A'B'\rangle$ are also eigenvectors of U with eigenvalues

$$U' = A' + iB' \qquad (A', B' = \text{real}) \qquad (14.105)$$

Furthermore,

$$A^2 + B^2 = \tfrac{1}{4}(U + U^\dagger)^2 - \tfrac{1}{4}(U - U^\dagger)^2 = U^\dagger U = 1$$

Hence, $A'^2 + B'^2 = 1$, from which it follows that *the eigenvalues of U have absolute value unity* and can be written as

$$U' = e^{iC'} = \frac{1 + i \tan \dfrac{C'}{2}}{1 - i \tan \dfrac{C'}{2}} \qquad (C' = \text{real}) \qquad (14.106)$$

where $A' = \cos C'$, and $B' = \sin C'$. We now define a Hermitian operator C with eigenvalues C' and eigenvectors $|C'\rangle = |A'B'\rangle$. Since these eigenvectors span the space, we have the simple result

$$U = \sum_{C'} e^{iC'} P_{C'} = e^{iC} = \frac{1 + i \tan \dfrac{C}{2}}{1 - i \tan \dfrac{C}{2}} \qquad (14.107)$$

Every unitary operator can be expressed in terms of a Hermitian operator by a formula like (14.107).

Exercise 14.16. Prove the property (14.106) of the eigenvalues and the orthogonality of the eigenvectors of a unitary operator directly from the eigenvalue equation $U|U'\rangle = U'|U'\rangle$.

6. The Continuous Spectrum. So far we have assumed the vector space to be of a finite number of dimensions, n. If n is allowed to grow to infinity, a number of important and difficult mathematical questions arise concerning the propriety of the limiting processes, the convergence of the infinite series into which the previously finite sums expand, and, above all, the problem of completeness of an orthogonal set of vectors.

A few examples will suffice to put the reader on guard against translating the earlier results uncritically into the language of infinitely dimensional vector space. The matrix representing a Hermitian operator can still be defined, but in general neither its trace (infinite sum of eigenvalues) nor its determinant (infinite product of eigenvalues) exists.[10] Whereas in n-dimensional space an orthonormal set of n vectors is automatically complete, so that every vector of the space can be expanded in terms of these n vectors, the condition of completeness must be added separately to the orthogonality condition when $n \to \infty$. In fact, care must be taken to give a meaningful definition of completeness, since the expansions in terms of infinitely many basis vectors require that suitable convergence criteria be specified. Finally, it is important to note that in the infinitely dimensional case an operator B can be said to be the inverse of A only if both

$$AB = I \quad \text{and} \quad BA = I$$

have been separately tested, although if n is finite one of these equations, and the uniqueness of the inverse, follows from the other. In particular, then, an operator U is unitary only if *both* $UU^\dagger = I$ and $U^\dagger U = I$ are satisfied.

A mathematically precise formulation of a suitable infinitely dimensional linear vector space, the so-called *Hilbert space*, can be given, but for our purposes it is sufficient to let the requirements of quantum mechanics dictate the introduction of the concepts to be used and to refer to the extensive literature[11] for the proof that these concepts are part of a mathematically consistent scheme. This section is thus devoted to an enumeration, rather than a derivation, of the mathematical tools of which quantum mechanics in its abstract formulation makes use.

---

[10] Of course, for many operators these quantities do exist. For example, the trace of a projection operator is always equal to unity.

[11] Easily the most important work is J. von Neumann, *Mathematical Foundations of Quantum Mechanics*, translated by R. T. Beyer, Princeton University Press, Princeton, 1955. A readable modern textbook is N. I. Akhiezer and I. M. Glazman, *Theory of Linear Operators in Hilbert Space*, Volumes I and II, translated by M. Nestell, F. Ungar Publishing Company, New York, 1961 and 1963.

A most important feature of the infinitely dimensional vector space is the possibility of using an *indenumerable* set of basis vectors and of replacing the discrete sums over n terms by integrals over some continuous variables. It will be seen that the mathematics of the n-dimensional vector space, as developed in the previous sections, serves extraordinarily well and needs but minor generalizations to allow for the appearance of continuous portions in the eigenvalue spectra of some important Hermitian operators.

The formulation of quantum mechanics in abstract vector space is based on the following *assumptions* which underlie the physical interpretation of the theory:

We associate with a physical system a linear vector space (of finite or infinite number of dimensions, depending on the nature of the system). To each (pure) state a of the system belongs a unit vector, $\hat{\Psi}_a$ or $|a\rangle$. The same state is described by all vectors $e^{i\alpha} |a\rangle$ (α = real), and any two vectors which differ other than by a phase factor describe different physical states. We speak of such normalized vectors, $\langle a | a \rangle = 1$, as *state vectors*. Further, it is assumed that to each physically observable quantity corresponds a Hermitian operator A such that

$$\langle a| A |b\rangle = \langle b| A |a\rangle^* \qquad (14.65)$$

for any two vectors in the space. The eigenvectors of each *observable*, as these operators are often called, are assumed to form a complete set; i.e., each vector in the space can be expressed as a linear combination of eigenvectors. The possible results of measuring the observable are its eigenvalues, and these are real numbers. The spectrum of eigenvalues consists of discrete points and continuous portions. The eigenvectors corresponding to discrete eigenvalues can be normalized to unity. In the continuous portion of the spectrum, we assume that the eigenvector is a continuous function of the eigenvalue. The physical interpretation of the theory demands that in the continuous part of the spectrum we admit only those solutions of the equation

$$A |A'\rangle = A' |A'\rangle$$

for which A' is real, and for which it is possible to adopt the normalization[12]

$$\langle A' | A'' \rangle = \delta(A' - A'') \qquad (14.108)$$

in analogy with

$$\langle A' | A'' \rangle = \delta_{A'A''} \qquad (14.109)$$

for the discrete eigenvalues.

With these normalizations all the formulas for the discrete and continuous cases are very similar, except that integrals in the latter replace sums in the

---

[12] It is easy to generalize the following equations if there are repeated eigenvalues.

former. Thus an arbitrary vector can be written as

$$|a\rangle = \sum_{K'} |K'\rangle\langle K' \mid a\rangle + \int |K''\rangle \, dK'' \, \langle K'' \mid a\rangle \tag{14.110}$$

the sum being over the discrete and the integral over the continuous eigenvalues of the complete set of commuting observables symbolized by K.

$|\langle K' \mid a\rangle|^2$ is the probability of finding the value K' for the observable K in the state $|a\rangle$ if we are in the discrete part of the spectrum. Similarly, $|\langle K'' \mid a\rangle|^2 \, dK''$ is the probability of finding a value between K'' and $K'' + dK''$ when K'' lies in the continuous part of the spectrum.

As a further example we formulate the eigenvalue problem of an operator A, assuming a purely continuous set of basis vectors. The equation

$$A \,|A'\rangle = A' \,|A'\rangle$$

becomes

$$\int A \,|K'\rangle \, dK'\langle K' \mid A'\rangle = A' \int |K'\rangle \, dK'\langle K' \mid A'\rangle$$

or

$$\int \langle K''| \, A \,|K'\rangle \, dK'\langle K' \mid A'\rangle = A'\langle K'' \mid A'\rangle \tag{14.111}$$

where $\langle K'' \mid K'\rangle = \delta(K' - K'')$ has been used. Equation (14.111), which is the analog of (14.102), is an integral equation for the unknown function $\langle K' \mid A'\rangle$.

7. *Application to Wave Mechanics in One Dimension.* The wave mechanics of a point particle, for simplicity restricted to one dimension, offers an instructive example. The state of the system is determined by a ket $|a\rangle$. Since we can always measure the particle's position along the x-axis, there must be a Hermitian operator x corresponding to this observable. The results of a measurement of x can be any real number between $-\infty$ and $+\infty$. Hence, the eigenvalues of x, denoted by x', form a continuum. The corresponding eigenvectors are denoted by $|x'\rangle$, and we have

$$x \,|x'\rangle = x' \,|x'\rangle \tag{14.112}$$

with the assumed normalization

$$\langle x'' \mid x'\rangle = \delta(x' - x'') \tag{14.113}$$

Evidently the vector $|a\rangle$ can be expanded as

$$|a\rangle = \int |x'\rangle \, dx'\langle x' \mid a\rangle \tag{14.114}$$

The components $\langle x' \mid a\rangle$ constitute a complex valued function of the real variable x'. We can now link the wave function with the state vector explicitly.

If we write

$$\psi_a(x') = \langle x' \mid a \rangle \tag{14.115}$$

the connection between the *state vector* $|a\rangle$ and the *wave function* $\psi_a(x')$ is established. x' is the continuously variable label of the components $\psi_a(x')$ of a state vector $\Psi_a = |a\rangle$ in an infinitely dimensional abstract space. From this point of view $\psi_a(x')$ is merely one of many possible ways of representing the state vector. The special representation used here, which is spanned by the eigenvectors of the position operator x, is called *coordinate representation*, and we might say that wave mechanics is quantum mechanics conducted in the coordinate representation.

We note that the scalar product of two states becomes

$$\langle b \mid a \rangle = \int\int \langle b \mid x'' \rangle \, dx'' \langle x'' \mid x' \rangle \, dx' \langle x' \mid a \rangle$$

$$= \int\int \langle b \mid x'' \rangle \, dx'' \delta(x'' - x') \, dx' \langle x' \mid a \rangle = \int_{-\infty}^{+\infty} \psi_b^*(x')\psi_a(x') \, dx' \tag{14.116}$$

The orthogonality of two states is expressed by the equation

$$\langle b \mid a \rangle = \int_{-\infty}^{+\infty} \psi_b^*(x')\psi_a(x') \, dx' = 0$$

This terminology is in agreement with our earlier conventions (Chapter 8).

The matrix which represents the operator x is now a matrix whose indices, instead of being denumerable, take on a continuum of values. Although such a matrix can no longer be written down as a square array, it is nevertheless appropriate to use the term *matrix elements* for the quantities

$$\langle x'' \mid x \mid x' \rangle = x' \delta(x' - x'') \tag{14.117}$$

and to say that this matrix is diagonal, as is the matrix representing any function of x:

$$\langle x'' \mid f(x) \mid x' \rangle = f(x')\delta(x' - x'')$$

The effect of an operator $f(x)$ on an arbitrary state $\Psi_a = |a\rangle$ with components $\langle x' \mid a \rangle = \psi_a(x')$ is given by

$$f(x)\Psi_a = f(x) \mid a \rangle = \int \mid x' \rangle \, dx' \langle x' \mid f(x) \mid x'' \rangle \, dx'' \langle x'' \mid a \rangle = \int \mid x' \rangle \, dx' f(x')\psi_a(x')$$

Hence

$$(\Psi_b, f(x)\Psi_a) = \langle b \mid f(x) \mid a \rangle = \int_{-\infty}^{+\infty} \psi_b^*(x')f(x')\psi_a(x') \, dx'' \tag{14.118}$$

An important linear operator in this vector space is the *translation operator* D_ξ, whose effect is to shift the coordinate by a constant distance ξ, i.e.,

$$D_\xi \mid x' \rangle = \mid x' + \xi \rangle \tag{14.119}$$

The matrix elements of the translation operator in the coordinate representation are

$$\langle x''| D_\xi |x'\rangle = \langle x'' | x' + \xi\rangle = \delta(x'' - x' - \xi) \qquad (14.120)$$

Evidently we have for two successive translations ξ and η

$$D_\eta D_\xi = D_\xi D_\eta = D_{\xi+\eta} \qquad (14.121)$$

expressing the simple fact that two successive translations are equivalent to one resultant translation.

Since D, according to its definition, preserves the orthonormality and the completeness of the basis vectors, it is a unitary operator; hence it can be written in the form

$$D_\xi = e^{iA(\xi)} \qquad (14.122)$$

where $A(\xi)$ is a Hermitian operator. The form of $A(\xi)$ can be determined from (14.121). Substituting $\eta = \xi$, we get

$$(D_\xi)^2 = D_{2\xi} \qquad \text{or} \qquad 2A(\xi) = A(2\xi)$$

More generally the reader may prove that

$$nA(\xi) = A(n\xi)$$

for any rational number n. Hence, if the operator is to be continuous,

$$A(n) = nA(1)$$

must hold for all real numbers n. It follows that $A(\xi) \propto \xi$ and that D_ξ has the form

$$D_\xi = e^{-i\xi k} \qquad (14.123)$$

where k is a Hermitian operator. The operator k is called the *generator of infinitesimal translations*.

The effect of D_ξ on an arbitrary ket can be exhibited as follows:

$$D_\xi |a\rangle = e^{-i\xi k} |a\rangle = D_\xi \int |x'\rangle \, dx' \langle x' | a\rangle = \int |x' + \xi\rangle \, dx' \langle x' | a\rangle$$

$$= \int |x'\rangle \, dx' \langle x' - \xi | a\rangle \qquad (14.124)$$

Thus in the displacement the state with the wave function $\psi_a(x') = \langle x' | a\rangle$ is changed to a new state with the wave function $\bar{\psi}_a(x') = \langle x' - \xi | a\rangle = \psi_a(x' - \xi)$.

Expanding (14.124) in a Taylor series in powers of ξ, we obtain by comparing the terms of nth degree:

$$k^n |a\rangle = \frac{1}{i^n} \int |x'\rangle \, dx' \frac{\partial^n}{\partial x'^n} \langle x' | a\rangle \qquad (14.125a)$$

or

$$k^n \Psi_a = \int |x'\rangle \, dx' \left(\frac{1}{i}\frac{\partial}{\partial x'}\right)^n \psi_a(x') \tag{14.125b}$$

From these equations we deduce the effect of any function of k which can be written as a power series:

$$f(k)\Psi_a = f(k) |a\rangle = \int |x'\rangle \, dx' f\left(\frac{1}{i}\frac{\partial}{\partial x'}\right)\psi_a(x')$$

and

$$(\Psi_b, f(k)\Psi_a) = \langle b| f(k) |a\rangle = \int_{-\infty}^{+\infty} \psi_b^*(x') f\left(\frac{1}{i}\frac{\partial}{\partial x'}\right)\psi_a(x') \, dx' \tag{14.126}$$

Exercise 14.17. Show that the matrix elements of k are given by

$$\langle x''| k |x'\rangle = i\frac{\partial}{\partial x'} \delta(x' - x'') \tag{14.127}$$

It should be pointed out that the matrix elements of the translation operator in the coordinate representation are not unique, because there is a fundamental ambiguity in the definition of a coordinate representation. This arises from the fact that a unitary transformation which changes merely the phase of every basis vector leads to a new and equivalent representation spanned by the vectors

$$|\bar{x}'\rangle = e^{i\varphi(x)} |x'\rangle = e^{i\varphi(x')} |x'\rangle$$

In terms of this new coordinate representation (14.125a) becomes for $n = 1$

$$k |a\rangle = \frac{1}{i}\int |\bar{x}'\rangle e^{-i\varphi(x')} \, dx' \frac{\partial}{\partial x'} [e^{i\varphi(x')}\langle \bar{x}' | a\rangle]$$

$$= \int |\bar{x}'\rangle \, dx' \left[\frac{1}{i}\frac{\partial}{\partial x'} + \frac{\partial \varphi(x')}{\partial x'}\right]\langle \bar{x}' | a\rangle \tag{14.128}$$

Hence, in the coordinate representation the effect of k on the state is found by letting the operator

$$\frac{1}{i}\frac{\partial}{\partial x'} + f(x')$$

act on the wave function. Since the arbitrary function $f(x')$ can be removed by a unitary transformation $\exp(i\int f(x) \, dx)$, there is no loss of generality if $f(x')$ is set equal to zero.

Using (14.125) or (14.128) it is easy to prove that

$$xk - kx = il \tag{14.129}$$

In connection with this familiar-looking equation it is clear that the linear displacement operator has a physical significance and that $\hbar k$ or $\hbar/i\ \partial/\partial x'$ corresponds to the x-component of linear momentum. This obvious identification will be discussed in detail in Chapter 15.

Exercise 14.18. Using (14.77), we obtain a contradiction when taking the trace of (14.129). Resolve this paradox.

The operators D_ξ and k have the same eigenvectors. Their eigenvalues are $e^{-ik'\xi}$ and k' respectively, where k' is a real number, and the eigenvectors are denoted by $|k'\rangle$. If we set $|a\rangle = |k'\rangle$ in (14.125a) for $n = 1$ and multiply on the left by $\langle x''|$, we obtain the differential equation

$$ik'\langle x'' \mid k'\rangle = \frac{\partial}{\partial x''}\langle x'' \mid k'\rangle$$

The solution of this equation is

$$\langle x'' \mid k'\rangle = g(k')e^{ik'x''} \tag{14.130}$$

There is no restriction on the values of k' other than that they must be real. Hence, the spectrum of k includes all real numbers from $-\infty$ to $+\infty$. Normalization of the eigenvectors of k requires

$$\langle k'' \mid k'\rangle = \delta(k'' - k') = \int \langle k'' \mid x'\rangle\, dx'\langle x' \mid k'\rangle$$

$$= g^*(k'')g(k')\int_{-\infty}^{+\infty} \exp\left[i(k' - k'')x'\right]dx' = 2\pi\,|g(k')|^2\,\delta(k' - k'')$$

Hence,

$$g(k') = \frac{1}{\sqrt{2\pi}}\,e^{i\alpha(k')}$$

Arbitrarily, but conveniently, the phase factor is chosen to be equal to unity so that finally

$$\langle x' \mid k'\rangle = \frac{1}{\sqrt{2\pi}}\,e^{ik'x'} \tag{14.131}$$

We may thus construct the normalized eigenkets of k,

$$|k'\rangle = \int |x'\rangle\, dx'\langle x' \mid k'\rangle = (2\pi)^{-1/2}\int e^{ik'x'}\,|x'\rangle\, dx' \tag{14.132}$$

Exercise 14.19. Verify that $|k'\rangle$ is an eigenvector of D_ξ. Show that all eigenvectors $|k', \xi\rangle$ of D_ξ corresponding to the eigenvalue $e^{-i\xi k'}$ have the form $\langle x' \mid k', \xi\rangle = e^{ik'x'}u_\xi(x')$, where $u_\xi(x')$ is an arbitrary periodic function of x' with period ξ. (Bloch waves, see Section 6.7.)

The eigenvectors $|k'\rangle$ may be used as a basis of a representation. In this case

$$|a\rangle = \int |k'\rangle \, dk' \langle k' \mid a\rangle$$

The connection between the k-representation and the coordinate representation is given by

$$\langle k' \mid a\rangle = \int \langle k' \mid x'\rangle \, dx' \langle x' \mid a\rangle = \frac{1}{\sqrt{2\pi}} \int e^{-ik'x'} \langle x' \mid a\rangle \, dx'$$

Denoting the wave function in the k-representation by

$$\varphi_a(k') = \langle k' \mid a\rangle$$

we obtain

$$\varphi_a(k') = \frac{1}{\sqrt{2\pi}} \int_{-\infty}^{+\infty} \psi_a(x') e^{-ik'x'} \, dx' \tag{14.133}$$

which is the familiar Fourier integral.

In view of its connection with momentum the k-representation is also called the *momentum representation*. The matrix elements of an operator which is a function of x only are given by the Fourier integral

$$\langle k'| \, V(x) \, |k''\rangle = \frac{1}{2\pi} \int \exp \, [ix'(k'' - k')] V(x') \, dx' \tag{14.134}$$

In conclusion let us consider a particular Hermitian operator which is a function of x and k,

$$H = \frac{k^2 + x^2}{2} \tag{14.135}$$

Such an operator characterizes the energy of a linear harmonic oscillator. Here only its eigenvalue problem in the coordinate representation will be formulated.

Following custom, we denote the eigenvalues of H by E:

$$H \, |E\rangle = E \, |E\rangle$$

With the notation

$$\psi_E(x') = \langle x' \mid E\rangle$$

the eigenvalue equation becomes

$$\tfrac{1}{2}(k^2 + x^2) \, |E\rangle \equiv \frac{1}{2} \int |x'\rangle \left(-\frac{\partial^2}{\partial x'^2} \right) \psi_E(x') \, dx' + \frac{1}{2} \int x'^2 \, |x'\rangle \, dx' \psi_E(x')$$

$$= E \int |x'\rangle \, dx' \psi_E(x')$$

From this we infer

$$\int |x'\rangle \, dx' \left(-\frac{1}{2}\frac{\partial^2}{\partial x'^2} + \frac{1}{2} x'^2 - E\right) \psi_E(x') = 0$$

Since the vectors $|x'\rangle$ are all linearly independent, this last equation requires that

$$\left(-\frac{1}{2}\frac{d^2}{dx'^2} + \frac{x'^2}{2} - E\right) \psi_E(x') = 0 \tag{14.136}$$

The eigenvalue problem for H of (14.135) thus leads to the Schrödinger equation for the harmonic oscillator. Only those solutions are admissible for which the normalization

$$\langle E' \mid E'' \rangle = \int \psi_{E'}{}^*(x') \psi_{E''}(x') \, dx' = \delta_{E', E''} \qquad \text{or} \qquad \delta(E' - E'')$$

can be achieved. It is easy to verify that for (14.136) this condition is equivalent to

$$\psi_E(+\infty) = \psi_E(-\infty) = 0$$

With these boundary conditions the problem becomes identical with the eigenvalue problem for the harmonic oscillator. From Chapter 5 we know that only discrete eigenvalues exist:

$$E = n + \tfrac{1}{2} \qquad (n = 0, 1, 2, \ldots) \tag{14.137}$$

and, by (5.24),

$$\langle x' \mid E \rangle = \psi_E(x') = 2^{-(n/2)}(n!)^{-1/2}(\pi)^{-1/4} \exp\left(-\frac{x'^2}{2}\right) H_n(x') \tag{14.138}$$

We may use the eigenvectors of H as the basis of a representation, which we designate as the (*harmonic oscillator*) *energy representation*. This basis is qualitatively very different from the basis of either the coordinate or the momentum representation. The latter two possess a continuum of basis vectors, whereas our new representation is spanned by a denumerable infinity of basis vectors. All three bases can be used to span the same space, although the dimensionality of the two continuous representations appears to be much greater than that of the discrete harmonic oscillator energy representation.

The example of the oscillator shows that we can go from a continuous basis to a discrete one, and vice versa. The transformation coefficients $\langle x' \mid E \rangle$ are subject to the unitarity conditions

$$\int \langle E' \mid x' \rangle \, dx' \langle x' \mid E'' \rangle = \delta_{E'E''} \tag{14.139a}$$

$$\sum_E \langle x' \mid E \rangle \langle E \mid x'' \rangle = \delta(x' - x'') \tag{14.139b}$$

which are satisfied by virtue of the orthonormality and completeness of the eigenfunctions (14.138).

Exercise 14.20. If $|\mathbf{r}'\rangle$ is the ket corresponding to a particle at position $\mathbf{r}'$, show that the displacement operator is $e^{-i\boldsymbol{\xi}\cdot\mathbf{k}}$ and that the generators of infinitesimal translations satisfy the commutation relations[13]

$$[k_x, k_y] = [k_y, k_z] = [k_z, k_x] = 0 \tag{14.140}$$

$$[x, k_x] = [y, k_y] = [z, k_z] = iI \tag{14.141}$$

$$[x, k_y] = [x, k_z] = [y, k_z] = [y, k_x] = [z, k_x] = [z, k_y] = 0 \tag{14.142}$$

[13] We again use the notation $AB - BA = [A, B]$.

CHAPTER 15

Quantum Dynamics

1. The Equation of Motion. According to the fundamental postulate of quantum mechanics, the state of a physical system at a time t is completely specified by a vector $|a, t\rangle$ in a vector space which is characterized by the nature of the system.

The basic question of nonrelativistic quantum dynamics is this: Given an initial state, $\Psi(t_0) = |a, t_0\rangle$, of the system, how is the state at time t, $\Psi(t) = |a, t\rangle$, determined from this—if indeed it is so determined at all? The assertion that $|a, t_0\rangle$ determines $|a, t\rangle$ is the quantum mechanical form of the principle of causality, and we shall assume it. In addition, we postulate an extension of the principle of superposition to include the temporal development of states. This states that, if $|a, t_0\rangle$ and $|b, t_0\rangle$ separately evolve into $|a, t\rangle$ and $|b, t\rangle$, then a superposition $c_1 |a, t_0\rangle + c_2 |b, t_0\rangle$ develops into $c_1 |a, t\rangle + c_2 |b, t\rangle$, i.e., each component of the state "moves" independently of all the others. This means that $|a, t\rangle$ can be obtained from an arbitrary state $|a, t_0\rangle$ by the application of a *linear* operator:

$$\Psi(t) = |a, t\rangle = T(t, t_0) |a, t_0\rangle = T(t, t_0)\Psi(t_0) \tag{15.1}$$

The operator $T(t, t_0)$ does not depend on $\Psi(t_0)$. It follows immediately that

$$\Psi(t_2) = T(t_2, t_1)\Psi(t_1) = T(t_2, t_1)T(t_1, t_0)\Psi(t_0) = T(t_2, t_0)\Psi(t_0)$$

Hence, the time development operators have the property,[1]

$$T(t_2, t_0) = T(t_2, t_1)T(t_1, t_0) \tag{15.2}$$

From the definition (15.1) it is also evident that

$$T(t, t) = 1 \tag{15.3}$$

Hence,

$$T(t, t_0)T(t_0, t) = T(t_0, t)T(t, t_0) = 1$$

---

[1] The property (15.2) is sometimes called the *group property*. The translation operators [(14.121)] and the rotation operators [(12.34)] exhibit a similar behavior. For the definition of a group see Chapter 16.

or

$$[T(t, t_0)]^{-1} = T(t_0, t) \tag{15.4}$$

For small ε we may write

$$T(t + \varepsilon, t) = I - \frac{i}{\hbar} \varepsilon H(t) \tag{15.5}$$

defining an operator $H(t)$. (The reason for introducing the factor $i/\hbar$ will become evident forthwith.) Since by (15.2)

$$T(t + \varepsilon, t_0) = T(t + \varepsilon, t)T(t, t_0)$$

we have with (15.5) the differential equation for T

$$\frac{dT(t, t_0)}{dt} = \lim_{\varepsilon \to 0} \frac{T(t + \varepsilon, t_0) - T(t, t_0)}{\varepsilon} = -\frac{i}{\hbar} H(t)T(t, t_0)$$

or

$$i\hbar \frac{dT(t, t_0)}{dt} = H(t)T(t, t_0) \tag{15.6}$$

with the initial condition $T(t_0, t_0) = I$.

The linear operator $H(t)$ is characteristic of the physical system under consideration. We shall see that it is analogous to the Hamiltonian function in classical mechanics. This analogy has led to the name *Hamiltonian operator* for $H(t)$, even when the system has no classical counterpart.

We also have

$$\Psi(t + \varepsilon) = T(t + \varepsilon, t)\Psi(t)$$

or, to first order in ε,

$$\Psi(t) + \varepsilon \frac{d\Psi(t)}{dt} = \left[I - \frac{i\varepsilon}{\hbar} H(t) \right]\Psi(t)$$

hence,

$$i\hbar \frac{d\Psi(t)}{dt} = H(t)\Psi(t) \tag{15.7}$$

This is the *equation of motion* for the state vector. In Dirac notation,

$$i\hbar \frac{d}{dt} |a, t\rangle = H(t) |a, t\rangle \tag{15.8}$$

where the derivative of the state vector is defined by

$$\frac{d}{dt} |a, t\rangle = \lim_{\varepsilon \to 0} \frac{|a, t + \varepsilon\rangle - |a, t\rangle}{\varepsilon}$$

Equations (15.7) and (15.8) give the general law of motion for any system. To specialize to a particular system, we must select a Hamiltonian operator.

The form of (15.7) is strongly reminiscent of the wave equation (8.79) and of the equation of motion for spinors, (13.1). This is no accident, for we have here merely reformulated in abstract language those fundamental assumptions which have already proved their worth in wave mechanics and in the spin theory. Of course, (15.7) is an equation for the state vector rather than for its components, but the difference is one of generality only. It now becomes clear why Planck's constant was introduced in (15.5).

In the two special cases in which we have already dealt with an equation of the type of (15.7), the differential operator or the 2×2 matrix H was Hermitian, as it must be whenever the Hamiltonian corresponds to the energy operator. We shall assume that H is always *Hermitian*.[2] By (15.5) this assumption implies that T is *unitary*, hence that the length of any state vector remains unaltered during the motion. If $|a, t_0\rangle$ is normalized to unity, $\langle a, t_0 | a, t_0 \rangle = 1$, then the normalization will be preserved in the course of time, and we have $\langle a, t | a, t \rangle = 1$. This circumstance makes the interpretation of the expansion coefficients in

$$|a, t\rangle = \sum_{K'} |K'\rangle\langle K' | a, t\rangle$$

particularly straightforward. $|\langle K' | a, t\rangle|^2$ is the probability of finding the system at time t with the value K' for the observable K.

Often H does not depend on the time, and then T can be obtained for finite time intervals by applying the rule (15.2) repeatedly to n intervals, each of length $\varepsilon = (t - t_0)/n$. Hence, by (15.5) we have, with $T(t_0, t_0) = 1$,

$$T(t, t_0) = \lim_{\substack{\varepsilon \to 0 \\ n \to \infty}} \left(1 - \frac{i}{\hbar} \varepsilon H\right)^n = \lim_{n \to \infty} \left[1 - \frac{i}{\hbar} \frac{(t - t_0)}{n} H\right]^n$$

In the limit we have by the definition of the exponential function

$$T(t, t_0) = \exp\left[-\frac{i}{\hbar} (t - t_0)H\right] \tag{15.9}$$

It is obvious that T is unitary if H is Hermitian.

Quantum dynamics contains no general prescription for the construction of the operator H whose existence it asserts. The Hamiltonian operator must be found on the basis of experience, using the clues provided by the classical description, if one is available. Physical insight is required to make a judicious choice of the operators to be used in the description of the system (such as

---

[2] A non-Hermitian Hamiltonian operator, like a complex index of refraction in optics, can be useful for the description of processes in which particles are absorbed (Problem 1 in Chapter 4) or in which systems undergo decay.

coordinates, momenta, spin variables, etc.) and to construct the Hamiltonian in terms of these variables.

Contact with measurable quantities and classical concepts can be established if we calculate the time development of the expectation value of an operator A, which does not itself vary with time:

$$i\hbar \frac{d}{dt} \langle a, t| A |a, t \rangle = i\hbar \frac{d}{dt} (\Psi(t), A\Psi(t))$$

$$= \left(\Psi(t), A i\hbar \frac{\partial \Psi(t)}{\partial t} \right) - \left(i\hbar \frac{\partial \Psi(t)}{\partial t}, A\Psi(t) \right)$$

$$= (\Psi(t), AH\Psi(t)) - (\Psi(t), HA\Psi(t))$$

The Hermitian property of H has been used in the last step. Hence,

$$i\hbar \frac{d}{dt} \langle A \rangle = \langle AH - HA \rangle \tag{15.10}$$

where the brackets indicate expectation values of the operators enclosed. We thus see that the commutators of H with the observables play an important role in the theory. If A commutes with H, the expectation value of A is constant, and A is said to be a *constant of the motion* (see Section 15.6).

If A depends on the time, as, for instance, an external field which acts on the system may, we have instead of (15.10)

$$i\hbar \frac{d}{dt} \langle A \rangle = \langle AH - HA \rangle + i\hbar \left\langle \frac{\partial A}{\partial t} \right\rangle \tag{15.11}$$

It is possible to define an operator dA/dt, and name it the *total time derivative of A*, by requiring the equality of the expectation values

$$\left\langle \frac{dA}{dt} \right\rangle = \frac{d}{dt} \langle A \rangle \tag{15.12}$$

for every state. Substituting this into (15.11), we may write the dynamical law in operator form as

$$\frac{dA}{dt} = \frac{AH - HA}{i\hbar} + \frac{\partial A}{\partial t} \tag{15.13}$$

2. The Quantization Postulates. Let us now apply these general equations to the special system of a mass point particle. We are concerned with the quantum behavior of this system, but it does have a classical analog— Newtonian mechanics, or its more sophisticated Lagrangian or Hamiltonian forms. The position operators x, y, z are assumed to form a complete set of commuting observables for this physical system.

Exclusively for the purposes of this chapter it is worthwhile to distinguish in the notation between the *classical* observables x_c, y_c, z_c, which are numbers, and the corresponding *quantum* observables x, y, z, which are operators.[3]

Setting $A = x$ in (15.10), we obtain

$$\frac{d\langle x \rangle}{dt} = \frac{\langle xH - Hx \rangle}{i\hbar} \qquad (15.14)$$

On the left-hand side there is a velocity, but if we wish to compare this equation with the classical equation for velocities we cannot simply let the operators go over into their classical analogs, because classical observables commute and we would have zero on the right-hand side. Hence, we must let $\hbar \to 0$ at the same time. Thus, we formulate the *correspondence principle* as follows:

If a quantum system has a classical analog, expectation values of operators behave, in the limit $\hbar \to 0$, like the corresponding classical quantities.[4] This principle provides us with a test which the quantum theory of any system with a classical analog must meet, but it does not give us an unambiguous prescription of how to construct the quantum version of any given classical theory. For this purpose the correspondence principle must be supplemented by a set of *quantization rules*. These rules have to be consistent with the correspondence principle, but their ultimate test lies in a comparison between the theoretical predictions and the experimental data.

We expect that (15.14) is the quantum analog of one of Hamilton's equations,

$$\frac{dx_c}{dt} = \frac{\partial H_c}{\partial p_{x_c}} \qquad (15.15)$$

where H_c is the classical Hamiltonian function of x_c, y_c, z_c, p_{x_c}, p_{y_c}, p_{z_c} characterizing the system. The correspondence principle requires that

$$\lim_{\hbar \to 0} \frac{\langle xH - Hx \rangle}{i\hbar} = \frac{\partial H_c}{\partial p_{x_c}} \qquad (15.16)$$

Similarly, for $A = p_x$ we have in quantum mechanics

$$\frac{d\langle p_x \rangle}{dt} = \frac{\langle p_x H - H p_x \rangle}{i\hbar} \qquad (15.17)$$

---

[3] Dirac distinguishes *c*-numbers (classical observables) and *q*-numbers (quantum operators).

[4] This does not mean that every valid classical equation can be turned into a correct quantum equation by replacing classical variables by expectation values. For example, $x_c p_c = \mu x_c (dx_c/dt) = \frac{1}{2}\mu(dx_c^2/dt)$ is a valid classical equation, if not a particularly useful one; yet, $\langle xp \rangle = \frac{1}{2}\mu(d/dt)\langle x^2 \rangle$ is generally wrong, although $\langle x \rangle \langle p \rangle = \mu\langle x \rangle(d/dt)\langle x \rangle$ and $\langle \frac{1}{2}(xp + px) \rangle = \frac{1}{2}\mu(d/dt)\langle x^2 \rangle$ are both correct. The "trouble" comes from the non-commutivity of x and p.

and classically,

$$\frac{dp_{x_c}}{dt} = -\frac{\partial H_c}{\partial x_c} \tag{15.18}$$

The correspondence principle requires that

$$\lim_{\hbar \to 0} \frac{\langle p_x H - H p_x \rangle}{i\hbar} = -\frac{\partial H_c}{\partial x_c} \tag{15.19}$$

Similar equations follow for y and z and their conjugate momenta.

All these conditions can be satisfied if we do the following:

(1) Let H be a Hermitian operator identical in form with H_c, but replacing all coordinates and momenta by their corresponding operators.

(2) Postulate the fundamental commutation relations between the Hermitian operators representing coordinates and momenta:

$$[x, p_x] = [y, p_y] = [z, p_z] = i\hbar \tag{15.20}$$

$$[x, p_y] = [x, p_z] = [y, p_z] = [y, p_x] = [z, p_x] = [z, p_y] = 0 \tag{15.21}$$

The three coordinates, x, y, z, commute with each other; the three momenta, p_x, p_y, p_z, also commute.

Prescription (1) must be applied with care if H_c contains terms such as $x_c p_{x_c}$, because x and p_x are noncommuting and would upon translation into operator language give rise to a non-Hermitian H. The symmetrized operator $\frac{1}{2}(x p_x + p_x x)$ can then be used instead. It is Hermitian and leads to the correct classical limit. Sometimes there may be several different ways of symmetrizing a term to make it Hermitian. Thus, the Hermitian operators $x^2 p_x{}^2 + p_x{}^2 x^2$ and $\frac{1}{2}(x p_x + p_x x)^2$ both have the classical limit $2 x_c{}^2 p_{x_c}{}^2$, but they are not identical. In practice it is usually possible to avoid such ambiguities.

Exercise 15.1. Show that the operators $x^2 p_x{}^2 + p_x{}^2 x^2$ and $\frac{1}{2}(x p_x + p_x x)^2$ differ only by terms of order $\hbar^2$.

The consistency of conditions (1) and (2) and their agreement with (15.16) and (15.19) can be verified for any H_c which can be expanded in powers of the coordinates and momenta. For instance, the commutation relation

$$x p_x - p_x x = i\hbar 1 \tag{15.20a}$$

agrees with (15.16) and (15.19), as can be seen if we choose $H = p_x$ and $H = x$ respectively. The consistency proof can be continued by letting

$H = x^2$. Then

$$\frac{1}{i\hbar} [p_x, x^2] = \frac{x}{i\hbar} [p_x, x] + \frac{1}{i\hbar} [p_x, x]x = -2x$$

by virtue of the quantum condition (15.20a). This is in agreement with the classical limit (15.19), because for $H_c = x_c^2$

$$-\frac{\partial x_c^2}{\partial x_c} = -2x_c$$

More generally, we can continue this type of argument to prove that for any two functions, F and G, of the coordinates and momenta, which can be expanded in a power series, the relation

$$\lim_{\hbar \to 0} \frac{\langle GF - FG \rangle}{i\hbar} = \frac{\partial G_c}{\partial x_c} \frac{\partial F_c}{\partial p_{x_c}} - \frac{\partial F_c}{\partial x_c} \frac{\partial G_c}{\partial p_{x_c}} + \frac{\partial G_c}{\partial y_c} \frac{\partial F_c}{\partial p_{y_c}}$$

$$- \frac{\partial F_c}{\partial y_c} \frac{\partial G_c}{\partial p_{y_c}} + \frac{\partial G_c}{\partial z_c} \frac{\partial F_c}{\partial p_{z_c}} - \frac{\partial F_c}{\partial z_c} \frac{\partial G_c}{\partial p_{z_c}} \qquad (15.22)$$

holds, where F_c and G_c are the same functions of the ordinary variables as F and G are of the corresponding operators. Equation (15.22) is assumed to be valid for any two functions of coordinates and momenta, even if a power series expansion cannot be made. Equations (15.16) and (15.19) are special cases of (15.22).

The expression on the right-hand side of (15.22) is abbreviated as $\{G_c, F_c\}$ and is known as the *Poisson bracket* of G and F in classical mechanics.[5] Dirac discovered that this is the classical analog of the commutator $(GF - FG)/i\hbar$.

An example will illustrate (15.22). Let $G = x^2$, $F = p_x^2$. Then

$$x^2 p_x^2 - p_x^2 x^2 = x[x, p_x]p_x + xp_x[x, p_x]$$
$$+ p_x[x, p_x]x + [x, p_x]xp_x = 4i\hbar xp_x + 2\hbar^2$$

or

$$\frac{x^2 p_x^2 - p_x^2 x^2}{i\hbar} = 4xp_x - 2i\hbar$$

Hence, as $\hbar \to 0$, this becomes $4x_c p_{x_c}$. But this is the same as

$$\{x_c^2, p_{x_c}^2\} = \frac{\partial x_c^2}{\partial x_c} \frac{\partial p_{x_c}^2}{\partial p_{x_c}} - \frac{\partial p_{x_c}^2}{\partial x_c} \frac{\partial x_c^2}{\partial p_{x_c}} = 4x_c p_{x_c}$$

[5] See H. C. Corben and P. Stehle, *Classical Mechanics*, second edition, John Wiley and Sons New York, 1960, Section 66.

Exercise 15.2. Invent some quantization rules which are different from (1) and (2) postulated above, but which satisfy the same requirements. Discuss the difference between the two sets of rules. (Consider this problem again in the light of the next section.)

3. Wave Mechanics Regained. In order to test the quantization procedure just outlined, we must show that the quantum equations obtained from it are identical with the equations of wave mechanics which we know to give an accurate description of many physical phenomena.

Thus we specialize to a classical Hamiltonian function of the form

$$H_c(\mathbf{r}_c, \mathbf{p}_c) = \frac{1}{2\mu}\, \mathbf{p}_c{}^2 + V(\mathbf{r}_c) \tag{15.23}$$

where we have assumed a conservative system. Comparing the commutation relations (15.20) with (14.140), (14.141), and (14.142) for the generators of linear displacements, we see that the former can be satisfied if we make the identification

$$\mathbf{p} = \hbar\mathbf{k} \tag{15.24}$$

The commutation relations permit the addition of the gradient of an arbitrary scalar function $\varphi(\mathbf{r})$ on the right-hand side of this equation. Since the interaction of a magnetic field $\mathbf{B} = \nabla \times \mathbf{A}$ with a charged particle is accomplished through replacing $\mathbf{p}$ by $\mathbf{p} - q\mathbf{A}/c$, the vector field $\nabla\varphi$ may be interpreted as the vector potential of zero magnetic field. As was shown in Section 14.7, it is possible to change the gauge by a unitary transformation $\exp[i\varphi(\mathbf{r})]$ and to return to the simple formula (15.24). (Also see Problem 7 in Chapter 8.)

It follows from (14.118) and (14.126) that for any function of $\mathbf{r}$ and $\mathbf{p}$, in which no questions of ordering arise, we have in the coordinate representation:

$$\langle \mathbf{r}'|\, F(\mathbf{r}, \mathbf{p})\, |a\rangle = F\left(\mathbf{r}', \frac{\hbar}{i}\nabla'\right)\langle \mathbf{r}' \,|\, a\rangle \tag{15.25}$$

Multiplying the equation of motion (15.8) on the left by $\langle \mathbf{r}'|$, we obtain

$$i\hbar\frac{\partial}{\partial t}\langle \mathbf{r}' \,|\, a, t\rangle = \langle \mathbf{r}'|\, H\, |a, t\rangle = H\left(\mathbf{r}', \frac{\hbar}{i}\nabla'\right)\langle \mathbf{r}' \,|\, a, t\rangle$$

For a Hamiltonian function of the form (15.23), we arrive at the result

$$i\hbar\frac{\partial\langle \mathbf{r}' \,|\, a, t\rangle}{\partial t} = \left[-\frac{\hbar^2}{2\mu}\nabla'^2 + V(\mathbf{r}')\right]\langle \mathbf{r}' \,|\, a, t\rangle \tag{15.26}$$

which is identical with the wave equation (4.1) if we identify the wave function as

$$\psi(\mathbf{r}', t) = \langle \mathbf{r}' \mid a, t \rangle \tag{15.27}$$

By separation of the variable t in (15.26), the Schrödinger equation is obtained in the usual manner:

$$H\left(\mathbf{r}', \frac{\hbar}{i} \nabla'\right) \langle \mathbf{r}' \mid E \rangle = E \langle \mathbf{r}' \mid E \rangle \tag{15.28}$$

The functions $\langle \mathbf{r}' \mid E \rangle$ are the components of the eigenkets of H in the coordinate representation, but they are also the transformation coefficients of the unitary transformation which takes us from the coordinate representation to the energy representation. In this representation the matrix H is diagonal, the diagonal elements being the allowed energy values of the stationary states.[6]

If we introduce the momentum representation, we obtain in a similar fashion

$$i\hbar \frac{\partial}{\partial t} \langle \mathbf{p}' \mid a, t \rangle = \frac{p'^2}{2\mu} \langle \mathbf{p}' \mid a, t \rangle + \int \langle \mathbf{p}' \mid V(\mathbf{r}) \mid \mathbf{p}'' \rangle \, d^3 p'' \langle \mathbf{p}'' \mid a, t \rangle \tag{15.29}$$

which is the same integral-differential equation as (8.15). The matrix element $\langle \mathbf{p}' \mid V \mid \mathbf{p}'' \rangle$ is a Fourier integral. If $V(\mathbf{r})$ has the form of a power series, we can simplify (15.29) and obtain

$$i\hbar \frac{\partial}{\partial t} \langle \mathbf{p}' \mid a, t \rangle = \frac{p'^2}{2\mu} \langle \mathbf{p}' \mid a, t \rangle + V\left(-\frac{\hbar}{i} \nabla_{\mathbf{p}'}\right) \langle \mathbf{p}' \mid a, t \rangle \tag{15.30}$$

The complete equivalence of these results with the earlier equations suggest that we are on the right track. In addition we see that the constant $\hbar$ introduced in this chapter is indeed Planck's constant, as anticipated.

4. Canonical Quantization. So far we have considered only descriptions of the physical system in terms of Cartesian coordinates. Yet, in Section 15.2 the connection between classical and quantum mechanics was established by the use of Hamilton's equations of classical mechanics, which are by no means restricted to Cartesian coordinates. Rather, these equations are well known to have the same general form for a very large class of variables, called *canonical coordinates and momenta*, and denoted by the symbols q_c and p_c. Since in the Hamiltonian form of mechanics the Cartesian coordinates do not occupy a unique position, we ask whether the quantization procedure of Section 15.2 could not equally well have been applied to more general canonical variables. Could we replace x by q and p_x by p (assuming for convenience only one degree of freedom) satisfying the more general

---

[6] Remember that H indicates the *matrix* representing the *operator* H.

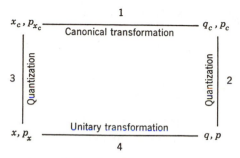

Figure 15.1. Canonical quantization.

commutation relations

$$qp - pq = i\hbar 1 \tag{15.31}$$

instead of (15.20), and could we still apply the same quantization rules?

To show that we are indeed at liberty to use canonical variables other than the Cartesian ones, we must prove that the same form of quantum mechanics results whether we use x_c, p_{x_c}, or q_c, p_c to make the transition to quantum mechanics. To prove that we can pass from the upper left to the lower right corner of Figure 15.1 equivalently by routes 1, 2 or 3, 4 we first consider an infinitesimal canonical transformation (step 1), i.e., a transformation which is generated by an infinitesimal function $\varepsilon G_c(x_c, p_c)$ from the relations[7]

$$q_c = x_c + \varepsilon \frac{\partial G_c}{\partial p_{x_c}}, \qquad p_c = p_{x_c} - \varepsilon \frac{\partial G_c}{\partial x_c} \tag{15.32}$$

The new Hamiltonian is

$$H_c'(q_c, p_c) = H_c(x_c, p_{x_c}) = H_c(q_c, p_c) - \varepsilon \frac{\partial H_c}{\partial x_c} \frac{\partial G_c}{\partial p_{x_c}} + \varepsilon \frac{\partial H_c}{\partial p_{x_c}} \frac{\partial G_c}{\partial x_c} \tag{15.33}$$

The canonical transformation is paralleled in quantum theory by step 4. Agreement with (15.32) and (15.33) in the correspondence limit is assured if we define the Hermitian operators

$$q = x + \frac{\varepsilon}{i\hbar} [x, G], \qquad p = p_x + \frac{\varepsilon}{i\hbar} [p_x, G] \tag{15.34}$$

where the Hermitian operator $G(x, p_x)$ is constructed from $G_c(x_c, p_{x_c})$ by letting x_c and p_{x_c} become operators.[8] The commutators are evaluated by applying the quantization rules of the last section for Cartesian coordinates.

[7] H. Goldstein, *Classical Mechanics*, Addison-Wesley Publishing Company, Cambridge, 1953, Section 8-6.

[8] It is again assumed that no ordering ambiguities arise in constructing $G(x, p_x)$.

To the first order in ε, (15.34) may be written as

$$q = \left(1 + \frac{i\varepsilon}{\hbar} G\right) x \left(1 - \frac{i\varepsilon}{\hbar} G\right) = U_\varepsilon x U_\varepsilon^{\;\dagger} \tag{15.35a}$$

$$p = \left(1 + \frac{i\varepsilon}{\hbar} G\right) p_x \left(1 - \frac{i\varepsilon}{\hbar} G\right) = U_\varepsilon p_x U_\varepsilon^{\;\dagger} \tag{15.35b}$$

showing that q and p are obtained from x and p_x by a unitary transformation $U_\varepsilon = 1 + (i\varepsilon/\hbar)G$. The Hamiltonian operator becomes in terms of the new variables

$$H'(q, p) = H(x, p_x) = H(U_\varepsilon^{\;\dagger} q U_\varepsilon, U_\varepsilon^{\;\dagger} p U_\varepsilon) = U_\varepsilon^{\;\dagger} H(q, p) U_\varepsilon$$

$$= H(q, p) + \frac{\varepsilon}{i\hbar} [G, H] \tag{15.36}$$

By (15.22) this agrees with (15.33) in the correspondence limit. [See Exercise 14.8 for the conditions under which (15.36) holds.]

Finally, the commutation relations are invariant under unitary transformations because

$$[q, p] = U_\varepsilon x U_\varepsilon^{\;\dagger} U_\varepsilon p_x U_\varepsilon^{\;\dagger} - U_\varepsilon p_x U_\varepsilon^{\;\dagger} U_\varepsilon x U_\varepsilon^{\;\dagger} = U_\varepsilon (x p_x - p_x x) U_\varepsilon^{\;\dagger} = i\hbar 1$$

and we have arrived at (15.31). This completes the proof that the quantization rules of Section 15.2 can be extended to new canonical variables which are infinitesimally close to Cartesian.

The quantization procedure based on rules (1) and (2) of Section 15.2 can now be immediately generalized to all those canonical variables which can be obtained from the Cartesian ones by a succession of infinitesimal canonical transformations. This is true because two classical canonical transformations made in succession can be replaced by a single direct one. Similarly, in quantum mechanical transformations, successive application of unitary operators is equivalent to the application of a single unitary operator. If we let $\varepsilon = \lambda/N$, and apply the same unitary operator N times, we obtain in the limit $N \rightarrow \infty$ the unitary operator

$$U = \lim_{N \to \infty} \left(1 + \frac{i\lambda}{\hbar N} G\right)^N = \exp\left(\frac{i\lambda}{\hbar} G\right) \tag{15.37}$$

This finite unitary transformation changes the Cartesian variables into

$$q = UxU^\dagger, \qquad p = Up_x U^\dagger \tag{15.38}$$

The commutation relations are, of course, also invariant under the finite transformations.

We note that if (15.38) holds the eigenvalue spectra of x and q are the same, as

$$x \, |x'\rangle = x' \, |x'\rangle$$

$$U x U^{\dagger} U \, |x'\rangle = q U \, |x'\rangle = x' U \, |x'\rangle$$

Hence

$$q' = x' \quad \text{and} \quad |q'\rangle = U \, |x'\rangle$$

We see that the quantization of the system can be carried through by the use of the general commutation relations (15.31) for any pair of canonical variables which can be obtained from x, p_x by a continuous succession of infinitesimal transformations. For more general canonical transformations than these the standard quantization procedure may or may not be valid. Clearly, it will be valid whenever the new operators can be obtained from the old ones by a unitary transformation. A simple example of a failure of the standard quantization is provided by the transition to spherical polar coordinates, r_c, φ_c, θ_c. The transformation to these from Cartesian coordinates is canonical, but it cannot be generated by a succession of infinitesimal transformations, because of the discontinuity of the spherical polar coordinates. Nor does a unitary transformation between x, y, z and r, φ, θ exist, for if it did the eigenvalues of the latter operators would have to be the same as those of x, y, z, and range from $-\infty$ to $+\infty$, contrary to the definition of r, φ, and θ.

Because of its close connection with the classical canonical formalism the quantization procedure described here is referred to as *canonical quantization*. The correspondence between canonical transformations and unitary operators has led to the frequent designation of the latter as quantum mechanical *canonical transformations*. This terminology has asserted itself even though there are unitary transformations which have no classical canonical analog, and vice versa.

Exercise 15.3. Construct the unitary operator which transforms the pair of variables ax, bp_x into bp_x, $-ax$, where a and b are real constants. Calculate the matrix elements of this operator in the coordinate representation. Noting that this transformation leaves the operator $a^2x^2 + b^2p_x^2$ invariant, rederive equation (8.18).

Exercise 15.4. Show that the reflection operator, defined by the relation $U \, |x'\rangle = |-x'\rangle$, gives rise to a unitary transformation which takes x into $-x$, and p_x into $-p_x$.

5. Systems of Two Distinguishable Particles. So far we have been concerned with systems in which only one particle was treated by quantum dynamics. Even when discussing the energy levels of a hydrogen atom, we assumed in accordance with classical physics that the system can be treated effectively as one body attracted by a fixed center of force, provided that the reduced rather than the real mass is used.

The general formulation of quantum mechanics enables us to generalize the theory unambiguously to systems composed of several particles. If the particles are identical, very important special peculiarities require considerations. Since Chapter 20 will deal with these exclusively, we confine ourselves in this section to the quantum mechanics of systems with two *distinguishable* particles. Examples are the hydrogen atom (electron and proton), the deuteron (proton and neutron), and positronium (electron and positron).

We denote the two particles by 1 and 2. We assume that each has its own complete set of dynamical variables, and that all variables belonging to particle 1 commute with those belonging to particle 2. For example, we suppose that x_1 and p_{2_x} are simultaneously measurable. The appropriate abstract vector space is the *direct product* space of the two vector spaces belonging to the two particles. The direct product space is defined as the space spanned by the basis vectors

$$|A_1'B_2''\rangle = |A_1'\rangle |B_2''\rangle \tag{15.39}$$

where both A_1' and B_2'' range through complete sets of eigenvalues. If n_1 and n_2 are the dimensions of the two factor spaces, the product space has $n_1 \times n_2$ dimensions. Any operator which pertains to only one of the two factor spaces is regarded as acting as an identity operator with respect to the other. More generally, if M_1 and N_2 are two operators belonging to the vector spaces 1 and 2 such that

$$M_1 |A_1'\rangle = \sum_{A_1''} |A_1''\rangle\langle A_1''| M_1 |A_1'\rangle$$

and

$$N_2 |B_2'\rangle = \sum_{B_2''} |B_2''\rangle\langle B_2''| N_2 |B_2'\rangle$$

we define the *direct product operator* $M_1 \times N_2$ by the equation

$$(M_1 \times N_2) |A_1'B_2'\rangle = \sum_{A_1'',B_2''} |A_1''B_2''\rangle\langle A_1''| M_1 |A_1'\rangle\langle B_2''| N_2 |B_2'\rangle \tag{15.40}$$

Hence, $M_1 \times N_2$ is represented by a matrix which is said to be the direct product of the two matrices representing M_1 and N_2 separately and which is defined by

$$\langle A_1''B_2''| M_1 \times N_2 |A_1'B_2'\rangle = \langle A_1''| M_1 |A_1'\rangle\langle B_2''| N_2 |B_2'\rangle \tag{15.41}$$

Exercise 15.5. If M_1 and P_1 are operators in space 1 and N_2 and Q_2 are operators in space 2, prove the identity

$$M_1P_1 \times N_2Q_2 = (M_1 \times N_2)(P_1 \times Q_2) \tag{15.42}$$

We have already made implicit use of the notion of a direct product when we added the spin to the space coordinates for the description of electron states. The state vector space for an electron with spin is the direct product of the infinitely dimensional space spanned by the coordinate eigenvectors, $|x'y'z'\rangle$, and the two-dimensional space spanned by $\alpha = |S_z' = +\tfrac{1}{2}\rangle$ and $\beta = |S_z' = -\tfrac{1}{2}\rangle$. It is useful to think in terms of direct products whenever two physically independent sets of variables are combined into a single set of dynamical variables.

As an example, consider the hydrogen atom as a two-body system but neglect the spin for the time being. If the six coordinates of electron and proton are used to define the basis $|\mathbf{r}_1\mathbf{r}_2\rangle = |\mathbf{r}_1\rangle\,|\mathbf{r}_2\rangle$, then in analogy with (14.115) we can introduce the two-particle wave function,

$$\psi(\mathbf{r}_1, \mathbf{r}_2) = \langle \mathbf{r}_1\mathbf{r}_2 \,|\, \Psi \rangle$$

The interpretation is of course the usual one: $|\psi(\mathbf{r}_1, \mathbf{r}_2)|^2\, d^3r_1\, d^3r_2$ is (proportional to) the probability that particle 1 will be found in volume element d^3r_1 centered at $\mathbf{r}_1$ and simultaneously particle 2 in volume element d^3r_2 centered at $\mathbf{r}_2$. If the integral of $|\psi|^2$ is quadratically integrable, we assume usually the normalization

$$\iint |\psi(\mathbf{r}_1, \mathbf{r}_2)|^2\, d^3r_1\, d^3r_2 = 1 \qquad (15.43)$$

Since ψ is now a function of two different points in space, it can no longer be pictured as a wave in the naïve sense which we found so fruitful in the early chapters of this book. Instead, ψ for two particles is sometimes considered as a wave in a six-dimensional *configuration space* of the coordinates $\mathbf{r}_1$ and $\mathbf{r}_2$.

The Hamiltonian of the hydrogen atom (without spin) is taken over from classical mechanics and has the general form

$$H = \frac{\mathbf{p}_1{}^2}{2\mu_1} + \frac{\mathbf{p}_2{}^2}{2\mu_2} + V(|\mathbf{r}_1 - \mathbf{r}_2|) \qquad (15.44)$$

In the coordinate representation this leads to the Schrödinger equation

$$\left[-\frac{\hbar^2}{2\mu_1}\nabla_1{}^2 - \frac{\hbar^2}{2\mu_2}\nabla_2{}^2 + V(|\mathbf{r}_1 - \mathbf{r}_2|) \right]\psi(\mathbf{r}_1, \mathbf{r}_2) = E\psi(\mathbf{r}_1, \mathbf{r}_2) \quad (15.45)$$

It is easily verified that the substitution

$$\mathbf{r} = \mathbf{r}_1 - \mathbf{r}_2 \qquad \mathbf{R} = \frac{\mu_1\mathbf{r}_1 + \mu_2\mathbf{r}_2}{\mu_1 + \mu_2} \qquad (15.46)$$

changes the Schrödinger equation to

$$-\frac{\hbar^2}{2(\mu_1 + \mu_2)}\left(\frac{\partial^2}{\partial X^2} + \frac{\partial^2}{\partial Y^2} + \frac{\partial^2}{\partial Z^2}\right)\psi$$

$$-\frac{\hbar^2}{2\,\dfrac{\mu_1\mu_2}{\mu_1 + \mu_2}}\left(\frac{\partial^2}{\partial x^2} + \frac{\partial^2}{\partial y^2} + \frac{\partial^2}{\partial z^2}\right)\psi + V(r)\psi = E\psi \quad (15.47)$$

where now $\psi = \psi(\mathbf{r}, \mathbf{R})$ is a function of the *relative coordinates*, $\mathbf{r}(x, y, z)$, and the coordinates of the center of mass, $\mathbf{R}(X, Y, Z)$. The new Hamiltonian is a sum

$$H = H_R + H_r \qquad (15.48)$$

and each of the two sub-Hamiltonians obviously possesses a complete set of eigenfunctions. Hence, all the eigenfunctions of (15.47) can be obtained by assuming that ψ is a product

$$\psi(\mathbf{r}, \mathbf{R}) = \psi(\mathbf{r})\varphi(\mathbf{R}) \qquad (15.49)$$

and the energy a sum, $E = E_R + E_r$ such that

$$-\frac{\hbar^2}{2(\mu_1 + \mu_2)}\nabla^2\varphi(\mathbf{R}) = E_R\varphi(\mathbf{R}) \qquad (15.50)$$

and

$$-\frac{\hbar^2}{2\mu_r}\nabla^2\psi(\mathbf{r}) + V(r)\psi(\mathbf{r}) = E_r\psi(\mathbf{r}) \qquad (15.51)$$

where μ_r is the *reduced mass*, $\mu_r = \mu_1\mu_2/(\mu_1 + \mu_2)$.

Equation (15.51) confirms that the relative motion of a system of two particles subject to central forces can be treated like a one-body problem if the reduced mass is used. Equation (15.50) whose solutions are plane waves, represents the quantum form of Newton's first law: the total momentum of an isolated system is constant. The most conspicuous manifestation of the effect of the reduced mass is the shift which is observed in a comparison of the spectral lines of hydrogen, deuterium, and positronium, which all have different reduced masses.

It is noteworthy that the transition to relative coordinates could equally well have been made before the quantization. The classical transformation

$$\mathbf{r} = \mathbf{r}_1 - \mathbf{r}_2 \qquad \mathbf{R} = \frac{\mu_1\mathbf{r}_1 + \mu_2\mathbf{r}_2}{\mu_1 + \mu_2}$$

$$\mathbf{p} = \frac{\mu_2\mathbf{p}_1 - \mu_1\mathbf{p}_2}{\mu_1 + \mu_2} \qquad \mathbf{P} = \mathbf{p}_1 + \mathbf{p}_2 \qquad (15.52)$$

is canonical and changes H into

$$\frac{\mathbf{P}^2}{2(\mu_1 + \mu_2)} + \frac{\mathbf{p}^2}{2\mu_r} + V(r)$$

Subsequent quantization and use of the coordinate representation leads again to (15.47).

Exercise 15.6. Prove that $\mathbf{r}$, $\mathbf{p}$ and $\mathbf{R}$, $\mathbf{P}$ defined in (15.52) satisfy the commutation relations for conjugate canonical variables.

6. Constants of the Motion and Invariance Properties. An important application of the ideas given in the last section can now be made. It concerns the finding of *constants of the motion*. These are observables which commute with H (see Section 15.1).

A useful way of obtaining constants of the motion for a Hamiltonian operator $H(q, p)$ consists in noting that, if a (finite or infinitesimal) canonical transformation to new variables q', p' is made, the new Hamiltonian H' is obtained from the old one by the equation

$$H'(q', p') = H(q, p) \tag{15.53}$$

i.e., by simply substituting q, p as functions of q', p' into $H(q, p)$. For a unitary transformation we have, on the other hand (see Exercise 14.8),

$$H'(q', p') = H'(UqU^\dagger, UpU^\dagger) = UH'(q, p)U^\dagger \tag{15.54}$$

Hence, comparing with (15.53),

$$H(q, p) = UH'(q, p)U^\dagger \tag{15.55}$$

If the canonical transformation leaves the Hamiltonian *invariant*, so that the new Hamiltonian is the same function of the q', p' as the old one was of the q, p, or,

$$H'(q, p) = H(q, p) \tag{15.56}$$

then

$$UH - HU = 0$$

Hence,

$$\frac{d}{dt} \langle U \rangle = \frac{1}{i\hbar} \langle UH - HU \rangle = 0$$

and thus the unitary operator U is a constant of the motion if the transformation leaves H invariant. If, in particular, H is invariant under an infinitesimal transformation,

$$U = 1 + \frac{i\varepsilon}{\hbar} G$$

then the (Hermitian) generator εG of the transformation commutes with H and is a constant of the motion. In this way physical observables are obtained which are constants of the motion.

Exercise 15.7. Show that if A and B are constants of the motion, then the commutator $i[A, B]$ is also a constant of the motion.

As an example consider a free particle whose Hamiltonian is

$$H = \frac{1}{2\mu} (p_x{}^2 + p_y{}^2 + p_z{}^2) \tag{15.57}$$

According to (15.34), the generator

$$\varepsilon G = -\varepsilon_x p_x - \varepsilon_y p_y - \varepsilon_z p_z = -\boldsymbol{\varepsilon} \cdot \mathbf{p}$$

produces no change in the momenta

$$p_x{}' = p_x, \qquad p_y{}' = p_y, \qquad p_z{}' = p_z \tag{15.58}$$

but changes the x-coordinates to

$$x' = x - \frac{1}{i\hbar} [x, \boldsymbol{\varepsilon} \cdot \mathbf{p}]$$

which by the fundamental commutation relations becomes

$$x' = x - \frac{1}{i\hbar} \varepsilon_x [x, p_x] = x - \varepsilon_x 1$$

Applying the same transformation to all three coordinates, we get

$$x' = x - \varepsilon_x 1, \qquad y' = y - \varepsilon_y 1, \qquad z' = z - \varepsilon_z 1 \tag{15.59}$$

i.e., the transformation describes a coordinate translation as expected from the connection between momentum and translation operators.

Evidently, any H which does not depend on the coordinates is invariant under the transformation (15.58), (15.59). Hence, $G = -\boldsymbol{\varepsilon} \cdot \mathbf{p}$ (for arbitrary $\boldsymbol{\varepsilon}$) and $\mathbf{p}$ itself are constants of the motion, and we conclude that *linear momentum is conserved.*

In Chapter 16 the connection between invariances and conservation laws will be taken up more fully and applied to less trivial examples.

7. *The Heisenberg Picture.* We saw that the equation of motion is of the form

$$i\hbar \frac{\partial \Psi(t)}{\partial t} = H\Psi(t) \tag{15.60}$$

where H is a Hermitian operator. This formulation of quantum dynamics is referred to as the *Schrödinger picture* because, as was shown in Section 15.3, (15.60) leads directly to the wave equation discovered by Schrödinger. Other equivalent formulations are available, and the present section will be devoted to the most important of these, known as the *Heisenberg picture*.

The possibility of formulating quantum dynamics in different pictures arises because the mathematical entities such as state vectors and operators are not the quantities which are directly accessible to physical measurement. Rather, the comparison with observation is made in terms of eigenvalues and of expansion coefficients which are scalar products in an abstract vector space. Measuring an observable A means finding one of its eigenvalues A', the probability of the particular result being given by $|\langle A' \mid a\rangle|^2$ if $|a\rangle$ denotes the state of the system. It follows that any formulation of quantum mechanics is equally acceptable as the Schrödinger picture if in the new picture

(a) the operators corresponding to the observables have the same eigenvalue spectrum as in the Schrödinger picture, and if

(b) the scalar products of the vector representing a state in the new picture with the new eigenvectors have the same values as the corresponding scalar products in the Schrödinger picture.[9]

These conditions are satisfied by any unitary transformation U applied to the vectors and operators in the Schrödinger picture. A bar over the vectors and operators will denote quantities in the new picture whenever the distinction is important. Thus, we write

$$|\bar{a}\rangle = U\,|a\rangle \qquad \text{and} \qquad \langle\bar{a}| = \langle a|\,U^\dagger \tag{15.61}$$

for vectors. Defining $\bar{A}$ as the new operator which maps $\bar{\Psi}$ into $\bar{\Phi}$, if A maps Ψ into Φ, we have

$$\bar{A} = UAU^\dagger \tag{15.62}$$

From the eigenvalue equation

$$A\,|A'\rangle = A'\,|A'\rangle$$

we obtain

$$UAU^\dagger U\,|A'\rangle = A'U\,|A'\rangle$$

or

$$\bar{A}\,|\bar{A}'\rangle = A'\,|\bar{A}'\rangle \tag{15.63}$$

This equation shows that the eigenvalues of $\bar{A}$ are the same as those of A in agreement with condition (a). The eigenvectors change from $|A'\rangle$ to $|\bar{A}'\rangle = U\,|A'\rangle$. Condition (b) is obviously satisfied if U is a unitary operator.

---

[9] It is sufficient to demand that the absolute values of scalar products remain unchanged, but a theorem to be stated in Section 16.1 shows that much greater generality cannot be gained by relaxing condition (b) in this way.

Infinitely many different choices of U are possible, leading to different pictures of quantum dynamics besides the Schrödinger picture, which we have used so far exclusively. The Heisenberg picture is obtained if we choose for the operator U at time t the inverse of the unitary time development operator introduced in Section 15.1:

$$U(t) = [T(t, t_0)]^{-1} = T^\dagger(t, t_0) = T(t_0, t) \qquad (15.64)$$

Hence, in the new picture the state of a system at time t is determined by the state vector

$$\overline{\Psi}(t) = U(t)\Psi(t) = T(t_0, t)T(t, t_0)\Psi(t_0)$$

by the definition of T in (15.1). It follows from (15.4) that

$$\overline{\Psi}(t) = \Psi(t_0) \qquad (15.65)$$

no matter what the value of t is. We thus see that in the new picture the state vector does not change in time at all. It simply retains its initial value, which it had at time t_0, and there is no equation of motion like (15.7). We may therefore write simply $\overline{\Psi}$ for the state vector in the Heisenberg picture. On the other hand, the operators, which in the Schrödinger picture were fixed in time (unless they happened to have an explicit time dependence), now vary with time, as exhibited by the equation

$$\overline{A}(t) = U(t)AU^\dagger(t) \qquad (15.66)$$

As a special case, note that if H is constant, $\overline{H}(t) = H$, i.e., if H is constant, the Hamiltonian operator does not change in time even in the Heisenberg picture.

The expectation value of an observable is, of course, the same in either picture, since

$$\langle \bar{a} |\ \overline{A}(t)\ | \bar{a} \rangle = (\overline{\Psi}(t), U(t)AU^\dagger(t)\overline{\Psi}(t)) = (\Psi, A\Psi) = \langle a|A|a \rangle$$

We can obtain a differential equation for $\overline{A}(t)$ by defining the derivative of $\overline{A}$ with respect to the time by

$$\frac{d\overline{A}}{dt} = \lim_{\varepsilon \to 0} \frac{\overline{A}(t + \varepsilon) - \overline{A}(t)}{\varepsilon} \qquad (15.67)$$

Differentiating (15.66) and noting that by (15.6) and (15.64)

$$i\hbar \frac{dU^\dagger(t)}{dt} = HU^\dagger(t) \qquad \text{and} \qquad -i\hbar \frac{dU(t)}{dt} = U(t)H$$

we get the equation of motion for any operator $\overline{A}$,

$$i\hbar \frac{d\overline{A}}{dt} = \overline{A}\overline{H} - \overline{H}\overline{A} \qquad (15.68)$$

provided that A does not depend on t explicitly. If the operator A has an explicit time dependence, this equation is generalized to

$$i\hbar \frac{d\bar{A}}{dt} = \bar{A}\bar{H} - \bar{H}\bar{A} + i\hbar \frac{\partial \bar{A}}{\partial t} \tag{15.69}$$

where the last term is defined by

$$\frac{\partial \bar{A}}{\partial t} = U(t) \frac{\partial A}{\partial t} U^{\dagger}(t) \tag{15.70}$$

Equation (15.69) is almost identical with (15.13), but two different pictures are employed.

Let us summarize the difference between the Schrödinger picture and the Heisenberg picture. Table 15.1 gives a schematic comparison. Since the

Table 15.1

	Schrödinger Picture	Heisenberg Picture
State vectors	Moving	Fixed
Operators	Fixed	Moving

operators, like the classical dynamical variables which they represent, change in time in the Heisenberg picture, this picture is more akin to classical dynamics than the Schrödinger picture. In the Schrödinger picture (15.12) *defines* the operator dA/dt.

The connection between the two pictures can be further clarified by noting that in the Heisenberg picture the eigenvectors of the observables change in time according to the equation

$$|\overline{A', t}\rangle = U(t) |A'\rangle$$

or

$$|A'\rangle = T(t, t_0) |\overline{A', t}\rangle \tag{15.71}$$

Differentiating this equation with respect to t, we obtain

$$\frac{dT(t, t_0)}{dt} |\overline{A', t}\rangle + T(t, t_0) \frac{d}{dt} |\overline{A', t}\rangle = 0$$

or, using (15.6),

$$i\hbar \frac{d}{dt} |\overline{A', t}\rangle = -H |\overline{A', t}\rangle \tag{15.72}$$

which is very similar to the equation of motion (15.8) in the Schrödinger picture, except for the all-important minus sign. Its appearance shows that, if in the Schrödinger picture we regard the state vectors as "rotating" in a certain direction in abstract vector space and the operators with their eigenvectors as fixed, then in the Heisenberg picture the state vectors stand still and the operators with their eigenvectors "rotate" in the opposite direction. But the mutual relation between state vectors and operators is the same in the two pictures. The two pictures are related to each other in much the same way as the two descriptions of the rotation of a rigid body with respect to a coordinate system: We can consider the body moving in a fixed system, or the body as being at rest with the coordinate system rotating backwards.

Exercise 15.8. Find the equation of motion for $\langle \overline{A', t} |$.

The canonical quantization procedure can be formulated in the Heisenberg picture as well as in the Schrödinger picture. Since the two pictures are related by a unitary operator U which changes every Schrödinger operator A into a Heisenberg operator

$$\bar{A} = UAU^\dagger$$

it is clear that every operator equation

$$f(A, B, C, \ldots) = 0$$

becomes in the Heisenberg picture

$$f(\bar{A}, \bar{B}, \bar{C}, \ldots) = 0$$

with the same function f.[10] In particular, the fundamental commutation relations for canonically conjugate observables are

$$qp - pq = i\hbar 1$$

or

$$UqU^\dagger UpU^\dagger - UpU^\dagger UqU^\dagger = i\hbar 1$$

Hence

$$\bar{q}(t)\bar{p}(t) - \bar{p}(t)\bar{q}(t) = i\hbar 1 \tag{15.73}$$

which is formally the same as in the Schrödinger picture.

The equations of motion for $\bar{q}(t)$ and $\bar{p}(t)$ are, according to (15.68),

$$i\hbar \frac{d\bar{q}(t)}{dt} = \bar{q}(t)\bar{H} - \bar{H}\bar{q}(t) \tag{15.74}$$

and

$$i\hbar \frac{d\bar{p}(t)}{dt} = \bar{p}(t)\bar{H} - \bar{H}\bar{p}(t) \tag{15.75}$$

---

[10] The class of functions for which this is true is delineated in Exercise 14.8.

It is consistent with the commutation relations to *define*

$$\frac{\partial \bar{H}}{\partial \bar{p}} = \frac{(\bar{q}\bar{H} - \bar{H}\bar{q})}{i\hbar} \tag{15.76}$$

and

$$\frac{\partial \bar{H}}{\partial \bar{q}} = \frac{(\bar{H}\bar{p} - \bar{p}\bar{H})}{i\hbar} \tag{15.77}$$

The equations of motion then become

$$\frac{d\bar{q}}{dt} = \frac{\partial \bar{H}}{\partial \bar{p}} \qquad \frac{d\bar{p}}{dt} = -\frac{\partial \bar{H}}{\partial \bar{q}} \tag{15.78}$$

These are formally identical with the classical equations of motion, except that all canonical variables are replaced by the corresponding operators. In the Heisenberg picture the transition from classical to quantum theory for a system which has a classical analog is made simply by replacing the canonical variables by operators which can change in time, subject to the commutation relations

$$\bar{q}_k(t)\bar{p}_l(t) - \bar{p}_l(t)\bar{q}_k(t) = i\hbar\delta_{kl}1 \tag{15.79a}$$

$$\bar{q}_k(t)\bar{q}_l(t) - \bar{q}_l(t)\bar{q}_k(t) = \bar{p}_k(t)\bar{p}_l(t) - \bar{p}_l(t)\bar{p}_k(t) = 0 \tag{15.79b}$$

and by postulating that with the definitions (15.76) and (15.77) the equations of motion in quantum mechanics shall have the same form as the corresponding classical equations—barring, as usual, complications that may arise from ambiguities in the ordering of operators.

The simple commutation relations (15.79) are not valid if the operators are taken at two different times. 'Thus, generally, $\bar{q}(t)$ and $\bar{q}(0)$ do not commute, nor is the commutator $[\bar{q}(0), \bar{p}(t)]$ equal to $i\hbar1$. For example, if the system is a free particle with $H = p^2/2\mu$, we have

$$\bar{q}(t) = \bar{q}(0) + \frac{\bar{p}(0)}{\mu} t$$

hence,

$$[\bar{q}(t), \bar{q}(0)] = -\frac{i\hbar}{\mu} t \tag{15.80}$$

Exercise 15.9. What conclusions follow from (15.80) for the coordinate uncertainty? Specialize to a state represented by a minimum uncertainty wave packet ($\Delta q \, \Delta p = \hbar/2$) at $t = 0$, and compare your result with (8.94).

If the close correspondence between the classical theory and quantum dynamics lends the Heisenberg picture some distinction, the Schrödinger picture is perhaps a more intuitive form of quantum mechanics, suited particularly for a discussion of scattering processes. Such processes are more naturally described by moving wave packets, albeit complex ones, than by operators changing in time.

From our present general point of view, however, it is a matter of taste whether we use the Heisenberg or the Schrödinger picture. The passage from one picture to the other can be made at will. The quantum mechanical concept of the transition probability will serve to illustrate this further.

Let us suppose that the system is at time t_0 in an eigenstate $|A'\rangle$ of the observable A. We ask: "What is the probability that at time t it will be in an eigenstate $|B''\rangle$ of the observable B?" Since at time t the system is in the state $T(t, t_0) |A'\rangle$, the answer, using the Schrödinger picture, is that the required probability is the square of the absolute value of

$$\langle B''| \, T(t, t_0) \, |A'\rangle \tag{15.81}$$

In the Heisenberg picture the state is assumed to be $\overline{|A', t_0\rangle} = |A'\rangle$, and we ask again for the probability that B has the value B'' at time t. Now the answer is even simpler: It is the square of the absolute value of

$$\overline{\langle B'', t \,|\, \overline{A', t_0}\rangle} \tag{15.81a}$$

Equation (15.71) shows that expressions (15.81) and (15.81a) are identical. The quantity

$$P(B'', t; A', t_0) = \overline{|\langle B'', t \,|\, \overline{A', t_0}\rangle|^2} = |\langle B''| \, T(t, t_0) \, |A'\rangle|^2 \tag{15.82}$$

is called the *transition probability* for a transition from state A' to B''. Expression (15.81) or (15.81a) is called the *transition amplitude*.[11]

In the following section the general operator formalism will be applied to the ubiquitous linear harmonic oscillator.

8. The Harmonic Oscillator. The generalized harmonic oscillator is a system whose Hamiltonian, expressed in terms of two canonical observables p and q is given by

$$H = \frac{p^2}{2\mu} + \tfrac{1}{2}\mu\omega^2 q^2 \tag{15.83}$$

where the Hermitian operators p and q satisfy the relation

$$qp - pq = i\hbar 1 \tag{15.84}$$

Both p and q have continuous spectra extending from $-\infty$ to $+\infty$.

---

[11] Transition amplitudes play a central role in an interesting alternative formulation of quantum dynamics. See R. P. Feynman and A. R. Hibbs, *Quantum Mechanics and Path Integrals*, McGraw-Hill Book Publishing Company, New York, 1965.

We first consider the eigenvalue problem of H, because it will give us the important stationary states of the system. It is convenient to introduce a new operator

$$a = \sqrt{\frac{\mu\omega}{2\hbar}} \left(q + i\frac{p}{\mu\omega} \right) \tag{15.85}$$

which is, of course, not Hermitian. By the use of the commutation relation, we prove easily that

$$H = \hbar\omega(a^\dagger a + \tfrac{1}{2}) = \hbar\omega(aa^\dagger - \tfrac{1}{2}) \tag{15.86}$$

where $a^\dagger$ is the Hermitian adjoint of a, i.e.,

$$a^\dagger = \sqrt{\frac{\mu\omega}{2\hbar}} \left(q - i\frac{p}{\mu\omega} \right) \tag{15.87}$$

The commutator of a and $a^\dagger$ is

$$aa^\dagger - a^\dagger a = 1 \tag{15.88}$$

Since by (15.86) H commutes with $a^\dagger a$, the eigenvectors of H and of $a^\dagger a$ can be assumed to be the same, and it is sufficient to solve the eigenvalue problem for $a^\dagger a$. Labeling the eigenvalues by λ_n $(n = 0, 1, 2, \ldots)$ and the corresponding eigenvectors by Ψ_n, we have

$$a^\dagger a \Psi_n = \lambda_n \Psi_n \tag{15.89}$$

This is the equation which we must solve.

First we prove that $\lambda_n \geqslant 0$. From (15.89) we get

$$(\Psi_n, a^\dagger a \Psi_n) = \lambda_n(\Psi_n, \Psi_n) \tag{15.90}$$

and, using the definition of the adjoint operator,

$$(a\Psi_n, a\Psi_n) = \lambda_n(\Psi_n, \Psi_n)$$

Since the norm of a vector is nonnegative, we conclude that

$$\lambda_n \geqslant 0 \tag{15.91}$$

If Ψ_n is an eigenvector of $a^\dagger a$, then $a^\dagger \Psi_n$ is also an eigenvector, as can be seen from the equation

$$(a^\dagger a)a^\dagger \Psi_n = a^\dagger(a^\dagger a + 1)\Psi_n = (\lambda_n + 1)a^\dagger \Psi_n$$

where the commutation relation (15.88) has been used. Hence, $a^\dagger \Psi_n$ is an eigenvector with eigenvalue $\lambda_{n'} = \lambda_n + 1$. Similarly, we can show that $a\Psi_n$ is also an eigenvector of $a^\dagger a$ with eigenvalue $\lambda_{n''} = \lambda_n - 1$. These properties justify the designation of $a^\dagger$ as the *raising operator*, and a as the *lowering operator*. By applying these operators repeatedly, we can generate, from any given eigenvector Ψ_n, new eigenvectors with different eigenvalues by what is graphically called a *ladder method*. However, condition (15.91) limits the number of times a lowering operator can be applied. When by successive downward steps an eigenvalue between 0 and 1 has been reached, by applying a again we do not obtain a new eigenvector, because that would be an eigenvector violating the restriction (15.91). Hence, we must have for the lowest step in the ladder (labeled $n = 0$)

$$a^\dagger a \Psi'_0 = \lambda_0 \Psi'_0, \qquad 1 > \lambda_0 \geqslant 0$$

and

$$a \Psi'_0 = 0 \tag{15.92}$$

Consequently,

$$\lambda_0 = 0 \tag{15.93}$$

and this is the only eigenvalue below unity.

Starting from Ψ'_0 and $\lambda_0 = 0$, we obtain all other eigenvectors and eigenvalues by repeated application of the raising operator $a^\dagger$. The eigenvalues increase in unit steps. Hence,

$$\Psi'_n \propto (a^\dagger)^n \Psi'_0 \qquad (n = 0, 1, 2, \ldots) \tag{15.94}$$

$$\lambda_n = n \tag{15.95}$$

There is no degeneracy as long as no dynamical variables other than p and q appear to characterize the system. The set of eigenvectors obtained is complete.

Combining (15.86), (15.89), and (15.95), we obtain

$$H\Psi'_n = \hbar\omega(n + \tfrac{1}{2})\Psi'_n \tag{15.96}$$

Hence, the eigenvalues of the Hamiltonian are

$$E_n = \hbar\omega(n + \tfrac{1}{2}) \tag{15.97}$$

in agreement with the discrete energy eigenvalues found in Chapter 5.

Since the Hamiltonian is the same in the Schrödinger and Heisenberg pictures, there has so far been no need to specify the picture in which we are working. It is instructive to consider now the dynamical equations in the Heisenberg picture. We set $t_0 = 0$, thus causing the two pictures to coalesce at time zero.

In order to simplify the notation we shall omit the bar introduced in Section 15.7 and characterize operators in the Heisenberg picture merely by writing them as functions of time, e.g., $A(t)$ instead of $\bar{A}(t)$.

According to (15.68) and (15.86), the equation of motion for the operator $a(t)$ is

$$i\hbar \frac{da(t)}{dt} = [a(t), H] = \hbar\omega[a(t)a^\dagger(t)a(t) - a^\dagger(t)a(t)a(t)]$$

Using the commutation relation (15.88), translated into the Heisenberg picture, we obtain the simple differential equation

$$\frac{da(t)}{dt} + i\omega a(t) = 0 \qquad (15.98)$$

Although the operator equations of motion are difficult to solve directly in most problems, thus necessitating passage to a representation in which these equations become conventional systems of linear differential and integral equations, the present example is an exception. Equation (15.98) can be solved immediately, even though a is an operator:

$$a(t) = a(0)e^{-i\omega t} \qquad (15.99)$$

Similarly,

$$a^{\dagger}(t) = a^{\dagger}(0)e^{i\omega t} \qquad (15.100)$$

The initial values, $a(0)$ and $a^{\dagger}(0)$, are any two operators which are adjoint to each other and which satisfy the condition (15.88). They are the constants of integration of the dynamical problem. p and q, and any function of these, can be expressed in terms of a and $a^{\dagger}$. Thus in principle the equations of motion have all been integrated.

Exercise 15.10. Work out $a(t) = \exp(+iHt/\hbar)a(0)\exp(-iHt/\hbar)$ directly as an application of (8.106). Determine $q(t)$ and $p(t)$ for the oscillator.

To study the problem in more detail, we now introduce a fixed representation. The most convenient one to use is the energy representation whose basis is spanned by the eigenvectors of the energy operator at $t = 0$ (see Section 14.7). The basis vectors will be characterized by the quantum numbers $n\ (= 0, 1, 2, \ldots)$ of the harmonic oscillator

$$\Psi_{n}(0) = |n, 0\rangle = \Psi_{n} = |n\rangle, \qquad E_{n} = \hbar\omega(n + \tfrac{1}{2}) \qquad (15.101)$$

At time t each eigenvector becomes multiplied by a phase factor

$$|n, t\rangle = \exp\left(\frac{i}{\hbar}Ht\right)|n\rangle = \exp[i\omega(n + \tfrac{1}{2})t]\,|n\rangle$$

but we shall retain for our representation the *fixed* basis vectors, $\Psi_{n} = |n\rangle = |n, 0\rangle$, which are assumed to be normalized:

$$\langle n\,|\,n\rangle = 1 \qquad (15.102)$$

In this representation H is, of course, diagonal at all times and is given by[12]

$$H = \frac{\hbar\omega}{2} \begin{pmatrix} 1 & 0 & 0 & \cdot & \cdot \\ 0 & 3 & 0 & \cdot & \cdot \\ 0 & 0 & 5 & \cdot & \cdot \\ \cdot & \cdot & \cdot & \cdot & \cdot \\ \cdot & \cdot & \cdot & \cdot & \cdot \end{pmatrix} \tag{15.103}$$

The matrices for $a(0) = a$ and $a^\dagger(0) = a^\dagger$ can be constructed easily if we recall that

$$a\Psi_n = C_n\Psi_{n-1} \tag{15.104}$$

The coefficient C_n has yet to be determined. Evidently, owing to the normalization of all eigenvectors,

$$(a\Psi_n, a\Psi_n) = |C_n|^2$$

or, by (15.89) and (15.95),

$$|C_n|^2 = (\Psi_n, a^\dagger a\Psi_n) = \lambda_n = n$$

Hence,

$$C_n = \sqrt{n}\, e^{i\alpha_n}$$

$\alpha_n = 0$ for all n is a possible and consistent choice. It follows that we may write

$$\Psi_n = |n\rangle = (n!)^{-1/2}(a^\dagger)^n\Psi_0 = (n!)^{-1/2}(a^\dagger)^n|0\rangle \tag{15.105}$$

and

$$a = \begin{pmatrix} 0 & \sqrt{1} & 0 & 0 & \cdot & \cdot \\ 0 & 0 & \sqrt{2} & 0 & \cdot & \cdot \\ 0 & 0 & 0 & \sqrt{3} & \cdot & \cdot \\ \cdot & \cdot & \cdot & \cdot & \cdot & \cdot \\ \cdot & \cdot & \cdot & \cdot & \cdot & \cdot \\ \cdot & \cdot & \cdot & \cdot & \cdot & \cdot \end{pmatrix} \tag{15.106a}$$

$$a^\dagger = \begin{pmatrix} 0 & 0 & 0 & \cdot & \cdot & \cdot \\ \sqrt{1} & 0 & 0 & \cdot & \cdot & \cdot \\ 0 & \sqrt{2} & 0 & \cdot & \cdot & \cdot \\ 0 & 0 & \sqrt{3} & \cdot & \cdot & \cdot \\ \cdot & \cdot & \cdot & \cdot & \cdot & \cdot \end{pmatrix} \tag{15.106b}$$

[12] Remember that the symbols H, a, and q indicate matrices representing certain operators.

The matrices representing $a(t)$ and $a^\dagger(t)$ are obtained by multiplying a and $a^\dagger$ by $e^{-i\omega t}$ and $e^{i\omega t}$ respectively. Finally we get for the coordinate operator of the harmonic oscillator in the Heisenberg picture:

$$q(t) = \sqrt{\frac{\hbar}{2\mu\omega}} \, [a(t) + a^\dagger(t)]$$

$$= \sqrt{\frac{\hbar}{2\mu\omega}} \begin{pmatrix} 0 & \sqrt{1}\,e^{-i\omega t} & 0 & \cdot & \cdot \\ \sqrt{1}\,e^{i\omega t} & 0 & \sqrt{2}\,e^{-i\omega t} & \cdot & \cdot \\ 0 & \sqrt{2}\,e^{i\omega t} & 0 & \cdot & \cdot \\ \cdot & \cdot & \cdot & \cdot & \cdot \\ \cdot & \cdot & \cdot & \cdot & \cdot \end{pmatrix} \qquad (15.107)$$

This matrix is evidently Hermitian, as it should be. Its elements are harmonic functions of the time—a general property of the matrix representing any Heisenberg operator in an energy representation which is spanned by the fixed eigenvectors of H:

$$H\,|E'\rangle = E'\,|E'\rangle$$

Indeed, for any operator which is not explicitly time-dependent, we obtain from

$$\bar{A}(t) = \exp\left(\frac{i}{\hbar} Ht\right) A \exp\left(-\frac{i}{\hbar} Ht\right)$$

the matrix element

$$\langle E'|\,\bar{A}(t)\,|E''\rangle = \langle E'|\,A\,|E''\rangle \exp\left[\frac{i}{\hbar}(E' - E'')t\right] \qquad (15.108)$$

The special feature of the harmonic oscillator is that all matrix elements oscillate with the same frequency.

An eigenstate of the coordinate q with eigenvalue q' is represented in the Schrödinger picture (or at $t = 0$ in the Heisenberg picture) by a column matrix, and the eigenvalue equation for q appears in the form

$$\sqrt{\frac{\hbar}{2\mu\omega}} \begin{pmatrix} 0 & \sqrt{1} & 0 & \cdot & \cdot \\ \sqrt{1} & 0 & \sqrt{2} & \cdot & \cdot \\ 0 & \sqrt{2} & 0 & \cdot & \cdot \\ \cdot & \cdot & \cdot & \cdot & \cdot \end{pmatrix} \begin{pmatrix} a_0 \\ a_1 \\ a_2 \\ \cdot \end{pmatrix} = q' \begin{pmatrix} a_0 \\ a_1 \\ a_2 \\ \cdot \end{pmatrix} \qquad (15.109)$$

where the components of the eigenvector of q are the transformation coefficients

$$a_k = \langle k\,|\,q'\rangle \qquad (15.110)$$

Equation (15.109) leads to a set of simultaneous linear equations:

$$\sqrt{1}\, a_1 \qquad\qquad\qquad = \sqrt{\frac{2\mu\omega}{\hbar}}\, q' a_0$$

$$\sqrt{2}\, a_2 + \sqrt{1}\, a_0 \qquad\quad = \sqrt{\frac{2\mu\omega}{\hbar}}\, q' a_1$$

$$. \quad . \quad . \quad . \quad . \quad . \quad . \quad . \quad . \quad . \quad . \quad .$$

$$\sqrt{n+1}\, a_{n+1} + \sqrt{n}\, a_{n-1} = \sqrt{\frac{2\mu\omega}{\hbar}}\, q' a_n$$

$$. \quad . \quad . \quad . \quad . \quad . \quad . \quad . \quad . \quad . \quad . \quad .$$

It is easy to verify that the function

$$a_n(q') \propto 2^{-(n/2)}(n!)^{-1/2} H_n\left(\sqrt{\frac{\mu\omega}{\hbar}}\, q'\right) \tag{15.111}$$

satisfies these conditions by virtue of the recurrence relation

$$H_{n+1}(x) - 2x H_n(x) + 2n H_{n-1}(x) = 0$$

for Hermite polynomials.

The closure condition (14.91),

$$\sum_{n=0}^{\infty} \langle q' \mid n \rangle \langle n \mid q'' \rangle = \delta(q' - q'')$$

finally determines the constant factor in (15.111). The result is

$$\langle n \mid q' \rangle = 2^{-(n/2)}(n!)^{-1/2}\left(\frac{\mu\omega}{\hbar\pi}\right)^{1/4} \exp\left(-\frac{\mu\omega}{2\hbar} q'^2\right) H_n\left(\sqrt{\frac{\mu\omega}{\hbar}}\, q'\right) \tag{15.112}$$

in agreement with (5.24) and (14.138). The complete equivalence of the two pictures should thus be apparent.

Exercise 15.11. Transcribe equations (15.92) and (15.105) in the coordinate representation and calculate $\langle q' \mid n \rangle$ from these differential relations. Using the mathematical tools of Section 5.3, verify equation (15.112).

9. *The Forced Linear Harmonic Oscillator.* For many applications, especially in field theory, it is desirable to consider the dynamical effects produced by the addition of an external time-dependent force which does not depend on q. Instead of (15.83) we assume a Hamiltonian of the form

$$H = \frac{p^2}{2\mu} + \tfrac{1}{2}\mu\omega^2 q^2 - q F(t)$$

where $F(t)$ is a real function of t. At no additional expense, we may generalize the Hamiltonian even further by introducing a velocity-dependent term:

$$H = \frac{p^2}{2\mu} + \tfrac{1}{2}\mu\omega^2 q^2 - qF(t) - pG(t) \tag{15.113}$$

where $G(t)$ is also a real function of t.

With the substitutions (15.85) and (15.87), the Hamiltonian (15.113) may be cast in the form

$$H = \hbar\omega(a^\dagger a + \tfrac{1}{2}) + f(t)a + f^*(t)a^\dagger \tag{15.114}$$

provided we define the complex-valued function $f(t)$:

$$f(t) = -\sqrt{\frac{\hbar}{2\mu\omega}}\, F(t) + i\sqrt{\frac{\hbar\mu\omega}{2}}\, G(t) \tag{15.115}$$

In most applications we shall be interested in the lasting, rather than the transient, changes produced by the time-dependent forces in an initially unperturbed linear harmonic oscillator. It is therefore reasonable to assume that before t_1 and after t_2 the Hamiltonian is that of a free oscillator and that the disturbance $f(t) \neq 0$ only during a finite time interval $t_1 < t < t_2$. The time development of the system is conveniently studied in the Heisenberg picture, in which the operators are subject to a unitary transformation as they change from the free oscillation regime before t_1 to a free oscillation regime after t_2.

The relevant commutation relations for a and $a^\dagger$, taken at equal time, are still

$$[a(t), a^\dagger(t)] = I \tag{15.116}$$

and the equation of motion for $a(t)$ is

$$i\hbar\frac{da(t)}{dt} = [a(t), H(t)] = \hbar\omega a(t) + f^*(t)$$

or

$$\frac{da(t)}{dt} + i\omega a(t) = -\frac{i}{\hbar}f^*(t) \tag{15.117}$$

This inhomogeneous differential equation is easily solved by standard methods. We choose the Green's function approach because it is illustrative of a method which has proved useful in many similar but more difficult problems.

A Green's function appropriate to equation (15.117) is a solution of the equation

$$\frac{dG(t - t')}{dt} + i\omega G(t - t') = \delta(t - t') \tag{15.118}$$

because such a function permits us to write a particular solution of equation (15.117) as

$$a(t) = -\frac{i}{\hbar} \int_{-\infty}^{+\infty} G(t - t')f^*(t') \, dt' \tag{15.119}$$

Obviously, for $t \neq t'$ the Green's function is proportional to $e^{-i\omega(t-t')}$, but at $t = t'$ there is a discontinuity. By integrating (15.118) over an interval which includes t', we derive the condition

$$\lim_{\varepsilon \to 0} [G(+\varepsilon) - G(-\varepsilon)] = 1$$

for $\varepsilon > 0$.

Two particular Green's functions are useful:

$$G_R(t - t') = \eta(t - t')e^{-i\omega(t-t')} \tag{15.120}$$

and

$$G_A(t - t') = -\eta(t' - t)e^{-i\omega(t-t')} \tag{15.121}$$

where the notation of (6.19) has been used for the step function η. These two particular solutions of (15.118) are known as the *retarded* and *advanced* Green's functions, respectively.

If we denote by $a_{\text{in}}(t)$ and $a_{\text{out}}(t)$ those solutions of the homogeneous equation

$$\frac{da(t)}{dt} + i\omega a(t) = 0$$

which coincide with the solution $a(t)$ of equation (15.117) for $t < t_1$ and $t > t_2$ respectively, we can write

$$a(t) = a_{\text{in}}(t) - \frac{i}{\hbar} \int_{-\infty}^{+\infty} G_R(t - t')f^*(t') \, dt'$$
$$= a_{\text{in}}(t) - \frac{i}{\hbar} \int_{-\infty}^{t} e^{-i\omega(t-t')}f^*(t') \, dt' \tag{15.122}$$

or, equivalently,

$$a(t) = a_{\text{out}}(t) - \frac{i}{\hbar} \int_{-\infty}^{+\infty} G_A(t - t')f^*(t') \, dt'$$
$$= a_{\text{out}}(t) + \frac{i}{\hbar} \int_{t}^{+\infty} e^{-i\omega(t-t')}f^*(t') \, dt' \tag{15.123}$$

By equating the right-hand sides of (15.122) and (15.123), we obtain the relation

$$a_{\text{out}}(t) = a_{\text{in}}(t) - \frac{i}{\hbar} \int_{-\infty}^{+\infty} e^{-i\omega(t-t')}f^*(t') \, dt' \tag{15.124}$$

As we saw in Section 15.8, the "free" operators a_{in} and a_{out} have the simple time dependence

$$a_{in}(t) = a_{in}e^{-i\omega t}$$

$$a_{out}(t) = a_{out}e^{-i\omega t}$$

Hence, from (15.124), we get

$$a_{out} = a_{in} - \frac{i}{\hbar}g^*(\omega) \tag{15.125}$$

where

$$g(\omega) = \int_{-\infty}^{+\infty} e^{-i\omega t'}f(t') \, dt' \tag{15.126}$$

is the Fourier transform of the "force" $f(t)$.

We now seek to determine the unitary operator S which transforms a_{in} into a_{out} such that

$$a_{out} = S^\dagger a_{in}S \tag{15.127}$$

The operator S provides the vital link between the description of the system before t_1 and after t_2. In particular, the Hamiltonian changes from

$$H_{in} = \hbar\omega(a_{in}^\dagger a_{in} + \tfrac{1}{2}) \qquad \text{for} \qquad t < t_1 \tag{15.128}$$

to

$$H_{out} = \hbar\omega(a_{out}^\dagger a_{out} + \tfrac{1}{2}) \qquad \text{for} \qquad t > t_2 \tag{15.129}$$

The operators $a_{in}^\dagger a_{in}$ and $a_{out}^\dagger a_{out}$ have eigenvalues $n = 0, 1, 2, \ldots$ and the corresponding eigenvectors may be denoted by $|n\rangle_{in}$ and $|n\rangle_{out}$. From equation (15.127) we infer that these eigenvectors are connected by the unitary transformation S according to

$$|n\rangle_{out} = S^\dagger |n\rangle_{in} \tag{15.130}$$

The operator S can be determined quite easily if we note the simple identity

$$e^{-\alpha a^\dagger + \alpha^* a}a e^{\alpha a^\dagger - \alpha^* a} = a + \alpha \tag{15.131}$$

for any complex number α. This equation follows from (8.105) and the commutation relation for a and $a^\dagger$.

Comparison of equations (15.125), (15.127), and (15.131) shows that the unitary transformation we seek is $S = \exp(\alpha a^\dagger - \alpha^* a)$ with $\alpha = -ig^*(\omega)/\hbar$, or

$$S = \exp\left[-\frac{i}{\hbar}g^*(\omega)a_{in}^\dagger - \frac{i}{\hbar}g(\omega)a_{in}\right] \tag{15.132}$$

We note that

$$\int_{-\infty}^{+\infty} [f(t)a_{\text{in}}(t) + f^*(t)a_{\text{in}}^\dagger(t)]\, dt = \int_{-\infty}^{+\infty} [f(t)e^{-i\omega t}a_{\text{in}} + f^*(t)e^{i\omega t}a_{\text{in}}^\dagger)\, dt$$

$$= g(\omega)a_{\text{in}} + g^*(\omega)a_{\text{in}}^\dagger$$

Hence we may express S as

$$S = \exp\left[-\frac{i}{\hbar}\int_{-\infty}^{+\infty} [f(t)a_{\text{in}}(t) + f^*(t)a_{\text{in}}^\dagger(t)]\, dt\right] \tag{15.133}$$

Exercise 15.12. Show that an equivalent expression for S is obtained by replacing a_{in} and $a_{\text{in}}^\dagger$ by a_{out} and $a_{\text{out}}^\dagger$.

The operator S allows us to determine all the physically relevant information about the system after t_2 from a knowledge of its condition before t_1. The simplest initial condition assumes that the system is in the ground state of H_{in}, so that its energy before the onset of the interaction is a minimum. In the Heisenberg picture the state does not change in time, and the system remains in the state

$$\Psi = |0\rangle_{\text{in}}$$

A quantity of considerable physical interest is the probability amplitude for finding the system in one of the eigenstates of H_{out}. According to equation (15.130), this amplitude, which measures the likelihood that the perturbing force induces permanent transitions to an excited state with quantum number n of the harmonic oscillator, is given by

$$_{\text{out}}\langle n\,|\,0\rangle_{\text{in}} = {}_{\text{out}}\langle n|\, S\,|0\rangle_{\text{out}} = {}_{\text{in}}\langle n|\, S\,|0\rangle_{\text{in}} \tag{15.134}$$

In evaluating $S\,|0\rangle$, it is convenient to make use of the property

$$S\,|0\rangle = e^{\alpha a^\dagger - \alpha^* a}\,|0\rangle = e^{-|\alpha|^2/2}e^{\alpha a^\dagger}e^{-\alpha^* a}\,|0\rangle = e^{-|\alpha|^2/2}e^{\alpha a^\dagger}\,|0\rangle \tag{15.135}$$

which follows from equation (8.107) and from $a\,|0\rangle = 0$. Hence,

$$\langle n|\, S\,|0\rangle = e^{-|\alpha|^2/2}\,\langle n|\, e^{\alpha a^\dagger}\,|0\rangle = e^{-|\alpha|^2/2}\frac{\alpha^n}{\sqrt{n!}} \tag{15.136}$$

In the last step, equation (15.105) and the orthonormality of the oscillator eigenvectors have been used.

If we apply the result (15.136) to the relation (15.134) and insert the explicit form (15.132) for S, we obtain the probability of finding the system at $t > t_2$ in the nth eigenstate of H_{out}, provided that it was certain to be in the ground state of H_{in} at $t < t_1$:

$$P_n(\omega) = |{}_{\text{out}}\langle n\,|\,0\rangle_{\text{in}}|^2 = e^{-|g(\omega)|^2/\hbar^2}\frac{1}{n!}\left|\frac{g(\omega)}{\hbar}\right|^{2n} \tag{15.137}$$

This is a *Poisson distribution* with an expectation value

$$_{\text{in}}\langle 0| \, a^\dagger_{\text{out}} \, a_{\text{out}} \, |0\rangle_{\text{in}} = |g(\omega)|^2/\hbar^2 \tag{15.138}$$

for the oscillator quantum number after the interaction has ceased.

An intriguing consequence of the formalism can be read off equation (15.125) if it is applied to the state $|0\rangle_{\text{in}} = S \, |0\rangle_{\text{out}}$. This vector is seen to satisfy the relation

$$a_{\text{out}} \, |0\rangle_{\text{in}} = \alpha \, |0\rangle_{\text{in}} \tag{15.139}$$

showing that $|0\rangle_{\text{in}}$ is an eigenvector of a_{out} with eigenvalue $\alpha = -ig^*(\omega)/\hbar$. Since a is not a normal operator, its eigenvectors are not subject to any orthogonality requirement, and since $g(\omega)$ is arbitrary, any complex number is an allowed eigenvalue of a. From (15.139) it follows immediately that

$$_{\text{in}}\langle 0| \, a^\dagger_{\text{out}} \, a_{\text{out}} \, |0\rangle_{\text{in}} = {}_{\text{in}}\langle 0| \, a^\dagger_{\text{out}} \, |0\rangle_{\text{in}} \, {}_{\text{in}}\langle 0| \, a_{\text{out}} \, |0\rangle_{\text{in}} \tag{15.140}$$

This simple factorization property, which testifies to the absence of certain quantum correlations in these special states, has led to their designation as *quasiclassical* states. Because of their role in optics and quantum electronics, these states are also known as *coherent*. A few elementary properties of coherent states are given in the next section.

Exercise 15.13. Show that in the state $|0\rangle_{\text{in}}$ the variance of the operator $a^\dagger_{\text{out}} a_{\text{out}}$ is equal to its expectation value.

Further insight is gained if we construct the time-dependent state vector in the Schrödinger picture for $t > t_2$. From (15.127) it follows that for all t

$$a_{\text{out}}(t) = S^\dagger a_{\text{in}}(t) S \tag{15.141}$$

since both $a_{\text{out}}(t)$ and $a_{\text{in}}(t)$ depend on the time through the same factor $e^{-i\omega t}$. Using the result of Exercise 15.10, we may write equivalently

$$a_{\text{out}}(t) = S^\dagger e^{iH_{\text{int}}t/\hbar} \, a_{\text{in}} e^{-iH_{\text{int}}t/\hbar} \, S$$

If we choose the initial time 0 to precede and the final time t to follow the external perturbation, $0 < t_1 < t_2 < t$, we may identify $a_{\text{in}} = a(0)$ and $a_{\text{out}}(t) = a(t)$ and conclude from comparison with (15.66) that the time development operator is (except for a constant phase factor)

$$T(t, 0) = e^{-iH_{\text{int}}t/\hbar} S \tag{15.142}$$

The Schrödinger operators a and $a^\dagger$ may be equated with the Heisenberg operators $a(t)$ and $a^\dagger(t)$ evaluated at $t = 0$. Hence, the notation $a = a(0) = a_{\text{in}}$ is appropriate.

For times $t > t_2$ the state vector is given by

$$\Psi(t) = e^{-i/\hbar H_0 t} S \Psi(0) \tag{15.143}$$

where $H_0 = \hbar\omega(a^\dagger a + \frac{1}{2})$ is the unperturbed Hamiltonian. If the initial state is the harmonic oscillator ground state, $\Psi(0) = |0\rangle$, after the interaction is terminated the state develops in time as

$$\Psi(t) = e^{-(|\alpha|^2 + i\omega t)/2} \exp(\alpha a^\dagger e^{-i\omega t}) |0\rangle \tag{15.144}$$

where, as before, $\alpha = -ig^*(\omega)/\hbar$.

Exercise 15.14. Derive equation (15.144), using formula (8.106). Show that $\Psi(t)$ is an eigenstate of a with eigenvalue $\alpha e^{-i\omega t}$.

The last exercise shows explicitly how the applied time-dependent force changes the state of the harmonic oscillator from its initial ground state into another coherent state. Although the detailed features of the force essentially affect the evolution of the state during the interaction, the state of the system after the complete cessation of the interaction is determined exclusively by the Fourier component $g(\omega)$ which is in resonance with the free harmonic oscillator.

In the coordinate representation the state $\Psi(t)$ is represented by the wave function $\psi(q', t) = \langle q' | \Psi(t)\rangle$. The equation which expresses the fact that (15.144) is an eigenstate of a appears transcribed as

$$\left(q' + \frac{\hbar}{\mu\omega} \frac{\partial}{\partial q'}\right) \psi(q', t) = \sqrt{\frac{2\hbar}{\mu\omega}} \, \alpha e^{-i\omega t} \psi(q', t) \tag{15.145}$$

where use has been made of the definition (15.85). This differential equation has precisely the same form as (8.77), and comparison of the two leads to the identification

$$(\Delta x)^2 = \frac{\hbar}{2\mu\omega}$$

and

$$\langle x\rangle + \frac{i}{\mu\omega} \langle p_x\rangle = \sqrt{\frac{2\hbar}{\mu\omega}} \, \alpha e^{-i\omega t} \tag{15.146}$$

The coherent state is thus represented by a minimum uncertainty wave packet which has constant width but oscillates harmonically in coordinate and momentum space.

Exercise 15.15. Show that $|\psi(q', t)|^2$ for the wave packet oscillates in time like the classical free harmonic oscillator.

Exercise 15.16. Prove the minimum uncertainty property for a coherent state directly by considering the expectation values of $a^\dagger a$ and $a^2 - (a^\dagger)^2$.

10. *Coherent States.* The coherent states were defined in the last section as eigenkets of the lowering operator a by the equation

$$a \, |\alpha\rangle = \alpha \, |\alpha\rangle \qquad (15.147)$$

From (15.131) we deduce that $S_\alpha = \exp(\alpha a^\dagger - \alpha^* a)$ has the properties $S_\alpha{}^\dagger = S_\alpha{}^{-1} = S_{-\alpha}$ and $aS_\alpha \, |0\rangle = \alpha S_\alpha \, |0\rangle$. The coherent state $|\alpha\rangle$ may therefore be written as

$$|\alpha\rangle = e^{\alpha a^\dagger - \alpha^* a} \, |0\rangle = e^{-|\alpha|^2/2} e^{\alpha a^\dagger} \, |0\rangle \qquad (15.148)$$

with the normalization

$$\langle \alpha \, | \, \alpha \rangle = \langle 0 \, | \, 0 \rangle = 1 \qquad (15.149)$$

There is an eigenstate $|\alpha\rangle$ of a for any complex number α, but the coherent states do not form an orthogonal set. The scalar product of two coherent states $|\alpha\rangle$ and $|\beta\rangle$ is

$$\langle \beta \, | \, \alpha \rangle = e^{-|\alpha|^2/2 - |\beta|^2/2} \langle 0| \, e^{\beta^* a} e^{\alpha a^\dagger} \, |0\rangle = e^{-|\alpha|^2/2 - |\beta|^2/2} \langle 0| \, e^{\alpha a^\dagger} e^{\beta^* a} e^{\alpha\beta^*} \, |0\rangle$$

$$= e^{-|\alpha|^2/2 - |\beta|^2/2 + \alpha\beta^*} \qquad (15.150)$$

and

$$|\langle \beta \, | \, \alpha \rangle|^2 = e^{-|\alpha - \beta|^2} \qquad (15.151)$$

Hence, the distance $|\alpha - \beta|$ in the complex eigenvalue plane measures the degree to which the two eigenstates are approximately orthogonal.

As we might expect from the lack of restrictions imposed on the eigenvalues and eigenstates of a, the latter form an *overcomplete* set, i.e., an arbitrary state can be expanded in terms of them in infinitely many different ways. Even so, an identity which bears a remarkable similarity to a closure relation can be proved:

$$\frac{1}{\pi} \int |\alpha\rangle \, d^2\alpha\langle\alpha| = \frac{1}{\pi} \int |\alpha\rangle \, dRe(\alpha) \, dIm(\alpha) \, \langle\alpha| = 1 \qquad (15.152)$$

Here, the integration is extended over the entire α-plane with a real element of area.

Exercise 15.17. Prove equation (15.152). This is most easily done by expanding the coherent states in terms of the harmonic oscillator eigenstates, using (15.136).

Exercise 15.18. Prove that the raising operator $a^\dagger$ has no normalizable eigenvectors.

Problems

1. A particle of charge q moves in a uniform magnetic field B which is directed along the z axis. Using a gauge in which $A_z = 0$, show that $q = (cp_x - qA_x)/qB$ and $p = (cp_y - qA_y)/c$ may be used as suitable canonically conjugate coordinate and momentum together with the pair z, p_z. Derive the energy spectrum and the eigenfunctions in the q-representation. Discuss the remaining degeneracy. Propose alternate methods for solving this eigenvalue problem.

2. If a linear harmonic oscillator in the ground state is exposed to a spatially uniform time-dependent external force $F(t)$, which acts in the time interval from t_1 to t_2, calculate the expected value of the total energy transfer both in classical and quantum mechanics.

3. A linear harmonic oscillator is subjected to a spatially uniform external force $F(t) = C\eta(t)e^{-\lambda t}$ where λ is a positive constant and $\eta(t)$ the step function (6.19). If the oscillator is in the ground state at $t < 0$, calculate the probability of finding the oscillator at t in an oscillator eigenstate with quantum number n. Assuming $C = (\hbar\mu\lambda^3)^{1/2}$, examine the variation of the transition probabilities with n and with the ratio λ/ω, ω being the natural frequency of the harmonic oscillator.

4. Two particles of equal mass are constrained to move on a straight line in a common harmonic oscillator potential and are coupled by a force which depends only on the distance between the particles. Construct the Schrödinger equation for the system and transform it into a separable equation by using relative coordinates and the coordinates of the center of mass. Show that the same equation is obtained by first constructing a separable classical Hamiltonian and subjecting it to canonical quantization.

5. Assuming the two particles of the preceding problem to be coupled by an elastic force, obtain the eigenvalues and eigenfunctions of the Schrödinger equation and show that the eigenfunctions are either symmetric or antisymmetric with respect to an interchange of the two particles.

CHAPTER 16

Rotations and other Symmetry Operations

1. The Euclidean Principle of Relativity and State Vector Transformations.
We now turn to a topic which has proved to be immensely fruitful in the
theory and practice of quantum mechanics: rotations in ordinary space and
their formal description in abstract vector space.

The fundamental assumption underlying all applications of quantum
mechanics is that ordinary space is subject to the laws of Euclidean geometry
and that it is physically homogeneous and isotropic. By this we mean that we
can move our entire physical apparatus from one place to another and we can
change its orientation in space without affecting the outcome of any experiment. We say that there is no preferred position or orientation in space.

A transformation which leaves the mutual relations of the physically
relevant aspects of a system unaltered is said to be a *symmetry operation*.
The assumption that geometrical translations and rotations of physical
systems are symmetry operations, or that space is homogeneous and isotropic, will be called the *Euclidean principle of relativity* because it denies that
spatial location and orientation have any absolute significance. In this
chapter we shall be concerned with rotations, since translations were already
discussed in Sections 14.7 and 15.6. Some preliminary remarks about
reflections will be found in Section 16.10. The symmetry operations associated
with the *Einstein principle of relativity* and with Lorentz transformations will
be taken up in Chapter 23.

Gravity seems at first sight to introduce inevitably a preferred direction,
the vertical, into any experiment performed on the surface of the earth,
but in quantum physics we are concerned with atomic and nuclear processes
in which gravitational effects play a negligible role. The apparent anisotropy
of space in an earthbound laboratory can then be ignored and the isotropy
of space for quantum processes tested directly by rotating the system at any
desired angle. No violation of the Euclidean principle of relativity in this
restricted area has ever been found.[1] We shall discuss some of the remarkable

[1] In classical, macroscopic mechanics, in which gravitation cannot be neglected, there is
no conflict with the Euclidean principle of relativity, because we can imagine the earth to be
part of the mechanical system and take its gravitational field into account when a rotation is
performed. On an even grander, cosmological scale there are legitimate serious questions
about the validity of the principle.

consequences which this principle has for the structure of quantum mechanics, and we shall find that it severely restricts the possible forms which the quantum description of a given system can take.

When a quantum system with a state vector Ψ is rotated in space to a new orientation, the state vector changes to Ψ''. The Euclidean principle of relativity requires that under rotation all probabilities be invariant, i.e., all scalar products remain invariant in absolute value. We thus have a mapping of the vector space onto itself, $\Psi \leftrightarrow \Psi''$, such that $|(\Psi'', \Phi')|^2 = |(\Psi, \Phi)|^2$ for every pair of vectors. The mapping must be reversible, because we could equally well have started from the new orientation and rotated the system back to its old orientation.

It should be noted that we do not require invariance of scalar products, which is the hallmark of unitary transformations, but only that the absolute values be invariant. Yet because of a remarkable theorem, we shall ultimately be able to confine our attention essentially to unitary and antiunitary transformations. The reasoning given here applies to any symmetry operation and not just to rotations.

Theorem. If a mapping of the vector space onto itself is given such that

$$\Psi \leftrightarrow \Psi', \qquad |(\Psi'', \Phi')|^2 = |(\Psi, \Phi)|^2 \tag{16.1}$$

then a second mapping, which is merely a phase change of every vector,

$$\Psi'' = e^{i\alpha(\Psi')}\Psi' \tag{16.2}$$

can be found such that

$$\Psi = \Psi_a + \Psi_b$$

is mapped into

$$\Psi'' = \Psi_a'' + \Psi_b''$$

For the proof of this theorem the reader is referred to the literature.[2] The theorem shows that by a rephasing of all vectors we can achieve a mapping which has one of the two fundamental properties of a linear operator: The transform of the sum of two vectors is equal to the sum of the transforms of the two vectors [see (14.19)]. It follows from this result and (16.1) that

$$|(\Psi_a'', \Psi_a'' + \Psi_b'')|^2 = |(\Psi_a, \Psi_a + \Psi_b)|^2$$

Hence, by applying (16.1) again,

$$(\Psi_a'', \Psi_b'') + (\Psi_a'', \Psi_b'')^* = (\Psi_a, \Psi_b) + (\Psi_a, \Psi_b)^*$$

[2] See E. P. Wigner, *Group Theory and Its Application to the Quantum Mechanics of Atomic Spectra*, translated by J. J. Griffin, Academic Press, New York, 1959, Appendix to Chapter 20, p. 233. See also V. Bargmann, *J. of Math. Phys*, **5**, 862 (1964).

or

$$\text{Re}\,(\Psi_a{}'', \Psi_b{}'') = \text{Re}\,(\Psi_a, \Psi_b)$$

Since the absolute value of the scalar product (Ψ_a, Ψ_b) is invariant, we must have either

$$\text{Im}\,(\Psi_a{}'', \Psi_b{}'') = \text{Im}\,(\Psi_a, \Psi_b) \tag{16.3a}$$

or

$$\text{Im}\,(\Psi_a{}'', \Psi_b{}'') = -\text{Im}\,(\Psi_a, \Psi_b) \tag{16.3b}$$

In the first case

$$(\Psi_a{}'', \Psi_b{}'') = (\Psi_a, \Psi_b) \tag{16.4}$$

and

$$(\lambda\Psi)'' = \lambda\Psi'' \tag{16.5}$$

whereas in the second case

$$(\Psi_a{}'', \Psi_b{}'') = (\Psi_a, \Psi_b)^* \tag{16.4a}$$

and

$$(\lambda\Psi)'' = \lambda^*\Psi'' \tag{16.5a}$$

Equation (16.5) expresses the second fundamental property of a *linear* operator [see (14.20)], and from condition (16.4) we infer that in the first case the transformation is *unitary*. Equation (16.5a), on the other hand, characterizes an *antilinear* operator [see equations (4.20a) and (14.20a)].

It is easy to see the profound implications of this theorem. State vectors which differ by phase factors represent the same state, and rephasing transformations have no physical significance.[3] It follows that in symmetry operations of a physical system we may confine ourselves to two simple types of transformations, linear and antilinear. Any more general transformation can be supplemented by a phase change and made to fall into one of these two simple categories which are mutually exclusive.

In particular, if the symmetry operation is a rotation, the antilinear case is excluded as a possibility because rotations can be generated continuously from the identity operation, which is inconsistent with complex conjugation. Antilinear transformations are important in describing the behavior of a system under time reversal (Sections 4.5, 16.11, and 23.5).

2. The Rotation Operator and Angular Momentum. The result of the last section is that, if the Euclidean principle of relativity holds, rotations in quantum mechanics are represented by unitary transformations. It was assumed that a one-to-one correspondence between state vectors and physical states can be set up such that two state vectors correspond to the same physical state if and only if they are equal or differ at most by a phase

---

[3] The rephasing operation is not unitary because each vector may be multiplied by a different phase factor.

factor.[4] It therefore follows from the Euclidean principle of relativity that, if there are several unitary operators describing the same rotation, they should differ only by a phase factor. Except for such a factor, the unitary rotation operator can depend only on the relative orientations of the system before and after the rotation, but not on its absolute orientation in space, nor on the manner in which the rotation is actually carried out. Consequently, if R_1 and R_2 are two rotations which are performed in succession, leading to the resultant rotation R_3, the unitary operators for these rotations must satisfy the condition

$$U_2 U_1 = e^{i\varphi(R_1 \cdot R_2)} U_3 \qquad (16.6)$$

The construction of a set of unitary operators (matrices) with these properties was already accomplished in Chapter 12 for the very special case of a two-dimensional state vector space. However, most of the discussion of Section 12.5 did not depend on the dimensionality of the vector space and may therefore be taken over by changing the notation only. Replacing the 2×2 matrices $\frac{1}{2}\boldsymbol{\sigma}$ by the Hermitian operators $\mathbf{J}/\hbar$, we see from (12.38) that a rotation operator U_R can be written in the form

$$U_R = \exp\left(-\frac{i}{\hbar}\,\hat{\mathbf{n}} \cdot \mathbf{J}\theta\right) \qquad (16.7)$$

where $\hat{\mathbf{n}}$ is the unit vector defining the axis of rotation and θ the angle of rotation. Planck's constant $\hbar$ has been inserted in anticipation of the identification of $\mathbf{J}$ with angular momentum.

Following the same arguments as in Section 12.5, we derive the fundamental commutation relations for the Hermitian generators, J_x, J_y, J_z, of the rotations in analogy with (12.43):

$$\begin{aligned} J_x J_y - J_y J_x &= i\hbar J_z \\ J_y J_z - J_z J_y &= i\hbar J_x \\ J_z J_x - J_x J_z &= i\hbar J_y \end{aligned} \qquad (16.8)$$

These are necessary conditions if U_R is to effect a rotation.

Exercise 16.1. Show that

$$\exp\left(\frac{i}{\hbar}\,\hat{\mathbf{n}} \cdot \mathbf{J}\theta\right) \mathbf{J} \exp\left(-\frac{i}{\hbar}\,\hat{\mathbf{n}} \cdot \mathbf{J}\theta\right)$$

$$= \hat{\mathbf{n}}(\hat{\mathbf{n}} \cdot \mathbf{J}) - \hat{\mathbf{n}} \times (\hat{\mathbf{n}} \times \mathbf{J}) \cos\theta + \hat{\mathbf{n}} \times \mathbf{J} \sin\theta \qquad (16.9)$$

---

[4] This assumption characterizes ordinary quantum mechanics of one-particle systems. It can and must be relaxed when the theory is generalized to systems with nontrivial Hermitian operators which commute with all observables (superselection rules).

and verify that the right-hand side of this equation is the vector obtained by rotating $\mathbf{J}$ about $\hat{\mathbf{n}}$ by an angle θ. [*Hint.* Differentiate the expression on the left-hand side of (16.9) twice with respect to θ to establish a simple differential equation. Compare also Exercise 12.3.]

As we know from Chapter 9, the operator $\mathbf{L} = \mathbf{r} \times \mathbf{p}$ satisfies the commutation relations (16.8) by virtue of the fundamental commutation relations between the components of $\mathbf{r}$ and $\mathbf{p}$. However, the operator $\mathbf{J}$ need not bear any relation to an *orbital* angular momentum, $\mathbf{L} = \mathbf{r} \times \mathbf{p}$, and it is of great interest to inquire into the general nature of the operators which satisfy the commutation relations (16.8).

To this end let us consider the eigenvalue problem of one of the components of $\mathbf{J}$, say J_z. We construct the operators

$$J_+ = J_x + iJ_y, \qquad J_- = J_x - iJ_y \tag{16.10}$$

and

$$\mathbf{J}^2 = J_x{}^2 + J_y{}^2 + J_z{}^2 \tag{16.11}$$

Of these three operators only $\mathbf{J}^2$ is Hermitian. J_- is the adjoint to J_+. From the commutation relations (16.8) we infer further commutation relations:

$$J_z J_+ - J_+ J_z = \hbar J_+ \tag{16.12}$$

$$J_- J_z - J_z J_- = \hbar J_- \tag{16.13}$$

$$J_+ J_- - J_- J_+ = 2\hbar J_z \tag{16.14}$$

$$\mathbf{J}^2 \mathbf{J} - \mathbf{J} \mathbf{J}^2 = 0 \tag{16.15}$$

We note the useful identity

$$\mathbf{J}^2 - J_z{}^2 \pm \hbar J_z = J_\pm J_\mp \tag{16.16}$$

Exercise 16.2. Prove (16.16).

Since, according to (16.15), J_z commutes with $\mathbf{J}^2$, it is possible to obtain simultaneous eigenvectors for these two operators. (This fact will help us in distinguishing between the various independent eigenvectors of J_z.) If we denote the eigenvalues of J_z by $m\hbar$, and those of $\mathbf{J}^2$ by $\lambda\hbar^2$, the eigenvalue problem can be written as

$$J_z \, |\lambda m\rangle = m\hbar \, |\lambda m\rangle \tag{16.17}$$

$$\mathbf{J}^2 \, |\lambda m\rangle = \lambda\hbar^2 \, |\lambda m\rangle \tag{16.18}$$

Theorem. The eigenvalues m and λ, belonging to the same eigenvector, satisfy the inequality

$$\lambda \geqslant m^2 \tag{16.19}$$

Proof

$$\mathbf{J}^2 - J_z^{\,2} = J_x^{\,2} + J_y^{\,2} = \tfrac{1}{2}(J_+J_- + J_-J_+) = \tfrac{1}{2}(J_+J_+^{\,\dagger} + J_+^{\,\dagger}J_+)$$

Since an operator of the form $AA^\dagger$ has only nonnegative expectation values, we conclude that

$$\langle \lambda m | \, \mathbf{J}^2 - J_z^{\,2} \, | \lambda m \rangle \geqslant 0$$

from which the inequality follows.

Next we develop again a ladder procedure similar to the method employed in Section 15.8 for the harmonic oscillator. Indeed, if we let (16.12) and (16.13) act on the ket $|\lambda m\rangle$, we obtain

$$J_z J_+ | \lambda m \rangle = (m + 1)\hbar J_+ | \lambda m \rangle \tag{16.20}$$

$$J_z J_- | \lambda m \rangle = (m - 1)\hbar J_- | \lambda m \rangle \tag{16.21}$$

Also,

$$\mathbf{J}^2 J_\pm \, | \lambda m \rangle = \lambda \hbar^2 J_\pm \, | \lambda m \rangle$$

Hence, if $|\lambda m\rangle$ is an eigenvector of J_z and $\mathbf{J}^2$ with eigenvalues $m\hbar$ and $\lambda\hbar^2$, then $J_\pm \, |\lambda m\rangle$ is likewise an eigenket of these same operators but with eigenvalues $(m \pm 1)\hbar$ and $\lambda\hbar^2$. We may therefore write

$$J_+ | \lambda m \rangle = C_+(\lambda m)\hbar \, | \lambda \, m{+}1 \rangle \tag{16.22}$$

$$J_- | \lambda m \rangle = C_-(\lambda m)\hbar \, | \lambda \, m{-}1 \rangle \tag{16.23}$$

where $C_+(\lambda m)$ are complex numbers yet to be determined.

For a given value of λ the inequality $\lambda \geqslant m^2$ limits the magnitude of m. Hence, there must be a greatest value of m, $m = j$, for any given λ. Application of the raising operator J_+ to the eigenket $|\lambda j\rangle$ should not lead to any new eigenket; hence,

$$J_+ | \lambda \, j \rangle = 0$$

Multiplying on the left by J_-, we obtain

$$J_- J_+ | \lambda \, j \rangle = (\mathbf{J}^2 - J_z^{\,2} - \hbar J_z) | \lambda \, j \rangle = (\lambda - j^2 - j)\hbar^2 | \lambda \, j \rangle = 0$$

from which the relation between j and λ follows:

$$\lambda = j(j + 1) \tag{16.24}$$

Similarly, there must be a lowest value of m, $m = j'$, such that

$$J_- | \lambda \, j' \rangle = 0$$

From this we deduce

$$\lambda = j'(j' - 1) \tag{16.25}$$

Equations (16.24) and (16.25) are consistent only if

$$j' = -j \quad \text{or} \quad j' = j + 1$$

The second solution is meaningless because it violates the assumption that j was the greatest and j' the smallest value of m. Hence, $j' = -j$.

Since the eigenvalues have both upper and lower bounds, it is clear that for a given value of λ, or j, it must be possible to reach $|\lambda j'\rangle = |\lambda, -j\rangle$ from $|\lambda j\rangle$ in a sufficient number of steps descending the ladder by application of the lowering operator J_-. In each downward step m decreases by unity; it follows that $j - j' = 2j$ must be a nonnegative integer. Hence, j must be either a *nonnegative integer or a half-integer*, i.e., the only possible values for j are

$$j = 0, \tfrac{1}{2}, 1, \tfrac{3}{2}, 2, \ldots \tag{16.26}$$

The corresponding values of $\lambda = j(j + 1)$ are

$$\lambda = 0, \tfrac{3}{4}, 2, \tfrac{15}{4}, 6, \ldots$$

For a given value of j the eigenvalues of J_z are

$$J_z' = m\hbar = j\hbar, (j - 1)\hbar, (j - 2)\hbar, \ldots, -(j - 1)\hbar, -j\hbar$$

These are $2j + 1$ in number.

With the aid of the identity (16.16) we can now determine the coefficients $C_\pm$ in (16.22) and (16.23). Note that from (16.22),

$$\langle \lambda\, m| J_- = \langle \lambda\, m+1| C_+{}^*(\lambda m)\hbar$$

Multiplying this and (16.22), we get

$$\langle \lambda\, m| J_- J_+ |\lambda\, m\rangle = |C_+(\lambda\, m)|^2 \hbar^2 \langle \lambda\, m+1 \mid \lambda\, m+1\rangle$$

Let us assume that all eigenvectors are normalized to unity. Then, since

$$\langle \lambda\, m| J_- J_+ |\lambda\, m\rangle = \langle \lambda\, m| \mathbf{J}^2 - J_z{}^2 - \hbar J_z |\lambda\, m\rangle$$

$$= [j(j + 1) - m^2 - m]\hbar^2 \langle \lambda\, m \mid \lambda\, m\rangle$$

we conclude that

$$|C_+(\lambda m)|^2 = j(j + 1) - m(m + 1) = (j - m)(j + m + 1)$$

The phases of C_+ are not determined and may be chosen arbitrarily. A usual choice is to make the phases equal to zero. We then have

$$J_+ |\lambda\, m\rangle = \sqrt{(j - m)(j + m + 1)}\,\hbar\, |\lambda\, m+1\rangle \tag{16.27}$$

Using the fact that J_- is the adjoint of J_+, we find

$$J_- |\lambda\, m\rangle = \sqrt{(j + m)(j - m + 1)}\,\hbar\, |\lambda\, m-1\rangle \tag{16.28}$$

By the use of these equations we can construct the matrices representing $J_x = (J_+ + J_-)/2$ and $J_y = (J_+ - J_-)/2i$ in a basis which consists of the common eigenvectors of J_z and $\mathbf{J}^2$.

We have thus completed the explicit construction of all the operators $\mathbf{J}$ which satisfy the commutation relations (16.8). The treatment of the eigenvalue problem given here has been a purely formal one. Only the commutation relations, the Hermitian nature of $\mathbf{J}$, and certain implicit assumptions about the existence of eigenvectors were utilized, but nothing else. In particular, no explicit use was made of the connection between $\mathbf{J}$ and spatial rotations. Our solution of the eigenvalue problem thus extends to any three operators which satisfy commutation relations like (16.8), e.g., the isospin operators, which are useful in the theory of elementary particles.

We summarize our results so far. If the state of a physical system is described by Ψ and if the system admits rotations as symmetry operations, a particular rotation R produces a new state Ψ'', related to Ψ by a unitary transformation U_R:

$$\Psi'' = U_R \Psi \tag{16.29}$$

The *rotation operator* U_R has the explicit form

$$U_R = \exp\left(-\frac{i}{\hbar}\,\hat{\mathbf{n}} \cdot \mathbf{J}\theta\right) \tag{16.7}$$

and the components of $\mathbf{J}$ are Hermitian operators satisfying the commutation relations (16.8).

In constructing the rotation operator explicitly we must take into account a further condition arising from the observation that the same rotation R is represented by all operators that have the form

$$\exp\left[-\frac{i}{\hbar}\,\hat{\mathbf{n}} \cdot \mathbf{J}(\theta + 2\pi k)\right] = \exp\left(-\frac{i}{\hbar}\,\hat{\mathbf{n}} \cdot \mathbf{J}\theta\right)\exp\left(-\frac{i}{\hbar}\,\hat{\mathbf{n}} \cdot \mathbf{J}2\pi k\right)$$

where k is an arbitrary integer. According to our formulation of the Euclidean principle of relativity in quantum mechanics, these operators must not differ from U_R by more than a phase factor[4]; hence, we must demand that $\exp\left(-i\hat{\mathbf{n}} \cdot \mathbf{J}2\pi k/\hbar\right)$ be a multiple of the identity operator and explore the consequences of this requirement. If the eigenvalue problem of the operator $\mathbf{J}$ is solved in the state vector space of the physical system, certain integral or half-integral quantum numbers $j_1, j_2, \ldots$ will be found. The effect of operating with $\exp\left(-i\hat{\mathbf{n}} \cdot \mathbf{J}2\pi k/\hbar\right)$ on an eigenstate of $\hat{\mathbf{n}} \cdot \mathbf{J}$ is to multiply the state by $\exp\left(-im2\pi k\right) = (-1)^{2km}$. For a given value of k all of these multiplicative constants must be the same, but this is possible only if either $j_1, j_2, \ldots$ are *all* integers or *all* half-integers. The Euclidean principle of relativity prohibits the mixing of these two classes, and the eigenstates of $\mathbf{J}^2$, which a given physical system may assume, are either all characterized by integral values of j or by half-integral values. In the integral case there is a unique operator U_R corresponding to a rotation R, whereas in the half-integral case each rotation is characterized by the operators U_R and $-U_R$, since a rotation by an odd multiple of 2π changes the sign of the rotation operator.

We thus see that since under a 2π rotation, which is tantamount to no rotation, all states of a system must transform uniformly, either all retaining the sign or all changing it, only states with either integral or half-integral j values may be superposed. In the framework of ordinary quantum mechanics the superposition of states with integral and half-integral j values is inadmissible. This inviolate *superselection rule* has important physical consequences: for instance, particles of half-integral spin cannot be created or destroyed singly (or, generally, in odd numbers).

We have considered the conditions which U_R must necessarily satisfy if it is to represent rotations in quantum mechanics. Conversely, it can be shown[5] that, if $R_1(\theta_1, \hat{\mathbf{n}}_1)$ and $R_2(\theta_2, \hat{\mathbf{n}}_2)$ are two successive rotations and if $R_3(\theta_3, \hat{\mathbf{n}}_3)$ is the composite single rotation, the commutation relations (16.8) alone suffice to insure the equality

$$\exp\left(-\frac{i}{\hbar}\,\hat{\mathbf{n}}_2 \cdot \mathbf{J}\theta_2\right) \exp\left(-\frac{i}{\hbar}\,\hat{\mathbf{n}}_1 \cdot \mathbf{J}\theta_1\right) = \pm \exp\left(-\frac{i}{\hbar}\,\hat{\mathbf{n}}_3 \cdot \mathbf{J}\theta_3\right) \quad (16.30)$$

if mixing of integral and half-integral j values is avoided. The composition of two rotations is thus represented by

$$U_2 U_1 = \pm U_3 \quad (16.31)$$

in agreement with the requirement (16.6).

The theory of this section presents us with all the possible ways in which state vectors may transform under rotation. It can of course not tell us which of these possibilities are realized in nature. We have already seen two actual and important examples: the orbital angular momentum, $\mathbf{J} = \mathbf{L}$, and the spin angular momentum, $\mathbf{J} = \mathbf{S} = \hbar\boldsymbol{\sigma}/2$, of electrons, protons, neutrons, etc. Both of these vector operators satisfy the commutation relations (16.8). They correspond to the values $j = l = 0, 1, 2, \ldots$ and $j = s = \frac{1}{2}$, respectively.

However, the notion of angular momentum can be generalized further. If we call *angular momentum* the observable corresponding to any operator $\mathbf{J}$, which generates infinitesimal rotations, $\mathbf{J}^2$ is capable of taking on the discrete set of eigenvalues $\lambda\hbar^2 = j(j+1)\hbar^2 = 0, \frac{3}{4}\hbar^2, 2\hbar^2, \ldots$. In order to make explicit use of the general theory we must know something about the nature of the particular physical system under consideration. We must know

---

[5] For the proof of sufficiency see W. I. Smirnov, *Lehrgang der Höheren Mathematik*, Teil 3/1, Sections 82 and 87, Deutscher Verlag der Wissenschaften, Berlin, 1954. For $j = \frac{1}{2}$ the proof is easy and can be obtained from Exercise 12.3.

the observables which describe it, and how they behave under rotation. Thus, in the case of orbital angular momentum (Chapter 9) we were dealing with the transformation of a function ψ of the position coordinates x, y, z, or r, φ, θ, and we were led to the study of spherical harmonics. In the case of the spin (Chapter 12) we deduced the behavior of two-component spinors under rotation from the physical connection between angular momentum and the magnetic moment, and from the vectorial character of these physical quantities. Other, more complex examples of angular momentum will appear shortly.

3. Symmetry Properties and Conservation Laws. In the last section we saw how a symmetry principle like the Euclidean principle of relativity circumscribes the geometric structure of quantum mechanics. There are equally important dynamical conclusions to be drawn from such a principle, leading to physical conservation laws by which the theory can be tested experimentally. Although the same ideas are applicable to any symmetry property, to be explicit we shall formulate the discussion in this section in terms of *conservation of angular momentum* which is a result of *rotational invariance*.

The dynamical system under consideration, which we shall assume to be characterized by a time-independent Hamiltonian operator, evolves in time from its initial state according to the formula

$$\Psi(t) = \exp\left[-\frac{i}{\hbar} H(\alpha)t\right]\Psi(0) \tag{16.32}$$

The Hamiltonian depends on the dynamical variables which describe the system, and by adding a parameter α we have explicitly allowed for the possibility that the system may be acted upon by external forces, constraints, or other agents which are not part of the system itself. The division of the world into "the system" and "the environment" in which the system finds itself is, of course, arbitrary. But, if the cut between the system and its surroundings is made suitably, the system may often be described to a highly accurate approximation by neglecting its action on the "rest of the world." It is in this spirit that the parameter α symbolizes the external fields acting on what, by an arbitrary but appropriate choice, we have delineated as the dynamical system under consideration.

We have seen that a rotation induces a unitary transformation U_R of the state vectors of the system. If at a time $t = 0$ we rotate the system but not its external environment, the state vector will in general follow a different course in time and become

$$\Psi''(t) = \exp\left[-\frac{i}{\hbar} H(\alpha)t\right]U_R\Psi(0) \tag{16.33}$$

On the other hand, we may also let the external fields participate in the rotation, thus giving rise to a new Hamiltonian, $H(\alpha_R)$. The Euclidean principle of relativity in conjunction with the principle of causality asserts that, if the system and the external fields acting on it are rotated together, the two arrangements, obtained from each other by rotation, must be equivalent *at all times*. Mathematically this implies

$$U_R \Psi(t) = \exp\left[-\frac{i}{\hbar} H(\alpha_R)t\right] U_R \Psi(0) \qquad (16.34)$$

From (16.32) and (16.34) we obtain the important connection,

$$H(\alpha_R)U_R = U_R H(\alpha) \qquad (16.35)$$

Now it frequently happens that the effect of the external parameters on the system is invariant under rotation. In mathematical terms, we then have the equality

$$H(\alpha_R) = H(\alpha) \qquad (16.36)$$

hence, U_R commutes with H. Since, according to (16.7), U_R is a function of the infinitesimal generator **J**, the latter becomes a constant of the motion. Conservation of angular momentum is thus seen to be a direct consequence of invariance under all rotations.

Note that the present discussion parallels that of Section 15.6, where the connection between invariance properties and conservation laws was discussed in general terms. The equivalence of the two approaches will impress itself on the reader.

As an important special case, (16.36) must obviously be valid for an isolated system which does not depend on any external parameters. We thus see that the isotropy of space, as expressed by the Euclidean principle of relativity, requires that the total angular momentum **J** of an isolated system be a constant of the motion.

Frequently it is possible to subject certain parts and variables of a system separately and independently to a rotation. For example, the spin of a particle can be rotated independently of its position coordinates. In the formalism this independence appears as the mutual commutivity of the operators **S** and **L** which describe rotations of the spin and the position coordinates respectively. If the Hamiltonian is such that no correlations are introduced in time between these two kinds of variables, then they may in effect be regarded as dynamical variables of two separate systems. Invariance under rotation then implies that both **S** and **L** commute separately with the Hamiltonian and that each constitutes a constant of the motion. The non-relativistic Hamiltonian of a particle with spin moving in a central force field actually couples **L** with **S**. As we saw in (13.42), this Hamiltonian may be written in the form

$$H = \frac{\mathbf{p}^2}{2\mu} + V(r) + W(r)\mathbf{L} \cdot \mathbf{S}$$

If the $\mathbf{L} \cdot \mathbf{S}$ term, which correlates spin and orbital motion, can be neglected in a first approximation, H commutes with both $\mathbf{L}$ and $\mathbf{S}$, and both of these are thus approximate constants of the motion. However, only the total angular momentum, $\mathbf{J} = \mathbf{L} + \mathbf{S}$, is rigorously conserved by this Hamiltonian, as was shown in Section 12.7. We shall see in Chapters 23 and 24 that in the relativistic theory of the electron even the free particle Hamiltonian does not commute with $\mathbf{L}$ and $\mathbf{S}$.

Exercise 16.3. How much rotational symmetry does a system possess, which consists of a spinless charged particle moving in a central field and a uniform magnetic field? What observable is conserved?

4. Symmetry Groups and Group Representations. Because of the great importance of rotational symmetry, the preceding sections of this chapter were devoted to a study of rotations in quantum mechanics. However, rotations are but one type of many different symmetry operations which play a role in physics. It is worthwhile to introduce the general notion of a group in this section, because symmetry operations are usually elements of certain groups, and group theory classifies and analyzes systematically a multitude of different symmetries which appear in nature.

A *group* is a set of distinct elements $a, b, c, \ldots$, subject to the following postulates:

1. To each ordered pair of elements a, b of the group belongs a product ab (usually not equal to ba), which is also an element of the group. We say that the law of group multiplication or the multiplication table of the group is given. The product of two symmetry operations, ba, is the symmetry operation which is equivalent to the successive application of a and b, performed in that order.

For example, rotations form a group in which the product ba of two elements b and a is defined as the single rotation which is equivalent to the successive rotations a and b. By a rotation we mean the mapping of a physical system, or of a Cartesian coordinate system, into a new physical or coordinate system obtainable from the old system by actually rotating it. The term *rotation* is, however, not to be understood as the physical motion which takes the system from one orientation to another. In fact, the intervening orientations which a system assumes during the motion are ignored, and two rotations are identified as equal if they lead from the same initial configuration to the same final configuration irrespective of the way in which the operation is performed. In the rotation group, generally, $ab \neq ba$. For instance, two successive rotations by 90°, one about the x-axis, the other about the y-axis, do not lead to the same over-all rotation when performed in reverse order. (Convince yourself of the truth of this statement by rotating this book successively about two different axes.)

2. $(ab)c = a(bc)$, i.e., the *associative law* holds. Since symmetry operations are usually motions or substitutions, this postulate is automatically satisfied.

3. The group contains a single *identity* element e, with the property $ea = ae = a$ for all elements a of the group. The operation "no rotation" is the identity element of the rotation group.

4. Each element has an *inverse*, denoted by a^{-1}, which is also an element of the group and has the property $aa^{-1} = a^{-1}a = e$. All symmetry operations are reversible and thus have inverses.

A symmetry operation a transforms the state Ψ of a system into a state Ψ_a. It was shown in Section 16.1 that under quite general conditions this transformation may be assumed to be either unitary, so that $\Psi_a = U_a\Psi$, or antilinear.

We assume here that the symmetry operations of interest belong to a group called a *symmetry group* of the system, which induces unitary linear transformations on the state vectors such that, if a and b are two elements of the group[6]

$$U_{ba} = U_b U_a \tag{16.37}$$

When (16.37) is translated into a matrix equation by introducing a complete set of basis vectors in the vector space of Ψ, each element a of the group becomes associated with a matrix $D(a)$ such that

$$D(ab) = D(a)D(b) \tag{16.38}$$

i.e., the matrices have the same multiplication table as the group elements to which they correspond. The set of matrices $D(a)$ is said to constitute a (matrix) *representation*[7] of the group. A change of basis will of course change the matrices of a representation according to the relation

$$D'(a) = S^{-1}D(a)S \tag{16.39}$$

---

[6] From the point of view of quantum mechanics we require only

$$U_b U_a = e^{i\varphi(b,a)} U_{ba}$$

[see (16.6)], but (16.37) is sufficiently general for our purposes, provided that the possibility of double-valued representations of the rotation group is not excluded. See below.

[7] The term *representation* has so far in this book been used mainly to describe a basis in an abstract vector space. In this chapter the same term will be used for the more specific *group representation*. The context usually establishes the intended meaning and misunderstandings are not likely to occur.

but from a group theoretical point of view two representations which can be transformed into each other by a similarity transformation S are not really different, and are called *equivalent*. A transformation S as in (16.39) is therefore also known as an *equivalence transformation*. Two representations are *inequivalent* if there is *no* transformation S which will take one into the other. Since the operators U_a were assumed to be unitary, the representation matrices are also unitary if the basis is orthonormal. In the following all representations $D(a)$ and all transformations S will be assumed in unitary form.

From an arbitrary given representation it is frequently possible to derive simpler representations by choosing a special basis in which all matrices of the representation simultaneously break up into a number of submatrices arrayed along the diagonal:

$$D(a) = \begin{pmatrix} D_1(a) & 0 & 0 & \cdot \\ 0 & D_2(a) & 0 & \cdot \\ 0 & 0 & D_3(a) & \cdot \\ \cdot & \cdot & \cdot & \cdot \end{pmatrix} \tag{16.40}$$

We suppose that each matrix of the representation acquires the same kind of block structure. If n is the dimensionality of D, each block D_1, D_2, ... is a matrix of dimension n_1, n_2, ... with $n_1 + n_2 + \cdots = n$. It is then obvious that the matrices D_1 by themselves constitute an n_1-dimensional representation. Similarly, D_2 gives an n_2-dimensional representation, etc. The original representation has thus been *reduced* to a number of simple representations. If no conceivable equivalence transformation can reduce all D matrices of the representation simultaneously to block structure, the representation is said to be *irreducible*. Otherwise it is called *reducible*. It is not difficult to prove that any reducible representation can be reduced in an essentially unique manner into irreducible components. The reduction is essentially unique because only an equivalence transformation remains unspecified.

Since, apart from possible equivalence transformations, the decomposition of a given representation into irreducible blocks is unique, there is a definite sense in stating the irreducible representations which make up a given reducible representation. Some of these may, of course, occur more than once.

The importance of these considerations lies in the fact that it is clearly sufficient to study all inequivalent irreducible representations of a group. All reducible representations are then built up from these. Group theory provides the means to construct systematically all irreducible representations from the group multiplication table.

The usefulness of the theory of group representations for quantum mechanics and notably of the idea of irreducibility will come into sharper focus if the Schrödinger equation $H\Psi = E\Psi$ is considered. A symmetry operation must leave the Schrödinger equation invariant so that the energies of the system are unaltered. The criterion for the invariance of the Schrödinger equation under the operations of the group is that the Hamiltonian commute with U_a for every element a of the group:

$$[H, U_a] = 0 \qquad (16.41)$$

In Section 16.3 the same condition was obtained by the application of symmetry requirements to the dynamical equations, and the connection with conservation laws and constants of the motion was established. By studying the symmetry group, which gives rise to these constants of the motion, it is possible to shed considerable light on the eigenvalue spectrum of the Hamiltonian and on the corresponding eigenfunctions.

If E is an n-fold degenerate eigenvalue of the Hamiltonian,

$$H\Psi_k = E\Psi_k \qquad (k = 1, 2, \ldots, n) \qquad (16.42)$$

then, owing to (16.41),

$$HU_a\Psi_k = U_aH\Psi_k = EU_a\Psi_k$$

Thus, if Ψ_k is an eigenvector of H with eigenvalue E, $U_a\Psi_k$ is also an eigenvector and belongs to the same eigenvalue. Hence, it must be equal to a linear combination of the degenerate eigenvectors,

$$U_a\Psi_k = \sum_{j=1}^{n} \Psi_j D_{jk}(a) \qquad (16.43)$$

where the D_{jk} are complex coefficients which depend on the group element. Repeated application of symmetry operations gives

$$U_bU_a\Psi_k = \sum_{j=1}^{n} U_b\Psi_j D_{jk}(a) = \sum_{j=1}^{n}\sum_{l=1}^{n} \Psi_l D_{lj}(b)D_{jk}(a) \qquad (16.44)$$

But we have also

$$U_{ba}\Psi_k = \sum_{l=1}^{n} \Psi_l D_{lk}(ba) \qquad (16.45)$$

By the assumption of (16.37) the left-hand sides of (16.44) and (16.45) are identical. Hence, comparing the right-hand sides, it follows that

$$D_{lk}(ba) = \sum_{j=1}^{n} D_{lj}(b)D_{jk}(a) \qquad (16.46)$$

This is the central equation of the theory. It shows that the coefficients D_{ij} define a unitary representation of the symmetry group. The n degenerate eigenvectors of H thus span an n-dimensional subspace of the state vector space of the system, and the operations of the group transform any vector which lies entirely in this subspace into another vector lying entirely in the same subspace, i.e., the symmetry operations leave the subspace *invariant*.

Since any representation D of the symmetry group can be characterized by the irreducible representations which it contains, it is possible to classify the stationary states of a system by the irreducible representations to which the eigenvectors of H belong. A partial determination of these eigenvectors can be accomplished thereby. The labels of the irreducible representations to which an energy eigenvalue belongs are the quantum numbers of the stationary state.

These considerations exhibit the mutual relationship between group theory and quantum mechanics: The eigenfunctions of (16.42) generate representations of the symmetry groups of the system described by H. Conversely, a knowledge of the appropriate symmetry groups and their irreducible representations can aid considerably in the solution of the Schrödinger equation for a complex system. If all symmetries of a system are recognized, much light can be shed on the eigenvalue spectrum and on the nature of the eigenstates. It must be apparent that the use of group theoretical methods is called for particularly in the study of the spectra of complex atoms and nuclei. The Schrödinger equation for such many-body systems is hopelessly complicated, but its complexity can be reduced very much and a great deal of information inferred from the various symmetry properties, such as rotational symmetry, reflection symmetry, and symmetry under exchange of identical particles.

The study of symmetries can be very fruitful even when details about forces and interactions are unknown. For instance, we know that the strong interactions between elementary particles satisfy certain general symmetry principles, such as invariance under rotations, Lorentz transformations, charge conjugation, interchange of identical particles, rotations in isospin space, and, at least approximately, under the operations of a special group of unitary transformations in a three-dimensional complex vector space [$SU(3)$]. By constructing all irreducible representations of the groups which correspond to these symmetries, we may obtain some of the basic quantum numbers and selection rules for the system without committing ourselves with regard to the ultimate form of a complete dynamical theory governing elementary particles.

5. *The Representations of the Rotation Group*. The representations of the rotation group, which is our prime example, are generated from the rotation operator U_R, as given in equation (16.7), by introducing a basis. The simplest basis to choose is one in which one component of the angular momentum operator $\mathbf{J}$ appears as a diagonal matrix. If we use the simultaneous eigenvectors of J_z and $\mathbf{J}^2$ as basis, ordered according to the eigenvalues j, the matrix of U_R breaks up into a number of blocks because none of the components of $\mathbf{J}$ has any nonvanishing matrix elements linking states of different j values. This follows immediately from the fundamental equations of Section 16.2, which are reproduced here for convenience in slightly altered but self-explanatory notation:

$$J_z \,|\alpha jm\rangle = m\hbar \,|\alpha jm\rangle \tag{16.47}$$

$$J_+ \,|\alpha jm\rangle = \sqrt{(j - m)(j + m + 1)}\hbar \,|\alpha j\, m{+}1\rangle \tag{16.48}$$

$$J_- \,|\alpha jm\rangle = \sqrt{(j + m)(j - m + 1)}\hbar \,|\alpha j\, m{-}1\rangle \tag{16.49}$$

α denotes all quantum numbers other than j and m which may be required to specify a basis vector. Since these quantum numbers are entirely unaffected by rotation, they may be omitted in some of the subsequent formulas. The vector space of the system at hand thus breaks up into a number of $(2j + 1)$-dimensional subspaces which are invariant under rotation. An arbitrary rotation R is represented in this subspace by the matrix

$$D^{(j)}_{m'm}(R) = \langle jm'|\, U_R \,|jm\rangle = \langle jm'|\exp\left(-\frac{i}{\hbar}\,\hat{\mathbf{n}}\cdot\mathbf{J}\theta\right)|jm\rangle \tag{16.50}$$

or, owing to (14.36) and (14.37), by

$$D^{(j)}(R) = \exp\left(-\frac{i}{\hbar}\,\hat{\mathbf{n}}\cdot\mathbf{J}\theta\right) \tag{16.51}$$

where $\mathbf{J}$ stands for the matrix $\langle jm'|\,\mathbf{J}\,|jm\rangle$. The simplicity of (16.50) and (16.51) is deceptive. For, the components of $\mathbf{J}$ other than J_z are represented by nondiagonal matrices, and the detailed dependence of D on the quantum numbers and on $\hat{\mathbf{n}}$ and θ is quite complicated.[8] For moderate values of j we can make use of equation (14.86) to construct the rotation matrices in terms of the first $2j$ powers of the matrix $\hat{\mathbf{n}}\cdot\mathbf{J}$:

$$D^{(j)}(R) = \sum_{k=-j}^{j} e^{-ik\theta} \prod_{l\neq k} \frac{l\hbar - \hat{\mathbf{n}}\cdot\mathbf{J}}{(l - k)\hbar} \tag{16.52}$$

Exceptional simplifications occur of course for rotations about the axis of quantization, the z-axis if J_z is chosen to be diagonal. For such special rotations

$$D^{(j)}_{m'm}(\theta) = e^{-im\theta}\,\delta_{m',m} \tag{16.53}$$

The representation matrices are also simple for infinitesimal rotations, i.e., when $\theta \ll 1$. In this case we have from (16.47), (16.48), (16.49), and (16.50),

---

[8] See, e.g., M. E. Rose, *Elementary Theory of Angular Momentum*, John Wiley and Sons, New York, 1957.

if we use the notation $\boldsymbol{\epsilon} = \theta\hat{\mathbf{n}}$,

$$D^{(j)}_{m'm}(\boldsymbol{\epsilon}) = \delta_{m',m} - \frac{i\varepsilon_x + \varepsilon_y}{2}\sqrt{(j - m)(j + m + 1)}\,\delta_{m',m+1}$$

$$- \frac{i\varepsilon_x - \varepsilon_y}{2}\sqrt{(j + m)(j - m + 1)}\,\delta_{m',m-1} - i\varepsilon_z m\,\delta_{m',m} \quad (16.54)$$

From this it is easy to see that a further subdiagonalization of all rotation matrices is impossible.[9] Hence, for a given value of j, the representations of the rotation group just obtained are irreducible. All continuous unitary irreducible representations of the rotation group are obtained by allowing j to assume the values $j = 0, \frac{1}{2}, 1, \ldots$.

A remark must be made concerning the half-integral values of j. From the point of view of the infinitesimal rotations these are on the same footing as the integral values, and (16.54) confirms this. However, when finite rotations are considered, an important distinction arises, for upon rotation about a fixed axis by 2π, see (16.53), the matrix $D^{(j)}(2\pi) = -1$ results. Yet, the system has been restored to its initial configuration, and the rotation is entirely equivalent to the operation "no rotation at all." Hence, for half-integral values of j a *double-valued representation* of the group is obtained: to any rotation R correspond two distinct matrices differing by a sign. As we have already seen in Section 16.2, owing to the usual phase ambiguities of quantum mechanics, there is no objection to the appearance of this kind of double-valued representation in the reduction of the matrix of U_R, provided that U_R contains then *only* double-valued representations.

Exercise 16.4. Show that the representation matrices $D^{(1/2)}(R)$ are equivalent to the matrices U_R introduced in Section 12.5 and that these matrices constitute the elements of the *unitary unimodular* (det $U_R = 1$) group in two dimensions. (This group is denoted as $SU(2)$ and said to be the *universal covering group* of the rotation group.)

If j is integral, $j = l$, a simple application of the representation theory developed here can be given. Expressed in terms of a (single particle) coordinate basis, the eigenvectors of J_z and $\mathbf{J}^2$ are then spherical harmonics:

$$\langle xyz \mid lm \rangle = Y_l{}^m(\theta, \varphi) \quad (16.55)$$

Under a rotation an eigenvector $|lm\rangle$ of J_z transforms into an eigenvector $|lm\rangle'$ of $J_{z'}$, z' being the axis obtained by the rotation from the z-axis. Hence,

$$|lm\rangle' = U_R |lm\rangle = \sum_{m'} |lm'\rangle D^{(l)}_{m'm}(R) \quad (16.56)$$

[9] The reason is this: In diagonalizing, for instance, $J_x = (J_+ + J_-)/2$ we would necessarily undiagonalize J_z.

and, if this is multiplied on the left by $\langle \mathbf{r}|$,

$$Y_l^m(\theta', \varphi') = \sum_{m'=-l}^{l} Y_l^{m'}(\theta, \varphi)D_{m'm}^{(l)}(R) \tag{16.57}$$

where θ, φ and θ', φ' are coordinates of the same physical point.

This equation can be inverted by using the unitary property of the representation:

$$Y_l^m(\theta, \varphi) = \sum_{m'=-l}^{l} D_{mm'}^{(l)*}(R)Y_l^{m'}(\theta', \varphi') \tag{16.58}$$

Consider in particular a point on the new z'-axis, $\theta' = 0$. Using (9.77), we find

$$D_{m0}^{(l)}(R) = \sqrt{\frac{4\pi}{2l+1}} \, Y_l^{m*}(\beta, \alpha) \tag{16.59}$$

where β, α are the spherical polar coordinates of the new z'-axis in the old coordinate system (see Figure 9.1). If this is substituted into (16.57) for $m = 0$, we get

$$Y_l^0(\theta', \varphi') = \sqrt{\frac{4\pi}{2l+1}} \sum_{m=-l}^{l} Y_l^m(\theta, \varphi)Y_l^{m*}(\beta, \alpha) \tag{16.60}$$

By (9.78) we finally obtain the addition theorem for spherical harmonics,

$$P_l(\cos \theta') = \frac{4\pi}{2l+1} \sum_{m-l}^{l} Y_l^{m*}(\beta, \alpha)Y_l^m(\theta, \varphi) \tag{16.61}$$

[see also (9.79)].

Exercise 16.5. Find the matrix which represents an infinitesimal rotation described by the Euler angles α, β, γ.

6. The Addition of Angular Momenta. If two distinct physical systems or two distinct sets of dynamical variables of one system, which are described in two different vector spaces, are merged, the states of the composite system are vectors in the *direct product space* of the two previously separate vector spaces. The mathematical procedure was outlined in Section 15.5. A common rotation of the composite system is represented by the direct product of the rotation operators for each subsystem and may, according to equation (15.42), be written as

$$e^{-(i/\hbar)\hat{n}\cdot\mathbf{J}_1\theta} \times e^{-(i/\hbar)\hat{n}\cdot\mathbf{J}_2\theta} = e^{-(i/\hbar)\hat{n}\cdot(\mathbf{J}_1\times I + I\times\mathbf{J}_2)\theta} \tag{16.62}$$

The operator

$$\mathbf{J} = \mathbf{J}_1 \times I + I \times \mathbf{J}_2 \tag{16.63}$$

often written more simply, if less accurately, as

$$\mathbf{J} = \mathbf{J}_1 + \mathbf{J}_2 \qquad (16.64)$$

is the *total angular momentum* of the entire system.

Exercise 16.6. Prove equation (16.62) and show that $\mathbf{J}_1 \times I$ commutes with $I \times \mathbf{J}_2$.

An important example of adding commuting angular momentum operators was already encountered in Section 12.7 and again in Section 16.3, where $\mathbf{L}$ and $\mathbf{S}$ were combined into $\mathbf{J}$. This section is devoted to the formal and general solution of the problem of adding any two commuting angular momenta.

Since each component of $\mathbf{J}_1$ commutes with each component of $\mathbf{J}_2$, which separately satisfy the usual commutation relations for angular momentum, the total angular momentum $\mathbf{J}$ also satisfies the angular momentum commutation relations:

$$[J_x, J_y] = i\hbar J_z, \qquad [J_y, J_z] = i\hbar J_x, \qquad [J_z, J_x] = i\hbar J_y \qquad (16.65)$$

The problem of the addition of two angular momenta consists of obtaining the eigenvalues of J_z and $\mathbf{J}^2$ and their eigenvectors in terms of the direct products of the eigenvectors of J_{1_z}, $\mathbf{J}_1^2$ and of J_{2_z}, $\mathbf{J}_2^2$. The normalized simultaneous eigenvectors of the four operators $\mathbf{J}_1^2$, $\mathbf{J}_2^2$, J_{1_z}, J_{2_z} can be symbolized by the direct product kets,

$$|j_1 j_2 m_1 m_2\rangle = |j_1 m_1\rangle \, |j_2 m_2\rangle \qquad (16.66)$$

These constitute a basis in the product space. It is desired to construct from this basis the eigenvectors of J_z, $\mathbf{J}^2$ which form a new basis.

$\mathbf{J}_1^2$ and $\mathbf{J}_2^2$ commute with every component of $\mathbf{J}$. Hence,

$$[J_z, \mathbf{J}_1^2] = [J_z, \mathbf{J}_2^2] = [\mathbf{J}^2, \mathbf{J}_1^2] = [\mathbf{J}^2, \mathbf{J}_2^2] = 0$$

and the eigenvectors of J_z and $\mathbf{J}^2$ can be required to be simultaneously eigenvectors of $\mathbf{J}_1^2$ and $\mathbf{J}_2^2$ also. (But $\mathbf{J}^2$ does not commute with J_{1_z} or J_{2_z}!). In the subspace of the simultaneous eigenvectors of $\mathbf{J}_1^2$ and $\mathbf{J}_2^2$ with eigenvalues j_1 and j_2 respectively we can thus write the transformation equation

$$|j_1 j_2 jm\rangle = \sum_{m_1 m_2} |j_1 j_2 m_1 m_2\rangle\langle j_1 j_2 m_1 m_2 | j_1 j_2 jm\rangle \qquad (16.67)$$

connecting the two sets of normalized eigenvectors. The summation needs to be carried out only over the eigenvalues m_1 and m_2, for j_1 and j_2 can be assumed to have fixed values. The problem of adding angular momenta is thus the problem of determining the transformation coefficients

$$\langle j_1 j_2 m_1 m_2 | j_1 j_2 jm\rangle.$$

These elements of the transformation matrix are called *vector addition* or *Clebsch-Gordan* or *Wigner coefficients*. Simplifying the notation, we may occasionally write

$$\langle j_1 j_2 m_1 m_2 \, | \, j_1 j_2 jm \rangle = \langle m_1 m_2 \, | \, jm \rangle$$

it being understood that j_1 and j_2 are the maximum values of m_1 and m_2.[10]

If the operator $J_z = J_{1_z} + J_{2_z}$ is applied to (16.67) and if the eigenvalue conditions

$$J_z \, | \, j_1 j_2 jm \rangle = m\hbar \, | \, j_1 j_2 jm \rangle$$

$$(J_{1_z} + J_{2_z}) \, | \, j_1 j_2 m_1 m_2 \rangle = (m_1 + m_2)\hbar \, | \, j_1 j_2 m_1 m_2 \rangle$$

are used, it can be concluded immediately that

$$\langle j_1 j_2 m_1 m_2 \, | \, j_1 j_2 jm \rangle = 0 \qquad \text{unless } m = m_1 + m_2 \qquad (16.68)$$

Applying J_+ and J_- to (16.67), we obtain readily the following recursion relations for the transformation coefficients:

$$\sqrt{(j \pm m)(j \mp m + 1)} \langle m_1 m_2 \, | \, j, m \mp 1 \rangle$$
$$= \sqrt{(j_1 \mp m_1)(j_1 \pm m_1 + 1)} \langle m_1 \pm 1, m_2 \, | \, jm \rangle$$
$$+ \sqrt{(j_2 \mp m_2)(j_2 \pm m_2 + 1)} \langle m_1, m_2 \pm 1 \, | \, jm \rangle \qquad (16.69)$$

To appreciate the usefulness of these equations, let us set $m_1 = j_1$ and $m = j$ in the upper part of (16.69). Since, following the instruction (16.68), m_2 can have only the value $m_2 = j - j_1 - 1$, we get

$$\sqrt{2j} \langle j_1, j - j_1 - 1 \, | \, j, j - 1 \rangle$$
$$= \sqrt{(j_2 - j + j_1 + 1)(j_2 + j - j_1)} \langle j_1, j - j_1 \, | \, jj \rangle \qquad (16.70)$$

Let us further set $m_1 = j_1$, $m_2 = j - j_1$, $m = j - 1$ in the lower part of (16.69). This results in

$$\sqrt{2j} \langle j_1, j - j_1 \, | \, jj \rangle = \sqrt{2j_1} \langle j_1 - 1, j - j_1 \, | \, j, j - 1 \rangle$$
$$+ \sqrt{(j_2 + j - j_1)(j_2 - j + j_1 + 1)} \, \langle j_1, j - j_1 - 1 \, | \, j, j - 1 \rangle \qquad (16.71)$$

According to condition (16.70), $\langle j_1, j - j_1 - 1 \, | \, j, j - 1 \rangle$ can be determined if $\langle j_1, j - j_1 \, | \, jj \rangle$ is known. Subsequently (16.71) can be used to compute

[10] For an alternate method of computing Wigner coefficients by using projection operators, see M. L. Goldberger and K. M. Watson, *Collision Theory*, John Wiley and Sons, New York, 1964, Section 1.2. Many different notations are in use for these coefficients such as $\langle j_1 j_2 m_1 m_2 \, | \, jm \rangle$ or $C(j_1 j_2 j; m_1 m_2)$. Related quantities are the 3*j*-symbols of Wigner and the *V* functions introduced by Racah. For a compendium, see A. R. Edmonds, *Angular Momentum in Quantum Mechanics*, Princeton University Press, Princeton, 1957. Also D. M. Brink and G. R. Satchler, *Angular Momentum*, 2nd ed., Clarendon Press, Oxford, 1968.

$\langle j_1 - 1, j - j_1 \mid j, j-1 \rangle$ from these two coefficients. Continuing in this manner, the recursion relations (16.69) can be used to give for fixed values of j_1, j_2, and j all the Clebsch-Gordan coefficients in terms of just one of them, namely

$$\langle j_1 j_2 j_1, j - j_1 \mid j_1 j_2 j\, j \rangle \tag{16.72}$$

The absolute value of this coefficient is determined by normalization (see below). The coefficient (16.72) is different from zero only if

$$-j_2 \leqslant j - j_1 \leqslant j_2$$

this being the range of the values of m_2. Hence, j is restricted to the range

$$j_1 - j_2 \leqslant j \leqslant j_1 + j_2 \tag{16.73a}$$

But we could equally well have expressed all these Clebsch-Gordan coefficients in terms of

$$\langle j_1 j_2, j - j_2, j_2 \mid j_1 j_2 j j \rangle$$

leading to the condition

$$j_2 - j_1 \leqslant j \leqslant j_1 + j_2 \tag{16.73b}$$

Hence, the three angular momentum quantum numbers must satisfy the so-called *triangular condition*

$$|j_1 - j_2| \leqslant j \leqslant j_1 + j_2 \tag{16.74}$$

i.e., the three numbers j_1, j_2, j must be such that they could constitute the three sides of a triangle. Since $m = m_1 + m_2$ ranges between $-j$ and j, it follows that j can assume only the values

$$j = j_1 + j_2, \ j_1 + j_2 - 1, \ldots, |j_1 - j_2| \tag{16.75}$$

Hence, either all three quantum numbers j_1, j_2, j are integers, or two of them are half-integral and one is an integer.

We observe that for fixed values of j_1 and j_2 (16.67) gives a complete new basis in the $(2j_1 + 1)(2j_2 + 1)$-dimensional vector space spanned by the kets $|j_1 j_2 m_1 m_2\rangle$. Indeed, the new kets $|j_1 j_2 j m\rangle$ with $j = j_1 + j_2$, $j_1 + j_2 - 1, \ldots, |j_1 - j_2|$ are again $(2j_1 + 1)(2j_2 + 1)$ in number, and being eigenkets of Hermitian operators they are also orthogonal. Hence, since the old and new bases are both normalized to unity, the Clebsch-Gordan coefficients must constitute a unitary matrix. From the recursion relations (16.69) it is clear that all Clebsch-Gordan coefficients are real numbers if one of them, say (16.72), is chosen real. If this is done, the Clebsch-Gordan coefficients satisfy the condition

$$\sum_{m_1, m_2} \langle j_1 j_2 m_1 m_2 \mid j_1 j_2 j m \rangle \langle j_1 j_2 m_1 m_2 \mid j_1 j_2 j' m' \rangle = \delta_{mm'} \, \delta_{jj'} \tag{16.76}$$

or inversely,

$$\sum_{j,m} \langle j_1 j_2 m_1 m_2 | j_1 j_2 jm \rangle \langle j_1 j_2 m_1' m_2' | j_1 j_2 jm \rangle = \delta_{m_1 m_1'} \delta_{m_2 m_2'} \quad (16.77)$$

The double sums in these two equations can be substantially simplified by the use of the selection rule (16.68). Condition (16.76) in conjunction with the recursion relations determines all Clebsch-Gordan coefficients except for a sign. It is conventional to choose the latter by demanding that the Clebsch-Gordan coefficient (16.72) be *real and positive*. Extensive numerical tables are available for the frequently needed Clebsch-Gordan coefficients.[11]

Some of the most useful symmetry relations for the Clebsch-Gordan coefficients are summarized by the equation

$$\langle j_1 j_2 m_1 m_2 | j_1 j_2 jm \rangle$$
$$= (-1)^{j-j_1-j_2} \langle j_2 j_1 m_2 m_1 | j_2 j_1 jm \rangle = \langle j_2 j_1, -m_2, -m_1 | j_2 j_1 j, -m \rangle \quad (16.78)$$

For spin $\frac{1}{2}$ particles the Clebsch-Gordan coefficients for $j_2 = \frac{1}{2}$ are particularly relevant. From the recursion relations (16.69), the normalization condition (16.76), and the convention that $\langle j_1 \frac{1}{2} j_1, j-j_1 | j_1 \frac{1}{2} j j \rangle$ be real positive, we readily obtain the values

$$\langle j_1 \frac{1}{2}, m-\frac{1}{2}, \frac{1}{2} | j_1 \frac{1}{2}, j_1 \pm \frac{1}{2}, m \rangle = \pm \sqrt{\frac{j_1 \pm m + \frac{1}{2}}{2j_1 + 1}} \quad (16.79a)$$

$$\langle j_1 \frac{1}{2}, m+\frac{1}{2}, -\frac{1}{2} | j_1 \frac{1}{2}, j_1 \pm \frac{1}{2}, m \rangle = \sqrt{\frac{j_1 \mp m + \frac{1}{2}}{2j_1 + 1}} \quad (16.79b)$$

since the allowed eigenvalues are $j = j_1 \pm \frac{1}{2}$.

If the coordinate representation is used for the eigenstates of orbital angular momentum $\mathbf{L}$ and if, as usual, the eigenstates of S_z are chosen as a basis to represent the spin, the total angular momentum $\mathbf{J} = \mathbf{L} + \mathbf{S}$ and its eigenvectors may be represented in the direct product spin-coordinate basis. We shall denote the common eigenspinor-functions of J_z and $\mathbf{J}^2$ by $\mathscr{Y}_l^{jm}$. Using the Clebsch-Gordan coefficients (16.79) with $j_1 = l$, we have

$$\mathscr{Y}_l^{l \pm (1/2), m} = \frac{1}{\sqrt{2l + 1}} \begin{pmatrix} \pm\sqrt{l \pm m + \frac{1}{2}} \; Y_l^{m-(1/2)} \\ \sqrt{l \mp m + \frac{1}{2}} \; Y_l^{m+(1/2)} \end{pmatrix} \quad (16.80)$$

for the eigenstates with $j = l \pm \frac{1}{2}$.

---

[11] The treatise of E. U. Condon and G. H. Shortley, *The Theory of Atomic Spectra*, University Press, Cambridge, 1935, contains some very useful tables. See also G. J. Nijgh, A. H. Wapstra, and R. van Lieshout, *Nuclear Spectroscopy Tables*, North-Holland Publishing Co., Section 9.7, Amsterdam, 1959, and M. Rotenberg, R. Bivins, N. Metropolis, and J. K. Wooten, Jr., *The 3-j and 6-j Symbols*. Technology Press, Cambridge, Mass., 1959.

Exercise 16.7. Apply

$$\mathbf{J}^2 = \mathbf{L}^2 + \mathbf{S}^2 + 2\mathbf{L} \cdot \mathbf{S} = \mathbf{L}^2 + \mathbf{S}^2 + L_+S_- + L_-S_+ + 2L_zS_z$$
$$(16.81)$$

to equation (16.80) and verify that $\mathscr{Y}_l^{jm}$ is an eigenstate of $\mathbf{J}^2$.

It is reasonable to anticipate that these eigenstates will play an important role in the quantum mechanics of single electron and single nucleon systems.

The formulas (16.79) are also useful when the spins of two particles are added to give the *total spin*

$$\mathbf{S} = \mathbf{S}_1 + \mathbf{S}_2 \qquad (16.82)$$

If the two particles have spin $\frac{1}{2}$, the direct product spin space of the composite system is four-dimensional, and a particular basis is spanned by the eigenvectors of S_{1_z} and S_{2_z}:

$$\alpha_1\alpha_2, \qquad \alpha_1\beta_2, \qquad \beta_1\alpha_2, \qquad \beta_1\beta_2 \qquad (16.83)$$

By letting $j_1 = \frac{1}{2}$ in the expressions (16.79), we obtain the appropriate Clebsch-Gordan coefficients, allowing us to write the simultaneous eigenstates of S_z and $\mathbf{S}^2$ in the form:

$$|\tfrac{1}{2}\tfrac{1}{2}00\rangle = \frac{1}{\sqrt{2}}(\alpha_1\beta_2 - \beta_1\alpha_2) \qquad (16.84)$$

$$\left.\begin{aligned} |\tfrac{1}{2}\tfrac{1}{2}11\rangle &= \alpha_1\alpha_2 \\[1mm] |\tfrac{1}{2}\tfrac{1}{2}10\rangle &= \frac{1}{\sqrt{2}}(\alpha_1\beta_2 + \beta_1\alpha_2) \\[1mm] |\tfrac{1}{2}\tfrac{1}{2}1, -1\rangle &= \beta_1\beta_2 \end{aligned}\right\} \qquad (16.85)$$

Usually we write the eigenvalue $\mathbf{S}^{2\prime}$ in the form

$$\mathbf{S}^{2\prime} = \hbar^2 S(S + 1)$$

and then the total spin quantum numbers $S = 0$ and 1 characterize the eigenstates of $\mathbf{S}^2$. The state (16.84) corresponds to $S = 0$ and is called a *singlet* state. The three states (16.85) correspond to $S = 1$ and are said to be the members of a *triplet*. The three states listed under (16.85) are, successively, eigenstates of S_z with eigenvalues $\hbar$, 0, and $-\hbar$.

Exercise 16.8. Work out the 4×4 matrix $\mathbf{S}^2$ in the direct product space with the basis (16.83) and show by explicit diagonalization that

$$\mathbf{S}^2(\alpha_1\beta_2 + \beta_1\alpha_2) = 2\hbar^2(\alpha_1\beta_2 + \beta_1\alpha_2)$$

and

$$\mathbf{S}^2(\alpha_1\beta_2 - \beta_1\alpha_2) = 0$$

Exercise 16.9. Show that

$$\langle j1j0 \,|\, j1jj \rangle = \sqrt{\frac{j}{j+1}} \qquad \text{and} \qquad \langle j2j0 \,|\, j2jj \rangle = \sqrt{\frac{j(2j-1)}{(j+1)(2j+3)}} \qquad (16.86)$$

7. The Clebsch-Gordan Series. The direct products of the matrices of two representations of a group constitute again a representation of the same group. The latter is usually reducible even if the original two representations are irreducible. The product representation can then be reduced, giving us generally new irreducible representations.

For the rotation group, equations (16.62) and (16.63) show that the problem of reducing the direct product representation $D^{(j_1)}(R) \times D^{(j_2)}(R)$ is intimately related to the problem of adding two angular momenta, $\mathbf{J} = \mathbf{J}_1 + \mathbf{J}_2$. Under rotations the state vectors in the $(2j_1 + 1)(2j_2 + 1)$ dimensional direct product space transforms according to $D^{(j_1)} \times D^{(j_2)}$, and the reduction consists of determining the invariant subspaces contained in this space. Since the irreducible representations are characterized by the eigenvalues of $\mathbf{J}^2$, the quantum number j, which can assume the values $j_1 + j_2$, $j_1 + j_2 - 1, \ldots, |j_1 - j_2|$, labels the several representations which the direct product representation contains. It follows from the last section that in the reduction each of these irreducible representations appears once and only once. Formally this fact is expressed by the equivalence

$$D^{(j_1)} \times D^{(j_2)} \rightarrow \begin{pmatrix} D^{(j_1+j_2)} & 0 & & & \\ 0 & D^{(j_1+j_2-1)} & & & \\ & & \cdot & & \\ & & & \cdot & \\ & & & & D^{|j_1-j_2|} \end{pmatrix} \qquad (16.87)$$

Exercise 16.10. Show by explicit counting that the matrices on both sides of (16.87) have the same dimensions.

The Clebsch-Gordan coefficients furnish the unitary transformation from the basis $|j_1j_2m_1m_2\rangle$, in which the matrices $D^{(j_1)}$ and $D^{(j_2)}$ represent rotations, to the basis $|j_1j_2jm\rangle$, in which $D^{(j)}$ represents rotations. Hence, noting also that the Clebsch-Gordan coefficients are real, we may write the equivalence (16.87) immediately and explicitly as

$$D^{(j_1)}_{m_1'm_1}(R)D^{(j_2)}_{m_2'm_2}(R)$$

$$= \sum_{j=|j_1-j_2|}^{j_1+j_2} \sum_{mm'} \langle j_1j_2m_1m_2 \,|\, j_1j_2jm\rangle\langle j_1j_2m_1'm_2' \,|\, j_1j_2jm'\rangle D^{(j)}_{m'm}(R) \qquad (16.88)$$

This expansion is called the *Clebsch-Gordan series*.

As a very useful application of the last equation, we use the identity (16.59) to obtain

$$
Y_{l_1}{}^{m_1}(\theta, \varphi) Y_{l_2}{}^{m_2}(\theta, \varphi) = \sum_l \sqrt{\frac{(2l_1 + 1)(2l_2 + 1)}{4\pi(2l + 1)}}
$$
$$
\times \langle l_1 l_2 00 \mid l_1 l_2 l 0 \rangle \langle l_1 l_2 m_1 m_2 \mid l_1 l_2 l, m_1 + m_2 \rangle Y_l^{m_1 + m_2}(\theta, \varphi) \quad (16.89)
$$

From this we find the value of the frequently used integral,

$$
\int Y_{l_3}^{m_3 *}(\theta, \varphi) Y_{l_2}{}^{m_2}(\theta, \varphi) Y_{l_1}{}^{m_1}(\theta, \varphi) \, d\Omega
$$
$$
= \sqrt{\frac{(2l_1 + 1)(2l_2 + 1)}{4\pi(2l_3 + 1)}} \langle l_1 l_2 00 \mid l_1 l_2 l_3 0 \rangle \langle l_1 l_2 m_1 m_2 \mid l_1 l_2 l_3 m_3 \rangle \quad (16.90)
$$

From (16.88) the following formula may be derived:

$$
\sum_{m_2'} \langle j_1 j_2 m_1' m_2' \mid j_1 j_2 j_3 m_3 \rangle D^{(j_2)}_{m_2' m_2}(R)
$$
$$
= \sum_{m_1 m} D^{(j_3)}_{m_3 m}(R) \langle j_1 j_2 m_1 m_2 \mid j_1 j_2 j_3 m \rangle D^{(j_1) *}_{m_1' m_1}(R) \quad (16.91)
$$

Exercise 16.11. Derive equation (16.91) by use of the unitarity condition for the *D* matrices and the orthogonality of the Clebsch-Gordan coefficients.

This identity will prove useful in the remainder of this chapter. It differs from (16.88) by being a linear homogeneous relation for the Clebsch-Gordan coefficients. For infinitesimal rotations this becomes identical with the recursion relations (16.69) for the Clebsch-Gordan coefficients.

8. Tensor Operators and the Wigner-Eckart Theorem. So far this chapter has been mainly concerned with the behavior under rotation of state vectors and wave functions. In the following, attention will be focused on the rotational transformation properties of the operators of quantum mechanics. The operators corresponding to various physical quantities will be characterized by their behavior under rotation as scalars, vectors, and tensors, and from a knowledge of this behavior alone much information will be inferred about the structure of the matrix elements. Such information is useful in many applications of quantum mechanics.

Let us suppose that a rotation R takes a state vector Ψ by a unitary transformation U_R into the state vector Ψ'':

$$
\Psi'' = U_R \Psi \quad (16.92)
$$

Three operators A_x, A_y, A_z are said to be the Cartesian components of a *vector operator* **A** if under every rotation the expectation values transform like the components of a vector. We thus require of a vector operator that for any Ψ the old and new expectation values of **A** be related by the equation

$$(\Psi', A_i\Psi') = \sum_{j=1}^{3} R_{ij} (\Psi, A_j\Psi) \qquad (i = 1, 2, 3) \qquad (16.93)$$

where R_{ij} represents the real orthogonal matrix which characterizes the rotation of the Cartesian coordinates x, y, z. The three components are conveniently characterized by subscripts 1, 2, 3.

For example, a rotation about the z-axis by an angle φ would, according to the rules of analytic geometry, be written as

$$R = \begin{pmatrix} \cos \varphi & -\sin \varphi & 0 \\ \sin \varphi & \cos \varphi & 0 \\ 0 & 0 & 1 \end{pmatrix} \qquad (16.94)$$

The matrices R constitute an irreducible representation of the rotation group; they are equivalent to $D^{(1)}$.

Substituting Ψ'' from (16.92) into (16.93) and noting that the resulting relation should hold for arbitrary states Ψ, we obtain the condition

$$U_R{}^{\dagger} A_i U_R = \sum_{j=1}^{3} R_{ij} A_j \qquad (16.95)$$

as the fundamental criterion for whether **A** is a vector operator or not.

A necessary (and, as we shall see, also sufficient) condition for a vector operator can be derived by applying (16.95) to the special case of an infinitesimal rotation. For such a rotation R is very close to the identity matrix, and U_R is very close to the identity operator. We thus write

$$R_{ij} = \delta_{ij} + \varepsilon_{ij} \qquad (16.96)$$

The orthogonality of this matrix implies to first order in ε_{ij},

$$\sum_{j} R_{ij} R_{kj} = \delta_{ik}$$

$$\sum_{j} (\delta_{ij} + \varepsilon_{ij})(\delta_{kj} + \varepsilon_{kj}) = \delta_{ik}$$

or

$$\varepsilon_{ik} + \varepsilon_{ki} = 0 \qquad (16.97)$$

i.e., the matrix ε_{ij} must be antisymmetric. Setting $\varepsilon_{12} = -\varepsilon_z$, $\varepsilon_{23} = -\varepsilon_x$, $\varepsilon_{31} = -\varepsilon_y$, we can write for the infinitesimal rotation:

$$R = \begin{pmatrix} 1 & -\varepsilon_z & \varepsilon_y \\ \varepsilon_z & 1 & -\varepsilon_x \\ -\varepsilon_y & \varepsilon_x & 1 \end{pmatrix} \qquad (16.98)$$

For the special case of an infinitesimal rotation about the z-axis this agrees with (16.94) if we set $\varphi = \varepsilon_z$. Generally, the rotation (16.98) takes place in a plane perpendicular to the vector $\boldsymbol{\epsilon}(\varepsilon_x, \varepsilon_y, \varepsilon_z)$; $|\boldsymbol{\epsilon}|$ is the angle of rotation, and the rotation is such that it forms a right-handed screw advancing in the direction of $\boldsymbol{\epsilon}$.

The unitary operator which corresponds to such a rotation is

$$U_R = 1 - \frac{i}{\hbar}\boldsymbol{\epsilon}\cdot\mathbf{J} \tag{16.99}$$

Expressions (16.96) and (16.99) are now substituted in (16.95) with the result

$$\left(1 + \frac{i}{\hbar}\boldsymbol{\epsilon}\cdot\mathbf{J}\right)A_i\left(1 - \frac{i}{\hbar}\boldsymbol{\epsilon}\cdot\mathbf{J}\right) = A_i + \varepsilon_{ij}A_j + \varepsilon_{ik}A_k = A_i - \varepsilon_k A_j + \varepsilon_j A_k$$

Here i, j, k represent the indices 1, 2, 3 in cyclic order. By comparing the two sides of this equation to first order in ε, we obtain the relations

$$A_x J_y - J_y A_x = i\hbar A_z, \qquad A_x J_z - J_z A_x = -i\hbar A_y$$
$$A_y J_z - J_z A_y = i\hbar A_x, \qquad A_y J_x - J_x A_y = -i\hbar A_z \tag{16.100}$$
$$A_z J_x - J_x A_z = i\hbar A_y, \qquad A_z J_y - J_y A_z = -i\hbar A_x$$

and

$$A_x J_x - J_x A_x = A_y J_y - J_y A_y = A_z J_z - J_z A_z = 0 \tag{16.101}$$

For infinitesimal rotations these commutation relations are equivalent to condition (16.95). They are also sufficient to ensure that condition (16.95) is satisfied for finite rotations. Indeed, we observe that, if (16.95) holds for two different rotations R and S,

$$U_R{}^\dagger A_i U_R = \sum_j R_{ij}A_j$$

and

$$U_S{}^\dagger A_i U_S = \sum_j S_{ij}A_j$$

it follows that for the compound rotation SR

$$U_{SR}{}^\dagger A_i U_{SR} = U_R{}^\dagger U_S{}^\dagger A_i U_S U_R = U_R{}^\dagger \sum_j S_{ij}A_j U_R$$

$$= \sum_j \sum_k S_{ij}R_{jk}A_k = \sum_k (SR)_{ik}A_k$$

substantiating again our earlier assertion that the matrices form a representation of the rotation group. Any finite rotation can be compounded of a large number of infinitesimal rotations. Hence, if (16.95) holds for all infinitesimal rotations, as is guaranteed by the commutation relations (16.100) and (16.101), then (16.95) will also hold for any finite rotation.

In all this it must not be forgotten that the operators J_x, J_y, J_z themselves are not arbitrary, but must satisfy the commutation relations (16.8). Comparing these relations with (16.100), it is evident that **J** itself is a vector operator. If **J** is to represent the angular momentum of the system, it is, of course, necessary that it be a vector operator.

Whether or not a given operator **A** constitutes a vector operator depends entirely on the definition of the physical system and the structure of its angular momentum operator **J**. An operator **A** can never be said to be a vector operator per se; it can only be a vector operator with respect to some particular dynamical system.

As an example consider the case where the coordinate operators x, y, z provide a complete description of the dynamical system, such as a particle without spin. In this case we can identify $\mathbf{J} = \mathbf{L} = \mathbf{r} \times \mathbf{p}$. The quantities **r**, **p**, **L** are all vector operators, as can be verified easily from (16.100) and (16.101) if the fundamental commutation relations between **r** and **p** are used. On the other hand, an external electric field **E** acting on the system does not in general make up a *vector operator* with respect to this system—even though **E** is, of course, a vector. Condition (16.95) is not satisfied by such a field, because **E** is external to the system and not subject to rotation together with it. But **E** would become a proper vector operator if the system were enlarged so as to include the sources of the electric field in the dynamical description. This would result in a much more complicated operator **J**, and the commutation relations (16.100) and (16.101) would then be satisfied by **E**.

If the dynamical system is a particle with spin, **S** becomes a vector operator provided that **J** is taken to be $\mathbf{L} + \mathbf{S}$, i.e., the spin wave function must be rotated together with the space wave function.

The generalization from vector to tensor operator is best made after the defining relation (16.95) for a vector operator has been rewritten slightly. Multiplying (16.95) from the left by U_R and from the right by $U_R^\dagger$, and using the orthogonality relation for the matrix R_{ij}, we obtain the condition

$$U_R A_k U_R^\dagger = \sum_{i=1}^{3} A_i R_{ik} \tag{16.102}$$

This is entirely equivalent to (16.95). It is convenient to regard the operator $U_R A_k U_R^\dagger$ as a new operator A_k', a *rotational transform* of A_k. A_k' is an operator which has the same expectation value with respect to the rotated state Ψ' as the old operator A_k had with respect to the unrotated state Ψ, for indeed by (16.92),

$$(\Psi', A_k'\Psi') = (\Psi, A_k\Psi)$$

A vector operator is thus a set of three operators A_x, A_y, A_z, whose rotational transforms A_x', A_y', A_z' are certain linear functions of A_x, A_y, A_z. This is the definition which can be generalized most readily.

A *tensor operator* is a set of n operators $T_1, T_2, \ldots, T_n$, such that their rotational transforms are linear functions of the n operators:

$$T_{k'} = U_R T_k U_R^\dagger = \sum_i T_i D_{ik}(R) \tag{16.103}$$

The coefficients $D_{ik}(R)$ depend on the rotation and are obviously representations of the rotation group. Every tensor induces a representation which, generally, is reducible. (In particular, the usual Cartesian tensors induce reducible representations if the rank of the tensors exceeds 1.)

From our discussion of group theory we expect that those tensors will be particularly simple which transform according to *irreducible* representations, i.e., such that the coefficients $D_{ik}(R)$ in (16.103) are matrix elements of an irreducible representation. Thus emerges the useful concept of the *irreducible spherical tensor operator* of rank k. This is a set of $2k + 1$ operators T_k^q, which satisfy the transformation equation

$$U_R T_k^q U_R^\dagger = \sum_{q'=-k}^{k} T_k^{q'} D_{q'q}^{(k)}(R) \tag{16.104}$$

Exercise 16.12. Show that an irreducible tensor operator of rank 1 is related to a vector operator by the equations

$$T_1^1 = -\frac{A_x + iA_y}{\sqrt{2}}, \quad T_1^0 = A_z, \quad T_1^{-1} = \frac{A_x - iA_y}{\sqrt{2}} \tag{16.105}$$

Armed with this definition we can now formulate and prove the *Wigner-Eckart theorem*, which answers the following question: If T_k^q ($q = -k, \ldots, k$) is a tensor operator, how much information about its matrix elements between simultaneous eigenstates of J_z and $\mathbf{J}^2$ can be inferred?

To obtain the answer, let us take the matrix element of the operator equation (16.104) between the states $\langle \alpha' j'm' |$ and $|\alpha jm\rangle$. Here j, j', m, and m' are angular momentum quantum numbers, as usual, and α and α' symbolize the totality of all other quantum numbers needed to specify the eigenstates of the system completely. We obtain

$$\langle \alpha' j'm' | U_R T_k^q U_R^\dagger |\alpha jm\rangle = \sum_{q'=-k}^{k} \langle \alpha' j'm' | T_k^{q'} |\alpha jm\rangle D_{q'q}^{(k)}(R)$$

Now we use the relations

$$U_R^\dagger |\alpha jm\rangle = \sum_\mu |\alpha j\mu\rangle D_{m\mu}^{(j)*}(R)$$

and

$$\langle \alpha' j'm' | U_R = \sum_{\mu'} D_{m'\mu'}^{(j')}(R) \langle \alpha' j'\mu' |$$

which follow directly from (16.7) and (16.50). Hence,

$$\sum_{\mu\mu'} D^{(j')}_{m'\mu'}(R)\langle\alpha'j'\mu'|\,T_k^q\,|\alpha j\mu\rangle D^{(j)*}_{m\mu}(R) = \sum_{q'} \langle\alpha'j'm'|\,T_k^{q'}\,|\alpha jm\rangle D^{(k)}_{q'q}(R) \quad (16.106)$$

A glance at this equation and (16.91) shows that the two have exactly the same structure. If the D are known, (16.91) determines the Clebsch-Gordan coefficients for given values of j_1, j_2, j_3, except for a common factor. Hence, (16.106) must similarly determine the matrix elements of the tensor operator. If we identify $j = j_1$, $k = j_2$, $j' = j_3$, we can conclude that the two solutions of these linear homogeneous recursion relations must be proportional:

$$\langle\alpha'j'm'|\,T_k^q\,|\alpha jm\rangle = \langle jkmq|jkj'm'\rangle\langle\alpha'j'\|\,T_k\,\|\alpha j\rangle \quad (16.107)$$

This important formula embodies the *Wigner-Eckart theorem.*

The constant of proportionality $\langle\alpha'j'\|\,T_k\,\|\alpha j\rangle$ is called the *reduced matrix element* of the irreducible spherical tensor operator T_k. It depends only on the nature of the tensor, on the total angular momentum quantum numbers j and j', and on the quantum numbers α, α', but not on the quantum numbers m, m', and q which specify the orientation of the system in space.

The Wigner-Eckart theorem provides us with a fundamental insight, because it separates the purely geometrical properties of the matrix element from the physical properties which are contained in the reduced matrix element. The theorem has also great practical value, since the Clebsch-Gordan coefficients have been tabulated.[11]

From the fundamental properties (16.68) and (16.74) of the Clebsch-Gordan coefficients, we infer the angular momentum *selection rules* for irreducible spherical tensor operators. The matrix element $\langle\alpha'j'm'|\,T_k^q\,|\alpha jm\rangle$ vanishes unless

$$q = m' - m \quad (16.108)$$

and

$$|j - j'| \leqslant k \leqslant j + j' \quad (16.109)$$

In particular, it follows that a scalar operator ($k = 0$) has nonvanishing matrix elements only if $m = m'$ and $j = j'$. The selection rules for a vector operator are $\Delta m \equiv m' - m = 0$, ± 1 and $\Delta j \equiv j' - j = 0$, ± 1, with $j = j' = 0$ excluded.

Although (16.104) defines a tensor operator, it is often not a simple matter to apply it when we wish to test a given set of $2k + 1$ operators T_k^q for its rotational behavior, because in general U_R is a complicated operator. In fact, it is out of the question to apply (16.104) to all possible rotations R, but, fortunately, it is not necessary to do this. Rather, it is entirely sufficient to check condition (16.104) for infinitesimal rotations, whose irreducible representations are given by (16.54). If this expression is substituted in (16.104) and if only terms of the first order are retained, (16.104) is found to be equivalent to the commutation relations

$$[J_z, T_k^q] = q\hbar T_k^q \tag{16.110a}$$

$$[J_+, T_k^q] = \sqrt{(k-q)(k+q+1)}\hbar T_k^{q+1} \tag{16.110b}$$

$$[J_-, T_k^q] = \sqrt{(k+q)(k-q+1)}\hbar T_k^{q-1} \tag{16.110c}$$

These relations serve to test whether a set of $2k+1$ operators T_k^q constitutes an irreducible spherical tensor operator with respect to the system whose rotations are generated by the angular momentum operator $\mathbf{J}$.

Exercise 16.13. Show that the trace of any irreducible spherical tensor operator vanishes, except those of rank 0 (scalar operators).

The static electric moments of a charge distribution, such as an atom or a nucleus, are examples of tensor operators. They are best defined in terms of the first-order perturbation energy of the particles in an external electric field $\mathbf{E} = -\nabla\phi$. A particle of charge q contributes

$$E = q\int \psi^*(\mathbf{r})\phi(\mathbf{r})\psi(\mathbf{r})d^3r$$

to the interaction energy. If the sources of the electric field are at large distances, ϕ satisfies Laplace's equation, and may be written in the form

$$\phi = \sum_{l=0}^{\infty}\sum_{m=-l}^{+l} A_l^m r^l Y_l^m(\theta, \varphi)$$

the origin being chosen at the center of mass of the atom or nucleus. The energy can therefore be expressed as the *electric multipole expansion*,

$$E = \int \rho(\mathbf{r})\phi(\mathbf{r})\,d^3r = \sum_{k,m}A_k^m\int \rho r^k Y_k^m\, d^3r$$

At each order of k, the field is characterized by $2k+1$ constants A_k^m, and, there is correspondingly a spherical tensor of rank k, $r^k Y_k^m$, the *electric 2^k-pole moment*. The components of these tensors pass the test (16.110) for irreducible spherical tensor operators. Their matrix elements may be partially evaluated by use of the Wigner-Eckart theorem.

Exercise 16.14. Prove that the static 2^k-pole moment of a charge distribution has zero expectation value in any state with angular momentum $j < (k/2)$. Verify this property explicitly for the quadrupole moment by use of (16.86) and the Wigner-Eckart theorem.

9. The Matrix Elements of Vector Operators. As an application of the Wigner-Eckart theorem it is useful to calculate the reduced matrix element of the special vector operator **J**. The particular linear combinations

$$J^1 = -\frac{1}{\sqrt{2}}(J_x + iJ_y), \qquad J^0 = J_z, \qquad J^{-1} = \frac{1}{\sqrt{2}}(J_x - iJ_y) \quad (16.111)$$

make up the components of an irreducible spherical tensor operator of rank 1 (Exercise 16.12).

By the Wigner-Eckart theorem,

$$\langle \alpha' jj | J_z | \alpha jj \rangle = \langle j1j0 | j1jj \rangle \langle \alpha' j \| \mathbf{J} \| \alpha j \rangle \quad (16.112)$$

The left-hand side of this equation equals $\hbar j \delta_{\alpha\alpha'}$, and the Clebsch-Gordan coefficient is given by (16.86). Hence,

$$\langle \alpha' j' \| \mathbf{J} \| \alpha j \rangle = \sqrt{j(j+1)}\hbar\, \delta_{jj'}\, \delta_{\alpha\alpha'} \quad (16.113)$$

The last equation can be used to derive a simple relation based on the fact that all vector operators have the same structure.

From the Wigner-Eckart theorem we have for any vector operator **A**

$$\langle \alpha' j' m' | A^q | \alpha jm \rangle = \frac{\langle \alpha' j' \| \mathbf{A} \| \alpha j \rangle}{\langle \alpha' j' \| \mathbf{J} \| \alpha j \rangle} \langle \alpha' j' m' | J^q | \alpha jm \rangle \quad (16.114)$$

For $j' = j$ this may be further developed by noting that by virtue of the simple properties of **J** we must have for the scalar operator $\mathbf{J} \cdot \mathbf{A}$:

$$\langle \alpha' jm' | \mathbf{J} \cdot \mathbf{A} | \alpha jm \rangle = c \langle \alpha' j \| \mathbf{A} \| \alpha j \rangle$$

c must be independent of the nature of **A** and of α and α'; hence, it can be evaluated by setting $\mathbf{A} = \mathbf{J}$. Then we have, using (16.113),

$$c = \sqrt{j(j+1)}\hbar\, \delta_{mm'},$$

hence,

$$\langle \alpha' j \| \mathbf{A} \| \alpha j \rangle = \frac{1}{\hbar\sqrt{j(j+1)}} \langle \alpha' jm | \mathbf{J} \cdot \mathbf{A} | \alpha jm \rangle$$

Substituting this into (16.114), we obtain the important result

$$\langle \alpha' jm' | A^q | \alpha jm \rangle = \frac{\langle \alpha' jm | \mathbf{J} \cdot \mathbf{A} | \alpha jm \rangle}{\hbar^2 j(j+1)} \langle jm' | J^q | jm \rangle \quad (16.115)$$

by which the j-diagonal matrix elements of a vector operator **A** are expressed in terms of the matrix element of the scalar operator $\mathbf{J} \cdot \mathbf{A}$ and of other known quantities. This formula contains the theoretical justification for the *vector model* of angular momenta and can be used as the starting point of the derivation of the matrix elements of magnetic moment operators which are important in spectroscopy.

Contributions to the magnetic moment of an atom (or nucleus) arise from the orbital motion of the charged particles and the intrinsic spins in the system. Generally, the magnetic moment operator of an atom may be assumed to have the structure

$$\mathbf{m} = -\frac{e}{2\mu c}(g_L\mathbf{L} + g_S\mathbf{S}) \tag{16.116}$$

Since both $\mathbf{L}$ and $\mathbf{S}$ are vector operators with respect to the total angular momentum $\mathbf{J} = \mathbf{L} + \mathbf{S}$, $\mathbf{m}$ is also a vector operator.

According to the Wigner-Eckart theorem, all matrix elements of $\mathbf{m}$ are proportional to each other. We therefore speak of *the magnetic moment* of the atom and, in so doing, have reference to the expectation value

$$m = \langle \alpha jj| \, m_z \, |\alpha jj\rangle = \langle j1j0 \, | \, j1jj\rangle\langle \alpha j\| \, \mathbf{m} \, \|\alpha j\rangle = \sqrt{\frac{j}{j+1}}\,\langle \alpha j\| \, \mathbf{m} \, \|\alpha j\rangle \tag{16.117}$$

From this we observe that in the classical limit ($j \to \infty$) the reduced matrix element becomes identical with the magnetic moment m.

The application of (16.115) to the magnetic moment vector operator (16.116) gives immediately

$$m = \langle \alpha jj|m_z|\alpha jj\rangle = \frac{-e}{2\mu\hbar c(j+1)}\langle \alpha jj| \, g_L\mathbf{L}\cdot\mathbf{J} + g_S\mathbf{S}\cdot\mathbf{J} \, |\alpha jj\rangle \tag{16.118}$$

But

$$\mathbf{L}\cdot\mathbf{J} = \tfrac{1}{2}(\mathbf{J}^2 + \mathbf{L}^2 - \mathbf{S}^2)$$

and

$$\mathbf{S}\cdot\mathbf{J} = \tfrac{1}{2}(\mathbf{J}^2 + \mathbf{S}^2 - \mathbf{L}^2)$$

Hence,

$$m = \frac{-e}{4\mu\hbar c(j+1)}\langle \alpha jj| \, (g_L + g_S)\mathbf{J}^2 + (g_L - g_S)(\mathbf{L}^2 - \mathbf{S}^2) \, |\alpha jj\rangle \tag{16.119}$$

If *L-S* coupling describes the ground state of the atom, $|\alpha jj\rangle$ is an approximate eigenstate of the total orbital and the total spin angular momenta with quantum numbers l and s. Hence, in this approximation, with $g_L = 1$ and $g_S = 2$,

$$m = -\frac{e\hbar j}{2\mu c}\left[1 + \frac{j(j+1) - l(l+1) + s(s+1)}{2j(j+1)}\right] \tag{16.120}$$

The expression in the bracket represents the general form for *L-S* coupling of the *Landé g-factor*.

10. Reflection Symmetry and Parity. The Euclidean principle of relativity may be supplemented by the further assumption that space has no intrinsic *chirality*, or handedness, by which is meant that processes take place in the same way in a physical system and its mirror image, obtained from one another by *reflection* with respect to a plane. We may call this assumption the *extended Euclidean principle of relativity*. By definition, a reflection with respect to the yz-plane changes x into $-x$, and leaves y and z unchanged; similarly, p_x goes over into $-p_x$, and p_y and p_z remain unchanged.

In any test of the extended Euclidean principle of relativity an important difference between rotational symmetry and reflection symmetry must be remembered. In rotating a system from one orientation to another we can proceed gradually and take it through a continuous sequence of rigid displacements, all of which are equivalent. Not so in the case of reflections, since it is impossible to transform a system into its mirror image without distorting it into some intermediate configurations which are physically very different from the original system. It is therefore not at all obvious how some quantities, seemingly unrelated to coordinate displacements, such as the electric charge of a particle, should be treated in a reflection if symmetry is to be preserved. Only experience can tell us whether it is possible to create a suitable mirror image of a complex physical system in accordance with the assumption of reflection symmetry.

In developing the mathematical formalism it is convenient to consider *inversions* through a fixed origin ($\mathbf{r} \to -\mathbf{r}$) instead of plane reflections. This is no limitation, since any reflection can be obtained by an inversion followed by a rotation. Conversely, an inversion is the same as three successive reflections with respect to perpendicular planes. Corresponding to an inversion there is an operator U_P with the following properties:

$$U_P^\dagger \mathbf{r} U_P = -\mathbf{r}, \qquad U_P^\dagger \mathbf{p} U_P = -\mathbf{p} \tag{16.121}$$

If the system is completely described by spatial coordinates, these requirements are met by the unitary operator U_P, defined by

$$U_P \, |\mathbf{r}'\rangle = |-\mathbf{r}'\rangle \tag{16.122}$$

If other dynamical variables, such as the spin, play a role, their transformation properties must also be suitably defined (see, e.g., Section 12.8.)

Since two successive inversions are the same as the identity operation, we may require that[12]

$$U_P^2 = 1 \tag{16.123}$$

---

[12] An arbitrary phase factor could be introduced in (16.122) with a consequent change of (16.123), but in ordinary quantum mechanics there is no need for this. The situation is different in the quantum theory of interacting fields which describe the creation and destruction of different particles. When the number of particles changes, the relative intrinsic parity of various particle species can be observed. See G. C. Wick, *Invariance Principles of Nuclear Physics in Annual Review of Nuclear Science*, Vol. 8, p. 1, Palo Alto, 1958. For further discussion of reflection and other discrete symmetries, see Chapter 23.

and this is in agreement with (16.122). The eigenvalues of U_P are ± 1 and are said to define the *parity* of the eigenstate. If the system is specified by $\mathbf{r}$, the eigenstates are the states with even and odd wave functions, since

$$U_P \int |\mathbf{r}'\rangle \, d\tau' \{\langle \mathbf{r}'| \rangle \pm \langle -\mathbf{r}'| \rangle\} = \pm \int |\mathbf{r}'\rangle \, d\tau' \{\langle \mathbf{r}'| \rangle \pm \langle -\mathbf{r}'| \rangle\}$$

Exercise 16.15. Show that $\exp(i\pi H/\hbar\omega)$ is a reflection operator, if H is the Hamiltonian of a linear harmonic oscillator with frequency ω.

A final implication of the extended Euclidean principle of relativity is that rotations and inversions commute. If this is assumed, we have

$$[U_P, \mathbf{J}] = 0 \tag{16.124}$$

and it follows that all the substates $|jm\rangle$ of a given angular momentum j have the same parity, because they are obtained by successive application of J_+.

If the Hamiltonian of a system is invariant under inversion, the *parity operator* U_P is a constant of the motion, and all eigenvectors of H may be assumed to have definite parity. Parity is conserved for a particle in a central field even in the presence of a uniform external magnetic field (Zeeman effect), but the presence of a uniform electric field causes states of opposite parities to be mixed (see also Chapter 17).

There are selection rules for the matrix elements of operators which transform simply under the parity operation. An *even* operator is characterized by the property

$$U_P A U_P^\dagger = A \tag{16.125}$$

and has nonvanishing matrix elements between states of definite parity only if the two states have the same parity. Similarly,

$$U_P B U_P^\dagger = -B \tag{16.126}$$

characterizes an *odd* operator, which has nonvanishing matrix elements between states of definite parity only if the two states have opposite parity. For instance, if $-e\mathbf{r}$ is the electric dipole moment operator, the expectation value of this operator is zero in any state of definite parity. More generally, an atom or nucleus in a state of definite parity has no electric 2^k-pole moment corresponding to odd values of k. If, as is known from the properties of the weak interactions, conservation of parity is only an approximate symmetry, small violations of this selection rule may be expected.

11. Time Reversal Symmetry. A system is said to exhibit symmetry under time reversal if, at least in principle, its time development may be reversed and all physical processes run backwards, with initial and final states interchanged. Symmetry between the two directions of motion in time implies that to every state Ψ there corresponds a time-reversed state $\Theta\Psi$ and that the transformation Θ preserves the values of all probabilities, thus leaving invariant the absolute value of any scalar product between two states.

From Section 16.1 we know that Θ may be assumed to be either a unitary or an antiunitary transformation. The physical significance of Θ as the time reversal operator requires that, while spatial relations must remain unchanged, all velocities must be reversed. Hence, we postulate the conditions

$$\Theta\mathbf{r}\Theta^{-1} = \mathbf{r} \tag{16.127}$$

$$\Theta\mathbf{p}\Theta^{-1} = -\mathbf{p} \tag{16.128}$$

$$\Theta\mathbf{J}\Theta^{-1} = -\mathbf{J} \tag{16.129}$$

If the time development of the system is given by

$$\Psi(t) = e^{-(i/\hbar)Ht}\Psi(0)$$

time reversal symmetry demands that the time-reversed initial state $\Theta\Psi(0)$ evolves into

$$\Theta\Psi(-t) = e^{-(i/\hbar)Ht}\Theta\Psi(0)$$

From the last two equations we obtain the condition

$$e^{-(i/\hbar)Ht}\Theta = \Theta e^{(i/\hbar)Ht} \tag{16.130}$$

if the theory is to be invariant under time reversal.

If Θ were unitary, condition (16.130) would be equivalent to

$$\Theta H + H\Theta = 0$$

If such an operator Θ existed, every stationary state Ψ_E of the system with energy E would be accompanied by another stationary state $\Theta\Psi_E$ with energy $-E$. This change of sign of the energy is in conflict with our classical notions about the behavior of the energy if all velocities are reversed, and it is inconsistent with the existence of a lower bound to the energy. Hence, Θ cannot be unitary.

If Θ is assumed to be *antiunitary* so that

$$\Theta\lambda\Psi = \lambda^*\Theta\Psi \tag{16.131}$$

$$(\Theta\Psi_a, \Theta\Psi_b) = (\Psi_b, \Psi_a) \tag{16.132}$$

for any two states, invariance under time reversal requires that

$$\Theta H - H\Theta = 0 \tag{16.133}$$

Although the operator Θ commutes with the Hamiltonian, it is not a constant of the motion because (15.10) holds only for linear operators.

Obviously a double reversal of time, corresponding to the application of Θ^2 to all states, has no physical consequences. If, as is the case in ordinary quantum mechanics, a one-to-one correspondence may be set up between physical states and state vectors, with only a phase factor remaining arbitrary, Θ^2 must satisfy

$$\Theta^2 \Psi = c \Psi$$

with the same constant c for all Ψ. From this condition the following chain of equalities flows as a consequence of the antiunitarity of Θ:

$$(\Theta \Psi_a, \Psi_b) = (\Theta \Psi_b, \Theta^2 \Psi_a)$$
$$= c(\Theta \Psi_b, \Psi_a) = c(\Theta \Psi_a, \Theta^2 \Psi_b) = c^2 (\Theta \Psi_a, \Psi_b) \quad (16.134)$$

Hence, either $c = 1$ or $c = -1$, depending on the nature of the system. If we set $\Psi_b = \Psi_a$, we also see from (16.134) that in case $c = -1$,

$$(\Theta \Psi_a, \Psi_a) = 0 \quad (16.135)$$

As a corollary of this result, we note that, if $c = -1$ and the Hamiltonian is invariant under time reversal, the energy eigenstates may be classified in degenerate time-reversed pairs. This property is known as *Kramers degeneracy*.

From the time reversal behavior of angular momentum, as defined in (16.129), and the equations of Section 16.2, it is easy to infer that the simultaneous eigenvectors of J_z and $\mathbf{J}^2$ must transform under time reversal according to the relation

$$\Theta | \alpha j m \rangle = e^{i\delta} (-1)^m | \alpha j, -m \rangle \quad (16.136)$$

where the real phase constant δ may depend on j and α but not on m. From (16.136) and the antilinear property of Θ, it follows by repeated application of Θ that

$$\Theta^2 | \alpha j m \rangle = (-1)^{2j} | \alpha j m \rangle \quad (16.137)$$

Hence, $\Theta^2 = I$ if j is integral, and $\Theta^2 = -I$ if j is half-integral. Kramers degeneracy implies that in atoms with an odd number of electrons the energy levels are doubly degenerate even in the presence of a static electric field, which does not destroy time reversal symmetry. A magnetic field violates time reversal symmetry and thus splits the degeneracy.

Exercise 16.16. Prove equations (16.136) and (16.137).

Although time reversal invariance does not lead to any conservation law, selection rules may be inferred just as from rotation and reflection invariance because the important tensor operators usually have simple transformation properties under time reversal. An irreducible spherical tensor operator T_k^q is said to be even or odd with respect to Θ if it satisfies the condition

$$\Theta T_k^q \Theta^{-1} = \pm (-1)^q T_k^{-q} \quad (16.138)$$

The + sign refers to tensors that are even under time reversal, and the − sign to tensors that are odd.

For matrix elements between states of sharp angular momentum, we can readily derive the equation

$$\langle \alpha' j'm'| T_k^q |\alpha jm\rangle = \pm e^{i(\delta - \delta')}\langle \alpha' j', -m'| T_k^{-q} |\alpha j, -m\rangle^* \quad (16.139)$$

If we confine our attention to matrix elements that are diagonal in α and j, it follows from (16.139), the Wigner-Eckart theorem, and (16.78) that

$$\langle \alpha j\| T_k \|\alpha j\rangle = \pm(-1)^k \langle \alpha j\| T_k \|\alpha j\rangle^* \quad (16.140)$$

For example, the electric dipole moment is an odd rank ($k = 1$) tensor with a real reduced matrix element, and it is even under time reversal. Hence, its expectation value vanishes as a consequence of time reversal symmetry. We may therefore conclude that the observation of a static electric dipole moment in a stationary state of definite angular momentum can be expected only if both space reflection and time reversal symmetries are violated by the dynamical interactions.

Exercise 16.17. Derive the results (16.139) and (16.140).

Problems

1. Show that the tensor operators $S_k^q = (-1)^q T_k^{-q\dagger}$ and T_k^q transform in the same way under rotations. Prove that

$$\langle \alpha' j\| S_k \|\alpha j\rangle = \langle \alpha j\| T_k \|\alpha' j\rangle^*$$

and

$$\langle jkmq \mid jkjm'\rangle = (-1)^q \langle jkm', -q \mid jkjm\rangle$$

2. If S_k^q and T_k^q are two irreducible spherical tensor operators of rank k, prove that

$$\sum_{q=-k}^{k} (-1)^q S_k^q T_k^{-q}$$

is a scalar operator.

3. The magnetic moment operator for a nucleon is $\mathbf{m} = e(g_l \mathbf{L} + g_s \mathbf{S})/2\mu_{nc}$, where $g_l = 1$ and $g_s = 5.587$ for a proton, $g_l = 0$ and $g_s = -3.826$ for a neutron. In a central field with an additional spin-orbit interaction the nucleons move in shells characterized by the quantum numbers l and $j = l \pm \frac{1}{2}$. Calculate the magnetic moment of a single nucleon as a function of j for the two kinds of nucleons, distinguishing the two cases $j = l + \frac{1}{2}$ and $j = l - \frac{1}{2}$. Plot j times the effective gyromagnetic ratio versus j, connecting in each case the points by straight line segments (*Schmidt lines*).

4. Show that, for a particle without spin in the coordinate representation, the time reversal operation may be represented as a complex conjugation of all wave functions. How is time reversal represented in the momentum representation?

Show that for a particle with spin $\frac{1}{2}$ in the usual basis α and β, which are the eigenspinors of σ_z, time reversal may be represented by $\sigma_y K$, where K stands for complex conjugation.

5. Show that the single-particle orbital angular momentum eigenfunctions with quantum numbers l, m in the momentum representation are spherical harmonics. Show that the choice of phase implied by $\langle \mathbf{p} \mid lm \rangle = i^l Y_l^m(\hat{\mathbf{p}})$ leads to correct and simple time reversal transformation properties for the angular momentum eigenfunctions in momentum space. Compare with the time reversal transformation properties of the orbital angular momentum eigenfunctions in the coordinate representation.

6. Show that, for a particle of spin $\frac{1}{2}$ and total angular momentum j, the eigenstates $\mathscr{Y}_l^{jm}$ defined in (16.80) transform under time reversal into $\pm(-1)^m \mathscr{Y}_l^{j,-m}$, the sign depending on whether $l = j - \frac{1}{2}$ or $l = j + \frac{1}{2}$.

7. The state of a spin $\frac{1}{2}$ particle with sharp total angular momentum j, m, is, in the notation of (16.80),

$$a \mathscr{Y}_{j-1/2}^{jm} + b \mathscr{Y}_{j+1/2}^{jm}$$

Assume this state to be an eigenstate of the Hamiltonian with no degeneracy other than that demanded by rotation invariance.

If H conserves parity, how are the coefficients a and b restricted? If H is invariant under time reversal, show that a/b must be imaginary. Verify explicitly that the expectation value of the electric dipole moment $-e\mathbf{r}$ vanishes if either parity is conserved or time reversal invariance holds (or both).

8. A particle (lambda hyperon) with spin $\frac{1}{2}$ decays at rest into two particles with spin $\frac{1}{2}$ (nucleon) and spin zero (pion). Show that in the representation in which the relative momentum of the decay products is diagonal the final state wave functions corresponding to $m = \pm\frac{1}{2}$ may be written in the form

$$\langle \mathbf{p} \mid \tfrac{1}{2}\tfrac{1}{2} \rangle = A_S \alpha + A_P(\cos \theta \, \alpha + e^{i\varphi} \sin \theta \, \beta)$$

$$\langle \mathbf{p} \mid \tfrac{1}{2}, -\tfrac{1}{2} \rangle = A_S \beta - A_P(\cos \theta \, \beta - e^{-i\varphi} \sin \theta \, \alpha)$$

where θ is the angle between the polarization vector of the decaying particle and the momentum vector of the spinless particle. (Neglect any interactions between the decay products in the final state.)

Show that the angular distribution of the spinless particles is of the form $1 + \lambda \cos \theta$, and evaluate λ in terms of A_S and A_P.

Prove that the polarization of the spin $\frac{1}{2}$ decay products can be written in the form

$$\langle \boldsymbol{\sigma} \rangle = \frac{1}{1 - \lambda \hat{\mathbf{p}} \cdot \mathbf{P}} [-(\lambda - \hat{\mathbf{p}} \cdot \mathbf{P})\hat{\mathbf{p}} + \mu \hat{\mathbf{p}} \times \mathbf{P} + \nu(\hat{\mathbf{p}} \times \mathbf{P}) \times \hat{\mathbf{p}}]$$

where $\hat{\mathbf{p}}$ is the unit vector in the direction of emission of the spin $\frac{1}{2}$ particle, and $\mathbf{P}$ denotes the initial polarization of the decaying particles. Determine μ and ν in terms of A_S and A_P, and show that $\lambda^2 + \mu^2 + \nu^2 = 1$.

Discuss the simplifications that occur in the expressions for the angular distribution and final state polarization if conservation of parity or invariance under time reversal is assumed for the decay-inducing interaction.

9. The Hamiltonian of the positronium atom in the $1S$ state in a magnetic field B along the z axis is to good approximation

$$H = A\mathbf{S}_1 \cdot \mathbf{S}_2 + \frac{eB}{\mu c}(S_{1_z} - S_{2_z})$$

if all higher energy states are neglected. The electron is labeled as particle 1 and the positron as particle 2. Using the *coupled representation* in which $\mathbf{S}^2 = (\mathbf{S}_1 + \mathbf{S}_2)^2$ and $S_z = S_{1_z} + S_{2_z}$ are diagonal, obtain the energy eigenvalues and eigenvectors and classify them according to the quantum numbers associated with constants of the motion.

   Empirically it is known that for $B = 0$ the frequency of the $1^3S \rightarrow 1^1S$ transition is 2.0338×10^5 Mc/sec and that the mean lifetimes against annihilation are 10^{-10} sec (two-photon decay) for the singlet and 10^{-7} sec (three-photon decay) for the triplet state. Estimate the magnetic field strength B which will cause the lifetime of the longer lived $m = 0$ state to be reduced to 10^{-8} sec.

10. An alternative to the usual representation for states of a particle with spin $\frac{1}{2}$, in which the simultaneous eigenstates of $\mathbf{r}$ and σ_z are used as a basis, is to employ a basis spanned by the simultaneous eigenstates of $\mathbf{r}$ and $\kappa = \frac{1}{2}\boldsymbol{\sigma} \cdot \hat{\mathbf{r}}$. Show that the operators $\mathbf{S}^2$, $\mathbf{J}^2$, J_z, κ commute and that their eigenfunctions may be represented as

$$\langle \hat{\mathbf{r}}\kappa' \mid jm\kappa'' \rangle \propto D_{m\kappa''}^{(j)*}(\varphi, \theta)\, \delta_{\kappa'\kappa''}$$

where φ, θ denotes the rotation which turns the z axis into the direction of $\mathbf{r}$. Can this representation be generalized to particles of higher spin? Can an analogous basis be constructed in the momentum representation?[13]

11. Let H be the Hamiltonian for the hydrogen atom

$$H = \frac{\mathbf{p}^2}{2\mu} - \frac{e^2}{r}$$

Prove that the *Runge-Lenz* vector

$$\mathbf{K} = \frac{1}{2\mu e^2}[\mathbf{L} \times \mathbf{p} - \mathbf{p} \times \mathbf{L}] + \frac{\mathbf{r}}{r}$$

is a vector operator which commutes with H and which has the properties $\mathbf{L} \cdot \mathbf{K} = \mathbf{K} \cdot \mathbf{L} = 0$ and

$$[K_x, K_y] = i\hbar\left(-\frac{2H}{\mu e^4}\right)L_z \qquad \text{et cycl.}$$

Show that the vector operator $\mathbf{A} = \sqrt{-\mu e^4/2E}\,\mathbf{K}$ defined in the subspace of hydrogen bound states with energy E ($\leqslant 0$) satisfies the commutation relations

$$[A_x, A_y] = i\hbar L_z \qquad \text{et cycl.}$$

---

[13] The *helicity operator* $\boldsymbol{\sigma} \cdot \mathbf{p}$ will be encountered in Chapter 23. See also Section 35.3 in K. Gottfried, *Quantum Mechanics*, W. A Benjamin, New York 1966. The usefulness of the helicity representation was emphasized by M. Jacob and G. C. Wick, *Ann. Physics*, **7**, 404 (1959).

and infer that the operators

$$\mathbf{J} = \tfrac{1}{2}(\mathbf{L} + \mathbf{A}) \qquad \text{and} \qquad \mathbf{J}' = \tfrac{1}{2}(\mathbf{L} - \mathbf{A})$$

obey the angular momentum commutation relations and the condition $\mathbf{J}^2 = \mathbf{J}'^2$. Derive the identity

$$\mathbf{J}^2 + \mathbf{J}'^2 = -\tfrac{1}{2}\hbar^2 - \frac{1}{2}\frac{\mu e^4}{2E}$$

and deduce from it the Balmer-Bohr formula for the energy levels of the hydrogen atom.

12. Use the fact that the time reversal operation commutes with rotations to prove that $D^{(j)*}_{m'm}(R) = (-1)^{m-m'} D^{(j)}_{-m',-m}(R)$.

CHAPTER 17

Bound State Perturbation Theory

1. The Perturbation Method. In this chapter we shall be concerned with *discrete* eigensolutions of the time-independent Schrödinger equation, i.e., with discrete eigenstates of H and notably with bound states. By the use of what is known as *Rayleigh-Schrödinger perturbation theory* we shall first obtain approximate energy eigenvalues and approximate stationary state wave functions. Sections 17.8 and 17.9 contain some comments of a more general nature.

The Rayleigh-Schrödinger perturbation theory applies to the discrete energy levels of a physical system whose Hamiltonian operator H can be broken up into two Hermitian parts:

$$H = H_0 + gV \tag{17.1}$$

Of these H_0 will be regarded as the unperturbed part, and gV as the perturbation. g is a real parameter, which will be used for bookkeeping purposes. We can let it become zero, in which case the Hamiltonian collapses into the unperturbed one, H_0, or we may let it grow to its full value, which may be chosen as $g = 1$. The eigenvalues and eigenfunctions of H are, of course, functions of g. Simple perturbation theory applies when these eigenvalues and eigenfunctions can be expanded in powers of g (at least in the sense of an asymptotic expansion) in the hope that for practical calculations only the first few terms of the expansion need be considered.

The eigenvalue problem which we wish to solve is

$$H\Psi_n = E_n\Psi_n \tag{17.2}$$

We suppose that the unperturbed eigenvalue problem,

$$H_0\Psi_n^{(0)} = E_n^{(0)}\Psi_n^{(0)} \tag{17.3}$$

has already been solved. Assuming, first, that no degeneracy has occurred here, let us inquire what happens to the energy eigenvalues and the corresponding eigenvectors as we allow g to grow continuously from zero to some finite value. In this process the energy will change to

$$E_n = E_n^{(0)} + \Delta E_n \tag{17.4}$$

and the eigenvector will change to

$$\Psi_n = \Psi_n^{(0)} + \Delta \Psi_n \tag{17.5}$$

Perturbation theory assumes that the changes ΔE_n and $\Delta \Psi_n$ are small, where "small" must yet be precisely defined. We substitute expressions (17.4) and (17.5) into (17.2) and make use of (17.3). The approximation consists of neglecting products of small changes (i.e., terms containing $\Delta E_n \, \Delta \Psi_n$), so that we obtain

$$H_0 \Delta \Psi_n + g V \Psi_n^{(0)} \approx E_n^{(0)} \Delta \Psi_n + \Delta E_n \Psi_n^{(0)}$$

The projection of this equation in the direction of $\Psi_n^{(0)}$ is obtained by multiplication from the left by the normalized vector $\Psi_n^{(0)}$:

$$(\Psi_n^{(0)}, H_0 \Delta \Psi_n) + (\Psi_n^{(0)}, g V \Psi_n^{(0)}) \approx E_n^{(0)}(\Psi_n^{(0)}, \Delta \Psi_n) + \Delta E_n$$

The first terms on the two sides of this equation cancel, because H_0 is Hermitian. Hence, we get in *first approximation*

$$\Delta E_n \approx (\Psi_n^{(0)}, g V \Psi_n^{(0)})$$

In words: The change in the nth eigenvalue is approximately equal to the expectation value of the perturbation in the nth unperturbed eigenstate. This result was already derived in Section 4.6.

For a systematic treatment we now return to the eigenvalue problem $H \Psi_n = E_n \Psi_n$ of the Hamiltonian $H = H_0 + g V$ under the assumption that the solutions of the unperturbed eigenvalue equation $H_0 \Psi_n^{(0)} = E_n^{(0)} \Psi_n^{(0)}$ are already known. We shall assume for the time being that the energy levels, $E_n^{(0)}$, of H_0 are *nondegenerate*.

The fundamental idea is to assume that both the eigenvalues and eigenvectors of H can be expanded in powers of the perturbation parameter g, and to determine the coefficients in the perturbation expansions:

$$E_n = E_n^{(0)} + g E_n^{(1)} + g^2 E_n^{(2)} + \cdots \tag{17.6}$$

and

$$\Psi_n = \Psi_n^{(0)} + g \Psi_n^{(1)} + g^2 \Psi_n^{(2)} + \cdots \tag{17.7}$$

Already here we should observe that the expansion of the eigenvector Ψ_n in powers of g is not unique, for if Ψ_n is a solution of (17.2), then $(a_0 + a_1 g + a_2 g^2 + \cdots) \Psi_n$ is an equally good solution, but its expansion in powers of g may be entirely different from (17.7), although the two solutions differ, of course, only by a constant factor. In spite of this apparent ambiguity in the eigenvector, all physically observable quantities, such as the energies (17.6) or expectation values of operators calculated from Ψ_n, have unique expansions in g.

Substituting the two expressions (17.6) and (17.7) into (17.2) and comparing systematically terms of equal power in g, we obtain the successive approximation equations:

$$H_0\Psi_n^{(0)} = E_n^{(0)}\Psi_n^{(0)} \tag{17.8}$$

$$H_0\Psi_n^{(1)} + V\Psi_n^{(0)} = E_n^{(0)}\Psi_n^{(1)} + E_n^{(1)}\Psi_n^{(0)} \tag{17.9}$$

$$H_0\Psi_n^{(2)} + V\Psi_n^{(1)} = E_n^{(0)}\Psi_n^{(2)} + E_n^{(1)}\Psi_n^{(1)} + E_n^{(2)}\Psi_n^{(0)} \tag{17.10}$$

Of these, (17.8) is identical with (17.3) and gives nothing new. Equation (17.9) can be rewritten as

$$(H_0 - E_n^{(0)})\Psi_n^{(1)} = (E_n^{(1)} - V)\Psi_n^{(0)} \tag{17.11}$$

The right-hand side of this equation is known except for the value of $E_n^{(1)}$; the unknown function $\Psi_n^{(1)}$ stands on the left. Thus (17.11) is an *inhomogeneous linear equation* for $\Psi_n^{(1)}$. Before considering this particular equation, it is worthwhile to review the properties of such equations, since all successive approximations and many other problems of quantum mechanics lead to the same type of equation.

2. Inhomogeneous Linear Equations. Assuming temporarily that the right-hand side is completely known, (17.11) is of the general type,

$$Au = v \tag{17.12}$$

where A is a given Hermitian operator with a complete set of eigenvectors, v is a given vector, and u is sought. For our purposes it is quite immaterial whether A is represented by a square matrix, in which case u and v are one-column matrices, or by a differential operator, so that u and v are ordinary functions. We can treat this equation in the general abstract vector space of Chapter 14, thus encompassing all special cases simultaneously. But to most students the theorems to be stated will be familiar from the theory of systems of simultaneous linear equations, where A is a finite dimensional square matrix.[1]

The alternatives to be distinguished are these:

(1) Either the homogeneous equation

$$Au' = 0 \tag{17.13}$$

possesses nontrivial solutions, i.e., A has zero eigenvalues, or

(2) Equation (17.13) has no nontrivial solution. In this latter case (which corresponds to det $A \neq 0$ for finite dimensional matrices) the operator A has

---

[1] See, for instance, R. Courant and D. Hilbert, *Methods of Mathematical Physics*, Volume I, Interscience Publishers, New York, 1953, Chapter 1, or P. R. Halmos, *Finite-Dimensional Vector Spaces*, 2nd ed., D. Van Nostrand Co., Princeton, 1958, and B. Friedman, *Principles and Techniques of Applied Mathematics*, John Wiley and Sons, New York, 1956.

a unique inverse A^{-1} (see Section 14.2). For any given v the solution of (17.12) is uniquely

$$u = A^{-1}v \qquad (17.14)$$

In case (1), on the other hand, (17.12) may have infinitely many solutions, for if f is a particular solution,

$$Af = v \qquad (17.15)$$

then any vector

$$u = u' + f \qquad (17.16)$$

will automatically also be a solution. By letting u' symbolize all solutions of the homogeneous equation (17.13), u of (17.16) represents all solutions of the inhomogeneous equation. However, in this case the existence of a solution depends on a further necessary (and sufficient) condition: v must have no component in the subspace spanned by the eigenvectors u' of A which correspond to *zero* eigenvalue.

Proof. Let P_0 be the projection operator which projects every vector into said subspace; then clearly

$$P_0 A = P_0 \sum_i A_i' P_i = 0 \qquad (17.17)$$

Hence,

$$P_0 v = P_0 A u = 0 \qquad (17.18)$$

which proves the assertion that v is orthogonal to the space of the solutions u'. We shall make essential use of the condition $P_0 v = 0$.

Instead of guessing a particular solution f, it is possible to construct one by the following procedure. By condition (17.18), we may write the equation to be solved, (17.15), as

$$Af = (1 - P_0)v \qquad (17.19)$$

seemingly a needless complication of the simple equation (17.15). Yet, although A, which has zero eigenvalues, has no inverse, there exists an operator K such that

$$AK = 1 - P_0 \qquad (17.20)$$

In fact, there are infinitely many operators K which satisfy this equation, because to any solution we may add an operator $P_0 B$ (B arbitrary) and still have a solution. It is convenient to remove this ambiguity and to select a unique solution of (17.20) by imposing the subsidiary conditions,

$$P_0 K = 0 \qquad (17.21)$$

The unique operator which is defined by (17.20) and (17.21) may symbolically be written as

$$K \equiv \frac{1 - P_0}{A} \qquad (17.22)$$

This expression is, however, not intended to imply that A has an inverse, and it is not permissible to apply the distributive law to it and write it as $A^{-1} - A^{-1}P_0$, or the like.

By eliminating $1 - P_0$ from (17.19) and (17.20) we get $Af = AKv$, and it follows that

$$f = Kv = \frac{1 - P_0}{A} v \qquad (17.23)$$

is a particular solution of $Af = v$. Because of (17.21), it is that particular solution which is orthogonal to the subspace of the u', i.e.,

$$P_0 f = 0 \qquad (17.24)$$

We thus conclude that, if $Au' = 0$ possesses nontrivial solutions, then $Au = v$ has the general solution

$$u = u' + \frac{1 - P_0}{A} v \qquad (17.25)$$

provided that $P_0 v = 0$. We shall now apply these results to perturbation theory.

3. *Solution of the Perturbation Equations.* The first approximation of the Rayleigh-Schrödinger perturbation procedure has led us to the inhomogeneous equation (17.11)

$$(H_0 - E_n^{(0)})\Psi_n^{(1)} = (E_n^{(1)} - V)\Psi_n^{(0)} \qquad (17.11)$$

Identifying the operator $H_0 - E_n^{(0)} = A$, we see that the homogeneous equation has indeed nontrivial solutions $\Psi_n^{(0)}$, which we assume *normalized to unity*. We have case (1) of the last section before us. Fixing our attention on a definite unperturbed state with a particular value of n, and denoting by $P_n^{(0)}$ the projection operator for the direction $\Psi_n^{(0)}$, we see that condition (17.18) appears as the equation

$$P_n^{(0)}(E_n^{(1)} - V)\Psi_n^{(0)} = 0 \qquad (17.26)$$

But

$$P_n^{(0)}\Psi_n^{(0)} = \Psi_n^{(0)} \qquad (17.27)$$

Hence,

$$E_n^{(1)}\Psi_n^{(0)} = P_n^{(0)}V\Psi_n^{(0)} \qquad (17.28)$$

and, assuming that $\Psi_n^{(0)}$ is normalized to unity,

$$E_n^{(1)} = (\Psi_n^{(0)}, P_n^{(0)}V\Psi_n^{(0)}) = (\Psi_n^{(0)}, V\Psi_n^{(0)}) \qquad (17.29)$$

because $P_n^{(0)}$ is Hermitian. The first-order correction to the energy is thus (omitting g)

$$E_n^{(1)} = (V)_{nn} = (\Psi_n^{(0)}, V\Psi_n^{(0)}) \tag{17.30}$$

or equal to the average value of the perturbation in the unperturbed state in agreement with the preliminary result of Sections 17.1 and 4.6.

Inserting the result (17.30) into (17.11) we can now solve the latter by comparing it with the general solution (17.25). Since it is assumed that boundary or other physical conditions limit the number of linearly independent eigensolutions for every $E_n^{(0)}$ to one (no degeneracy in H_0), it is evident that the solution of (17.11) is

$$\Psi_n^{(1)} = C_n^{(1)}\Psi_n^{(0)} - \frac{1 - P_n^{(0)}}{E_n^{(0)} - H_0}(E_n^{(1)} - V)\Psi_n^{(0)}$$

We must remember that the operator $(1 - P_0)/(E_n^{(0)} - H_0)$ is merely a symbol for K, and, by (17.21) gives zero when it acts on any solution of the homogeneous equation. Hence, by definition we have

$$\frac{1 - P_n^{(0)}}{E_n^{(0)} - H_0}\Psi_n^{(0)} = 0 \tag{17.31}$$

and thus,

$$\Psi_n^{(1)} = C_n^{(1)}\Psi_n^{(0)} + \frac{1 - P_n^{(0)}}{E_n^{(0)} - H_0}V\Psi_n^{(0)} \tag{17.32}$$

where $C_n^{(1)}$ is an arbitrary constant.

Although it is far from easy to see this in detail to all orders of the perturbation expansion, it should be clear that the choice of $C_n^{(1)}$ and all further arbitrary constants can have no physical consequences. Sometimes one chooses these constants such that at the kth stage of approximation the vector

$$\Psi_n = \Psi_n^{(0)} + g\Psi_n^{(1)} + \cdots + g^k\Psi_n^{(k)} + O(g^{k+1}) \tag{17.33}$$

is normalized to unity in the sense that

$$(\Psi_n, \Psi_n) = 1 + O(g^{k+1}) \tag{17.34}$$

This condition reduces but does not entirely remove the arbitrariness of the constants. For instance, if we require for $k = 1$ that

$$(\Psi_n^{(0)} + g\Psi_n^{(1)}, \Psi_n^{(0)} + g\Psi_n^{(1)}) = 1 + O(g^2)$$

it follows by substituting (17.32) into this equation that $C_n^{(1)}$ must be purely imaginary, but otherwise it is still undetermined. However, it can be shown that the remaining arbitrariness corresponds merely to the option we have of multiplying Ψ_n by a phase factor without destroying the normalization. The phase factor may be a function of g.

In practice it is simplest to set all arbitrary constants $C_n^{(k)}$ equal to zero and, if desired, to normalize the approximate eigenvectors at the end of the calculation. The normalization constant is, of course, a function of g.

With $C_n^{(1)} = 0$, i.e., $P_n^{(0)} \Psi_n^{(1)} = 0$, we have

$$\Psi_n^{(1)} = \frac{1 - P_n^{(0)}}{E_n^{(0)} - H_0} V \Psi_n^{(0)} \tag{17.35}$$

if we insert for the identity operator I its closure equivalent, $I = \sum_k P_k^{(0)}$, and note that by the definition of a projection operator

$$P_k^{(0)} V \Psi_n^{(0)} = \Psi_k^{(0)} (\Psi_k^{(0)}, V \Psi_n^{(0)}) = \Psi_k^{(0)} V_{kn}$$

we obtain from (17.35) explicitly

$$\Psi_n^{(1)} = \sum_{k \neq n} \Psi_k^{(0)} \frac{V_{kn}}{E_n^{(0)} - E_k^{(0)}} \tag{17.36}$$

giving us the first correction to the nth eigenvector in terms of the unperturbed eigenvectors. It is to be noted that, if (17.36) is used, $\Psi_n^{(0)} + g\Psi_n^{(1)}$ is correctly normalized to unity to first order in g.

The same procedure can be continued systematically as the perturbation theory is carried to higher orders. In second order we must solve (17.10) or

$$(H_0 - E_n^{(0)})\Psi_n^{(2)} = (E_n^{(1)} - V)\Psi_n^{(1)} + E_n^{(2)}\Psi_n^{(0)} \tag{17.37}$$

which is also of the inhomogeneous type. The homogeneous equation, obtained by replacing the right-hand side by zero, has again nontrivial solutions. Hence, we must require that the inhomogeneity have no component in the direction of $\Psi_n^{(0)}$, or that its scalar product with $\Psi_n^{(0)}$ vanish:

$$(\Psi_n^{(0)}, (E_n^{(1)} - V)\Psi_n^{(1)} + E_n^{(2)}\Psi_n^{(0)}) = 0$$

$\Psi_n^{(0)}$ is orthogonal to $\Psi_n^{(1)}$, and $\Psi_n^{(0)}$ is normalized to unity. Hence, we obtain the simple relation

$$E_n^{(2)} = (\Psi_n^{(0)}, V\Psi_n^{(1)}) \tag{17.38}$$

Substituting (17.35) or (17.36) into this formula, we get finally for the second-order correction to the energy,

$$E_n^{(2)} = \left(\Psi_n^{(0)}, V \frac{1 - P_n^{(0)}}{E_n^{(0)} - H_0} V \Psi_n^{(0)} \right) = \sum_{k \neq n} \frac{V_{nk} V_{kn}}{E_n^{(0)} - E_k^{(0)}} = \sum_{k \neq n} \frac{|V_{nk}|^2}{E_n^{(0)} - E_k^{(0)}} \tag{17.39}$$

It is evident from (17.36) and (17.39) that for the Rayleigh-Schrödinger perturbation expansion to converge rapidly it is necessary that the quantities $|V_{nk}/(E_n^{(0)} - E_k^{(0)})|$ be small.

In turn we can now calculate the eigenvector correction $\Psi_n^{\prime(2)}$. Applying the same method as before to (17.37), we find

$$\Psi_n^{\prime(2)} = C_n^{(2)}\Psi_n^{\prime(0)} - \frac{1 - P_n^{(0)}}{E_n^{(0)} - H_0}(E_n^{(1)} - V)\Psi_n^{\prime(1)}$$

Again we may set $C_n^{(2)} = 0$, i.e., $P_n^{(0)}\Psi_n^{\prime(2)} = 0$. Substituting also (17.35) for $\Psi_n^{\prime(1)}$, we obtain

$$\Psi_n^{\prime(2)} = -\frac{1 - P_n^{(0)}}{E_n^{(0)} - H_0}(E_n^{(1)} - V)\frac{1 - P_n^{(0)}}{E_n^{(0)} - H_0}V\Psi_n^{\prime(0)} \qquad (17.40)$$

The relation (17.28) finally allows us to write $\Psi_n^{\prime(2)}$ entirely in terms of the unperturbed system as

$$\Psi_n^{\prime(2)} = -\frac{1 - P_n^{(0)}}{(E_n^{(0)} - H_0)^2}VP_n^{(0)}V\Psi_n^{\prime(0)} + \frac{1 - P_n^{(0)}}{E_n^{(0)} - H_0}V\frac{1 - P_n^{(0)}}{E_n^{(0)} - H_0}V\Psi_n^{\prime(0)} \qquad (17.41)$$

Exercise 17.1. Show that if $C_n^{(1)} = C_n^{(2)} = 0$,

$$(\Psi_n', V\Psi_n') = E_n^{(1)} + 2gE_n^{(2)} + 0(g^2) \qquad (17.42)$$

Also evaluate $(\Psi_n', H_0\Psi_n')$ to second order in g. Does the sum of $(\Psi_n', H_0\Psi_n')$ and $(\Psi_n', gV\Psi_n')$ give E_n to second order in g?

Exercise 17.2. Calculate the normalization factor for the perturbed eigenvector to second order in g.

The perturbation theory can be further developed in this way to any desired order.[2] For practical purposes it is rarely necessary to go beyond the second order.

Exercise 17.3. Obtain expressions for the mth-order correction to the energy and the eigenvectors in terms of the corrections of lower order. Show that with the choice $C_n^{(k)} = 0$, for all $k > 0$, the perturbed eigenvector, instead of being normalized to unity, satisfies the condition

$$(\Psi_n^{\prime(0)}, \Psi_n') = 1$$

Derive the formula for the energy shift,

$$\Delta E_n = E_n - E_n^{(0)} = (\Psi_n^{\prime(0)}, V\Psi_n') \qquad (17.43)$$

4. *Electrostatic Polarization and the Dipole Moment.* As an important example we consider an electron bound in an atom and placed in a weak

---

[2] T. Kato, *Progr. of Theor. Phys.*, **4**, 514 (1949); K. A. Brueckner, *Phys. Rev.*, **100**, 36 (1955). As shown by R. M. Sternheimer in *Phys. Rev.* **84**, 244 (1951) and *Phys. Rev.* **95**, 736 (1954), it is sometimes practicable to solve the inhomogeneous perturbation equations directly without making an expansion in terms of unperturbed eigenfunctions.

uniform external electric field **E**. The field can be derived from an electro-static potential

$$\phi(\mathbf{r}) = -\mathbf{E} \cdot \mathbf{r}$$

where the coordinate origin is most conveniently chosen at the position of the nucleus, and the perturbation potential is

$$gV = -e\phi = e\mathbf{E} \cdot \mathbf{r}$$

The energy of the system to second order is given by the formulas of the last section as

$$E_n = E_n^{(0)} + e\mathbf{E} \cdot \mathbf{r}_{nn} + e^2 \sum_{k \neq n} \frac{(\mathbf{E} \cdot \mathbf{r}_{nk})(\mathbf{E} \cdot \mathbf{r}_{kn})}{E_n^{(0)} - E_k^{(0)}} \qquad (17.44)$$

where all matrix elements are to be taken with respect to the unperturbed eigenstates.

The shift of energy levels in an electric field is known as the *Stark effect*. The first two terms of the perturbation expansion give accurate results for applied fields which are small compared to the internal electric field of the atom. The latter is in order of magnitude given by $E^{(0)}/ea \simeq 10^{10}$ volts/meter. In practice this condition is always well satisfied and successive terms in the perturbation expansion decrease rapidly and uniformly, except that some terms may vanish owing to certain symmetry properties of the system. The most important instance of this is conservation of parity which results in the absence of the first-order term in almost all atomic states, with the important exception of the excited states in hydrogenic atoms. If the unperturbed electron is in a central field, H_0 is invariant under coordinate inversion through the center of force, and the energy eigenstates may be taken to have definite parity. We saw in Section 16.10 that the expectation value of the operator **r**, which is odd under reflection, vanishes for states of definite parity; hence, the external electric field can, in general, produce no first-order, or linear, Stark effect. An exception arises if the central field is a pure Coulomb field (hydrogenic atoms) because the excited states of such atoms exhibit degeneracy of states with opposite parity. Superposition of such states yields energy eigenstates which have no definite parity, and the expectation value of **r** need no longer vanish. We shall resume discussion of the linear Stark effect of the first excited state of hydrogen as an example of degenerate perturbation theory in Section 17.6. The inevitable degeneracy of the magnetic substates for states of nonzero angular momentum, on the other hand, does not affect our conclusion concerning the absence of the linear Stark effect, because all these substates have the same parity (Section 16.10).

Usually, then, the first-order term in (17.44) is absent. The second-order term gives rise to the so-called *quadratic Stark effect*. If the electric field is along the z-axis, the quadratic Stark effect is given by the formula

$$E_n = E_n^{(0)} + e^2 \mathbf{E}^2 \sum_{k \neq n} \frac{|z_{nk}|^2}{E_n^{(0)} - E_k^{(0)}} \tag{17.45}$$

The parity selection rule insures that the trivial degeneracy of magnetic substates does not interfere with the applicability of this formula, because two states which differ only by their magnetic quantum number have the same parity. By inspection of the work of the preceding section we see that such apparently indeterminate (0/0) terms may simply be omitted from the sum in (17.45).

Perturbation theory may also be used to calculate the expectation value of the static electric dipole moment, $-e\mathbf{r}$, in a stationary state of the one-electron atom. In the lowest approximation,

$$\mathbf{p_0} = -e\mathbf{r}_{nn} = -e \int \mathbf{r} \, |\psi_n^{(0)}(\mathbf{r})|^2 \, d\tau \tag{17.46}$$

This is called the *permanent electric dipole moment* of the system, because it represents a vector which is determined by the unperturbed state of the system and entirely independent of the applied field. It vanishes, of course, for all states which possess definite parity.

A better approximation is obtained by using the correction (17.36):

$$\rho \approx |\psi_n^{(0)} + g\psi_n^{(1)}|^2 \approx |\psi_n^{(0)}|^2 + e\psi_n^{(0)*} \sum_{k \neq n} \psi_k^{(0)} \frac{\mathbf{E} \cdot \mathbf{r}_{kn}}{E_n^{(0)} - E_k^{(0)}}$$

$$+ e\psi_n^{(0)} \sum_{k \neq n} \psi_k^{(0)*} \frac{\mathbf{E} \cdot \mathbf{r}_{kn}}{E_n^{(0)} - E_k^{(0)}} \tag{17.47}$$

The last two terms describe the polarization of the atom by the applied field.

In this approximation we obtain for the dipole moment of the one-electron atom

$$\mathbf{p} = -e \int \rho \mathbf{r} \, d\tau = \mathbf{p_0} - e^2 \sum_{k \neq n} \frac{\mathbf{r}_{nk}\mathbf{r}_{kn} + \mathbf{r}_{kn}\mathbf{r}_{nk}}{E_n^{(0)} - E_k^{(0)}} \cdot \mathbf{E}$$

where the last term represents the *induced dipole moment* in the state n,

$$\mathbf{p_1} = e^2 \sum_{k \neq n} \frac{\mathbf{r}_{nk}\mathbf{r}_{kn} + \mathbf{r}_{kn}\mathbf{r}_{nk}}{E_k^{(0)} - E_n^{(0)}} \cdot \mathbf{E} = \boldsymbol{\alpha} \cdot \mathbf{E} \tag{17.48}$$

This equation defines a *tensor* (or dyadic) *of polarizability* for the state n,

$$\boldsymbol{\alpha} = e^2 \sum_{k \neq n} \frac{\mathbf{r}_{nk}\mathbf{r}_{kn} + \mathbf{r}_{kn}\mathbf{r}_{nk}}{E_k^{(0)} - E_n^{(0)}} \tag{17.49}$$

It is of interest to note that this tensor is symmetric.[3] In many applications we find $\alpha_{xy} = \alpha_{yz} = \alpha_{zx} = 0$, and $\alpha_{xx} = \alpha_{yy} = \alpha_{zz}$, so that the polarizability is a scalar.

Exercise 17.4. Calculate the polarizability of an isotropic harmonic oscillator from (17.49), and verify that the result agrees with an exact calculation of the induced dipole moment.

Note that

$$\mathbf{E} \cdot \mathbf{p} = -\int \psi_n^* g V \psi_n \, d\tau = \mathbf{E} \cdot \mathbf{p_0} + \mathbf{E} \cdot \boldsymbol{\alpha} \cdot \mathbf{E} + 0(\mathbf{E}^3)$$

Comparing this with (17.42), we obtain

$$E_n = E_n^{(0)} - \mathbf{E} \cdot \mathbf{P_0} - \tfrac{1}{2}\mathbf{E} \cdot \boldsymbol{\alpha} \cdot \mathbf{E} + \cdots$$

which upon substitution of $\mathbf{p_0}$ and $\boldsymbol{\alpha}$ can easily be seen to be identical with (17.44). The factor of $\tfrac{1}{2}$ which appears in the energy owing to the induced dipole moment is the same as that customarily found when "stress" ($\mathbf{E}$) and "strain" ($\mathbf{p_1}$) are proportional, as is the case in the approximation leading to (17.48).

The rigorous evaluation of the sums over unperturbed states, which are encountered in all higher order perturbation calculations, is usually a difficult problem. However, sometimes special techniques may allow such sums to be performed. As an example consider the quadratic Stark effect or the polarizability of the ground state of the hydrogen atom. According to (17.44) and (17.49), this requires the evaluation of

$$\sum_{k \neq 0} \frac{|z_{0k}|^2}{E_0^{(0)} - E_k^{(0)}} = \sum_{k \neq 0} \frac{z_{0k} z_{k0}}{E_0^{(0)} - E_k^{(0)}}$$

where the subscript 0 labels the ground state of hydrogen $|0\rangle$ and k labels all other states of hydrogen.

Let us suppose that it is possible to find, by whatever procedure, an operator F which satisfies the equation

$$z \, |0\rangle = (FH_0 - H_0 F) \, |0\rangle \tag{17.50}$$

Then we have

$$z_{k0} = \langle k| \, z \, |0\rangle = \langle k| \, FH_0 \, |0\rangle - \langle k| \, H_0 F \, |0\rangle = (E_0^{(0)} - E_k^{(0)})\langle k| \, F \, |0\rangle$$

[3] M. Born and E. Wolf, *Principles of Optics*, 3rd ed., Pergamon Press, New York, 1965, p. 366. See also W K. H. Panofsky and M. Phillips, *Classical Electricity and Magnetism*, 2nd ed., Addison-Wesley Publishing Company, Reading, 1962, p. 30 and Section 6-2.

and

$$\sum_{k \neq 0} \frac{|z_{0k}|^2}{E_0^{(0)} - E_k^{(0)}} = \sum_{k \neq 0} \langle 0| \, z \, |k\rangle\langle k| \, F \, |0\rangle = \langle 0| \, zF \, |0\rangle - \langle 0| \, z \, |0\rangle\langle 0| \, F \, |0\rangle$$

(17.51)

By the use of closure in the last step the sum over states has thus been transformed into the calculation of expectation values in a single state. Of course, the usefulness of (17.51) hinges on our ability to determine the operator F.[4]

If H_0 stands for the Hamiltonian of the unperturbed hydrogen atom and $|0\rangle$ for the ground state of hydrogen, it is easy to determine F by writing (17.50) explicitly in the coordinate representation. Assuming that F is a function of the coordinates only (and not of the momenta), a differential equation for F is obtained, which is conveniently expressed in terms of spherical polar coordinates and may be solved by separation of variables. The details of the calculation are left to the reader, who may also verify by direct substitution that

$$F = -\frac{\mu a}{\hbar^2} \left(\frac{r}{2} + a\right) z$$

(17.52)

(where $a = $ Bohr radius) satisfies (17.50) in our example.

Since the expectation value of z in the ground state of hydrogen vanishes, it follows from (17.51) that

$$\sum_{k \neq 0} \frac{|z_{0k}|^2}{E_0^{(0)} - E_k^{(0)}} = -\frac{\mu a}{\hbar^2} \langle 0| \left(\frac{r}{2} + a\right) z^2 |0\rangle$$

The remaining expectation value is easily evaluated by noting that by virtue of the spherical symmetry of the ground state (S-state)

$$\langle 0| f(r)z^2 |0\rangle = \langle 0| f(r)x^2 |0\rangle = \langle 0| f(r)y^2 |0\rangle = \tfrac{1}{3}\langle 0| f(r)r^2 |0\rangle$$

Hence,

$$\sum_{k \neq 0} \frac{|z_{0k}|^2}{E_0^{(0)} - E_k^{(0)}} = -\frac{\mu a}{3\hbar^2} (\tfrac{1}{2}\langle r^3\rangle_0 + a\langle r^2\rangle_0)$$

But

$$\langle r^n\rangle_0 = \frac{1}{\pi a^3} \int d\Omega \int_0^\infty r^{n+2} \exp\left(-\frac{2r}{a}\right) dr = \frac{a^n}{2^{n+1}} (n + 2)!$$

We thus finally obtain

$$E_0 = -\frac{e^2}{2a} - \frac{9}{4} a^3 |\mathbf{E}|^2$$

(17.53)

[4] Ingenious use was made of this method by A. Dalgarno and J. T. Lewis, *Proc. Roy. Soc.*, A **233**, 70 (1955). For earlier polarizability calculations, based on direct solutions of the inhomogeneous equations arising in perturbation theory, see H. M. Foley, O. M. Sternheimer, and D. Tycko, *Phys. Rev.* **93**, 734 (1954), and R. M. Sternheimer, *Phys. Rev.* **96**, 951 (1954) and *Phys. Rev.* **127**, 1220 (1962).

for the ground state of the hydrogen atom to second order in the applied electric field.[5] The presence of the field causes a lowering of the ground state energy. This was to be expected from (17.39), since every term in the sum may be regarded as a repulsion of the nth level by the kth level.

5. Degenerate Perturbation Theory. We must now supplement our perturbation methods by admitting the possibility that the nth unperturbed state may be degenerate, usually as the result of certain symmetries. Thus, if the system is in a central force field, the magnetic substates of a given angular momentum all have the same energy, owing to rotational symmetry. If this symmetry is disturbed, as by the application of a magnetic field, the degeneracy is usually removed.

The perturbation procedure developed in Sections 17.1 and 17.3 cannot be applied without modification, because the expansion (17.7) of the eigenvector was based of the assumption that we know into which unperturbed eigenvector $\Psi_n^{(0)}$ the exact perturbed eigenvector Ψ_n collapses as g approaches zero. This assumption breaks down when the unperturbed state is degenerate and we have no prior knowledge which would allow us to predict what particular linear combination of the given degenerate substates the eigenvector Ψ_n will go into as $g \to 0$. (However, frequently symmetry properties can be used to avoid this ambiguity. For the resulting simplifications see below.)

The breakdown of the simple Rayleigh-Schrödinger theory in the case of degenerate unperturbed states appears formally as the vanishing of some of the energy denominators in (17.36) and (17.39). When this happens, the perturbation expansions become meaningless (except if $V_{kn} = 0$ as a result of some symmetry). For practical applications it is important to realize that these difficulties arise not only if the unperturbed states are strictly degenerate, but also if they are merely so close in energy that $|V_{kn}/(E_n^{(0)} - E_n^{(0)})|$ is large and causes large-scale mixing of unperturbed states in (17.36).

In order to keep the notation uncluttered, let us suppose that the unperturbed eigenvalue $E_n^{(0)}$ is only doubly degenerate, i.e., that two linearly independent eigenvectors $\Psi_{n1}^{(0)}$ and $\Psi_{n2}^{(0)}$ belong to it. We may assume these two eigenvectors to be orthonormal. When the perturbation is "turned on," the level usually splits into two components, and we have the expansions,

$$E_{n1} = E_n^{(0)} + g E_{n1}^{(1)} + g^2 E_{n1}^{(2)} + \cdots \qquad (17.54a)$$

$$E_{n2} = E_n^{(0)} + g E_{n2}^{(1)} + g^2 E_{n2}^{(2)} + \cdots \qquad (17.54b)$$

---

[5] An alternative method for obtaining the result (17.53) consists of using parabolic coordinates in which the Schrödinger equation for the hydrogen atom is separable even in the presence of a uniform electric field.

with the eigenvectors,

$$\Psi_{n1} = c_{11}\Psi_{n1}^{(0)} + c_{21}\Psi_{n2}^{(0)} + g\Psi_{n1}^{(1)} + \cdots \tag{17.55a}$$

$$\Psi_{n2} = c_{12}\Psi_{n1}^{(0)} + c_{22}\Psi_{n2}^{(0)} + g\Psi_{n2}^{(1)} + \cdots \tag{17.55b}$$

The problem of degenerate perturbation theory is to determine the coefficients c_{ij} and to find the correct linear combinations of the unperturbed eigenvectors which may serve as zero-order approximations to the actual perturbed eigenvectors.

We now use these expansions in (17.2) and obtain a set of successive approximation equations, but they are slightly more complicated than in the nondegenerate case. Proceeding only to the first order in g, we have

$$(H_0 - E_n^{(0)})\Psi_{n1}^{(0)} = 0 \tag{17.56a}$$

$$(H_0 - E_n^{(0)})\Psi_{n2}^{(0)} = 0 \tag{17.56b}$$

$$(H_0 - E_n^{(0)})\Psi_{n1}^{(1)} = (E_{n1}^{(1)} - V)(c_{11}\Psi_{n1}^{(0)} + c_{21}\Psi_{n2}^{(0)}) \tag{17.56c}$$

$$(H_0 - E_n^{(0)})\Psi_{n2}^{(1)} = (E_{n2}^{(1)} - V)(c_{12}\Psi_{n1}^{(0)} + c_{22}\Psi_{n2}^{(0)}) \tag{17.56d}$$

These equations are of the same inhomogeneous type as before. The criterion for the solubility of (17.56c) and (17.56d) is that the inhomogeneous term must have no component in the subspace spanned by the solutions, $\Psi_{n1}^{(0)}$ and $\Psi_{n2}^{(0)}$, of the homogeneous equations (17.56a) and (17.56b). Introducing the projection operator,

$$P_n^{(0)} = P_{n1}^{(0)} + P_{n2}^{(0)} = |E_{n1}^{(0)}\rangle\langle E_{n1}^{(0)}| + |E_{n2}^{(0)}\rangle\langle E_{n2}^{(0)}| \tag{17.57}$$

we demand by application to (17.56c) that

$$c_{11}(E_{n1}^{(1)}\Psi_{n1}^{(0)} - P_n^{(0)}V\Psi_{n1}^{(0)}) + c_{21}(E_{n1}^{(1)}\Psi_{n2}^{(0)} - P_n^{(0)}V\Psi_{n2}^{(0)}) = 0 \tag{17.58}$$

But

$$P_n^{(0)}V\Psi_{n1}^{(0)} = \Psi_{n1}^{(0)}V_{11} + \Psi_{n2}^{(0)}V_{21}$$

$$P_n^{(0)}V\Psi_{n2}^{(0)} = \Psi_{n1}^{(0)}V_{12} + \Psi_{n2}^{(0)}V_{22}$$

Hence, we obtain from (17.58) two linear homogeneous equations for determining the as yet unknown coefficient c_{11} and c_{21},

$$(V_{11} - E_{n1}^{(1)})c_{11} + V_{12}c_{21} = 0$$
$$V_{21}c_{11} + (V_{22} - E_{n1}^{(1)})c_{21} = 0 \tag{17.59}$$

or, in matrix form,

$$\begin{pmatrix} V_{11} & V_{12} \\ V_{21} & V_{22} \end{pmatrix} \begin{pmatrix} c_{11} \\ c_{21} \end{pmatrix} = E_{n1}^{(1)} \begin{pmatrix} c_{11} \\ c_{21} \end{pmatrix} \tag{17.60}$$

We thus see that the determination of the unknown coefficients has led us back to the familiar problem of matrix mechanics: the diagonalization, or the finding of the eigenvectors, of a Hermitian matrix. However, instead of having to solve this problem for matrices of infinitely many rows and columns, we only have to diagonalize explicitly a matrix whose dimensionality equals the degree of degeneracy of the level $E_n^{(0)}$—two in our example.

Not only does (17.60) determine the correct linear combinations of the unperturbed zero-order eigenvectors to be used, but in the process we obtain the first-order corrections to the energy as well. Indeed, (17.60) has nontrivial solutions if and only if

$$\begin{vmatrix} V_{11} - E_{n1}^{(1)} & V_{12} \\ V_{21} & V_{22} - E_{n1}^{(1)} \end{vmatrix} = 0 \qquad (17.61)$$

One of the two roots of the secular equation is $E_{n1}^{(1)}$. The other is $E_{n2}^{(1)}$, as we see if from (17.56d) we derive a set of equations for c_{12}, c_{22}:

$$\begin{pmatrix} V_{11} & V_{12} \\ V_{21} & V_{22} \end{pmatrix} \begin{pmatrix} c_{12} \\ c_{22} \end{pmatrix} = E_{n2}^{(1)} \begin{pmatrix} c_{12} \\ c_{22} \end{pmatrix} \qquad (17.62)$$

This leads to the same secular equation (17.61).

In our simple case, with $d = 2$, it is clear that (17.61) has distinct roots, and thus the degeneracy is removed by the perturbation in first order, unless $V_{11} = V_{22}$ and $V_{12} = V_{21} = 0$. If this happens, the degeneracy may still be removable in a higher approximation, but the general formulas become complicated.

Summarizing: If $E_n^{(0)}$ is d-fold degenerate, we construct the matrix of the perturbation V with respect to the orthonormal degenerate eigenstates, $\Psi_{n1}^{(0)}$, $\Psi_{n2}^{(0)}$, ... The d roots of the corresponding secular equation are the first-order corrections to the energy.

Plainly, the program of first-order perturbation theory requires the calculation of a d-dimensional Hamiltonian submatrix in the space of the degenerate eigenstates corresponding to the eigenvalue $E_n^{(0)}$. At this level of approximation, all other unperturbed eigenstates are simply ignored and the eigenvalue problem for the truncated Hamiltonian is solved exactly.

Such a procedure of partial exact diagonalization of the energy matrix H, by including in the calculation a limited set of eigenstates of a suitably chosen zero-order Hamiltonian H_0 and excluding all other eigenstates, is frequently resorted to in obtaining approximate solutions of the eigenvalue problem of H for complex systems. The included unperturbed eigenstates of H_0 need not be strictly degenerate, but reasonably good results can be expected only if they are well separated in energy from the excluded states or if the matrix elements of the perturbation between included and excluded states are small, or—even better—if both of these requirements are satisfied. Under favorable conditions, accurate eigenvectors of H can be constructed as superpositions of included states only, and it is then said that the *interaction* between included and excluded states is negligible.

In general, the perturbation calculation for a *d*-fold degenerate unperturbed state requires us to compute the eigenvalues of a $d \times d$ matrix. Although modern computing techniques have made it possible to solve secular equations which previously were considered forbidding, nevertheless time, and incidentally physical insight, can be gained if we take full advantage of the symmetry properties of the system. If certain symmetries of the unperturbed system survive in the presence of the perturbation, there exist commuting Hermitian operators A which commute both with H_0 and H, hence also with their difference gV:

$$H_0 A - A H_0 = HA - AH = VA - AV = 0 \qquad (17.63)$$

For example, if we have an electron in a central field and place this system in a uniform external magnetic or electric field which is parallel to the *z*-direction, the rotational symmetry about the *z*-axis is preserved, and as a consequence L_z commutes with H and V as well as with H_0. (L_x or L_y commute also with H_0, but not with H.) If an operator A satisfying (17.63) exists, the eigenvectors of H may be selected to be also eigenvectors of A, and this can be required for all values of g. Consequently, in constructing the correct linear combinations of unperturbed eigenvectors we need to include only those eigenvectors which belong to the same eigenvalue, A', of A. Formally, the simplification comes about, because if A and V commute,

$$\langle A' | V | A'' \rangle = 0 \qquad (17.64)$$

unless $A' = A''$ (*selection rule*).

If the Hamiltonian of a system depends on a variable parameter B, such as a magnetic field, the energy eigenvalues are functions of B. The theory of this section may be used to discuss the conditions under which *level crossing* as a function of B may occur. Let us tentatively assume that for a particular value of the parameter, $B = B_0$, two energy eigenvalues become degenerate. For values of B near B_0 we can write the Hamiltonian as

$$H(B) = H(B_0) + (B - B_0)V \qquad (17.65)$$

For sufficiently small deviations of B from B_0 we may use degenerate perturbation theory. The rules require that the matrix of V with respect to the correct unperturbed eigenfunctions be diagonal, i.e.,

$$V_{12} = (\overline{\Psi}_1^{(0)}, V\overline{\Psi}_2^{(0)}) = 0 \qquad (17.66)$$

if $\overline{\Psi}_1^{(0)}$ and $\overline{\Psi}_2^{(0)}$ are the eigenfunctions of $H(B)$, evaluated at $B = B_0$:

$$\overline{\Psi}_1^{(0)} = \lim_{B \to B_0} \Psi_1, \qquad \overline{\Psi}_2^{(0)} = \lim_{B \to B_0} \Psi_2$$

Generally, the complex equation (17.66) represents two equations for the real parameter B_0. The equality of the eigenvalues

$$E_1(B_0) = E_2(B_0) \tag{17.67}$$

is a third condition for B_0. Hence, there are three conditions which must be satisfied by the single unknown B_0. This is generally impossible without contradiction. An exception arises if, owing to some symmetry properties, condition (17.66) holds identically for all values of B_0. Then the two energy levels may be made to intersect, but otherwise the two levels can in general not cross.

6. Applications to Atoms. We are now in a position to solve a great number of realistic physical problems, and this section contains a sampling from atomic spectroscopy.[6]

The theory of the *linear Stark effect* in the hydrogen atom can be used to illustrate degenerate perturbation theory. The Hamiltonian is

$$H = \frac{\mathbf{p}^2}{2\mu} - \frac{e^2}{r} + e\,|\mathbf{E}|\,z \tag{17.68}$$

where the last term shall be regarded as a perturbation. The ground state (1S) of hydrogen is nondegenerate and has even parity; hence no linear Stark effect occurs, and there is no permanent dipole moment. The situation is different for the excited states, of which we shall treat only the lowest one. The 2S state and the three 2P states are degenerate, the former being of even parity, whereas the latter three are odd. At first sight degenerate perturbation theory requires that we allow the unperturbed eigenfunctions to be linear combinations of all four unperturbed eigenstates. However, H shares cylindrical symmetry with H_0, and L_z commutes with H. Hence, the perturbed eigenstates can still be required to be eigenstates of L_z with the eigenvalues $m\hbar$, where $m = -1$, 0, or $+1$. If we denote the several eigenstates of the level n by $|nlm\rangle$, we see that of the degenerate states only $|2,0,0\rangle$ and $|2,1,0\rangle$ are mixed by the perturbation. The states $|2,1,1\rangle$ and $|2,1,-1\rangle$ are single and remain so. They do not exhibit any linear Stark effect, because they have definite parity. What remains to solve is the secular equation,

$$\begin{vmatrix} e|\mathbf{E}|\langle 2,0,0|\,z\,|2,0,0\rangle - \lambda & e|\mathbf{E}|\langle 2,0,0|\,z\,|2,1,0\rangle \\ e|\mathbf{E}|\langle 2,1,0|\,z\,|2,0,0\rangle & e|\mathbf{E}|\langle 2,1,0|\,z\,|2,1,0\rangle - \lambda \end{vmatrix} = 0$$

[6] The article by Bethe and Salpeter in Volume 35 of the *Encyclopedia of Physics* is an outstanding review of the applications of quantum mechanics in atomic physics. Entitled *Quantum Mechanics of One- and Two-Electron Systems*, it has been reprinted separately by Academic Press, New York, 1957.

Here again, because of conservation of parity, the diagonal elements $\langle 2,0,0| z |2,0,0\rangle$ and $\langle 2,1,0| z |2,1,0\rangle$ vanish. Hence, the first-order change in energy is

$$\lambda = \pm e\,|\mathbf{E}|\,|\langle 2,0,0| z |2,1,0\rangle|$$

Thus, only one matrix element has to be evaluated. For this purpose it is necessary to use the unperturbed eigenfunctions explicitly. They are (Chapter 10)

$$2S\ (m = 0)\colon \psi_{2S}^{(0)} = \frac{1}{\sqrt{4\pi}}\left(\frac{1}{2a}\right)^{3/2}\left(2 - \frac{r}{a}\right)\exp\left(-\frac{r}{2a}\right)$$

$$2P\ (m = 0)\colon \psi_{2P}^{(0)} = \frac{1}{\sqrt{4\pi}}\left(\frac{1}{2a}\right)^{3/2}\frac{r}{a}\exp\left(-\frac{r}{2a}\right)\cos\theta$$

We calculate

$$\langle 2,0,0| z |2,1,0\rangle = \langle 2,0,0| r\cos\theta |2,1,0\rangle$$

$$= \frac{1}{4\pi}\left(\frac{1}{2a}\right)^{3}\frac{1}{a}\int_{0}^{\infty}\int_{0}^{\pi}\int_{0}^{2\pi} r^{4}\left(2 - \frac{r}{a}\right)\exp\left(-\frac{r}{a}\right)\cos^{2}\theta\,\sin\theta\,dr\,d\theta\,d\varphi = -3a$$

Hence, the linear Stark effect splits the degenerate $m = 0$ level into two components, the shifts being

$$\Delta E = \pm 3ae\,|\mathbf{E}| \tag{17.69}$$

The corresponding eigenfunctions are easily seen to be

$$\frac{1}{\sqrt{2}}\left(\psi_{S}^{(0)} \mp \psi_{P}^{(0)}\right) \tag{17.70}$$

mixing the two components in equal proportions.

The degeneracy of the 2S and 2P states in the hydrogenic atom is removed by any perturbation of the pure Coulomb field. Thus the 2s and 2p states are no longer degenerate in the lithium atom, nor in the heavy atoms, where they constitute components of the inner L-shell. In both cases there are central screening fields which modify the pure Coulomb field.[7] On the other hand, the degeneracy of the $2l + 1$ magnetic substates survives the addition of such perturbations. But, in a field along the z-direction the energy eigenstates are eigenstates of L_z, and this fact allows us to avoid the use of degenerate perturbation theory, once levels of different l are split appreciably.

Exercise 17.5. Calculate the linear Stark effect for the n = 3 level of hydrogen.

---

[7] The $2S - 2P$ degeneracy in the hydrogen atom also disappears when the electron is treated as a relativistic particle with spin and its interaction with the electromagnetic field is fully taken into account (Lamb shift).

Next we shall deal more accurately than before with the motion of an electron in a central field with the inclusion of spin effects. We saw in Section 13.3 that the Schrödinger equation of a spinning electron in a central field has the general form

$$\left[-\frac{\hbar^2}{2\mu} \nabla^2 + V(r) + W(r) \mathbf{L} \cdot \boldsymbol{\sigma} \right] \begin{pmatrix} \psi_1 \\ \psi_2 \end{pmatrix} = E \begin{pmatrix} \psi_1 \\ \psi_2 \end{pmatrix} \qquad (17.71)$$

The *spin-orbit interaction* $W(r) \mathbf{L} \cdot \boldsymbol{\sigma}$ will be treated as a small perturbation. Actually, equation (17.71) is not sufficiently complex to yield precise values for the fine structure of the hydrogen energy levels because these are affected by relativistic corrections to the kinetic energy operator, $-(\hbar^2/2\mu)\nabla^2$, as much as by the specific spin-orbit interaction. Indeed, there is no advantage in using a fundamentally nonrelativistic Schrödinger equation like (17.71) for the hydrogen atom. This simple system requires and can be accorded a much more accurate treatment by the use of the fully relativistic Dirac equation of the electron. Such a treatment will be given in Section 24.4.

However, (17.71) can be regarded as a useful guide to an understanding of the role which the electron spin plays in general in so-called one-electron systems. For example, an investigation of (17.71) can give us insight into the qualitative features of the alkali spectra. An alkali atom may indeed for many purposes be regarded as a one-electron atom, with all electrons in closed shells contributing merely to an effective screening of the electrostatic potential in which the single valence electron moves.

The unperturbed Hamiltonian,

$$H_0 = -\frac{\hbar^2}{2\mu} \nabla^2 + V(r) \qquad (17.72)$$

represents a familiar central force problem. We know from Chapter 9 that its eigenfunctions may be assumed to be also eigenfunctions of L_z and $\mathbf{L}^2$. H_0, which contains no reference to the spin variables, commutes of course with S_z and $\mathbf{S}^2$, and it is thus possible to make the unperturbed eigenfunctions of H_0 eigenfunctions of these operators also. The unperturbed eigenfunctions may then be written as

$$R_{ln}(r) Y_l^{m_l}(\theta, \varphi) \, \alpha, \qquad R_{ln}(r) Y_l^{m_l}(\theta, \varphi) \, \beta$$

Here n denotes a radial quantum number that characterizes the unperturbed energy eigenvalues which are in general $2(2l + 1)$-fold degenerate. Straightforward application of the methods of Section 16.5 would require us first to diagonalize the submatrix of the perturbation with respect to these degenerate eigenstates. However, a great simplification results if it is recognized that, although neither $\mathbf{L}$ nor $\mathbf{S}$ commutes with the complete Hamiltonian $H = H_0 + W(r) \mathbf{L} \cdot \boldsymbol{\sigma}$, the *total angular momentum* vector operator, $\mathbf{J} = \mathbf{L} + \mathbf{S}$, does commute with H (see Section 12.7). Hence, the simultaneous eigenfunctions of the mutually commuting operators $\mathbf{L}^2$, $\mathbf{S}^2$, $\mathbf{J}^2$, and J_z (instead of $\mathbf{L}^2$, L_z, $\mathbf{S}^2$, and S_z) make up a suitable set of correct unperturbed eigenfunctions, the perturbation matrix being automatically diagonal with respect to them. We have already calculated these eigenfunctions, and they are given explicitly in formula (16.80). However, on account of the identity $\mathbf{L} \cdot \mathbf{S} = \frac{1}{2}(\mathbf{J}^2 - \mathbf{L}^2 - \mathbf{S}^2)$, we do not need to know the detailed form of the eigenfunctions if we only wish to evaluate the first-order change in energy due to the spin-orbit interaction:

$$\Delta E = \frac{1}{\hbar} \langle E_n^{(0)}, l\tfrac{1}{2}jm|\ W(r)(\mathbf{J}^2 - \mathbf{L}^2 - \mathbf{S}^2)\ |E_n^{(0)}, l\tfrac{1}{2}jm\rangle$$

$$= [j(j + 1) - l(l + 1) - \tfrac{3}{4}]\hbar \int |R_{ln}(r)|^2\ W(r)r^2\ dr \qquad (17.73)$$

or

$$\Delta E = \left\{ \begin{matrix} l\hbar \\ -(l + 1)\hbar \end{matrix} \right\} \int |R_{ln}(r)|^2\ W(r)r^2\ dr \quad \text{for} \quad \begin{matrix} j = l + \tfrac{1}{2} \\ j = l - \tfrac{1}{2} \end{matrix} \qquad (17.74)$$

From this formula, *Landé's interval rule*, the so-called *fine structure splitting* can be evaluated if $W(r)$ is known.

As was pointed out in Section 12.1, a spin-orbit interaction in an atom arises from the interaction between the Coulomb field of the nucleus and the intrinsic magnetic moment of the moving electron. The energy associated with this interaction is

$$H_{\mathrm{magn}} = \mathbf{m} \cdot \frac{\mathbf{v}}{c} \times \mathbf{E} \approx \frac{e}{\mu c^2} \mathbf{S} \cdot \mathbf{v} \times \nabla\phi \qquad (17.75)$$

and the potential energy due to the central forces is $V(r) = -e\phi$.
Hence,

$$e\mathbf{E} = -e\nabla\phi = \frac{\mathbf{r}}{r} \frac{dV}{dr}$$

and

$$H_{\mathrm{magn}} = \frac{1}{\mu^2 c^2} \frac{1}{r} \frac{dV}{dr} \mathbf{L} \cdot \mathbf{S} \qquad (17.76)$$

When the actual calculation is made with the proper Lorentz transformation for the fields, it is found that owing to purely kinematic effects we must add a term to the energy, which has the same form as (17.76) but a different coefficient. Known as the *Thomas term*, this contribution to the Hamiltonian is

$$H_{\mathrm{Thomas}} = -\frac{1}{2\mu^2 c^2} \frac{1}{r} \frac{dV}{dr} \mathbf{L} \cdot \mathbf{S} \qquad (17.77)$$

Such a term is expected whether the central potential is of electromagnetic origin or not. The total spin-orbit interaction in atoms is the sum of expressions (17.76) and (17.77),

$$H_{\text{spin-orbit}} = \frac{1}{2\mu^2 c^2} \frac{1}{r} \frac{dV}{dr} \mathbf{L} \cdot \mathbf{S} \tag{17.78}$$

Hence, the splitting of a level of given l and n into two components with $j = l + \frac{1}{2}$ and $j = l - \frac{1}{2}$ according to (17.74) and (17.78) is determined by the interaction potential

$$W(r) = \frac{\hbar}{4\mu^2 c^2} \frac{1}{r} \frac{dV}{dr} \tag{17.79}$$

For an attractive central potential this quantity is positive, and consequently the level with $j = l - \frac{1}{2}$ lies below that with $j = l + \frac{1}{2}$.

In order to get an idea of the magnitudes involved, we note that if $V(r)$ were a pure Coulomb potential,

$$\langle W \rangle = \frac{\hbar}{4\mu^2 c^2} Ze^2 \left\langle \frac{1}{r^3} \right\rangle$$

A rigorous evaluation of $\langle 1/r^3 \rangle$ would lead us into difficulties which a proper relativistic treatment can avoid. For our purpose it is sufficient to estimate for the nth orbit

$$\left\langle \frac{1}{r^3} \right\rangle \simeq \frac{Z^3}{n^3 a^3} = \left(\frac{Ze^2 \mu}{\hbar^2 n} \right)^3$$

where a is the Bohr radius of hydrogen. Hence, owing to (10.72),

$$\langle \hbar W \rangle \simeq \left(\frac{e^2}{\hbar c} \right)^2 \frac{Z^2}{n} |E_n^{(0)}|$$

The dimensionless constant,

$$\alpha = \frac{e^2}{\hbar c} = \frac{1}{137.037} \tag{17.80}$$

is called the *fine structure constant*, and it is clear that its small value is responsible for the relative smallness of the fine structure splitting, $\Delta E_{l+1/2} - \Delta E_{l-1/2}$, compared with the gross structure energies, $E_n^{(0)} (\simeq eV)$:

$$\frac{\Delta E_{l+1/2} - \Delta E_{l-1/2}}{E_n^{(0)}} \simeq \frac{(Z\alpha)^2}{n}$$

Although the fine structure constant α was introduced here in a somewhat special context, it is of much more general significance. As a pure number which is independent of the units chosen, it has an absolute meaning. It measures the relative magnitudes of the Bohr radius, the Compton wavelength, and the classical radius of the electron:

$$\frac{\hbar^2}{\mu e^2} :: \frac{\hbar}{\mu c} :: \frac{e^2}{\mu c^2} = 1 :: \alpha :: \alpha^2$$

If $\hbar$ and c are considered to be more fundamental than e, the fine structure constant becomes a measure of the electronic charge. This point of view is dominant in current quantum field theories in which e plays the role of an interaction strength between charged matter and the electromagnetic field.

When an atom is exposed to a constant uniform magnetic field, the energy levels split further. Neglecting terms which are quadratic in the field,[8] the central force Hamiltonian (17.72) is perturbed by an interaction energy

$$H' = \frac{2}{\hbar} W \mathbf{L} \cdot \mathbf{S} + \frac{eB}{2\mu c} (g_L L_z + g_S S_z) \qquad (17.81)$$

where the z axis has conveniently been chosen in the direction of the external field, and equation (16.116) has been used for the magnetic moment of the atom, with $g_L = 1$ and $g_S = 2$ for a one-electron atom.

If the matrix elements of H' are small compared with the energy level separations of H_0, first-order perturbation theory requires, according to the discussion of Section 17.5, that we solve the eigenvalue problem of H' in a space of degenerate eigenstates of H_0. For a general atom, the zero-order energies are labeled by a set of quantum numbers of which l and s only are relevant for the present purpose, because the corresponding operators $\mathbf{L}^2$ and $\mathbf{S}^2$ commute with H'. In the $(2l + 1)(2s + 1)$-dimensional space characterized by the quantum numbers l and s, the perturbation may be written in the form

$$H' = \lambda \mathbf{L} \cdot \mathbf{S} + \mu L_z + \nu S_z \qquad (17.82)$$

which is sufficiently general to render the present discussion applicable to a large class of problems involving complex as well as one-electron atoms, and a variety of other types of angular momentum couplings, such as hyperfine interactions, etc. The coefficients λ, μ, ν in (17.82) are real constants.

The eigenvalue problem posed by the effective Hamiltonian H' may be solved in any representation, but a judicious choice of basis may save a considerable amount of labor. Evidently, in order to minimize the work of computing matrix elements, it is desirable to use as many constants of the motion as possible for the specification of the complete set of operators whose

---

[8] The neglect of changes in the energy which are quadratic in the field strength **B** implies that we ignore such induced effects as diamagnetism. (See Problem 11 at the end of this chapter.)

eigenvectors are to serve as basis. We thus see the advisability of selecting $\mathbf{L}^2$, $\mathbf{S}^2$, $\mathbf{J}^2$, J_z as the appropriate complete set spanning the coupled representation $|lsjm\rangle$. Only $\mathbf{J}^2$ fails to commute with H', and the only nonvanishing off-diagonal matrix elements are, therefore, of the form $\langle lsj'm|\,H'\,|lsjm\rangle$. In fact, since H' can be rewritten as

$$H' = \frac{\lambda}{2}(\mathbf{J}^2 - \mathbf{L}^2 - \mathbf{S}^2) + \mu J_z + (\nu - \mu)S_z \tag{17.83}$$

we only have to evaluate $\langle lsj'm|\,S_z\,|lsjm\rangle$.

The evaluation of the matrix elements of S_z is easily accomplished by using the defining relation (16.67) of the Clebsch-Gordan coefficients, and we obtain

$$\langle lsj'm|\,S_z\,|lsjm\rangle = \hbar \sum_{m_s} m_s \langle m-m_s, m_s | j'm\rangle\langle m-m_s, m_s | jm\rangle \tag{17.84}$$

We now specialize to the case $s = \tfrac{1}{2}$, which includes one-electron atoms. From (16.79) and (17.84) we derive the values

$$\langle l\tfrac{1}{2}, l\pm\tfrac{1}{2}, m|\,S_z\,|l\tfrac{1}{2}, l\pm\tfrac{1}{2}, m\rangle = \pm\frac{m\hbar}{2l+1} \tag{17.85}$$

$$\langle l\tfrac{1}{2}, l\pm\tfrac{1}{2}, m|\,S_z\,|l\tfrac{1}{2}, l\mp\tfrac{1}{2}, m\rangle = -\frac{\hbar}{2l+1}\sqrt{(l+\tfrac{1}{2})^2 - m^2} \tag{17.86}$$

As an important example we consider a ^2P state ($l = 1$) with total angular momentum $j = \tfrac{1}{2}$ and $\tfrac{3}{2}$. The operator S_z has nonvanishing off-diagonal matrix elements only between the states $|jm\rangle = |\tfrac{1}{2}\,\tfrac{1}{2}\rangle$ and $|\tfrac{3}{2}\,\tfrac{1}{2}\rangle$ and between the states $|\tfrac{1}{2}, -\tfrac{1}{2}\rangle$ and $|\tfrac{3}{2}, -\tfrac{1}{2}\rangle$. Hence, the states $|\tfrac{3}{2}\,\tfrac{3}{2}\rangle$ and $|\tfrac{3}{2}, -\tfrac{3}{2}\rangle$ are eigenvectors of H', and the remaining eigenvectors are of the form

$$a\,|\tfrac{3}{2}\,\tfrac{1}{2}\rangle + b\,|\tfrac{1}{2}\,\tfrac{1}{2}\rangle$$

and

$$c\,|\tfrac{3}{2}, -\tfrac{1}{2}\rangle + d\,|\tfrac{1}{2}, -\tfrac{1}{2}\rangle$$

The eigenvalues of H' are the first-order corrections to the energy. Denoted by E_m, they are

$$E_{\pm 3/2} = \tfrac{1}{2}\lambda\hbar^2 \pm \tfrac{1}{2}(2\mu + \nu)\hbar \tag{17.87}$$

and the roots of the secular equation

$$\begin{vmatrix} \tfrac{1}{2}\lambda\hbar^2 \pm \tfrac{1}{6}(2\mu + \nu)\hbar - E_{\pm 1/2} & \dfrac{\sqrt{2}}{3}(\mu - \nu)\hbar \\[2mm] \dfrac{\sqrt{2}}{3}(\mu - \nu)\hbar & -\lambda\hbar^2 \pm \tfrac{1}{6}(4\mu - \nu)\hbar - E_{\pm 1/2} \end{vmatrix} = 0 \tag{17.88}$$

There are two roots for $m = \tfrac{1}{2}$ and two roots for $m = -\tfrac{1}{2}$.

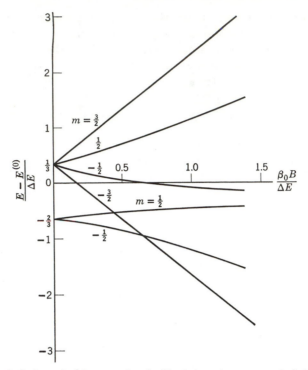

Figure 17.1. Splitting of a ^2P energy level of hydrogen in a magnetic field **B** (Zeeman effect). $E^{(0)}$ is the unperturbed energy, ΔE is the fine structure splitting.

For the case of a one-electron atom ($g_L = 1, g_S = 2$) the perturbed ^2P energies are plotted in Figure 17.1 as a function of the applied magnetic field. The resulting splitting of spectral lines is, of course, known as the *Zeeman effect*. The operator J_z commutes with H' for all values of the magnetic field, and m is thus a *"good quantum number"* throughout. In conformity with the discussion in the last section, only levels belonging to different values of m are seen to cross.

Exercise 17.6. Carry out the calculations of the Zeeman effect in detail. Show that, if the external field is either very small or very large compared with the internal magnetic field of the atom, the energy depends linearly on B. Show that the internal field which produces the fine structure is of the order of 10^4 Gauss.

7. *General Perturbation Methods.* The perturbation expansion with which we have been concerned so far in this chapter is basically simple, but its practical applicability is in many cases limited by poor convergence. Moreover, the range of values of the perturbation parameter g for which the series converges at all, even in an asymptotic sense, is frequently very restricted. Owing to this circumstance the Rayleigh-Schrödinger expansion has its greatest usefulness when the perturbation gV is produced by an external field whose magnitude can be controlled experimentally. For instance, by adjusting an applied electric or magnetic field, we can conduct the experiment at a perturbation level which is sufficiently low to warrant applicability of the power series expansion.

Frequently, however, perturbation methods are called for in the treatment of systems with interactions gV over which we have no control. For example, the Coulomb interaction between electrons in an atom or the force between nucleons in a nucleus may or may not be weak, but we certainly cannot influence it. Many iteration and successive approximation schemes have been devised in the search for perturbation methods more suitable than the Rayleigh-Schrödinger procedure. Here we cannot go into details, but a brief and general discussion of the fundamental equations on which all perturbation theories for discrete eigenstates are based is in order.[9]

We write the Hamiltonian as usual as the sum of H_0 and the perturbation gV: $H = H_0 + gV$. We assume that the spectra of both H and H_0 are discrete, but we shall not exclude degeneracy of eigenvalues. Let us suppose that the degree of degeneracy of a particular unperturbed energy level $E_n^{(0)}$ is d_n. Its eigenvectors span a d_n-dimensional subspace $\Omega_n^{(0)}$ of the general state vector space. The operator $P_n^{(0)}$ projects any vector onto this subspace. As g grows from $g = 0$ to a finite value, $\Omega_n^{(0)}$ changes to a subspace $\Omega_n^{(g)}$ which is generally different from $\Omega_n^{(0)}$ but is assumed to have the same dimensionality, d_n. It is further assumed that every vector in $\Omega_n^{(g)}$ has a nonvanishing projection on $\Omega_n^{(0)}$, i.e., $\Omega_n^{(g)}$ contains no vector which is orthogonal to the subspace $\Omega_n^{(0)}$. All this is assured for sufficiently small values of g.[10]

We shall denote by $E_{n1} \cdots E_{nd_n}$ the eigenvalues of H which grow out of $E_n^{(0)}$. In this section the eigenvectors $|E_{ni}\rangle$ are assumed to be normalized to unity. Since it is the goal of perturbation theory to work in $\Omega_n^{(0)}$, rather than $\Omega_n^{(g)}$, it is convenient to introduce the projection of $|E_{ni}\rangle$ onto $\Omega_n^{(0)}$,

$$P_n^{(0)} |E_{ni}\rangle = |\alpha_{ni}\rangle \tag{17.89}$$

The $|\alpha_{ni}\rangle$ form a set of d_n linearly independent eigenvectors of H_0, but they

---

[9] See Chapter 9 in P. M. Morse and H. Feshbach, *Methods of Theoretical Physics*, McGraw-Hill Book Company, New York, 1953. An elegant formulation of perturbation theory by C. Bloch, *Nucl. Phys.*, **6**, 329 (1958) was helpful in the preparation of this section. See also C. H. Wilcox, editor, *Perturbation Theory and its Applications in Quantum Mechanics*, John Wiley and Sons, New York, 1966.

[10] We assume g to be small enough so that $\Omega_n^{(g)}$ is a continuous function of g. This may cease to be the case when g is so large that perturbed energy levels cross.

are not necessarily orthogonal or normalized. The norm,

$$\langle \alpha_{ni} | \alpha_{ni} \rangle = \langle E_{ni} | P_n^{(0)} | E_{ni} \rangle \tag{17.90}$$

of the projection of $|E_{ni}\rangle$ is the probability of obtaining the value $E_n^{(0)}$ if a measurement of H_0 is made on the state $|E_{ni}\rangle$. The fundamental equation to be solved is

$$(H_0 - E_n^{(0)}) |E_{ni}\rangle = (\Delta E_{ni} - V) |E_{ni}\rangle \tag{17.91}$$

A simple formula is immediately obtained from this equation by multiplying on the left by $P_n^{(0)}$:

$$P_n^{(0)} V |E_{ni}\rangle = \Delta E_{ni} P_n^{(0)} |E_{ni}\rangle \tag{17.92}$$

This can be used to write for the level shift,

$$\Delta E_{ni} = \frac{\langle a | P_n^{(0)} V |E_{ni}\rangle}{\langle a | P_n^{(0)} |E_{ni}\rangle} \tag{17.93}$$

provided that $\langle a |$ is a ket which is arbitrary, except that it must not be orthogonal to $\Omega_n^{(0)}$. (Compare with Exercise 17.3.)

We can now "solve" (17.91) by regarding the right-hand side as an inhomogeneous term and using the technique of Section 17.2. The result is, of course, merely another equation for $|E_{ni}\rangle$:

$$|E_{ni}\rangle = |x\rangle + \frac{1 - P_n^{(0)}}{E_n^{(0)} - H_0} (V - \Delta E_{ni}) |E_{ni}\rangle \tag{17.94}$$

Here $|x\rangle$ is a solution of the (homogeneous) equation $(H_0 - E_n^{(0)}) |x\rangle = 0$. Multiplying both sides of (17.94) by $P_n^{(0)}$, and comparing with (17.89), we see that $|x\rangle = |\alpha_{ni}\rangle$. Hence,

$$|E_{ni}\rangle = |\alpha_{ni}\rangle + \frac{1 - P_n^{(0)}}{E_n^{(0)} - H_0} (V - \Delta E_{ni}) |E_{ni}\rangle \tag{17.95}$$

If H_0 is a differential operator, and thus (17.91) a differential equation, then (17.95) is the corresponding integral equation. It contains all sorts of unknown quantities, but if the perturbation is sufficiently small, as stipulated, we may invert this equation and write

$$|E_{ni}\rangle = \left[1 + \frac{1 - P_n^{(0)}}{E_n^{(0)} - H_0} (\Delta E_{ni} - V) \right]^{-1} |\alpha_{ni}\rangle \tag{17.96}$$

Substituting this expression finally into (17.92), we obtain

$$P_n^{(0)} V \left[1 + \frac{1 - P_n^{(0)}}{E_n^{(0)} - H_0} (\Delta E_{ni} - V) \right]^{-1} |\alpha_{ni}\rangle = \Delta E_{ni} |\alpha_{ni}\rangle \tag{17.97}$$

as a new eigenvalue equation in the subspace $\Omega_n^{(0)}$. If an expansion in powers of V is made, we are led back to the Rayleigh-Schrödinger scheme. However, other iteration techniques can be employed to obtain approximate, and possibly more rapidly convergent, solutions to this equation.

Exercise 17.7. Prove that

$$|E_{ni}^{(0)}\rangle = \lim_{g \to 0} |\alpha_{ni}\rangle$$

is an eigenvector of the operator $P_n^{(0)} V$ and that the eigenvalues are the first-order energy shifts.

Equation (17.97) is not the only useful alternative form of the eigenvalue problem. Starting from

$$(H_0 - E_{ni}) |E_{ni}\rangle = -V |E_{ni}\rangle \qquad (17.98)$$

another calculation method may be developed. Depending on whether the level shift is zero or finite, we must distinguish two separate cases. If $\Delta E_{ni} = 0$, there is no difference between (17.98) and (17.91), and thus nothing new results.[11]

Here we assume $\Delta E_{ni} \neq 0$. Hence, the operator $H_0 - E_{ni}$ has an inverse, and we have

$$|E_{ni}\rangle = - \frac{1}{H_0 - E_{ni}} V |E_{ni}\rangle \qquad (17.99)$$

Separating out on the right the projection onto $\Omega_n^{(0)}$, we have an equation similar to our earlier result:

$$|E_{ni}\rangle = |\alpha_{ni}\rangle + \frac{1 - P_n^{(0)}}{E_{ni} - H_0} V |E_{ni}\rangle \qquad (17.100)$$

and solving for $|E_{ni}\rangle$,

$$|E_{ni}\rangle = \left(1 - \frac{1 - P_n^{(0)}}{E_{ni} - H_0} V\right)^{-1} |\alpha_{ni}\rangle \qquad (17.101)$$

If we substitute this into (17.92), we finally obtain the set of equations:

$$P_n^{(0)} V \left(1 - \frac{1 - P_n^{(0)}}{E_{ni} - H_0} V\right)^{-1} |\alpha_{ni}\rangle = \Delta E_{ni} |\alpha_{ni}\rangle \qquad (17.102)$$

This can also be made the starting point of useful iteration procedures. It is important to note that the perturbed energy E_{ni}, rather than the unperturbed $E_n^{(0)}$, appears in the denominator.

8. Variational Methods. In many actual applications of quantum mechanics the Schrödinger equation cannot be solved rigorously, nor can a neighboring

---

[11] $\Delta E = 0$ occurs infrequently in bound state perturbation theory. On the other hand, negligibly small level shifts are common in scattering theory, for a spatially confined perturbing potential cannot alter the energy of a free particle at great distances. See Chapter 19.

unperturbed Hamiltonian be found which affords a good approximation and a suitable starting point for a perturbation treatment. One then often falls back upon a general approximation method based on the equivalence of the eigenvalue problem of H with a *variational problem*. For the standard one-particle form of the Schrödinger equation this equivalence was demonstrated in Section 4.6, and the general case was taken up in Section 14.5. Here, we repeat a shorthand derivation of this fundamental proposition.

First, we show that the eigenvectors of H cause the variation of the expectation value

$$\langle H \rangle = \frac{(\Psi', H\Psi')}{(\Psi', \Psi')} \tag{17.103}$$

to vanish. Indeed, for a small change $\delta\Psi'$, we have

$$(\Psi', \Psi')^2 \, \delta\langle H \rangle = (\Psi', \Psi')[(\delta\Psi', H\Psi') + (\Psi', H\,\delta\Psi')]$$
$$- (\Psi', H\Psi')[(\delta\Psi', \Psi') + (\Psi', \delta\Psi')] + 0[(\delta\Psi')^2] \tag{17.104}$$

Since H is Hermitian, it is easily seen that the vectors $\Psi' = \Psi'_k$ which make $\delta\langle H \rangle = 0$ to first order in $\delta\Psi'$ must satisfy an equation

$$H\Psi'_k = E_k\Psi'_k \tag{17.105}$$

Conversely, every eigenvector of H makes the variation of the expression (17.103) zero. The stationary values of $\langle H \rangle$ are the eigenvalues E_k.

Since we cannot extract an infinite amount of energy from any physical system, the expectation value of the energy must have a lower bound. Hence, if H represents the energy, the absolute minimum of $\langle H \rangle$ is the lowest energy eigenvalue E_0, and this value is reached if the vector Ψ'_0, used to calculate $\langle H \rangle$, is an eigenvector corresponding to E_0. Upper bounds to the ground state energy of a system are thus obtained by evaluating $\langle H \rangle$ for various trial states. This technique was discussed in Section 4.6 and utilized in several problems included in earlier chapters.

As an almost trivial application of the variational method consider a Rayleigh-Schrödinger perturbation calculation of the ground state energy of a system with a Hamiltonian H. If the kth approximation,

$$\Psi' = \Psi'^{(0)}_0 + g\Psi'^{(1)}_0 + g^2\Psi'^{(2)}_0 + \cdots + g^k\Psi'^{(k)}_0 \tag{17.106}$$

which differs from the correct eigenstate Ψ'_0 by terms of order g^{k+1}, is used in (17.103), the value of $\langle H \rangle$ will differ from E_0 by terms of order g^{2k+2} only [see (17.104)]. It follows that $\langle H \rangle$, computed with the trial vector (17.106), gives E_0 correctly to order g^{2k+1}.

Exercise 17.8. Illustrate the connection between the variational and the Rayleigh-Schrödinger perturbation methods by applying a trial vector

$$\Psi_0 = \Psi_0^{(0)} + g\Psi_0^{(1)}$$

to approximate the ground state.

In principle, the variational method can be used for states other than the ground state. We already saw in Chapter 14 how it can be applied to states of successively higher energy by gradually narrowing the vector space in which H operates. If the eigenvector Ψ_0 has been obtained, its subspace is split off and the variational procedure repeated in the remaining subspace of all vectors which are orthogonal to Ψ_0. Obviously, the method will give more and more inaccurate results if Ψ_0 is known only approximately. This difficulty can be avoided if it is possible to construct a subset of trial vectors which are known to be rigorously orthogonal to the exact Ψ_0. For instance, in the case of a particle in a central field, where the ground state is an S-state, we use only spherically symmetric trial functions to estimate E_0. Similarly, by using P-state wave functions we can obtain an upper limit for the lowest P-state, etc.

If we know some constant of the motion A, such that

$$HA - AH = 0 \tag{17.107}$$

the projection operator $P_{A_i'}$, which projects any vector into the subspace spanned by the eigenvectors of A with eigenvalue A_i', is also a constant of the motion. Hence,

$$HP_{A_i'} - P_{A_i'}H = 0 \tag{17.108}$$

Applying $P_{A_i'}$ to (17.105) and noting that a projection operator is idempotent, we see that

$$HP_{A_i'}(P_{A_i'}\Psi_k) = E_k(P_{A_i'}\Psi_k) \tag{17.109}$$

We now set $HP_{A_i'} = H_{A_i'}$ and obtain a new set of eigenvalue problems,

$$H_{A_i'}\Psi_k' = E_k\Psi_k' \tag{17.110}$$

The eigenvalues clearly belong to the eigenvalue spectrum of H. We can approximate the lowest energy eigenvalue for each value A_i' by minimizing the expectation value $\langle H_{A_i'}\rangle$ of the new Hamiltonian.

A generalization of the variational principle can be formulated if we vary not only the state vector Ψ but also the Hamiltonian H. Denoting such generalized variations by the symbol Δ, and neglecting terms which are quadratic in Δ, we have by the usual rule concerning the variation of a product $[\Delta(ab) = (\Delta a)b + a(\Delta b)]$

$$\Delta\langle H\rangle \equiv \Delta\frac{(\Psi, H\Psi)}{(\Psi, \Psi)} = \frac{(\Psi, \Delta H\Psi)}{(\Psi, \Psi)} + \delta\frac{(\Psi, H\Psi)}{(\Psi, \Psi)} \tag{17.111}$$

where the δ-variation is one in which the operator is kept fixed. If the variation is performed near an eigenvector Ψ'_k of the operator H, the last term vanishes, and we are left with

$$\Delta \frac{(\Psi'_k, H\Psi'_k)}{(\Psi'_k, \Psi'_k)} = \frac{(\Psi'_k, \Delta H\Psi'_k)}{(\Psi'_k, \Psi'_k)} \tag{17.112}$$

Conversely, any vector Ψ'_k for which relation (17.112) holds is an eigenvector of H. For $\Delta H = 0$ the Δ-variation goes over into the δ-variation, and the extremum property of the eigenvectors is recovered.

So far the variations of Ψ' and H have been independent. If we now restrict them by requiring that Ψ' shall remain an eigenvector of H during the variation, $\Delta\langle H \rangle$ becomes the change ΔE_k of the eigenvalue E_k, and we get

$$\Delta E_k = \frac{(\Psi'_k, \Delta H\Psi'_k)}{(\Psi'_k, \Psi'_k)} \tag{17.113}$$

which is, of course, identical with the result of first-order perturbation theory.

From (17.112) and (17.113) the variational theorem may be generalized as follows: Ψ' is an eigenvector of H if and only if $\Delta\langle H \rangle$ depends only on the change in H, and not on the change in Ψ'. The value of $\Delta\langle H \rangle$ is then equal to the change of the corresponding eigenvalue.

Exercise 17.9. If $H(\lambda)$ depends on a parameter λ, and $\Psi'_k(\lambda)$ is an eigenvector which is normalized to unity, prove that

$$\frac{dE_k(\lambda)}{d\lambda} = \left(\Psi'_k(\lambda), \frac{\partial H(\lambda)}{\partial \lambda} \Psi'_k(\lambda) \right) \tag{17.114}$$

(Hellmann-Feynman theorem).

Exercise 17.10. If $\Psi'^{(0)}_{n1}$ and $\Psi'^{(0)}_{n2}$ are two orthogonal degenerate eigenvectors of an unperturbed Hamiltonian H_0, calculate the energy splitting produced by a perturbation V by using a variational trial vector $\alpha\Psi'^{(0)}_{n1} + \beta\Psi'^{(0)}_{n2}$, where α and β are variable parameters subject to the constraint $|\alpha|^2 + |\beta|^2 = 1$. Compare with the degenerate perturbation theory of Section 17.5.

9. The Helium Atom. The neutral helium atom with a fixed nucleus is described by the Schrödinger equation in configuration space:

$$\left[-\frac{\hbar^2}{2\mu}(\nabla'^2 + \nabla''^2) - \frac{2e^2}{r'} - \frac{2e^2}{r''} + \frac{e^2}{|\mathbf{r}' - \mathbf{r}''|} \right] \psi(\mathbf{r}', \mathbf{r}'') = E\psi(\mathbf{r}', \mathbf{r}'') \tag{17.115}$$

The coordinates of the two electrons are labeled $\mathbf{r}'$ and $\mathbf{r}''$ under the provisional assumption that the particles are in principle distinguishable. Of course, we know this assumption to be false but, since electrons have spin $\frac{1}{2}$, with two possible substates, and to the extent that spin-orbit interactions may be neglected, the fiction of two (but no more!) distinguishable electrons can be maintained in solving equation (17.115). We may think of one electron as having spin "up" and the other spin "down," so that the two particles belong in effect to two different, distinguishable electron species. We shall see that with this assumption we can obtain the entire spectrum of a two-electron system. The corrections to this picture, imposed by the identity of two electrons in the same spin state and by the Pauli exclusion principle, will be found in Chapter 21.

Owing to the symmetry of the differential operator in (17.115), the solutions of the Schrödinger equation,

$$H(\mathbf{r}', \mathbf{r}'')\psi(\mathbf{r}', \mathbf{r}'') = E\psi(\mathbf{r}', \mathbf{r}'') \tag{17.116}$$

fall naturally into two classes: Every solution can be assumed to be either symmetric or antisymmetric in the space coordinate

$$\psi(\mathbf{r}'', \mathbf{r}') = \pm\psi(\mathbf{r}', \mathbf{r}'')$$

If a particular solution $f(\mathbf{r}', \mathbf{r}'')$ fails to have this property, we observe that $f(\mathbf{r}'', \mathbf{r}')$ is also a solution (a phenomenon which is referred to as *exchange degeneracy*, because both solutions belong to the same energy eigenvalue). Hence, owing to the linearity of the Schrödinger equation, the linear combinations

$$f(\mathbf{r}', \mathbf{r}'') + f(\mathbf{r}'', \mathbf{r}') = \psi_+(\mathbf{r}', \mathbf{r}'') \quad \text{and} \quad f(\mathbf{r}', \mathbf{r}'') - f(\mathbf{r}'', \mathbf{r}') = \psi_-(\mathbf{r}', \mathbf{r}'')$$

are also solutions. Since conversely every solution can be written as a superposition of solutions with definite symmetry,

$$f(\mathbf{r}', \mathbf{r}'') = \tfrac{1}{2}\psi_+(\mathbf{r}', \mathbf{r}'') + \tfrac{1}{2}\psi_-(\mathbf{r}', \mathbf{r}''),$$

we need be concerned only with the *symmetric* and *antisymmetric* eigensolutions of the Schrödinger equation (17.115).

It is tempting to neglect in lowest approximation the repulsive interaction between the electrons and treat the neglected part of the potential energy as a perturbation. Since the repulsive term is clearly of the same order of magnitude as the Coulomb attraction due to the nucleus, there is no justification for this procedure other than the simplification which results from it. We permit ourselves to neglect this term because we shall thus obtain a qualitative understanding of the level scheme of helium.

The simplification results because the approximate Schrödinger equation,

$$\left[-\frac{\hbar^2}{2\mu}(\nabla'^2 + \nabla''^2) - 2e^2\left(\frac{1}{r'} + \frac{1}{r''}\right) \right]\psi^{(0)}(\mathbf{r}', \mathbf{r}'') = E^{(0)}\psi^{(0)}(\mathbf{r}', \mathbf{r}'') \tag{17.117}$$

is further separable into two hydrogenic equations

$$\left(-\frac{\hbar^2}{2\mu}\nabla'^2 - \frac{2e^2}{r'}\right)g(\mathbf{r}') = E'g(\mathbf{r}')$$

$$\left(-\frac{\hbar^2}{2\mu}\nabla''^2 - \frac{2e^2}{r''}\right)h(\mathbf{r}'') = E''h(\mathbf{r}'')$$

if we assume that

$$E^{(0)} = E' + E''$$

and

$$\psi^{(0)}(\mathbf{r}', \mathbf{r}'') = g(\mathbf{r}')h(\mathbf{r}'')$$

The lowest unperturbed energy level of the atom corresponds to $E' = E'' = -4 \times 13.6\,\text{eV} = -54.4\,\text{eV}$ with the wave function ($Z = 2$)

$$\psi_+^{(0)} = \frac{Z^3}{\pi a^3} \exp\left[-\frac{Z(r' + r'')}{a}\right] \tag{17.118}$$

which is a product of two hydrogenic $1s$ wave functions with $Z = 2$. This is symmetric under an exchange of r' and r''.

The first-order correction to the energy is

$$\Delta E = \int |\psi_+^{(0)}(\mathbf{r}', \mathbf{r}'')|^2 \frac{e^2}{|\mathbf{r}' - \mathbf{r}''|} d^3r'\, d^3r'' = \frac{5}{4}\frac{e^2}{a} = \frac{5}{2} \times 13.6 = 34\,\text{eV}$$

The total energy in first approximation is then

$$E_{\text{cal}} = -2 \times 54.4 + 34 = -74.8\,\text{eV}$$

The measured ionization potential of helium is 24.46 Volts. Hence the total energy of the atom is

$$E_{\text{obs}} = -54.4 - 24.5 = -78.9\,\text{eV}$$

$E_{\text{cal}} > E_{\text{obs}}$ in agreement with the variational principle.

The agreement between theory and experiment can be further improved by using a better trial function than (17.118). The design of suitable trial functions has been the subject of ceaseless investigation ever since quantum mechanics was invented. The problem offers a challenge to the theoretician because the two-electron system is mathematically manageable, although solutions in closed form cannot be found. Notable efforts have been made in devising better and better variational trial functions for the ground states of helium, the test of their quality being how closely $\langle H \rangle$ lies above the observed value of the energy.

We shall content ourselves with describing the very simplest variational method. It uses trial functions of the form (17.118) but with Z replaced by an effective nuclear charge, $Z_{\text{eff}} = Z - \sigma$. This is physically reasonable, because each electron is partially screened from seeing the full charge of the nucleus by the presence of the other electron. If σ is left arbitrary, we can calculate the expectation value of the energy

$$E(\sigma) = \langle H \rangle \tag{17.119}$$

where H is the complete Hamiltonian for the helium atom, including the interaction between the electrons. According to the variational theorem, $E(\sigma) \geqslant E$ for any value of σ. The best estimate with a trial function of the form (17.118) is therefore obtained by minimizing $E(\sigma)$ with respect to σ. The result for any two-electron atom with nuclear charge Z is

$$E(\sigma) = - \frac{e^2}{a}(Z^2 - \tfrac{5}{8}Z + \tfrac{5}{8}\sigma - \sigma^2)$$

Exercise 17.11. Work out this result.

The minimum of $E(\sigma)$ is obtained for $\sigma = \tfrac{5}{16}$, and has the value

$$E(\tfrac{5}{16}) = -(Z - \tfrac{5}{16})^2 \frac{e^2}{a} \tag{17.120}$$

Exercise 17.12. Show that, if an unperturbed Hamiltonian,

$$H_0 = -\frac{\hbar^2}{2\mu}\nabla'^2 - \frac{\hbar^2}{2\mu}\nabla''^2 - \frac{(2 - \tfrac{5}{16})e^2}{r'} - \frac{(2 - \tfrac{5}{16})e^2}{r''} \tag{17.121}$$

is chosen, the first-order perturbation energy of the ground state of helium vanishes.

Numerically, according to (17.120), the ground state energy for helium is

$$E_{\text{cal}} = -\tfrac{729}{128} \times 13.6 = -77.46 \text{ eV}$$

which is quite close to the measured value of -78.86 eV. Trial functions with many more adjustable parameters have been employed and have led to extremely satisfactory agreement between theory and experiment.[12] The excited states of helium are obtained by lifting one electron to an excited hydrogenic level, and leaving the other electron in the ground state. The corresponding zero-order wave functions are

$$\psi_{\pm}^{(0)} = \frac{1}{\sqrt{2}}[\psi_{100}(\mathbf{r}')\psi_{nlm_l}(\mathbf{r}'') \pm \psi_{nlm_l}(\mathbf{r}')\psi_{100}(\mathbf{r}'')] \tag{17.122}$$

with an energy $E^{(0)} = E_0 + E_n$. The unperturbed levels are degenerate

[12] C. L. Pekeris, *Phys. Rev.*, **115**, 1216 (1959), has used electronic computation to arrive at a theoretical value of 198,310.687 cm^{-1} for the ionization potential of helium. The best experimental value is 198,310.8$_2$ $\pm$ 0.15 cm^{-1}.

exactly as in the hydrogen atom. The interaction between the two electrons removes the l-degeneracy. The unperturbed eigenfunctions (17.122) are just the correct linear combinations to begin a degenerate perturbation treatment: The perturbation potential is diagonal with respect to any two states of the type (17.122) because the total orbital angular momentum $\mathbf{L}$ is a constant of the motion and because two states of different symmetry cannot combine.

The excited states thus split into components according to their symmetry under exchange of the particle coordinates. The energy correction due to the symmetric perturbation $V(\mathbf{r}', \mathbf{r}'')$ is

$$\Delta E = \int \psi_{100}^*(\mathbf{r}')\psi_{nlm_l}^*(\mathbf{r}'')V(\mathbf{r}', \mathbf{r}'')\psi_{100}(\mathbf{r}')\psi_{nlm_l}(\mathbf{r}'')\, d^3r'\, d^3r''$$

$$\pm \int \psi_{100}^*(\mathbf{r}')\psi_{nlm_l}^*(\mathbf{r}'')V(\mathbf{r}', \mathbf{r}'')\psi_{nlm_l}(\mathbf{r}')\psi_{100}(\mathbf{r}'')\, d^3r'\, d^3r'' \quad (17.123)$$

The upper sign holds for the symmetric and the lower sign for the antisymmetric states of the form (17.122). We can write ΔE as

$$\Delta E = I \pm J \quad (17.124)$$

Here the so-called *direct integral*,

$$I = \int \psi_{100}^*(\mathbf{r}')\psi_{nlm_l}^*(\mathbf{r}'')V(\mathbf{r}', \mathbf{r}'')\psi_{100}(\mathbf{r}')\psi_{nlm_l}(\mathbf{r}'')\, d^3r'\, d^3r'' \quad (17.125)$$

is the expectation value of the perturbation which would obtain if exchange degeneracy were absent from the unperturbed Hamiltonian, or if ψ_{100} and ψ_{nlm_l} did not overlap. The other term,

$$J = \int \psi_{100}^*(\mathbf{r}')\psi_{nlm_l}^*(\mathbf{r}'')V(\mathbf{r}', \mathbf{r}'')\psi_{nlm_l}(\mathbf{r}')\psi_{100}(\mathbf{r}'')\, d^3r'\, d^3r'' \quad (17.126)$$

bears the descriptive name *exchange integral*.

Exercise 17.13. To give an interpretation of the two nonstationary states

$$\psi_{100}(\mathbf{r}')\psi_{nlm_l}(\mathbf{r}'') \qquad \text{and} \qquad \psi_{nlm_l}(\mathbf{r}')\psi_{100}(\mathbf{r}'')$$

show that, if the atom is initially in one of these, it shuttles back and forth with an exchange frequency $J/\hbar$. [See (5.62).]

For our perturbation, $V = e^2/|\mathbf{r}' - \mathbf{r}''|$, I is of course positive. Calculation shows that J is also positive. Hence, all symmetric terms lie slightly higher than the corresponding antisymmetric terms. This behavior is understandable, since in the spatially antisymmetric state the probability is small for the electrons to be found near each other, while in the symmetric state they have a greater opportunity to repel each other, thereby raising the energy.

Since helium atoms in spatially symmetric states are physically quite different from those in spatially antisymmetric states, a special terminology is often used. Helium atoms in spatially symmetric states are said to form *parahelium*, those in spatially antisymmetric states form *orthohelium*. It is clear that the normal ground state of helium is a para state.

Problems

1. The Hamiltonian of a rigid rotator in a magnetic field perpendicular to the x-axis is of the form $AL^2 + BL_z + CL_y$, if the term which is quadratic in the field is neglected. Obtain the exact energy eigenvalues and eigenfunctions of this Hamiltonian. Then, assuming $B \gg C$, use second-order perturbation theory to get approximate eigenvalues and compare these with the exact answers.

2. A charged particle is constrained to move on a spherical shell in a weak uniform electric field. Obtain the energy spectrum to second order in the field strength.

3. Apply perturbation theory to the elastically coupled harmonic oscillators of Problems 4 and 5 in Chapter 15, assuming that the interaction between the two particles is weak, and compare with the rigorous solutions.

4. Use second-order perturbation theory to calculate the change in energy of a linear harmonic oscillator when a constant force is added, and compare with the exact result.

5. Solve the Schrödinger problem for a particle which is confined in a two-dimensional square box whose sides have length L and are oriented along the x and y coordinate axes with one corner at the origin. Find the eigenvalues and eigenfunctions, and calculate the number of eigenstates per unit energy interval for high energies.

   A small perturbation $V = Cxy$ is now introduced. Find the energy change of the ground state and the first excited state in the lowest nonvanishing order. Construct the appropriate zero-order wave function for the perturbed problem in the case of the first excited state.

6. Set up an equation for the first-order correction to the energies of two almost, but not quite, degenerate unperturbed levels, when a small perturbation is applied.

7. A slightly anisotropic three-dimensional harmonic oscillator has $\omega_z \approx \omega_x = \omega_y$. A charged particle moves in the field of this oscillator and is at the same time exposed to a uniform magnetic field in the x-direction. Assuming that the Zeeman splitting is comparable to the splitting produced by the anisotropy, but small compared to $\hbar\omega$, calculate to first order the energies of the components of the first excited state. Discuss various limiting cases.

8. Prove that if $\psi_0 = e^{-\varphi_0}$ is a positive bounded function satisfying appropriate boundary conditions, it represents the ground state for a particle moving in a potential

$$V = \frac{\hbar^2}{2\mu} [(\nabla \varphi_0)^2 - \nabla^2 \varphi_0]$$

and the corresponding energy is zero. Verify the theorem for (a) the isotropic harmonic oscillator and (b) the hydrogen atom.

9. Prove that the trace of the direct product of two matrices equals the product of the traces of the matrices. Apply this result to show that the "center of gravity" of a multiplet split by the spin-orbit interaction is at the position of the unperturbed energy level.

10. Obtain the relativistic correction $\propto p^4$ to the nonrelativistic kinetic energy of an electron and, using first-order perturbation theory, evaluate the energy shift which it produces in the ground state of hydrogen.

11. Using the Hamiltonian for an atomic electron in a magnetic field, determine, for a state of zero angular momentum, the energy change to order $\mathbf{B}^2$, if the system is in a uniform magnetic field represented by the vector potential $\mathbf{A} = \frac{1}{2}\mathbf{B} \times \mathbf{r}$.

   Defining the atomic diamagnetic susceptibility χ by $E = -\frac{1}{2}\chi \mathbf{B}^2$, calculate χ for a helium atom in the ground state and compare the result with the measured value.

12. The energy of the lowest eigenstate of a double harmonic oscillator with fixed ω (Section 5.6) depends on the distance a and has a minimum at $a = a_0$ (see Figure 5.4). Use Exercise 17.9 to show that if $a = a_0$, the expectation value of $|x|$ is equal to a_0.

13. Apply second-order perturbation theory to a one-dimensional periodic perturbing potential

$$V(x) = \sum_{n=-\infty}^{+\infty} V_n e^{2\pi i n x/l}$$

with period l. Assume that the entire "crystal" has N periods and use the periodic boundary condition $\psi(x + \frac{1}{2}Nl) = \psi(x - \frac{1}{2}Nl)$, where N is an even number. Assume that the zero-order Hamiltonian is that of a free particle. Show that nondegenerate perturbation theory breaks down at the band edges.

   Assuming that the zero-order free particle energy is large compared with g/l, apply the results to the special example of a Kronig-Penney potential with δ-functions of strength g, and verify the result of perturbation theory by expanding the exact eigenvalue condition at $kl = \pi/2$.

14. Apply degenerate perturbation theory to estimate the widths of the forbidden gaps in the energy bands of the periodic potential shown in Figure 6.2. Consider the entire potential as a perturbation on an unperturbed free particle state. Verify the result by obtaining approximate solutions of equation (6.74) at the band edges, or of equation (6.78).

15. For Bloch wave functions in a periodic potential, prove by the use of Exercises 6.11 and 17.9 that the expectation value of the velocity is equal to the gradient of the energy in $\hbar k$-space.

16. Let ψ be the variational trial function for the ground state ψ_0 of a system with nondegenerate energy eigenvalues. Assume that ψ and ψ_0 are real, normalized to unity, and that $\int \psi\psi_0 \, d\tau$ is positive. Show that

$$\frac{1}{2}\int |\psi - \psi_0|^2 \, d\tau \leqslant 1 - \left(1 - \frac{\langle H \rangle - E_0}{E_1 - E_0}\right)^{1/2} \approx \frac{1}{2}\frac{\langle H \rangle - E_0}{E_1 - E_0}$$

where E_0 and E_1 are the exact energies of the ground and first excited states, and

$\langle H \rangle$ is the expectation value of the Hamiltonian in the state ψ. Estimate the accuracy of the trial functions used in Section 17.9 for the ground state of the helium atom.

17. A rotator whose orientation is specified by the angular coordinates θ and φ performs a *hindered rotation* described by the Hamiltonian

$$H = AL^2 + B\hbar^2 \cos 2\varphi$$

with $A \gg B$. Calculate the S, P, and D energy levels of this system in first-order perturbation theory, and work out the corresponding unperturbed energy eigenfunctions.

18. Assume that n unperturbed, but not necessarily degenerate, eigenstates $|k\rangle$ of H_0 (with $k = 1, 2, \ldots, n$) all interact with one of them, say $|1\rangle$, but not otherwise so that $\langle k| V |k'\rangle \neq 0$ only if either $k = k'$ or $k = 1$ or $k' = 1$. Solve the eigenvalue problem in the n-dimensional vector space exactly and derive an implicit equation for the perturbed energies. Using a graphic method, discuss the solutions of the eigenvalue problem for various assumed values of the non-vanishing matrix elements of V, and exhibit the nature of the perturbed eigenstates.

CHAPTER 18

Time-Dependent Perturbation Theory

1. The Equation of Motion. When the nature of a physical system, i.e., its Hamiltonian, is known, we are often confronted with the problem of predicting the time development of the system from its initial conditions. If H is a time-independent Hamiltonian, this problem has a straightforward answer when all solutions of the eigenvalue problem,

$$H\Psi_n = E_n\Psi_n \tag{18.1}$$

are known. Indeed, as we have seen repeatedly, the state vector

$$\Psi(t) = \sum_n c_n \exp\left(-\frac{i}{\hbar}E_nt\right)\Psi_n \tag{18.2}$$

represents the general solution of the equation of motion,

$$i\hbar\frac{d\Psi(t)}{dt} = H\Psi(t) \tag{18.3}$$

and can be adapted to coincide with a given initial state $\Psi(t_0)$, if the coefficients are chosen to be

$$c_n = (\Psi_n, \Psi(t_0)) \exp\left(\frac{i}{\hbar}E_nt_0\right) \tag{18.4}$$

In Section 11.2 this procedure was applied to the scattering of a wave packet by a fixed potential.

Usually, H is a complicated operator, and rigorous solutions of (18.1) are not available except in special cases. Stationary state perturbation theory could be used to obtain approximate eigenvalues and eigenfunctions of H for substitution into (18.2). Although this procedure is in principle perfectly suitable if H is time-independent, we shall in this chapter take up a mathematically simpler approach, which has the additional advantage of lending itself to immediate physical interpretation.

Since a perturbation method will be applied directly to the *time-dependent* equation (18.3), H need not be time-independent. Indeed, perturbations which are explicitly dependent on the time are of considerable practical importance. The excitation or ionization of an atom or nucleus by a varying electric field is an obvious example, and we shall work out the cross sections for excitation by a passing charged particle and for the absorption of light. When H depends on t there are, of course, no strictly stationary states, and the energy of the system is not conserved. Hence the algorithm (18.1) through (18.4) cannot be used.

It is characteristic of the physical processes mentioned—scattering of a particle by a fixed potential, and absorption of energy by a system from a passing disturbance—that the perturbing interaction is limited in space and time. The system is unperturbed before it enters the scattering region or before it is hit by the wave, and it is again free from perturbation after sufficient time has elapsed. This sequence of events suggests that the total Hamiltonian be again considered as the sum of two terms

$$H = H_0 + V \tag{18.5}$$

where the time-independent operator H_0 describes the unperturbed system, and V is the perturbation which may be explicitly time-dependent (as in the case of excitation by a transient electromagnetic field) or not (as in the case of scattering). In either case the initial state is, at least approximately, an eigenstate of H_0. After the perturbation has acted, we observe the system again and determine the probabilities with which the system is found in the various eigenstates of H_0. The forced linear harmonic oscillator, for which exact solutions were worked out in Section 15.9, may serve to test any approximation methods to be developed.

The basic idea of time-dependent perturbation theory is simple. First it is recognized that, if the perturbation V were absent, the eigenfunctions appropriate to the problem would be given by the equation

$$H_0 \Psi_n^{(0)} = E_n^{(0)} \Psi_n^{(0)} \tag{18.6}$$

If the initial state is expanded in terms of these unperturbed eigenfunctions,

$$\Psi(t_0) = \sum_n c_n \exp\left(-\frac{i}{\hbar} E_n^{(0)} t_0\right) \Psi_n^{(0)} \tag{18.7}$$

and we would, in the absence of a perturbation, have for all times

$$\Psi(t) = \sum_n c_n \exp\left(-\frac{i}{\hbar} E_n^{(0)} t\right) \Psi_n^{(0)} \tag{18.8}$$

If the perturbation V is present, (18.8) is no longer a solution of the wave equation, but it is still legitimate to expand $\Psi(t)$ at every instant in terms of the eigenfunctions $\Psi_n^{(0)}$ of the unperturbed problem, provided that the coefficients are now regarded as depending on the time:

$$\Psi(t) = \sum_n c_n(t) \exp\left(-\frac{i}{\hbar} E_n^{(0)} t\right) \Psi_n^{(0)} \tag{18.9}$$

Evidently,

$$c_n(t) = c_n = (\Psi_n^{(0)}, \Psi(t)) \exp\left(\frac{i}{\hbar} E_n^{(0)} t\right)$$

is the probability amplitude for finding the system in the nth unperturbed state. If (18.9) is substituted into the equation of motion,

$$i\hbar \frac{d\Psi}{dt} = (H_0 + V)\Psi \tag{18.3}$$

equations of motion are obtained for the amplitudes $c_n(t)$, using the linear independence of the eigenstates $\Psi_n^{(0)}$; we get readily

$$i\hbar \frac{dc_k}{dt} = \sum_n V_{kn} c_n e^{i\omega_{kn}t} \tag{18.10}$$

where

$$\omega_{kn} = \frac{E_k^{(0)} - E_n^{(0)}}{\hbar} \tag{18.11}$$

and

$$V_{kn} = (\Psi_k^{(0)}, V\Psi_n^{(0)}) \tag{18.12}$$

Thus, V_{kn} is the matrix element of the perturbation between the unperturbed eigenstates.

Equation (18.10) is a system of simultaneous linear homogeneous differential equations. No approximation has yet been made. Equation (18.10) expresses the equation of motion in a particular representation spanned by the moving eigenvectors of the unperturbed Hamiltonian H_0. In matrix notation we can write:

$$i\hbar \frac{d}{dt} \begin{pmatrix} c_1 \\ c_2 \\ \cdot \\ \cdot \\ \cdot \end{pmatrix} = \begin{pmatrix} V_{11} & V_{12} e^{i\omega_{12}t} \cdots \\ V_{21} e^{-i\omega_{12}t} & V_{22} \cdots \\ & \cdot \\ \cdot & \cdot \end{pmatrix} \begin{pmatrix} c_1 \\ c_2 \\ \cdot \\ \cdot \\ \cdot \end{pmatrix}$$

It is in the solution of this complicated linear system that the perturbation approximation is invoked. The method takes its name, *time-dependent perturbation theory*, from the fact that it solves directly the time-dependent equation (18.3) rather than the time-independent Schrödinger equation. The perturbation may or may not depend on the time.

2. The Perturbation Method. The solution of (18.10) depends critically on the initial conditions. For simplicity let us first suppose that the system is at the initial time $t_0 = -\infty$ definitely in one of the stationary states of the unperturbed Hamiltonian. Such a state is in general not an eigenstate of the perturbed Hamiltonian; hence its development in time is not simply oscillatory. How this kind of an initial state is adapted to the real physical conditions which prevail in an experiment will be shown later in several examples and applications.

Assuming that H_0 possesses only discrete energy levels, the initial conditions

$$c_s(-\infty) = 1, \qquad c_k(-\infty) = 0 \qquad \text{if } k \neq s \tag{18.13}$$

are introduced, and a successive approximation method is instituted to solve (18.10) subject to these initial conditions. The initial values of the coefficients c_k are substituted on the right-hand side of (18.10). For $k \neq s$ the approximate equations

$$i\hbar \frac{dc_k}{dt} = V_{ks} e^{i\omega_{ks}t} \qquad (k \neq s) \tag{18.14}$$

are thus obtained. Obviously, these equations are valid only for values of t such that $c_k(t) \ll c_s(t) \approx 1$, if $k \neq s$. Equation (18.14) can be integrated if use is made of the initial conditions (18.13):

$$c_k(t) = -\frac{i}{\hbar} \int_{-\infty}^{t} V_{ks} e^{i\omega_{ks}t'} dt' \qquad (k \neq s) \tag{18.15}$$

If the perturbation is a transient one and sufficiently small in magnitude, the amplitudes c_k may remain small throughout. After the perturbation has ceased, the system settles down again to constant values for the coefficients c_k, and these are given by (18.15) evaluated at $t = +\infty$:

$$c_k(+\infty) = -\frac{i}{\hbar} \int_{-\infty}^{+\infty} V_{ks} e^{i\omega_{ks}t'} dt' \tag{18.16}$$

This formula shows that under the influence of the time-dependent perturbation the system makes transitions to other eigenstates of H_0. The transition probability $|c_k(+\infty)|^2$ to state k is proportional to the square of the absolute value of the Fourier component of the perturbation matrix element V_{ks} evaluated at the transition frequency ω_{ks}.

Frequently, the approximation (18.14) is too drastic to be practically useful but can be significantly improved by taking the interaction of a selected number of states more accurately into account. As in all perturbation theories, the chief criterion of success is the consistency of the solution with the simplifying assumptions made in arriving at the approximate equations. If a proper choice of *closely coupled states* is made, the perturbed amplitudes may turn out to be sufficiently accurate for times long enough to have physical relevance.

If, for example, r and s label two states which interact strongly with each other but only weakly with all other states k, and if the initial state is a linear combination of $\Psi_r^{(0)}$ and $\Psi_s^{(0)}$ only, the equations (18.14) are replaced by the set of first-order perturbation equations:

$$i\hbar \frac{dc_k}{dt} = V_{kr}e^{i\omega_{kr}t}c_r + V_{ks}e^{i\omega_{ks}t}c_s \qquad (k \neq r, s)$$

$$i\hbar \frac{dc_r}{dt} = V_{rr}c_r + V_{rs}e^{i\omega_{rs}t}c_s \qquad\qquad (18.17)$$

$$i\hbar \frac{dc_s}{dt} = V_{sr}e^{i\omega_{sr}t}c_r + V_{ss}c_s$$

An example of the use of these coupled equations will be given in Section 18.5.

3. *Coulomb Excitation.* We shall now apply time-dependent perturbation theory to study the effect of a transient Coulomb field on a compact bound system of charged particles, which we may call the *target*.

As a concrete example consider the excitation of a *nucleus* by a particle with charge $Z_1 e$ which sweeps by. When charged particles, e.g., alpha particles of moderate energy, bombard a nucleus of atomic number Z_2, the particles are deflected by the mutual Coulomb repulsion, and the familiar Rutherford scattering takes place (Section 11.8). However, in the process the nucleus sees a rapidly varying electric field which can cause transitions from the ground state to some excited nuclear state. When this happens, we speak of *Coulomb excitation* in contrast to other inelastic processes which occur if the projectile is a nucleus and fast enough to allow penetration of the Coulomb barrier, so that the two nuclei can interact strongly through forces of origin other than electromagnetic.

A second example is the excitation of an *atom* by the passage of a charged particle. This is the dominant mechanism by which charged particles are slowed down in matter.

In lowest approximation it is frequently legitimate to treat the projectile as a classical particle moving with a definite initial momentum and impact parameter in the field of the target and to neglect altogether the energy loss which this particle must suffer if excitation results. The perturbation interaction is simply of the form

$$V = \pm Z_1 e^2 \sum_{i=1}^{Z_2} \frac{1}{|\mathbf{r}_i - \mathbf{R}(t)|} \qquad (18.18)$$

where $\mathbf{r}_1, \mathbf{r}_2, \ldots, \mathbf{r}_{Z_2}$ denote the position operators for the Z_2 charged particles in the target, and $\mathbf{R}(t)$ is the position vector of the projectile with charge Z_1e. The $+$ or $-$ sign is to be used depending on whether the charges are of like sign or not. It is convenient to choose the center of mass of the target as the coordinate origin.

If the projectile never enters the target, an electric multipole expansion of the form

$$\frac{1}{|\mathbf{r}_i - \mathbf{R}|} = \sum_{l=0}^{\infty} \frac{r_i^l}{R^{l+1}} P_l(\hat{\mathbf{r}}_i \cdot \hat{\mathbf{R}}) \tag{18.19}$$

is appropriate. Substituting this into (18.16), we obtain

$$c_k(+\infty) = \mp \frac{iZ_1e^2}{\hbar} \sum_{l=0}^{\infty} \int_{-\infty}^{+\infty} e^{i\omega_{ks}t'} \frac{\langle k| \sum_{i=1}^{Z_2} r_i^l P_l(\hat{\mathbf{r}}_i \cdot \hat{\mathbf{R}}(t')) |s\rangle}{[R(t')]^{l+1}} \, dt' \tag{18.20}$$

If the distance between the projectile and the target is always large compared with the size of the target the multipole expansion converges rapidly, and only the first nonvanishing term in the power series in l in (18.20) need be retained. For a given initial state s and a final state k, there is always a smallest value of l for which the matrix element $\langle k| \sum_{i=1}^{Z_2} r_i^l P_l(\hat{\mathbf{r}}_i \cdot \hat{\mathbf{R}}(t')) |s\rangle$ will not vanish. Since $P_0 = 1$, and k and s are orthogonal states, the matrix elements for $l = 0$ (monopole transitions) always vanish. $l = 1$ corresponds to electric dipole transitions, $l = 2$ to quadrupole transitions, and generally we speak of 2^l-pole transitions.

The calculation of the matrix elements is in general complicated but becomes fairly simple if we assume that the transitions are of the *electric dipole* type ($l = 1$). If the plane of the orbit is taken to be the xz-coordinate plane, we have ($\varphi = 0$)

$$r_i P_1(\hat{\mathbf{r}}_i \cdot \hat{\mathbf{R}}(t)) = \mathbf{r}_i \cdot \hat{\mathbf{R}}(t) = x_i \sin \theta(t) + z_i \cos \theta(t) \tag{18.21}$$

where $\theta(t)$ is the instantaneous polar angle of the projectile (see Figure 18.1). If we assume that the projectile moves in a central field originating from the target, conservation of angular momentum may be invoked in the classical description of this motion, and we have

$$[R(t)]^2 \frac{d\theta}{dt} = \rho v \tag{18.22}$$

where v is the initial velocity of the incident particle and ρ its impact parameter. Substituting (18.21) and (18.22) into (18.20), we obtain for the dipole transition probability amplitude

$$c_k(+\infty) = \mp \frac{iZ_1e^2}{\hbar v\rho} \int_{-\pi}^{\theta(\rho)} e^{i\omega_{ks}t'} (x_{ks} \sin \theta + z_{ks} \cos \theta) \, d\theta \tag{18.23}$$

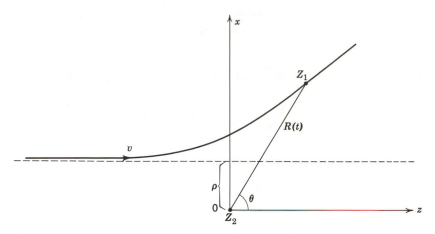

Figure 18.1. The Coulomb field of a charged particle Z_1 passing near a charged target Z_2 causes internal excitation. ρ is the impact parameter, and the coordinates θ and $R(t)$ determine the instantaneous position of the projectile.

where ex_{ks} and ez_{ks} are two components of the *dipole matrix element*

$$e\mathbf{r}_{ks} = e\langle k| \sum_{i=1}^{Z_2} \mathbf{r}_i\, |s\rangle \qquad (18.24)$$

The integration extends to the final scattering angle $\theta(\rho)$, which corresponds to the prescribed impact parameter ρ. Although the integrand in (18.23) looks simple, it should be remembered that t' is usually a complicated function of θ, and the integral can be evaluated only if the classical motion of the projectile is known.

In the case of nuclear Coulomb excitation the projectile performs a Kepler motion under the influence of the Coulomb force, and the integrals can be evaluated in terms of Hankel functions of imaginary argument.[1] The motion of a charged particle passing near an atom is more complicated. However, for some purposes it is a good approximation to assume that the exponent $\omega_{ks}t'$ in the integrand of (18.23) is negligible. This approximation is valid if the collision time during which the projectile is close enough to the target to be effective in exciting the latter is short compared with $1/\omega_{ks}$, which is a typical period associated with the proper oscillations of the target system. In

[1] A comprehensive review of nuclear Coulomb excitation was written by K. Alder, A. Bohr, T. Huus, B. Mottelson, and A. Winther, *Rev. Mod. Phys.*, **28**, 432 (1956). In actual nuclei electric dipole excitation is rarer than quadrupole excitation. See also L. C. Biedenharn and P. J. Brussaard, *Coulomb Excitation*, Clarendon Press, Oxford, 1965.

the limit $\omega_{ks}t' \to 0$, (18.23) can be integrated easily and yields

$$c_k(+\infty) = \pm \frac{iZ_1e^2}{\hbar v \rho} \{x_{ks}[\cos\theta(\rho) + 1] - z_{ks}\sin\theta(\rho)\}$$

Niels Bohr has shown[2] under what conditions this is a good approximation in the treatment of collisions of heavy charged particles, say protons, with atoms leading to energy loss by excitation and ionization. If we neglect the relatively rare collisions in which the projectile is sharply deflected and make the approximation $\theta(\rho) = 0$, a very simple expression is obtained for the transition probability amplitude:

$$c_k(+\infty) = \pm \frac{2iZ_1e^2}{\hbar v \rho} x_{ks} \tag{18.25}$$

The average energy loss T in such inelastic collisions of impact parameter ρ is

$$T = \sum_k (E_k - E_s) |c_k(+\infty)|^2 = \frac{4Z_1^2 e^4}{\rho^2 v^2 \hbar^2} \sum_k |x_{ks}|^2 (E_k - E_s) \tag{18.26}$$

where the sum is to be taken over all excited states of the target atom.

Equation (18.26) can be simplified by the use of an important identity. Consider the expression

$$\tfrac{1}{2}[[x, H_0], x] = xH_0x - \tfrac{1}{2}H_0x^2 - \tfrac{1}{2}x^2H_0$$

Its expectation value in the initial state s is

$$\tfrac{1}{2}\langle s | [[x, H_0], x] | s \rangle = \langle s | xH_0x | s \rangle - E_s\langle s | x^2 | s \rangle$$

$$= \sum_k (\langle s | x | k \rangle E_k \langle k | x | s \rangle - E_s \langle s | x | k \rangle \langle k | x | s \rangle)$$

$$= \sum_k (E_k - E_s) |x_{sk}|^2$$

which is precisely the sum that appears in (18.26). If the Hamiltonian H_0 contains the momentum only in the kinetic energy operator, we have

$$[x, H_0] = \frac{i\hbar}{\mu} p_x \quad \text{and} \quad \tfrac{1}{2}[[x, H_0], x] = \frac{\hbar^2}{2\mu} 1$$

hence, the identity

$$\frac{2\mu}{\hbar^2} \sum_k (E_k - E_s) |x_{sk}|^2 = 1 \tag{18.27}$$

---

[2] For a thorough discussion of this and other approximations and their mutual consistency see N. Bohr, *Kgl. Danske Videnskab Selskab, Mat.-fys. Medd*, **18**, No. 8 (1948). Note that collisions with the nuclei do not contribute appreciably to the energy loss of charged particles traversing matter.

by virtue of the normalization of the wave functions. If x is replaced by the x-component of the total dipole moment operator, $x_1 + x_2 + \cdots + x_Z$, for Z particles this identity can be generalized to

$$\frac{2\mu}{\hbar^2} \sum_k (E_k - E_s) \, |x_{sk}|^2 = Z \tag{18.28}$$

The quantities $f_{ks} = 2\mu(E_k - E_s) |x_{sk}|^2/\hbar^2$ are known as *oscillator strengths*, and (18.28) is the *Thomas-Reiche-Kuhn sum rule*.

If we neglect the effect of collisions with the nuclei, the average energy loss of a fast charged particle in a distant collision with an atom containing Z_2 electrons is according to (18.26) and (18.28) given by

$$T = \frac{2Z_1^{\,2}Z_2 e^4}{\mu v^2 \rho^2}$$

which is essentially the classical result. An approximate formula for the *stopping power* of charged particles can be derived from the semiclassical considerations of this section, but a more accurate stopping power formula was obtained by Bethe in a completely quantal treatment, which is based on the Born approximation, but which avoids the radical assumptions made in this section.[3]

4. The Atom in a Radiation Field. As a further illustration of the use of (18.16), it is instructive to calculate the absorption cross section of an atom for light. The electron spin will be neglected, and for simplicity it will be assumed that only one atomic electron is involved in the interaction with the incident radiation and that the nucleus is infinitely heavy, but the results can be generalized easily to other systems (molecules, nuclei) which can absorb radiation.

The Hamiltonian of the atomic electron (charge $-e$) in an electromagnetic field is given by (8.80) as

$$H = \frac{1}{2\mu}\left(\mathbf{p} + \frac{e}{c}\mathbf{A}\right)^2 - e\phi \tag{18.29}$$

In the coordinate representation this becomes the operator

$$H = \frac{\mathbf{p}^2}{2\mu} - e\phi + \frac{e}{\mu c}\mathbf{A}\cdot\frac{\hbar}{i}\nabla - \frac{ie\hbar}{2\mu c}(\nabla\cdot\mathbf{A}) + \frac{e^2}{2\mu c^2}\mathbf{A}^2 \tag{18.30}$$

We shall apply this Hamiltonian only under conditions that allow the $\mathbf{A}^2$ term to be neglected.

It is convenient to separate the electromagnetic field which represents the incident light wave from all other contributions to the Hamiltonian, whether

[3] H. A. Bethe, *Ann. Phys.*, **5**, 325 (1930); see also the reference of Footnote 2.

they are of electromagnetic origin, such as the Coulomb interactions within the atom, or not. Only the effect of the pure radiation will be treated as a perturbation and all other forces which act on the electron are incorporated into the V_0 term of the unperturbed Hamiltonian,

$$H_0 = \frac{\mathbf{p}^2}{2\mu} + V_0 \qquad (18.31)$$

According to Maxwell's theory the pure radiation field which perturbs the atom can be described in terms of a vector potential $\mathbf{A}$ alone. As this field has no sources in the vicinity of the atom, it may be derived from potentials which in Cartesian coordinates satisfy the equations

$$\nabla^2 \mathbf{A} - \frac{1}{c^2} \frac{\partial^2 \mathbf{A}}{\partial t^2} = 0 \qquad (18.32)$$

$$\phi = 0 \quad \text{and} \quad \nabla \cdot \mathbf{A} = 0 \qquad (18.33)$$

With this choice the complete Hamiltonian simplifies to

$$H = \frac{\mathbf{p}^2}{2\mu} + V_0 + \frac{e}{\mu c} \mathbf{A} \cdot \mathbf{p} \qquad (18.34)$$

where the last term will be regarded as an external perturbation. If V_0 represents the potential energy of a harmonic oscillator and if the spatial variation of $\mathbf{A}$ can be neglected, (18.34) is the Hamiltonian of a forced harmonic oscillator which was treated in detail in Section 15.9.

Exercise 18.1. Prove that the perturbation term in (18.34) is Hermitian.

It is convenient, and corresponds to actual experimental conditions, to suppose that the incident radiation can be described as a very broad plane wave packet moving in a given direction. Any such wave can be written as a superposition of harmonic plane waves,

$$\mathbf{A}(\mathbf{r}, t) = \int_{-\infty}^{+\infty} \mathbf{A}(\omega) \exp\left[-i\omega\left(t - \frac{\hat{\mathbf{n}} \cdot \mathbf{r}}{c}\right)\right] d\omega \qquad (18.35)$$

where $\hat{\mathbf{n}}$ is the unit vector pointing in the direction of propagation. The vector potential must be a real quantity, which requires that

$$\mathbf{A}^*(\omega) = \mathbf{A}(-\omega) \qquad (18.36)$$

On the other hand, the solenoidal character of the vector potential

$(\nabla \cdot \mathbf{A} = 0)$ implies that the plane wave is transverse:

$$\hat{\mathbf{n}} \cdot \mathbf{A}(\omega) = 0 \tag{18.37}$$

The perturbation

$$V = \frac{e}{\mu c} \int_{-\infty}^{+\infty} \exp\left[-i\omega\left(t - \frac{\hat{\mathbf{n}} \cdot \mathbf{r}}{c}\right)\right] \mathbf{A}(\omega) \cdot \mathbf{p} \, d\omega \tag{18.38}$$

is now substituted into (18.16) to give the transition probability amplitude,

$$c_k(+\infty)$$

$$= -\frac{ie}{\hbar\mu c} \int_{-\infty}^{+\infty} \int_{-\infty}^{+\infty} \langle k| \exp\left(i\frac{\omega}{c} \hat{\mathbf{n}} \cdot \mathbf{r}\right) \mathbf{p} \, |s\rangle \cdot \mathbf{A}(\omega) \, d\omega \exp\left[i(\omega_{ks} - \omega)t'\right] dt'$$

$$= -\frac{2\pi ie}{\hbar\mu c} \int_{-\infty}^{+\infty} \langle k| \exp\left(i\frac{\omega}{c} \hat{\mathbf{n}} \cdot \mathbf{r}\right) \mathbf{p} \, |s\rangle \cdot \mathbf{A}(\omega) \, \delta(\omega - \omega_{ks}) \, d\omega \tag{18.39}$$

hence,

$$c_k(+\infty) = -\frac{2\pi ie}{\hbar\mu c} \langle k| \exp\left(i\frac{\omega_{ks}}{c} \hat{\mathbf{n}} \cdot \mathbf{r}\right) \mathbf{p} \, |s\rangle \cdot \mathbf{A}(\omega_{ks}) \tag{18.40}$$

The last equation shows that the only Fourier component of the incident radiation which is effective in the absorption processes leading to the final state k corresponds to the frequency

$$\omega_{ks} = \frac{E_k^{(0)} - E_s^{(0)}}{\hbar} \tag{18.41}$$

in agreement with the frequency condition originally postulated by Bohr. The energy difference $\hbar\omega_{ks}$ is given up by the light pulse to the absorbing system, but the change which the radiation field undergoes in this energy transfer can only be described if the electromagnetic field, instead of being treated as an external prescribed driving force, is made part of the dynamical system and thus is itself amenable to quantization. The quantum theory of the radiation field will be developed in Chapter 22. It should, of course, always be remembered that (18.40) holds only if the light wave is sufficiently weak so that first-order perturbation theory is valid for the entire duration of the pulse.

The light pulse also induces inverse transitions if the atom is initially in state k. The transition probability amplitude to the state s of lower energy is evidently given by

$$c_s'(+\infty) = -\frac{i}{\hbar} \int_{-\infty}^{+\infty} V_{sk} e^{i\omega_{sk}t'} \, dt'$$

Since V is a Hermitian perturbation operator (see Exercise 18.1) and

$\omega_{sk} = -\omega_{ks}$, it follows that

$$c_s'(+\infty) = -c_k^*(+\infty) \tag{18.42}$$

showing that the two transition probabilities are equal. This property is known as the condition of *detailed balancing*. Since $E_k^{(0)} > E_s^{(0)}$, the excess energy $\hbar\omega_{ks}$ is transferred to the radiation field. A proper quantum treatment of the radiation field in Chapter 22 will confirm that this energy appears as light of frequency ω_{ks}. The transition from k to s, induced by the applied radiation pulse, is known as *stimulated emission*. If the radiating system is near thermal equilibrium, stimulated emission between two energy levels is usually much less intense than absorption, despite the equality of the transition probabilities, because the initial population of the upper level is generally smaller. In masers and lasers, however, a population inversion is achieved, and stimulated emission predominates.

For actual calculations of transition probabilities using (18.40) we may assume the pulse (18.35) to be linearly polarized, which implies that all frequency components $A(\omega)$ have the same direction independent of ω. There is no loss of generality involved in this assumption, if the cross section for absorption of light is to be calculated, since in an actual experiment a large number of incoherent pulses of the form (18.35) are directed at the system. Incoherence here simply means that the phases of the pulses are arbitrary and independent of each other. According to the wave equation each Fourier component is elliptically polarized.[4] Each elliptical oscillation can be decomposed into two linear ones directed along two fixed perpendicular axes, both of which are orthogonal to the propagation vector $\hat{\mathbf{n}}$. Because of the uniform randomness of the phases, the interference terms which arise between the two linear oscillations upon squaring the amplitudes (18.40) will cancel when an average over many pulses is performed. Hence, the total absorption probability is the sum of two expressions of the form $|c_k(+\infty)|^2$, as obtained by squaring (18.40), one for each direction of polarization.

If we adopt a definite direction of linear polarization $\hat{\mathbf{e}}$ and write

$$\mathbf{A}(\omega) = A(\omega)\hat{\mathbf{e}}$$

the transition probability becomes

$$|c_k(+\infty)|^2 = \frac{4\pi^2 e^2}{\hbar^2 \mu^2 c^2} |A(\omega_{ks})|^2 \left| \langle k| \exp\left(i\frac{\omega_{ks}}{c}\hat{\mathbf{n}}\cdot\mathbf{r}\right)\mathbf{p}\cdot\hat{\mathbf{e}}\,|s\rangle \right|^2 \tag{18.43}$$

This result can be related to the intensity of the incident pulse by comparing

---

[4] M. Born and E. Wolf, *Principles of Optics*, 3rd ed., Pergamon Press, New York, 1965, p. 33.

it with the Poynting vector,

$$\mathbf{N} = c\,\frac{\mathbf{E} \times \mathbf{H}}{4\pi} = -\frac{1}{4\pi}\frac{\partial \mathbf{A}}{\partial t} \times (\nabla \times \mathbf{A})$$

$$= -\frac{1}{4\pi c}\int_{-\infty}^{+\infty} \omega A(\omega)\hat{\mathbf{e}}\,\exp\left[-i\omega\left(t - \frac{\hat{\mathbf{n}} \cdot \mathbf{r}}{c}\right)\right]\,d\omega$$

$$\times \int_{-\infty}^{+\infty} \omega' A(\omega')(\hat{\mathbf{n}} \times \hat{\mathbf{e}})\,\exp\left[-i\omega'\left(t - \frac{\hat{\mathbf{n}} \cdot \mathbf{r}}{c}\right)\right]\,d\omega'$$

The triple product, $\hat{\mathbf{e}} \times (\hat{\mathbf{n}} \times \hat{\mathbf{e}}) = \hat{\mathbf{n}} - \hat{\mathbf{e}}(\hat{\mathbf{e}} \cdot \hat{\mathbf{n}})$, simplifies because the wave is transverse, whence $\hat{\mathbf{e}} \cdot \hat{\mathbf{n}} = 0$, giving the equation

$$\mathbf{N} = -\frac{1}{4\pi c}\,\hat{\mathbf{n}}\int_{-\infty}^{+\infty} d\omega \int_{-\infty}^{+\infty} d\omega'\,\omega\omega'\,A(\omega)A(\omega')\,\exp\left[-i(\omega + \omega')\left(t - \frac{\hat{\mathbf{n}} \cdot \mathbf{r}}{c}\right)\right]$$

(18.44)

It follows that the energy flows in the direction of propagation, and that the total energy which the pulse carries through a unit area placed perpendicular to the direction of propagation is given by

$$\int_{-\infty}^{+\infty} \mathbf{N} \cdot \hat{\mathbf{n}}\,dt = \frac{1}{2c}\int_{-\infty}^{+\infty} d\omega\,\omega^2\,|A(\omega)|^2 = \frac{1}{c}\int_0^\infty d\omega\,\omega^2|A(\omega)|^2 = \int_0^\infty N(\omega)\,d\omega$$

(18.45)

Hence, the energy carried through a unit area in the frequency interval between ω and $\omega + d\omega$ is

$$N(\omega) = \frac{\omega^2}{c}|A(\omega)|^2$$

(18.46)

Combining this result with (18.43), we obtain

$$|c_k(+\infty)|^2 = \frac{4\pi^2 e^2}{c\mu^2\hbar^2\omega_{ks}{}^2}\left|\langle k|\exp\left(i\frac{\omega_{ks}}{c}\hat{\mathbf{n}} \cdot \mathbf{r}\right)\mathbf{p} \cdot \hat{\mathbf{e}}\,|s\rangle\right|^2 N(\omega_{k_s}) \quad (18.47)$$

Energy transfer from the incident radiation pulse to the atom takes place only at those frequencies which the system can absorb in accordance with the Bohr frequency condition (18.41). The average energy loss of the pulse at the frequency ω_{ks} is

$$\hbar\omega_{ks}\,|c_k(+\infty)|^2 = \frac{4\pi^2\alpha}{\mu^2\omega_{ks}}\left|\langle k|\exp\left(i\frac{\omega_{ks}}{c}\hat{\mathbf{n}} \cdot \mathbf{r}\right)\mathbf{p} \cdot \hat{\mathbf{e}}\,|s\rangle\right|^2 N(\omega_{ks}) \quad (18.48)$$

where α denotes the fine structure constant, $\alpha = e^2/\hbar c \approx 1/137$.

It is customary to base the definition of an *absorption cross section* for linearly polarized electromagnetic radiation of a certain frequency on (18.48).

The absorption cross section is a fictitious area which, when placed perpendicular to the incident pulse, would be exposed to and traversed by an amount of radiation energy equal to that absorbed by the atom or nucleus. The notion of the cross section is a useful one only if it is independent of the detailed shape of the incident pulse; it should depend only on the frequency, direction of incidence, and state of polarization of the incident beam as well as on the nature and initial condition of the absorbing system. Under the usual experimental conditions these goals can be achieved and a cross section $\sigma(\omega)$ defined such that the energy absorbed in the frequency interval $d\omega$ is

$$\sigma(\omega)N(\omega)\, d\omega \tag{18.49}$$

Absorption cross sections corresponding to two diametrically opposite situations will now be evaluated:

1. The initial and final states are discrete, and the absorption spectrum consists of sharp lines (Section 18.5).
2. The initial state is discrete, but the final state is in the continuum. This is the photoelectric effect, and $\sigma(\omega)$ is a slowly varying function of the frequency (Section 18.6).

5. *The Absorption Cross Section.* If the states s and k are discrete, radiation can be absorbed only if its frequency lies very close to ω_{ks}. Hence, $\sigma(\omega)$ is a strongly peaked function of ω. The total energy absorbed is obtained by integrating the expression (18.49) and this quantity can then be equated to the expression (18.48):

$$\int_{\Delta\omega} \sigma(\omega)N(\omega)\, d\omega = \frac{4\pi^2\alpha}{\mu^2\omega_{ks}} \left| \langle k| \exp\left(i\,\frac{\omega_{ks}}{c}\,\hat{\mathbf{n}}\cdot\mathbf{r}\right)\mathbf{p}\cdot\hat{\mathbf{e}}|s\rangle \right|^2 N(\omega_{ks}) \tag{18.50}$$

Here the integral on the left is to be extended over an interval $\Delta\omega$ which completely contains the line (but no other absorption frequencies). Since $\sigma(\omega)$ has a steep maximum at $\omega = \omega_{ks}$, a reasonable definition of the cross section can be obtained if $N(\omega)$ varies relatively slowly. If we restrict ourselves to pulses broad in frequency compared with the line, the left-hand side of (18.50) can be written as

$$N(\omega_{ks})\int_{\Delta\omega} \sigma(\omega)\, d\omega$$

and an integrated cross section can be obtained:

$$\int_{\Delta\omega} \sigma(\omega)\, d\omega = \frac{4\pi^2\alpha}{\mu^2\omega_{ks}} \left| \langle k| \exp\left(i\,\frac{\omega_{ks}}{c}\,\hat{\mathbf{n}}\cdot\mathbf{r}\right)\mathbf{p}\cdot\hat{\mathbf{e}}\,|s\rangle \right|^2 \tag{18.51}$$

Further progress in the evaluation of this cross section depends on specific assumptions about the structure of the unperturbed Hamiltonian H_0 and its eigenstates s and k. The coordinate origin is chosen at the center of mass of the atom. Generally it is possible to make an important approximation based on the fact that the wavelength of the incident light, $\lambda = 2\pi c/\omega_{ks}$, is large compared with the linear dimensions of the absorbing system. In this case the exponential operator in the matrix element of (18.51) is advantageously expanded in a power series

$$\exp\left(i\,\frac{\omega_{ks}}{c}\,\hat{\mathbf{n}}\cdot\mathbf{r}\right) = 1 + i\,\frac{\omega_{ks}}{c}\,\hat{\mathbf{n}}\cdot\mathbf{r} + \cdots \tag{18.52}$$

and only the first nonvanishing term in the corresponding series of matrix elements,

$$\langle k|\,\mathbf{p}\cdot\hat{\mathbf{e}}\,|s\rangle + \langle k|\,i\,\frac{\omega_{ks}}{c}\,\hat{\mathbf{n}}\cdot\mathbf{r}\mathbf{p}\cdot\hat{\mathbf{e}}\,|s\rangle + \cdots \tag{18.53}$$

is retained. The ratio of two successive nonvanishing terms in this series can be estimated by assuming that r is effectively of the order of magnitude of the radius of the system. For an atom of radius a/Z (a: Bohr radius) the expansion parameter is equal to

$$\frac{\omega a}{cZ} \simeq \frac{Ze^2}{\hbar c} = Z\alpha = \frac{Z}{137}$$

and thus the series (18.53) converges rapidly for small Z.

Selection rules for the wave functions of the unperturbed states determine which term in the expansion (18.53) is the first nonvanishing matrix element. If the first term in the expansion, $\langle k|\,\mathbf{p}\cdot\hat{\mathbf{e}}\,|s\rangle$ is different from zero, the transition is said to be of the *electric dipole* type. Since in this approximation we effectively replace $\exp(i\omega\hat{\mathbf{n}}\cdot\mathbf{r}/c)$ by unity, the electric dipole approximation also implies a complete neglect of retardation across the atom. Owing to the inherent weakness of transitions of higher multipolarity ("forbidden" transitions), electric dipole (or "allowed") transitions are the most important ones in atoms. In nuclei the ratio $r\omega/c$ is not quite so small. $\hbar\omega$ is of the order of $\hbar^2/\mu_n R^2$, where μ_n is the nucleon mass, and R is the nuclear radius. Hence, the expansion parameter is

$$\frac{\omega}{c}R \simeq \frac{\hbar}{\mu_n cR} = \frac{\text{Compton wavelength of nucleon}}{\text{Nuclear radius}}$$

which for light nuclei can be as large as $1/10$. The powerful detection techniques of nuclear physics make it easy to observe transitions of higher multipolarity. When such transitions are considered, it is no longer permissible to neglect the interaction of the electromagnetic field with the spin magnetic moment.

In electric dipole transitions the matrix element $\langle k| \mathbf{p} |s \rangle$ is the decisive quantity which must be evaluated. This can be related to the matrix element of the position operator $\mathbf{r}$ if the unperturbed Hamiltonian is of the form $H_0 = (\mathbf{p}^2/2\mu) + V_0$, and if V_0 commutes with $\mathbf{r}$. Under these conditions

$$\mathbf{r}H_0 - H_0\mathbf{r} = i\hbar \frac{\mathbf{p}}{\mu} \tag{18.54}$$

hence, by taking the matrix element of both sides of this equation,

$$\langle k| \mathbf{p} |s \rangle = \frac{\mu}{i\hbar} (E_s^{(0)} - E_k^{(0)})\langle k| \mathbf{r} |s \rangle = i\mu\omega_{ks}\langle k| \mathbf{r} |s \rangle \tag{18.55}$$

Making the dipole approximation, and substituting (18.55) into (18.51), we obtain the integrated absorption cross section,

$$\int_{\Delta\omega} \sigma(\omega) \, d\omega = 4\pi^2\alpha\omega_{ks}|\langle k| \mathbf{r} \cdot \hat{\mathbf{e}} |s \rangle|^2 \tag{18.56}$$

For many purposes, it is useful to rewrite the transition matrix element as follows:

$$\langle k| \mathbf{r} \cdot \hat{\mathbf{e}} |s \rangle = \frac{4\pi}{3} \sum_q (-1)^q \langle k| r Y_1^q(\hat{\mathbf{r}}) |s \rangle Y_1^{-q}(\hat{\mathbf{e}}) \tag{18.57}$$

If the absorbing system is a one-electron atom, and if the fine structure may be neglected, the quantum numbers nlm and $n'l'm'$ characterize the initial and final states. Equation (16.90) may then be applied to carry out the integration over the angle coordinates in the evaluation of the matrix element, giving us

$\langle n'l'm'| \mathbf{r} \cdot \hat{\mathbf{e}} |nlm \rangle$

$$= \sqrt{\frac{4\pi}{3} \frac{2l+1}{2l'+1}} \sum_q \langle l1mq \mid l1l'm' \rangle \langle l100 \mid l1l'0 \rangle [Y_1^q(\hat{\mathbf{e}})]^* R_{nl}^{n'l'} \tag{18.58}$$

where

$$R_{nl}^{n'l'} = \int_0^\infty R_{n'l'}(r)R_{nl}(r)r^3 \, dr \tag{18.59}$$

and $R_{nl}(r)$ are the real normalized radial wave functions. The values of the relevant Clebsch-Gordan coefficients are

$$\langle l100 \mid l1, l+1, 0 \rangle = \sqrt{\frac{l+1}{2l+1}} \tag{18.60}$$

$$\langle l100 \mid l1, l-1, 0 \rangle = -\sqrt{\frac{l}{2l+1}} \tag{18.61}$$

The electric dipole selection rules

$$\Delta l = l' - l = \pm 1, \qquad \Delta m = m' - m = 0, \pm 1$$

are an immediate consequence of (18.58). Using (18.58) and the selection rules, we may write

$$|\langle n'l'm'| \mathbf{r} \cdot \hat{\mathbf{e}} |nlm\rangle|^2$$
$$= \frac{4\pi}{3} \frac{2l+1}{2l'+1} \langle l1m, m'-m \mid l1l'm'\rangle^2 \langle l100 \mid l1l'0\rangle^2 |Y_1^{m'-m}(\hat{\mathbf{e}})|^2 (R_{nl}^{n'l'})^2$$

$$(18.62)$$

for the last factor in (18.56).

We may also calculate the average rate of absorption for a system (atom or molecule) which is exposed to a continuous frequency distribution of incoherent electromagnetic radiation from all directions and with random polarization. These conditions may, for example, represent the absorption of blackbody radiation. Usually only electric dipole transitions are important, and we shall restrict our formulas to these.[5]

Applying the dipole approximation to (18.50), we find that the energy absorbed from a single polarized unidirectional pulse is

$$4\pi^2 \alpha \omega_{ks} |\langle k| \mathbf{r} \cdot \hat{\mathbf{e}}_i |s\rangle|^2 N_i(\omega_{ks}) \qquad (18.63)$$

where $\hat{\mathbf{e}}_i$ denotes the polarization vector of the ith pulse, and $N_i(\omega_{ks})$ the intensity of its frequency component ω_{ks}. Expression (18.63) must be summed over the n light pulses which on the average strike the atom per unit time. Since in blackbody radiation there is no correlation between $\hat{\mathbf{e}}_i$ and $N_i(\omega)$, the average of the product is the product of the averages; hence, the *rate of absorption* from the radiation field is given by

$$R = 4\pi^2 \alpha \omega_{ks} \overline{|\langle k| \mathbf{r} \cdot \hat{\mathbf{e}} |s\rangle|^2} \, \overline{N}(\omega_{ks})n \qquad (18.64)$$

The average intensity is related to the energy density per unit frequency interval $u(\omega)$ by

$$n\overline{N}(\omega) = cu(\omega)$$

The average over the polarization directions $\hat{\mathbf{e}}$ can be performed easily, since for Cartesian components of $\hat{\mathbf{e}}$,

$$\int e_i e_j \, d\Omega_{\hat{\mathbf{e}}} = \frac{4\pi}{3} \delta_{ij} \qquad (18.65)$$

Hence,

$$\overline{|\langle k| \mathbf{r} \cdot \hat{\mathbf{e}} |s\rangle|^2} = \tfrac{1}{3} |\langle k| \mathbf{r} |s\rangle|^2 \qquad (18.66)$$

---

[5] Forbidden atomic lines are important in stellar spectra.

and the average rate of absorption is

$$R = \frac{4\pi^2}{3} \alpha \omega_{ks} c \; |\langle k| \mathbf{r} \,|s\rangle|^2 \, u(\omega_{ks}) \tag{18.67}$$

Alternatively, the average

$$\overline{|Y_1^{m'-m}(\hat{\mathbf{e}})|^2} = \frac{1}{4\pi}$$

may be used to calculate from (18.62) and (18.64) the average rate of absorption by initial states of sharp angular momentum. The total transition probability from a given state nlm to all the degenerate m-substates of the level $n'l'$ is obtained by summing over the allowed values of m'. Since the Clebsch-Gordan coefficients satisfy the relation

$$\sum_{m'} \langle lkm, m'-m \,|\, lkl'm'\rangle^2 = \frac{2l'+1}{2l+1} \tag{18.68}$$

it follows from (18.62) that for $l' = l + 1$

$$\sum_{m'} \overline{|\langle n'l'm'| \mathbf{r} \cdot \hat{\mathbf{e}} \,|nlm\rangle|^2} = \frac{1}{3} \frac{l+1}{2l+1} (R_{nl}^{n'l'})^2 \tag{18.69}$$

and for $l' = l - 1$

$$\sum_{m'} \overline{|\langle n'l'm'| \mathbf{r} \cdot \hat{\mathbf{e}} \,|nlm\rangle|^2} = \frac{1}{3} \frac{l}{2l+1} (R_{nl}^{n'l'})^2 \tag{18.70}$$

Substituting these values in (18.64), we obtain the absorption rates. It is important to note that these rates do not depend on the m quantum number of the particular initial substate. (*Principle of spectroscopic stability.*)

A cross section, $\sigma_{\text{s.e.}}(\omega)$, for stimulated emission (s.e.) can be defined, in analogy with the absorption cross section, σ_{abs}, by equating to $N(\omega)\sigma_{\text{s.e.}}(\omega)\, d\omega$ the energy given off by an atom as a consequence of the passage of the light wave in the frequency interval $d\omega$ around ω. Since the transition probabilities for stimulated emission and absorption are the same, we see from (18.56) that for any initial state s,

$$\int_0^\infty [\sigma_{\text{abs}}(\omega) - \sigma_{\text{s.e.}}(\omega)] \, d\omega$$

$$= 4\pi^2\alpha \left[\sum_{E_k > E_s} \omega_{ks} \, |\langle k| \mathbf{r} \cdot \hat{\mathbf{e}} \,|s\rangle|^2 - \sum_{E_k < E_s} \omega_{sk} \, |\langle s| \mathbf{r} \cdot \hat{\mathbf{e}} \,|k\rangle|^2 \right]$$

provided that the sum is taken over all eigenstates of H_0. From the sum rule (18.27), we then deduce for electric dipole transitions in one-electron atoms the general formula

$$\int_0^\infty [\sigma_{\text{abs}}(\omega) - \sigma_{\text{s.e.}}(\omega)] \, d\omega = 4\pi^2\alpha \frac{\hbar}{2\mu} = 2\pi^2 \frac{e^2}{\mu c^2} c = 2\pi^2 r_0 c \quad (18.71)$$

where r_0 is the classical electron radius. This relation is independent of the direction of polarization and of the particular initial state and holds for any system for which the Thomas-Reiche-Kuhn sum rule is valid.

For an atom with Z electrons, the sum rule (18.28) holds and gives

$$\int_0^\infty [\sigma_{\text{abs}}(\omega) - \sigma_{\text{s.e.}}(\omega)] \, d\omega = 2\pi^2 Z r_0 c \quad (18.72)$$

provided that all possible electric dipole transitions of the atom are included. Equation (18.72) is identical with the result of a similar classical calculation.[6]

Sum rules for electric dipole cross sections can also be derived for nuclei which are exposed to gamma radiation, but care is required because nuclei are composed of charged and neutral particles. Since dipole absorption comes about as a result of the relative displacement between the center of charge of the nucleus and its center of mass, the dipole sum rule for nuclei depends critically on the relative number of protons and neutrons.[7]

So far only the integrated cross section $\int \sigma(\omega) \, d\omega$ has been discussed, but the detailed frequency dependence of $\sigma(\omega)$ was left unspecified. All that could be said about it was that it has a very pronounced peak at $\omega = \omega_{ks}$. In fact, in the approximation used in this chapter no absorption occurs unless the exact frequency component ω_{ks} is represented in the incident beam. Hence, $\sigma(\omega)$ must be proportional to a delta function, and according to (18.50) must be given by

$$\sigma(\omega) = \frac{4\pi^2\alpha}{\mu^2\omega_{ks}} \left| \langle k | \exp\left(i\frac{\omega_{ks}}{c} \hat{\mathbf{n}} \cdot \mathbf{r}\right) \mathbf{p} \cdot \hat{\mathbf{e}} \, |s\rangle \right|^2 \delta(\omega - \omega_{ks}) \quad (18.73)$$

Equation (18.73) is of course an approximation. Observed spectral absorption lines are not infinitely sharp but possess a finite width which the simple theory neglects. In actual fact, the excited state k is not, as was tacitly assumed, a stable and therefore strictly discrete energy level. Rather, it decays with a characteristic lifetime τ back to the original level s, and possibly to other lower lying levels if such are available. Besides, an atomic level is depleted not only by emission of radiation but also, and often predominantly, by collisions with other atoms. In a more accurate form of perturbation theory these interactions, which also cause transitions, must be included. The shape

[6] Compare equation (17.74) in J. D. Jackson, *Classical Electrodynamics*, John Wiley and Sons, New York, 1962.

[7] J. M. Blatt and V. F. Weisskopf, *Theoretical Nuclear Physics*, John Wiley and Sons, New York, 1952, pp. 640–643.

of the absorption line with its finite width can thus only be predicted if the subsequent return of the atom to its ground state is taken into account. It will be shown in Section 18.9 how interactions may lead to the exponential decay of excited states and to a more realistic expression for the line shape than (18.73).

In retrospect it may be well to stress some tacit assumptions which underlie the treatment of absorption given in this section.

We have assumed that the incident radiation is weak, so that no atom is ever hit by several light pulses simultaneously; and in fact we have assumed that the atom could settle down to an undisturbed existence for comparatively long periods between pulses. A sufficient number of collisions in these intervals insure that almost certainly the atom is in the ground state when the next pulse arrives, or that at least thermal equilibrium has been restored, and each atom is in a definite unperturbed state. Also we have assumed that each pulse is sufficiently weak so that the probability of absorption per single pulse is numerically small. If, in violation of these assumptions, the incident radiation were intense, we would have to use a more accurate form of perturbation theory with the result that, in the case of such *saturation* of the absorption line, additional broadening of the line would occur. This broadening is due to the fact that the incident radiation stimulates emissive transitions appreciably, and furthermore there is a measurable depletion of the ground state. The additional broadening of the absorption line depends on the intensity of the incident radiation.

Masers and lasers produce conditions that are totally incompatible with the assumptions outlined above. The radiation in these devices is coherent, sharply monochromatic, and extremely intense, and it is essential that excited atomic states of the active substance are appreciably populated. If, under these circumstances, only two levels r and s of an atomic system, which interacts with the radiation, are important, equations (18.17) may be employed with the perturbation

$$V = \frac{e}{\mu c} (e^{i\omega t} \mathbf{A}_0 + e^{-i\omega t} \mathbf{A}_0{}^*) \cdot \mathbf{p} \tag{18.74}$$

Here, $\mathbf{A}_0$ is a complex constant vector, and the electric dipole approximation has been made. If selection rules cause the diagonal matrix elements of V in the states r and s to vanish and if, furthermore, the off-diagonal matrix element $(\mathbf{A}_0 \cdot \mathbf{p})_{rs} \neq 0$, the equations (18.17) assume the simple form

$$i\hbar \frac{d}{dt}\begin{pmatrix} c_r \\ c_s \end{pmatrix} = \begin{pmatrix} 0 & \alpha e^{i[(\omega_0 - \omega)t + \delta]} \\ \alpha e^{-i[(\omega_0 - \omega)t + \delta]} & 0 \end{pmatrix}\begin{pmatrix} c_r \\ c_s \end{pmatrix} \tag{18.75}$$

where $\hbar\omega_0 = \hbar\omega_{rs} = E_r^{(0)} - E_s^{(s)}$. The positive constant α is proportional to the strength of the electromagnetic field.

Equation (18.75) is formally the same as the equation of motion of a spatially fixed spin $\frac{1}{2}$ particle with magnetic moment in a rotating magnetic field. Hence, the solutions to this problem, worked out in Section 13.2, may be transcribed appropriately.

Exercise 18.2. Apply the results of Exercise 13.6 to the problem of a two-level atom interacting with a classical monochromatic electromagnetic wave in the limit of long wavelength. Discuss the resonance behavior of the transition probability, and examine the dependence of the resonance width on the physical parameters.

Finally, it should be remembered that in this chapter the field is regarded merely as a prescribed external agent, perturbing the quantum system such as an atom, but itself incapable of being reacted upon. This approximation is evidently quite restrictive and ignores all effects in the interaction between light and matter which go beyond simple absorption, notably scattering of light. Even the process of *spontaneous emission*, i.e., emission from an excited atom in the absence of all applied electromagnetic fields, cannot be described properly without including the field as a dynamical system that can be found in various quantum states of excitation, quite similar to the material systems with which it interacts. The elements of the *quantum theory of radiation* will be taken up in Chapter 22.

6. The Photoelectric Effect. In the photoelectric effect the absorption of light by an atom (or nucleus) leads to the emission of an electron (or nucleon) into a continuum state. In contradistinction to the behavior of $\sigma(\omega)$ for jumps from discrete to discrete levels, the cross section for the photoelectric effect is a smooth function of the frequency. Hence, by comparing (18.48) with (18.49), we obtain readily

$$d\sigma(\omega) = \frac{4\pi^2\alpha}{\mu^2\omega} \left| \langle k| \exp\left(i\frac{\omega}{c}\,\hat{\mathbf{n}}\cdot\mathbf{r} \right) \mathbf{p}\cdot\hat{\mathbf{e}}\,|s\rangle \right|^2 \frac{\Delta n}{\Delta\omega} \qquad (18.76)$$

Here ω is the frequency of the incident radiation and of the quantum jump. s is the discrete initial state, but k, the final state, lies in the continuum. $d\sigma(\omega)$ denotes the differential cross section corresponding to emission of the electron into a sharply defined solid angle $d\Omega$ with respect to the linearly polarized incident beam. Finally, Δn is the number of electron eigenstates in the frequency interval $\Delta\omega$ with an average energy $E_k = \hbar\omega + E_s$ and corresponding to the given solid angle.

For the evaluation of (18.76) the extreme assumption will be made that $\hbar\omega$ is much larger than the ionization potential of the atom, so large in fact that the final electron states can be satisfactorily approximated by plane waves. If we use plane waves which are normalized in a very large cube of length L containing the atom, we have

$$\langle \mathbf{r} \mid k \rangle = \frac{1}{L^{3/2}} \exp{(i\mathbf{k} \cdot \mathbf{r})}, \qquad E_k = \frac{\hbar^2 k^2}{2\mu}$$

and according to (10.10)

$$\frac{\Delta n}{\Delta \omega} = \frac{d\Omega}{4\pi} \frac{\mu^{3/2}\sqrt{E_k}L^3}{\sqrt{2}\pi^2\hbar^2}$$

If the electron is initially in the K-shell, ejection from which is comparatively probable, the initial wave function is

$$\langle \mathbf{r} \mid s \rangle = \frac{1}{\sqrt{\pi}}\left(\frac{Z}{a}\right)^{3/2} \exp\left(-\frac{Zr}{a}\right)$$

Substituting all these quantities into (18.76), we obtain for the differential cross section

$$\frac{d\sigma(\omega)}{d\Omega} = \frac{e^2 k}{2\mu c\hbar^2\pi^2\omega}\frac{Z^3}{a^3}$$

$$\times \left| \int \exp{(-i\mathbf{k} \cdot \mathbf{r})} \exp\left(i\frac{\omega}{c}\hat{\mathbf{n}} \cdot \mathbf{r}\right) \hat{\mathbf{e}} \cdot \frac{\hbar}{i}\nabla \exp\left(-\frac{Zr}{a}\right) d^3r \right|^2$$

Owing to the Hermitian property of the momentum operator, and the transverse character of the light waves ($\hat{\mathbf{e}} \cdot \hat{\mathbf{n}} = 0$), we can write

$$\int \exp\left[i\left(-\mathbf{k} \cdot \mathbf{r} + \frac{\omega}{c}\hat{\mathbf{n}} \cdot \mathbf{r}\right)\right] \hat{\mathbf{e}} \cdot \frac{\hbar}{i}\nabla \exp\left(-\frac{Zr}{a}\right) d^3r$$

$$= \hat{\mathbf{e}} \cdot \int \left\{\frac{\hbar}{i}\nabla \exp\left[i\left(\mathbf{k} - \frac{\omega}{c}\hat{\mathbf{n}}\right) \cdot \mathbf{r}\right]\right\}^* \exp\left(-\frac{Zr}{a}\right) d^3r$$

$$= \hbar\hat{\mathbf{e}} \cdot \mathbf{k} \int \exp\left[i\left(\frac{\omega}{c}\hat{\mathbf{n}} - \mathbf{k}\right) \cdot \mathbf{r} - \frac{Zr}{a}\right] d^3r = \hat{\mathbf{e}} \cdot \mathbf{k}\frac{8\pi Z\hbar}{a}\left(\frac{Z^2}{a^2} + q^2\right)^{-2}$$

where $\mathbf{q}$ is the momentum transfer in units of $\hbar$, i.e.,

$$\mathbf{q} = \mathbf{k} - \frac{\omega}{c}\hat{\mathbf{n}} \tag{18.77}$$

Hence, finally,

$$\frac{d\sigma(\omega)}{d\Omega} = \frac{32e^2 k(\hat{\mathbf{e}} \cdot \mathbf{k})^2}{\mu c\omega}\frac{Z^5}{a^5}\left(\frac{Z^2}{a^2} + q^2\right)^{-4} \tag{18.78}$$

which is in good agreement with experimental observations.

If $\hbar\omega$ is comparable with the ionization potential, a better approximation to the final state electron wave function must be used. Ideally, exact continuum eigenfunctions could be used which represent the emission of an electron into a state approaching asymptotically the plane wave $e^{i\mathbf{k}\cdot\mathbf{r}}$. Such a state, although with a different normalization, is $\psi_{\mathbf{k}}^{(-)}$ defined by (11.31). The state $\psi_{\mathbf{k}}^{(-)}$ with asymptotically *incoming* waves rather than $\psi_{\mathbf{k}}^{(+)}$ with *outgoing* waves must be chosen, because a wave packet of approximate momentum $\mathbf{k}$ moving *away* from the origin is constructed as a superposition of neighboring functions $\psi_{\mathbf{k}}^{(-)}$, just as a wave packet moving *toward* the origin was shown in Section 11.2 to be a simple superposition of $\psi_{\mathbf{k}}^{(+)}$ functions.[8]

7. The Interaction Picture. In the previous sections we saw that the simplest formulas of time-dependent perturbation theory have a wide range of applicability. Yet the formalism of Section 18.2 is only the first step in a systematic approximation procedure which can be carried to higher orders if necessary. This section will be devoted to a more general treatment of time-dependent perturbations. The general expansion can be written compactly and elegantly if use is made of the operator formalism of quantum mechanics, and we shall use this alternative approach here; the procedure is entirely equivalent to a conventional successive approximation scheme applied to (18.10).

The starting point is the observation that in Section 18.1 it was found practical to separate out the oscillating time dependence which is due to the unperturbed Hamiltonian, thus letting the amplitudes c_k respond only to the perturbation V. The equation of motion in the Schrödinger picture is

$$i\hbar \frac{d\Psi(t)}{dt} = (H_0 + V)\Psi(t) \tag{18.79}$$

H_0 will be assumed to be a constant operator, but V may depend on the time. We define a state vector

$$\Psi''(t) = \exp\left(\frac{i}{\hbar} H_0 t\right)\Psi(t) \tag{18.80}$$

which at $t = 0$ coalesces with $\Psi(0)$ and evolves from the latter by a time-dependent unitary transformation. If Ψ'' is differentiated and (18.79) used, we

---

[8] If bound states are included, both $\psi^{(+)}$ and $\psi^{(-)}$ are separately complete sets, but if a nearly plane wave packet moving away from the origin were expressed in terms of $\psi_{\mathbf{k}}^{(+)}$, momenta $\mathbf{k}$ of all directions would have to be used in the superposition in order to produce the desired wave packet by judicious destructive interference. For a more accurate treatment near threshold, see Section 58 in K. Gottfried, *Quantum Mechanics*, Volume I, W. A. Benjamin, Inc., New York, 1966.

obtain the equation of motion for Ψ'',

$$i\hbar \frac{d\Psi''(t)}{dt} = V'\Psi''(t) \tag{18.81}$$

where

$$V'(t) = \exp\left(\frac{i}{\hbar} H_0 t\right) V \exp\left(-\frac{i}{\hbar} H_0 t\right) \tag{18.82}$$

is the effective Hamiltonian for the transformed state vector. Equation (18.81) shows that Ψ'' changes in time only if there is a perturbation. Using the terminology of Section 15.7, we may say that Ψ'' describes the same quantum state as Ψ but in a different picture, known as the *interaction picture*. An arbitrary operator A in the Schrödinger picture is transformed into

$$A'(t) = \exp\left(\frac{i}{\hbar} H_0 t\right) A \exp\left(-\frac{i}{\hbar} H_0 t\right) \tag{18.83}$$

by the unitary transformation (18.80). We may also write

$$\frac{dA'}{dt} = \frac{\partial A'}{\partial t} + \frac{1}{i\hbar} (A'H_0 - H_0 A') \tag{18.84}$$

Equations (18.81) and (18.84) exhibit the fact that in the interaction picture both state vectors and operators "move," but their time developments are governed by different portions of the total Hamiltonian. The motion of the state vector is dictated by V, that of the operators by H_0. Thus the interaction picture is intermediate between the Schrödinger picture, in which only the state vectors change in time, and the Heisenberg picture, in which all operators are subject to time development.

If, as a special case, (18.83) is applied to the operator $A = H_0$, we obtain $H_0'(t) = H_0$, i.e., the unperturbed Hamiltonian in the interaction picture is the same as in the Schrödinger picture.

The equation of motion (18.81) can be formally solved by a linear relation

$$\Psi''(t) = U(t, t_0)\Psi''(t_0) \tag{18.85}$$

Combining equation (18.80) with (15.1), we see that the time development operators in the Schrödinger and interaction pictures are related by

$$U(t, t_0) = e^{(i/\hbar)H_0 t} T(t, t_0) e^{-(i/\hbar)H_0 t_0} \tag{18.86}$$

If H_0 is Hermitian and $T(t, t_0)$ is unitary, $U(t, t_0)$ is also unitary. From the equations defining U, it follows that this operator satisfies the differential equation

$$i\hbar \frac{dU(t, t_0)}{dt} = V'(t)U(t, t_0) \tag{18.87}$$

subject to the initial condition

$$U(t_0, t_0) = 1 \qquad (18.88)$$

and that it possesses also the group property,

$$U(t, t'') = U(t, t')U(t', t'') \qquad (18.89)$$

Exercise 18.3. Prove the statements made in the last paragraph.

For calculations it is often useful to replace the differential equation for U by an integral equation, obtained by integrating (18.87) from t_0 to t:

$$U(t, t_0) = 1 - \frac{i}{\hbar} \int_{t_0}^{t} V'(t')U(t', t_0) \, dt' \qquad (18.90)$$

If the perturbation V is small, this equation can be made the starting point of an iteration procedure. First, $U = 1$ is substituted under the integral; the approximate U so obtained is substituted again in the integrand, etc. The result of this repeated iteration is very simple and appears as a power series in terms of V:

$$U(t, t_0) = 1 - \frac{i}{\hbar} \int_{t_0}^{t} V'(t') \, dt' + \left(\frac{-i}{\hbar}\right)^2 \int_{t_0}^{t} V'(t') \, dt' \int_{t_0}^{t'} V'(t'') \, dt'' + \cdots \qquad (18.91)$$

Approximate solutions of the equation of motion are obtained if this power series is arbitrarily terminated.

In particular, if the experimental arrangement is such that the system is known to be in an eigenstate s of the unperturbed Hamiltonian at $t = t_0$, then the transition amplitude to some eigenstate k of H_0 is given by the scalar product

$$(\Psi_k, \Psi'(t)) = (\Psi_k, U(t, t_0)\Psi_s) \qquad (18.92)$$

where Ψ_s and Ψ_k are normalized eigenvectors of $H_0'(t) = H_0$. If the final state k is different from the initial state s, substitution of the expansion (18.91) into (18.92) gives

$$(\Psi_k, \Psi'(t)) = -\frac{i}{\hbar} \int_{t_0}^{t} \langle k| \, V'(t') \, |s\rangle \, dt'$$
$$+ \left(-\frac{i}{\hbar}\right)^2 \int_{t_0}^{t} dt' \int_{t_0}^{t'} dt'' \langle k| \, V'(t')V'(t'') \, |s\rangle + \cdots \qquad (18.93)$$

Remembering the definition of V', (18.82), the transition amplitude can be written in the first approximation as

$$(\Psi_k, \Psi'(t)) \approx -\frac{i}{\hbar} \int_{t_0}^{t} \langle k| \, V \, |s\rangle e^{i\omega_{ks}t'} \, dt' \qquad (18.94)$$

If $t_0 \to -\infty$, this expression becomes identical with the amplitude $c_k(t)$ in (18.15). Higher approximations can be calculated from (18.93) when needed.

For some purposes it is convenient to write the series expansion (18.91) in a more symmetric form. This is achieved by noting that the second term can be transformed by a change in the order of integration as follows:

$$\int_{t_0}^{t} V'(t')\, dt' \int_{t_0}^{t'} V'(t'')\, dt''$$

$$= \int_{t_0}^{t} dt'' \int_{t''}^{t} dt'\, V'(t')V'(t'') = \int_{t_0}^{t} dt' \int_{t'}^{t} dt''\, V'(t'')V'(t')$$

$$= \frac{1}{2} \int_{t_0}^{t} dt' \int_{t_0}^{t'} dt''\, V'(t')V'(t'') + \frac{1}{2} \int_{t_0}^{t} dt' \int_{t'}^{t} dt''\, V'(t'')V'(t')$$

If $t > t_0$ this may also be written concisely in the form

$$\frac{1}{2} \int_{t_0}^{t} \int_{t_0}^{t} dt'\, dt''\, P[V'(t')V'(t'')] \tag{18.95}$$

if we define the *time-ordered product* $P[V'(t')V'(t'')]$ by requiring that

$$P[V'(t')V'(t'')] = \begin{cases} V'(t')V'(t'') & \text{if } t'' \leqslant t' \\ V'(t'')V'(t') & \text{if } t' \leqslant t'' \end{cases}$$

Generalizing this convention to products of any number of time-dependent operators, it can be shown that $U(t, t_0)$ may be written in the form

$$U(t, t_0) = 1 + \sum_{n=1}^{\infty} \frac{1}{n!} \left(-\frac{i}{\hbar} \right)^n \int_{t_0}^{t}\int_{t_0}^{t} \cdots \int_{t_0}^{t} dt_1\, dt_2 \cdots dt_n P[V'(t_1)V'(t_2) \cdots V'(t_n)] \tag{18.96}$$

or formally and compactly as

$$U(t, t_0) = P \exp\left[-\frac{i}{\hbar} \int_{t_0}^{t} V'(t')\, dt' \right] \tag{18.97}$$

Exercise 18.4. Verify that (18.96) is a solution of (18.87) by constructing the time derivative of U as $\lim_{\varepsilon \to 0} [U(t + \varepsilon, t_0) - U(t, t_0)]/\varepsilon$ explicitly.

8. The Golden Rule for Constant Transition Rates. In many cases of practical interest the perturbation V does not itself depend on the time, but the time development of the system with a Hamiltonian $H = H_0 + V$ is nevertheless conveniently described in terms of transitions between eigenstates of the unperturbed Hamiltonian H_0. For example, a fixed potential scatters a beam of particles from an initial momentum state into various possible final momentum states, thus causing transitions between the eigenstates of the free particle Hamiltonian. Scattering will be discussed from this point of view in Chapter 19.

The theory is again based on the approximate equations (18.14), which read

$$i\hbar \frac{dc_k}{dt} = V_{ks}e^{i\omega_{ks}t} \qquad (k \neq s)$$

If V does not depend on the time, these equations can be immediately integrated. It is now convenient to choose the initial time as $t_0 = 0$. We find (for $k \neq s$)

$$c_k(t) = \frac{\langle k| V |s\rangle}{E_k^{(0)} - E_s^{(0)}} (1 - e^{i\omega_{ks}t}) \tag{18.98}$$

if $c_k(0) = 0$ and $c_s(0) = 1$. The probability that the system, known to have been in the initial discrete state s at $t = 0$, will be in the unperturbed final eigenstate $k \neq s$ at time t is then given by

$$|c_k(t)|^2 = 2 |\langle k| V |s\rangle|^2 \frac{1 - \cos \omega_{ks}t}{(E_k^{(0)} - E_s^{(0)})^2} \tag{18.99}$$

This is an oscillating function of t with a period equal to $2\pi/|\omega_{ks}|$. Its amplitude has a pronounced peak at $\omega_{ks} = 0$. It remains a valid expression only as long as $c_s(t)$ can be legitimately approximated by $c_s(t) \approx 1$. During such times the transition probability to states for which $E_k^{(0)} \neq E_s^{(0)}$ remains small, if the perturbation is weak. We know from the Rayleigh-Schrödinger perturbation theory that the eigenstate Ψ_s of H_0 is very nearly an eigenstate also of the actual Hamiltonian H if

$$\left| \frac{\langle k| V |s\rangle}{E_k^{(0)} - E_s^{(0)}} \right| \ll 1$$

This same inequality ensures that the transition probabilities to states with an unperturbed energy appreciably different from the initial value can never become large. On the other hand, transitions to states with unperturbed energies which are very nearly equal to $E_s^{(0)}$ can be very important, and if such states exist, there may be a good chance of finding the system in them after some time has elapsed. Since transitions to states with an $E_k^{(0)}$ radically different from $E_s^{(0)}$ are rare, we may say that, if the perturbation is weak, the unperturbed energy is nearly conserved, although the *total energy* is, of course, strictly conserved and $\langle H \rangle$ is a rigorous constant of the motion.

Transitions to states with $E_k^{(0)} \approx E_s^{(0)}$ have an important property. As long as

$$|\omega_{ks}| t \ll 1 \tag{18.100}$$

(18.99) becomes approximately

$$|c_k(t)|^2 \approx \frac{1}{\hbar^2} |\langle k|V|s\rangle|^2 t^2 \tag{18.101}$$

i.e., the transition probability increases quadratically with the time. This circumstance has interesting consequences if there are very many states k available having an energy $E_k^{(0)} \approx E_s^{(0)}$, as will happen if H_0 has a near continuum of eigenstates in the vicinity of $E_s^{(0)}$.

As a concrete example upon which we can fix our ideas, let us consider the helium atom again, regarding the electrostatic repulsion between the electrons as a perturbation (and neglecting radiative transitions altogether):

$$H_0 = \frac{\mathbf{p_1}^2}{2\mu} + \frac{\mathbf{p_2}^2}{2\mu} - \frac{2e^2}{r_1} - \frac{2e^2}{r_2} \tag{18.102}$$

$$V = \frac{e^2}{|\mathbf{r}_1 - \mathbf{r}_2|} \tag{18.103}$$

The excited state in which both electrons occupy the hydrogenic $2s$ level has an unperturbed energy of $6 \times 13.6 = 81.6$ eV above the ground state. Since this energy is greater than the energy required (54.4 eV) to lift one electron into the continuum with zero kinetic energy, the $(2s)^2$ level coincides in energy with a continuum state formed by placing one electron in the bound $1s$ state and the other electron in the continuum. But the wave functions of these two types of states are very different. The first is a discontinuous function of E, the second a continuous one. Figure 18.2 is an energy level diagram for the unperturbed two-electron system. If at $t = t_0$ the atom has been excited to the $(2s)^2$ state, the electrostatic perturbation (18.103) will cause so-called *radiationless* (autoionizing or Auger) transitions to the continuum. Radiative transitions to lower levels with the emission of light also take place, but are much less probable. The effects of radiationless transitions from autoionizing states, like the $(2s)^2$ state in the helium atom, are observable in electron-atom collisions. Auger transitions are observed in heavy atoms which have been excited by the removal of an electron from an inner shell. From such states the emission into the continuum of so-called *Auger electrons* competes with the emission of X rays. Internal conversion of nuclear gamma rays is another example of radiationless transitions.[9]

Since the formulas in this section were developed for quadratically integrable eigenstates of H_0, it is convenient to imagine the entire system to be enclosed in a very large box, or better still, to enforce discreteness of the

---

[9] For a review of resonant scattering of electrons by atomic systems see K. Smith, *Reports on Progress in Physics* **29**, 373 (1966). Radiationless transitions are discussed in E. H. S. Burhop, *The Auger Effect and Other Radiationless Transitions*, Cambridge University Press, Cambridge, 1952.

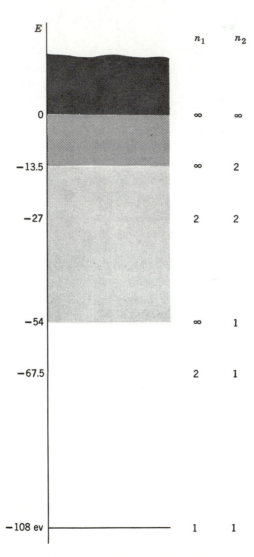

Figure 18.2. Partial energy level diagram for the unperturbed states of two electrons in the Coulomb field of a helium nucleus. n_1 and n_2 are the principal quantum numbers for the two electrons. The continuous portions of the spectrum are shaded. Note that the discrete state $n_1 = n_2 = 2$ is imbedded in the continuum.

unperturbed eigenstates by imposing on the unbound states periodic boundary conditions in such a large box. (In our example these states are approximately plane waves, although at the lower energies they must be taken to be appropriately screened Coulomb wave functions.) In order to be able to replace sums by integrals, the number of these *quasi-continuum* states per unit energy interval is introduced. This density of final unperturbed states is denoted by $\rho_f(E)$.

According to (18.99), the total transition probability to all final states under consideration, labeled by f, is given by

$$\sum_{k \in f} |c_k(t)|^2 = 2 \int |V_{ks}|^2 \frac{1 - \cos \omega_{ks} t}{(E_k^{(0)} - E_s^{(0)})^2} \rho_f(E_k^{(0)}) \, dE_k^{(0)} \qquad (18.104)$$

where we have again set $V_{ks} = \langle k| \, V \, |s \rangle$ as an abbreviation. Of particular interest is the time rate of change of the total transition probability:

$$w = \frac{d}{dt} \sum_k |c_k(t)|^2 = \frac{2}{\hbar^2} \int |V_{ks}|^2 \frac{\sin \omega_{ks} t}{\omega_{ks}} \rho_f(E_k^{(0)}) \, dE_k^{(0)} \qquad (18.105)$$

In usual practice both $|V_{ks}|^2$ and $\rho_f(E_s^{(0)})$ are sensibly constant over an energy range ΔE in the neighborhood of E_s. But, $\sin \omega_{ks} t / \omega_{ks}$ oscillates rapidly as a function of $E_k^{(0)}$ in this same interval for all t satisfying

$$t \gg \hbar/\Delta E \qquad (18.106)$$

and has a pronounced peak at $E_k^{(0)} = E_s^{(0)}$. Again it is seen that transitions which tend to conserve the unperturbed energy are dominant. ΔE is generally comparable in magnitude to $E_s^{(0)}$ itself. If this is so, $\hbar/\Delta E$ is a very short time and there is a considerable range of t such that (18.106) is fulfilled and yet the initial state s not appreciably depleted. During such times (18.105) can be simplified and gives to good approximation

$$\frac{d}{dt} \sum_k |c_k(t)|^2 = \frac{2}{\hbar} |V_{ks}|^2 \rho_f(E_s^{(0)}) \int_{-\infty}^{+\infty} \frac{\sin \omega_{ks} t}{\omega_{ks}} \, d\omega_{ks}$$

where condition (18.106) has been invoked in replacing the limits of the integral by $-\infty$ and $+\infty$. Under the conditions stated we have for the transition probability per unit time (for $t > 0$):

$$w = \frac{d}{dt} \sum_k |c_k(t)|^2 = \frac{2\pi}{\hbar} |\langle k| \, V \, |s \rangle|^2 \rho_f(E_s^{(0)}) \qquad (18.107)$$

Formula (18.107) represents a *constant rate of transition*. This remarkable result comes about because we have summed over transitions which conserve the unperturbed energy strictly ($E_k^{(0)} = E_s^{(0)}$) and transitions that violate this energy conservation ($E_k^{(0)} \neq E_s^{(0)}$). According to (18.101), the probability of the former increases quadratically with time, but the probability of the latter oscillates periodically. The sum is a compromise between these two tendencies: the transition rate is constant. Fermi has called formula (18.107) the *Golden Rule of time-dependent perturbation theory* because it plays a fundamental part in many applications.

If $\langle k | V | s \rangle = 0$, the probability of a transition from the unperturbed state s to a state k may still be different from zero, since it becomes now necessary to retain the second-order term in the transition amplitude (18.93), and possibly terms of even higher order. It is easily verified that a second-order calculation, based on the same initial conditions, $c_k(0) = \delta_{ks}$, as before, gives for $k \neq s$

$$c_k(t) = \sum_m{}' V_{km} V_{ms} \left[\frac{1 - e^{i\omega_{ks}t}}{(E_k^{(0)} - E_s^{(0)})(E_s^{(0)} - E_m^{(0)})} - \frac{1 - e^{i\omega_{km}t}}{(E_k^{(0)} - E_m^{(0)})(E_s^{(0)} - E_m^{(0)})} \right]$$

$$(18.108)$$

in place of (18.98). The sum must be carried out over the complete set of unperturbed states m, but we must omit those eigenstates of H_0 for which either $V_{km} = 0$ or $V_{ms} = 0$. The prime on the summation sign is intended as a reminder of this restriction.

Again the only transitions with appreciable probability are those for which $E_k^{(0)} \approx E_s^{(0)}$. If there are no states m to which first-order transitions can take place, i.e., if $V_{ms} = 0$ for all $E_m^{(0)} \approx E_s^{(0)}$, the second term in (18.108) can be neglected because it represents oscillations of small amplitude. If the final state k is again assumed to lie in a continuum, the first term in (18.108) leads to a constant transition probability per unit time, and the Golden Rule (18.107) still holds provided that the first-order transition matrix element $\langle k | V | s \rangle$ is replaced by the expression

$$\sum_m{}' \frac{V_{km} V_{ms}}{E_s^{(0)} - E_m^{(0)}} \tag{18.109}$$

This effective transition matrix element may be interpreted by describing the transition as a two-step process, in which a *virtual* transition from the initial state s to an *intermediate state m* is followed by a second virtual transition from the state m to the final state k. The energy of the intermediate statë can be entirely different from the energy of the initial and final states, but such transitory violations of the conservation of energy are of no moment, and in the over-all transition from the initial to the final states energy is, of course, conserved.

If there are any states m to which first-order transitions are possible (i.e., $V_{ms} \neq 0$, $E_m^{(0)} \approx E_s^{(0)}$) but which are not among the final states k of interest, the application of (18.108) requires some care. In this case the vanishing energy denominators signal the need for a more accurate treatment, and the second term in (18.108) can no longer be neglected.[10]

9. Exponential Decay. Suppose that a system, which is perturbed by a constant V as described in the last section, is known to be in the initial state s at time t. The probability that it will make a transition in the ensuing interval between t and $t + dt$ is equal to $w\, dt$, if the conditions under which (18.107) was derived obtain. *Stochastic processes* with constant w are very familiar in probability theory.[11] The probability $P_s(t + dt)$ of finding the system in state s at time $t + dt$ can be obtained easily if we argue as follows: The system will be in state s at $t + dt$ only if (a) it was in s at time t, and (b) it has not decayed from this state during the interval dt. Since the probability for not decaying in dt is $(1 - w\, dt)$, we have

$$P_s(t + dt) = P_s(t)(1 - w\, dt) \tag{18.110}$$

with the initial condition $P_s(t_0) = 1$. Solving (18.110), we infer the probability of finding the system at any time t still undecayed in the initial state:

$$P_s(t) = e^{-wt} \tag{18.111}$$

This is the famous *exponential decay law*.

However, caution is required in making the preceding argument because it assumes that P_s changes only by virtue of transitions *out of* state s *into* other states but disregards the possibility that the state s may be replenished by reverse transitions from the final states. Even so, the conditional probability $P_s(t + dt)$ can be equated to the product of $P_s(t)$ and $(1 - w\, dt)$ only if the actual determination of whether at time t the system is in state s or not does not influence its future development. In general, this condition for the validity of (18.110) is emphatically not satisfied in quantum mechanics. Such a measurement does in fact interfere with the course of events and alters the chances of finding the system in s at time $t + dt$ from what they would have been had we refrained from attempting the measurement.

Starting with

$$P(t_0) = |c_s(t_0)|^2 = 1$$

quantum mechanics requires that the probability at all times is to be calculated from the *amplitudes*, which are the primary concepts, hence,

$$P_s(t + dt) = |c_s(t + dt)|^2 \tag{18.112}$$

---

[10] For an example of discrete energy-conserving intermediate states see the discussion of resonance fluorescence in W. Heitler, *The Quantum Theory of Radiation*, 3rd ed., Section 5.20, Clarendon Press, Oxford, 1954. For energy-conserving intermediate states which lie in the continuum see L. I. Schiff, *Quantum Mechanics*, 2nd ed., McGraw-Hill Book Company, New York, 1955, pp. 202–204.

[11] W. Feller, *An Introduction to Probability Theory and Its Applications*, Volume I, 2nd ed., pp. 397–402, John Wiley & Sons, New York, 1957.

There is in general no reason why this expression should be equal to

$$P_s(t)(1 - w\,dt) = |c_s(t)|^2\,(1 - w\,dt) \qquad (18.113)$$

as was assumed in writing (18.110). However, it can be shown that the very conditions which insure that there is a constant transition rate also allow us to equate (18.112) with (18.113). From (18.10) we recall that the motion is rigorously determined by the coupled linear equations

$$i\hbar\frac{dc_k}{dt} = \sum_l V_{kl}c_l e^{i\omega_{kl}t} \qquad (18.114)$$

According to these equations, a state l feeds a state k if $V_{kl} = \langle k|\,V\,|l\rangle \neq 0$. Hence, transitions from the initial state s to the various available final states k occur, and at the same time these final states contribute by a feedback process to the amplitude of the initial state. As the amplitudes of the final states k increase from their initial value zero, they must grow at the expense of the initial state, since probability is conserved. We might expect, however, that as the amplitudes of the states k increase, they would begin to feed back into state s. Indeed, this is what happens, but because of the different frequencies ω_{ks} with which this feedback occurs, the contributions from the many amplitudes c_k to c_s are all of different phases. Hence, if there are very many states k, forming essentially a continuum, these contributions tend to cancel. It is this destructive interference which causes the gradual exponential depletion of the initial state without corresponding regeneration.

In order to derive the exponential decay law, we must solve the equations of motion under the same assumptions as before (i.e., constant V, transitions from a discrete initial state s to a quasi continuum of final states), but we must remove the uncomfortable restriction to times which are short compared with the lifetime of the initial state.[12] This means, of course, that it is no longer legitimate to replace $c_s(t)$ on the right-hand side of (18.114) by $c_s(t) = 1$. However—and this is the fundamental assumption—we continue to neglect all other contributions to the change in $c_k(t)$ and use the approximate equations for $t \geqslant 0$,

$$i\hbar\frac{dc_k}{dt} = V_{ks}c_s e^{i\omega_{ks}t} \qquad \text{for} \qquad k \neq s \qquad (18.115)$$

The justification for this assumption is essentially physical and a posteriori. To a certain extent it is based on our previous experience with the short-term approximation. If the perturbation is weak, c_k will remain small for

---

[12] The fundamental paper on the quantum theory of exponential decay and the natural line width is by V. F. Weisskopf and E. P. Wigner, *Z. Physik*, **63**, 54 (1930).

those transitions for which ω_{ks} is appreciably different from zero. Hence, only those amplitudes c_k are likely to be important for which $\omega_{ks} \approx 0$. On the other hand, the matrix elements $\langle k| V |l \rangle$ which connect two possible final unperturbed states $k \neq s$ and $l \neq s$ for which $E_k^{(0)} \approx E_l^{(0)}$ are usually small or vanish and will be neglected.[13]

Equation (18.115) can be integrated,

$$c_k(t) = -\frac{i}{\hbar} V_{ks} \int_0^t c_s(t')e^{i\omega_{ks}t'} \, dt' \qquad (k \neq s) \tag{18.116}$$

since $c_k(0) = 0$ for all $k \neq s$. The equation of motion for $c_s(t)$ is rigorously

$$i\hbar \frac{dc_s}{dt} = \sum_{k \neq s}' V_{sk}c_k(t)e^{i\omega_{sk}t} + V_{ss}c_s(t) \tag{18.117}$$

where the prime on the summation sign is to remind us that the term for $k = s$ has been omitted from the sum; it appears separately in the equation. (In decay problems we have frequently $\langle s| V |s \rangle = 0$, but in any case this term produces only a shift in the unperturbed energy level s, as is already known from the Rayleigh-Schrödinger perturbation theory.)

If (18.116) is substituted into (18.117), we have for $t \geq 0$,

$$\frac{dc_s}{dt} = -\frac{1}{\hbar^2} \sum_{k \neq s}' |V_{ks}|^2 \int_0^t c_s(t')e^{i\omega_{ks}(t'-t)} \, dt' - \frac{i}{\hbar} V_{ss}c_s(t) \tag{18.118}$$

It is convenient to supplement this equation arbitrarily by the assumption

$$c_s(t) = 0 \qquad \text{for} \qquad t < 0$$

and we require, of course, $c_s(0) = 1$. Equation (18.118) may be solved for $t \geq 0$ by use of Fourier transforms, which we define as

$$c_s(t) = \int_{-\infty}^{+\infty} f(\omega)e^{-i\omega t} \, d\omega \tag{18.119}$$

with its inverse,

$$f(\omega) = \frac{1}{2\pi} \int_0^{+\infty} c_s(t)e^{i\omega t} \, dt \tag{18.120}$$

We multiply equation (18.118) by $e^{i\omega t}$ and integrate from $t = 0$ to $t = +\infty$, assuming tentatively, and subject to eventual verification by the solution, that $c_s(+\infty) = 0$. Hence, the left-hand side of the transformed equation becomes

$$\int_0^{+\infty} \frac{dc_s}{dt} e^{i\omega t} \, dt = c_s(t)e^{i\omega t}\Big|_0^{+\infty} - i\omega \int_0^{+\infty} c_s(t)e^{i\omega t} \, dt = -c_s(0) - 2\pi i\omega f(\omega)$$

[13] Such transitions are basic to the scattering process; see Chapter 19.

and (18.118) is transformed into

$$2\pi i \omega f(\omega) + 1 = \frac{1}{\hbar^2} \sum' |V_{ks}|^2 \int_0^{+\infty} dt \; e^{i(\omega_{sk}+\omega)t - \varepsilon t} \int_0^t e^{-i\omega_{sk}t'} c_s(t') \, dt' + \frac{2\pi i}{\hbar} V_{ss} f(\omega)$$

The factor $e^{-\varepsilon t}$ has been inserted to render the integral convergent, and the limit $\varepsilon \to 0$ will be taken at the end of the calculation. Integration by parts on the right-hand side yields

$$2\pi i \omega f(\omega) + 1 = \frac{1}{\hbar^2} \sum' |V_{ks}|^2 \frac{2\pi f(\omega)}{-i(\omega_{sk} + \omega) + \varepsilon} + \frac{2\pi i}{\hbar} V_{ss} f(\omega)$$

An important approximation is made at this stage by setting $\omega = 0$ in the denominator on the right-hand side of this equation. This step can be justified if we are not interested in values of $c_s(t)$ within the first few natural periods of the radiating system after $t = 0$. From the last equation, we then obtain[14]

$$2\pi i f(\omega) = \left[-\omega + \frac{1}{\hbar} V_{ss} + \frac{1}{\hbar^2} \sum' |V_{ks}|^2 \frac{1}{\omega_{sk} + i\varepsilon} \right]^{-1} \qquad (18.121)$$

and, with the use of (18.119) and the same integration technique which gave us (7.54) and (11.79),

$$c_s(t) = \begin{cases} \exp \left\{ -\frac{i}{\hbar} \left[V_{ss} + \frac{1}{\hbar} \sum' |V_{ks}|^2 \frac{1}{\omega_{sk} + i\varepsilon} \right] t \right\}, & t \geqslant 0 \\ 0 & t < 0 \end{cases} \qquad (18.122)$$

The exponent may be converted into a more manageable form by applying equation (6.18):

$$c_s(t) = \eta(t) \exp \left\{ -\frac{i}{\hbar} \left[V_{ss} + \sum' \frac{|V_{ks}|^2}{E_s^{(0)} - E_k^{(0)}} - \frac{i}{\hbar} \pi \sum' |V_{ks}|^2 \delta(\omega_{sk}) \right] t \right\} \quad (18.123)$$

Equation (18.123) is the anticipated result. If ΔE_s denotes the shift of the unperturbed energy level $E_s^{(0)}$ due to the perturbation in second order, we may write

$$c_s(t) = \exp \left(-\frac{w}{2} t - \frac{i}{\hbar} t \, \Delta E \right) \qquad (18.124)$$

describing the exponential decay of the unstable state.

To obtain nonreversing transitions and a progressive depletion of the initial state it is essential that the discrete initial state be coupled to a very large number of states with similar frequencies. However, the fact remains

---

[14] For more details, see Chapter 8 in M. L. Goldberger and K. M. Watson, *Collision Theory*, John Wiley and Sons, New York, 1964.

that the exponential decay law, for which we have so much empirical support in radioactive processes, is not a rigorous consequence of quantum mechanics but the result of somewhat delicate approximations.

If (18.124) is substituted back into (18.116), the integration can be carried out and we obtain for $t \geqslant 0$,

$$
c_k(t) = -\frac{i}{\hbar} V_{ks} \int_0^t \exp\left[-\frac{i}{\hbar}\left(E_s^{(0)} + \Delta E_s - E_k^{(0)} - i\hbar \frac{w}{2}\right)t'\right] dt'
$$

$$
= V_{ks} \frac{1 - \exp\left(-\frac{w}{2}t\right)\exp\left[-\frac{i}{\hbar}(E_s^{(0)} + \Delta E - E_k^{(0)})t\right]}{E_k^{(0)} - (E_s^{(0)} + \Delta E_s) + i\hbar \frac{w}{2}} \tag{18.125}
$$

Hence, the probability that the system has decayed into state k is

$$
|c_k(t)|^2 = |V_{ks}|^2 \frac{1 - 2\exp\left(-\frac{\Gamma}{2\hbar}t\right)\cos\left(\frac{E_s^{(0)} + \Delta E_s - E_k^{(0)}}{\hbar}t\right) + \exp\left(-\frac{\Gamma}{\hbar}t\right)}{(E_k^{(0)} - E_s^{(0)} - \Delta E_s)^2 + \frac{\Gamma^2}{4}} \tag{18.126}
$$

where we have set $\Gamma = \hbar w$. After a time very long compared with the life-time $\hbar/\Gamma$ we obtain the distribution

$$
|c_k(+\infty)|^2 = \frac{|V_{ks}|^2}{(E_k^{(0)} - E_s^{(0)} - \Delta E_s)^2 + \frac{\Gamma^2}{4}} \tag{18.127}
$$

exhibiting the typical bell-shaped resonance behavior with a peak at $E_s^{(0)} + \Delta E_s$ and a width equal to Γ.

Exercise 18.5. Prove from (18.126) that the sum $\sum_{k \neq s} |c_k(t)|^2$ over all final states is equal to $1 - \exp(-\Gamma t/\hbar)$ as required by the conservation of probability.

In a somewhat imprecise manner, the results of this section may be interpreted as implying that the interaction causes the state s to change from a strictly stationary state of H_0 with energy $E_s^{(0)}$ into a decaying state with probability amplitude

$$
\propto \left(E - E_s^{(0)} - \Delta E_s + i\frac{\Gamma}{2}\right)^{-1} \tag{18.128}
$$

for having the energy E.

A remark about the shape of the absorption cross section calculated in Section 18.5 may now be made. If natural decay is the mechanism responsible for the depletion of a state k excited by absorption of radiation from a stable state s, the absorption probability for the frequency $\omega = (E - E_s^{(0)})/\hbar$ must, according to (18.128), be weighted by the factor

$$p(\omega)\, d\omega = \frac{(\Gamma/2\pi\hbar)d\omega}{(\omega - \omega_{ks})^2 + (\Gamma/2\,\hbar)^2} \tag{18.129}$$

(Note the interchange of the labels k and s, because the final state in absorption is the initial state in emission.) The absorption cross section is obtained from (18.73) by replacing the delta function by the distribution (18.129), and hence,

$$\sigma(\omega) = \frac{4\pi^2\alpha}{\mu^2\omega_{ks}} \left| \langle k| \exp\left(i\,\frac{\omega_{ks}}{c}\,\hat{\mathbf{n}}\cdot\mathbf{r}\right)\mathbf{p}\cdot\hat{\mathbf{e}}\,|s\rangle \right|^2 \frac{\Gamma/2\pi\hbar}{(\omega - \omega_{ks})^2 + \frac{1}{4}\Gamma^2/\hbar^2} \tag{18.130}$$

showing the characteristic resonance (or Lorentz) shape of the absorption "line."

Problems

1. Show that if the term V in the Hamiltonian changes suddenly (by an "impulse"), i.e., in a time Δt short compared with all relevant periods, the change of the state vector is given by

$$\Psi(t + \Delta t) = \exp\left[-i\int_t^{t+\Delta t} V(t')\, dt'/\hbar\right]\Psi(t)$$

assuming only that $[V(t'), V(t'')] = 0$ during the impulse.

   Note especially that the wave function remains unchanged during a sudden change of V by a finite amount.

   If H is constant in time up to $t = 0$ and suddenly changes to H' (also constant) at $t = 0$, show that if the system was in an eigenstate Ψ_n of H for $t < 0$, the probability of finding it at $t > 0$ in an eigenstate Ψ_m' of H' is $|(\Psi_m', \Psi_n)|^2$.

2. A linear harmonic oscillator in its ground state is exposed to a constant force which at $t = 0$ is suddenly removed. Compute the transition probabilities to the first and second excited states of the oscillator. Use the generating function (5.18) to obtain a general formula.

3. Calculate the cross section for the emission of a photoelectron ejected when linearly polarized monochromatic light of frequency ω is incident on a complex atom. Describe the initial state of the electron by the ground state wave function of an isotropic three-dimensional harmonic oscillator and the final state by a plane wave. Obtain the angular distribution as a function of the angle of emission and sketch it on a polar graph for suitable assumed values of the parameters.

4. Calculate the total cross section for photoemission from the K-shell as a function of the frequency of the incident light and the frequency of the K-shell absorption edge, assuming that $\hbar\omega$ is much larger than the ionization potential but that nevertheless the photon momentum is much less than the momentum of the ejected electron.

5. Obtain the hydrogenic ground state wave function in momentum space and show that it plays an important role in the cross-section calculation for the photoelectric effect.

6. By considering the double commutator

$$[[H, e^{i\mathbf{k}\cdot\mathbf{r}}], e^{-i\mathbf{k}\cdot\mathbf{r}}]$$

obtain as a generalization of the Thomas-Reiche-Kuhn sum rule the formula

$$\sum_n (E_n - E_s)\,|\langle n|\,e^{i\mathbf{k}\cdot\mathbf{r}}\,|s\rangle|^2 = \hbar^2 k^2/2\mu$$

Specify the conditions on the Hamiltonian H required for the validity of this sum rule.

7. Apply the methods of time-dependent perturbation theory to a forced linear harmonic oscillator which is initially in the ground state, and compare the results with the exact calculation of Section 15.8. Prove that the first-order perturbation value of the energy transfer to the oscillator is equal to the exact result (Problem 2 in Chapter 15). Explain the agreement.

8. A charged particle moving in a linear harmonic oscillator potential is exposed to electromagnetic radiation. Initially, the particle is in the oscillator ground state. Discuss the conditions under which the electric dipole—no retardation approximation is good. In this approximation, show that the first-order perturbation value of the integrated absorption cross section is equal to the sum of dipole absorption cross sections, calculated exactly.

   Derive the selection rules for transitions in the next (electric quadrupole) approximation, which corresponds to retaining the second term in the expansion (18.53). Calculate the absorption rate for quadrupole transitions and compare with the rate for dipole transitions.

CHAPTER 19

The Formal Theory of Scattering

1. *Introduction.* There are several reasons for taking up the theory of scattering once more. In the first place it is very natural to look at a scattering process as a transition from one unperturbed state to another. If the scattering region is of finite extent, the initial and final states are simply plane wave eigenstates of definite momentum of the unperturbed Hamiltonian, $H_0 = \mathbf{p}^2/2\mu$, and the scattering potential causes transitions from an initial state with propagation vector $\mathbf{k}$ to the final states characterized by propagation vectors $\mathbf{k}'$.

This description of the scattering process, in which the infinite plane waves serve as idealizations of very broad and long wave packets, replaces the formulation of Chapter 11 in terms of finite wave packets. At first sight it may seem strange that an incident wave can be represented by an infinite plane wave which is equally as intense in front of the scatterer as behind it. In Chapter 11 the unphysical appearance of incident waves behind the scatterer was avoided by superposing waves of different $\mathbf{k}$ and canceling the unwanted portion of the wave by destructive interference. The resulting theory, involving Fourier integrals at every step, was correct but clumsy. In the present chapter precisely the same results will be achieved in a more elegant fashion by the use of suitable mathematical limiting procedures, but if mistakes are to be avoided it is advisable always to keep in mind the physical picture of the scattering of particles, which first impinge upon and subsequently move away from the scatterer.

The methods with which we shall deal in this chapter are formal and may seem to sacrifice ease of physical interpretation for mathematical compactness. It is true that when scattering from a potential is discussed there is no essential difference between the formal theory of scattering and the more pedestrian wave packet method of Chapter 11, and this equivalence is useful in setting up the ground rules of the formal theory. But the formal theory of scattering is a powerful tool in the application to processes which are more complex than the elastic deflection of particles by a fixed potential. Examples of such processes are the following: collisions between complex systems, such as are encountered in nuclear reactions; inelastic collisions; processes involving photons, such as the Compton effect or pair production; collisions between pions and nucleons, etc. Indeed, almost every laboratory experiment of atomic and nuclear physics can be described as a generalized scattering process with an initial incident state, an interaction between all the components of the system, and a final scattered state. But whereas the wave packet analysis becomes prohibitively difficult for any but the simplest scattering problems, the methods which will be developed in this chapter are capable of enormous generalization and can with suitable but often minor emendation be applied to many complicated systems and interactions.[1]

Finally, the formal scattering theory permits the quick utilization of the symmetry properties of the system, enabling us to predict from these symmetry properties the general form of observable quantities such as transition probabilities or cross sections. We have already encountered examples of this kind in Chapters 6 and 13. There we found the scattering matrix to be a useful concept which contains all relevant dynamical information independent of the special choice of the incident wave. Evidently, then, we should expect the discussion of this chapter to center around a generalization of this concept, a scattering matrix or scattering operator.

2. The Equations of Motion and the Transition Matrix. As in Chapter 11, elastic scattering of a particle without spin from a fixed potential will be considered, since this process, to which we shall refer as *simple scattering*, is the prototype of all more complex processes. The Hamiltonian is

$$H = H_0 + V = \frac{\mathbf{p}^2}{2\mu} + V \tag{19.1}$$

The solution of the equation of motion may be written in terms of the eigenvectors of H_0 as[2]

$$\Psi(t) = \sum_n c_n(t) \exp\left(-\frac{i}{\hbar} E_n t\right) \Psi_n \tag{19.2}$$

where

$$H_0 \Psi_n = E_n \Psi_n \tag{19.3}$$

Although we assume that H_0 is simply the kinetic energy operator, in more sophisticated applications H_0 may include part of the interaction, provided

---

[1] A comprehensive treatise on scattering processes is M. L. Goldberger and K. M Watson, *Collision Theory*, John Wiley and Sons, New York, 1964.

[2] Equation (19.2) is the same as (18.9), except for a slight change in notation: the superscripts on the eigenvectors and eigenvalues of the unperturbed Hamiltonian have for simplicity been omitted. Compare also (19.4) and (18.10).

only that the solutions to the eigenvalue problem (19.3) are known. The equation of motion is now expressible as

$$i\hbar \frac{dc_r}{dt} = \sum_n V_{rn} c_n e^{i\omega_{rn} t} \tag{19.4}$$

where

$$\hbar \omega_{rn} = E_r - E_n \tag{19.5}$$

and

$$V_{rn} = (\Psi_r, V\Psi_n) \tag{19.6}$$

and where it has been assumed that the unperturbed eigenvectors are normalized to unity:

$$(\Psi_r, \Psi_n) = \delta_{rn} \tag{19.7}$$

This means that, if the unperturbed states are plane waves, periodic boundary conditions must be imposed in a large box of volume L^3. Eventually the limit $L \to \infty$ may be taken.

The initial conditions of the problem are as follows:

$$c_s(-\infty) = 1, \qquad c_r(-\infty) = 0 \qquad \text{for} \qquad r \neq s \tag{19.8}$$

if s denotes the incident state, namely, e^{ikz} for a spinless particle incident along the z-axis. If perturbation theory were now used to solve (19.4), we would write

$$c_r(t) = -\frac{i}{\hbar} V_{rs} \int_{-\infty}^{t} e^{i\omega_{rs} t'} \, dt' + \delta_{rs} \tag{19.9}$$

since V is constant. However, the integral in (19.9) does not exist, so this cannot be a meaningful solution. The same difficulty did not arise for the constant perturbations of Chapter 18, because the initial time was chosen as $t_0 = 0$. Here we could do the same thing and let t_0 be a finite value, but this would bring into the calculation confusing transients which are unphysical because they have their origin in the totally unrealistic assumption that somehow at time t_0 the scattering region has been embedded in a perfect plane wave which is now being released.

There are two possible ways out of this dilemma: a wave packet may be formed which represents the physically sensible condition at time t_0 of a particle moving toward the scatterer; alternatively, $t_0 = -\infty$ may be maintained but the integral in (19.9) rendered meaningful by the introduction of a limiting process. The first approach is that of Chapter 11; here the second road will be taken.

There are, of course, a great many possible ways of altering the integral in (11.9) "slightly" so that it will converge. Fortunately, the most obvious proposal works. It consists of merely inserting a factor $e^{\alpha t}$ in the integrand

and writing

$$c_r(t) = -\frac{i}{\hbar} T_{rs} \int_{t_0}^{t} e^{i\omega_{rs}t' + \alpha t'} \, dt' + \delta_{rs} \tag{19.10}$$

with the understanding that α is positive and that the limit $\alpha \to 0$ must be taken after the limit $t_0 \to -\infty$. Equation (19.10) will be assumed to give $c_k(t)$ correctly only for times t which satisfy the relation

$$|t| \ll (1/\alpha) \tag{19.11}$$

It is essential to keep these restrictions in mind. If they are disregarded, the equations of the formal theory may lead to painful contradictions. Such contradictions arise easily because the formal theory is designed to be a shorthand notation in which conditions like (19.11) are implied but never spelled out. The formal theory thus operates with a set of conventions. From these it derives its conciseness and flexibility. Those who consider the absence of explicit mathematical instructions at every step too high a price to pay can always return to the wave packet form of the theory. The connection between the two points of view is never lost if it is noted that $1/\alpha$ measures crudely the length of time during which the wave packet strikes, envelops, and passes the scattering region. If v is the mean particle velocity, v/α is thus roughly the length of the wave packet.

One other change has been made in generalizing (19.9) to (19.10): the known matrix V_{rs} has been changed into an unknown matrix T_{rs} in the hope that by doing this we might possibly be able to avoid making the perturbation approximation upon which (19.9) was based.

Having thus given some motivation for the form (19.10), we may now simply regard this as an *Ansatz* for solving (19.4) and ask if the matrix T_{rs} can be determined so that (19.10) is the solution of (19.4). In the next section it will be shown that under the conditions prevailing in scattering problems such a solution does indeed exist. Moreover, it is rigorous and not approximate.

Assuming the existence of this solution, an important conclusion can be drawn immediately. Upon integrating (19.10) we obtain

$$c_r(t) = \frac{T_{rs} e^{i\omega_{rs}t + \alpha t}}{\hbar(-\omega_{rs} + i\alpha)} + \delta_{rs} \tag{19.12}$$

as $\lim_{\substack{\alpha \to 0 \\ t_0 \to -\infty}} e^{\alpha t_0} = 0$. For states $r \neq s$ we thus have

$$|c_r(t)|^2 = \frac{|T_{rs}|^2 e^{2\alpha t}}{\hbar^2(\omega_{rs}^2 + \alpha^2)} \tag{19.13}$$

hence, for the rate of transition into state r:

$$\frac{d}{dt}|c_r(t)|^2 = \frac{2\alpha}{\omega_{rs}^2 + \alpha^2} e^{2\alpha t} \frac{1}{\hbar^2} |T_{rs}|^2 \tag{19.14}$$

In the limit $\alpha \to 0$, which must always be taken but at finite values of t, this becomes [see (6.11b)]

$$\frac{d}{dt}|c_r(t)|^2 = \frac{2\pi}{\hbar^2} \delta(\omega_{rs})|T_{rs}|^2 = \frac{2\pi}{\hbar} \delta(E_r - E_s)|T_{rs}|^2 \tag{19.15}$$

if $r \neq s$ and if it is assumed that T_{rs} has no singularity as a function of energy at $E_r = E_s$. Evidently, the solution thus implies a constant transition rate—precisely what we expect to be the effect of the scatterer causing transitions from state s to r. This makes it clear why T_{sr} is called the *transition matrix*.

Of course, (19.15) is meaningful only if there is a quasi-continuum of unperturbed states with energies $E_r \approx E_s$. If V_{rs} had been used instead of T_{rs}, (19.15) would be simply equivalent to the Golden Rule of time-dependent perturbation theory, (18.107).

The result (19.15) will be used in the calculation of scattering cross sections. But first it is imperative to demonstrate that the matrix T_{rs} exists and that in scattering problems (19.4) has solutions of the form (19.12).

3. *The Integral Equations of Scattering Theory.* We now substitute $c_r(t)$ from (19.12) into (19.4) and immediately set $\alpha t = 0$ in accordance with the restriction (19.11). The matrix T_{rs} must then satisfy the system of simultaneous linear equations,

$$T_{rs} = \frac{1}{\hbar} \sum_n \frac{V_{rn} T_{ns}}{-\omega_{ns} + i\alpha} + V_{rs} \tag{19.16}$$

If T_{rs} satisfies this equation, the expression (19.12) is a solution of the equation of motion (19.4) for times $|t| \ll 1/\alpha$. The solution (19.12) also satisfies the initial conditions (19.8) as $t \to -\infty$; yet it should be realized that it is not a solution of (19.4) for times $t \ll -1/\alpha$.

The summation in (19.16) extends over all unperturbed eigenstates, hence, includes states for which $E_n = E_s$. The corresponding denominator, $-\omega_{ns} + i\alpha$, vanishes in the limit $\alpha \to 0$, and we would thus be led to a meaningless equation if the spectrum of H_0 were truly discrete. The unmanageable singularity can be averted and the transition matrix expected to exist only if the unperturbed states form a quasi-continuum with energies very close to the energy E_s of the initial state. This is precisely the situation which prevails in scattering, where the initial and final states have the same, or very nearly the same, energies. T_{rs} is then proportional to $1/L^3$. Hence, the amplitudes of the final states, $r \neq s$, will remain small, at least for

$t \ll 1/\alpha$. It follows that the initial state is not appreciably depleted over a period of the order $1/\alpha$, and we can have a *constant rate of transition.*[3]

Equation (19.16) can be used to connect the formal theory with the more explicit description of scattering in Chapter 11. For this purpose it is convenient to define a set of vectors $\Psi_s^{\prime(+)}$ by the linear equations,

$$T_{rs} = \sum_j (\Psi_r^{\prime}, V\Psi_j^{\prime})(\Psi_j^{\prime}, \Psi_s^{\prime(+)}) = (\Psi_r^{\prime}, V\Psi_s^{\prime(+)}) \tag{19.17}$$

Substituting this scalar product in (19.16), we obtain an equation which $\Psi_s^{\prime(+)}$ must satisfy:

$$(\Psi_r^{\prime}, V\Psi_s^{\prime(+)}) = \sum_n (\Psi_r^{\prime}, V\Psi_n^{\prime}) \frac{(\Psi_n^{\prime}, V\Psi_s^{\prime(+)})}{E_s - E_n + i\hbar\alpha} + (\Psi_r^{\prime}, V\Psi_s^{\prime})$$

or, since this must be true for all r,

$$\Psi_s^{\prime(+)} = \Psi_s^{\prime} + \sum_n \Psi_n^{\prime} \frac{(\Psi_n^{\prime}, V\Psi_s^{\prime(+)})}{E_s - E_n + i\hbar\alpha}$$

$$= \Psi_s^{\prime} + \sum_n \frac{1}{E_s - H_0 + i\hbar\alpha} \Psi_n^{\prime}(\Psi_n^{\prime}, V\Psi_s^{\prime(+)}) \tag{19.18}$$

However, by completeness,

$$\sum_n \Psi_n^{\prime}(\Psi_n^{\prime}, V\Psi_s^{\prime(+)}) = V\Psi_s^{\prime(+)} \tag{19.19}$$

Hence, we obtain as a final result the *implicit equation*

$$\Psi_s^{\prime(+)} = \Psi_s^{\prime} + \frac{1}{E_s - H_0 + i\hbar\alpha} V\Psi_s^{\prime(+)} \tag{19.20}$$

This is a fundamental equation of the formal theory of scattering. The problem of obtaining the transition matrix has thus been reduced to solving (19.20), known as the *Lippmann–Schwinger equation.*

By applying the operator $H_0 - E_s + i\hbar\alpha$ to (19.20), it is immediately established that in the limit $\alpha \to 0$,

$$(H_0 - E_s)\Psi_s^{\prime(+)} = -V\Psi_s^{\prime(+)} \tag{19.21}$$

i.e., $\Psi_s^{\prime(+)}$ becomes an eigenvector of $H = H_0 + V$, and E_s the corresponding eigenvalue. Hence, the procedure of solving the equation of motion outlined in this chapter can succeed only if in the limit $\alpha \to 0$ E_s is an eigenvalue of *both* H_0 and H. In simple scattering systems where $V \to 0$ as $r \to \infty$ both Hamiltonian operators have the same continuous spectrum: The energy

---

[3] If the initial state s were a truly discrete state, the method of solution outlined here would fail. Such a state decays exponentially, and an altogether different solution to the equation of motion is appropriate. See Chapter 18.

can have any value between 0 and ∞, and the presence of the potential does not change this fact. It will be assumed throughout this chapter that the continuous portions of the spectra of H_0 and H coincide and extend from $E = 0$ to ∞, although for complicated systems this is a severe restriction which cannot and need not be maintained. In addition, H may possess discrete eigenvalues which have no counterpart in the spectrum of H_0, corresponding to bound states.

Actually, the periodic boundary conditions on the cube of length L cause the entire spectra of both H_0 and H to be discrete. Even the quasi-continuous energy levels of H_0 and H will then generally not coincide, because the two operators differ by V. But the shift, ΔE_r, of each level goes to zero as $L \rightarrow \infty$. The limit $L \rightarrow \infty$ must be taken together with the limit $\alpha \rightarrow 0$, since v/α signifies approximately the length of the incident wave packet and we must require that $L > v/\alpha$, or else the normalization box could not contain the wave packet.[4] If the limit $L \rightarrow \infty$ is accompanied by a change of normalization of positive energy states from unity to k-normalization, as defined in (10.6), the formal identification $L = 2\pi$ allows the same equations to be used after the limit has been taken as before.

In retrospect it is instructive to rederive (19.20) by making the eigenvalue equation (19.21) the starting point, and to ask for a solution of the eigenvalue problem of H, assuming that the fixed eigenvalue E_s lies in the continuous spectrum which is common to both H and H_0. Equation (19.21) may then be regarded as an inhomogeneous equation to be solved. Its homogeneous counterpart,

$$(H_0 - E_s)\Psi_s = 0 \tag{19.22}$$

does have solutions (by assumption). Hence, according to the theorems of Section 17.2 the inhomogeneous equation has a solution if and only if the inhomogeneous term $-V\Psi_s^{(+)}$ is orthogonal to the subspace spanned by the solutions of the homogeneous equation (19.22). This condition is generally not satisfied. We can nevertheless ignore this apparent obstacle in solving (19.21), because, when the spectrum is quasi-continuous, the component of $-V\Psi_s^{(+)}$ in the troublesome subspace can be neglected without any appreciable alteration.

According to Section 17.2, the general "solution" of the "inhomogeneous" equation (19.21) is

$$\Psi_s^{(+)} = \Psi_s + G_+(E_s)V\Psi_s^{(+)} \tag{19.23}$$

---

[4] In the formal scattering theory the order in which the several limiting processes are executed is very important. No simple mathematical notation has yet been invented for operators which depend on parameters whose limiting values are to be taken after the operation, although the *theory of distributions*, or *generalized functions*, is a beginning in this direction (see Chapter 6, Footnote 2).

where $G_+(E_s)$ is a particular solution of the operator equation

$$(E_s - H_0)G(E_s) = 1 - P_s \qquad (19.24)$$

This equation has infinitely many different solutions. In order to obtain the solutions $\Psi_s^{\prime(+)}$ we must obviously make the choice

$$G_+(E_s) = \frac{1}{E_s - H_0 + i\hbar\alpha} \qquad (19.25)$$

since then (19.20) and (19.23) become identical. The operator (19.25) is indeed a particular solution of (19.24), since $(E_s - H_0)G_+(E_s)$ gives zero when it acts on an eigenvector of H_0 with eigenvalue E_s and has the effect of the identity operation when it acts on an eigenvector Ψ_r of H_0 with $E_r \neq E_s$.

The implicit equation (19.20) for $\Psi_s^{\prime(+)}$ looks like the integral equation (11.20). Indeed, they are equivalent if the unperturbed eigenstates Ψ_s are momentum eigenstates. To prove this it is best to go back to (19.18) and write it in the coordinate representation. We set

$$\langle \mathbf{r} \mid \Psi_s^{\prime(+)} \rangle = \psi_{\mathbf{k}}^{(+)}(\mathbf{r}) \qquad (19.26)$$

and

$$\langle \mathbf{r} \mid \Psi_s \rangle = \frac{1}{(2\pi)^{3/2}} e^{i\mathbf{k}\cdot\mathbf{r}}$$

and obtain for the local potential $\langle \mathbf{r}' \mid V \mid \mathbf{r}'' \rangle = V(\mathbf{r})\,\delta(\mathbf{r}' - \mathbf{r}'')$,

$$\psi_{\mathbf{k}}^{(+)}(\mathbf{r}) = \frac{1}{(2\pi)^{3/2}} e^{i\mathbf{k}\cdot\mathbf{r}} - \frac{1}{(2\pi)^3} \iint \frac{e^{i\mathbf{k}'\cdot(\mathbf{r}-\mathbf{r}')}}{k'^2 - (k^2 + i\varepsilon)}\, d^3k'\, \frac{2\mu}{\hbar^2}\, V(\mathbf{r}')\psi_{\mathbf{k}}^{(+)}(\mathbf{r}')\, d^3r'$$

$$(19.27)$$

By comparison with the equations of Section 11.3, we see that this equation is identical with the integral equation (11.31).

The preceding discussion shows that the Green's function $G_+(\mathbf{r}, \mathbf{r}')$ is proportional to the matrix element of the operator

$$G_+(E) = (E - H_0 + i\hbar\alpha)^{-1}$$

in the coordinate representation:

$$G_+(\mathbf{r}, \mathbf{r}') = -4\pi\, \frac{\hbar^2}{2\mu}\, \langle \mathbf{r} \mid G_+(E) \mid \mathbf{r}' \rangle \qquad (19.28)$$

It was shown in Chapter 11 that this particular Green's function ensures that the solutions $\psi^{(+)}$ of the integral equation (11.20) asymptotically represent *outgoing* spherical waves in addition to incident plane waves. Hence, $\Psi_s^{\prime(+)}$ is properly called an *outgoing* eigenstate of H, if Ψ_s is a momentum eigenvector.

Exercise 19.1. Work out the Green's function $\langle x | G_+(E) | x' \rangle$ for a particle in one dimension.

The operator G_+ is only one particular solution of (19.24). The particular solution

$$G_-(E_s) = \frac{1}{E_s - H_0 - i\hbar\alpha} \tag{19.29}$$

of (19.24) leads to the replacement of (19.21) by the equation

$$\Psi_s^{(-)} = \Psi_s + G_-(E_s) V \Psi_s^{(-)} \tag{19.30}$$

If Ψ_s is a momentum eigenvector, the eigenstate $\Psi_s^{(-)}$ of H describes spherically *incoming* waves.

Another solution of (19.24) which is often useful is the operator

$$G_1(E_s) = \tfrac{1}{2}G_+(E_s) + \tfrac{1}{2}G_-(E_s) = \frac{1}{2}\left(\frac{1}{E_s - H_0 + i\hbar\alpha} + \frac{1}{E_s - H_0 - i\hbar\alpha} \right)$$

The expression on the right may be simplified by using the identity (6.18), and we are led to write G_1 as

$$G_1(E) = \mathrm{P}\left(\frac{1}{E - H_0} \right) \tag{19.31}$$

With this operator we can define a set of eigenstates $\Psi^{(1)}$ by the equation

$$\Psi_s^{(1)} = \Psi_s + \mathrm{P}\left(\frac{1}{E - H_0} \right) V \Psi_s^{(1)} \tag{19.32}$$

In the coordinate representation the operator $G_1(E)$ is related to the standing wave Green's function of Exercise 11.2.

It should always be remembered that the solutions of equations like (19.23), (19.30), and (19.32) depend on the choice of Ψ_s.

4. *The Scattering Cross Section.* In this section the formula for the scattering cross section will be derived from the transition rate, (19.15).

The unperturbed states are again assumed to be normalized momentum eigenstates. Since

$$\delta(E_k - E_{k'}) = \frac{\mu}{\hbar^2 k}\, \delta(k - k')$$

we obtain for the total transition rate from an incident momentum state **k** into a solid angle $d\Omega$,

$$w = \sum_r \frac{d}{dt} |c_r(t)|^2 = \frac{2\pi}{\hbar} d\Omega \int_0^\infty \frac{\mu}{\hbar^2 k} \delta(k - k') |T_{k'k}|^2 k'^2 \frac{L^3}{(2\pi)^3} dk'$$

$$= \frac{\mu k L^3}{(2\pi)^2 \hbar^3} |T_{k'k}|^2 d\Omega \tag{19.33}$$

where $\mathbf{k}'$ is the momentum of the scattered particle $(k' = k)$. Hence, if $v = \hbar k/\mu$ is the velocity of the incident particles,

$$w = \frac{\mu^2 v L^3}{(2\pi)^2 \hbar^4} |T_{k'k}|^2 d\Omega \tag{19.34}$$

The probability of finding a particle in a unit volume of the incident beam is $1/L^3$. Hence, v/L^3 is the probability that a particle is incident on a unit area perpendicular to the beam per unit time. If this probability current density is multiplied by the differential cross section $d\sigma$, as defined in Section 11.1, the transition rate w is obtained; hence,

$$d\sigma = \frac{w}{v/L^3} = \left(\frac{\mu L^3}{2\pi\hbar^2}\right)^2 |T_{k'k}|^2 d\Omega \tag{19.35}$$

The transition matrix can, by (19.17), be expressed as

$$T_{k'k} = \frac{1}{L^{3/2}} \int e^{-i\mathbf{k}'\cdot\mathbf{r}} V(\mathbf{r}) \psi_k^{(+)}(\mathbf{r}) \, d^3r \tag{19.36}$$

Comparing this with (11.35), we find a simple relation between the *transition matrix* and the *scattering amplitude*:

$$T_{k'k} = -\frac{2\pi\hbar^2}{\mu L^3} f_k(\hat{\mathbf{k}}') \tag{19.37}$$

Substituting $T_{k'k}$ from here into (19.35), we obtain

$$d\sigma = |f_k(\hat{\mathbf{k}}')|^2 \, d\Omega \tag{19.38}$$

This result is identical with (11.15).

5. *Properties of the Scattering States.* The fundamental problem of scattering theory is to solve the equation[5]

$$\Psi_s^{(+)} = \Psi_s + \frac{1}{E_s - H_0 + i\hbar\alpha} V \Psi_s^{(+)} \tag{19.20}$$

[5] Note the formal similarity between (19.20) and (17.95) and (17.100) of bound state perturbation theory, except that in scattering theory we have assumed (quasi-) continuous rather than discrete eigenvalues and zero level shift.

The solutions can then be used to determine the transition matrix which, according to the last section, is directly related to the cross section.

Formally we may solve (19.20) by multiplying it by $E_s - H_0 + i\hbar\alpha$, and adding and subtracting $-V\Psi_s$ on the right-hand side of the equation. Thus we obtain

$$(E_s - H + i\hbar\alpha)\Psi_s^{(+)} = (E_s - H + i\hbar\alpha)\Psi_s + V\Psi_s$$

or

$$\Psi_s^{(+)} = \Psi_s + \frac{1}{E_s - H + i\hbar\alpha} V\Psi_s \qquad (19.39)$$

The important distinction between this equation and (19.20) is the appearance of H rather than H_0 in the denominator.

If the solution (19.39) is substituted in (19.17) for the transition matrix, we get

$$T_{rs} = (\Psi_r, V\Psi_s) + \left(\Psi_r, V\frac{1}{E_s - H + i\hbar\alpha} V\Psi_s\right) \qquad (19.40)$$

In this way the cross section for a scattering process can in principle be calculated. However, for practical purposes not much is gained, because the effect of the operator $(E_s - H + i\hbar\alpha)^{-1}$ is not known unless the eigenvectors of H have already been determined—but this is the problem which we wish to solve. It is therefore usually necessary to resort to approximation methods to solve (19.20).

Exercise 19.2. Show that, if $E_r = E_s$,

$$T_{rs} = (\Psi_r^{(-)}, V\Psi_s) \qquad (19.41)$$

The crudest approximation is obtained if in (19.20) the term proportional to V on the right-hand side of the equation is neglected altogether:

$$\Psi_s^{(+)} \simeq \Psi_s \qquad (19.42)$$

This is simply the first term in the solution of (19.20) obtained by systematic iteration:

$$\Psi_s^{(+)} = \Psi_s + G_+(E_s)V\Psi_s + G_+(E_s)VG_+(E_s)V\Psi_s + \cdots \qquad (19.43)$$

where $G_+(E_s) = (E_s - H_0 - i\hbar\alpha)^{-1}$. This form is also arrived at by rewriting (19.23) as

$$\Psi_s^{(+)} = \frac{1}{1 - G_+(E_s)V} \Psi_s \qquad (19.44)$$

and expanding $(1 - G_+V)^{-1}$ as a power series. The *n*th *Born approximation* consists of terminating the expansion (19.43) arbitrarily after *n* terms.

The convergence of the *Born series* (19.43) is often difficult to ascertain, but it is easy to see that it will certainly *not* converge, if the equation

$$\lambda_0 G_+(E_s) V \Psi = \Psi$$

has an eigenvalue λ_0 whose absolute value is less than 1. For the operator $(1 - \lambda G_+ V)^{-1}$ has a singularity at $\lambda = \lambda_0$; consequently, the radius of convergence of the series expansion of this operator in powers of λ must be less than $|\lambda_0|$. If $|\lambda_0| < 1$, the Born series, which corresponds to $\lambda = 1$, is divergent. If, as happens frequently in cases of practical interest, the Born series fails to converge or converges too slowly to be useful, more powerful approximation techniques may be employed for the determination of $\Psi'^{(+)}$.

If the first Born approximation (19.42) is substituted into the transition matrix, we obtain from (19.36) and (19.37)

$$f_{\text{Born}} = -\frac{\mu}{2\pi\hbar^2} \int e^{-i\mathbf{k}'\cdot\mathbf{r}} V e^{i\mathbf{k}\cdot\mathbf{r}} \, d^3r = -\frac{\mu L^3}{2\pi\hbar^2} V_{\mathbf{k}'\mathbf{k}} \qquad (19.45)$$

in agreement with (11.36). The first Born approximation is the result of a first-order perturbation treatment of scattering, in which the accurate equation (19.10) is replaced by the approximate equation (19.9).

The formal solution (19.39) can be used to demonstrate the orthonormality of the eigenvectors $\Psi'^{(+)}$. This is seen by the following simple manipulations:

$$(\Psi_r'^{(+)}, \Psi_s'^{(+)}) = \left(\Psi_r + \frac{1}{E_r - H + i\hbar\alpha} V\Psi_r, \Psi_s'^{(+)} \right)$$

$$= \left(\Psi_r, \Psi_s'^{(+)} + V \frac{1}{E_r - H - i\hbar\alpha} \Psi_s'^{(+)} \right)$$

$$= \left(\Psi_r, \Psi_s'^{(+)} + V \frac{1}{E_r - E_s - i\hbar\alpha} \Psi_s'^{(+)} \right)$$

$$= \left(\Psi_r, \Psi_s'^{(+)} - \frac{1}{E_s - H_0 + i\hbar\alpha} V\Psi_s'^{(+)} \right)$$

If we finally use (19.20), we get the result[6]

$$(\Psi_r'^{(+)}, \Psi_s'^{(+)}) = (\Psi_r, \Psi_s) = \delta_{rs} \qquad (19.46)$$

Entirely analogous arguments can be made to show that

$$(\Psi_r'^{(-)}, \Psi_s'^{(-)}) = \delta_{rs} \qquad (19.47)$$

Corresponding to an orthonormal set of Ψ_s we thus obtain two sets, $\Psi_s'^{(+)}$ and $\Psi_s'^{(-)}$, of orthonormal eigenvectors of the total Hamiltonian H. The

[6] Equation (19.46) is true only in the limit $L \rightarrow \infty$ when E_r becomes an eigenvalue of both H and H_0.

question arises whether these sets are complete. It would appear that each set by itself is a complete set, because the vectors Ψ'_s form a complete set, and $\Psi'^{(+)}_s$ (or $\Psi'^{(-)}_s$) goes over into Ψ'_s as $V \to 0$. However, one reservation is called for: H may have discrete energy eigenvalues corresponding to bound states produced by the interaction V. These discrete states, which have no counterpart in the spectrum of H_0 and are never found among the solutions of (19.20), are orthogonal to the scattering states and must be added to all the $\Psi'^{(+)}_s$ (or $\Psi'^{(-)}_s$) to complete the set of eigenvectors.

6. *The Scattering Matrix.* From the preceding discussion it follows that the continuum eigenstates $\Psi'^{(+)}$ must be expressible as linear combinations of the $\Psi'^{(-)}$:

$$\Psi'^{(+)}_q = \sum_q \Psi'^{(-)}_r S_{rq} \tag{19.48}$$

No discrete bound state eigenvectors appear on the right, because these are orthogonal to $\Psi'^{(+)}_q$. From the orthonormality of the scattering states we obtain

$$S_{rq} = (\Psi'^{(-)}_r, \Psi'^{(+)}_q) \tag{19.49}$$

This matrix is called a *scattering matrix* or simply an *S-matrix*. Its detailed form depends, of course, on the choice of the basis Ψ'_r.[7]

Since two eigenvectors of H belonging to different energy eigenvalues are orthogonal, any scattering matrix is always diagonal with respect to the energy. Hence, if in the limit $L \to \infty$ the energy is regarded as a continuous variable, the transformation matrix must be of the form

$$S_{rq} = \delta_{rq} + \delta(E_r - E_q)U_{rq} \tag{19.50}$$

where U_{rq} is not singular at $E_r = E_q$. It is hoped that the reader will not be confused by the seemingly inconsistent use of both Kronecker delta and delta function on the right-hand side of this equation.

Using the identity

$$\delta(x) = \lim_{\varepsilon \to 0} \frac{1}{2\pi i}\left(\frac{1}{x - i\varepsilon} - \frac{1}{x + i\varepsilon}\right)$$

we may write

$$S_{rq} = \delta_{rq} + \frac{1}{2\pi i}\left(\frac{1}{E_r - E_q - i\hbar\alpha} - \frac{1}{E_r - E_q + i\hbar\alpha}\right)U_{rq} \tag{19.51}$$

To relate U_{rq} to the transition matrix T_{rq} we substitute (19.20), (19.30), and (19.51) into (19.48). By a simple rearrangement and use of the definition

---

[7] Some authors reserve the term *scattering matrix* for the special case where the states labeled by q and r correspond to eigenstates of momentum. See (19.55) and (19.66).

(19.17), we obtain

$$\frac{1}{E_q - H_0 + i\hbar\alpha} \sum_r \Psi_r T_{rq}$$

$$= \frac{1}{2\pi i} \left(\frac{1}{E_q - H_0 - i\hbar\alpha} - \frac{1}{E_q - H_0 + i\hbar\alpha} \right) \sum_r \Psi_r U_{rq}$$

$$+ \sum_r \frac{1}{E_r - H_0 - i\hbar\alpha} V\Psi_r^{(-)} S_{rq} \quad (19.52)$$

It is now legitimate to compare terms with $(E - H_0 + i\hbar\alpha)^{-1}$ and terms with $(E - H_0 - i\hbar\alpha)^{-1}$. These terms are linearly independent, and in the coordinate representation give rise to asymptotically outgoing and ingoing spherical waves. From a comparison of the terms proportional to $(E - H_0 + i\hbar\alpha)^{-1}$ we obtain immediately

$$U_{rq} = -2\pi i T_{rq} \quad (19.53)$$

hence, by (19.50), the fundamental relation,

$$S_{rq} = \delta_{rq} - 2\pi i\, \delta(E_r - E_q)T_{rq} \quad (19.54)$$

If plane waves are chosen for the set of eigenvectors Ψ_s, the scattering matrix becomes [see (19.37)]

$$S_{\mathbf{k}'\mathbf{k}} = \delta_{\mathbf{k}',\mathbf{k}} + \frac{4\pi^2 i}{kL^3}\, \delta(k - k') f_{\mathbf{k}}(\hat{\mathbf{k}}') \quad (19.55)$$

From this simple relation between the scattering amplitude and the scattering matrix it becomes obvious that the scattering cross section must be proportional to $|S_{\mathbf{k}'\mathbf{k}} - \delta_{\mathbf{k}',\mathbf{k}}|^2$.

If in (19.52) we compare the terms which are proportional to $(E - H_0 - i\hbar\alpha)^{-1}$, the simple formula

$$T_{jq} = \sum_r (\Psi_j, V\Psi_r^{(-)}) S_{rq} \quad (19.56)$$

results. This follows also directly from the definition of the scattering matrix and may if $E_j = E_q$, on account of (19.41), be written as

$$T_{jq} = \sum_r T_{rj}^* S_{rq} \quad (19.57)$$

The scattering matrix owes its central importance to the fact that it is *unitary*. To prove the unitary property we must show that

$$\sum_n S_{nl}^* S_{nj} = \delta_{lj} \quad (19.58)$$

and

$$\sum_n S_{ln} S_{jn}^* = \delta_{lj} \quad (19.59)$$

The first of these equations follows immediately from the definition (19.48) and the orthonormality of the $\Psi'^{(+)}$ and $\Psi'^{(-)}$. The second equation is proved by using (19.49) to construct

$$\sum_n S_{ln}S_{jn}{}^* = \sum_n (\Psi_l'^{(-)}, \Psi_n'^{(+)})(\Psi_n'^{(+)}, \Psi_j'^{(-)})$$

If $\Psi_i'^{(b)}$ denotes the bound states, we have as a result of completeness the closure relation

$$\sum_n (\Phi_a, \Psi_n'^{(+)})(\Psi_n'^{(+)}, \Phi_b) + \sum_i (\Phi_a, \Psi_i'^{(b)})(\Psi_i'^{(b)}, \Phi_b) = (\Phi_a, \Phi_b)$$

for any two states Φ_a and Φ_b. Applying this to the previous equation, we get

$$\sum_n S_{ln}S_{jn}{}^* = (\Psi_l'^{(-)}, \Psi_j'^{(-)}) - \sum_i (\Psi_l'^{(-)}, \Psi_i'^{(b)})(\Psi_i'^{(b)}, \Psi_j'^{(-)}) = (\Psi_l'^{(-)}, \Psi_j'^{(-)}) = \delta_{lj}$$

because the bound states are orthogonal to the scattering states.

A very important relation is discovered if the fundamental equation (19.10) is re-examined. In the limit $t_0 \to -\infty$, and $\alpha \to 0$, and $t \to +\infty$ this equation reduces to

$$c_r(+\infty) = -\frac{2\pi i}{\hbar} T_{rq}\, \delta(\omega_{rq}) + \delta_{rq} \tag{19.60}$$

if the initial state is denoted by q. Comparing this with (19.54), we obtain the all-important result:

$$c_r(+\infty) = S_{rq} \tag{19.61}$$

According to (18.92), the transition amplitude $c_r(+\infty)$ is equal to the matrix element of $U(+\infty, -\infty)$ between the initial unperturbed state Ψ_q and the final unperturbed state Ψ_r. Hence, we have

$$S_{rq} = (\Psi_r, U(+\infty,-\infty)\,\Psi_q) \tag{19.62}$$

The element S_{rq} of the S-matrix is thus the probability amplitude for finding the system at $t = +\infty$ in state Ψ_r if it was known to have been in state Ψ_q at $t = -\infty$. This interpretation is consistent with the following equivalent expressions for the scattering state:

$$\Psi_q'^{(+)} = \Psi_q + G_+(E_q)V\Psi_q'^{(+)} \tag{19.63}$$

$$\Psi_q'^{(+)} = \sum_r [\Psi_r S_{rq} + G_-(E_r)V\Psi_r'^{(-)} S_{rq}] \tag{19.64}$$

Equation (19.64) is obtained by substituting (19.30) into (19.48). If we imagine that wave packets are constructed from these equations by superposition, we may relate them to the time-dependent description of the scattering process. At $t = -\infty$ only the first term in (19.63) contributes, and at $t = +\infty$, only the first terms in (19.64) contribute, since the retarded

(G_+) wave vanishes before the scattering and the advanced (G_-) vanishes after the scattering. Hence, the matrix S_{rq} connects the free initial state q with the free final states r, as described by (19.62). Thus, the elements of the scattering matrix can be regarded as the matrix elements of a *unitary scattering operator S*, defined by

$$S = U(+\infty, -\infty) \qquad (19.65)$$

in the representation spanned by the unperturbed eigenstates.

In self-explanatory notation an element of the scattering matrix can be written as

$$S_{\mathbf{k'k}} = \langle \mathbf{k'}|\, S\, |\mathbf{k}\rangle \qquad (19.66)$$

The definition of the scattering operator is a useful one because the operator S depends only on the nature of the system and its Hamiltonian but not on the particular incident state. A simple application to scattering from a central force field will illustrate the advantages of working with the scattering operator.

7. Rotational Invariance and the S-Matrix. If a spatial rotation is applied to all the states of a physical system in which scattering occurs, the initial and final momentum states are rotated rigidly. According to Section 16.2, this is accomplished by applying a unitary operator U_R. If the forces are central, $V = V(r)$, the scattering matrix will be the same before and after the rotation, and we have

$$\langle U_R\mathbf{k'}|\, S\, |U_R\mathbf{k}\rangle = \langle \mathbf{k'}|\, S\, |\mathbf{k}\rangle$$

Hence, the scattering matrix cannot depend on the absolute orientation of the vectors $\mathbf{k}$ and $\mathbf{k'}$ in space. It can be only a function of the energy and of the angle between the initial and final momenta. If the particles have no spin, the scattering matrix can thus be written in the form

$$\langle \mathbf{k'}|\, S\, |\mathbf{k}\rangle = \delta(k - k') \sum_{l=0}^{\infty} F_l(k) P_l(\hat{\mathbf{k}} \cdot \hat{\mathbf{k}'}) \qquad (19.67)$$

with undetermined coefficients $F_l(k)$. The delta function has been included as a separate factor, because we already know that the S-matrix has non-vanishing elements only "on the energy shell," i.e., between two states of the same energy.

The coefficients $F_l(k)$ can be determined to within a phase factor by invoking the unitarity of the scattering matrix:

$$\int \langle \mathbf{k'}|\, S\, |\mathbf{k''}\rangle\langle \mathbf{k}|\, S\, |\mathbf{k''}\rangle^* \, d^3k'' = \delta(\mathbf{k} - \mathbf{k'})$$

Substituting (19.67) here and carrying out the integration in $\mathbf{k''}$-space, we obtain

$$k^2\,\delta(k - k')\sum_{l=0}^{\infty}\frac{4\pi}{2l + 1}|F_l(k)|^2\,P_l(\hat{\mathbf{k}}\cdot\hat{\mathbf{k}}') = \delta(\mathbf{k} - \mathbf{k}')$$

We now use the following identity:

$$\delta(\mathbf{k} - \mathbf{k}') = \frac{\delta(k - k')}{k^2}\sum_{l=0}^{\infty}\frac{2l + 1}{4\pi}\,P_l(\hat{\mathbf{k}}\cdot\hat{\mathbf{k}}') \tag{19.68}$$

Exercise 19.3. Derive (19.68) from the completeness of the Legendre polynomials.

From the last two equations we find immediately that the coefficients $F_l(k)$ must be of the form

$$F_l(k) = \frac{2l + 1}{4\pi k^2}\,e^{2i\delta_l(k)} \tag{19.69}$$

where the $\delta_l(k)$ are real functions of the momentum (or energy). Inserting this result in (19.67), we conclude that the scattering matrix is expressible as

$$\langle \mathbf{k}'|\,S\,|\mathbf{k}\rangle = \delta(k - k')\sum_{l=0}^{\infty}\frac{2l + 1}{4\pi k^2}\,e^{2i\delta_l(k)}\,P_l(\hat{\mathbf{k}}\cdot\hat{\mathbf{k}}') \tag{19.70}$$

On the other hand, according to (19.55) and (19.68) this same matrix element can also be written in terms of the scattering amplitude as

$$\langle \mathbf{k}'|\,S\,|\mathbf{k}\rangle = \delta(k - k')\left[\sum_{l=0}^{\infty}\frac{2l + 1}{4\pi k^2}\,P_l(\hat{\mathbf{k}}\cdot\hat{\mathbf{k}}') + \frac{i}{2\pi k}\,f_{\mathbf{k}}(\hat{\mathbf{k}}')\right] \tag{19.71}$$

Here we have chosen the length of the normalization cube $L = 2\pi$, so that the unperturbed eigenstates are normalized as $\langle \mathbf{k}'\,|\,\mathbf{k}\rangle = \delta(\mathbf{k} - \mathbf{k}')$ throughout. Comparing (19.70) and (19.71), we get

$$f_{\mathbf{k}}(\hat{\mathbf{k}}') = \frac{1}{k}\sum_{l=0}^{\infty}(2l + 1)e^{i\delta_l(k)}\sin\delta_l(k)P_l(\hat{\mathbf{k}}\cdot\hat{\mathbf{k}}') \tag{19.72}$$

We have thus rederived the main result of the partial wave analysis (11.59) directly from the rotational invariance of the scattering operator.

Exercise 19.4. Transform the matrix element (19.70) into the orbital angular momentum representation (see Problem 5, Chapter 16), and show that[8]

$$\langle \alpha\,l'm'|\,S\,|\alpha lm\rangle = e^{2i\delta_l(k)}\,\delta_{m'm}\,\delta_{l'l} \tag{19.73}$$

in agreement with equation (11.67).

---

[8] The same symbol (δ) is used for delta functions and phase shifts in this section, but the context always determines the meaning unambiguously.

8. The Optical Theorem. From the unitary property of the scattering matrix we can derive an important theorem for the scattering amplitudes. If we substitute the expression (19.54) for the S-matrix in (19.58) and work out the result, we get

$$2\pi \sum_n \delta(E_n - E_i)T_{ni}^* T_{nj} = i(T_{ij} - T_{ji}^*) \tag{19.74}$$

By (19.37) this formula can also be written in terms of the scattering amplitudes. Replacing the summation by an integration, we obtain by use of the appropriate density factor

$$2\pi \frac{2\pi\hbar^2}{\mu L^3} \frac{L^3}{(2\pi)^3} \int f_{\mathbf{k}}^*(\hat{\mathbf{k}}'') f_{\mathbf{k}'}(\hat{\mathbf{k}}'') \frac{\mu}{\hbar^2 k''} \delta(k'' - k)k''^2 \, dk'' \, d\Omega''$$

$$= -i[f_{\mathbf{k}'}(\hat{\mathbf{k}}) - f_{\mathbf{k}}^*(\hat{\mathbf{k}}')]$$

or

$$\int f_{\mathbf{k}}^*(\hat{\mathbf{k}}'')f_{\mathbf{k}'}(\hat{\mathbf{k}}'') \, d\Omega'' = \frac{4\pi}{k} \frac{f_{\mathbf{k}'}(\hat{\mathbf{k}}) - f_{\mathbf{k}}^*(\hat{\mathbf{k}}')}{2i} \tag{19.75}$$

As a special case of this relation we may identify $\mathbf{k}' = \mathbf{k}$ and then obtain by comparison with (19.38):

$$\sigma = \int d\sigma = \int |f_{\mathbf{k}}(\hat{\mathbf{k}}'')|^2 \, d\Omega'' = \frac{4\pi}{k} \operatorname{Im} f_{\mathbf{k}}(\hat{\mathbf{k}}) \tag{19.76}$$

This formula shows that the imaginary part of the forward scattering amplitude $f_{\mathbf{k}}(\hat{\mathbf{k}})$ measures the loss of intensity which the incident beam suffers because of the scattering. It therefore expresses the conservation of probability. Since the latter is a consequence of the Hermitian nature of the Hamiltonian, we see that the unitarity of S is linked with the Hermitian property of H. Equation (19.76) is known as the *optical theorem*.[9]

Exercise 19.5. Derive the optical theorem from the conservation of probability and (19.12).

Exercise 19.6. Verify that the scattering amplitude (19.72) is consistent with the optical theorem.

Exercise 19.7. Show that the first Born approximation violates the optical theorem. Explain this failure and show how it can be remedied by including the second Born approximation for the forward scattering amplitude.

[9] So called because of the analogy with light which passes through a medium. The imaginary part of the complex index of refraction is related to the absorption coefficient, i.e., the total absorption cross section. See L. Rosenfeld, *Theory of Electrons*, North-Holland Publishing Co., Amsterdam, 1951.

9. Time-Reversal Symmetry. We finally consider the properties of the scattering states under the time reversal operation. To this end we write the fundamental integral equations with momentum eigenstates as the unperturbed states:

$$\Psi_{\mathbf{k}}^{'(+)} = \Psi_{\mathbf{k}}' + \frac{1}{E_k - H_0 + i\hbar\alpha} \, V\Psi_{\mathbf{k}}^{'(+)} \tag{19.77}$$

and

$$\Psi_{\mathbf{k}}^{'(-)} = \Psi_{\mathbf{k}}' + \frac{1}{E_k - H_0 - i\hbar\alpha} \, V\Psi_{\mathbf{k}}^{'(-)} \tag{19.78}$$

If we now apply the operator Θ defined in Section 16.11, choosing the phases of the momentum eigenstates such that $\Theta\Psi_{\mathbf{k}}' = \Psi_{-\mathbf{k}}'$, we obtain

$$\Theta\Psi_{\mathbf{k}}^{'(+)} = \Psi_{-\mathbf{k}}' + \frac{1}{E_{\mathbf{k}} - H_0 - i\hbar\alpha} \, \Theta V\Theta^{-1}\Theta\Psi_{\mathbf{k}}^{'(+)} \tag{19.79}$$

where use has been made of the invariance of H_0 under time reversal. Comparing this relation with (19.78), we observe that $\Psi_{\mathbf{k}}^{'(+)}$ and $\Psi_{-\mathbf{k}}^{'(-)}$ are mutually time reversed states,

$$\Theta\Psi_{\mathbf{k}}^{'(+)} = \Psi_{-\mathbf{k}}^{'(-)} \tag{19.80}$$

if the interaction V is invariant under time reversal:

$$\Theta V\Theta^{-1} = V \tag{19.81}$$

In this case the S-matrix satisfies the condition

$$\langle \mathbf{k}'| \, S \, |\mathbf{k} \rangle = (\Psi_{\mathbf{k}'}^{'(-)}, \Psi_{\mathbf{k}}^{'(+)}) = (\Theta\Psi_{\mathbf{k}'}^{'(-)}, \Theta\Psi_{\mathbf{k}}^{'(+)})^*$$

$$= (\Psi_{-\mathbf{k}}^{'(-)}, \Psi_{-\mathbf{k}'}^{'(+)}) = \langle -\mathbf{k}| \, S \, |-\mathbf{k}' \rangle \tag{19.82}$$

owing to the antiunitary property of Θ. For the scattering amplitude this implies by (19.55) the relation

$$f_{\mathbf{k}}(\hat{\mathbf{k}}') = f_{-\mathbf{k}'}(-\hat{\mathbf{k}}) \tag{19.83}$$

This equation, derived from very general symmetry properties, expresses the equality of two scattering processes obtained by reversing the path of the particle and is known as the *reciprocity relation*.[10]

---

[10] The implications of invariance under time reversal for more general collision processes are discussed in J. M. Blatt and V. F. Weisskopf, *Theoretical Nuclear Physics*, John Wiley and Sons, New York, 1952, p. 528.

Problems

1. For formal manipulations in scattering theory, it is sometimes convenient to use the operators T, $\Omega^{(+)}$, and $\Omega^{(-)}$ defined by the equations

$$T_{rs} = (\Psi_r, T\Psi_s) = (\Psi_r, V\Psi_s^{(+)})$$

and

$$\Omega^{(+)}\Psi_r = \Psi_r^{(+)}, \qquad \Omega^{(-)}\Psi_r = \Psi_r^{(-)}$$

Prove that

$$T = V\Omega^{(+)} \qquad \text{and} \qquad S = [\Omega^{(-)}]^\dagger\Omega^{(+)}$$

2. Obtain the "scattering states" (energy eigenstates with $E \geqslant 0$) for a one-dimensional delta function potential, $g\,\delta(x)$. Calculate the matrix elements $\langle k'| S |k \rangle$ and verify the unitarity of the S-matrix. Obtain the transmission coefficient, and compare with the result of Problem 3 in Chapter 6. Perform the calculations in both the coordinate and momentum representations.

3. Use the Born approximation to calculate the differential and total cross sections for the elastic scattering of electrons by a hydrogen atom which is in its ground state. Neglect exchange phenomena.

4. The cross section for two-quantum annihilation of positrons of velocity v with an electron at rest has, for $v \ll c$, the "classical" value

$$\sigma = \pi\left(\frac{e^2}{\mu c^2}\right)^2 \frac{c}{v}$$

Use this information to derive the annihilation probability per unit time from a plane wave state and (assuming that annihilation occurs only if the two particles are at the same place) estimate the decay probability in the singlet ground state of positronium.

5. Using the Born approximation, and neglecting relativistic effects, express the differential cross section for scattering of an electron from a spherically symmetric charge distribution $\rho(r)$ as the product of the Rutherford scattering cross section for a point charge and the square of a *form factor* F. Obtain an expression for the form factor and evaluate it as a function of the momentum transfer for (a) a uniform charge distribution of radius R, (b) a Gaussian charge distribution with the same root-mean-square radius.

6. If the nonlocal *separable* scattering potential

$$\langle \mathbf{r}'| V |\mathbf{r}'' \rangle = \lambda v(r)v(r')$$

is given, work out explicitly and solve the integral equation for $\Psi^{(+)}$. Obtain the scattering amplitude, and discuss the Born series for this potential.

CHAPTER 20

Identical Particles

1. The Indistinguishability of Identical Particles. When we endeavor to describe the quantum behavior of systems containing several identical particles, altogether new, strange, and unclassical features are encountered because there is no way of keeping track of each particle separately when the wave functions of two identical particles overlap. The indistinguishability of the two particles makes it impossible to follow them individually in the region where they both may be found simultaneously. Similar complications do not arise in classical mechanics from the identity of two bodies because their wave packets do not overlap, and the particles move in separate, distinguishable, continuous orbits.

Even if they are not followed as they move along their orbits, two objects that are identical can, classical mechanics assumes, always be made distinguishable by marking them in such a way that there is no measurable influence on the physical process under consideration. For instance, we suppose that coloring the balls in billiards differently has no influence on their motion; yet it serves to distinguish them individually and to identify them after any number of collisions.

In quantum mechanics the state of a system of n identical particles is described in terms of some complete set of dynamical variables K, appropriate for each individual particle. A particular state of the system is specified by stating that of the n particles n' have the value K', n'' have the value K'', etc. More general states are obtained by superposition of such particular states. It is impossible to tell which particle has the value K', which one the value K'', etc. For example, the coordinates x, y, z and the spin component s_z constitute a complete set of dynamical variables for electrons. By a measurement it can be determined that there is one electron at position $\mathbf{r}'$ with spin component $\hbar\sigma/2$ (where $\sigma = +1$ or -1), another electron at position $\mathbf{r}''$ with spin component $\hbar\sigma/2$, etc., but it is not possible to make any further identification of the electrons.

We shall see that the indistinguishability of identical particles has important, experimentally verifiable consequences. For instance, the spectra of many-electron atoms would be altogether different if the atomic electrons were distinguishable. Furthermore, the exclusion principle, which explains the periodic system of elements and the stability of complex atoms, presupposes the indistinguishability of electrons.[1]

It should be emphasized that there is a basic arbitrariness inherent in the definition of the term *particle* as it is used here and that the identity of two particles is to a certain extent a matter of convention. For example, protons and neutrons may be regarded as two distinct species, distinguishable by differences in mass, charge, magnetic moment, and decay properties. Yet frequently, it is convenient to describe them as two different states of the same species, the *nucleon*. All nucleons are then considered as identical, and proton and neutron states are characterized by different values of a new dynamical variable, the *isobaric "spin,"* or *isospin*. Current views of the nature of fundamental particles suggest that it will be fruitful to regard nucleons in turn as particular quantum states of the baryon species.

In certain problems, on the other hand, it may be convenient to regard entire nuclei, atoms, or molecules as particles. Their composite nature gives rise to some new internal degrees of freedom, which must be included among the dynamical variables used to describe the state of the system.[2]

The latitude in the definition of what is meant by a particle is no obstacle in the development of the quantum mechanics for identical particles. On the contrary, the general principles of the theory are applicable to any species of particle that is characterized by a complete set of dynamical variables.

2. The State Vector Space for a System of Identical Particles. In defining the quantum state vector space for a system of n *identical* particles, we make the basic assumption that any complete set of dynamical variables K, which describes the behavior of a *single particle*, can also be employed for n particles of the same kind. This assumption, postulated to be true even in the presence of interactions between the particles, is a bold one, since it implies that the composite system retains the properties of the individual particles to a considerable extent. Mathematically it means that to each eigenvalue K_i' of K corresponds an *occupation number operator N_i* whose eigenvectors characterize the states in which a definite number, n_i, of particles has the value K_i'. The eigenvalues of N_i are the occupation numbers n_i. As a fundamental postulate, we assume that the totality of the operators N_i forms a *complete set* of commuting Hermitian operators for the system of identical

---

[1] There are important consequences of the indistinguishability of identical particles even in classical statistical mechanics. See a discussion of the *Gibbs paradox* in D. ter Haar, *Elements of Statistical Mechanics*, Rinehart and Co., New York, 1954, p. 145.

[2] *Examples.* In the description of the rotation and vibration of molecules (deformed nuclei), the motion of the constituent electrons (nucleons) is only indirectly taken into account.

particles. In the next section we shall see that this postulate is responsible primarily for the unique roles played by Bose-Einstein and Fermi-Dirac statistics. However, we must first construct the state vector space of the many-body system by a suitable generalization of one-particle quantum mechanics, paying due attention to the indistinguishability of the particles.

The fundamental postulate implies that, in the state vector space of the many-body system, also known as *Fock space*, the basis vectors

$$|n_1, n_2, n_3, \ldots\rangle \tag{20.1}$$

which allot the eigenvalue K_1' to n_1 particles, the eigenvalue K_2' to n_2 particles, etc., constitute a complete set of orthonormal basis vectors for the system of identical particles. The most general state of the system is a linear combination of the kets (20.1).

In particular, we have the "no-particle" (or vacuum) state

$$\Psi^{(0)} = |0, 0, \ldots\rangle \tag{20.2}$$

Then there are the one-particle states

$$\Psi_i^{(1)} = |0, 0, \ldots, n_i{=}1, 0, 0, \ldots\rangle \equiv |K_i'\rangle \tag{20.3}$$

spanning the one-particle subspace of the much larger vector space of the many-body system, which is composed of states with zero, one, two, ... particles and their linear combinations. The principles of quantum mechanics in the one-particle subspace were the subject of Chapters 14 and 15. In this chapter, these principles are generalized to systems containing an arbitrary number of identical particles.

Does the postulate adopted in this section provide the necessary ingredients for a quantum theory of interacting particles? To answer this question it must be shown that a consistent theory can be constructed in this framework and that the theoretical calculations are in satisfactory agreement with experimental observations. The assumption that the states of a system of interacting particles can be expanded in terms of the states of noninteracting particles has its roots in the doctrine of perturbation theory and finds broad support in many successful calculations. Especially when the relatively weak electromagnetic interactions predominate, as in atoms, molecules, and solids, this formulation affords an account of almost unlimited precision, despite the unsettling fact that the theory of interacting fields and particles, even in its most respectable version (quantum electrodynamics) contains some serious inconsistencies. Except for a few simple applications of first-order perturbation theory to the interaction between particles and fields, we shall confine our attention to problems in which it appears legitimate to suppose that the behavior of a system of identical particles (electrons, nucleons, photons) is governed to sufficient approximation by an external field or source, or by an interaction potential that describes the effect of the particles on each other. This program, which originates from a general conception of the correspondence principle, is consistent with the fundamentally inductive approach followed in this book.[3]

3. Creation and Annihilation Operators. It is useful to define *creation operators* $a_i^\dagger$ with the property

$$a_i^\dagger \, |n_1, n_2, \ldots n_{i-1}, n_i, n_{i+1}, \ldots\rangle \propto |n_1, n_2, \ldots n_{i-1}, n_i+1, n_{i+1}, \ldots\rangle$$

$$(20.4)$$

adding to a basis state one particle with quantum number K_i'. From the definition of a Hermitian adjoint operator, it follows conversely that an *annihilation operator* a_i has the property

$$a_i \, |n_1, n_2, \ldots n_{i-1}, n_i, n_{i+1}, \ldots\rangle \propto |n_1, n_2, \ldots n_{i-1}, n_i-1, n_{i+1}, \ldots\rangle \quad (20.5)$$

and thus in effect removes one particle with quantum number K_i'. The constants of proportionality in these defining relations are yet to be chosen. In the interest of simplicity we begin this determination by requiring

$$a_i^\dagger \Psi'^{(0)} = \Psi_i'^{(1)}, \qquad a_i \Psi_i'^{(1)} = \Psi'^{(0)} \tag{20.6}$$

Since the vacuum contains no particle to be destroyed, we also demand that

$$a_i \Psi'^{(0)} = 0 \quad \text{and} \quad a_j \Psi_i'^{(1)} = 0 \quad (j \neq i) \tag{20.7}$$

Exercise 20.1. If $\Psi'^{(1)}$ is an arbitrary one-particle state, prove that the probability $|(a_i^\dagger \Psi'^{(0)}, \Psi'^{(1)})|^2$ equals the expectation value $(\Psi'^{(1)}, a_i^\dagger a_i \Psi'^{(1)})$.

In order to establish how a unitary transformation from one basis to another in the one-particle theory can be expressed in this new notation, we introduce a second complete set of one-particle observables L with eigenvalues L_q'. The corresponding occupation number operators will be denoted by $\tilde{N}_q$ with eigenvalues $\tilde{n}_q$. The transformation between the two representations is effected by

$$|K_i'\rangle = \sum_q |L_q'\rangle \langle L_q' \, | \, K_i'\rangle = \sum_q |L_q'\rangle c_{qi} \tag{20.8}$$

The complex transformation coefficients $c_{qi} = \langle L_q' \, | \, K_i'\rangle$ form a unitary matrix [see equations (14.97)].

The vacuum state $\Phi^{(0)}$ is the same in the new as in the old basis: $\Phi^{(0)} = \Psi'^{(0)}$. The one-particle states in the new basis are

$$\Phi_q^{(1)} = |0, 0, \ldots, \tilde{n}_q=1, 0, 0, \ldots\rangle = |L_q'\rangle \tag{20.9}$$

[3] For a full account of the contemporary quantum theory of interacting fields, see J. D. Bjorken and S. D. Drell, *Relativistic Quantum Fields*, McGraw-Hill Book Company, New York, 1965.

We now introduce creation operators $b_q^\dagger$ and annihilation operators b_q in complete analogy with (20.4) and (20.5):

$$b_q^\dagger \, |\tilde{n}_1, \tilde{n}_2, \ldots \tilde{n}_q, \ldots\rangle \propto |\tilde{n}_1, \tilde{n}_2, \ldots \tilde{n}_q+1, \ldots\rangle \qquad (20.10)$$

$$b_q \, |\tilde{n}_1, \tilde{n}_2, \ldots \tilde{n}_q, \ldots\rangle \propto |\tilde{n}_1, \tilde{n}_2, \ldots \tilde{n}_q-1, \ldots\rangle \qquad (20.11)$$

Evidently, for the one-particle states,

$$a_i^\dagger \Psi^{(0)} = \Psi_i^{(1)} = \sum_q \Phi_q^{(1)} c_{qi} = \sum_q b_q^\dagger \, c_{qi} \Psi^{(0)}$$

This equation is satisfied by the transformation condition

$$a_i^\dagger = \sum_q b_q^\dagger \, c_{qi} \qquad (20.12)$$

The Hermitian adjoint of this equation

$$a_i = \sum_q b_q c_{qi}{}^* \qquad (20.13)$$

is trivially satisfied when acting on the vacuum state. Applying a_i to a one-particle state we have, by the use of equations (20.6) and (20.7) and their analogs,

$$a_i \Psi_j^{(1)} = \delta_{ij} \Psi^{(0)} = \sum_q c_{qi}{}^* c_{qj} \Psi^{(0)} = \sum_{q,r} c_{qi}{}^* c_{rj} \, \delta_{qr} \Psi^{(0)}$$

$$= \sum_q b_q c_{qi}{}^* \sum_r \Phi_r^{(1)} c_{rj} = \sum_q b_q c_{qi}{}^* \Psi_j^{(1)}$$

This relation again is satisfied by equation (20.13). Hence, we may adopt the operator equations (20.12) and (20.13) in the subspace of no particles and one particle.

It is a basic notion of quantum mechanics that the creation of a particle with quantum number K_i' is equivalent to the creation of a particle with any one of the quantum numbers L_q', each contributing an amplitude $c_{qi} = \langle L_q' \mid K_i' \rangle$ in a linear superposition. Equation (20.12) is consistent with this general requirement. The change to yet another one-particle basis M_s' is effected by the relation

$$a_i^\dagger = \sum_q b_q^\dagger \, \langle L_q' \mid K_i' \rangle = \sum_s \left(\sum_q b_q^\dagger \, \langle L_q' \mid M_s' \rangle \right) \langle M_s' \mid K_i' \rangle$$

explicitly exhibiting the formal equivalence of all one-particle bases or representations if (20.12) is valid.

We therefore *assume* the transformation equations (20.12) and (20.13) to hold beyond the no-particle and one-particle subspaces as operator equations in the entire state vector space of the system of any number of identical particles. This assumption is technical rather than fundamental and determines the form of the ensuing theory and not its physical content. We shall see that it leads to an almost unambiguous specification of the constants of proportionality entering equations (20.4), (20.5), (20.10), and (20.11).

The first step in deriving the remaining properties of the creation and annihilation operators consists of observing that two creation operators $a_i^\dagger$ and $a_j^\dagger$, when applied successively to a basis state, produce the same physical state, although the normalization may offhand be thought to depend on the order in which the two particles are created. Hence, for any basis state Ψ,

$$a_i^\dagger a_j^\dagger \Psi = \lambda a_j^\dagger a_i^\dagger \Psi \tag{20.14}$$

To show that the constant λ is, in fact, independent of the state and of the subscripts i and j, it is advantageous to consider what happens in changing to another representation. Substituting $a^\dagger$ from equation (20.12) into equation (20.14), we get

$$(a_i^\dagger a_j^\dagger - \lambda a_j^\dagger a_i^\dagger)\Psi = \sum_{k,l} c_{ki} c_{lj} (b_k^\dagger b_l^\dagger - \lambda b_l^\dagger b_k^\dagger)\Psi = 0 \tag{20.15}$$

The coefficients c_{ki} are arbitrary complex numbers except for the constraint which unitarity imposes:

$$\sum_k c_{ki}^* c_{kj} = \delta_{ij} \tag{20.16}$$

If the theory is to have the same form in any representation, equation (20.15) can be satisfied for all basis vectors Ψ only if, for all values of k and l,

$$b_k^\dagger b_l^\dagger - \lambda b_l^\dagger b_k^\dagger = 0$$

Hence, also,

$$b_l^\dagger b_k^\dagger - \lambda b_k^\dagger b_l^\dagger = 0$$

and consequently

$$\lambda^2 = 1 \qquad \text{or} \qquad \lambda = \pm 1$$

It follows that there are two classes of relations for creation operators of various particle species. They satisfy either the *commutation relations*

$$a_i^\dagger a_j^\dagger - a_j^\dagger a_i^\dagger = 0 \tag{20.17}$$

or the *anticommutation relations*

$$a_i^\dagger a_j^\dagger + a_j^\dagger a_i^\dagger = 0 \tag{20.18}$$

The possibility that for a specific kind of particle (20.17) holds for some creation operators and (20.18) for others need not be contemplated because it is assumed that there are no superselection rules operative that would restrict the unitary transformations c_{qi} in the one-particle theory to certain subspaces.

By taking the Hermitian adjoint of equations (20.17) and (20.18), we obtain

$$a_i a_j \mp a_j a_i = 0 \tag{20.19}$$

A similar argument can be made for two operators a_i and $a_j^\dagger$, with $i \neq j$. Consider

$$a_i a_j^\dagger \Psi = \mu a_j^\dagger a_i \Psi$$

The unitary transformation (20.12) and (20.13) gives for $i \neq j$,

$$(a_i a_j^\dagger - \mu a_j^\dagger a_i)\Psi = \sum_{k,l} c_{ki}{}^* c_{lj}(b_k b_l^\dagger - \mu b_l^\dagger b_k)\Psi = 0 \qquad (20.20)$$

In satisfying this condition, we must take the constraint (20.16) into account. It can be inferred from (20.20), which must be valid for any basis vector, that for all values of k and l, but subject to the restriction $k \neq l$, we must have

$$b_k b_l^\dagger - \mu b_l^\dagger b_k = 0 \qquad (20.21)$$

If we substitute this back into equation (20.20), we get $(i \neq j)$

$$\sum_k c_{ki}{}^* c_{kj}(b_k b_k^\dagger - \mu b_k^\dagger b_k)\Psi = 0$$

This equation is compatible with (20.16) only if

$$b_k b_k^\dagger - \mu b_k^\dagger b_k = A \qquad (20.22)$$

where the operator A is independent of the subscript k. If we now expand the expression $a_i a_i^\dagger - \mu a_i^\dagger a_i$ by use of the transformation (20.12) and (20.13), the conditions (20.21) and (20.16) insure that $a_i a_i^\dagger - \mu a_i^\dagger a_i = A$; hence, the operator A is the same for all pairs of creation and annihilation operators in any one-particle basis.

To complete the theoretical framework we define the Hermitian operators N_i whose eigenvalues give the occupation numbers, n_i, of the particles for which K has the value K_i'. The sum of these,

$$N = \sum_i N_i \qquad (20.23)$$

must be the operator measuring the total number of particles, and this must be invariant under a change of the one-particle basis:

$$N = \sum_i N_i = \sum_k \tilde{N}_k$$

There are only three such operators which are invariant under the transformation (20.12), (20.13). One of these is a constant independent of i and k. The other two,

$$\sum_i a_i a_i^\dagger = \sum_k b_k b_k^\dagger \qquad (20.24)$$

and

$$\sum_i a_i^\dagger a_i = \sum_k b_k^\dagger b_k \qquad (20.25)$$

are invariant owing to the unitarity of the transformation c_{qi}. Thus we expect N_i to have the form

$$N_i = xa_i^\dagger a_i + ya_i a_i^\dagger + z1$$

where x, y, and z are constants. We must, of course, require that

$$N_i \Psi'^{(0)} = 0$$

which implies that $y + z = 0$, or

$$N_i = xa_i^\dagger a_i + y(a_i a_i^\dagger - 1) \tag{20.26}$$

If the basis vectors (20.1) are to be eigenstates of N_i, it is easy from the defining relations (20.4) and (20.5) for the creation and annihilation operators to derive the commutation relations

$$N_i a_k - a_k N_i = N_i a_k^\dagger - a_k^\dagger N_i = 0 \qquad i \neq k \tag{20.27}$$

and

$$N_i a_i - a_i N_i = -a_i \tag{20.28}$$

and

$$N_i a_i^\dagger - a_i^\dagger N_i = a_i^\dagger \tag{20.29}$$

If the expression (20.26) is now substituted into these commutation relations, consistency with conditions (20.21) and (20.22) requires that $\mu = +1$ if (20.17) is valid, and $\mu = -1$ if (20.18) is valid. Furthermore, it follows that

$$Aa_k^\dagger - a_k^\dagger A = 0$$

showing that A must be a multiple of the identity operator. By letting (20.22) act on the vacuum, we see that $A\Psi'^{(0)} = \Psi'^{(0)}$, hence $A = 1$.

We therefore arrive at the conclusion that there are two and only two forms of quantum mechanics for identical particles, consistent with our fundamental assumptions:

Bose-Einstein Case

$$a_k^\dagger a_l^\dagger - a_l^\dagger a_k^\dagger = 0$$
$$a_k a_l - a_l a_k = 0 \tag{20.30}$$
$$a_k a_l^\dagger - a_l^\dagger a_k = \delta_{kl} 1$$

Fermi Dirac Case

$$a_k^\dagger a_l^\dagger + a_l^\dagger a_k^\dagger = 0$$
$$a_k a_l + a_l a_k = 0 \tag{20.31}$$
$$a_k a_l^\dagger + a_l^\dagger a_k = \delta_{kl} 1$$

The designations (B.E. and F.D.) associated with the two classes of commutation relations[4] will be seen to be justified when the statistical behavior

---

[4] When no confusion is likely, *anticommutation* relations, in which a plus rather than a minus sign appears, will, in the interest of brevity, also be referred to as *commutation* relations.

of identical particles is examined. It is a characteristic property of each kind of particle that it belongs to one or the other of these two classes. We call the former *bosons*, since they obey Bose-Einstein statistics, and the latter *fermions*, since they obey Fermi-Dirac statistics, and we characterize a particle by its *statistics*. A deep physical connection between the statistics of a particle and transformation of states under spatial rotations is found in relativistic quantum field theory and leads to the conclusion that all particles with integral spin are bosons, and all particles with half-integral spin are fermions. One immediate consequence of equation (20.31) is that, for fermions, $a_k^\dagger a_k^\dagger = 0$, i.e., there are no fermion states in which two or more particles share the same quantum numbers. This is, of course, the traditional expression of the *Pauli exclusion principle*.

With the commutation relations (20.30) and (20.31), we obtain from (20.26),

$$N_i = (x \pm y)a_i^\dagger a_i$$

for the two kinds of statistics. Finally, from (20.6) we see that for a one-particle state the requirements that

$$N_i \Psi^{(1)} = \Psi^{(1)} \tag{20.32}$$

gives us the value $x \pm y = 1$. Hence, for bosons as well as fermions, the occupation number operator N_i is unambiguously determined to be

$$N_i = a_i^\dagger a_i \tag{20.33}$$

The expectation value of N_i in any state can never be negative, for $a_i^\dagger a_i$ is a positive definite operator.

Exercise 20.2. From the commutation relations for a_i and $a_i^\dagger$, deduce that the operator $N_i = a_i^\dagger a_i$ has as its eigenvalues all nonnegative integers in the case of Bose statistics, and 0 and 1 in the case of Fermi statistics.

The operators $N_i \equiv a_i^\dagger a_i$ determine the constants of proportionality in equations (20.4) and (20.5) except for a phase factor. Indeed

$$\langle n_1, n_2 \cdots n_i \cdots | \, N_i \, | n_1, n_2 \cdots n_i \cdots \rangle$$
$$= n_i = \langle n_1, n_2 \cdots n_i \cdots | \, a_i^\dagger a_i \, | n_1, n_2 \cdots n_i \cdots \rangle$$

Hence,

$$a_i \, | n_1 n_2 \cdots n_i \cdots \rangle = e^{i\alpha} \sqrt{n_i} \, | n_1 n_2 \cdots n_i - 1 \cdots \rangle \tag{20.34}$$

For Bose statistics we may set $\alpha = 0$. For Fermi statistics, it is consistent with the anticommutation relations to chose $e^{i\alpha}$ to be equal to 1 if the number of occupied one-particle states with index less than i is even, and equal to -1 if this number is odd.

Exercise 20.3. Prove the feasibility of these phase assignments.

4. *Dynamical Variables.* Useful dynamical variables for a system of an arbitrary number of identical particles can now be constructed. The operator that measures the total value of an *additive one-particle quantity K*, like the kinetic energy, is

$$\mathcal{K} = \sum_i K_i' N_i = \sum_i K_i' a_i^\dagger a_i \tag{20.35}$$

It is desirable to express this operator in terms of an arbitrary set of creation and annihilation operators $b_q^\dagger$, b_q. The transformations (20.12) and (20.13), applied to the expression (20.35), yield

$$\mathcal{K} = \sum_{q,r} b_q^\dagger b_r \langle L_q' | K | L_r' \rangle \tag{20.36}$$

since

$$\langle L_q' | K | L_r' \rangle = \sum_i \langle L_q' | K_i' \rangle K_i' \langle K_i' | L_r' \rangle$$

$\mathcal{K}$ is the most general Hermitian operator which is bilinear in annihilation and creation operators.

An *additive two-particle operator* like the mutual potential energy can be used to define

$$\mathcal{V} = \tfrac{1}{2} \sum_{i \ne j} N_i N_j V_{ij}' + \tfrac{1}{2} \sum_i N_i(N_i - 1)V_{ii}' = \tfrac{1}{2} \sum_{i,j} (N_i N_j - N_i \delta_{ij})V_{ij}' \tag{20.37}$$

where we may assume, without loss of generality, that the numbers V_{ij}' form a real symmetric matrix. It is easy to see that the *pair distribution operator* $P_{ij} = N_i N_j - \delta_{ij} N_i$ may be written as

$$P_{ij} = N_i N_j - \delta_{ij} N_j = a_i^\dagger a_j^\dagger a_j a_i \tag{20.38}$$

for either the Bose-Einstein or Fermi-Dirac case. Hence, equation (20.37) may be written more concisely as

$$\mathcal{V} = \tfrac{1}{2} \sum_{i,j} a_i^\dagger a_j^\dagger a_j a_i V_{ij}' \tag{20.39}$$

After the transformation (20.12) and (20.13) is applied, this operator acquires its generalized form

$$\mathcal{V} = \tfrac{1}{2} \sum_{qrst} b_q^\dagger b_r^\dagger b_s b_t \langle qr | V | ts \rangle \tag{20.40}$$

since

$$\langle qr| V |ts \rangle = \sum_{ij} \langle L_q' | K_i' \rangle \langle K_i' | L_t' \rangle \langle L_r' | K_j' \rangle \langle K_j' | L_s' \rangle V_{ij}' \tag{20.41}$$

is the general two-particle matrix element. Here q and t belong to one particle r and s to the other.[5] From the last equation it is easily verified that

---

[5] Some authors use a different ordering of the quantum numbers in the two-particle matrix element. They would write the matrix element in (20.40) as $\langle qr| V |st \rangle$.

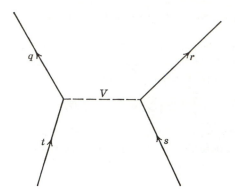

Figure 20.1. Diagram representation of the two-particle matrix element $\langle qr| \, V \, |ts\rangle$.

the condition V_{ij}' be real, and $V_{ji}' = V_{ij}'$ implies for the two-particle operator matrix element that

$$\langle qr| \, V \, |ts\rangle = \langle rq| \, V \, |st\rangle \qquad (20.42)$$

Figure 20.1 shows a diagram representation of the matrix element $\langle qr| \, V \, |ts\rangle$, which needs no explanation.

Operators involving more than two particles can be defined in a similar manner.

5. *The Continuous One-Particle Spectrum.* So far we have implicitly assumed that the eigenvectors of the complete set of commuting one-particle operators are normalizable to unity, corresponding to a discrete set of eigenvalues. Now we must consider the case that the one-particle basis used belongs to the *continuous* spectrum of a complete set of commuting operators symbolized by x, such that the one-particle eigenvectors are normalized as

$$\langle x', \sigma' \, | \, x'', \sigma''\rangle = \delta(x' - x'') \, \delta_{\sigma'\sigma''}$$

The continuous eigenvalues x' of x and the discrete eigenvalues σ' of σ label the basis vectors. The most common example of a continuous basis occurs, of course, if x is the ordinary position coordinate $\mathbf{r}$ of a particle. The discrete variable σ may represent the z-component of the spin of the particle, as explained in Section 12.3.

The formalism of the preceding sections can be modified and transcribed without difficulty to be applicable to this case, but the occupation number representation is no longer useful, unless the $\mathbf{r}$-space is artifically partitioned into a large number of small but finite cells.

It is customary to write $\psi_{\sigma'}(\mathbf{r}')$ for the annihilation operator, instead of using the notation a_i which is employed in the discrete case. An operator like $\psi_{\sigma'}(\mathbf{r}')$, which depends on the position coordinates, is generally referred to as a *field operator* or, simply, a *field*. With this new notation, we now generalize the commutation or anticommutation relations (20.30) and (20.31) and assume

$$\psi_{\sigma'}(\mathbf{r}')\psi_{\sigma''}(\mathbf{r}'') \mp \psi_{\sigma''}(\mathbf{r}'')\psi_{\sigma'}(\mathbf{r}') = 0$$

$$\psi_{\sigma'}^{\dagger}(\mathbf{r}')\psi_{\sigma''}^{\dagger}(\mathbf{r}'') \mp \psi_{\sigma''}^{\dagger}(\mathbf{r}'')\psi_{\sigma'}^{\dagger}(\mathbf{r}') = 0 \qquad (20.43)$$

$$\psi_{\sigma'}(\mathbf{r}')\psi_{\sigma''}^{\dagger}(\mathbf{r}'') \mp \psi_{\sigma''}^{\dagger}(\mathbf{r}'')\psi_{\sigma'}(\mathbf{r}') = \delta(\mathbf{r}' - \mathbf{r}'')\,\delta_{\sigma'\sigma''}$$

The operator $\psi_{\sigma'}^{\dagger}(\mathbf{r}')\psi_{\sigma'}(\mathbf{r}')$ measures the density of particles for which the discrete variable has the value σ' at the coordinate $\mathbf{r}'$, and

$$N = \sum_{\sigma'} \int \psi_{\sigma'}^{\dagger}(\mathbf{r}')\psi_{\sigma'}(\mathbf{r})\, d^3r' \qquad (20.44)$$

is the operator whose eigenvalues measure the total number of particles.

For an additive one-particle operator, we have

$$\mathcal{K} = \sum_{\sigma'\sigma''} \iint \psi_{\sigma'}^{\dagger}(\mathbf{r}') \langle \mathbf{r}'\sigma'|\, K\, |\mathbf{r}''\sigma'' \rangle\, \psi_{\sigma''}(\mathbf{r}'')\, d^3r'\, d^3r'' \qquad (20.45)$$

and, for the common kind of additive two-particle operator,

$$\mathcal{V} = \tfrac{1}{2} \sum_{\sigma'\sigma''\sigma'''\sigma''''} \iiiint \psi_{\sigma'}^{\dagger}(\mathbf{r}')\psi_{\sigma''}^{\dagger}(\mathbf{r}'')\psi_{\sigma'''}(\mathbf{r}''')\psi_{\sigma''''}(\mathbf{r}'''')$$

$$\times \langle \mathbf{r}'\sigma', \mathbf{r}''\sigma''|\, V\, |\mathbf{r}''''\sigma'''', \mathbf{r}'''\sigma''' \rangle\, d^3r'\, d^3r''\, d^3r'''\, d^3r'''' \qquad (20.46)$$

An arbitrary *n*-particle state Ψ can be conveniently represented by the function

$$\psi(\mathbf{r}_1\sigma_1, \mathbf{r}_2\sigma_2 \cdots \mathbf{r}_n\sigma_n) = \frac{1}{\sqrt{n!}} (\psi_{\sigma_n}^{\dagger}(\mathbf{r}_n) \cdots \psi_{\sigma_1}^{\dagger}(\mathbf{r}_1)\Psi^{(0)}, \Psi) \qquad (20.47)$$

Evidently, with respect to permutations of the subscripts $1, 2, \dots n$, ψ is *symmetric* in the Bose-Einstein case, and *antisymmetric* in the Fermi-Dirac case. In the trivial case of a one-particle state, $n = 1$, equation (20.47) reduces to

$$\psi(\mathbf{r}\sigma) = (\psi_{\sigma}^{\dagger}(\mathbf{r})\Psi^{(0)}, \Psi) \equiv \langle \mathbf{r}\sigma\,|\,\Psi \rangle$$

which is simply the ordinary Schrödinger wave function with spin, representing the state Ψ.

It is now essential to prove that, if the operators $\mathbf{r}$, σ constitute a complete set of observables for the single particle, ψ as given by equation (20.47) completely determines the state of a system of n identical particles Ψ. A proof of the important formula

$$\Psi = \sum_{\sigma_1\sigma_2\cdots} \int \frac{1}{\sqrt{n!}}\, \psi_{\sigma_n}^{\dagger}(\mathbf{r}_n) \cdots \psi_{\sigma_1}^{\dagger}(\mathbf{r}_1)\Psi^{(0)}\psi(\mathbf{r}_1\sigma_1, \mathbf{r}_2\sigma_2 \cdots \mathbf{r}_n\sigma_n)\, d^3r_1 \cdots d^3r_n$$

$$(20.48)$$

which is the inverse of equation (20.47), will be given in Section 20.7. This representation of an n-particle state is convenient for many purposes; the function $\psi(\mathbf{r}_1\sigma_1, \mathbf{r}_2\sigma_2, \ldots \mathbf{r}_n\sigma_n)$ is said to be the *wave function* of the n-particle system in the Cartesian $3n$-dimensional *configuration space* of the coordinates $\mathbf{r}_1, \mathbf{r}_2, \ldots \mathbf{r}_n$.

By multiplying equation (20.47) by $\psi(\mathbf{r}_1\sigma_1 \cdots \mathbf{r}_n\sigma_n)^*$, integrating over the entire configuration space, and summing over the discrete variables, we get

$$\sum_{\sigma_1\sigma_2\cdots} \int |\psi(\mathbf{r}_1\sigma_1 \cdots \mathbf{r}_n\sigma_n)|^2 \, d^3r_1 \cdots d^3r_n$$

$$= \sum \int \frac{1}{\sqrt{n!}} (\psi(\mathbf{r}_1\sigma_1 \cdots \mathbf{r}_n\sigma_n)\psi_{\sigma_n}^\dagger(\mathbf{r}_n) \cdots \psi_{\sigma_1}^\dagger(\mathbf{r}_1) \, \Psi^{(0)}, \Psi) \, d^3r_1 \cdots d^3r_n$$

$$= (\Psi, \Psi) \tag{20.49}$$

where equation (20.48) has been used. Since equation (20.49) establishes the normalization of ψ, the wave function in configuration space can be made the basis of a straightforward probability interpretation, and this has been the traditional approach to the quantum mechanics of many particles.

Exercise 20.4. Show that if $K(\mathbf{r}, \mathbf{p})$ is a one-particle operator, which is a function of the coordinate $\mathbf{r}$ and the conjugate momentum $\mathbf{p}$, the expectation value of the sum operator $\mathscr{K}$ in a normalized n-particle state Ψ is

$$(\Psi, \mathscr{K}\Psi)$$

$$= \left(\Psi, \sum_{\sigma'} \int \psi_{\sigma'}^\dagger (\mathbf{r}')K\left(\mathbf{r}', \frac{\hbar}{i}\nabla'\right)\psi_{\sigma'}(\mathbf{r}') \, d^3r' \, \Psi\right)$$

$$= \sum_{j=1}^n \sum_{\sigma_1\sigma_2\cdots} \int \psi^*(\mathbf{r}_1\sigma_1, \ldots \mathbf{r}_n\sigma_n)K\left(\mathbf{r}_j, \frac{\hbar}{i}\nabla_j\right)\psi(\mathbf{r}_1\sigma_1, \ldots \mathbf{r}_n\sigma_n) \, d^3r_1 \cdots d^3r_n \tag{20.50}$$

Exercise 20.5. Show that, if the two-particle interaction is spin-independent and *local*, such that

$$\langle r'\sigma', r''\sigma''|\, V \,|r'''\sigma''', r''''\sigma''''\rangle = V(\mathbf{r}', \mathbf{r}'') \, \delta(\mathbf{r}' - \mathbf{r}''') \, \delta(\mathbf{r}'' - \mathbf{r}'''') \, \delta_{\sigma'\sigma'''}\delta_{\sigma''\sigma''''}$$

we may write in an n-particle state,

$$(\Psi, \mathscr{V}\Psi) = \sum_{i>j=1}^n \sum_{\sigma_1\sigma_2\cdots} \int |\psi(\mathbf{r}_1\sigma_1, \ldots, \mathbf{r}_n\sigma_n)|^2 \, V(\mathbf{r}_i, \mathbf{r}_j) \, d^3r_1 \cdots d^3r_n \tag{20.51}$$

6. *Quantum Dynamics of Identical Particle Systems.* Up to this point, nothing has been said about changes in *time*. Following the program of Chapter 15, we now assume that the temporal development of a system is governed by a Hermitian operator $\mathscr{H}$, the *Hamiltonian* of the system. For simplicity we also assume that $\mathscr{H}$ is not explicitly dependent on the time t.

We must decide which picture to use, such as Schrödinger, Heisenberg, or interaction. In the remainder of this section we shall use the *Heisenberg picture* exclusively because it is instructive to see the similarity of the equation of motion for the operator $\psi(\mathbf{r}, t)$ with the wave equation in the one-particle theory.

All physical observables can be constructed from creation and annihilation operators. It is, therefore, of particular interest to determine the temporal behavior of these fundamental entities and to derive the equation of motion for some arbitrary destruction operator $a_j(t)$. Since, in what follows, all operators will be evaluated at time t, we shall, for brevity, omit the reference to the time dependence in some of the ancillary equations. The commutation relations are, of course, equally valid at all times, provided that the two factors are taken *at the same time*.

In order to obtain an equation for the time development of a_j, it is helpful to make explicit assumptions about the structure of the Hamiltonian operator. We shall cover many important applications by making the physical assumption that $\mathscr{H}$ preserves the total number of particles and that it is composed only of one- and two-particle operators. The most general form of $\mathscr{H}$ is thus assumed to be

$$\mathscr{H} = \sum_{k,l} a_k^\dagger \langle k|H_0|l\rangle\, a_l + \tfrac{1}{2} \sum_{q,r,s,t} a_q^\dagger a_r^\dagger \langle qr|V|ts\rangle\, a_s a_t \qquad (20.52)$$

where H_0 is a one-particle operator, such as the kinetic energy plus the potential energy in an external field, and V is a two-particle interaction operator.

To obtain the equation of motion for a_j, we must evaluate the commutators

$$[a_j, a_k^\dagger a_l] \qquad \text{and} \qquad [a_j, a_q^\dagger a_r^\dagger a_s a_t]$$

This is easily done by using the commutation or anticommutation relations for creation and annihilation operators. The results, after some straight-forward algebra, are

$$[a_j, a_k^\dagger a_l] = a_l\, \delta_{kj} \qquad (20.53)$$

and

$$[a_j, a_q^\dagger a_r^\dagger a_s a_t] = a_r^\dagger a_s a_t\, \delta_{qj} + a_q^\dagger a_t a_s\, \delta_{rj} \qquad (20.54)$$

It is remarkable that these relations hold for both Bose-Einstein and Fermi-Dirac statistics.

Using these identities, we obtain in the Heisenberg picture:

$$i\hbar \frac{da_j}{dt} = [a_j, \mathcal{H}]$$

$$= \sum_l \langle j| H_0 |l\rangle a_l + \tfrac{1}{2}\sum_{r,s,t} a_r^\dagger a_s a_t \langle jr| V |ts\rangle + \tfrac{1}{2}\sum_{q,s,t} a_q^\dagger a_t a_s \langle qj| V |ts\rangle$$

$$= \sum_l \langle j| H_0 |l\rangle a_l + \tfrac{1}{2}\sum_{r,s,t} a_r^\dagger a_s a_t (\langle jr| V |ts\rangle + \langle rj| V |st\rangle)$$

The symmetry relation (20.42) finally gives us, for both statistics, the desired equation of motion:

$$i\hbar \frac{da_j(t)}{dt} = \sum_l \langle j| H_0 |l\rangle a_l(t) + \sum_{r,s,t} a_r^\dagger(t) a_s(t) a_t(t) \langle jr| V |ts\rangle \qquad (20.55)$$

A case of special interest arises if we work in a representation in which the two-particle interaction matrix element is diagonal:

$$\langle qr| V |ts\rangle = V_{qr}\delta_{qt}\delta_{rs} \qquad (20.56)$$

The equation of motion for the annihilation operator is then

$$i\hbar \frac{da_j(t)}{dt} = \sum_l \langle j| H_0 |l\rangle a_l(t) + \sum_r a_r^\dagger(t) a_r(t) a_j(t) V_{jr} \qquad (20.57)$$

It is important to examine the form that this equation takes if the one-particle basis used is that provided by the eigenvectors of the coordinate $\mathbf{r}$ and the discrete variable σ as described in Section 20.5. The interaction energy is assumed to be spin-independent and local as in Exercise 20.5. Equation (20.57) is then transcribed unambiguously and becomes

$$i\hbar \frac{\partial \psi_{\sigma'}(\mathbf{r}', t)}{\partial t} = \sum_{s''} \int \langle \mathbf{r}'s'| H_0 |\mathbf{r}''s''\rangle \psi_{\sigma''}(\mathbf{r}'', t)\, d^3r''$$

$$+ \sum_{\sigma''} \int V(\mathbf{r}', \mathbf{r}'') \psi_{\sigma''}^\dagger(\mathbf{r}'', t) \psi_{\sigma''}(\mathbf{r}'', t)\, d^3r'' \psi_{\sigma'}(\mathbf{r}', t) \qquad (20.58)$$

If, as is often the case, the one-particle operator H_0 is a function only of coordinates and conjugate momenta so that

$$\langle \mathbf{r}'\sigma'| H_0(\mathbf{r}, \mathbf{p}) |\mathbf{r}''\sigma''\rangle = H_0\!\left(\mathbf{r}', \frac{\hbar}{i}\nabla'\right)\delta(\mathbf{r}' - \mathbf{r}'')\,\delta_{\sigma'\sigma''} \qquad (20.59)$$

equation (20.58) can be rewritten as an integral-differential equation in ordinary space and time:

$$i\hbar \frac{\partial \psi_{\sigma'}(\mathbf{r}', t)}{\partial t} = H_0\!\left(\mathbf{r}', \frac{\hbar}{i}\nabla'\right)\psi_{\sigma'}(\mathbf{r}', t)$$

$$+ \sum_{\sigma''} \int V(\mathbf{r}', \mathbf{r}'') \psi_{\sigma''}^\dagger(\mathbf{r}'', t) \psi_{\sigma''}(\mathbf{r}'', t)\, d^3r'' \psi_{\sigma'}(\mathbf{r}', t) \qquad (20.60)$$

Again, the universality of this equation, applicable to both bosons and fermions, deserves to be emphasized.

If there is no interaction between the particles, (20.60) reduces to

$$i\hbar \frac{\partial \psi_{\sigma'}(\mathbf{r}', t)}{\partial t} = H_0\left(\mathbf{r}', \frac{\hbar}{i}\nabla'\right)\psi_\sigma(\mathbf{r}', t) \tag{20.61}$$

which, in form, is completely identical with the one-particle wave equation (8.79). However, it must be remembered that $\psi_{\sigma'}(\mathbf{r}', t)$ here is a *field operator* and not an ordinary *wave function*. Equation (20.61) can be regarded as the *quantized* version of the wave equation (3.16) when the latter is interpreted as a *classical* field equation. The annihilation operator corresponding to the coordinate $\mathbf{r}'$ was, of course, denoted by $\psi(\mathbf{r}')$, rather than some other symbol, in the foreknowledge of its ultimate identification with the quantized Schrödinger field. This is the point of contact between *quantum field theory* and the many-body formulation developed in this chapter and traditionally known as *second quantization*.

The interaction term in equation (20.60) causes the equation to be non-linear and can be interpreted simply as describing the effect of all the particles on a single one. The effective potential

$$\sum_{\sigma''} \int V(\mathbf{r}', \mathbf{r}'')\psi_{\sigma''}^\dagger(\mathbf{r}'', t)\psi_{\sigma''}(\mathbf{r}'', t)\, d^3r'' = \sum_{\sigma''} \int V(\mathbf{r}', \mathbf{r}'')\rho_{\sigma''}(\mathbf{r}'', t)\, d^3r''$$

can be calculated only if the solution of the equation is already known, thus suggesting an iteration procedure to generate a self-consistent solution. Such techniques for solving the many-body problem are, indeed, frequently applied (see Chapter 21).

7. Mathematical Detail. Finally, the proof of equation (20.48) will be sketched. This is accomplished by use of the following lemma:

$$F \equiv (\psi_{\sigma_n}^\dagger(\mathbf{r}_n) \cdots \psi_{\sigma_1}^\dagger(\mathbf{r}_1)\Psi^{(0)},\ \psi_{\sigma_n'}^\dagger(\mathbf{r}_n') \cdots \psi_{\sigma_1'}^\dagger(\mathbf{r}_1')\Psi^{(0)})$$
$$= \sum_{P_n} sgn(P_n)\, \delta(\mathbf{r}_1 - \mathbf{r}_1')\, \delta(\mathbf{r}_2 - \mathbf{r}_2') \cdots \delta(\mathbf{r}_n - \mathbf{r}_n')\, \delta_{\sigma_n, \sigma_n'} \cdots \delta_{\sigma_1, \sigma_1'} \tag{20.62}$$

The sum is to be taken over all $n!$ permutations P_n of the coordinates $\mathbf{r}_1\sigma_1,\ \mathbf{r}_2\sigma_2 \cdots \mathbf{r}_n\sigma_n$ (but not the primed coordinates). For the Bose-Einstein case $sgn(P_n) = 1$; for the Fermi-Dirac case $sgn(P_n) = +1$ if the permutation P_n is even, and $sgn(P_n) = -1$ if it is odd.

This formula is proved by induction. It is obviously valid for $n = 1$. Use of the commutation relations (20.43) gives

$$F = (\psi_{\sigma_{n-1}}^\dagger(\mathbf{r}_{n-1}) \cdots \psi_{\sigma_1}^\dagger(\mathbf{r}_1)\Psi^{(0)},\ \psi_{\sigma_{n-1}}^\dagger(\mathbf{r}_{n-1}') \cdots \psi_{\sigma_1}^\dagger(\mathbf{r}_1')\Psi^{(0)})\, \delta(\mathbf{r}_n - \mathbf{r}_n')$$
$$\pm (\psi_{\sigma_{n-1}}^\dagger(\mathbf{r}_{n-1}) \cdots \psi_{\sigma_1}^\dagger(\mathbf{r}_1)\Psi^{(0)},\ \psi_{\sigma_n'}^\dagger(\mathbf{r}_n')\psi_{\sigma_n}(\mathbf{r}_n)\psi_{\sigma_{n-1}'}^\dagger(\mathbf{r}_{n-1}') \cdots \psi_{\sigma_1'}^\dagger(\mathbf{r}_1')\Psi^{(0)})$$

By continuing to move the operator $\psi_{\sigma_n}(\mathbf{r}_n)$ to the right, step by step, and by applying the equation (20.62) for $n - 1$, we eventually obtain[6]

$$F = \sum_{P_{n-1}} sgn(P_{n-1})\, \delta(\mathbf{r}_1 - \mathbf{r}_1') \cdots \delta(\mathbf{r}_{n-1} - \mathbf{r}_{n-1}')\, \delta(\mathbf{r}_n - \mathbf{r}_n')$$

$$\pm \sum_{P_{n-1}} sgn(P_{n-1})\, \delta(\mathbf{r}_1 - \mathbf{r}_1') \cdots \delta(\mathbf{r}_{n-2} - \mathbf{r}_{n-2}')\, \delta(\mathbf{r}_{n-1} - \mathbf{r}_n')\, \delta(\mathbf{r}_n - \mathbf{r}_{n-1}')$$

$$+ \sum_{P_{n-1}} sgn(P_{n-1})\, \delta(\mathbf{r}_1 - \mathbf{r}_1') \cdots \delta(\mathbf{r}_{n-2} - \mathbf{r}_{n-1}')\, \delta(\mathbf{r}_{n-1} - \mathbf{r}_n')\, \delta(\mathbf{r}_n - \mathbf{r}_{n-2}')$$

$$\pm \cdots$$

which can obviously be contracted into the right-hand side of equation (20.62).

Armed with the generalized orthogonality relation (20.62), we can now deduce equation (20.48) from (20.47). To this end we note that a general n-particle state can surely be represented by

$$\Psi = \frac{1}{\sqrt{n!}} \sum_{\sigma_1' \cdots \sigma_n'} \int \psi_{\sigma_n'}^\dagger(\mathbf{r}_n') \cdots \psi_{\sigma_1'}^\dagger(\mathbf{r}_1')\Psi^{(0)}$$

$$\times f(\mathbf{r}_1'\sigma_1', \mathbf{r}_2'\sigma_2' \cdots \mathbf{r}_n'\sigma_n')\, d^3\mathbf{r}_1'\, d^3\mathbf{r}_2' \cdots d^3\mathbf{r}_n' \quad (20.63)$$

If this expression is substituted for Ψ in equation (20.47) and the integrations are carried out,

$$\psi(\mathbf{r}_1\sigma_1 \cdots \mathbf{r}_n\sigma_n) = \frac{1}{n!} \sum_{P_n} sgn(P_n) f(\mathbf{r}_1\sigma_1, \mathbf{r}_2\sigma_2 \cdots \mathbf{r}_n\sigma_n)$$

results. On the other hand, owing to the (anti) commutivity of the creation operators, (20.63) can also be written as

$$\Psi = \frac{1}{\sqrt{n!}} \sum_{\sigma_1\sigma_2 \cdots \sigma_n} \int d^3\mathbf{r}_1 \cdots d^3\mathbf{r}_n \psi_{\sigma_n}^\dagger(\mathbf{r}_n) \cdots \psi_{\sigma_n}^\dagger(\mathbf{r}_1)\Psi^{(0)}$$

$$\times \frac{1}{n!} \sum_{P_n} sgn(P_n) f(\mathbf{r}_1\sigma_1 \cdots \mathbf{r}_n\sigma_n)$$

Hence, equation (20.48) follows.

Problem

1. Two identical bosons or fermions in a state

$$\Psi^{(2)} = A \sum c_i\, d_j a_j^\dagger a_i^\dagger \Psi^{(0)}$$

are said to be *uncorrelated* (except for the effect of statistics). If $\sum |c_i|^2 = \sum |d_i|^2 = 1$, determine the normalization constant A in terms of the sum $S = \sum c_i^* d_i$. In this state, work out the expectation value of an additive one-particle operator in terms of the one-particle amplitudes c_i and d_i and the matrix elements $\langle i|\, K\, |j\rangle$. Show that if $S = 0$, the expectation value is the same as if the two particles with amplitudes c_i and d_i were distinguishable. Work out the expectation value of a diagonal interaction operator in terms of c_i, d_i, and $\langle ij|\, V\, |kl\rangle = V_{ij}\, \delta_{ik}\, \delta_{jl}$. Show that the result is the same as for distinguishable particles if the states of the two particles do not overlap, i.e., if $c_i d_i = 0$ for all i.

[6] The Kronecker deltas carrying the spin variable have been suppressed for brevity.

CHAPTER 21

Applications of Second Quantization

1. Angular Momentum in a System of Identical Particles. An important example of an observable in a system of identical particles is the angular momentum operator, which according to (20.36) is the additive one-particle operator

$$\mathcal{J} = \sum_\alpha \sum_j \sum_{m,m'} a^\dagger_{jm'\alpha} a_{jm\alpha} \langle jm' | \mathbf{J} | jm \rangle \tag{21.1}$$

if the one-particle basis is characterized by the angular momentum quantum numbers j and m, as defined in Section 16.5, and α stands for all remaining quantum numbers needed to specify the basis. The total angular momentum operator (21.1) owes its simple structure to the absence of offdiagonal matrix elements of $\mathbf{J}$ with respect to j and α.

The operator $\mathcal{J}^2 = \mathcal{J} \cdot \mathcal{J}$ is not just the sum of the $\mathbf{J}^2$ for the individual particles but contains terms that couple two particles, and thus it serves as an example of an additive *two-particle* operator. Since it conserves the number of particles, annihilating one and creating one, the operator $\mathcal{J}$ commutes with the total number-of-particles operator, N.

Exercise 21.1. Exhibit the two-particle matrix elements of the square of the total angular momentum explicitly.

The *two-particle* states in which $\mathcal{J}_z$ and $\mathcal{J}^2$ have the sharp values $M\hbar$ and $J(J + 1)\hbar^2$ are readily constructed by the use of the Clebsch-Gordan coefficients defined in (16.67):

$$\Psi^{(2)}_{JM} = C \sum_{m_1 m_2} a^\dagger_{j_2 m_2 \alpha_2} a^\dagger_{j_1 m_1 \alpha_1} \langle j_1 j_2 m_1 m_2 | j_1 j_2 JM \rangle \Psi^{(0)} \tag{21.2}$$

Since the one-particle state $a^\dagger_{jm\alpha}\Psi^{(0)}$ is normalized to unity, (21.2) completely parallels expression (16.67), and the normalization constant $C = 1$, unless $\alpha_1 = \alpha_2$ and $j_1 = j_2$.

The expression (21.2) remains an eigenvector of $\mathcal{J}_z$ and $\mathcal{J}^2$ even if $\alpha_1 = \alpha_2$ and $j_1 = j_2$, but the normalization is altered. The symmetry relation (16.78) permits us to rewrite (21.2) in the form

$$\Psi_{JM}^{'(2)} = \tfrac{1}{2}[1 \pm (-1)^{J-2j}]C \sum_{m_1 m_2} a_{j m_2 \alpha}^{\dagger} a_{j m_1 \alpha}^{\dagger} \langle j j m_1 m_2 \,|\, j j J M \rangle \Psi^{'(0)} \quad (21.3)$$

with the upper sign applicable to bosons and the lower sign to fermions. Hence, the angular momenta of two identical bosons (fermions), which share all one-particle quantum numbers except m, cannot couple to a state for which $J - 2j$ is an odd (even) number. If the usual connection between spin and statistics is assumed and bosons have integral spin, odd J values of the total angular momentum cannot occur for two alike bosons with the same α and j. Similarly, two identical fermions, which have half-integral spin, cannot couple to odd values of J if they have the same α and j.

If $\Psi_{JM}^{'(2)} \neq 0$, the value of the normalization constant C may be determined by requiring $(\Psi_{JM}^{'(2)}, \Psi_{JM}^{'(2)}) = 1$. The unitarity condition (16.76) yields readily the value $C = 1/\sqrt{2}$, so that if $\alpha_1 = \alpha_2 = \alpha$ and $j_1 = j_2 = j$,

$$\Psi_{JM}^{'(2)} = \frac{1}{\sqrt{2}} \sum_{m_1 m_2} a_{j m_2 \alpha}^{\dagger} a_{j m_1 \alpha}^{\dagger} \langle j j m_1 m_2 \,|\, j j J M \rangle \Psi^{'(0)} \quad (21.4)$$

Exercise 21.2. Verify the normalization (21.4).

Exercise 21.3. Construct explicitly in terms of states of the form $a_{j m_2 \alpha}^{\dagger} a_{j m_1 \alpha}^{\dagger} \Psi^{'(0)}$ the total angular momentum eigenstates for two neutrons in the configurations $(p_{1/2})^2$ and $(p_{3/2})^2$. How would the angular momentum eigenstates look if the two particles were a neutron and a proton but otherwise had the same quantum numbers as before?

Exercise 21.4. Show that if two identical particles with the same quantum numbers α and with angular momentum j couple to zero total angular momentum, the resulting *pair state* is, in an obviously simplified notation,

$$\Psi_{00}^{'(2)} = [2(2j + 1)]^{-1/2} \sum_{m=-j}^{j} (-1)^m a_m^{\dagger} a_{-m}^{\dagger} \Psi^{'(0)} \quad (21.5)$$

2. Angular Momentum and Spin $\tfrac{1}{2}$ Boson Operators. If we postulate a *boson* with spin $\tfrac{1}{2}$ and with no other dynamical properties, the total angular momentum operator for a system of identical bosons of this kind takes the form

$$\mathscr{J} = \frac{\hbar}{2} (a_{1/2}^{\dagger} \; a_{-1/2}^{\dagger}) \boldsymbol{\sigma} \begin{pmatrix} a_{1/2} \\ a_{-1/2} \end{pmatrix} \quad (21.6)$$

if the creation operators for the two spin states, $m = +\tfrac{1}{2}$ and $-\tfrac{1}{2}$ are denoted simply by $a^\dagger_{1/2}$ and $a^\dagger_{-1/2}$. Equation (21.6) may be decomposed into

$$\mathscr{J}_+ = \hbar a^\dagger_{1/2} a_{-1/2} \qquad \mathscr{J}_- = \hbar a^\dagger_{-1/2} a_{1/2}$$

$$\mathscr{J}_z = \frac{\hbar}{2}(a^\dagger_{1/2} a_{1/2} - a^\dagger_{-1/2} a_{-1/2}) \tag{21.7}$$

in agreement with our expectations for $\mathscr{J}_\pm$ as the raising (lowering) operator which changes the state $|JM\rangle$ into $|J\,M+1\rangle$. From (21.7) we derive, using the boson commutation relations,

$$\mathscr{J}^2 = \frac{\hbar^2}{4}(a^\dagger_{1/2} a_{1/2} + a^\dagger_{-1/2} a_{-1/2})^2 + \frac{\hbar^2}{2}(a^\dagger_{1/2} a_{1/2} + a^\dagger_{-1/2} a_{-1/2})$$

$$= \frac{\hbar^2}{4} N^2 + \frac{\hbar^2}{2} N = \hbar^2 \frac{N}{2}\left(\frac{N}{2}+1\right) \tag{21.8}$$

Hence, the state with a total number $n = 2J$ of identical spin $\frac{1}{2}$ bosons is an eigenstate of $\mathscr{J}^2$ with eigenvalue $J(J+1)\hbar^2$, where J is either integral or half-integral. The simultaneous eigenstates of the occupation number operators $N_+ = a^\dagger_{1/2} a_{1/2}$ (number of "spin up" bosons) and $N_- = a^\dagger_{-1/2} a_{-1/2}$ (number of "spin down" bosons) are simultaneous eigenstates of $\mathscr{J}^2$ and $\mathscr{J}_z = \hbar(N_+ - N_-)/2$. The eigenvalues of the occupation number operators are determined by the relations

$$n_+ + n_- = 2J \qquad n_+ - n_- = 2M \tag{21.9}$$

or

$$n_+ = J + M \qquad n_- \equiv J - M$$

Hence, by (20.34), the eigenstates, normalized to unity, are

$$|JM\rangle = \frac{(a^\dagger_{1/2})^{J+M}(a^\dagger_{-1/2})^{J-M}}{\sqrt{(J+M)!\,(J-M)!}}\,\Psi^{(0)} \tag{21.10}$$

In terms of the vector model of angular momentum, this representation of the state $|JM\rangle$ may be recognized as the projection $M\hbar$ of the resultant of $2J$ spin $\frac{1}{2}$ vectors combined to produce the "stretched" vector polygon with all spin $\frac{1}{2}$ vectors "parallel." The requirements of Bose-Einstein statistics for the spins which make up this resultant cause this state to be uniquely defined.

The violation of the connection between spin and statistics implied by the use of spin $\frac{1}{2}$ bosons in this section does not vitiate the mathematical procedure which we have outlined. The bosons defined here are not particles in the usual sense, since they have no momentum or energy. Rather, they are somewhat abstract carriers of spin, allowing a particularly transparent description of angular momentum states. As auxiliary entities, these bosons may be used for a relatively easy evaluation of the Clebsch-Gordan coefficients and of the more complicated structures that arise in the coupling of more than two angular momenta.[1] An illustration of the usefulness of the spin $\frac{1}{2}$ boson operator formalism will be found in the next section.

3. Elements of the Theory of Spin Waves and Magnons. The magnetic behavior insulators is frequently described in terms of an effective spin Hamiltonian, known as the *Heisenberg Hamiltonian*, which in the simplest case takes the form

$$\mathscr{H} = -\sum_{jl} J_{jl} \mathbf{S}_j \cdot \mathbf{S}_l + Bg\beta_0 \sum_j (S_j)_z \qquad (21.11)$$

In this expression $\mathbf{S}_j$ is a spin operator corresponding to a localized spin (usually of an ion) at the site labeled by an integer j. The interaction constants J_{jl} are assumed to be nonnegative numbers, appropriate to ferromagnetic interactions which favor a lowering of the energy when the spins tend to be lined up parallel. Without loss of generality, we may assume $J_{jl} = J_{lj}$ and $J_{jj} = 0$. (See Problem 4 at the end of this chapter for a derivation of a Hamiltonian of the form (21.11) for spin $\frac{1}{2}$ in a special example.)

The second term on the right-hand side of equation (21.11) takes into account the effect of a uniform external magnetic field, directed along the positive z axis and of magnitude B, if $g\beta_0$ is the gyromagnetic ratio for the localized spins. We assume that all spins are identical and have maximum projection sh, where s may be integral or half-integral. For definiteness we also assume $g > 0$.

It is easy to see by inspection of (21.11) that the ground state of the spin system is the state in which the z component of each spin is $-sh$ and that the ground state energy is

$$E_0 = -\sum_{jl} J_{jl} s^2 \hbar^2 - nBg\beta_0 sh \qquad (21.12)$$

where n is the total number of spins or orbital sites occupied by spins. Assuming n to be an odd number, the summations may be extended from $j, l = -(n-1)/2$ to $(n-1)/2$.

It is much less easy to determine the excited states of the system and the energy spectrum, especially if n is large, because all spins are coupled by the interaction, even if only neighboring spins interact. It is convenient to transform the problem by introducing the spin $\frac{1}{2}$ boson representation of the preceding section. A distinct boson is defined for each spin site, and the appropriate site address j is affixed as subscript to the creation and annihilation operators. Operators pertaining to different sites are assumed to commute.

---

[1] For a full treatment, see J. Schwinger, *On Angular Momentum* in L. C. Biedenharn and H. Van Dam, editors, *Quantum Theory of Angular Momentum*, Academic Press, New York, 1965, p. 229.

Equation (21.9) imposes a constraint on the boson operators:

$$a^\dagger_{1/2,j}a_{1/2,j} + a^\dagger_{-1/2,j}a_{-1/2,j} = 2s \qquad (21.13)$$

Using this condition and the relations (21.7) for each spin, we obtain for the scalar product of any two spins:

$$S_j \cdot S_l = \frac{\hbar^2}{2}\left[a^\dagger_{1/2,j}a_{-1/2,j}a^\dagger_{-1/2,l}a_{1/2,l} + a^\dagger_{-1/2,j}a_{1/2,j}a^\dagger_{1/2,l}a_{-1/2,l} \right.$$

$$\left. + 2(a^\dagger_{1/2,j}a_{1/2,j} - s)(a^\dagger_{1/2,l}a_{1/2,l} - s) \right] \qquad (21.14)$$

The form of the Hamiltonian makes it evident that, in the low-lying excited states of the system, the spins remain pointed downward with high probability and the excitations are due to small departures from this perfect downward polarization of all the spins. The approximation to be made is based on these considerations and is characteristic of quantal many-body theories. It assumes that, with regard to their effect on low-lying energy eigenstates of the system, the operators $a^\dagger_{1/2,j}$ and $a_{1/2,j}$ are "small" compared with the operators $a^\dagger_{-1/2,j}$ and $a_{-1/2,j}$ and that, in the Hamiltonian, terms which are of order higher than second in $a^\dagger_{1/2,j}$ and $a_{1/2,j}$ may be neglected if n is large. Using the expansion

$$a^\dagger_{-1/2,j} \approx a_{-1/2,j} \approx \sqrt{2s} - \frac{a^\dagger_{1/2,j}a_{1/2,j}}{\sqrt{8s}}$$

dictated by (21.13), we obtain from (21.14) and (21.11) the *approximate* Hamiltonian

$$\mathcal{H}_{\mathrm{app}} = E_0 + 2\hbar^2 s \sum_j \sum_l J_{jl}a^\dagger_{1/2,j}a_{1/2,j} - 2\hbar^2 s \sum_{jl} J_{jl}a^\dagger_{1/2,j}a_{1/2,l}$$

$$+ Bg\beta_0\hbar \sum_j a^\dagger_{1/2,j}a_{1/2,j} \qquad (21.15)$$

Further progress toward solving the eigenvalue problem for this approximate Hamiltonian can be made if the spin sites exhibit a periodicity such as is encountered in a crystalline solid. Since the essence of the method emerges clearly even in a very simple example, we shall consider only a one-dimensional periodic lattice in which all sites are equivalent so that

$$J_{j+q,l+q} = J_{jl} \qquad (21.16)$$

If n, the number of spin sites, is large, mathematical simplicity is achieved, without any significant alteration of the energy spectrum, by ignoring end effects and taking the limit $n \to \infty$. In the infinite periodic one-dimensional lattice, equation (21.16) must hold for any integer q, and both $\mathcal{H}$ and $\mathcal{H}_{\mathrm{app}}$ are invariant under any translation of the spins from site j to site $j + q$.

This symmetry suggests that the Hamiltonian be transformed from the site (or coordinate) representation to the displacement (or momentum) representation.

The unitary operator D_q which effects such a displacement is defined by the relations

$$D_q a^\dagger_{1/2, j} D^\dagger_q = a^\dagger_{1/2, j+q} \tag{21.17}$$

and by the property $D_q \Psi^{(0)} = \Psi^{(0)}$, if $\Psi^{(0)}$ denotes the *ground state* of the system in which there are no "spin up" bosons.

In analogy with the derivation of Section 14.7, we find that the operators

$$\alpha^\dagger_\kappa = \frac{1}{\sqrt{2\pi}} \sum_{j=-\infty}^{+\infty} a^\dagger_{1/2, j} e^{i\kappa j} \tag{21.18}$$

which are defined in the interval $0 \leqslant \kappa < 2\pi$ have the simple property

$$D_q \alpha^\dagger_\kappa D^\dagger_q = e^{-i\kappa q} \alpha^\dagger_\kappa \tag{21.19}$$

This equation shows that $\alpha^\dagger_\kappa \Psi^{(0)}$ is a one-boson eigenstate of D_q with eigenvalue $e^{-i\kappa q}$. Since D_q is unitary, κ must be real.

Equation (21.18) defines a continuum of new creation operators $\alpha^\dagger_\kappa$ characterized by the variable κ, and these operators may be used instead of the discrete set of operators $a^\dagger_{1/2, j}$. Since

$$\int_0^{2\pi} \alpha^\dagger_\kappa \alpha_\kappa \, d\kappa = \sum_{j=-\infty}^{+\infty} a^\dagger_{1/2, j} a_{1/2, j} \tag{21.20}$$

the transformation conserves the number of "spin ups" and is precisely of the form (20.12). If it is remembered that α_κ is defined only for $0 \leqslant \kappa < 2\pi$, it is easily seen that the new operators satisfy the Bose-Einstein commutation relations

$$[\alpha_\kappa, \alpha_{\kappa'}] = [\alpha^\dagger_\kappa, \alpha^\dagger_{\kappa'}] = 0, \qquad [\alpha_\kappa, \alpha^\dagger_{\kappa'}] = \delta(\kappa - \kappa') \, I$$

The bosons which the operators $\alpha_\kappa^\dagger$ create are quanta associated with the collective motion of all the spins in the system. Such quanta, or *quasiparticles*, are called *magnons*. In the limit of large s spin operators may be treated as classical vectors. Their motion is governed by the interaction (21.11) which generates *spin waves*. Magnons may be regarded as the quanta corresponding to classical spin waves.[2]

Equation (21.18) may be inverted:

$$a^\dagger_{1/2, j} = \frac{1}{\sqrt{2\pi}} \int_0^{2\pi} \alpha^\dagger_\kappa e^{-ij\kappa} \, d\kappa \tag{21.21}$$

---

[2] For details and applications to solids, see C. Kittel, *Quantum Theory of Solids*, John Wiley and Sons, New York, 1963, Chapter 4.

Since translation invariance causes

$$J = \sum_l J_{jl} \tag{21.22}$$

to be independent of j, the second and last terms in (21.15) are readily transformed into the α representation by the use of (21.20). The third term in the Hamiltonian is transformed as follows:

$$\sum_{jl} J_{jl} a_{1/2,j}^\dagger a_{1/2,l} = \frac{1}{\sqrt{2\pi}} \sum_{jl} \int_0^{2\pi} d\kappa \int_0^{2\pi} d\kappa' \, J_{jl} \, \alpha_\kappa^\dagger \alpha_{\kappa'} e^{i(\kappa'l - \kappa j)}$$

$$= \frac{1}{\sqrt{2\pi}} \int_0^{2\pi} d\kappa \int_0^{2\pi} d\kappa' \sum_l \alpha_\kappa^\dagger \alpha_{\kappa'} e^{i(\kappa'-\kappa)l} \sum_j J_{jl} \, e^{i\kappa(l-j)}$$

By the assumption (21.16), the sum

$$\varphi(\kappa) = \sum_{j=-\infty}^{+\infty} J_{jl} \, e^{i\kappa(l-j)} \tag{21.23}$$

is independent of l. Using the *lattice sum* $(0 \leqslant \kappa, \kappa' < 2\pi)$

$$\sum_{l=-\infty}^{+\infty} e^{i(\kappa'-\kappa)l} = 2\pi\delta(\kappa - \kappa') \tag{21.24}$$

we finally obtain the Hamiltonian in terms of the magnon creation and annihilation operators:

$$\mathscr{H}_{app} = E_0 + \int_0^{2\pi} d\kappa \, \alpha_\kappa^\dagger \alpha_\kappa [2\hbar^2 sJ - 2\hbar^2 s\varphi(\kappa) + Bg\beta_0\hbar] \tag{21.25}$$

Equation (21.25) is the desired result of our calculation and has a straightforward interpretation. Since $\alpha_\kappa^\dagger \alpha_\kappa$ is the number density of magnons in κ-space, the quantity in brackets,

$$E_\kappa = 2\hbar^2 sJ - 2\hbar^2 s\varphi(\kappa) + Bg\beta_0\hbar \tag{21.26}$$

is the energy of a magnon with *displacement quantum number* κ. If d is the distance between adjacent spin lattice sites, κ/d becomes the wave number, and $p = \hbar\kappa/d$ may be identified as the linear momentum of a magnon. Its range is $0 \leqslant p < h/d$, in harmony with the fact that the shortest wavelength sustained by the lattice is equal to the spacing d. Equation (21.26) is called the *spin wave dispersion* formula.

Since $\mathscr{H}_{app}$ is expressible as a linear combination of occupation number operators $\alpha_\kappa^\dagger \alpha_\kappa$, the approximation made in deriving $\mathscr{H}_{app}$ from $\mathscr{H}$ consists of neglecting any interactions between magnons. $\mathscr{H}_{app}$ describes noninteracting magnons, and the excited eigenstates of the approximate Hamiltonian are simply the states obtained from the magnon vacuum $\Psi^{(0)}$ (ground state of the system) by creating one or more magnons in states with definite values of κ. The energy of such a state is the sum of the individual magnon energies. The one-magnon eigenstate with definite κ is

$$\alpha_\kappa^\dagger \Psi'^{(0)} = \frac{1}{\sqrt{2\pi}} \sum_j e^{i\kappa j} a_{1/2,j}^\dagger \Psi'^{(0)} \tag{21.27}$$

Exercise 21.5. Prove that $\alpha(\kappa)$ is an even function. Show that in the low-momentum limit the energy-momentum relation for a magnon is similar to that for a nonrelativistic particle. Identify the effective mass of a magnon.

Exercise 21.6. Derive the spin wave dispersion relation for a lattice in which each spin interacts only with its nearest neighbor.

Exercise 21.7. Prove that while (21.27) is an eigenstate of $\mathcal{H}_{\text{app}}$, the state

$$\frac{1}{\sqrt{2\pi}} \sum_j e^{i\kappa j} a_{1/2,j}^\dagger a_{-1/2,j} \Psi'^{(0)}$$

is an eigenstate of the exact Hamiltonian $\mathcal{H}$, and calculate the corresponding eigenvalue if the spin lattice is one-dimensional and all spins are equivalent.

Exercise 21.8. From the Hamiltonian (21.11) derive the equations of motion for the spin operators $\mathbf{S}_k$ in the Heisenberg picture. Assuming that the operators may be replaced approximately by classical vectors and that, as is the case near the ground state, each $\mathbf{S}_k(t)$ differs only slightly from a fixed common vector $\mathbf{S}$, obtain linear equations of motion for the *spin deviation vectors* $\mathbf{s}_k(t) = \mathbf{S}_k(t) - \mathbf{S}$, if $|\mathbf{s}_k(t)| \ll |\mathbf{S}|$. Show that the equations of motion for the spin deviation vectors describe the propagation of spin waves through a periodic lattice, each vector $\mathbf{s}_k(t)$ precessing about $\mathbf{S}$. For a one-dimensional lattice in which all spins are equivalent, show that the relation between the precession frequency ω and the wave number κ/d is given by the dispersion formula (21.26), provided that the magnon energy is identified as $E_\kappa = \hbar\omega$.

4. First-Order Perturbation Theory in Many-Body Systems. A simple and important illustration of the use of two-particle operators is afforded by a first-order perturbation calculation of the energy eigenvalues of a Hamiltonian which describes a system of interacting identical particles:

$$\mathcal{H} = \sum_i \varepsilon_i a_i^\dagger a_i + \tfrac{1}{2} \sum_{qrst} a_q^\dagger \, a_r^\dagger \, a_s a_t \langle qr| \, V \, |ts\rangle \tag{21.28}$$

It is assumed that the eigenstates of the unperturbed Hamiltonian of non-interacting particles,

$$\mathcal{H}_0 = \sum_i \varepsilon_i a_i^\dagger a_i \tag{21.29}$$

are known and characterized as $|n_1, n_2, \ldots n_i \ldots\rangle$ by the eigenvalues n_i of the occupation number operators $a_i^\dagger a_i$. If the eigenvalues of $\mathcal{H}_0$ are non-degenerate, first-order perturbation theory gives for the energies the approximate values

$$E_{n_1 n_2 \cdots} = \sum_i n_i \varepsilon_i + \tfrac{1}{2} \sum_{qrst} \langle n_1 n_2 \cdots | a_q^\dagger a_r^\dagger a_s a_t | n_1 n_2 \cdots \rangle \langle qr | V | ts \rangle \tag{21.30}$$

In evaluating the matrix element of the operator $a_q^\dagger a_q^\dagger a_s a_t$, it is helpful to recognize that, owing to the orthogonality of the unperturbed eigenstates, nonvanishing contributions to the interaction energy are obtained only if $q \neq r$ and either $s = r$ and $t = q$ or $s = q$ and $t = r$, or if $q = r = s = t$. Equation (21.30) is therefore reducible to

$$E_{n_1 n_2 \cdots} = \sum_i n_i \varepsilon_i + \tfrac{1}{2} \sum_{q \neq r} n_q n_r [\langle qr | V | qr \rangle \pm \langle qr | V | rq \rangle]$$

$$+ \tfrac{1}{2} \sum_q n_q (n_q - 1) \langle qq | V | qq \rangle \tag{21.31}$$

The $+$ sign holds for Bose-Einstein statistics and the $-$ sign for Fermi-Dirac statistics. The two matrix elements $\langle qr | V | qr \rangle$ and $\langle qr | V | rq \rangle$, connecting the two one-particle states q and r, are said to have *direct* and *exchange* character, respectively. The last term, which accounts for the interaction of particles occupying the same one-particle state, vanishes for fermions.

The evaluation of a matrix element of the product of several creation and annihilation operators carried out here is typical of most calculations in many-body theories. The labor involved in such computations is significantly reduced if the operators in a product are arranged in *normal ordering*, i.e., with all annihilation operators standing to the right of all creation operators. The operators in the Hamiltonian (21.28) are already normally ordered. If a product is not yet normally ordered, it may, by repeated application of the commutation relations, be transformed into a sum of normally ordered products. A set of simple manipulative rules may be formulated[3] which permit the expansion of an operator of arbitrary complexity into terms with normal ordering.

As an example, we briefly consider the fundamental problem of *atomic spectroscopy*, the determination of energy eigenvalues and eigenstates of an atom with n electrons.[4] If all spin-dependent interactions are neglected, only

---

[3] The operator algebra of normal ordering, and especially its relation to time ordering, was developed by G. C. Wick. For a lucid description of the use of these concepts and related diagrammatic techniques in the Brueckner-Goldstone theory of nuclear matter see B. D. Day, *Rev. Mod. Phys.* **39**, 719 (1967).

[4] A useful introduction to atomic, molecular, and solid state applications of quantum mechanics is M. Tinkham, *Group Theory and Quantum Mechanics*, McGraw-Hill Book Company, New York, 1964.

electrostatic potentials are effective. In this approximation both the total orbital and the total spin angular momentum commute with the Hamiltonian. As was suggested in Section 17.6, it is practical to require the eigenvectors of $\mathscr{H}_0$, on which the perturbation theory is based, to be also eigenvectors of the total orbital and the total spin angular momentum. A level with quantum numbers L and S is split by the spin-orbit interaction into a multiplet of eigenstates with definite J values ranging from $|L - S|$ to $L + S$. This scheme of building approximate energy eigenstates for an atom is known as *L-S* (or Russell-Saunders) *coupling*.

If $\mathscr{H}_0$ is a central force Hamiltonian for noninteracting particles, the unperturbed eigenstates are characterized by the set of occupation numbers for the one-particle states, or *shells*, or *orbitals* with radial and orbital quantum numbers n_i, l_i. Such a set of occupation numbers is said to define a *configuration*. A particular configuration usually contains many distinct states of the product form

$$\prod a^{\dagger}_{n_i l_i m_{l_i} m_s} \Psi^{(0)}$$

Eigenstates of $\mathscr{H}_0$ that are represented by a product of n creation operators are said to be *independent particle states*.

Although generally knowledge of the atomic configuration and the quantum numbers L and S, M_L and M_S is not sufficient to specify the state of an atom unambiguously, in simple cases, e.g., near closed shells, these specifications may determine the state uniquely. The states of the *helium atom* may be fully classified in this way, and we shall discuss these in some detail.

If the two electrons are in different shells, the states of any two-electron configuration $(n_1 l_1)(n_2 l_2)$ which are simultaneously eigenstates of the total orbital and the total spin angular momentum are, according to the formula (21.2)

$$\Psi^{(2)}_{n_1 l_1 n_2 l_2}(LSM_L M_S) = \sum_{m_1 m_2} \langle l_1 l_2 m_1 m_2 \mid l_1 l_2 L M_L \rangle \sum_{m_1' m_2'} \langle \tfrac{1}{2}\tfrac{1}{2} m_1' m_2' \mid \tfrac{1}{2}\tfrac{1}{2} S M_S \rangle$$

$$\times \, a^{\dagger}_{n_2 l_2 m_2 m_2'} a^{\dagger}_{n_1 l_1 m_1 m_1'} \Psi^{(0)} \tag{21.32}$$

If the two electrons are in the same shell and the configuration is $(nl)^2$, it is legitimate to set $n_1 = n_2 = n$ and $l_1 = l_2 = l$ in (21.32) provided that a normalization factor of $1/\sqrt{2}$ is furnished.

The ground state of the neutral helium atom is described by the configuration $(1s)^2$ and has the spectroscopic character $^1S_0$. In our notation this state may be expressed as

$$\Psi^{(2)}_{1010}(0000) = a^{\dagger}_{100,-1/2} a^{\dagger}_{100,1/2} \Psi^{(0)} \tag{21.33}$$

The configuration of the excited states is $(1s)(nl)$ with $n > 1$. Since the two spins may couple to 0 or 1, the excited states are classified as singlet $(S = 0)$ and triplet $(S = 1)$ states. With the appropriate values for the Clebsch-Gordan coefficients substituted in (21.32), we obtain for the triplet states:

$$\Psi^{\prime(2)}_{10nl}(l\ 1\ m,\ 1) = a^{\dagger}_{nlm,1/2}a^{\dagger}_{100,1/2}\Psi^{\prime(0)}$$

$$\Psi^{\prime(2)}_{10nl}(l\ 1\ m,\ 0) = \frac{1}{\sqrt{2}}(a^{\dagger}_{nlm,-1/2}a^{\dagger}_{100,1/2} + a^{\dagger}_{nlm,1/2}a^{\dagger}_{100,-1/2})\Psi^{\prime(0)} \quad (21.34)$$

$$\Psi^{\prime(2)}_{10nl}(l\ 1\ m,\ -1) = a^{\dagger}_{nlm,-1/2}a^{\dagger}_{100,-1/2}\Psi^{\prime(0)}$$

and for the singlet states:

$$\Psi^{\prime(2)}_{10nl}(l0m0) = \frac{1}{\sqrt{2}}(a^{\dagger}_{nlm,-1/2}a^{\dagger}_{100,1/2} - a^{\dagger}_{nlm,1/2}a^{\dagger}_{100,-1/2})\Psi^{\prime(0)} \quad (21.35)$$

Owing to the anticommutation properties of the creation operators, the triplet states are symmetric under an exchange of the spin quantum numbers of the two particles and antisymmetric under exchange of the set of radial and orbital quantum numbers. The situation is reversed for the singlet states.

The perturbation interaction, arising from the Coulomb repulsion of the electrons, is diagonal with respect to all the unperturbed states which we have constructed, and the first-order corrections to the energy are the expectation values of the interaction in these states. These energies were already worked out in terms of direct and exchange integrals in Section 17.9. We now see that the identity of the electrons, manifested in their statistics, results in a definite correlation between the spatial, or orbital, symmetry and the total spin S of the system. The states of parahelium are singlet states, and the states of orthohelium are triplet states. In complex atoms the connection between S and the spatial symmetry of the state is less simple and not necessarily unique, but S remains instrumental in classifying the orbital symmetry of the states and thus serves as a quantum number on which the energy levels depend, even though the interaction depends only on the position coordinates of the electrons.[5]

5. The Hartree-Fock Method. One of the most useful methods for approximating the ground state of a system of n interacting *fermions* is based on the variational property of the Hamiltonian

$$\mathcal{H} = \sum_{\alpha\alpha'} b^{\dagger}_{\alpha}\langle\alpha|\ H_0\ |\alpha'\rangle b_{\alpha'} + \tfrac{1}{2}\sum_{\alpha\beta\alpha'\beta'} b^{\dagger}_{\alpha}b^{\dagger}_{\beta}\langle\alpha\beta|\ V\ |\alpha'\beta'\rangle b_{\beta'}b_{\alpha'} \quad (21.36)$$

[5] For a compact treatment of the theory of atomic spectra in terms of the second quantization formalism, see B. R. Judd, *Second Quantization and Atomic Spectroscopy*, Johns Hopkins Press, Baltimore, 1967.

The essence of the *Hartree-Fock method* is to seek a new one-particle basis with creation operators $a_k^\dagger$ such that the individual particle state

$$\Psi_v = a_n^\dagger a_{n-1}^\dagger \cdots a_2^\dagger a_1^\dagger \Psi^{(0)} \tag{21.37}$$

renders the expectation value of $\mathscr{H}$ stationary. In this new basis the Hamiltonian appears as

$$\mathscr{H} = \sum_{k,l} a_k^\dagger \langle k | H_0 | l \rangle a_l + \tfrac{1}{2} \sum_{qrst} a_q^\dagger a_r^\dagger \langle qr | V | ts \rangle a_s a_t \tag{21.38}$$

The exact ground state of the Hamiltonian is, of course, usually not as simple as Ψ_v but can be thought of as a linear combination of individual particle states, with the expression (21.37) as the leading term.

The variation to be considered is a basis change, which is a unitary transformation and expressible as

$$a_k^\dagger + \delta a_k^\dagger = \sum_j a_j^\dagger (\delta_{jk} + i \varepsilon_{jk}) \tag{21.39}$$

or

$$\delta a_k^\dagger = i \sum_j a_j^\dagger \varepsilon_{jk}$$

with transformation coefficients ε_{jk} such that

$$|\varepsilon_{jk}| \ll 1$$

The general variation of the state Ψ_v can be built up as a linear combination of independent variations of the form

$$\delta \Psi_{jk} = \varepsilon_{jk} a_j^\dagger a_k \Psi_v \tag{21.40}$$

where a_k must annihilate a fermion in one of the *occupied* one-particle states $1 \cdots n$, and $a_i^\dagger$ must create a particle in one of the previously *unoccupied* one-particle states $n+1, \ldots$. The unitarity of the transformation coefficients in (21.39) requires only that the ε_{jk} form a Hermitian matrix. Since the variation $\delta \Psi_{kj}$ vanishes owing to the exclusion principle, the condition $\varepsilon_{kj} = \varepsilon_{jk}{}^*$ implies no restriction on the permissible variations, and the independence of the ε-variations is assured. (Variations with $j = k$ do not change the state and are therefore irrelevant.)

We may thus confine our attention to variations $\delta \Psi$ of the form (21.40) which are orthogonal to the "best" state Ψ_v of the structure (21.37). The variational theorem,

$$\delta \langle \mathscr{H} \rangle = 0$$

in conjunction with the Hermitian property of $\mathscr{H}$, requires that

$$(\Psi_v, \Psi_v)(\delta \Psi, \mathscr{H} \Psi_v) - (\Psi_v, \mathscr{H} \Psi_v)(\delta \Psi, \Psi_v) = 0$$

The orthogonality of $\delta \Psi$ and Ψ_v guarantees that the variation preserves the normalization of the state and, according to the last equation, makes it necessary that $\delta\Psi$ also be orthogonal to $\mathscr{H}\Psi_v$. Hence, the variational condition is (*Brillouin's* theorem)

$$(\delta\Psi, \mathscr{H}\Psi_v) = 0 \qquad (21.41)$$

If the Hamiltonian (21.38) and the variation (21.40) are substituted into this condition, we obtain

$$(a_j^\dagger a_k \Psi_v, \sum_{n,l} a_n^\dagger a_l \langle n| H_0 |l\rangle \Psi_v) + (a_j^\dagger a_k \Psi_v, \tfrac{1}{2} \sum_{qrst} a_q^\dagger a_r^\dagger a_s a_t \langle qr| V |ts\rangle \Psi_v) = 0$$

This relation is easily seen to be equivalent to the equation

$$\langle j| H_0 |k\rangle + \sum_{t=1}^{n} [\langle jt| V |kt\rangle - \langle jt| V |tk\rangle] = 0 \qquad (21.42)$$

The sum over t is to be taken only over the occupied one-particle states; j denotes any unoccupied, k any occupied one-particle state.

If in the original one-particle basis $b_\alpha^\dagger$ the interaction between the fermions is diagonal,

$$\langle \alpha\beta| V |\alpha'\beta'\rangle = V_{\alpha\beta}\, \delta_{\alpha\alpha'}\, \delta_{\beta\beta'} \qquad (21.43)$$

condition (21.42) can be construed as expressing the orthogonality of the eigenkets of an *effective one-particle Hamiltonian*, H_{HF}, corresponding to the eigenvalue problem

$$H_{HF} |k\rangle \equiv \left[H_0 + \sum_{t=1}^{n} \sum_{\alpha\beta} (|\alpha\rangle\langle t | \beta\rangle V_{\alpha\beta}\langle\beta | t\rangle\langle\alpha| \right.$$

$$\left. - |\alpha\rangle\langle t | \beta\rangle V_{\alpha\beta}\langle\alpha | t\rangle\langle\beta|) \right] |k\rangle = \varepsilon_k |k\rangle \quad (21.44)$$

The summation over the Greek indices extends over the complete set of one-particle states. Equations (21.44) are known as the *Hartree-Fock equations*. From them we immediately infer that

$$\langle k| H_0 |k\rangle + \sum_{t=1}^{n} [\langle kt| V |kt\rangle - \langle kt| V |tk\rangle] = \varepsilon_k \qquad (21.45)$$

Exercise 21.9. Verify that H_{HF}, as defined by (21.44) is Hermitian.

The occupied states $|t\rangle$ in equations (21.44) and (21.45) are not at our discretion; they must be chosen from among the eigenkets of (21.44) in a manner which will minimize the expectation value of the Hamiltonian. Frequently, the best choice corresponds to the use of those eigenkets which belong to the n lowest eigenvalues ε_k, although, perhaps contrary to expectations, the variationally minimal value of $\langle \mathcal{H} \rangle$ is not just the sum of the Hartree-Fock one-particle energies ε_k. Rather, the Hartree-Fock approximation E_v to the ground state energy is

$$
E_v = \langle \mathcal{H} \rangle = (\Psi_v, \mathcal{H}\Psi_v) = \sum_{k=1}^{n} \langle k| H_0 |k \rangle + \tfrac{1}{2} \sum_{i,k=1}^{n} [\langle ik| V |ik \rangle - \langle ik| V |ki \rangle]
$$

$$
= \tfrac{1}{2} \sum_{k=1}^{n} [\varepsilon_k + \langle k| H_0 |k \rangle] = \sum_{k=1}^{n} \varepsilon_k - \tfrac{1}{2} \sum_{i,k=1}^{n} [\langle ik| V |ik \rangle - \langle ik| V |ki \rangle] \quad (21.46)
$$

The state $a_j^{\dagger} a_k \Psi_v$ (with k and j labeling occupied and unoccupied single-particle states, respectively), is orthogonal to the approximate ground state Ψ_v and may thus be regarded as an approximation to an excited state of the system. In this state we have

$$
\langle \mathcal{H} \rangle = (a_j^{\dagger} a_k \Psi_v, \mathcal{H} a_j^{\dagger} a_k \Psi_v) = E_v + \varepsilon_j - \varepsilon_k - \langle jk| V |jk \rangle + \langle jk| V |kj \rangle
$$
$$
(21.47)
$$

If the last two terms can be neglected, $\varepsilon_j - \varepsilon_k$ represents an excitation energy of the system.

Exercise 21.10. Verify equation (21.47).

Exercise 21.11. Prove that the expectation value of $\mathcal{H}$ in the "ionized" state $a_k \Psi_v$ is

$$
\langle \mathcal{H} \rangle = E_v - \varepsilon_k \qquad (Koopmans' \text{ theorem}) \qquad (21.48)
$$

The practical task of solving the Hartree-Fock equations is far from being straightforward. The equations have the appearance of a common eigenvalue problem, but the matrix elements of the interaction V, which enter the construction of the effective one-particle Hamiltonian H_{HF}, cannot be computed without a foreknowledge of the appropriate n eigensolutions $|t\rangle$ of equations (21.44). The coupled equations (21.44) are thus thoroughly nonlinear and require an iteration technique for their solution. One starts out by guessing a set of occupied one-particle states $|t\rangle$; using these, one calculates the matrix elements of V; and one then solves the Hartree-Fock equations (21.44). If the initial guess was fortuitously good, n of the eigensolutions of (21.44) will be similar to the initially chosen kets. If, as is more likely, the eigensolutions of the Hartree-Fock equations do not reproduce the starting kets, the eigensolutions corresponding to the lowest n eigenvalues ε_k are used to recalculate the matrix elements of V, and this procedure is repeated until a self-consistent set of solutions is obtained. Sufficiently good initial guesses of the one-particle states are usually available, so that fairly rapid convergence of the iteration process is the rule rather than the exception in actual practice.

The Hartree-Fock equations can, in the representation which diagonalizes V, be rewritten as

$$\sum_{\beta} [\langle \alpha | H_0 | \beta \rangle \langle \beta | k \rangle + \sum_{t=1}^{n} \langle t | \beta \rangle V_{\alpha\beta} \langle \beta | t \rangle \langle \alpha | k \rangle$$

$$- \langle t | \beta \rangle V_{\alpha\beta} \langle \alpha | t \rangle \langle \beta | k \rangle] = \varepsilon_k \langle \alpha | k \rangle \quad (21.49)$$

As an application of these equations, we consider an atom with a nuclear charge Ze and with n electrons. Then

$$H_0 = \frac{\mathbf{p}^2}{2\mu} - \frac{Ze^2}{r}$$

The interaction V is diagonal in the coordinate representation and has the form

$$V(\mathbf{r}\sigma, \mathbf{r}'\sigma') = \frac{e^2}{|\mathbf{r} - \mathbf{r}'|}$$

We choose the coordinate representation with spin as the basis $|\alpha\rangle$ and $|\beta\rangle$, and we denote the Hartree-Fock eigenfunctions as

$$\langle \mathbf{r}\sigma | k \rangle = \psi_k(\mathbf{r}\sigma)$$

The Hartree-Fock equations (21.49) are then immediately transcribed into the form

$$-\frac{\hbar^2}{2\mu} \nabla^2 \psi_k(\mathbf{r}\sigma) - \frac{Ze^2}{r} \psi_k(\mathbf{r}\sigma) + e^2 \sum_{t=1}^{n} \sum_{\sigma'} \int \psi_t^*(\mathbf{r}'\sigma') \frac{1}{|\mathbf{r} - \mathbf{r}'|} \psi_t(\mathbf{r}'\sigma') \psi_k(\mathbf{r}\sigma) \, d^3r'$$

$$-e^2 \sum_{t=1}^{n} \sum_{\sigma'} \int \psi_t^*(\mathbf{r}'\sigma') \frac{1}{|\mathbf{r} - \mathbf{r}'|} \psi_t(\mathbf{r}\sigma) \psi_k(\mathbf{r}'\sigma') \, d^3r' = \varepsilon_k \psi_k(\mathbf{r}\sigma) \quad (21.50)$$

These coupled nonlinear differential-integral equations constitute the most familiar realization of the Hartree-Fock theory. If the last sum on the left-hand side, due to the exchange matrix elements of the interaction, is neglected and the term corresponding to $t = k$ in the sum over the direct matrix elements is omitted, the equations (21.50) reduce to the simpler *Hartree equations*, which before the advent of fast computers were of greater practical interest than the more accurate Hartree-Fock equations.

Exercise 21.12. Show that the configuration space wave function corresponding to the individual particle state (21.37) can be expressed as the *Slater* *determinant*

$$\psi(\mathbf{r}_1\sigma_1, \ldots, \mathbf{r}_n\sigma_n) = \frac{1}{\sqrt{n!}} \begin{vmatrix} \psi_1(\mathbf{r}_1\sigma_1)\psi_1(\mathbf{r}_2\sigma_2) \cdots \psi_1(\mathbf{r}_n\sigma_n) \\ \psi_2(\mathbf{r}_1\sigma_1)\psi_2(\mathbf{r}_2\sigma_2) \cdots \psi_2(\mathbf{r}_n\sigma_n) \\ \cdot \quad \cdot \quad \cdots \cdot \\ \cdot \quad \cdot \quad \cdots \cdot \\ \cdot \quad \cdot \quad \cdots \cdot \\ \psi_n(\mathbf{r}_1\sigma_1)\psi_n(\mathbf{r}_2\sigma_2) \cdots \psi_n(\mathbf{r}_n\sigma_n) \end{vmatrix}$$

6. Pairing Interactions. To illustrate further the utility of the operator formalism in the treatment of many-body problems, we now consider a system of identical fermions whose one-particle energy eigenstates may be grouped in pairs by use of suitably chosen quantum numbers and which are subject to a peculiar pairing interaction rather than the most general possible two-particle interaction. In the present context, a *pair* of fermions is defined as two particles occupying two paired one-particle states. These are conveniently labeled by the quantum numbers m and $-m$. An interaction between the members of a pair of the form

$$\mathscr{V} = \tfrac{1}{2} \sum_{q,m} a_q^\dagger a_{-q}^\dagger a_{-m} a_m \langle q, -q | V | m, -m \rangle \tag{21.51}$$

is said to be a *pairing interaction*. For example, the effective Hamiltonian for electrons in a solid appears to contain a term like (21.51), provided that an electron (or *Cooper*) *pair* is defined by two electrons which have equal energies but opposite linear momenta, $\mathbf{p}$ and $-\mathbf{p}$, and opposite spins. In complex nuclei, effective pairing occurs between nucleons in the same subshell but with opposite values of the component of angular momentum along the nuclear symmetry axis.

Generally, in the absence of the interaction, the one-particle states m and $-m$, which constitute a pair, are degenerate in energy, denoted as $\varepsilon_m' = \varepsilon_{-m}'$, and we shall assume this to be the case. Our assumptions make it possible and convenient to restrict all summations over the one-particle quantum numbers to positive values only and to write the Hamiltonian of the system of fermions as

$$\mathscr{H} = \sum_{m>0} \varepsilon_m'(a_m^\dagger a_m + a_{-m}^\dagger a_{-m}) + \sum_{q,m>0} a_q^\dagger a_{-q}^\dagger a_{-m} a_m$$

$$\times [\langle q, -q | V | m, -m \rangle - \langle q, -q | V | -m, m \rangle] \tag{21.52}$$

Of primary interest here is a method for dealing with this system which is capable of producing nonperturbative solutions of the eigenvalue problem for $\mathscr{H}$. The particular states to be considered are superpositions with appreciable amplitudes for a large number of pair configurations and differ significantly from perturbative solutions, in which one unperturbed configuration usually dominates. The so-called BCS states,[6] which are fundamental to an understanding of superconductivity in solids, are such collective states produced by pairing; in nuclei, a pairing interaction is responsible for a conspicuous gap in the energy spectrum of the system near its ground state.

The basic ideas of the theory are preserved, but the mathematics is much simplified if we assume that the pairing interaction in (21.52) is either equal to a constant $-G$ or zero, depending on whether q and m belong to a specified subset S of the quantum numbers or not. In the case of superconductivity, for instance, the electron pairing force is effective only for electrons near the Fermi energy. We thus assume that the Hamiltonian has the form

$$\mathscr{H} = \sum_{m>0} \varepsilon_m{}'(a_m^\dagger a_m + a_{-m}^\dagger a_{-m}) - G \sum_{q,m \in S} a_q^\dagger a_{-q}^\dagger a_{-m} a_m \qquad (21.53)$$

As usual in many-body theories, we now seek a transformation to a new set of creation and annihilation operators that will make the Hamiltonian correspond as nearly as possible to a system of noninteracting particles or quasiparticles. Here this objective is approached by defining for each value of m the new annihilation operators

$$\begin{aligned} \alpha_m &= u_m a_m - v_m a_{-m}^\dagger \\ \beta_m &= u_m a_{-m} + v_m a_m^\dagger \end{aligned} \qquad (21.54)$$

u_m and v_m are numbers which we choose to be real. If we further require the normalization

$$u_m{}^2 + v_m{}^2 = 1 \qquad (21.55)$$

but otherwise leave u_m and v_m temporarily unspecified, the fermion anticommutation relations for $a_m^\dagger$ and a_m can be easily seen to imply that

$$\begin{aligned} \alpha_m \alpha_q + \alpha_q \alpha_m &= \alpha_m^\dagger \alpha_q^\dagger + \alpha_q^\dagger \alpha_m^\dagger \\ &= \beta_m \beta_q + \beta_q \beta_m = \beta_m^\dagger \beta_q^\dagger + \beta_q^\dagger \beta_m^\dagger = 0 \\ \alpha_m \alpha_q^\dagger + \alpha_q^\dagger \alpha_m &= \beta_m \beta_q^\dagger + \beta_q^\dagger \beta_m = \delta_{qm} I \end{aligned} \qquad (21.56)$$

Exercise 21.13. Verify the anticommutation relations (21.56).

The canonical transformation defined by equations (21.54) is quite different in kind from a transformation like (20.12) and (20.13) of the one-particle basis, as applied to spin $\frac{1}{2}$ bosons in discussing the dynamics of

---

[6] The symbols BCS stand for Bardeen, Cooper, and Schrieffer, who recognized the role of the pairing interaction in producing the ground state of a superconductor.

magnons and to electrons in the Hartree-Fock theory. In the present trans-formation the new creation operators are linear combinations of the old creation *and* annihilation operators, and the number of particles is not conserved. We shall refer to the fermions which the operators $\alpha_m^\dagger$ and $\beta_m^\dagger$ create as *quasiparticles* associated with the given pairing interaction.[7] From (21.54) and (21.55) we derive the inverse transformation equations:

$$a_m = u_m \alpha_m + v_m \beta_m^\dagger$$
$$a_{-m} = u_m \beta_m - v_m \alpha_m^\dagger \tag{21.57}$$

The "vacuum" state Ψ_0 with no *quasiparticles*, which is defined by

$$\alpha_m \Psi_0 = \beta_m \Psi_0 = 0 \tag{21.58}$$

is obviously not identical with the vacuum state $\Psi'^{(0)}$ of the *particles*. It is easy to see that the two vacuum states are related by the equations

$$\Psi_0 = \prod_m (u_m + v_m a_m^\dagger a_{-m}^\dagger)\Psi'^{(0)}, \qquad \Psi'^{(0)} = \prod_m (u_m + v_m \beta_m^\dagger \alpha_m^\dagger)\Psi_0 \tag{21.59}$$

Exercise 21.14. Prove the relations between the vacuum state $\Psi'^{(0)}$ and the no-quasiparticle state Ψ_0.

The operator that measures the number of particles in the two paired one-particle states m, $-m$ can easily be transformed into the quasiparticle representation. With normal ordering of the operators,

$$a_m^\dagger a_m + a_{-m}^\dagger a_{-m}$$
$$= 2v_m^2 + (u_m^2 - v_m^2)(\alpha_m^\dagger \alpha_m + \beta_m^\dagger \beta_m) + 2u_m v_m(\alpha_m^\dagger \beta_m^\dagger + \beta_m \alpha_m) \tag{21.60}$$

In our effort to transform the Hamiltonian (21.53) into the quasiparticle representation, we also require the expansion of the pairing interaction in terms of the quasiparticle operators:

$$a_q^\dagger a_{-q}^\dagger a_{-m} a_m = (u_q \alpha_q^\dagger + v_q \beta_q)(u_q \beta_q^\dagger - v_q \alpha_q)(u_m \beta_m - v_m \alpha_m^\dagger)(u_m \alpha_m + v_m \beta_m^\dagger)$$

If we arrange the expansion in normal product form and *neglect* all terms

---

[7] In honor of N. N. Bogoliubov, who invented the transformation described in this section, this particular species of quasiparticle was christened *bogolon* in a lighthearted introduction to many-body theories, R. D. Mattuck, *A Guide to Feynman Diagrams in the Many-Body Problem*, McGraw-Hill Book Company, New York, 1967. For a more advanced treatment of many-body problems, see A. A. Abrikosov, L. P. Gorkov, and I. E. Dzyalo-shinski, *Methods of Quantum Field Theory in Statistical Physics*, translated by R. A. Silverman, Prentice-Hall, Inc., Englewood Cliffs, 1963.

which contain more than two quasiparticle operators, we arrive eventually at the result:

$$\sum_{q,m\in S} a_q^\dagger a_{-q}^\dagger a_{-m} a_m \simeq \sum_{q,m\in S} u_q v_q u_m v_m + \sum_{m\in S} v_m^4 +$$

$$+ \sum_{m\in S} \left[(u_m^2 - v_m^2) v_m^2 - 2 u_m v_m \sum_{q\in S} u_q v_q \right] (\alpha_m^\dagger \alpha_m + \beta_m^\dagger \beta_m)$$

$$+ \sum_{m\in S} \left[(u_m^2 - v_m^2) \sum_{q\in S} u_q v_q + 2 u_m v_m^3 \right] (\alpha_m^\dagger \beta_m^\dagger + \beta_m \alpha_m)$$

The quasiparticle Hamiltonian $\mathscr{H}_{app}$ which we obtain in this approximation does not commute with N, the operator corresponding to the total number of particles. In seeking the eigenstates of the approximate Hamiltonian it is, therefore, not possible to work in a subspace of Hilbert space corresponding to a fixed number of particles. When the full Hilbert space is available, the lowest energy state is the particle vacuum $\Psi^{(0)}$. Since we are not interested in this trivial solution, it is necessary to impose on the variational problem of $\mathscr{H}_{app}$ the constraint

$$\langle N \rangle \equiv (\Psi, N\Psi) = n \qquad (21.61)$$

which insures that at least the mean value of N equals the number of fermions present in the system. We shall take this constraint into account by solving the eigenvalue problem for the operator $\mathscr{H} - \lambda N$ and determining the Lagrangian multiplier λ subsequently from condition (21.61).

In the approximation described above which neglects all interactions between quasiparticles, we obtain

$$\mathscr{H}_{app} - \lambda N = 2 \sum_m (\varepsilon_m - \lambda) v_m^2 - G \left(\sum_{m\in S} u_m v_m \right)^2 + G \sum_{m\in S} v_m^4$$

$$+ \sum_m (\varepsilon_m - \lambda)(u_m^2 - v_m^2)(\alpha_m^\dagger \alpha_m + \beta_m^\dagger \beta_m)$$

$$+ 2G \sum_{q\in S} u_q v_q \sum_{m\in S} u_m v_m (\alpha_m^\dagger \alpha_m + \beta_m^\dagger \beta_m)$$

$$+ 2 \sum_m (\varepsilon_m - \lambda) u_m v_m (\alpha_m^\dagger \beta_m^\dagger + \beta_m \alpha_m)$$

$$- G \sum_{q\in S} u_q v_q \sum_{m\in S} (u_m^2 - v_m^2)(\alpha_m^\dagger \beta_m^\dagger + \beta_m \alpha_m) \qquad (21.62)$$

where

$$\varepsilon_m = \varepsilon_m' \qquad \text{if } m \notin S$$

and $\qquad (21.63)$

$$\varepsilon_m = \varepsilon_m' - G v_m^2 \qquad \text{if } m \in S$$

The eigenvalue problem posed by the operator (21.62) can be solved exactly and simply by a suitable disposition with regard to the choice of values for the as yet undetermined coefficients u_m and v_m. These are selected in such a manner that the terms containing $\alpha^\dagger\beta^\dagger$ and $\beta\alpha$ vanish identically:

$$u_m v_m = 0 \qquad \text{if} \qquad m \notin S \tag{21.64}$$

$$u_m{}^2 - v_m{}^2 = \frac{2u_m v_m(\varepsilon_m - \lambda)}{G \sum\limits_{q \in S} u_q v_q} \qquad \text{if} \qquad m \in S \tag{21.65}$$

It is convenient to introduce the abbreviation

$$\Delta = G \sum_m u_m v_m \tag{21.66}$$

From equations (21.55) and (21.65) we obtain[8]

$$2u_m v_m = \frac{\Delta}{\sqrt{(\varepsilon_m - \lambda)^2 + \Delta^2}}, \qquad m \in S \tag{21.67}$$

and

$$u_m{}^2 - v_m{}^2 = \frac{\varepsilon_m - \lambda}{\sqrt{(\varepsilon_m - \lambda)^2 + \Delta^2}}, \qquad m \in S \tag{21.68}$$

The equations for u_m and v_m are thus solved by the choice

$$u_m{}^2 = \frac{1}{2}\left(1 + \frac{\varepsilon_m - \lambda}{E_m}\right) \tag{21.69}$$

$$v_m{}^2 = \frac{1}{2}\left(1 - \frac{\varepsilon_m - \lambda}{E_m}\right) \tag{21.70}$$

where

$$E_m = |\varepsilon_m - \lambda| \qquad\qquad \text{if } m \notin S \tag{21.71}$$

and

$$E_m = \sqrt{(\varepsilon_m - \lambda)^2 + \Delta^2} \qquad \text{if } m \in S \tag{21.72}$$

With this notation the effective Hamiltonian can now be written as

$$\mathscr{H}_{app} - \lambda N = 2 \sum_m (\varepsilon_m - \lambda)v_m{}^2 - \frac{\Delta^2}{G} + G \sum_{m \in S} v_m{}^4 + \sum_m E_m(\alpha_m^\dagger \alpha_m + \beta_m^\dagger \beta_m) \tag{21.73}$$

The quantity Δ is seen to be determined from equations (21.66) and (21.67) by the condition

$$\frac{G}{2} \sum_{m \in S} \frac{\Delta}{\sqrt{(\varepsilon_m - \lambda)^2 + \Delta^2}} = \Delta \tag{21.74}$$

---

[8] If we replace u_m by v_m and v_m by $-u_m$, a second solution is obtained, but this does not correspond to a minimum of the energy.

A trivial solution of these equations is given by $\Delta = 0$ and

$$u_m^{(0)} = 1, \qquad v_m^{(0)} = 0 \qquad \text{if} \qquad \varepsilon_m > \lambda_0$$
$$u_m^{(0)} = 0, \qquad v_m^{(0)} = 1 \qquad \text{if} \qquad \varepsilon_m < \lambda_0$$

The no-quasiparticle state which this solution defines is

$$\Psi_0 = \prod_{\varepsilon_m < \lambda_0} a_m^\dagger a_{-m}^\dagger \Psi^{(0)} \tag{21.75}$$

If the pairing interaction is weak, $\varepsilon_m \approx \varepsilon_m{}'$, and (21.75) is the ground state of the unperturbed Hamiltonian

$$\mathcal{H}_0 = \sum_m \varepsilon_m{}'(a_m^\dagger a_m + a_{-m}^\dagger a_{-m})$$

In this state the one-particle energy levels are filled in pairs by n particles from the lowest energy level up to the *Fermi level* λ_0, and the expectation value of the energy is

$$E_0 = (\Psi_0, \mathcal{H}\Psi_0) = 2 \sum_{\varepsilon_m < \lambda_0} \varepsilon_m + 0(G) \tag{21.76}$$

The term $0(G)$ accounts for the first-order effect of the pairing interaction beyond the corrections which are already included in ε_m.

Excited states are obtained by creating one or more quasiparticles. The state

$$\alpha_q^{(0)\dagger}\Psi_0 = (u_q^{(0)}a_q^\dagger - v_q^{(0)}a_{-q})\Psi_0 \tag{21.77}$$

differs from Ψ_0, which represents the simply filled Fermi sea, by the addition of a particle if $\varepsilon_q > \lambda_0$, and the annihilation of a particle, or the creation of a hole, if $\varepsilon_q < \lambda_0$. In either case, an amount $|\varepsilon_q - \lambda_0|$ is added to the energy E_0.

The quasiparticle approximation defined by $\Delta = 0$ is thus equivalent to an ordinary perturbation procedure based on the noninteracting particles. The states obtained in this manner are called the *normal* states of the fermion system.

When G is different from zero and positive, the equations for u_m and v_m admit in addition of a qualitatively different solution, with $\Delta \neq 0$ determined by the condition (21.74), or

$$\frac{G}{2} \sum_{m \in S} \frac{1}{\sqrt{(\varepsilon_m - \lambda)^2 + \Delta^2}} = 1 \tag{21.78}$$

The "chemical potential" λ, which is the analog of the Fermi energy, is defined by equation (21.61) which, for the new quasiparticle vacuum state, reduces to

$$2 \sum_m v_m{}^2 = n \tag{21.79}$$

or

$$\sum_m \left(1 - \frac{\varepsilon_m - \lambda}{E_m}\right) = n \qquad (21.80)$$

It should be noted that equation (21.78) has a solution only if

$$\frac{G}{2} \sum_{m \in S} \frac{1}{|\varepsilon_m - \lambda|} > 1 \qquad (21.81)$$

The lowest energy state for $\Delta \neq 0$ is the state with no quasiparticles defined by (21.69) and (21.70). Great interest attaches to this particular state, denoted by

$$\Psi_{BCS} = \prod_m (u_m + v_m a_m^\dagger a_{-m}^\dagger) \Psi^{(0)} \qquad (21.82)$$

if its energy expectation value E_{BCS} lies below E_0, the energy of the lowest normal state. We find from (21.73) and (21.79),

$$E_{BCS} = 2 \sum_m \varepsilon_m v_m^2 - \frac{\Delta^2}{G} + G \sum_{m \in S} v_m^4 \qquad (21.83)$$

In applications to extended systems the last term in this expression may usually be neglected; also, in such systems it is a good approximation to make the identification $\lambda \approx \lambda_0$. Under these assumptions the *BCS* state differs from the lowest normal state by

$$\Delta E = E_{BCS} - E_0 = 2 \sum_{\varepsilon_m > \lambda_0} (\varepsilon_m - \lambda_0) v_m^2 + 2 \sum_{\varepsilon_m < \lambda_0} (\lambda_0 - \varepsilon_m) u_m^2 - \frac{\Delta^2}{G}$$

At first sight it may be confusing that in the description given in this section both the state Ψ_0 of (21.75) and the state Ψ_{BCS} of (21.82) are no-quasiparticle states, or quasiparticle vacuum states. This paradox is resolved when it is noted that the quasiparticles defined by $\Delta = 0$ are quite different from those defined by $\Delta \neq 0$.

Exercise 21.15. Show that the solution (21.69), and (21.70) is recovered if the expression (21.83) is minimized with respect to u_m and v_m, subject to the constraint $u_m^2 + v_m^2 = 1$.

The physical significance of the coefficients u_m and v_m is readily appreciated by considering the *BCS* ground state (21.82). The quantity v_m^2 measures the probability that a pair of particles occupies the one-particle levels m and $-m$. As Figure 21.1 shows, v_m^2 drops from the value 1, indicating certain occupancy of levels deep within the Fermi sea, to zero, high above the Fermi surface. The transition takes place in an energy interval of the

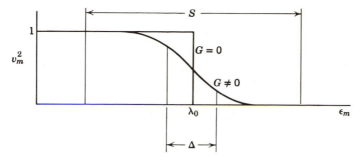

Figure 21.1

order of Δ, if the pairing interaction region S extends over a broad band of one-particle energy levels surrounding the Fermi level.

As equation (21.73) shows, the excited states are characterized by the creation of one or more quasiparticles, and the energy contributed by each quasiparticle is E_m. The minimum quasiparticle energy, corresponding to $\varepsilon_m \simeq \lambda_0$, is Δ, which is thus the *energy gap* between the quasiparticle vacuum and states with one quasiparticle.

In order to evaluate the energy difference between the *BCS* and the normal state for a case of practical interest, we assume that the energy range S, over which pairing interactions are effective, extends symmetrically about the Fermi energy λ_0, so that $G \neq 0$ if $\lambda_0 + \hbar\omega > \varepsilon_m > \lambda_0 - \hbar\omega$, and $G = 0$ otherwise. If we further assume the single particle energy levels to be dense enough and the range $2\hbar\omega$ to be small enough compared to λ_0 so that the sums can be replaced by integrals with a constant density of energy levels, we have

$$
\begin{aligned}
\Delta E &= 2\rho(\lambda_0)\int_0^{\hbar\omega} \varepsilon\left(1 - \frac{\varepsilon}{\sqrt{\varepsilon^2 + \Delta^2}}\right) d\varepsilon - \frac{\Delta^2}{G} \\
&= \rho(\lambda_0)(\hbar\omega)^2\left[1 - \sqrt{1 + \left(\frac{\Delta}{\hbar\omega}\right)^2}\right] + \rho(\lambda_0)\,\Delta^2\,\text{arcsinh}\,\frac{\hbar\omega}{\Delta} - \frac{\Delta^2}{G} \quad (21.84)
\end{aligned}
$$

where $\rho(\lambda_0)$ is the density of energy levels at the Fermi surface, counting each degenerate level of pair states only once. On the other hand, Δ is determined from equation (21.78) in the same approximation as

$$
1 = G\int_0^{\hbar\omega} \frac{\rho(\varepsilon + \lambda_0)\,d\varepsilon}{\sqrt{\varepsilon^2 + \Delta^2}} \simeq G\rho(\lambda_0)\int_0^{\hbar\omega} \frac{d\varepsilon}{\sqrt{\varepsilon^2 + \Delta^2}}
$$

or

$$
\frac{1}{\rho(\lambda_0)G} = \text{arcsinh}\,\frac{\hbar\omega}{\Delta}
$$

or

$$\Delta = \hbar\omega / \sinh \frac{1}{\rho(\lambda_0)G} \tag{21.85}$$

Substituting this result into (21.84), we find that the last two terms cancel and that the remainder reduces to

$$\Delta E = - \frac{2\rho(\lambda_0)(\hbar\omega)^2}{e^{2/\rho(\lambda_0)G} - 1} \tag{21.86}$$

The analysis carried out here is tailored to an undertanding of *super-conductivity*.[9] The Hamiltonian then describes the electrons in the solid and the attractive $(G > 0)$ pairing interaction between electrons of opposite momenta and opposite spins, resulting from the basic electron-phonon interaction through the elimination of virtual phonons. The range over which the pairing interaction is effective is defined by $\hbar\omega \approx k\theta$ (*Debye* temperature). The strength G of the interaction is inversely proportional to the volume of the solid; hence, $\rho(\lambda_0)G$ is independent of the volume. The energy gap (21.85) is therefore also independent of the volume. Since the interaction is partly compensated by the Coulomb repulsion of the electrons, G is weak, so that $\rho(\lambda_0)G \ll 1$ and

$$\Delta \approx 2\hbar\omega e^{-1/\rho(\lambda_0)G}$$

showing that the gap is much smaller than $\hbar\omega$. The energy gap and the distinctive, isolated *BCS* ground state, which the gap separates from all other states of the electron system, provide a basis for an explanation of many features of superconductivity.

Exercise 21.16. Discuss the limit of strong pairing interactions.

Exercise 21.17. Evaluate the variance of the total number of particles for the *BCS* ground state in terms of u_m and v_m.

7. *Elements of Quantum Statistics.* The operator formalism of Chapter 20 is ideally suited for treating large systems of identical particles in thermal equilibrium. If $\bar{\rho}$ denotes the average *density* or *statistical* operator for a grand canonical ensemble with fixed values for the averages of $\mathcal{H}$ and N, statistical thermodynamics requires that the entropy, defined as

$$S = -k \,\text{trace}\, (\bar{\rho} \log \bar{\rho}) \tag{21.87}$$

---

[9] Many important papers are included as reprints in D. Pines, *The Many-Body Problem*, W. A. Benjamin, New York, 1961.

be made a maximum subject to the constraints

$$\langle \bar{N} \rangle = \text{trace } (\bar{\rho} N) = n, \qquad \langle \overline{\mathscr{H}} \rangle = \text{trace } (\bar{\rho} \mathscr{H}) = E, \qquad \text{trace } \bar{\rho} = 1$$

in generalization of the formulas of Section 13.4. Using the Lagrangian multipliers α and β, and the variational principle

$$\delta(S - k\alpha \langle \bar{N} \rangle - k\beta \langle \overline{\mathscr{H}} \rangle) = 0$$

it is easy to derive the central formula

$$\bar{\rho} = e^{-\alpha N - \beta \mathscr{H}} / Z \qquad (21.88)$$

where

$$Z = \text{trace } e^{-\alpha N - \beta \mathscr{H}} \qquad (21.89)$$

is the grand canonical partition function.

For a system of noninteracting identical particles with one-particle energies ε_i, known as a *perfect gas* in thermodynamic parlance,

$$\mathscr{H} = \sum_i \varepsilon_i a_i^\dagger a_i = \sum_i \varepsilon_i N_i \qquad (21.90)$$

The quantum statistical average of any physical quantity Q may be computed by application of the formula

$$\langle \bar{Q} \rangle = \text{trace } \bar{\rho} Q \qquad (21.91)$$

We apply this relation to the evaluation of the average occupation numbers N_i:

$$\langle \bar{N}_i \rangle = \langle a_i^\dagger a_i \rangle = \text{trace } (e^{-\alpha N - \beta \mathscr{H}} a_i^\dagger a_i) / Z$$

Using equations (20.27), (20.28), and (20.29), and the identity (8.106), we find that

$$\text{trace } (e^{-\alpha N - \beta \mathscr{H}} a_i^\dagger a_i) = e^{-(\alpha + \beta \varepsilon_i)} \text{trace } (e^{-\alpha N - \beta \mathscr{H}} a_i a_i^\dagger) \qquad (21.92)$$

Exercise 21.18. Verify the equation (21.92).

If the commutation relations for bosons or fermions are used, we obtain (upper sign for bosons, lower sign for fermions)

$$\text{trace } (e^{-\alpha N - \beta \mathscr{H}} a_i^\dagger a_i) = e^{-(\alpha + \beta \varepsilon_i)} \text{trace } [e^{-\alpha N - \beta \mathscr{H}} (1 \mp a_i^\dagger a_i)]$$

and hence,

$$\langle \bar{N}_i \rangle = \langle a_i^\dagger a_i \rangle = (e^{\alpha + \beta \varepsilon_i} \mp 1)^{-1} \qquad (21.93)$$

which is the familiar formula for the distributions of particles in Bose-Einstein and Fermi-Dirac statistics, respectively. The derivation given here is intended to exhibit as plainly as possible the connection between the commutation or anticommutation relations for the creation and annihilation operators and

-1 or $+1$ which characteristically appears in the denominator of the distribution law.

Exercise 21.19. Use a similar technique to show that the square of the fractional deviation from the mean occupation number is

$$\frac{\overline{\langle(N_i - \langle \overline{N}_i \rangle)^2\rangle}}{\langle \overline{N}_i \rangle^2} = \frac{\overline{\langle N_i^2 \rangle} - \langle \overline{N}_i \rangle^2}{\langle \overline{N}_i \rangle^2} = e^{\alpha + \beta \varepsilon_i} = \frac{1}{\langle \overline{N}_i \rangle} \pm 1$$

Problems

1. Show that if $V(r)$ is a two-particle interaction which depends only on the distance r between the particles, the matrix element of the interaction in the **k**-representation may be written as

$$\langle \mathbf{k}_3 \mathbf{k}_4 | \, V \, | \mathbf{k}_1 \mathbf{k}_2 \rangle = \delta(\mathbf{k}_1 + \mathbf{k}_2 - \mathbf{k}_3 - \mathbf{k}_4) \frac{1}{(2\pi)^3} \int V(r) e^{-i\mathbf{q} \cdot \mathbf{r}} \, d^3r$$

where $\hbar \mathbf{q}$ is the momentum transfer $\hbar(\mathbf{k}_3 - \mathbf{k}_1)$.

   Work out the matrix elements for the screened Coulomb potential $V_0 e^{-\alpha r}/(\alpha r)$, and construct the corresponding two-particle interaction operator $\mathscr{V}$ for identical particles in terms of the creation and annihilation operators in **k**-space. Exhibit the direct ($\langle \mathbf{k}_1 \mathbf{k}_2 | \, V \, | \mathbf{k}_1 \mathbf{k}_2 \rangle$) and exchange ($\langle \mathbf{k}_2 \mathbf{k}_1 | \, V \, | \mathbf{k}_1 \mathbf{k}_2 \rangle$) terms in the interaction.

2. Consider a system of identical bosons with only two one-particle basis states, $a^\dagger_{1/2} \Psi^{(0)}$ and $a^\dagger_{-1/2} \Psi^{(0)}$. Define the Hermitian operators x, p_x, y, p_y by the relations

$$a_{1/2} = \frac{1}{\sqrt{2\hbar}} \left(cx + i \frac{p_x}{c} \right), \qquad a_{-1/2} = \frac{1}{\sqrt{2\hbar}} \left(cy + i \frac{p_y}{c} \right)$$

   where c is an arbitrary real constant, and derive the commutation relations for these Hermitian operators. Express the angular momentum operator (21.6) in terms of these "coordinates" and "momenta," and also evaluate $\mathscr{J}^2$. Relate $\mathscr{J}^2$ to the square of the Hamiltonian of an isotropic two-dimensional harmonic oscillator by making the identification $c = \sqrt{\mu \omega}$, and show the connection between the eigenvalues of these operators.

3. Using the fermion creation operators $a^\dagger_{jm}$, appropriate to particles with angular momentum j, form the closed-shell state in which all one-particle states $m = -j$ to j are occupied. Prove that the closed shell has zero total angular momentum.

   If a fermion with magnetic quantum number m is missing from a closed shell of particles with angular momentum j, show that the hole state may be treated like a one-particle state with magnetic quantum number $-m$ and an effective creation operator $(-1)^{j-m} a_{jm}$ in the coupling of angular momenta.

4. Consider the unperturbed states $a^\dagger_{n m_n} \cdots a^\dagger_{k m_k} \cdots a^\dagger_{1 m_1} \Psi^{(0)}$ of n electrons, each occupying one of n degenerate orthogonal orbitals labeled by the quantum number k, and with $m_k = \pm \frac{1}{2}$ denoting the spin quantum number associated

with the orbital k. Show that in the space of the 2^n unperturbed states a spin-independent two-body interaction may, in first-order perturbation theory, be replaced by the effective Hamiltonian

$$\mathscr{H}_{\text{eff}} = -\frac{1}{\hbar^2} \sum_{kl} \langle kl | \ V \ | lk \rangle \mathbf{S}_k \cdot \mathbf{S}_l + \text{const.}$$

where $\mathbf{S}_k$ is the *localized* spin operator

$$\mathbf{S}_k = \frac{\hbar}{2} \sum_{m_k m_k'} a^\dagger_{k m_k} a_{k m_k'} \langle m_k | \ \boldsymbol{\sigma} \ | m_k' \rangle$$

5. Consider a system of identical fermions with $2p$ degenerate one-particle states $a^\dagger_{\pm k} \Psi^{(0)}$ ($k = 1, 2, \ldots, p$) and with a pairing interaction

$$\mathscr{H} = -G \sum_{l=1}^{p} \sum_{k=1}^{p} a^\dagger_l a^\dagger_{-l} a_{-k} a_k$$

Define the fermion quasiparticles created by the operators

$$b^\dagger_{1/2, k} = a^\dagger_k, \qquad b^\dagger_{-1/2, k} = a_{-k}$$

and show that $\mathscr{H}$ can be expressed simply in terms of the *quasispin* operators

$$\mathbf{S}_k = \tfrac{1}{2} \sum_{\sigma' \sigma''} b^\dagger_{\sigma'/2, k} b_{\sigma''/2, k} \langle \sigma' | \ \boldsymbol{\sigma} \ | \sigma'' \rangle$$

where σ', σ'' are the eigenvalues ± 1 of σ_z. Show that the eigenvalues of the z component of each quasispin are $\frac{1}{2}$, $-\frac{1}{2}$, and zero, depending on whether the states k, $-k$ are occupied by a pair of particles, vacant, or occupied by one particle (*"broken pair"*). Show that the eigenvalues of the square of the total quasispin operator $\sum_k \mathbf{S}_k$ are expressible in terms of the total quasispin quantum number $S = \frac{1}{2}(p - v)$, where v (the *seniority*) is the number of broken pairs.

Derive the formula for the energy spectrum of $\mathscr{H}$:

$$E_v^{(n)} = -\frac{G}{4} (n - v)(2p - n - v + 2)$$

if n is the total number of particles in the state. Calculate the ground state ($v = 0$) energy approximately by the Bogoliubov quasiparticle method (Section 21.6), and compare the results for $p \gg 1$.

6. Apply the Hartree-Fock method to a system of two "electrons" which are attracted to the coordinate origin by an isotropic harmonic oscillator potential $V(r) = \frac{1}{2} \mu \omega^2 r^2$ and which interact with each other through a potential $c(\mathbf{r}' - \mathbf{r}'')^2$. Solve the Hartree-Fock equations for the ground state and compare with the exact result and with first-order perturbation theory. (See Problem 5 in Chapter 15.)[10]

[10] M. Moshinsky, *Am. J. Phys.*, **36**, 52 (1968).

7. For a Fermi gas of free particles with Fermi momentum p_F, calculate the ground state expectation value of the pair density operator

$$\sum_{\sigma, \sigma'} \psi_\sigma^\dagger (\mathbf{r}') \psi_{\sigma''}^\dagger (\mathbf{r}'') \psi_{\sigma''}(\mathbf{r}'') \psi_{\sigma'}(\mathbf{r}')$$

in coordinate space and show that there is a repulsive correlation which would be absent if the particles were not identical. Show that there is no spatial correlation between fermions of opposite spin.

8. In a superconductor the pairs are defined in terms of free electron momentum eigenstates with momenta $\mathbf{p}$ and $-\mathbf{p}$ and opposite z components of the spins. If the operator which creates an electron with momentum $\mathbf{p}$ ($-\mathbf{p}$) and spin up (spin down) is denoted by $a_{\mathbf{p}1/2}^\dagger(a_{-\mathbf{p}-1/2}^\dagger)$, the pairing interaction takes the form

$$-G \sum_{\mathbf{p}, \mathbf{p}'} a_{\mathbf{p}'1/2}^\dagger a_{-\mathbf{p}'-1/2}^\dagger a_{-\mathbf{p}-1/2} a_{\mathbf{p}1/2}$$

Show that the expectation value of the pair density operator

$$\psi_+^\dagger(\mathbf{r}')\psi_-^\dagger(\mathbf{r}'')\psi_-(\mathbf{r}'')\psi_+(\mathbf{r}') + \psi_-^\dagger(\mathbf{r}')\psi_+^\dagger(\mathbf{r}'')\psi_+(\mathbf{r}'')\psi_-(\mathbf{r}')$$

for electrons with opposite spins in coordinate space exhibits spatial correlations which decrease with increasing distance between the two electrons. (**Do not attempt a complete evaluation of the integrals in momentum space.**)[11]

9. Calculate in first order the energies of the ^1S, ^3P, and ^1D states arising from the atomic configuration p^2. Use the expansion

$$\frac{e^2}{|\mathbf{r}' - \mathbf{r}''|} = e^2 \sum_{k=0}^\infty \frac{4\pi}{2k+1} \, \gamma_k(r', r'') \sum_{q=-k}^k (-1)^q \, Y_k^q(\hat{\mathbf{r}}') Y_k^{-q}(\hat{\mathbf{r}}'')$$

for the interaction energy between the electrons, and show that the term energies may be expressed as

$$E(^1S) = E_0 + \langle \gamma_0 \rangle + \tfrac{10}{25}\langle \gamma_2 \rangle$$
$$E(^3P) = E_0 + \langle \gamma_0 \rangle - \tfrac{5}{25}\langle \gamma_2 \rangle$$
$$E(^1D) = E_0 + \langle \gamma_0 \rangle + \tfrac{1}{25}\langle \gamma_2 \rangle$$

where $\langle \gamma_k \rangle$ is the radial integral

$$\langle \gamma_k \rangle = e^2 \iint \gamma_k(r', r'')[R(r')]^2 [R(r'')]^2 r'^2 r''^2 \, dr' \, dr''$$

(*Note.* This is a somewhat difficult problem.)

---

[11] For details on estimating the integrals, see J. Bardeen, L. N. Cooper, and J. R. Schrieffer, *Phys. Rev.*, **108**, 1189 (1956), Appendix D.

CHAPTER 22

Photons and the Electromagnetic Field

1. Fundamental Notions. The formalism developed in Chapter 20 for the quantal description of identical particles is sufficiently general to be applicable to *photons*, the particles associated with the electromagnetic field. The theory is based on these empirical facts: photons are bosons with zero mass; they can carry linear momentum $\hbar\mathbf{k}$ and energy $\hbar\omega_k = c\hbar k$; and simultaneously with sharp linear momentum, they may have a definite value $+\hbar$ or $-\hbar$ for the component $\mathbf{J} \cdot \hat{\mathbf{k}}$ of angular momentum along the direction of propagation.[1] Photons with a sharp value for $\mathbf{J} \cdot \hat{\mathbf{k}}$ are said to possess definite, positive, or negative *helicity*. Since the component of orbital angular momentum parallel to $\mathbf{k}$ is necessarily zero, *spin* 1 may be attributed to the photon, provided that the eigenvalue zero of the projection $\mathbf{J} \cdot \hat{\mathbf{k}}$ is excluded. The consistent omission from the theory of the eigenstate for which $\mathbf{J} \cdot \hat{\mathbf{k}}$ has the value zero is possible only because the photon has vanishing rest mass and cannot be brought to rest in any Lorentz frame of reference for the purpose of measuring its intrinsic spin separately from the orbital angular momentum that it carries.

By definition, the operators $a_R^\dagger(\mathbf{k})$ and $a_L^\dagger(\mathbf{k})$ correspond to the creation of a photon with momentum $\hbar\mathbf{k}$ and $\mathbf{J} \cdot \hat{\mathbf{k}} = +\hbar$ and $-\hbar$. For convenience, the momentum states, or photon *modes*, will be assumed to form a discrete set defined by $\mathbf{k}$ vectors with allowed components $2\pi n_x/L$, $2\pi n_y/L$, $2\pi n_z/L$, as in (10.8). The transition to the continuum may be made by letting the arbitrary length $L \to \infty$, or formally by simply setting $L = 2\pi$ and replacing all sums over $\mathbf{k}$ by integrals and Kronecker deltas by delta functions. All spatial integrations extend over the cube of volume L^3.

The commutation relations for the photon creation and annihilation operators are

$$[a_R(\mathbf{k}), a_R^\dagger(\mathbf{k}')] = [a_L(\mathbf{k}), a_L^\dagger(\mathbf{k}')] = \delta_{\mathbf{kk}'} \tag{22.1}$$

All other commutators vanish.

---

[1] Throughout this chapter $\hat{\mathbf{k}}$ denotes the unit vector in the direction of the photon momentum, except in the equation that follows (22.34), where it retains its usual meaning as the unit vector in the z direction.

The important physical operators for a system of free photons are the total energy, the total linear momentum, and the total number of photons:

$$\mathscr{H} = \sum_{\mathbf{k}} c\hbar k [a_R^\dagger(\mathbf{k})a_R(\mathbf{k}) + a_L^\dagger(\mathbf{k})a_L(\mathbf{k})] \tag{22.2}$$

$$\mathbf{P} = \sum_{\mathbf{k}} \hbar\mathbf{k} [a_R^\dagger(\mathbf{k})a_R(\mathbf{k}) + a_L^\dagger(\mathbf{k})a_L(\mathbf{k})] \tag{22.3}$$

$$N = \sum_{\mathbf{k}} [a_R^\dagger(\mathbf{k})a_R(\mathbf{k}) + a_L^\dagger(\mathbf{k})a_L(\mathbf{k})] \tag{22.4}$$

The operator $\mathscr{J} \cdot \hat{\mathbf{k}}$, representing the total angular momentum component along the direction $\mathbf{k}$, is an additive one-particle operator and thus a bilinear combination of $a_R^\dagger(\mathbf{k}')$, $a_L^\dagger(\mathbf{k}')$, $a_R(\mathbf{k}'')$, and $a_L(\mathbf{k}'')$. Owing to the helicity property of photons, the operators creating and annihilating photons of momentum $\hbar\mathbf{k}$ occur in this linear combination only in the diagonal terms $a_R^\dagger(\mathbf{k})a_R(\mathbf{k})$ and $a_L^\dagger(\mathbf{k})a_L(\mathbf{k})$. These terms, which represent the contribution to $\mathscr{J} \cdot \hat{\mathbf{k}}$ of the photons with linear momentum $\mathbf{k}$, may be written as

$$[\mathscr{J} \cdot \hat{\mathbf{k}}, a_R^\dagger(\mathbf{k})]a_R(\mathbf{k}) + [\mathscr{J} \cdot \hat{\mathbf{k}}, a_L^\dagger(\mathbf{k})]a_L(\mathbf{k}) = \hbar[a_R^\dagger(\mathbf{k})a_R(\mathbf{k}) - a_L^\dagger(\mathbf{k})a_L(\mathbf{k})]$$
$$\tag{22.5}$$

This operator generates rotations of the states with definite $\mathbf{k}$ about the direction $\mathbf{k}$. According to Sections 16.5 and 20.3, we expect that such a rotation changes $a_R^\dagger(\mathbf{k})$ into $a_R^\dagger(\mathbf{k})e^{-i\alpha}$ and $a_L^\dagger(\mathbf{k})$ into $a_L^\dagger(\mathbf{k})e^{i\alpha}$, if α is the angle of rotation.

Exercise 22.1. Using the commutation relations, prove that

$$e^{-i\alpha[a_R^\dagger(\mathbf{k})a_R(\mathbf{k}) - a_L^\dagger(\mathbf{k})a_L(\mathbf{k})]} a_{R,L}^\dagger(\mathbf{k}) e^{i\alpha[a_R^\dagger(\mathbf{k})a_R(\mathbf{k}) - a_L^\dagger(\mathbf{k})a_L(\mathbf{k})]} = e^{\mp i\alpha} a_{R,L}^\dagger(\mathbf{k})$$

It is apparent that under the simple rotation considered, $a_R^\dagger(\mathbf{k})$ and $a_L^\dagger(\mathbf{k})$ transform like the T_1^1 and T_1^{-1} components of an irreducible tensor of rank one. Using equations (16.105), we introduce the new operators

$$a_1^\dagger(\mathbf{k}) = \frac{1}{\sqrt{2}} [-a_R^\dagger(\mathbf{k}) + a_L^\dagger(\mathbf{k})] \tag{22.6}$$

$$a_2^\dagger(\mathbf{k}) = \frac{i}{\sqrt{2}} [a_R^\dagger(\mathbf{k}) + a_L^\dagger(\mathbf{k})] \tag{22.7}$$

Under a rotation about $\mathbf{k}$ by an angle α, these operators are transformed into

$$a_1^\dagger(\mathbf{k}) \cos \alpha + a_2^\dagger(\mathbf{k}) \sin \alpha \qquad \text{and} \qquad -a_1^\dagger(\mathbf{k}) \sin \alpha + a_2^\dagger(\mathbf{k}) \cos \alpha$$

just as two orthogonal components of a vector in a plane perpendicular to $\mathbf{k}$. If we introduce two perpendicular unit vectors $\hat{\mathbf{e}}_{\mathbf{k}}^{(1)}$ and $\hat{\mathbf{e}}_{\mathbf{k}}^{(2)}$ in the plane perpendicular to k, such that

$$\hat{\mathbf{e}}_k^{(1)} \times \hat{\mathbf{e}}_k^{(2)} = \hat{\mathbf{k}}, \qquad \hat{\mathbf{e}}_k^{(2)} \times \hat{\mathbf{k}} = \hat{\mathbf{e}}_k^{(1)}, \qquad \hat{\mathbf{k}} \times \hat{\mathbf{e}}_k^{(1)} = \hat{\mathbf{e}}_k^{(2)},$$

it follows that

$$a_1^\dagger(\mathbf{k})\hat{\mathbf{e}}_k^{(1)} + a_2^\dagger(\mathbf{k})\hat{\mathbf{e}}_k^{(2)} = \frac{1}{\sqrt{2}} a_R^\dagger(\mathbf{k})(-\hat{\mathbf{e}}_k^{(1)} + i\hat{\mathbf{e}}_k^{(2)}) + \frac{1}{\sqrt{2}} a_L^\dagger(\mathbf{k})(\hat{\mathbf{e}}_k^{(1)} + i\hat{\mathbf{e}}_k^{(2)})$$

transforms like a three-vector which is perpendicular to $\mathbf{k}$.

The corresponding Hermitian conjugate operator is

$$a_1(\mathbf{k})\hat{\mathbf{e}}_k^{(1)} + a_2(\mathbf{k})\hat{\mathbf{e}}_k^{(2)} = -\frac{1}{\sqrt{2}} a_R(\mathbf{k})(\hat{\mathbf{e}}_k^{(1)} + i\hat{\mathbf{e}}_k^{(2)}) + \frac{1}{\sqrt{2}} a_L(\mathbf{k})(\hat{\mathbf{e}}_k^{(1)} - i\hat{\mathbf{e}}_k^{(2)})$$

$$(22.8)$$

It is tempting to consider now a unitary transformation from the *momentum* representation, which we have used to introduce photons, to a *coordinate* representation by the application of the one-particle transformation coefficients

$$\langle \mathbf{r} \mid \mathbf{k} \rangle = \frac{1}{L^{3/2}} e^{i\mathbf{k}\cdot\mathbf{r}} \tag{22.9}$$

In this way it is possible to define, as in Section 20.5, operators that describe the creation (and annihilation) of a photon at a position $\mathbf{r}$, such as

$$\psi^\dagger(\mathbf{r}) = \frac{1}{L^{3/2}} \sum_{\mathbf{k}} [a_1^\dagger(\mathbf{k})\hat{\mathbf{e}}_k^{(1)} + a_2^\dagger(\mathbf{k})\hat{\mathbf{e}}_k^{(2)}]e^{-i\mathbf{k}\cdot\mathbf{r}}$$

In conformity with equation (20.47) one could then also introduce a photon wave function in configuration space. However, such a representation—straightforward for the case of nonrelativistic particles—is of limited usefulness for photons. This is most easily seen by examining an observable like the energy of the system which, after transformation to coordinate space, will appear in the form

$$\mathcal{H} = \frac{\hbar c}{L^3} \sum_{\mathbf{k}} \iint \psi^\dagger(\mathbf{r}') \cdot \psi(\mathbf{r}) \, k \, e^{i\mathbf{k}\cdot(\mathbf{r}-\mathbf{r}')} \, d^3r \, d^3r'$$

Since $k = \sqrt{k_x^2 + k_y^2 + k_z^2}$, the integrand cannot be transformed into a local expression involving a finite number of field derivatives and interpretable as an energy *density*. As we shall see in Chapter 23, for particles of *half-integral* spin such a transformation is possible, i.e., it succeeds for electrons and neutrinos but not for photons.

For photons this deficiency is remedied by introducing a slightly modified set of *field operators;*

$$\mathbf{A}^{(+)}(\mathbf{r}) = \sqrt{4\pi\hbar c^2}\,\frac{1}{L^{3/2}}\sum_{\mathbf{k}}\frac{1}{\sqrt{2\omega_k}}\,[a_1(\mathbf{k})\hat{\mathbf{e}}_{\mathbf{k}}^{(1)} + a_2(\mathbf{k})\hat{\mathbf{e}}_{\mathbf{k}}^{(2)}]e^{i\mathbf{k}\cdot\mathbf{r}} \qquad (22.10)$$

The significant difference between this vector operator and $\boldsymbol{\psi}(\mathbf{r})$ is the appearance of the coefficient $(2\omega_k)^{-1/2}$, and we shall see that its insertion permits us to write the energy and other important observables as simple bilinear functions of the field operators. The constants in the definition (22.10) were chosen with the foreknowledge of the ultimate identification of this operator with the *vector potential* of the electromagnetic field. Equation (22.10) has a Hermitian adjoint companion:

$$\mathbf{A}^{(-)}(\mathbf{r}) = [\mathbf{A}^{(+)}(\mathbf{r})]^\dagger$$

$$= \sqrt{4\pi\hbar c^2}\,\frac{1}{L^{3/2}}\sum_{\mathbf{k}}\frac{1}{\sqrt{2\omega_k}}\,[a_1^\dagger(\mathbf{k})\hat{\mathbf{e}}_{\mathbf{k}}^{(1)} + a_2^\dagger(\mathbf{k})\hat{\mathbf{e}}_{\mathbf{k}}^{(2)}]e^{-i\mathbf{k}\cdot\mathbf{r}} \qquad (22.11)$$

All field operators are periodic functions of the coordinates x, y, z with period L.

If we work in the Heisenberg picture, the operators $a(\mathbf{k})$ have a simple time dependence, which can be inferred from the equation of motion for the free photon system:

$$i\hbar\,\frac{da(\mathbf{k}, t)}{dt} = [a(\mathbf{k}, t), \mathscr{H}\,]$$

Here $\mathscr{H}$ is the energy operator (22.2). Hence,

$$i\,\frac{da(\mathbf{k}, t)}{dt} = \omega_k a(\mathbf{k}, t)$$

and

$$a(\mathbf{k}, t) = a(\mathbf{k})e^{-i\omega_k t} \qquad a^\dagger(\mathbf{k}, t) = a^\dagger(\mathbf{k})e^{i\omega_k t} \qquad (22.12)$$

It follows that $\mathbf{A}^{(+)}(\mathbf{r}, t)$ is obtained from (22.10) if we replace $e^{i\mathbf{k}\cdot\mathbf{r}}$ by $e^{i(\mathbf{k}\cdot\mathbf{r}-\omega_k t)}$. Similarly, to obtain $\mathbf{A}^{(-)}(\mathbf{r}, t)$, replace $e^{-i\mathbf{k}\cdot\mathbf{r}}$ in (22.11) by $e^{-i(\mathbf{k}\cdot\mathbf{r}-\omega_k t)}$. These forms explain why $\mathbf{A}^{(+)}$ is called the *positive frequency part* and $\mathbf{A}^{(-)}$ the *negative frequency part* of the Hermitian operator $\mathbf{A}(\mathbf{r}, t)$ which is defined as the sum of the two:

$$\mathbf{A}(\mathbf{r}, t) = \sqrt{4\pi\hbar c^2}\,\frac{1}{L^{3/2}}\sum_{\mathbf{k}}\frac{1}{\sqrt{2\omega_k}}\,\{[a_1(\mathbf{k})\hat{\mathbf{e}}_{\mathbf{k}}^{(1)} + a_2(\mathbf{k})\hat{\mathbf{e}}_{\mathbf{k}}^{(2)}]e^{i(\mathbf{k}\cdot\mathbf{r}-\omega_k t)}$$
$$+ [a_1^\dagger(\mathbf{k})\hat{\mathbf{e}}_{\mathbf{k}}^{(1)} + a_2^\dagger(\mathbf{k})\hat{\mathbf{e}}_{\mathbf{k}}^{(2)}]e^{-i(\mathbf{k}\cdot\mathbf{r}-\omega_k t)}\} \quad (22.13)$$

Owing to the orthogonality of the *polarization* vectors $\hat{\mathbf{e}}_{\mathbf{k}}$ to $\mathbf{k}$, this operator satisfies the *transversality condition*

$$\nabla \cdot \mathbf{A}(\mathbf{r}, t) = 0 \qquad (22.14)$$

In Cartesian coordinates it also obeys the equation

$$\left(\nabla^2 - \frac{1}{c^2}\frac{\partial^2}{\partial t^2}\right)\mathbf{A}(\mathbf{r}, t) = 0 \tag{22.15}$$

because $\omega_k = ck$. The Hermitian operators $\mathbf{E}$ and $\mathbf{B}$ defined by

$$\mathbf{B} = \nabla \times \mathbf{A} \quad \text{and} \quad \mathbf{E} = -\frac{1}{c}\frac{\partial \mathbf{A}}{\partial t}$$

are thus local field operators which satisfy the *Maxwell equations* for the free radiation field in the so-called *Coulomb gauge*, excluding the electrostatic contribution $-\nabla\phi$ to the electric field which arises from the Coulomb interaction between charges.

2. Energy, Momentum, and Angular Momentum of the Radiation Field. The identification of $\mathbf{A}$, $\mathbf{B}$, and $\mathbf{E}$ will be complete only if we can verify that the expressions for physical observables, such as the energy and momentum of the field, have the correct classical form in the correspondence limit.

To check this, we consider

$$\mathbf{E}^{(+)}(\mathbf{r}, t) = -\frac{1}{c}\frac{\partial \mathbf{A}^{(+)}(\mathbf{r}, t)}{\partial t}$$

$$= i\sqrt{2\pi\hbar}\frac{1}{L^{3/2}}\sum_{\mathbf{k}}\sqrt{\omega_k}\,[a_1(\mathbf{k})\hat{\mathbf{e}}_{\mathbf{k}}^{(1)} + a_2(\mathbf{k})\hat{\mathbf{e}}_{\mathbf{k}}^{(2)}]e^{i(\mathbf{k}\cdot\mathbf{r}-\omega_k t)} \tag{22.16}$$

and, hence,

$$a_1(\mathbf{k})\hat{\mathbf{e}}_{\mathbf{k}}^{(1)} + a_2(\mathbf{k})\hat{\mathbf{e}}_{\mathbf{k}}^{(2)} = -\frac{i}{\sqrt{2\pi\hbar}}\frac{1}{\sqrt{\omega_k}}\frac{1}{L^{3/2}}\int e^{-i(\mathbf{k}\cdot\mathbf{r}-\omega_k t)}\mathbf{E}^{(+)}(\mathbf{r}, t)\,d^3r \tag{22.17}$$

The operator $\mathbf{E}^{(-)}$ is similarly defined.

Using (22.8) and (22.17), we may now transform the expression (22.2) for the total energy as follows:

$$\mathcal{H} = \sum_{\mathbf{k}}\hbar\omega_k[a_R^\dagger(\mathbf{k})a_R(\mathbf{k}) + a_L^\dagger(\mathbf{k})a_L(\mathbf{k})]$$

$$= \sum_{\mathbf{k}}\hbar\omega_k[a_1^\dagger(\mathbf{k})a_1(\mathbf{k}) + a_2^\dagger(\mathbf{k})a_2(\mathbf{k})]$$

$$= \sum_{\mathbf{k}}\hbar\omega_k[a_1^\dagger(\mathbf{k})\hat{\mathbf{e}}_{\mathbf{k}}^{(1)} + a_2^\dagger(\mathbf{k})\hat{\mathbf{e}}_{\mathbf{k}}^{(2)}]\cdot[a_1(\mathbf{k})\hat{\mathbf{e}}_{\mathbf{k}}^{(1)} + a_2(\mathbf{k})\hat{\mathbf{e}}_{\mathbf{k}}^{(2)}]$$

$$= \frac{1}{2\pi L^3}\sum_{\mathbf{k}}\int\int e^{i\mathbf{k}\cdot(\mathbf{r}-\mathbf{r}')}\mathbf{E}^{(-)}(\mathbf{r}, t)\cdot\mathbf{E}^{(+)}(\mathbf{r}', t)\,d^3r\,d^3r'$$

and finally,

$$\mathcal{H} = \frac{1}{2\pi}\int\mathbf{E}^{(-)}(\mathbf{r}, t)\cdot\mathbf{E}^{(+)}(\mathbf{r}, t)\,d^3r \tag{22.18}$$

But this is precisely the form which the energy of the classical electromagnetic radiation field,

$$\mathscr{H}_c = \frac{1}{8\pi} \int (\mathbf{E}_c^2 + \mathbf{B}_c^2)\, d^3r \qquad (22.19)$$

takes when a Fourier decomposition is made and the field is written as the
sum of a positive and negative frequency part. As is well known, the magnetic
and electric fields contribute equally to the total energy.

Exercise 22.2. Exhibit the connection between (22.18) and (22.19). Show
similarly that the total linear momentum (22.3) may be transformed into

$$\mathbf{P} = \frac{1}{2\pi c} \int \mathbf{E}^{(-)}(\mathbf{r}, t) \times \mathbf{B}^{(+)}(\mathbf{r}, t)\, d^3r \qquad (22.20)$$

and establish the correspondence with the classical expression for the
momentum of the field.

The equations (22.18) and (22.20) may be cast into a compact formula for
the energy-momentum four-vector

$$P^\mu = \left(\frac{E}{c}, \mathbf{P} \right) \qquad (22.21)$$

if the contravariant time-space coordinate four-vector

$$x^\mu = (x^0, x^1, x^2, x^3) = (ct, x, y, z) = (ct, \mathbf{r})$$

and the metric[2]

$$g^{11} = g^{22} = g^{33} = -g^{00} = -1, \qquad g^{\mu\nu} = 0 \text{ if } \mu \neq \nu$$

are introduced. The covariant four-vector x_μ is

$$x_\mu = g_{\mu\nu} x^\nu = (ct, -x, -y, -z)$$

Using the relation $\nabla \cdot \mathbf{E} = 0$, the integrals (22.18) and (22.20) may be
transformed into

$$P^\mu = -\frac{1}{2\pi c} \int \mathbf{E}^{(-)}(\mathbf{r}, t) \frac{\partial}{\partial x_\mu} \mathbf{A}^{(+)}(\mathbf{r}, t)\, d^3r \qquad (22.22)$$

This formula has the usual structure of an additive one-particle operator like
(20.50), except that the differential operator stands between $\mathbf{E}^{(-)}$ and $\mathbf{A}^{(+)}$
rather than between $\psi^\dagger$ and ψ.

---

[2] In Chapters 22–24 on relativistic quantum mechanics, we generally use the same
conventions as J. D. Bjorken and S. D. Drell in *Relativistic Quantum Mechanics*, McGraw-
Hill Book Company, New York, 1964. However, we retain h and c and employ unration-
alized electromagnetic units so that the fine structure constant is $e^2/hc \approx 1/137$.

The classical expression for the *angular momentum* of the electromagnetic field,

$$\mathcal{J}_c = \frac{1}{4\pi c} \int \mathbf{r} \times (\mathbf{E}_c \times \mathbf{B}_c)\, d^3r \tag{22.23}$$

may be shown similarly to be the classical limit of the operator equation

$$\mathcal{J} = \frac{1}{2\pi c} \int \sum_{i=1}^{3} E_i^{(-)}(\mathbf{r} \times \nabla)A_i^{(+)}\, d^3r + \frac{1}{2\pi c} \int \mathbf{E}^{(-)} \times \mathbf{A}^{(+)}\, d^3r \tag{22.24}$$

Exercise 22.3. Prove the connection between (22.23) and (22.24) by using the identity

$$[\mathbf{r} \times (\mathbf{E} \times \mathbf{B})]_x = \sum_{i=1}^{3} E_i(\mathbf{r} \times \nabla)_x A_i + (\mathbf{E} \times \mathbf{A})_x + \nabla \cdot [\mathbf{E}(zA_y - yA_z)]$$

The first term on the right-hand side of (22.24) has the form of an *orbital angular momentum* operator, and the second term accounts for the *spin* of the photons. If the expansion (22.10) and the Hermitian adjoint of (22.16) are substituted, the second term is seen to contribute precisely the value given by equation (22.5) to the component of total angular momentum along **k** for each photon momentum state **k**. The photon spin operator may also be written in the form

$$\mathcal{S} = \frac{i}{2\pi \hbar c} \int (E_x^{(-)}\ E_y^{(-)}\ E_z^{(-)})\mathbf{S} \begin{pmatrix} A_x^{(+)} \\ A_y^{(+)} \\ A_z^{(+)} \end{pmatrix} d^3r \tag{22.25}$$

The 3×3 matrices **S** satisfy the angular momentum commutation relations and have eigenvalues 0 and $\pm\hbar$, thus confirming that photons are particles with spin one.

Exercise 22.4. Construct the matrices **S** explicitly and verify their commutation relations.

From the behavior of the various physical quantities associated with the field, we infer that the annihilation and creation operators $a_1(\mathbf{k})$ and $a_1^\dagger(\mathbf{k})$ become in the classical limit proportional to the Fourier amplitudes of the electric radiation field linearly polarized in one direction, while the operators $a_2(\mathbf{k})$ and $a_2^\dagger(\mathbf{k})$ become proportional to the component with perpendicular linear polarization. According to the relation (22.8), the operators $a_R(\mathbf{k})$ and $a_R^\dagger(\mathbf{k})$ correspond in the classical limit to the amplitudes of an electric field with right-handed circular polarization; similarly, $a_L(\mathbf{k})$ and $a_L^\dagger(\mathbf{k})$

correspond to left-handed polarization. These identifications explain the notation R and L adopted for the subscripts on the operators.

Exercise 22.5. Work out the equal-time commutation relations between the components of $\mathbf{A}(\mathbf{r}, t)$ and $\mathbf{E}(\mathbf{r}, t)$.

3. Interactions With Charged Particles. So far, only free photons have been considered, and it has been shown that these particles are the quanta of the free electromagnetic field. With this background, we can now introduce the interaction between photons and charged material particles, for example, electrons. Although in many applications it is necessary to use a proper relativistic treatment of the massive particles, a nonrelativistic approximation will be used in this section for describing the electron.

We postulate that the Hamiltonian operator is obtained from the Hamiltonian for the free particle system by the substitution

$$\mathbf{p} \rightarrow \mathbf{p} - \frac{q}{c}\mathbf{A}$$

if the particles have charge q. We thus assume the Hamiltonian

$$\mathscr{H} = \frac{1}{2\pi}\int \mathbf{E}^{(-)}\cdot\mathbf{E}^{(+)}\, d^3r + \frac{1}{2\mu}\int \psi^\dagger\left(\frac{\hbar}{i}\nabla - \frac{q}{c}\mathbf{A}\right)^2\psi\, d^3r + \int V\psi^\dagger\psi\, d^3r \quad (22.26)$$

The first term refers to the electromagnetic field alone, and the last term describes the external forces as well as the static interactions between the particles, such as the Coulomb repulsion between electrons. The middle term, when expanded, contributes one term which solely refers to the electrons, and the remaining terms represent explicitly the interaction between radiation and matter fields. The Hamiltonian is therefore naturally divided into two parts, $\mathscr{H} = \mathscr{H}_0 + \mathscr{H}_1$, with $\mathscr{H}_0$ including the pure radiation and matter terms, while $\mathscr{H}_1$ symbolizes the interaction terms:

$$\mathscr{H}_0 = \frac{1}{2\pi}\int \mathbf{E}^{(-)}\cdot\mathbf{E}^{(+)}\, d^3r - \frac{\hbar^2}{2\mu}\int \psi^\dagger\nabla^2\psi\, d^3r + \int V\psi^\dagger\psi\, d^3r \quad (22.27)$$

$$\mathscr{H}_1 = -\frac{q\hbar}{\mu c i}\int \psi^\dagger\mathbf{A}\cdot\nabla\psi\, d^3r + \frac{q^2}{2\mu c^2}\int \mathbf{A}\cdot\mathbf{A}\psi^\dagger\psi\, d^3r \quad (22.28)$$

The dynamics of *interacting fields* is governed by the Hamiltonian (22.26) and described by field operators whose structure is far more complicated than the free fields like (22.13). The theory of interacting quantum fields, with its impressive achievements in the high-precision calculations of radiative corrections in quantum electrodynamics, lies outside the scope of this book.[3] We restrict ourselves to two simplified approaches: problems for which a first-order approximation is appropriate, and problems in which the prescribed motion of the charges may be treated by classical methods.

Since the electromagnetic interaction between the radiation field and a charged particle is comparatively weak, the use of perturbation theory is indicated in many practical problems. As an illustration, we shall derive the formulas for the intensity of *emission* and *absorption* of photons by a one-electron atom, neglecting the spin of the electron. The operator $\mathscr{H}_0$ describing the free photons and the atom is the unperturbed Hamiltonian, and the interaction $\mathscr{H}_1$ is the perturbation to which the general methods of Chapter 18 will be applied.

In first approximation the first term in the interaction (22.28), which is linear in photon creation and annihilation operators, is responsible for transitions in which *one* photon is emitted or absorbed. If the electron, with charge $q = -e$, is initially in an atomic state defined by the Schrödinger wave function $\psi_i(\mathbf{r})$, the initial state of the electron-photon system is, in a somewhat mixed notation, given by

$$\Psi_i = \int \psi_i(\mathbf{r})\psi^\dagger(\mathbf{r})\, d^3r \, \Psi_{el}^{(0)}|\cdots n_i(\mathbf{k})\cdots\rangle \qquad (22.29)$$

Here $\Psi_{el}^{(0)}$ denotes the no-electron state (electron vacuum), and $|\cdots n_i(\mathbf{k})\cdots\rangle$ symbolizes the state of the electromagnetic field in terms of photon occupation numbers. In the final state there must be one photon more or one photon less than in the initial state, and the electron is annihilated in state $\psi_i(\mathbf{r})$ and recreated in state $\psi_f(\mathbf{r})$.

In order to evaluate the matrix element of the perturbation term for the transition from state i to f, we merely have to substitute the expression (22.13) for the operator $\mathbf{A}$, evaluated at $t = 0$, into the interaction (22.28) and to calculate the scalar product $(\Psi_f, \mathscr{H}_1 \Psi_i)$ for the initial state (22.29) and the final state

$$\Psi_f = \int \psi_f(\mathbf{r})\psi^\dagger(\mathbf{r})\, d^3r \, \Psi_{el}^{(0)}|\ldots, n_i(\mathbf{k})\mp 1, \ldots\rangle \qquad (22.30)$$

The result is

$$(\Psi_f, \mathscr{H}_1\Psi_i) = \frac{e}{\mu c}\sqrt{4\pi\hbar c^2}\,\frac{1}{L^{3/2}}\,\frac{1}{\sqrt{2\omega_k}}\left\{\begin{matrix}\sqrt{n_i}\\\sqrt{n_i+1}\end{matrix}\right\}\int \psi_f{}^*(\mathbf{r})e^{\pm i\mathbf{k}\cdot\mathbf{r}}\hat{\mathbf{e}}_{\mathbf{k}}^{(\lambda)}\cdot\frac{\hbar}{i}\,\nabla\psi_i(\mathbf{r})\, d^3r$$

$$(22.31)$$

[3] A contemporary set of textbooks on quantum electrodynamics is J. D. Bjorken and S. D. Drell, *Relativistic Quantum Mechanics*, McGraw-Hill Book Company, New York, 1964 and J. D. Bjorken and S. D. Drell, *Relativistic Quantum Fields*, McGraw-Hill Book Company, New York, 1965.

The upper option in brackets applies to absorption of a photon and the lower option to emission. If E_i and E_f denote the initial and final energy of the unperturbed atom, transitions are, of course, appreciable only if the photon energy $\hbar\omega_k = E_f - E_i$ or $E_i - E_f$, depending on whether absorption or emission of a photon takes place.

Under the usual conditions of observation, the state of the electromagnetic field corresponds to a superposition of occupied photon states in a quasi-continuum and it is appropriate to characterize this state by an initial average[4] photon number $\bar{n}_i^{(\lambda)}(\mathbf{k})$ for photons in the polarization mode λ and with approximate momentum $\hbar\mathbf{k}$ directed within a solid angle $d\Omega$. Since the number of photon modes for each polarization per unit energy interval is given by

$$\rho(E) = \frac{\omega^2 L^3}{8\pi^3 \hbar c^3} \, d\Omega$$

the transition probability per unit time is according to the Golden Rule (18.107):

$$w = \frac{2\pi}{\hbar} \rho(\hbar\omega_k) \, |(\Psi_f, \mathcal{H}_1 \Psi_i)|^2$$

$$= \frac{\alpha}{2\pi c^2} \, \omega_k \, \bar{n}_{i,f}^{(\lambda)}(\mathbf{k}) \, |\langle f| \, e^{i\mathbf{k}\cdot\mathbf{r}}\hat{\mathbf{e}}_{\mathbf{k}}^{(\lambda)} \cdot \mathbf{v} \, |i\rangle|^2 \, d\Omega \qquad (22.32)$$

where $\alpha = e^2/\hbar c$ is the fine structure constant, and

$$\langle f| \, e^{i\mathbf{k}\cdot\mathbf{r}}\hat{\mathbf{e}}_{\mathbf{k}}^{(\lambda)} \cdot \mathbf{v} \, |i\rangle = \frac{1}{\mu} \int \psi_f{}^*(\mathbf{r}) e^{i\mathbf{k}\cdot\mathbf{r}}\hat{\mathbf{e}}_{\mathbf{k}}^{(\lambda)} \cdot \frac{\hbar}{i} \nabla \psi_i(\mathbf{r}) \, d^3r \qquad (22.33)$$

is the one-particle matrix element that determines the intensity of the transition. In formula (22.32) $\bar{n}_i$ is to be used for absorption and $\bar{n}_f$ for emission of a photon.

The photon flux per unit frequency is

$$I_0 = \frac{\bar{n}}{L^3} \, c\hbar\rho(\hbar\omega) = \frac{\bar{n}\omega^2}{8\pi^3 c^2} \, d\Omega$$

If we define a cross section $\sigma(\omega)$ for photon absorption (or stimulated emission) by the relation

$$w = \int \sigma(\omega) I_0 \, d\omega$$

---

[4] If in the initial state the photon modes are very selectively occupied, as in masers and lasers, application of the Golden Rule, which presupposes a weak dependence of the matrix element on the occupation number, may be wholly inappropriate. See also the remarks in Section 18.5.

the expressions obtained in Sections 18.4, 18.5, and 18.6 by treating the radiation field as classical and prescribed are recovered exactly.

If $n_i = 0$, (22.32) accounts for the rate of *spontaneous* emission of photons in the transition from an excited state of the radiating atom. If the transition is assumed to be of the *electric dipole* type ($e^{i\mathbf{k}\cdot\mathbf{r}} \approx 1$) and the substitution (18.55) is made for the matrix element, the transition rate for spontaneous emission of light into the solid angle $d\Omega$ with a polarization vector $\mathbf{e}_{\mathbf{k}}^{(\lambda)}$ and frequency ω becomes

$$w = \frac{\alpha}{2\pi c^2}\,\omega^3\,|\langle f|\,\hat{\mathbf{e}}_{\mathbf{k}}^{(\lambda)}\cdot\mathbf{r}\,|i\rangle|^2 = \frac{\alpha}{2\pi c^2}\,\omega^3\,\langle f|\,\hat{\mathbf{e}}_{\mathbf{k}}^{(\lambda)}\cdot\mathbf{r}\,|i\rangle\langle i|\,\hat{\mathbf{e}}_{\mathbf{k}}^{(\lambda)}\cdot\mathbf{r}\,|f\rangle$$

The total integrated emission rate is obtained by summing over the two polarizations and integrating over all angles of emission. Noting that in dyadic notation,

$$\int [\hat{\mathbf{e}}_{\mathbf{k}}^{(1)}\hat{\mathbf{e}}_{\mathbf{k}}^{(1)} + \hat{\mathbf{e}}_{\mathbf{k}}^{(2)}\hat{\mathbf{e}}_{\mathbf{k}}^{(2)}]\,d\Omega_{\mathbf{k}} = \int [I - \hat{\mathbf{k}}\hat{\mathbf{k}}]\,d\Omega_{\mathbf{k}} = \frac{8\pi}{3}\,I$$

we obtain

$$\sum w = \frac{4\alpha}{3c^2}\,\omega^3\,|\langle f|\,\mathbf{r}\,|i\rangle|^2 \tag{22.34}$$

If the initial state of the radiating one-electron atom is characterized by the quantum numbers *nlm*, the intensity may be calculated from equations (18.62) and (18.68) by observing that

$$|Y_1^{\,m}(\hat{\mathbf{i}})|^2 + |Y_1^{\,m}(\hat{\mathbf{j}})|^2 + |Y_1^{\,m}(\hat{\mathbf{k}})|^2 = \frac{3}{4\pi}$$

The selection rules for the emission of electric dipole radiation are, of course, the same as derived in Section 18.5. The emission rates for transitions of higher multipolarity may be computed by similar methods.

Proper higher-order perturbation calculations on interacting radiation and matter fields belong to the domain of quantum electrodynamics. If, however, the matter field that creates and annihilates photons can be approximated by a *prescribed* classical current distribution, the dynamics of the radiation field is considerably simplified, and it is not necessary to resort to perturbation expansions. The Hamiltonian of the radiation field coupled to a prescribed classical current density $\mathbf{j}(\mathbf{r}, t)$ is obtained from (22.26) by letting the material carrier of the current become very massive ($\mu \to \infty$) while keeping $\mathbf{j}(\mathbf{r}, t)$ finite:

$$\mathscr{H} = \frac{1}{2\pi}\int \mathbf{E}^{(-)}\cdot\mathbf{E}^{(+)}\,d^3r - \frac{1}{c}\int \mathbf{j}(\mathbf{r}, t)\cdot\mathbf{A}(\mathbf{r})\,d^3r \tag{22.35}$$

This expression is transformed in the momentum representation as:

$$\mathscr{H} = \sum_{\mathbf{k}} \hbar \omega_k [a_1^\dagger(\mathbf{k})a_1(\mathbf{k}) + a_2^\dagger(\mathbf{k})a_2(\mathbf{k})]$$

$$- \sqrt{4\pi\hbar} \sum_{\mathbf{k}} \frac{1}{\sqrt{2\omega_k}} \mathbf{j}(\mathbf{k}, t) \cdot [a_1(\mathbf{k})\hat{\mathbf{e}}_{\mathbf{k}}^{(1)} + a_2(\mathbf{k})\hat{\mathbf{e}}_{\mathbf{k}}^{(2)}]$$

$$- \sqrt{4\pi\hbar} \sum_{\mathbf{k}} \frac{1}{\sqrt{2\omega_k}} \mathbf{j}(\mathbf{k}, t) \cdot [a_1^\dagger(\mathbf{k})\hat{\mathbf{e}}_{\mathbf{k}}^{(1)} + a_2^\dagger(\mathbf{k})\hat{\mathbf{e}}_{\mathbf{k}}^{(2)}] \qquad (22.36)$$

where $\mathbf{j}(\mathbf{k}, t)$ is the Fourier transform of the applied current:

$$\mathbf{j}(\mathbf{k}, t) = \frac{1}{L^{3/2}} \int \mathbf{j}(\mathbf{r}, t)e^{i\mathbf{k}\cdot\mathbf{r}} \, d^3r \qquad (22.37)$$

The Hamiltonian (22.36) is a sum of independent contributions from each photon mode with wave vector $\mathbf{k}$ and a definite polarization. Omitting, for brevity, all reference to the particular mode, each addend is of the form

$$\hbar \omega a^\dagger a + f(t)a + f^*(t)a^\dagger$$

which, except for the missing zero-point energy, is the same as the Hamiltonian (15.114) for the forced linear harmonic oscillator. The complex-valued function of time $f(t)$ stands for

$$f(t) = -\sqrt{\frac{2\pi\hbar}{\omega_k}} \mathbf{j}(\mathbf{k}, t) \cdot \hat{\mathbf{e}}_{\mathbf{k}}^{(\lambda)} \qquad (22.38)$$

Our interest will be focused on the changes produced by a current distribution which is effective only during a finite time interval but vanishes in the distant past and the remote future as $t \rightarrow \pm\infty$. Indeed, it is convenient to assume that $f(t) \neq 0$ only for $t_2 > t > t_1 > 0$. Therefore, the electromagnetic field is free before t_1 and after t_2.

If at $t = 0$ there are no photons present, the initial state is the vacuum. Since, with all reactions on the sources of the electromagnetic field neglected, each photon mode develops independently in time, the results of Section 15.9 are directly applicable here. According to equations (15.143) and (15.133), in the Schrödinger picture the state of the radiation field at time $t > t_2$, i.e., after all currents have ceased, is given by

$$\Psi(t) = e^{-(i/\hbar)\mathscr{H}t} S\Psi^{(0)} \qquad (22.39)$$

where

$$S = \exp\left[\frac{i}{\hbar c} \int_{t_1}^{t_2} dt \int \mathbf{j}(\mathbf{r}, t) \cdot \mathbf{A}_{in}(\mathbf{r}, t) \, d^3r\right] \qquad (22.40)$$

The state $S\Psi^{(0)}$ may, by use of relation (15.135) and the definitions of

Section 15.9, be expressed in the form

$$S\Psi^{(0)} = \exp\left[-\pi \sum_{\mathbf{k},\lambda} \frac{1}{\hbar\omega_k} |j(\mathbf{k},\omega_k,\lambda)|^2\right] \exp\left[i \sum_{k,\lambda} \sqrt{\frac{2\pi}{\hbar\omega_k}} j(\mathbf{k},\omega_k,\lambda)a_\lambda^{\dagger}(\mathbf{k})\right]\Psi^{(0)}$$

(22.41)

where

$$j(\mathbf{k},\omega_k,\lambda) = \frac{1}{L^{3/2}} \int_{-\infty}^{+\infty} dt \int d^3r \, e^{i(\omega t - \mathbf{k}\cdot\mathbf{r})} \mathbf{j}(\mathbf{r},t) \cdot \mathbf{e}_{\mathbf{k}}^{(\lambda)}$$

(22.42)

From the discussion in Sections 15.9 and 15.10 on *coherent states* it follows that for $t > t_2$ the state (22.39) is an eigenstate of the operators $\mathbf{A}^{(+)}(\mathbf{r})$ and $\mathbf{E}^{(+)}(\mathbf{r})$ and that each photon mode with specified $\mathbf{k}$ and polarization is populated by photons according to a Poisson distribution. The successive emissions of photons from a prescribed current distribution may, therefore, be regarded as statistically independent events, and this behavior is relevant in quantum electronics.[5]

It can be verified that the expectation value of $\mathbf{A}(\mathbf{r})$ in the state (22.39) is identical with the classical retarded vector potential generated, in the Coulomb gauge, by the prescribed current distribution. This is, of course, as it should be according to the correspondence principle. The most characteristic feature of any coherent state Ψ_{coh} like (22.39) is the factorization law

$$\langle \Psi_{coh}| \prod_i \mathbf{E}^{(-)}(\mathbf{r}_i) \prod_j \mathbf{E}^{(+)}(\mathbf{r}_j) |\Psi_{coh}\rangle$$

$$= \prod_i \langle \Psi_{coh}| \mathbf{E}^{(-)}(\mathbf{r}_i) |\Psi_{coh}\rangle \prod_j \langle \Psi_{coh}| \mathbf{E}^{(+)}(\mathbf{r}_j) |\Psi_{coh}\rangle$$

for any normally ordered product of field operators. In particular, the energy density of the radiation field for a coherent state is proportional to

$$\langle \Psi_{coh}| \mathbf{E}^{(-)}(\mathbf{r}) \cdot \mathbf{E}^{(+)}(\mathbf{r}) |\Psi_{coh}\rangle = \langle \Psi_{coh}| \mathbf{E}^{(-)}(\mathbf{r}) |\Psi_{coh}\rangle \cdot \langle \Psi_{coh}| \mathbf{E}^{(+)}(\mathbf{r}) |\Psi_{coh}\rangle$$

showing that in a coherent, or quasiclassical, state the field intensity may be calculated by treating the expectation values of $\mathbf{E}^{(+)}$ and $\mathbf{E}^{(-)}$ like classical fields.[6]

If the radiation field is in thermal equilibrium, the theory of Section 21.7 may be applied provided that the Lagrangian multiplier α is set equal to

[5] For more information on coherent states and their role in quantum electronics and optics, see R. J. Glauber, *Optical Coherence and Photon Statistics* in C. DeWitt, A. Blandin, C. Cohen-Tannoudji, editors, *Quantum Optics and Electronics*, Gordon and Breach, New York, 1965, p. 63. Also, J. R. Klauder and E. C. G. Sudarshan, *Fundamentals of Quantum Optics*, W. A. Benjamin, New York, 1968, and W. H. Louisell, *Radiation and Noise in Quantum Electronics*, McGraw-Hill Book Company, New York, 1965.

[6] Coherent states of the radiation field produced by classical currents are also helpful in explaining, and removing, the "infrared catastrophe" of quantum electrodynamics; see J. D. Bjorken and S. D. Drell, *Relativistic Quantum Fields*, McGraw-Hill Book Company, New York, 1965, Section 17.10.

zero. This is to take account of the fact that it is inappropriate to require the total number of photons to be fixed because photons have zero mass and are created and annihilated singly and without energy threshold. Since photons are bosons, equation (21.92) gives the average number of photons in the mode **k** in thermal equilibrium as

$$\langle N_{\mathbf{k}} \rangle = [e^{\hbar\omega/kT} - 1]^{-1} \tag{22.43}$$

This is, of course, Planck's familiar *black-body radiation* formula with which the quantum theory was inaugurated.

Problems

1. Calculate the emission rate for the first line in the Balmer series of the hydrogen spectrum.
2. Prove that in a hydrogen atom the radiative transition from the 2S excited to the 1S ground state cannot occur by emission of one photon. Outline (but do not attempt to carry through in detail) the calculation of the transition rate for two-photon emission. In the dipole approximation, show that the two photons are preferentially emitted in the same direction or in opposite directions and that the angular correlation function is proportional to $1+\cos^2\theta$ if θ is the angle between the photons and if the polarization of the emitted light is not observed. Estimate the order of magnitude of the lifetime of the metastable 2S state.
3. Evaluate the peak value of the cross section for electric dipole absorption by a linear harmonic oscillator in its ground state, assuming that the excited state is depleted only by spontaneous emission. Use equation (18.130).
4. Show that the total probability for emitting from a prescribed classical current distribution n photons into a specified set R of photon modes, and none into any other mode, is given by

$$P_n(R) = e^{-\bar{n}} \frac{r^n}{n!}$$

where $\bar{n}$ is the mean number of photons emitted into all modes. Evaluate $\bar{n}$ and r in terms of the Fourier components of the impressed current density.
5. Show that the average density operator for a particular photon mode in thermal equilibrium may be written in the forms

$$\bar{\rho} = \frac{e^{-(\hbar\omega/kT)a^{\dagger}a}}{Z} = \frac{e^{-\hbar\omega/kT}}{\langle N \rangle} \sum_m e^{-m(\hbar\omega/kT)} |m\rangle\langle m| = \frac{1}{\pi\langle N \rangle} \int e^{-|\alpha|^2/\langle N \rangle} |\alpha\rangle \, d^2\alpha \langle\alpha|$$

where Z is the partition function and $\langle N \rangle$ is given by equation (22.43).

CHAPTER 23

Relativistic Electron Theory

1. The Electron-Positron Field. In developing the relativistic quantum theory of electrons, it is again appropriate to list the pertinent empirical facts first. These particles have mass μ and spin $\frac{1}{2}$, and they are fermions. In the interaction with the electromagnetic field, positrons of charge e are created simultaneously with electrons of charge $-e$, and it is a compelling feature of the theory that it can provide a unified description of both particles and antiparticles in terms of a common *electron-positron field*. A free electron or positron is most easily characterized by its linear momentum $\mathbf{p}$ and energy E_p, related as

$$E_p = \sqrt{c^2 p^2 + (\mu c^2)^2} \tag{23.1}$$

As in the case of the photon, only a measurement of the component of the particle's spin in the direction of the momentum is compatible with a sharp energy-momentum vector. Hence, the electron or positron may have definite positive (R) or negative (L) *helicity*.

We introduce creation and annihilation operators for electrons ($a^\dagger$ and a) and positrons ($b^\dagger$ and b) in the two helicity states, subject to the *anti-commutation* relations:

$$
\begin{aligned}
a_R(\mathbf{p})a_R^\dagger(\mathbf{p}') + a_R^\dagger(\mathbf{p}')a_R(\mathbf{p}) &= a_L(\mathbf{p})a_L^\dagger(\mathbf{p}') + a_L^\dagger(\mathbf{p}')a_L(\mathbf{p}) \\
&= b_R(\mathbf{p})b_R^\dagger(\mathbf{p}') + b_R^\dagger(\mathbf{p}')b_R(\mathbf{p}) \\
&= b_L(\mathbf{p})b_L^\dagger(\mathbf{p}') + b_L^\dagger(\mathbf{p}')b_L(\mathbf{p}) = \delta(\mathbf{p} - \mathbf{p}')
\end{aligned}
\tag{23.2}
$$

All other anticommutators of these eight operators are set equal to zero, partly as a consequence of the fermion theory developed in Chapter 20, and partly (namely, for anticommutators of a or $a^\dagger$ with b or $b^\dagger$) as an assumption that will be seen to be consistent with the formulation of a unified electron-positron theory.[1]

---

[1] In Chapter 22 the photon momentum was restricted to discrete values by the imposition of periodic boundary conditions on the radiation field. For the electron-positron field it is convenient to let $L \to \infty$ from the beginning and allow all of momentum space for $\mathbf{p}$.

The operators for the energy, linear momentum, and charge of a system of free electrons and positrons are easily written down:

$$\mathscr{H} = \int E_p[a_R^\dagger(\mathbf{p})a_R(\mathbf{p}) + a_L^\dagger(\mathbf{p})a_L(\mathbf{p})$$
$$+ b_R^\dagger(\mathbf{p})b_R(\mathbf{p}) + b_L^\dagger(\mathbf{p})b_L(\mathbf{p})]\, d^3p \quad (23.3)$$

$$\mathbf{P} = \int \mathbf{p}[a_R^\dagger(\mathbf{p})a_R(\mathbf{p}) + a_L^\dagger(\mathbf{p})a_L(\mathbf{p})$$
$$+ b_R^\dagger(\mathbf{p})b_R(\mathbf{p}) + b_L^\dagger(\mathbf{p})b_L(\mathbf{p})]\, d^3p \quad (23.4)$$

$$Q = e\int [-a_R^\dagger(\mathbf{p})a_R(\mathbf{p}) - a_L^\dagger(\mathbf{p})a_L(\mathbf{p})$$
$$+ b_R^\dagger(\mathbf{p})b_R(\mathbf{p}) + b_L^\dagger(\mathbf{p})b_L(\mathbf{p})]\, d^3p \quad (23.5)$$

Also,

$$\mathscr{J}\cdot\hat{\mathbf{p}},\, a_R^\dagger(\mathbf{p})]a_R(\mathbf{p}) + [\mathscr{J}\cdot\hat{\mathbf{p}},\, a_L^\dagger(\mathbf{p})]a_L(\mathbf{p})$$
$$+ [\mathscr{J}\cdot\hat{\mathbf{p}},\, b_R^\dagger(\mathbf{p})]b_R(\mathbf{p}) + [\mathscr{J}\cdot\hat{\mathbf{p}},\, b_L^\dagger(\mathbf{p})]b_L(\mathbf{p})$$
$$= \frac{\hbar}{2}[a_R^\dagger(\mathbf{p})a_R(\mathbf{p}) - a_L^\dagger(\mathbf{p})a_L(\mathbf{p}) + b_R^\dagger(\mathbf{p})b_R(\mathbf{p}) - b_L^\dagger(\mathbf{p})b_L(\mathbf{p})] \quad (23.6)$$

is the spin component along the direction of the particle momentum $\mathbf{p}$ per unit volume in momentum space.

It is the objective of *local quantum field theory* to seek ways of expressing these physical quantities as volume integrals of local (density) operators so that the operators for the total energy, momentum, etc. eventually appear in the form

$$\mathscr{H} = \int \psi^\dagger(\mathbf{r})K\psi(\mathbf{r})\, d^3r \quad (23.7)$$

where K is an appropriate one-particle operator. The field operators $\psi(\mathbf{r})$ are defined in the usual manner by a transformation from the momentum to the coordinate representation as Fourier integrals, but particular care is required in the construction of the Fourier coefficients as well as in the choice of the one-particle operators representing physical quantities.

For example, it is formally possible to write the energy of the system of free electrons and positrons as

$$\mathscr{H} = \int \psi^\dagger(\mathbf{r})\sqrt{-\hbar^2c^2\nabla^2 + (\mu c^2)^2}\,\psi(\mathbf{r})\, d^3r$$

Such a choice was seen to be unsatisfactory in the case of photons because it implies a nonlocal expression for the energy density. For photons, this impasse led to the inference that a reasonable definition of a one-photon probability density in ordinary space cannot be given.

In the case of relativistic particles with mass, the same conclusion holds (although the expansion of $\sqrt{-\hbar^2 c^2 \nabla^2 + (\mu c^2)^2}$ in powers of ∇^2 shows that the nonlocal effects, which arise from the presence of arbitrarily high derivatives, disappear in the nonrelativistic approximation[2]), and the goal of formulating a strictly *relativistic one-particle theory* must be sacrificed.

It is, of course, *possible* to produce a sensible field theory for particles with mass along similar lines as was done for photons, and this is customarily done for bosons such as pions (spin zero). However, Dirac discovered that for electrons (and positrons), which are fermions of spin one-half, the field theory may be developed in a form that is strongly reminiscent of one-particle quantum mechanics. A straightforward relativistic one-particle *approximation* thus becomes feasible (see Chapter 24).

In the language of quantum field theory, the essence of Dirac's discovery may be said to lie in the observation that the physical quantities (22.3)–(22.5) may be reexpressed in alternate form by the use of the anticommutation relations and some simple changes of variables of integration, resulting in:

$$\mathscr{H} = \int E_p [a_R^\dagger(\mathbf{p}) a_R(\mathbf{p}) + a_L^\dagger(\mathbf{p}) a_L(\mathbf{p})$$
$$- b_R(-\mathbf{p}) b_R^\dagger(-\mathbf{p}) - b_L(-\mathbf{p}) b_L^\dagger(-\mathbf{p}) + C] \, d^3 p \quad (23.8)$$

$$\mathbf{P} = \int \mathbf{p} [a_R^\dagger(\mathbf{p}) a_R(\mathbf{p}) + a_L^\dagger(\mathbf{p}) a_L(\mathbf{p})$$
$$+ b_R(-\mathbf{p}) b_R^\dagger(-\mathbf{p}) + b_L(-\mathbf{p}) b_L^\dagger(-\mathbf{p}) + C] \, d^3 p \quad (23.9)$$

$$Q = e \int [-a_R^\dagger(\mathbf{p}) a_R(\mathbf{p}) - a_L^\dagger(\mathbf{p}) a_L(\mathbf{p})$$
$$- b_R(-\mathbf{p}) b_R^\dagger(-\mathbf{p}) - b_L(-\mathbf{p}) b_L^\dagger(-\mathbf{p}) + C] \, d^3 p \quad (23.10)$$

and

$$-[\mathscr{J} \cdot \hat{\mathbf{p}}, a_R^\dagger(\mathbf{p})] a_R(\mathbf{p}) + [\mathscr{J} \cdot \hat{\mathbf{p}}, a_L^\dagger(\mathbf{p})] a_L(\mathbf{p})$$
$$- [\mathscr{J} \cdot \hat{\mathbf{p}}, b_R^\dagger(-\mathbf{p})] b_R(-\mathbf{p}) - [\mathscr{J} \cdot \hat{\mathbf{p}}, b_L^\dagger(-\mathbf{p})] b_L(-\mathbf{p})$$
$$= \frac{\hbar}{2} [a_R^\dagger(-\mathbf{p}) a_R(-\mathbf{p}) - a_L^\dagger(-\mathbf{p}) a_L(-\mathbf{p})$$
$$- b_R(-\mathbf{p}) b_R^\dagger(-\mathbf{p}) + b_L(-\mathbf{p}) b_L^\dagger(-\mathbf{p})] \quad (23.11)$$

If we momentarily disregard the singular additive terms symbolized by the infinite constant C, these expressions show that the *annihilation* operator for a *positron* $b_R(-\mathbf{p})$ can also be interpreted as an operator *creating* an *electron* of momentum $\mathbf{p}$ and positive helicity but negative energy, $-E_p$.

---

[2] The *Foldy-Wouthuysen* version of relativistic electron theory is based on the use of the square root operator for the energy and can be put in local form only by successive approximations. See M. E. Rose, *Relativistic Electron Theory*, John Wiley and Sons, New York, 1961.

Such negative energies appear quite naturally in a relativistic theory which relates energy and momentum by the equation

$$E_p{}^2 = c^2 p^2 + (\mu c^2)^2 \tag{23.12}$$

allowing in addition to equation (23.1) the solution $-\sqrt{c^2 p^2 + (\mu c^2)^2}$.

In accordance with these clues we construct a field operator

$$\psi(\mathbf{r}) = \psi^{(+)}(\mathbf{r}) + \psi^{(-)}(\mathbf{r}) \tag{23.13}$$

as the sum of positive and negative frequency (energy) parts defined as

$$\psi^{(+)}(\mathbf{r}) = \frac{1}{(2\pi\hbar)^{3/2}} \int [u^{(R)}(\mathbf{p}) a_R(\mathbf{p}) + u^{(L)}(\mathbf{p}) a_L(\mathbf{p})] e^{(i/\hbar)\mathbf{p}\cdot\mathbf{r}} \, d^3 p \tag{23.14}$$

$$\psi^{(-)}(\mathbf{r}) = \frac{1}{(2\pi\hbar)^{3/2}} \int [v^{(R)}(-\mathbf{p}) b_R^\dagger(\mathbf{p}) + v^{(L)}(-\mathbf{p}) b_L^\dagger(\mathbf{p})] e^{-(i/\hbar)\mathbf{p}\cdot\mathbf{r}} \, d^3 p$$

$$= \frac{1}{(2\pi\hbar)^{3/2}} \int [v^{(R)}(\mathbf{p}) b_R^\dagger(-\mathbf{p}) + v^{(L)}(\mathbf{p}) b_L^\dagger(-\mathbf{p})] e^{(i/\hbar)\mathbf{p}\cdot\mathbf{r}} \, d^3 p \tag{23.15}$$

The coefficients $u^{(R)}(\mathbf{p})$, $u^{(L)}(\mathbf{p})$, $v^{(R)}(\mathbf{p})$, and $v^{(L)}(\mathbf{p})$ are one-column matrices which must be orthogonal to each other, such that for a fixed momentum $\mathbf{p}$,

$$u^{(R)\dagger} u^{(L)} = (u_1^{(R)*} \; u_2^{(R)*} \cdots) \begin{pmatrix} u_1^{(L)} \\ u_2^{(L)} \\ \cdot \\ \cdot \\ \cdot \end{pmatrix} = 0$$

and similarly,

$$u^{(R)\dagger} u^{(L)} = u^{(R)\dagger} v^{(R)} = u^{(R)\dagger} v^{(L)} = u^{(L)\dagger} v^{(R)} = u^{(L)\dagger} v^{(L)} = v^{(R)\dagger} v^{(L)} = 0 \tag{23.16}$$

Generalizing the terminology introduced in Section 12.4, we shall call these one-column matrices, with an as yet unspecified number of rows, *spinors*, and *Dirac spinors* on occasion when it is essential to avoid confusion with the two-component matrices of Chapter 12. We shall assume these spinors to be normalized according to the relations[3]

$$u^{(R)\dagger} u^{(R)} = u^{(L)\dagger} u^{(L)} = v^{(R)\dagger} v^{(R)} = v^{(L)\dagger} v^{(L)} = 1 \tag{23.17}$$

If such spinors can be found, the total linear momentum and the total charge

---

[3] *Warning.* A variety of different normalizations for Dirac spinors are current in the literature. Often the right-hand side of (23.17) is chosen equal to $E_p/\mu c^2$.

of the system can be written as

$$\mathbf{P} = \int \psi^\dagger(\mathbf{r}) \frac{\hbar}{i} \nabla \psi(\mathbf{r}) \, d^3r \tag{23.18}$$

$$Q = -\frac{e}{2} \int [\psi^\dagger(\mathbf{r})\psi(\mathbf{r}) - \tilde{\psi}(\mathbf{r})\tilde{\psi}^\dagger(\mathbf{r})] \, d^3r \tag{23.19}$$

where the symbol $\sim$ indicates *matrix transposition*. Three comments are in order: (a) The constant term in the integrand of (23.9) has been omitted because it merely insures that the vacuum has zero momentum and is not needed if all momenta are measured relative to the vacuum. (b) Expression (23.18) for the linear momentum has the same form in the relativistic as in the nonrelativistic theory because, as indicated by equation (2.25), $(\hbar/i)\nabla$ represents the three spatial components of a four-vector. Angular momentum is made relativistic in a similar straightforward manner (Section 23.3). (c) The peculiar form of equation (23.19) arises from rewriting (23.5) as

$$Q = -\frac{e}{2} \int [a_R^\dagger(\mathbf{p})a_R(\mathbf{p}) - a_R(\mathbf{p})a_R^\dagger(\mathbf{p}) + a_L^\dagger(\mathbf{p})a_L(\mathbf{p})$$
$$- a_L(\mathbf{p})a_L^\dagger(\mathbf{p}) - b_R^\dagger(\mathbf{p})b_R(\mathbf{p}) + b_R(\mathbf{p})b_R^\dagger(\mathbf{p})$$
$$- b_L^\dagger(\mathbf{p})b_L(\mathbf{p}) + b_L(\mathbf{p})b_L^\dagger(\mathbf{p})] \, d^3p \tag{23.20}$$

2. The Dirac Equation. It remains to show that the energy of the system can also be written in the form.

$$\mathscr{H} = \int \psi^\dagger(\mathbf{r})H\psi(\mathbf{r}) \, d^3r + \text{const.} \tag{23.21}$$

Substitution of the fields (23.14) and (23.15) in this integral shows that this goal can be accomplished if we require that

$$Hu^{(R,L)}(\mathbf{p})e^{(i/\hbar)\mathbf{p}\cdot\mathbf{r}} = E_p u^{(R,L)}(\mathbf{p})e^{(i/\hbar)\mathbf{p}\cdot\mathbf{r}} \tag{23.22}$$

$$Hv^{(R,L)}(\mathbf{p})e^{(i/\hbar)\mathbf{p}\cdot\mathbf{r}} = -E_p v^{(R,L)}(\mathbf{p})e^{(i/\hbar)\mathbf{p}\cdot\mathbf{r}} \tag{23.23}$$

If (23.21) is to be an integral over a localized energy density, the requirements of Lorentz invariance make it mandatory to seek a Hamiltonian which is *linear* in the differential operator ∇. Therefore we attempt to construct H in the form

$$H = c\boldsymbol{\alpha} \cdot \frac{\hbar}{i} \nabla + \beta\mu c^2 \tag{23.24}$$

leaving the constant square *matrices* α_x, α_y, α_z, and β as yet undetermined. With this choice for H, equations (23.22) and (23.23) reduce to

$$(c\boldsymbol{\alpha} \cdot \mathbf{p} + \beta\mu c^2)u^{(R,L)}(\mathbf{p}) = E_p u^{(R,L)}(\mathbf{p}) \tag{23.25}$$

$$(c\boldsymbol{\alpha} \cdot \mathbf{p} + \beta\mu c^2)v^{(R,L)}(\mathbf{p}) = -E_p v^{(R,L)}(\mathbf{p}) \tag{23.26}$$

Since the eigenvalues $\pm E_p$ are real and the eigenspinors orthogonal, the operator-matrix

$$H_p = c\boldsymbol{\alpha} \cdot \mathbf{p} + \beta u c^2 \qquad (23.27)$$

must be Hermitian. Thus $\boldsymbol{\alpha}$ and β are four Hermitian matrices. They must be *at least* four dimensional (four rows and four columns), so that H_p will have four orthogonal eigenspinors, and they should be *no more* than four dimensional if the description of electrons and positrons in terms of momentum, energy, and helicity is complete.

Since the eigenvalues of H_p are to be E_p and $-E_p$, and each of these is to be doubly degenerate, all four eigenvalues of $(H_p)^2$ must be equal to $E_p{}^2$, hence $(H_p)^2 = E_p{}^2 I$ and trace $H_p = 0$ is required. If we take the square of (23.27) and use the relation (23.12), we thus obtain the conditions

$$\alpha_x{}^2 = \alpha_y{}^2 = \alpha_z{}^2 = \beta^2 = I \qquad (23.28)$$

and

$$\alpha_x\alpha_y + \alpha_y\alpha_x = \alpha_y\alpha_z + \alpha_z\alpha_y = \alpha_z\alpha_x + \alpha_x\alpha_z$$
$$= \alpha_x\beta + \beta\alpha_x = \alpha_y\beta + \beta\alpha_y = \alpha_z\beta + \beta\alpha_z = 0 \quad (23.29)$$

Our problem thus reduces to a purely algebraic one of finding *four-dimensional* Hermitian matrices with the properties (23.28) and (23.29). Pauli has proven that all matrix solutions to these equations for $\boldsymbol{\alpha}$ and β are reducible by unitary transformation to one another; hence, it is sufficient to determine one particular 4×4 solution and show that all traces vanish.[4]

Exercise 23.1. Noting that $\alpha_x = -\alpha_y\alpha_x\alpha_y$, prove that the trace of α_x, α_y, α_z and β vanishes, and show that each of these matrices has n eigenvalues $+1$ and n eigenvalues -1, where $2n$ is the dimension of the matrices.

Exercise 23.2. Using only the conditions (23.28) and (23.29), prove that $\boldsymbol{\alpha}$ and β are at least four-dimensional.

The most widely used representation of the $\boldsymbol{\alpha}$ and β matrices is specified in terms of the familiar 2×2 Pauli matrices:

$$\alpha = \begin{pmatrix} 0 & \sigma \\ \sigma & 0 \end{pmatrix}, \qquad \beta = \begin{pmatrix} I & 0 \\ 0 & -I \end{pmatrix} \qquad (23.30)$$

Every element in these 2×2 matrices is itself to be understood as a 2×2 matrix. We shall refer to (23.30) as the *standard representation*.

---

[4] For an exposition of the algebra of $\boldsymbol{\alpha}$ and β matrices and for further references, see M. E. Rose, *Relativistic Electron Theory*, John Wiley and Sons, New York, 1961.

Exercise 23.3. Verify the validity of the solutions (23.30) to the problem posed by conditions (23.28) and (23.29).

The discussion of this section so far leaves unidentified the Hermitian matrix which represents the helicity. Such a matrix must commute with H_p and distinguish, by its eigenvalues, the two helicity states R and L. The matrix can be readily derived after the angular momentum operator is obtained (Sections 23.3 and 23.4).

The anticommutation relations for the field operators can now be derived from equation (23.2) and the remarks following this equation. The four eigenspinors $u^{(R)}(\mathbf{p})$, $u^{(L)}(\mathbf{p})$, $v^{(R)}(\mathbf{p})$, and $v^{(L)}(\mathbf{p})$ of the 4×4 matrix H_p are orthonormal; hence, they form a complete set of spinors, and the *closure* relation

$$u^{(R)}u^{(R)\dagger} + u^{(L)}u^{(L)\dagger} + v^{(R)}v^{(R)\dagger} + v^{(L)}v^{(L)\dagger} = I \qquad (23.31)$$

holds. Using this relation, it is easy to verify that

$$\psi_\alpha(\mathbf{r})\psi_\beta(\mathbf{r}') + \psi_\beta(\mathbf{r}')\psi_\alpha(\mathbf{r}) = 0$$
$$\psi_\alpha^\dagger(\mathbf{r})\psi_\beta^\dagger(\mathbf{r}') + \psi_\beta^\dagger(\mathbf{r}')\psi_\alpha^\dagger(\mathbf{r}) = 0 \qquad (23.32)$$
$$\psi_\alpha(\mathbf{r})\psi_\beta^\dagger(\mathbf{r}') + \psi_\beta^\dagger(\mathbf{r}')\psi_\alpha(\mathbf{r}) = \delta(\mathbf{r} - \mathbf{r}')\delta_{\alpha\beta}$$

Exercise 23.4. Verify equations (23.32).

Exercise 23.5. Using (23.18) and (23.32), prove that

$$[\psi(\mathbf{r}), \mathbf{P}] = \frac{\hbar}{i}\nabla\psi(\mathbf{r}) \qquad (23.33)$$

From the equations of motion for the creation and annihilation operators, the *time development* of the *free Dirac field* is readily deduced by use of the Hamiltonian (23.3) and we obtain in the *Heisenberg picture*

$$\psi^{(+)}(\mathbf{r}, t) = \frac{1}{(2\pi\hbar)^{3/2}}\int [u^{(R)}(\mathbf{p})a_R(\mathbf{p}) + u^{(L)}(\mathbf{p})a_L(\mathbf{p})]e^{(i/\hbar)(\mathbf{p}\cdot\mathbf{r} - E_p t)} d^3p \quad (23.34)$$

$$\psi^{(-)}(\mathbf{r}, t) = \frac{1}{(2\pi\hbar)^{3/2}}\int [v^{(R)}(\mathbf{p})b_R^\dagger(-\mathbf{p}) + v^{(L)}(\mathbf{p})b_L^\dagger(-\mathbf{p})]e^{(i/\hbar)(\mathbf{p}\cdot\mathbf{r} + E_p t)} d^3p$$

$$(23.35)$$

If equations (23.22), (23.23), and (23.24) are applied, we see that both frequency components of ψ and the total field itself satisfy the *field equation*

$$i\hbar\frac{\partial\psi(\mathbf{r}, t)}{\partial t} = c\left(\frac{\hbar}{i}\boldsymbol{\alpha}\cdot\nabla + \beta\mu c\right)\psi(\mathbf{r}, t) \qquad (23.36)$$

This equation, which is the analog of the time-dependent Schrödinger equation of nonrelativistic-quantum mechanics and of Maxwell's equations for the electromagnetic field, is known as the *Dirac equation of the electron.* It can be cast in a more appealing form, particularly suitable for discussion of Lorentz covariance, by the introduction of a new set of 4 × 4 matrices:

$$\gamma^0 = \beta, \qquad \gamma^1 = \beta\alpha_x, \qquad \gamma^2 = \beta\alpha_y, \qquad \gamma^3 = \beta\alpha_z \qquad (23.37)$$

Using the relativistic notation and the metric introduced in Section 22.2 and the summation convention, with Greek indices running from 0 to 3, equation (23.36) may be rewritten in the compact form

$$\gamma^\mu \frac{\partial \psi}{\partial x^\mu} + i\kappa\psi = 0 \qquad (23.38)$$

where we have abbreviated the inverse of the Compton wavelength of the electron as

$$\kappa = \mu c/\hbar$$

and denoted

$$\gamma^\mu \frac{\partial}{\partial x^\mu} = \gamma^0 \frac{\partial}{\partial x^0} + \boldsymbol{\gamma} \cdot \nabla = \beta \frac{1}{c} \frac{\partial}{\partial t} + \beta\boldsymbol{\alpha} \cdot \nabla$$

The relations (23.28) and (23.29) may be summarized as anticommutation relations for the *Dirac γ matrices;*

$$\gamma^\mu \gamma^\nu + \gamma^\nu \gamma^\mu = 2g^{\mu\nu}1 \qquad (23.39)$$

The one-particle energy-momentum operator is given by

$$p^\mu = \left(\frac{E}{c}, \mathbf{p} \right) = i\hbar \frac{\partial}{\partial x_\mu} = \left(\frac{i\hbar}{c} \frac{\partial}{\partial t}, \frac{\hbar}{i} \nabla \right)$$

The presence of an external electromagnetic field, acting on the matter field, is as usual taken into account by the replacement

$$p^\mu \to p^\mu - \frac{q}{c} A^\mu$$

This prescription defines a *minimal interaction* with the field

$$A^\mu = (\phi, \mathbf{A})$$

With q = −e (e > 0) for electrons, this substitution changes the Dirac equation from its free field form into

$$\gamma^\mu \left(\frac{\partial}{\partial x^\mu} - \frac{ie}{\hbar c} A_\mu \right) \psi + i\kappa\psi = 0 \qquad (23.40)$$

or in the noncovariant form, analogous to equation (23.36)

$$i\hbar \frac{\partial \psi}{\partial t} = \left[c\boldsymbol{\alpha} \cdot \left(\frac{\hbar}{i} \nabla + \frac{e}{c} \mathbf{A} \right) - e\phi + \beta \mu c^2 \right] \psi \tag{23.41}$$

It is useful to define an *adjoint Dirac field* operator by the relation

$$\bar{\psi} = \psi^\dagger \gamma^0$$

Since $\boldsymbol{\gamma}$ is antihermitian and γ^0 Hermitian, Hermitian conjugation of equation (23.40) and multiplication on the right by γ^0 leads to

$$\left(\frac{\partial}{\partial x^\mu} + \frac{ie}{\hbar c} A_\mu \right) \bar{\psi} \gamma^\mu - i\kappa \bar{\psi} = 0 \tag{23.42}$$

If this equation is multiplied on the right by ψ and equation (23.40) on the left by $\bar{\psi}$, and if the resulting equations are added to one another, the *continuity equation*

$$\frac{\partial}{\partial x^\mu} (\bar{\psi} \gamma^\mu \psi) = 0 \tag{23.43}$$

is obtained.

It is easy to prove similarly the further continuity equation

$$\frac{\partial}{\partial x^\mu} (\tilde{\psi} \tilde{\gamma}^\mu \tilde{\bar{\psi}}) = 0 \tag{23.44}$$

Exercise 23.6. Derive the continuity equation (23.44).

Comparing these expressions with the total charge operator (23.19), we infer that the *electric current density* four vector of the electron-positron system is defined by

$$j^\mu = (c\rho, \mathbf{j}) = -\frac{ec}{2} (\bar{\psi} \gamma^\mu \psi - \tilde{\psi} \tilde{\gamma}^\mu \tilde{\bar{\psi}}) \tag{23.45}$$

Conservation of charge is insured by the continuity equations (23.43) and (23.44), or

$$\frac{\partial j^\mu}{\partial x^\mu} = 0 \tag{23.46}$$

3. Relativistic Invariance. While the relativistic invariance of Maxwell's equations for the free radiation field, even in quantized form, needs no proof, since the Lorentz transformations were designed to accomplish just this aim, it is necessary to demonstrate that the Dirac theory is in consonance with the demands of special relativity. Specifically, the requirement of invariance of the theory under inhomogeneous Lorentz (or Poincaré) transformations will serve as a guide in establishing the transformation properties of the electron-positron field. The general theory of the irreducible representations of the *Lorentz group* contains the relevant information, but if nothing more than the transformation properties of a special field is desired, the mathematical structure may be deduced from simple physical considerations.

Einstein's restricted principle of special relativity postulates the equivalence of physical systems which are obtained from each other by geometrical translation or rotation or which differ from one another only by being in uniform relative motion. According to Section 16.1, such equivalent systems can be connected by a unitary transformation of the respective state vectors.

The principle of relativity is implemented by constructing the coordinate transformation

$$x'^{\mu} = a^{\mu}{}_{\nu} x^{\nu} + b^{\mu} \tag{23.47}$$

with real coefficients $a^{\mu}{}_{\nu}$ and b^{μ}, subject to the orthogonality condition

$$dx'^{\mu} dx_{\mu}' = dx^{\nu} dx_{\nu} \qquad \text{or} \qquad a^{\mu}{}_{\lambda} a_{\mu}{}^{\nu} = \delta_{\lambda}{}^{\nu} \tag{23.48}$$

In addition to the *proper orthochronous* Lorentz transformations for which

$$\det a^{\mu}{}_{\nu} = 1, \qquad a^{0}{}_{0} \geq 1$$

the orthogonality condition allows *improper* Lorentz transformations such as space reflections and time reversal, as well as combinations of these with proper orthochronous transformations. Although there is no compelling reason to expect that the coverage of the principle of relativity extends to the improper Lorentz transformations and those reversing the sense of time, it is important to investigate if the proposed theory is invariant under the totality of the transformations licensed by the orthogonality condition (23.48).

It is a fundamental assertion of local quantum field theory that if a Lorentz transformation takes the point $(\mathbf{r}, t)$ into $(\mathbf{r}', t')$ and changes the state Ψ into a state $U\Psi$, the components of $\psi(\mathbf{r}', t')U\Psi$ must be related by a linear transformation to the components of $U\psi(\mathbf{r}, t)\Psi$. Hence, the field must transform as

$$\psi(\mathbf{r}', t')U\Psi = SU\psi(\mathbf{r}, t)\Psi$$

or

$$U^{\dagger}\psi(\mathbf{r}', t')U = S\psi(\mathbf{r}, t) \tag{23.49}$$

where S is a 4×4 matrix that defines the geometrical transformation properties of the spinor whose components, like those of a vector or tensor, are reshuffled in this symmetry operation. [Compare equation (16.104).] It is assumed that the vacuum state is left unchanged by a symmetry transformation: $U\Psi^{(0)} = \Psi^{(0)}$.

We shall first consider three-dimensional rotations as a subgroup of the Lorentz transformations. From the definition of rotations, it follows that we must expect the relations

$$Ua_{R,L}(\mathbf{p}) = a_{R,L}(\mathbf{p}')U, \qquad Ub_{R,L}(\mathbf{p}) = b_{R,L}(\mathbf{p}')U \qquad (23.50)$$

to hold, with $\mathbf{p}'$ being the momentum vector which is obtained from $\mathbf{p}$ by the rotation. Since $\mathbf{p} \cdot \mathbf{r} = \mathbf{p}' \cdot \mathbf{r}'$ and since the integral over the entire momentum space is invariant under rotations, it follows from equations (23.13), (23.14), (23.15), and (23.50) that condition (23.49) will be satisfied if we determine the matrix S such that

$$u^{(R,L)}(\mathbf{p}') = Su^{(R,L)}(\mathbf{p}) \qquad (23.51)$$

$$v^{(R,L)}(\mathbf{p}') = Sv^{(R,L)}(\mathbf{p}) \qquad (23.52)$$

Since E_p is invariant under rotations, the last two equations in conjunction with (23.25) and (23.26) imply the condition

$$H_p S = SH_p \qquad (23.53)$$

where

$$H_p = c\boldsymbol{\alpha} \cdot \mathbf{p} + \beta\mu c^2 = \beta(c\boldsymbol{\gamma} \cdot \mathbf{p} + \mu c^2)$$

If we write

$$p'^k = a^k{}_j p^j, \qquad p_k = p_l{}' a^l{}_k$$

with summations over repeated Latin indices extending from 1 to 3 only, substitution into (23.53) produces the conditions

$$\beta\gamma^l S = S\beta\gamma^k a^l{}_k, \qquad \beta S = S\beta$$

or

$$a^l{}_k \gamma^k = S^{-1}\gamma^l S \qquad (23.54)$$

and

$$\gamma^0 = S^{-1}\gamma^0 S \qquad (23.55)$$

The conditions (23.54) and (23.55) for the matrix S are included as special cases in the general condition that S must satisfy if the electron-positron field theory is to be invariant under all (homogeneous) Lorentz transformations:

$$a^\lambda{}_\mu \gamma^\mu = S^{-1}\gamma^\lambda S \qquad (23.56)$$

Although this condition may be obtained by generalizing the argument that we have given for spatial rotations, it is easier to derive (23.56) by requiring that the Dirac equation (23.38) must be invariant under the transformation:

$$x'^\mu = a^\mu{}_\nu x^\nu, \qquad \frac{\partial}{\partial x^\mu} = \frac{\partial}{\partial x'^\nu} a^\nu{}_\mu$$

and

$$\psi'(\mathbf{r}', t') = S\psi(\mathbf{r}, t) \qquad (23.57)$$

Exercise 23.7. Derive condition (23.56) from the Lorentz invariance of the Dirac equation.

The demonstration of the Lorentz invariance of the theory will be complete if the matrix S can be exhibited for each possible Lorentz transformation. The explicit construction of S for proper orthochronous Lorentz transformations, which can be obtained continuously from the identity operation, is most easily accomplished by considering the condition (23.56) in an *infinitesimal* neighborhood of the identity. We may write

$$a^\mu{}_\nu = \delta^\mu{}_\nu + \varepsilon^\mu{}_\nu \qquad (23.58)$$

with the condition $\varepsilon^\mu{}_\nu = -\varepsilon_\nu{}^\mu$ as an immediate consequence of the orthogonality condition (23.48). For the case of spatial rotations, see (16.97).

If an arbitrary Lorentz transformation is followed by an infinitesimal one, the infinitesimal transformation is represented by

$$(S + dS)S^{-1} = 1 + dS \cdot S^{-1}$$

Applying this to equation (23.56), we get

$$(1 - dS \cdot S^{-1})\gamma^\lambda(1 + dS \cdot S^{-1}) = \gamma^\lambda + \varepsilon^\lambda{}_\mu\gamma^\mu$$

or

$$[\gamma^\lambda, dS \cdot S^{-1}] = \varepsilon^\lambda{}_\mu\gamma^\mu \qquad (23.59)$$

for all values of λ. The solution of this commutation relation is readily seen to be

$$dS \cdot S^{-1} = \varepsilon_{\mu\nu}\gamma^\mu\gamma^\nu/4 \qquad (23.60)$$

A three-dimensional rotation by an angle $\delta\theta$ about an axis along the unit vector $\hat{\mathbf{n}}$ takes the position vector $\mathbf{r}$ into

$$\mathbf{r}' = \mathbf{r} + \delta\theta\hat{\mathbf{n}} \times \mathbf{r} \qquad (23.61)$$

By comparison with equation (23.58), the identification

$$\varepsilon^1{}_2 = -\varepsilon^2{}_1 = -\delta\theta\, n_3 = -\varepsilon_{12} = \varepsilon_{21}$$

$$\varepsilon^2{}_3 = -\varepsilon^3{}_2 = -\delta\theta\, n_1 = -\varepsilon_{23} = \varepsilon_{32}$$

$$\varepsilon^3{}_1 = -\varepsilon^1{}_3 = -\delta\theta\, n_2 = -\varepsilon_{31} = \varepsilon_{13}$$

emerges. If we define the matrix

$$\Sigma^{\mu\nu} = \frac{i}{2}[\gamma^\mu, \gamma^\nu] \qquad (23.62)$$

equation (23.60) reduces to

$$dS \cdot S^{-1} = -\frac{i}{2}\delta\theta(n_1\Sigma^{23} + n_2\Sigma^{31} + n_3\Sigma^{12}]$$

and this differential equation has the simple unitary solution

$$S = \exp\left[-\frac{i}{2}\theta(n_1\Sigma^{23} + n_2\Sigma^{31} + n_3\Sigma^{12})\right] = e^{-(i/2)\theta\hat{n}\cdot\mathbf{\Sigma}} \quad (23.63)$$

where we have used the notation

$$\Sigma_x = \Sigma^{23}, \qquad \Sigma_y = \Sigma^{31}, \qquad \Sigma_z = \Sigma^{12} \quad (23.64)$$

for the four-dimensional analogs of the *Pauli spin matrices.*

Exercise 23.8. Show that the 4×4 matrices $\mathbf{\Sigma}$ defined by (23.62) and (23.64) satisfy the usual commutation relations for Pauli spin matrices. Show that in the standard representation (23.30),

$$\mathbf{\Sigma} = \begin{pmatrix} \mathbf{\sigma} & 0 \\ 0 & \mathbf{\sigma} \end{pmatrix}$$

If the inverse of (23.61),

$$\mathbf{r} = \mathbf{r}' - \delta\theta\hat{n} \times \mathbf{r}'$$

is substituted into (23.49) and the integration performed, the behavior of the spinor field under finite rotations is obtained:

$$U^{\dagger}\psi(\mathbf{r}, t)U = \exp\left[-i\theta\hat{n}\cdot\left(\mathbf{r} \times \frac{1}{i}\nabla + \tfrac{1}{2}\mathbf{\Sigma}\right)\right]\psi(\mathbf{r}, t) \quad (23.65)$$

If the unitary operator U is expressed as

$$U = \exp\left(-\frac{i}{\hbar}\theta\hat{n}\cdot\mathbf{J}\right) \quad (23.66)$$

it follows from equation (23.65) that the Hermitian operator $\mathbf{J}$ must satisfy the commutation relations

$$[\psi(\mathbf{r}, t), \mathbf{J}] = \left(\mathbf{r} \times \frac{\hbar}{i}\nabla + \frac{\hbar}{2}\mathbf{\Sigma}\right)\psi(\mathbf{r}, t) \quad (23.67)$$

The *angular momentum operator*

$$\mathbf{J} = \int \psi^{\dagger}(\mathbf{r})\left(\mathbf{r} \times \frac{\hbar}{i}\nabla + \frac{\hbar}{2}\mathbf{\Sigma}\right)\psi(\mathbf{r}) \, d^3r \quad (23.68)$$

satisfies this equation.

Exercise 23.9. Verify that expression (23.68) is consistent with equation (23.67) and with the defining relation for helicity, (23.6).

Since any proper orthochronous homogeneous Lorentz transformation may be obtained as a succession of spatial rotations and special Lorentz transformations, it suffices for the invariance proof to show the existence of S for special Lorentz transformations.

Exercise 23.10. For a special Lorentz transformation corresponding to uniform motion with velocity $v = c \, arctanh \, \chi$ along the x-axis, show that

$$S = \exp\left(\frac{i}{2} \chi \Sigma^{01}\right) = \exp\left(-\tfrac{1}{2}\chi\alpha_x\right) \tag{23.69}$$

Note that since Σ^{01} is antihermitian, S is not unitary in this case. [The *unitary* operator U, which effects this transformation in accordance with equation (23.49), can again be constructed by starting from the infinitesimal transformation.]

Exercise 23.11. Discuss coordinate translations in the theory of the Dirac field.

From (23.63) and (23.69) it is easy to deduce that

$$S^\dagger \gamma^0 S = \gamma^0 \tag{23.70}$$

Exercise 23.12. Verify equation (23.70).

Combining (23.56) and (23.70), we obtain

$$S^\dagger \gamma^0 \gamma^\mu S = a^\mu{}_\nu \gamma^0 \gamma^\nu \tag{23.71}$$

If the unitary operator U, induced by a Lorentz transformation, is applied to the current density (23.45), use of (23.49) and (23.71) shows that the current density is a four-vector operator and satisfies the relation

$$U^\dagger j^\mu(\mathbf{r}', t') U = a^\mu{}_\nu j^\nu(\mathbf{r}, t) \tag{23.72}$$

in generalization of the concept of a vector operator defined in Section 16.8.

The study of proper orthochronous Lorentz transformations must be supplemented by consideration of the fundamental *improper transformations*. Spatial reflections will be discussed in the remainder of this section, but the study of time reversal is left to Section 23.5.

If it is assumed that *spatial reflection* of all three coordinates, or inversion, is a symmetry operation for the Dirac theory, condition (23.56) may be invoked to yield

$$S^{-1}\gamma^0 S = \gamma^0, \; S^{-1}\gamma^1 S = -\gamma^1, \; S^{-1}\gamma^2 S = -\gamma^2, \; S^{-1}\gamma^3 S = -\gamma^3 \tag{23.73}$$

From these equations and (23.49) for $\mathbf{r}' = -\mathbf{r}$ it follows that the current density (23.45) behaves as a four-vector under the action of the unitary inversion operator U_P:

$$U_P^\dagger j^\mu(-\mathbf{r}, t)U_P = j_\mu(\mathbf{r}, t) \tag{23.74}$$

only if S is unitary, $S^\dagger S = I$. Except for an arbitrary phase factor, the conditions on S are solved by

$$S = \gamma^0 = \beta \tag{23.75}$$

and the inversion is thus accomplished by the relation

$$U_P\psi(\mathbf{r}, t)U_P^\dagger = \gamma^0\psi(-\mathbf{r}, t) \tag{23.76}$$

The unitary operator U_P defined by this equation is the *parity operator*. It is conventional to assume that the vacuum state is an eigenstate of U_P with even parity.

Exercise 23.13. Attempt an explicit construction of the parity operator in terms of the field operators.

It is convenient to define an additional Hermitian 4×4 Dirac matrix,

$$\gamma^5 \equiv \gamma_5 = i\gamma^0\gamma^1\gamma^2\gamma^3 \tag{23.77}$$

which has the properties

$$\gamma^\mu\gamma^5 + \gamma^5\gamma^\mu = 0, \quad (\gamma^5)^2 = I \tag{23.78}$$

Exercise 23.14. Prove equation (23.78) and also the property

$$[\gamma^5, \Sigma^{\mu\nu}] = 0 \tag{23.79}$$

Exercise 23.15. Verify the following transformation properties for the designated bilinear functions of the field operators under proper orthochronous Lorentz transformations and under reflections:

$\overline{\psi}(\mathbf{r}, t)\psi(\mathbf{r}, t)$	scalar
$\overline{\psi}(\mathbf{r}, t)\gamma^\mu\psi(\mathbf{r}, t)$	vector
$\overline{\psi}(\mathbf{r}, t)\gamma^5\gamma^\mu\psi(\mathbf{r}, t)$	pseudovector
$\overline{\psi}(\mathbf{r}, t)\Sigma^{\mu\nu}\psi(\mathbf{r}, t)$	tensor of rank two
$\overline{\psi}(\mathbf{r}, t)\gamma^5\psi(\mathbf{r}, t)$	pseudoscalar

It can be shown that the set I, γ^μ, $\gamma^5\gamma^\mu$, $\Sigma^{\mu\nu}$, γ^5 is complete in the sense that any arbitrary 4×4 matrix can be expanded in terms of these 16 matrices.

4. *Solutions of the Free Field Dirac Equation.* In this section explicit solutions of the equations (23.25) and (23.26) will be given. For this purpose we shall employ the standard representation (23.30) of the Dirac matrices.

Exercise 23.16. Write out the four linear homogeneous equations implied by (23.25) and (23.26) in full detail, and show that the vanishing of their determinant is assured by the relation (23.12). Prove that all 3×3 minors of the secular determinant also vanish (but not all 2×2 minors), and interpret this result.

The simplest solutions are obtained if the momentum vector $\mathbf{p}$ points in the direction of the positive z-axis. In this case, equation (23.25) reduces to

$$
\begin{pmatrix}
\mu c^2 & 0 & cp & 0 \\
0 & \mu c^2 & 0 & -cp \\
cp & 0 & -\mu c^2 & 0 \\
0 & -cp & 0 & -\mu c^2
\end{pmatrix}
\begin{pmatrix}
u_1 \\ u_2 \\ u_3 \\ u_4
\end{pmatrix}
= E_p
\begin{pmatrix}
u_1 \\ u_2 \\ u_3 \\ u_4
\end{pmatrix}
$$

or

$$
\begin{aligned}
\mu c^2 u_1 + cp u_3 &= E_p u_1 \\
cp u_1 - \mu c^2 u_3 &= E_p u_3 \\
\mu c^2 u_2 - cp u_4 &= E_p u_2 \\
-cp u_2 - \mu c^2 u_4 &= E_p u_4
\end{aligned}
\tag{23.80}
$$

Evidently, this system of equations possesses two linearly independent and, in fact, orthogonal solutions:

$$
u^{(R)} \propto
\begin{pmatrix}
1 \\ 0 \\ \dfrac{cp}{E_p + \mu c^2} \\ 0
\end{pmatrix}
\quad \text{and} \quad
u^{(L)} \propto
\begin{pmatrix}
0 \\ 1 \\ 0 \\ \dfrac{-cp}{E_p + \mu c^2}
\end{pmatrix}
\tag{23.81}
$$

The labels R and L have been affixed to these spinors because they are eigenspinors of the one-particle *helicity operator* $\mathbf{\Sigma} \cdot \hat{\mathbf{p}}$ (here reduced to Σ_z) with eigenvalues $+1$ and -1, respectively. According to formula (23.68), this is the component of angular momentum along $\mathbf{p}$ for the particles of linear momentum $\mathbf{p}$, since orbital angular momentum contributes nothing to this projection (Exercise 23.9).

The corresponding solutions for the eigenvalue $-E_p$ are

$$
v^{(R)} \propto
\begin{pmatrix}
\dfrac{-cp}{E_p + \mu c^2} \\ 0 \\ 1 \\ 0
\end{pmatrix},
\quad
v^{(L)} \propto
\begin{pmatrix}
0 \\ \dfrac{cp}{E_p + \mu c^2} \\ 0 \\ 1
\end{pmatrix}
\tag{23.82}
$$

Exercise 23.17. Determine the multiplicative constants for each of the four solutions, insuring the normalization (23.17).

The eigenspinors with definite helicity but arbitrary linear momentum vector **p** are easily found by rotating the states described by (23.81) and (23.82) by an angle $\theta = arccos\,(p_z/p)$ about the axis determined by the vector $(-p_y, p_x, 0)$. Such a rotation takes the z-axis into the direction of **p**. The matrix operator which carries out this rotation is

$$S = \exp\left[-i\frac{\theta}{2}(-p_y\Sigma_x + p_x\Sigma_y)\Big/\sqrt{p_x^{\,2} + p_y^{\,2}}\right] \tag{23.83}$$

Using the generalization of the identity (12.57) to the 4×4 Pauli matrices, we may write this as

$$S = I\cos\frac{\theta}{2} + i\frac{(p_y\Sigma_x - p_x\Sigma_y)}{\sqrt{p_x^{\,2} + p_y^{\,2}}}\sin\frac{\theta}{2} \tag{23.84}$$

Hence, if the components of $\hat{\mathbf{p}}$ are denoted as (n_x, n_y, n_z)

$$u^{(R)}(\mathbf{p}) \propto \begin{pmatrix} \cos\dfrac{\theta}{2} \\[2mm] \dfrac{(p_x + ip_y)}{\sqrt{p^2 - p_z^{\,2}}}\sin\dfrac{\theta}{2} \\[2mm] \dfrac{cp}{E_p + \mu c^2}\cos\dfrac{\theta}{2} \\[2mm] \dfrac{cp}{E_p + \mu c^2}\dfrac{p_x + ip_y}{\sqrt{p^2 - p_z^{\,2}}}\sin\dfrac{\theta}{2} \end{pmatrix} \propto \begin{pmatrix} n_z + 1 \\[2mm] n_x + in_y \\[2mm] \dfrac{cp}{E_p + \mu c^2}(n_z + 1) \\[2mm] \dfrac{cp}{E_p + \mu c^2}(n_x + in_y) \end{pmatrix} \tag{23.85}$$

Similarly

$$u^{(L)}(\mathbf{p}) \propto \begin{pmatrix} \dfrac{-p_x + ip_y}{\sqrt{p^2 - p_z^{\,2}}}\sin\dfrac{\theta}{2} \\[2mm] \cos\dfrac{\theta}{2} \\[2mm] -\dfrac{cp}{E_p + \mu c^2}\dfrac{-p_x + ip_y}{\sqrt{p^2 - p_z^{\,2}}}\sin\dfrac{\theta}{2} \\[2mm] -\dfrac{cp}{E_p + \mu c^2}\cos\dfrac{\theta}{2} \end{pmatrix} \propto \begin{pmatrix} -n_x + in_y \\[2mm] n_z + 1 \\[2mm] -\dfrac{cp}{E_p + \mu c^2}(-n_x + in_y) \\[2mm] -\dfrac{cp}{E_p + \mu c^2}(n_z + 1) \end{pmatrix} \tag{23.86}$$

Exercise 23.18. Work out similar expressions for $v^{(R)}$ and $v^{(L)}$.

Exercise 23.19. Verify the closure relation (23.31).

The matrix

$$B_+(\mathbf{p}) = u^{(R)}(\mathbf{p})u^{(R)}(\mathbf{p})^\dagger + u^{(L)}(\mathbf{p})u^{(L)}(\mathbf{p})^\dagger \qquad (23.87)$$

constructed from the normalized eigenspinors, gives zero when applied to an eigenspinor of $c\boldsymbol{\alpha} \cdot \mathbf{p} + \beta\mu c^2$ with eigenvalue $-E_p$; and applied to an eigenspinor of $c\boldsymbol{\alpha} \cdot \mathbf{p} + \beta\mu c^2$ with eigenvalue E_p, it acts like the unit matrix. Hence, we have the (*Casimir*) projection operator

$$B_+(\mathbf{p}) = \frac{E_p + c\boldsymbol{\alpha} \cdot \mathbf{p} + \beta\mu c^2}{2E_p} \qquad (23.88)$$

Similarly, the matrix

$$B_-(p) = v^{(R)}(\mathbf{p})v^{(R)}(\mathbf{p})^\dagger + v^{(L)}(\mathbf{p})v^{(L)}(\mathbf{p})^\dagger = \frac{E_p - c\boldsymbol{\alpha} \cdot \mathbf{p} - \beta\mu c^2}{2E_p} \quad (23.89)$$

acts as a projection operator for the eigenspinors with eigenvalue $-E_p$.

Exercise 23.20. Show that for the eigenspinors of fixed momentum $\mathbf{p}$ and an arbitrary 4×4 matrix A,

$$u^{(R)\dagger}Au^{(R)} + u^{(L)\dagger}Au^{(L)} + v^{(R)\dagger}Av^{(R)} + v^{(L)\dagger}Av^{(L)} = \text{trace } A \qquad (23.90)$$

5. Charge Conjugation, Time Reversal, and the PCT Theorem. It is useful to record a simple relationship that follows from the similarity of the equations satisfied by $v^{(R,L)}$ ($\mathbf{p}$) and $u^{(R,L)}(\mathbf{p})$. If the complex conjugate of equation (23.26) is taken and $\mathbf{p}$ replaced by $-\mathbf{p}$, we obtain

$$(-c\boldsymbol{\alpha}^* \cdot \mathbf{p} + \mu c^2\beta^*)v^{(R,L)}(-\mathbf{p})^* = -E_p v^{(R,L)}(-\mathbf{p})^*$$

which is to be compared with (23.25),

$$(c\boldsymbol{\alpha} \cdot \mathbf{p} + \mu c^2\beta)u^{(R,L)}(\mathbf{p}) = E_p u^{(R,L)}(\mathbf{p})$$

If we can find a matrix C with the properties

$$C\boldsymbol{\alpha}^* = \boldsymbol{\alpha}C, \qquad C\beta^* = -\beta C \qquad (23.91)$$

it is seen that $Cv^{(R,L)}(-\mathbf{p})^*$ satisfies the same equation as $u^{(R,L)}(\mathbf{p})$.

Exercise 23.21. Establish that $C^{-1}\gamma^1 C = \bar{\gamma}^1$, $C^{-1}\gamma^2 C = \bar{\gamma}^2$, $C^{-1}\gamma^3 C = \bar{\gamma}^3$, as well as $\boldsymbol{\Sigma}C = -C\boldsymbol{\Sigma}^*$ and $C^{-1}\gamma^0 C = -\bar{\gamma}^0$.

Helicity is preserved under this transformation of solutions of the Dirac equation. Indeed, from the equations

$$\boldsymbol{\Sigma} \cdot \mathbf{p}Cv^{(R)}(-\mathbf{p})^* = -C(\boldsymbol{\Sigma} \cdot \mathbf{p})^*v^{(R)}(-\mathbf{p})^* = Cv^{(R)}(-\mathbf{p})^*$$

it follows that

$$u^{(R)}(\mathbf{p}) = Cv^{(R)}(-\mathbf{p})^* \tag{23.92}$$

and

$$u^{(L)}(\mathbf{p}) = Cv^{(L)}(-\mathbf{p})^* \tag{23.93}$$

may be postulated. By using the same matrix C in both of these equations, a partial choice of the previously undetermined relative phases of the spinors u and v is made. The normalization (23.17) requires that

$$u^{(R)}(\mathbf{p})^\dagger u^{(R)}(\mathbf{p}) = \tilde{v}^{(R)}(-\mathbf{p})C^\dagger Cv^{(R)}(-\mathbf{p})^* = v^{(R)}(-\mathbf{p})^\dagger \tilde{C}C^* v^{(R)}(-\mathbf{p}) = 1$$

or

$$\tilde{C}C^* = I \tag{23.94}$$

i.e., C must be unitary. In the standard representation (23.30) the conditions (23.91) and (23.94) are satisfied by the matrix

$$C = \beta\alpha_y = \gamma^2$$

Equations (23.14) and (23.15) show that $u^{(R,L)}(\mathbf{p})$ is associated with the annihilation of an electron and $v^{(R,L)}(-\mathbf{p})^*$ with the annihilation of a positron. The connection (23.92) and (23.93) between these two amplitudes suggests that the unitary transformation which takes electrons into positrons and vice versa, without changing either momentum or helicity, may have a simple local formulation in terms of the fields. We define the unitary operator $\mathscr{C}$, known as *charge conjugation* or *particle-antiparticle conjugation* operator, by the equations

$$\begin{aligned}
\mathscr{C}a_R(\mathbf{p})\mathscr{C}^\dagger &= b_R(\mathbf{p}) \\
\mathscr{C}b_R(\mathbf{p})\mathscr{C}^\dagger &= a_R(\mathbf{p}) \\
\mathscr{C}a_L(\mathbf{p})\mathscr{C}^\dagger &= b_L(\mathbf{p}) \\
\mathscr{C}b_L(\mathbf{p})\mathscr{C}^\dagger &= a_L(\mathbf{p})
\end{aligned} \tag{23.95}$$

From equations (23.14), (23.15), (23.92), and (23.93), it is easily seen that

$$\mathscr{C}\psi(\mathbf{r})\mathscr{C}^\dagger = C\tilde{\gamma}^0\tilde{\bar{\psi}}(\mathbf{r}) = C\tilde{\psi}^\dagger(\mathbf{r}) \tag{23.96}$$

Exercise 23.22. Verify equation (23.96) and show conversely that

$$\mathscr{C}\psi^\dagger(\mathbf{r})\mathscr{C}^\dagger = \tilde{\psi}(\mathbf{r})C^{-1} \tag{23.97}$$

The definition of $\mathscr{C}$ is supplemented by requiring that the vacuum state remain unchanged under charge conjugation: $\mathscr{C}\Psi^{(0)} = \Psi^{(0)}$.

As time develops, the relations (23.96) and (23.97) remain applicable if the electron-positron field is free. This follows from the definition of charge conjugation and can be verified by showing that if $\psi(\mathbf{r}, t)$ and $\psi^\dagger(\mathbf{r}, t)$ are connected at all times by (23.96), the two Dirac equations (23.40) and (23.42) with $A_\mu = 0$ imply one another.

Exercise 23.23. Prove the last statement.

In the presence of an external electromagnetic field ($A_\mu \neq 0$), the Dirac equation (23.40) is no longer invariant under charge conjugation as defined by relation (23.96). Applying this operation to equation (23.40), we obtain

$$\gamma^\mu \left(\frac{\partial}{\partial x^\mu} - \frac{ie}{\hbar c} A_\mu \right) C \tilde{\psi}^\dagger + i\kappa C \tilde{\psi}^\dagger = 0$$

If, by using the commutation properties of the matrix C, we reduce this equation, we find

$$\left(\frac{\partial}{\partial x^\mu} - \frac{ie}{\hbar c} A_\mu \right) \overline{\psi} \gamma^\mu - i\kappa \overline{\psi} = 0 \tag{23.98}$$

Exercise 23.24. Reproduce the steps leading to equation (23.98), using the results of Exercise 23.21 and the properties of the Dirac matrices.

This equation is the same as equation (23.42) except for the important change in sign in front of the vector potential. The presence of an external field thus destroys the invariance of the theory under charge conjugation. At the same time, it is apparent that the invariance is restored if the electromagnetic field is regarded as part of the dynamical system and is reversed ($A_\mu \rightarrow -A_\mu$) when charge conjugation is undertaken.

Exercise 23.25. Show that under charge conjugation the current density operator, defined in equation (23.45), changes into its negative if the anticommutation properties of the field are used.

We now return briefly to the parity operator U_P, defined in Section 23.3. We note that together with $u^{(R,L)}(\mathbf{p})$ the spinors $\gamma^0 u^{(R,L)}(-\mathbf{p})$, obtained by reflection, are also solutions of equation (23.25). Since $\mathbf{\Sigma} \cdot \hat{\mathbf{p}}$ changes sign under reflection, the helicity is reversed, and $\gamma^0 u^{(R)}(-\mathbf{p})$ must be proportional to $u^{(L)}(\mathbf{p})$. Similarly, $\gamma^0 v^{(R)}(-\mathbf{p})$ must be proportional to $v^{(L)}(\mathbf{p})$. With the phases of the Dirac spinors controlled by the relations (23.92) and (23.93), the constants of proportionality are partially determined by the condition (Exercise 23.21):

$$\gamma^0 C = -C \tilde{\gamma}^0$$

If we choose to set

$$\gamma^0 u^{(R)}(-\mathbf{p}) = u^{(L)}(\mathbf{p}) \tag{23.99}$$

and recall that γ^0 is Hermitian, the equations

$$u^{(L)}(\mathbf{p}) = \gamma^0 u^{(R)}(-\mathbf{p}) = \gamma^0 C v^{(R)}(\mathbf{p})^* = -C\bar{\gamma}^0 v^{(R)}(\mathbf{p})^* = -C[\gamma^0 v^{(R)}(\mathbf{p})]^*$$
$$u^{(L)}(\mathbf{p}) = C v^{(L)}(-\mathbf{p})^*$$

lead to the conclusion that we must have

$$\gamma^0 v^{(R)}(-\mathbf{p}) = -v^{(L)}(\mathbf{p}) \tag{23.100}$$

From (23.99), (23.100), and (23.76), we deduce the transformation properties of the electron and positron annihilation operators under spatial reflection as

$$\begin{aligned} U_P a_R(\mathbf{p}) U_P^\dagger &= a_L(-\mathbf{p}) \\ U_P a_L(\mathbf{p}) U_P^\dagger &= a_R(-\mathbf{p}) \\ U_P b_R(\mathbf{p}) U_P^\dagger &= -b_L(-\mathbf{p}) \\ U_P b_L(\mathbf{p}) U_P^\dagger &= -b_R(-\mathbf{p}) \end{aligned} \tag{23.101}$$

The difference in sign between the first two and the last two of these equations has important physical consequences, since it shows that an electron and a positron in the same orbital states have opposite parities.[5]

We conclude this chapter with some remarks about a third discrete symmetry operation: time reversal. The general concepts needed for the discussion were already contained in Section 16.11.

The *antiunitary* time reversal operator Θ is defined to reverse the sign of all momenta and spins. We therefore require that

$$\begin{aligned} \Theta a_R(\mathbf{p}) \Theta^{-1} &= e^{i\alpha_R(\mathbf{p})} a_R(-\mathbf{p}) \\ \Theta a_L(\mathbf{p}) \Theta^{-1} &= e^{i\alpha_L(\mathbf{p})} a_L(-\mathbf{p}) \\ \Theta b_R(\mathbf{p}) \Theta^{-1} &= e^{i\beta_R(\mathbf{p})} b_R(-\mathbf{p}) \\ \Theta b_L(\mathbf{p}) \Theta^{-1} &= e^{i\beta_L(\mathbf{p})} b_L(-\mathbf{p}) \end{aligned} \tag{23.102}$$

Although the phases in (23.102) are arbitrary, it is possible to choose them in such a manner that the fields undergo simple transformations under time reversal. From the antiunitary property of Θ one may derive the transformation properties of the creation operators:

$$\begin{aligned} \Theta a_R^\dagger(\mathbf{p}) \Theta^{-1} &= e^{-i\alpha_R(\mathbf{p})} a_R^\dagger(-\mathbf{p}) \\ \Theta a_L^\dagger(\mathbf{p}) \Theta^{-1} &= e^{-i\alpha_L(\mathbf{p})} a_L^\dagger(-\mathbf{p}) \\ \Theta b_R^\dagger(\mathbf{p}) \Theta^{-1} &= e^{-i\beta_R(\mathbf{p})} b_R^\dagger(-\mathbf{p}) \\ \Theta b_L^\dagger(\mathbf{p}) \Theta^{-1} &= e^{-i\beta_L(\mathbf{p})} b_L^\dagger(-\mathbf{p}) \end{aligned} \tag{23.103}$$

[5] For illustrations of the effectiveness of the selection rules that can be derived for interactions invariant under reflection and charge conjugation, see Section 4-4 of J. J. Sakurai, *Advanced Quantum Mechanics*, Addison-Wesley Publishing Company, Reading, 1967.

Exercise 23.26. Derive (23.103) from (23.102).

If we apply Θ to the fields defined in equations (23.34) and (23.35), and make some trivial substitutions in the integrand, we obtain

$$\Theta \psi^{(+)}(\mathbf{r}, t)\Theta^{-1} = \frac{1}{(2\pi\hbar)^{3/2}} \int [u^{(R)}(-\mathbf{p})^* e^{i\alpha_R(-\mathbf{p})} a_R(\mathbf{p})$$
$$+ u^{(L)}(-\mathbf{p})^* e^{i\alpha_L(-\mathbf{p})} a_L(\mathbf{p})] e^{(i/\hbar)(-\mathbf{p}\cdot\mathbf{r} - E_p t)} d^3p$$

In arriving at this equation, the antiunitary nature of Θ has been exploited and has resulted in complex conjugation. It is easily seen that the right-hand side of this equation becomes a local expression for the field at time $-t$, $T\psi^{(+)}(\mathbf{r}, -t)$, if a 4×4 matrix T can be found such that

$$u^{(R)}(-\mathbf{p})^* e^{i\alpha_R(-\mathbf{p})} = T u^{(R)}(\mathbf{p}) \tag{23.104}$$

and

$$u^{(L)}(-\mathbf{p})^* e^{i\alpha_L(-\mathbf{p})} = T u^{(L)}(\mathbf{p}) \tag{23.105}$$

The normalization (23.31) implies that T must be unitary:

$$TT^\dagger = I \tag{23.106}$$

The relations (23.104) and (23.105) are consistent with the Dirac equation (23.25) only if T satisfies the conditions

$$\boldsymbol{\alpha}^* T = -T\boldsymbol{\alpha}, \qquad \beta^* T = T\beta \tag{23.107}$$

The unitary solution to these equations is unique except for an arbitrary phase factor.[6] In the standard representation (23.30), the imaginary matrix

$$T = -i\alpha_z\alpha_x = \Sigma_y \tag{23.108}$$

is a solution. It has the important property

$$TT^* = -I \tag{23.109}$$

which can be proved to be independent of the representation.

Exercise 23.27. Apply the time reversal operator to the negative frequency part of the field, and show that the same matrix T may be used to transform $\psi^{(-)}$ as $\psi^{(+)}$.

It follows that the complete electron-positron field is transformed under time reversal according to

$$\Theta\psi(\mathbf{r}, t)\Theta^{-1} = T\psi(\mathbf{r}, -t), \qquad \Theta\psi^\dagger(\mathbf{r}, t)\Theta^{-1} = \psi^\dagger(\mathbf{r}, -t)T^\dagger \tag{23.110}$$

[6] If T_1 and T_2 are two different unitary matrices which satisfy (23.107), construct $T_1 T_2^{-1}$. Show that this commutes with all Dirac matrices and hence must be a multiple of the unit matrix.

Exercise 23.28. Show that $T\boldsymbol{\Sigma} \cdot \hat{\mathbf{p}} = -\boldsymbol{\Sigma}^* \cdot \hat{\mathbf{p}}T$ and that helicity is preserved under time reversal, substantiating (23.104) and (23.105).

It is easy to see that the properties of T impose some restrictions on the phases α and β in (23.102) and (23.103), since iteration of equations (23.104) and (23.105), in conjunction with the requirement (23.109), gives the result

$$e^{i\alpha_R(-\mathbf{p})-i\alpha_R(\mathbf{p})} = e^{i\alpha_L(-\mathbf{p})-i\alpha_L(\mathbf{p})} = e^{i\beta_R(-\mathbf{p})-i\beta_R(\mathbf{p})} = e^{i\beta_L(-\mathbf{p})-i\beta_L(\mathbf{p})} = -1$$

$$(23.111)$$

The effect of two successive time reversals can now be established. Owing to the antilinearity of Θ:

$$\Theta^2 a_R(\mathbf{p})\Theta^{-2} = e^{-i\alpha_R(\mathbf{p})}\Theta a_R(-\mathbf{p})\Theta^{-1} = e^{-i\alpha_R(\mathbf{p})+i\alpha_R(-\mathbf{p})}a_R(\mathbf{p}) = -a_R(\mathbf{p})$$

$$(23.112)$$

Thus, application of Θ^2 merely changes the sign of the annihilation operator. The same conclusion holds for all other annihilation and creation operators. Hence, Θ^2 acts like $+1$ on states with an even number of Dirac particles (and, more generally, fermions), and like -1 on states with an odd number of such particles. This conclusion conforms with the discussion of Section 16.11.

Since Θ^2 is a unitary operator that commutes with all observables, state vectors which are obtained by superposition of states with even and odd numbers of Dirac particles cannot be physically realized. This statement is a *superselection rule*, and it is consistent with a superselection rule inferred in Section 16.2 from the commutivity of observables with rotations by 2π.

In the presence of an external electromagnetic field, the time reversal operation is generally no longer a symmetry operation. However, the invariance of the Dirac equation (23.41) under time reversal as defined by (23.110) is restored if $\mathbf{A}$ is changed into $-\mathbf{A}$, while ϕ remains unchanged.

Exercise 23.29. Determine the transformation properties of the Dirac current density operator under time reversal.

In addition to the angular momentum operators and other generators of the proper Lorentz group, we have discussed in this chapter three discrete symmetry operations corresponding to reflection, charge conjugation, and time reversal. Originally defined for free fields, these operations may remain symmetry operations when interactions are introduced. For instance, quantum electrodynamics is invariant under each of these three operations. In more subtle cases interactions may no longer be invariant under these three operations separately; however, invariance still holds for the product (i.e., the successive application) of the three discrete operations (*PCT theorem*), provided only that the restricted principle of relativity is valid.[7]

[7] For further discussion of the fundamental questions related to the PCT theorem, the connection between spin and statistics, and superselection rules, see R. F. Streater and A. S. Wightman, *PCT, Spin and Statistics, and all that*, W. A. Benjamin, New York, 1963. Applications are found in S. Gasiorowicz, *Elementary Particle Physics*, John Wiley and Sons, New York, 1966.

Problems

1. If **A** and **B** are proportional to the unit 4×4 matrix, derive expansion formulas for the matrix products $(\boldsymbol{\alpha} \cdot \mathbf{A})(\boldsymbol{\alpha} \cdot \mathbf{B})$ and $(\boldsymbol{\alpha} \cdot \mathbf{A})(\boldsymbol{\Sigma} \cdot \mathbf{B})$ in terms of $\boldsymbol{\alpha}$ and $\boldsymbol{\Sigma}$ matrices in analogy with formula (12.55).

2. If a field theory of massless spin $\frac{1}{2}$ particles (neutrinos) is developed, so that the β matrix is absent, show that the conditions (23.28) and (23.29) are solved by 2×2 Pauli matrices, $\boldsymbol{\alpha} = \pm\boldsymbol{\sigma}$. Work out the details of the resulting *two-component* theory with particular attention to the helicity properties. Is this theory invariant under spatial reflection?

3. Develop the outlines of relativistic quantum field theory for neutral spinless bosons with mass. What modifications are indicated when the particles are charged?

One-Electron Dirac Theory

1. The One-Particle Approximation. In Chapter 20 quantum field theory was developed into a consistent description of systems of identical particles from the concepts of nonrelativistic quantum mechanics for a single particle; but it was emphasized in Chapters 22 and 23 that this procedure could not be used to construct a *relativistic* form of one-particle quantum mechanics in a rigorous manner. Inevitably such a theory is an approximation to a proper many-body theory.

It is, of course, tempting to identify the state

$$\psi_\alpha^\dagger(\mathbf{r})\Psi'^{(0)} = |\mathbf{r}, \alpha\rangle \qquad (24.1)$$

in analogy to equations (20.3) and (20.6) as the one-electron state which corresponds to a sharp position of the particle with quantum number α. The inadequacy of this identification in the relativistic theory is seen from the fact that, owing to the anticommutation relations, the state (24.1) cannot be normalized properly. The trouble stems, of course, from the properties of the field ψ whose expansion contains electron annihilation as well as positron creation operators, so that

$$\psi_\alpha(\mathbf{r})\Psi'^{(0)} \neq 0$$

An obvious possibility for remedying this difficulty is to use, instead of equation (24.1), the identification

$$\psi_\alpha^{(+)\dagger}(\mathbf{r})\Psi'^{(0)} = |\mathbf{r}, \alpha\rangle \qquad (24.2)$$

for a one-electron state. Such a theory, if pursued, would contain one-electron wave functions which do not correspond to any state in the original field theory (e.g., the negative energy eigenstates of H). This difficulty would make its appearance whenever we encountered an operator that connects the "physical" with the "unphysical" states. Arbitrary exclusion of the "unphysical" states would violate the completeness requirements and lead to incorrect results in calculations which involve virtual intermediate states. On the other hand, their inclusion would be embarrassing, since the theory then permits transitions to "unphysical" states if a strongly fluctuating perturbation is applied.

In view of these reservations, we choose a slightly more satisfactory approach by introducing a new *"electron vacuum"* Ψ_{0e} such that

$$\psi_\alpha(\mathbf{r})\Psi_{0e} = 0 \tag{24.3}$$

holds. Obviously, this is a state in which there are no electrons but in which all available positron one-particle states are occupied. It can thus hardly be called the physical vacuum, since it has obviously infinite charge and infinite energy. Nevertheless, departures from this state by the addition of one electron or subtraction of one positron can be treated effectively as one-particle states. We thus define a *one-electron state* as

$$\Psi_e = \int \psi^\dagger(\mathbf{r}, t)\psi_e(\mathbf{r}, t)\, d^3r\, \Psi_{0e} \tag{24.4}$$

where ψ_e is a *spinor wave function* with four components. From the anti-commutation relation,

$$\psi_\alpha^\dagger(\mathbf{r}, t)\psi_\beta(\mathbf{r}', t) + \psi_\beta(\mathbf{r}', t)\psi_\alpha^\dagger(\mathbf{r}, t) = \delta_{\alpha\beta}\,\delta(\mathbf{r} - \mathbf{r}')$$

and equation (24.3), it follows immediately that

$$(\Psi_e, \Psi_e) = \int \psi_e^\dagger(\mathbf{r}, t)\psi_e(\mathbf{r}, t)\, d^3r = 1$$

and that

$$\psi_e(\mathbf{r}, t) = (\Psi_{0e}, \psi(\mathbf{r}, t)\Psi_e) \tag{24.5}$$

Exercise 24.1. Verify that the total charge of the system in state Ψ_e of equation (24.4) differs from the charge in state Ψ_{0e} by $-e$.

Similarly, a *"positron vacuum"* Ψ_{0p} and a one-positron state Ψ_p are defined by the equations

$$\psi^+(\mathbf{r})\Psi_{0p} = 0 \tag{24.6}$$

$$\Psi_p = \int \tilde{\psi}(\mathbf{r}, t)C^{-1}\psi_p(\mathbf{r}, t)\, d^3r\, \Psi_{0p} \tag{24.7}$$

From the normalization of these states to unity, the *positron wave function* $\psi_p(\mathbf{r}, t)$ is obtained as

$$\psi_p(\mathbf{r}, t) = (\Psi_{0p}, C\tilde{\psi}^\dagger(\mathbf{r}, t)\Psi_p) \tag{24.8}$$

It is easily shown from equations (23.40) and (23.42) that in the presence of an external electromagnetic field the spinor *wave functions* ψ_e and ψ_p

satisfy the equations[1]

$$\gamma^\mu \left(\frac{\partial}{\partial x^\mu} - \frac{ie}{\hbar c} A_\mu \right) \psi_e + i\kappa \psi_e = 0 \tag{24.9}$$

$$\gamma^\mu \left(\frac{\partial}{\partial x^\mu} + \frac{ie}{\hbar c} A_\mu \right) \psi_p + i\kappa \psi_p = 0 \tag{24.10}$$

The usual probability interpretation of one-particle quantum mechanics is retrieved for the electron and positron if we consider an additive Hermitian one-particle operator, like the linear or angular momentum, which can be written in the form

$$\mathscr{K} = \int \psi^\dagger(\mathbf{r}) K \psi(\mathbf{r})\, d^3 r \tag{24.11}$$

where $K(\mathbf{r}, -i\hbar \nabla)$ is a function of the position and momentum vectors and, in addition, may be a 4×4 matrix acting on a spinor.

If $\mathscr{K}$ represents a physical quantity which is invariant under charge conjugation, the one-particle operator-matrix K must satisfy the condition

$$C^{-1} K C = -K^* \tag{24.12}$$

where C is the matrix defined in Section 23.5. The minus sign appearing in (24.12) is directly connected with the anticommutation properties of the Dirac field.

For a charge-independent operator $\mathscr{K}$, the one-electron and one-positron expectation values are simply

$$(\Psi_e, \mathscr{K} \Psi_e) = \int \psi_e^\dagger(\mathbf{r}) K \psi_e(\mathbf{r})\, d^3 r \tag{24.13}$$

$$(\Psi_p, \mathscr{K} \Psi_p) = \int \psi_p^\dagger(\mathbf{r}) K \psi_p(\mathbf{r})\, d^3 r \tag{24.14}$$

Exercise 24.2. Derive condition (24.12) and verify that linear and angular momentum operators satisfy it, as does the free-particle energy operator.

If forces are absent, the wave function of an electron or positron with momentum $\mathbf{p}$ and definite helicity is

$$\frac{1}{(2\pi\hbar)^{3/2}} u^{(R,L)}(\mathbf{p})\, e^{(i/\hbar)(\mathbf{p}\cdot\mathbf{r} - E_p t)} = \frac{1}{(2\pi\hbar)^{3/2}} C v^{(R,L)}(-\mathbf{p})^* \, e^{(i/\hbar)(\mathbf{p}\cdot\mathbf{r} - E_p t)} \tag{24.15}$$

---

[1] Dirac originally proposed the equation bearing his name essentially in the one-particle form (24.9).

The approximate nature of the one-electron or one-positron theory is visible in many ways. For example, the equation of motion for a free one-electron wave function,

$$\gamma^{\mu} \frac{\partial \psi_e}{\partial x^{\mu}} + i\kappa \psi_e = 0$$

has, for a given momentum $\mathbf{p}$, four linearly independent solutions. Two of these, for positive energy, correspond to the two spin states of the electron. The remaining two are eigenstates of H with negative eigenvalues and represent, according to the definition (24.4), the removal of a positron from rather than the addition of an electron to the vacuum Ψ_{0e}. These solutions cannot be ignored since, for a spin $\frac{1}{2}$ particle with mass, Lorentz invariance requires that ψ_e have four components, so that four linearly independent spinors are needed for specifying an arbitrary initial state. And even if initially the wave function were a superposition of only electron eigenstates, the amplitudes of the positron components may, under the influence of forces, eventually become appreciable.

As a simple example, we imagine that the free Dirac electron is subjected to a perturbation during a time interval of order τ. The transition amplitude for an exponential perturbation is according to equation (18.16)

$$-\frac{i}{\hbar} V_{ks} \int_{-\infty}^{+\infty} e^{i\omega_{ks}t'} e^{-|t'|/\tau} \, dt' = -\frac{2i}{\hbar} \frac{\tau}{1 + \omega_{ks}^2 \tau^2} V_{ks}$$

The optimum value of τ is evidently $\tau = 1/\omega_{ks}$. For this value the transition amplitude becomes $-2iV_{ks}/(\hbar\omega_{ks})$. Since $\hbar\omega_{ks} \simeq \mu c^2$ for transitions between positive and negative eigenstates of H, it is apparent that the one-particle approximation breaks down when interaction energies of strength $\simeq \mu c^2$ fluctuate in times of the order $\simeq \hbar/\mu c^2$. Translating these considerations from time into space language, we can say that if the potential energy changes by μc^2 over a distance of the order $\hbar/\mu c$, an initial one-electron state may lead to *pair annihilation* of the electron with one of the positrons present in the state Ψ_{0e}. This chapter thus may be said properly to deal with a one-charge rather than a one-particle theory.

While we view the one-electron Dirac theory as an approximation to a more accurate description involving interacting fields, mathematically it is possible to ignore the many-body aspects and to consider on its own merits the theory of the relativistic wave equation

$$i\hbar \frac{\partial \psi_e}{\partial t} = \left[c\boldsymbol{\alpha} \cdot \left(\frac{\hbar}{i} \nabla + \frac{e}{c} \mathbf{A} \right) - e\phi + \beta \mu c^2 \right] \psi_e \tag{24.16}$$

where ψ_e is a spinor function rather than a field operator.

Just as in Section 23.2, we can derive from this equation readily the continuity equation

$$\frac{\partial(\psi_e^\dagger \psi_e)}{\partial t} + \nabla \cdot (c\psi_e^\dagger \boldsymbol{\alpha} \psi_e) = 0 \tag{24.17}$$

Equation (24.16) has the usual form

$$i\hbar \frac{\partial \psi_e}{\partial t} = H\psi_e \tag{24.18}$$

familiar from ordinary quantum mechanics with H being a Hermitian operator. The only unusual feature of H is, of course, the fact that, unlike the nonrelativistic one-particle Hamiltonian and unlike the total field energy operator $\mathscr{H}$, H as defined by (24.16) and (24.18) is not a positive definite operator. We shall see in Section 24.3 that there is, nevertheless, a simple correspondence between the relativistic and nonrelativistic Hamiltonians.

The stationary state solutions of the Dirac equation for a *free electron* (or positron) need not be discussed in detail here, because this was in effect already done in Section 23.4. It should be pointed out, however, that if $E_p \simeq \mu c^2$, i.e., in the nonrelativistic limit, in the standard representation (23.81) the third and fourth components of $u^{(R,L)}$ are small in the ratio of v/c compared to the first two components. The converse is true for the spinors $v^{(R,L)}$ as (23.82) shows. It is, therefore, customary to speak of "large" and "small" components of the Dirac wave functions, but this terminology is dependent on the representation used for the Dirac matrices. The stationary state solutions of (24.16) for a static *central potential* will be discussed in Section 24.4.

2. Dynamical Variables in the Dirac Theory. In the quantum field theory, the time development of physical quantities was formulated in terms of the *Heisenberg* picture. When the transition to a one-electron theory is made, equation (24.16) shows that a formulation in the *Schrödinger* picture emerges. It is convenient to study the one-electron theory also in the Heisenberg picture, where state vectors are constant while operators move according to the formula (15.68), or

$$i\hbar \frac{dA(t)}{dt} = A(t)H(t) - H(t)A(t)$$

With the understanding that we are working in the Heisenberg picture, we may simplify the notation by writing A for $A(t)$ for the purposes of this section only.

The Hamiltonian operator has the form

$$H = c\boldsymbol{\alpha} \cdot \left(\mathbf{p} + \frac{e}{c} \mathbf{A} \right) - e\phi + \beta \mu c^2 \tag{24.19}$$

where not only **r** and **p** but also the matrices $\boldsymbol{\alpha}$ and β must be regarded as dynamical variables.

The *velocity operator* in this theory is, by definition,

$$\mathbf{v} = \frac{d\mathbf{r}}{dt} = \frac{1}{i\hbar} [\mathbf{r}, H] = c\boldsymbol{\alpha} \tag{24.20}$$

Furthermore

$$\frac{d\mathbf{p}}{dt} = \frac{1}{i\hbar} [\mathbf{p}, H] = e\nabla\phi - e\nabla(\boldsymbol{\alpha} \cdot \mathbf{A})$$

and

$$\frac{d\mathbf{A}}{dt} = \frac{\partial\mathbf{A}}{dt} + \frac{1}{i\hbar} [\mathbf{A}, H] = \frac{\partial\mathbf{A}}{dt} + c\boldsymbol{\alpha} \cdot \nabla\mathbf{A}$$

Hence,

$$\frac{d}{dt}\left(\mathbf{p} + \frac{e}{c}\mathbf{A}\right) = e\nabla\phi + \frac{e}{c}\frac{\partial\mathbf{A}}{\partial t} + e\boldsymbol{\alpha} \cdot \nabla\mathbf{A} - e\nabla(\boldsymbol{\alpha} \cdot \mathbf{A})$$

$$= -e\left(-\frac{1}{c}\frac{\partial\mathbf{A}}{\partial t} - \nabla\phi\right) - e\boldsymbol{\alpha} \times (\nabla \times \mathbf{A})$$

or,

$$\frac{d}{dt}\left(\mathbf{p} + \frac{e}{c}\mathbf{A}\right) = -e\mathbf{E} - e\boldsymbol{\alpha} \times \mathbf{B} = -e\mathbf{E} - e\frac{\mathbf{v}}{c} \times \mathbf{B} \tag{24.21}$$

Exercise 24.3. Carry out the details of the calculation leading to (24.21), evaluating all requisite commutators.

Equations (24.21) can be written in a form even more reminiscent of classical physics if we note the identity

$$\boldsymbol{\alpha}(H + e\phi) + (H + e\phi)\boldsymbol{\alpha} = 2c\left(\mathbf{p} + \frac{e}{c}\mathbf{A}\right) \tag{24.22}$$

Combining (24.21), (24.22), and (22.20), we thus obtain the quantum mechanical analog of the *Lorentz equation*,

$$\frac{d}{dt}\left[\frac{1}{2}\left(\mathbf{v}\frac{H + e\phi}{c^2} + \frac{H + e\phi}{c^2}\mathbf{v}\right)\right] = -e\left(\mathbf{E} + \frac{\mathbf{v}}{c} \times \mathbf{B}\right) \tag{24.23}$$

If expectation values are taken, $H + e\phi$ can be replaced in the lowest approximation by μc^2 or $-\mu c^2$, depending on whether the state is made up of positive or negative energy solutions of the Dirac equation. Effectively, therefore,

$$\mu\frac{d\mathbf{v}}{dt} \simeq \mp e\left(\mathbf{E} + \frac{\mathbf{v}}{c} \times \mathbf{B}\right) \tag{24.24}$$

The upper sign corresponds to a single electron with charge $-e$ and positive energy moving in the given external field; the lower sign corresponds to the motion of a single *hole* in the "sea of positrons," represented by the "electron vacuum" state Ψ_{0e}, in the presence of the same external field. Such a hole is equivalent to a particle of the opposite charge, i.e., an electron, with negative mass or energy, and equation (24.24) is consistent with this interpretation.

The result embodied in equation (24.20) is peculiar. In the relativistic one-electron theory the components of the velocity operator $c\boldsymbol{\alpha}$ do not commute with one another, but $v_x = c\alpha_x$ commutes with x. These properties are just the opposite of the nonrelativistic regime. Furthermore, since the eigenvalues of α_x are ± 1, the eigenvalues of any component of the velocity operator are $\pm c$, indicating that any velocity determination invariably leads to the conclusion that the particle, in spite of having a mass, moves with the speed of light. This result does not violate relativity, and it is understandable in terms of quantum mechanics, since a velocity measurement requires accurate successive position determinations, with a consequent infinite uncertainty in momentum. From the relativistic connection between velocity and momentum, it follows that $p_x \to \infty$ is only possible if $v \to \pm c$.

More insight into the significance of various operators can be gained if the particle is moving freely, such that $\mathbf{A} = 0$ and $\phi = 0$ everywhere. The Heisenberg equations of motion may then be solved explicitly. Equation (24.21) reduces to

$$\frac{d\mathbf{p}}{dt} = 0 \qquad \text{and} \qquad \mathbf{p} = \text{const.}$$

The free particle Hamiltonian is

$$H = c\boldsymbol{\alpha} \cdot \mathbf{p} + \beta \mu c^2$$

and the equation of motion for the operator $\boldsymbol{\alpha}$ becomes

$$\frac{d\boldsymbol{\alpha}}{dt} = \frac{1}{i\hbar}[\boldsymbol{\alpha}, H] = \frac{2}{i\hbar}(c\mathbf{p} - H\boldsymbol{\alpha}) \tag{24.25}$$

Since $H = \text{const.}$, this equation has a simple solution:

$$\frac{\mathbf{v}(t)}{c} = \boldsymbol{\alpha}(t) = cH^{-1}\mathbf{p} + e^{(2i/\hbar)Ht}[\boldsymbol{\alpha}(0) - cH^{-1}\mathbf{p}] \tag{24.26}$$

The last equation can readily be integrated:

$$\mathbf{r}(t) = \mathbf{r}(0) + \frac{c^2\mathbf{p}}{H}t + \frac{\hbar c}{2iH}(e^{(2i/\hbar)Ht} - 1)\left[\boldsymbol{\alpha}(0) - \frac{c\mathbf{p}}{H}\right] \tag{24.27}$$

The first two terms on the right-hand side describe simply the uniform motion of a free particle. The last term is a feature of relativistic quantum mechanics and connotes a high-frequency vibration ("*Zitterbewegung*") of the particle with frequency $\simeq \mu c^2/\hbar$ and amplitude $\hbar/\mu c$, the Compton wavelength of the particle. Since for a free particle,

$$\boldsymbol{\alpha} H + H \boldsymbol{\alpha} = 2c\mathbf{p} \tag{24.28}$$

it follows that in a representation based on momentum and energy, the operator $\boldsymbol{\alpha} - c\mathbf{p}/H$ has nonvanishing matrix elements only between states of equal momentum and *opposite* energies. The last term in equations (24.26) and (24.27) thus is intimately connected with the appearance of the negative energy states in a relativistic theory which simultaneously describes particles and antiparticles.

It is of interest to note a few further operator identities. For a free-particle Hamiltonian,

$$\beta H + H\beta = 2\mu c^2 \tag{24.29}$$

and

$$\gamma^5 H + H\gamma^5 = 2c\boldsymbol{\Sigma} \cdot \mathbf{p} \tag{24.30}$$

Hence, in a state of energy E, β has the expectation value

$$\langle \beta \rangle = \frac{\mu c^2}{E} = \pm\sqrt{1 - \frac{c^2 p^2}{E^2}} \tag{24.31}$$

so that $\langle \beta \rangle$ approaches ± 1 in the nonrelativistic approximation and vanishes as the speed of light is approached.

Similarly,

$$\langle \gamma^5 \rangle = \langle \boldsymbol{\Sigma} \cdot \mathbf{p} \rangle \frac{c}{E} \tag{24.32}$$

showing that the expectation value of γ^5 is v/c times the expectation value of the helicity operator.

Exercise 24.4. Verify equations (24.28) to (24.32) for free relativistic electrons.

The role of the spin in the one-electron Dirac theory is brought into focus if we evaluate the time derivative of $\boldsymbol{\Sigma}$ for an electron exposed to a vector potential $\mathbf{A}$ but no ϕ, so that

$$H = c\boldsymbol{\alpha} \cdot \left(\mathbf{p} + \frac{e}{c}\mathbf{A}\right) + \beta\mu c^2 \tag{24.33}$$

By a sequence of algebraic manipulations, we obtain

$$\frac{d\boldsymbol{\Sigma}}{dt} = \frac{1}{i\hbar}[\boldsymbol{\Sigma}, H] = \frac{2c}{\hbar}\left(\mathbf{p} + \frac{e}{c}\mathbf{A}\right) \times \boldsymbol{\alpha} \tag{24.34}$$

and

$$H \frac{d\Sigma}{dt} + \frac{d\Sigma}{dt} H = \frac{2c^2}{\hbar} \left\{ \left[\boldsymbol{\alpha} \cdot \left(\mathbf{p} + \frac{e}{c} \mathbf{A} \right) \right] \left[\boldsymbol{\alpha} \times \left(\mathbf{p} + \frac{e}{c} \mathbf{A} \right) \right] \right.$$

$$\left. + \left[\boldsymbol{\alpha} \times \left(\mathbf{p} + \frac{e}{c} \mathbf{A} \right) \right] \left[\boldsymbol{\alpha} \cdot \left(\mathbf{p} + \frac{e}{c} \mathbf{A} \right) \right] \right\}$$

The contents of the brace on right-hand side of the last equation may be reduced to $(e\hbar/c)\Sigma \times (\nabla \times \mathbf{A}) = (e\hbar/c)\Sigma \times \mathbf{B}$. Hence, the simple relation

$$H \frac{d\Sigma}{dt} + \frac{d\Sigma}{dt} H = 2ec\Sigma \times \mathbf{B} \tag{24.35}$$

is rigorously valid. In the nonrelativistic approximation $H \simeq \mu c^2$, this equation becomes the equation of motion for the one-electron spin operator $\mathbf{S} = \hbar \Sigma / 2$,

$$\frac{d\mathbf{S}}{dt} = -\frac{e}{\mu c} \mathbf{B} \times \mathbf{S} \tag{24.36}$$

A straightforward interpretation may be given this equation. The time rate of change of intrinsic angular momentum (spin) is a torque produced by the applied magnetic field. If a magnetic moment $\mathbf{m}$ is associated with the spin, the torque is $\mathbf{B} \times \mathbf{m}$. Comparison with (24.36) shows that in this approximation the magnetic moment operator for an electron is

$$\mathbf{m} = -\frac{e}{\mu c} \mathbf{S} = -\frac{e}{2\mu c} g_S \mathbf{S} \tag{24.37}$$

with g_S as defined in equation (16.116). The value $g_S = 2$, derived here from relativistic quantum mechanics for a charged Dirac particle, is, except for radiative corrections, in agreement with the experimental measurements.[2]

3. The Nonrelativistic Limit. In this section we return to the *Schrödinger picture* for the one-particle Dirac equation which we treat in its simple form

$$\gamma^{\mu} \left(\frac{\partial}{\partial x^{\mu}} - \frac{ie}{\hbar c} A_{\mu} \right) \psi_e + i\kappa \psi_e = 0 \tag{24.38}$$

A convenient second-order differential equation for ψ_e is obtained by iterating this equation as follows:

$$\left[\gamma^{\mu} \left(\frac{\partial}{\partial x^{\mu}} - \frac{ie}{\hbar c} A_{\mu} \right) - i\kappa \right] \left[\gamma^{\nu} \left(\frac{\partial}{\partial x^{\nu}} - \frac{ie}{\hbar c} A_{\nu} \right) + i\kappa \right] \psi_e = 0 \tag{24.39}$$

---

[2] Quantum electrodynamics gives the value $g_S = 2(1 + \alpha/2\pi)$ to first order in the fine structure constant. See Sakurai, *Advanced Quantum Mechanics*, Addison-Wesley Publishing Co., Reading, 1967.

whence,

$$\left[\gamma^\mu\gamma^\nu\left(\frac{\partial}{\partial x^\mu} - \frac{ie}{\hbar c}A_\mu\right)\left(\frac{\partial}{\partial x^\nu} - \frac{ie}{\hbar c}A_\nu\right) + \kappa^2\right]\psi_e = 0$$

or, separating the terms with $\mu = \nu$ from those with $\mu \neq \nu$,

$$\left[-\left(\nabla + \frac{ie}{\hbar c}\mathbf{A}\right)^2 + \frac{1}{c^2}\left(\frac{\partial}{\partial t} - \frac{ie}{\hbar}\phi\right)^2 + \frac{e}{\hbar c}\boldsymbol{\Sigma}\cdot\mathbf{B} - \frac{ie}{\hbar c}\boldsymbol{\alpha}\cdot\mathbf{E} + \kappa^2\right]\psi_e = 0$$

$$(24.40)$$

or, in more elegant relativistic notation:

$$\left[\left(\frac{\partial}{\partial x_\mu} - \frac{ie}{\hbar c}A^\mu\right)\left(\frac{\partial}{\partial x^\mu} - \frac{ie}{\hbar c}A_\mu\right) + \frac{e}{2\hbar c}\Sigma^{\mu\nu}F_{\mu\nu} + \kappa^2\right]\psi_e = 0 \quad (24.41)$$

where

$$F_{\mu\nu} = \frac{\partial A_\nu}{\partial x^\mu} - \frac{\partial A_\mu}{\partial x^\mu}$$

Exercise 24.5. Derive equations (24.40) and (24.41).

This second-order equation is similar to the (Klein-Gordon) equation governing the wave function for a relativistic particle with spin zero but differs from it by the terms containing the Dirac matrices and coupling the wave function directly to the electromagnetic field $\mathbf{B}$ and $\mathbf{E}$.

The second-order equation (24.39) has, of course, more solutions than the Dirac equation, from which it was obtained by iteration, and it is therefore necessary to select among its solutions only those which also satisfy the Dirac equation.

It is worth noting that there is another way of using equation (24.39) in the description of spin $\frac{1}{2}$ particles. If χ_e is an arbitrary solution of equation (24.39), the wave function

$$\psi_e = \frac{i}{\kappa}\left[\gamma^\mu\left(\frac{\partial}{\partial x^\mu} - \frac{ie}{\hbar c}A_\mu\right) - i\kappa\right]\chi_e \quad (24.42)$$

is evidently a solution of the Dirac equation (24.9). Since γ^5 is Hermitian and anticommutes with all γ^μ, the solutions of equation (24.39) can be assumed to be simultaneously eigenspinors of γ^5 with eigenvalues $+1$ or -1, so that

$$\gamma^5\chi_e^{(\pm)} = \pm\chi_e^{(\pm)}$$

From equation (24.42) we find readily that

$$\tfrac{1}{2}(1 \pm \gamma^5)\psi_e^{(\pm)} = \chi_e^{(\pm)} \quad (24.43)$$

if $\psi_e^{(\pm)}$ is generated from $\chi_e^{(\pm)}$.

Exercise 24.6. Prove that $\frac{1}{2}(1 \pm \gamma^5)$ is a projection operator and that $\frac{1}{2}(1 \pm \gamma^5)\psi_e$ is an eigenspinor of γ^5.

Equation (24.43) associates with every solution of the Dirac equation two solutions of equation (24.39) with sharp values (± 1) of γ^5. It is therefore possible to establish a one-to-one correspondence between the solutions of the first-order Dirac equation and those solutions of the second-order equation (24.39) which are simultaneously eigenfunctions of γ^5 with eigenvalue $+1$ (or, alternatively, -1). The important point is that a particular solution $\chi_e^{(+)}$ and a partner solution

$$\chi_e^{(-)} = \frac{i}{\kappa} \gamma^\mu \left(\frac{\partial}{\partial x^\mu} - \frac{ie}{\hbar c} A_\mu \right) \chi_e^{(+)}$$

yield the same Dirac wave function ψ_e.

Exercise 24.7. Find a representation of the Dirac matrices in which

$$\gamma^5 = \begin{pmatrix} 1 & 0 & 0 & 0 \\ 0 & 1 & 0 & 0 \\ 0 & 0 & -1 & 0 \\ 0 & 0 & 0 & -1 \end{pmatrix}$$

and

$$\Sigma_z = \begin{pmatrix} 1 & 0 & 0 & 0 \\ 0 & -1 & 0 & 0 \\ 0 & 0 & 1 & 0 \\ 0 & 0 & 0 & -1 \end{pmatrix}$$

Show that, in this representation, equations (24.39), (24.40), or (24.41) break up into two separate two-component equations, one for the spinor $\chi^{(+)}$, the other for $\chi^{(-)}$, each being separately equivalent to the Dirac equation. For a free Dirac particle, contrast this separation with the separation into positive and negative energy or "large" and "small" components effected by the use of the standard representation given by equation (23.30) (Feynman and Gell-Mann).

In order to interpret equation (24.40), we assume that the external field is time-independent and we consider a stationary state solution

$$\psi_e(\mathbf{r}, t) = e^{-(i/\hbar)Et} u(\mathbf{r})$$

Substitution into equation (24.40) gives

$$\left[\frac{1}{c^2}(E + e\phi)^2 - \left(\frac{\hbar}{i}\nabla + \frac{e}{c}\mathbf{A}\right)^2 - \frac{e\hbar}{c}(\mathbf{\Sigma}\cdot\mathbf{B} - i\boldsymbol{\alpha}\cdot\mathbf{E}) - \mu^2 c^2\right]u = 0$$

This equation is still exact. For a nonrelativistic electron, for which $E \simeq \mu c^2$, we approximate

$$\frac{1}{c^2}(E + e\phi)^2 - \mu^2 c^2 \simeq 2\mu(E + e\phi - \mu c^2)$$

Hence, we obtain

$$(E + e\phi - \mu c^2)u = \left[\frac{1}{2\mu}\left(\frac{\hbar}{i}\nabla + \frac{e}{c}\mathbf{A}\right)^2 + \frac{e\hbar}{2\mu c}(\mathbf{\Sigma}\cdot\mathbf{B} - i\boldsymbol{\alpha}\cdot\mathbf{E})\right]u \qquad (24.44)$$

which is very similar to the nonrelativistic Schrödinger equation. In the absence of an electric field, this equation describes the motion of the electron in an external magnetic field and shows again that an intrinsic magnetic moment as given by (24.37) must be ascribed to the electron.

The physical appreciation of the Dirac theory is further enhanced by rewriting the current density

$$j^\mu = -ec\bar{\psi}_e\gamma^\mu\psi_e \qquad (24.45)$$

in terms of operators which have a nonrelativistic interpretation. To accomplish this goal, we write the Dirac equations for ψ_e and $\bar{\psi}_e$ as

$$\psi_e = \frac{i}{\kappa}\gamma^\lambda\left(\frac{\partial}{\partial x^\lambda} - \frac{ie}{\hbar c}A_\lambda\right)\psi_e \qquad (24.46)$$

$$\bar{\psi}_e = -\frac{i}{\kappa}\left(\frac{\partial}{\partial x^\lambda} + \frac{ie}{\hbar c}A_\lambda\right)\bar{\psi}_e\gamma^\lambda \qquad (24.47)$$

Substituting (24.46) in one half of the current density and (24.47) in the other half, we get

$$j^\mu = -\frac{eci}{2\kappa}\left(\bar{\psi}_e\gamma^\mu\gamma^\lambda\frac{\partial\psi_e}{\partial x^\lambda} - \frac{\partial\bar{\psi}_e}{\partial x^\lambda}\gamma^\lambda\gamma^\mu\psi_e\right)(1 - \delta_{\mu\lambda})$$

$$-\frac{eci}{2\kappa}g^{\mu\lambda}\left\{\bar{\psi}_e\left(\frac{\partial}{\partial x^\lambda} - \frac{ie}{\hbar c}A_\lambda\right)\psi_e - \left[\left(\frac{\partial}{\partial x^\lambda} + \frac{ie}{\hbar c}A_\lambda\right)\bar{\psi}_e\right]\psi_e\right\}$$

where the terms with $\mu \neq \lambda$ have been separated from those with $\mu = \lambda$. Using the definition (23.62), j^μ is finally transformed into a sum of *polarization* and *convection* terms,

$$j^\mu = j^\mu_{\text{pol}} + j^\mu_{\text{conv}} \qquad (24.48)$$

where

$$j^\mu_{\text{pol}} = \frac{ec}{2\kappa} \frac{\partial}{\partial x^\lambda} (\bar{\psi}_e \Sigma^{\lambda\mu} \psi_e) \tag{24.49}$$

$$j^\mu_{\text{conv}} = -\frac{eci}{2\kappa} g^{\mu\lambda} \left\{ \bar{\psi}_e \left(\frac{\partial}{\partial x^\lambda} - \frac{ie}{\hbar c} A_\lambda \right) \psi_e - \left[\left(\frac{\partial}{\partial x^\lambda} + \frac{ie}{\hbar c} A_\lambda \right) \bar{\psi}_e \right] \psi_e \right\} \tag{24.50}$$

This procedure is known as the *Gordon decomposition* of the current density.

Exercise 24.7. Prove that both the polarization and convection currents are conserved, and show the relation of the convection current to the non-relativistic current density (Exercise 8.10).

4. Central Forces and the Hydrogen Atom. In an electrostatic central field, the one-particle Dirac Hamiltonian is

$$H = c\boldsymbol{\alpha} \cdot \frac{\hbar}{i} \nabla + \beta\mu c^2 - e\phi(r) \tag{24.51}$$

Since **L** and **S** do not separately commute with the free particle Hamiltonian, they certainly will not commute with the Hamiltonian (24.51) either. However, the components of the total angular momentum,

$$\mathbf{J} = \mathbf{r} \times \frac{\hbar}{i} \nabla + \frac{\hbar}{2} \boldsymbol{\Sigma} \tag{24.52}$$

evidently do commute with H, and we may therefore seek to find simultaneous eigenspinors of H, $\mathbf{J}^2$, and J_z. Parity will join these as a useful constant of the motion.

At this point it is convenient (although not unavoidable) to introduce the standard representation (23.30) of the Dirac matrices and to write all equations in two-component form. If we define

$$\psi_e = \begin{pmatrix} \varphi \\ \chi \end{pmatrix} \tag{24.53}$$

the Dirac equation decomposes according to Exercise 23.8 into the coupled equations

$$(E - \mu c^2 + e\phi)\varphi - \frac{\hbar}{i} c\boldsymbol{\sigma} \cdot \nabla\chi = 0 \tag{24.54}$$

$$(E + \mu c^2 + e\phi)\chi - \frac{\hbar}{i} c\boldsymbol{\sigma} \cdot \nabla\varphi = 0 \tag{24.55}$$

where all $\boldsymbol{\sigma}$ matrices are 2×2 Pauli matrices. The operators J_z and

$$\mathbf{J}^2 = \mathbf{L}^2 + \hbar \mathbf{L} \cdot \boldsymbol{\Sigma} + \tfrac{3}{4}\hbar^2$$

decompose similarly, and it is clear that we must seek to make φ and χ two-component eigenspinors satisfying the conditions

$$\left(L_z + \frac{\hbar}{2}\sigma_z\right)\binom{\varphi}{\chi} = m\hbar \binom{\varphi}{\chi} \tag{24.56}$$

and

$$\left(\mathbf{L}^2 + \hbar \mathbf{L} \cdot \boldsymbol{\sigma} + \tfrac{3}{4}\hbar^2\right)\binom{\varphi}{\chi} = j(j+1)\hbar^2 \binom{\varphi}{\chi} \tag{24.57}$$

Equation (16.80) contains the answer to this problem and shows that for a given value of j the spinors φ and χ must be proportional to $\mathscr{Y}^{jm}_{j\mp(1/2)}$. The two-component spinors $\mathscr{Y}$ are normalized as $\int \mathscr{Y}^\dagger \mathscr{Y} \, d\Omega = 1$ and they have the useful property

$$\boldsymbol{\sigma} \cdot \hat{\mathbf{r}} \mathscr{Y}^{jm}_{j\mp(1/2)} = -\mathscr{Y}^{jm}_{j\pm(1/2)} \tag{24.58}$$

Exercise 24.8. Prove equation (24.58) by using the facts that $\boldsymbol{\sigma} \cdot \hat{\mathbf{r}}$ is a pseudo-scalar under rotation and that $\mathscr{Y}$ has a simple value for $\theta = 0$.

Exercise 24.9. Prove that

$$\boldsymbol{\sigma} \cdot \mathbf{L}\mathscr{Y}^{jm}_{j-(1/2)} = (j - \tfrac{1}{2})\mathscr{Y}^{jm}_{j-(1/2)} \tag{24.59}$$

$$\boldsymbol{\sigma} \cdot \mathbf{L}\mathscr{Y}^{jm}_{j+(1/2)} = (-j - \tfrac{3}{2})\mathscr{Y}^{jm}_{j+(1/2)} \tag{24.60}$$

Since the *parity operator* also commutes with the other available constants of the motion, it may be chosen as a further "good quantum number" and we may require that

$$\beta\binom{\varphi(-\mathbf{r})}{\chi(-\mathbf{r})} = \pm \binom{\varphi(\mathbf{r})}{\chi(\mathbf{r})} \qquad \begin{array}{l}\text{(even)}\\\text{(odd)}\end{array}$$

The parity of the eigenfunction clearly dictates how the spinors $\mathscr{Y}^{jm}_{j\pm\frac{1}{2}}$ are associated with φ and χ. It is easily seen that the two solutions must have the form

$$\psi_e = \binom{F(r)\mathscr{Y}^{jm}_{j-(1/2)}}{-if(r)\mathscr{Y}^{jm}_{j+(1/2)}} \tag{24.61}$$

or

$$\psi_e = \binom{G(r)\mathscr{Y}^{jm}_{j+(1/2)}}{-ig(r)\mathscr{Y}^{jm}_{j-(1/2)}} \tag{24.62}$$

The former has even or odd parity depending on the parity of $j - \tfrac{1}{2}$. The latter has even or odd parity depending on the parity of $j + \tfrac{1}{2}$. The factors $-i$ have been introduced so that the radial equations will be real.

In order to derive the radial equations, we employ the following identity

$$\boldsymbol{\sigma} \cdot \mathbf{p} = \frac{1}{r^2}(\boldsymbol{\sigma} \cdot \mathbf{r})(\boldsymbol{\sigma} \cdot \mathbf{r})(\boldsymbol{\sigma} \cdot \mathbf{p}) = \boldsymbol{\sigma} \cdot \hat{\mathbf{r}}\left(\hat{\mathbf{r}} \cdot \mathbf{p} + \frac{i\boldsymbol{\sigma} \cdot \mathbf{L}}{r}\right) \tag{24.63}$$

which follows from (12.55). Here,

$$\hat{\mathbf{r}} \cdot \mathbf{p} = \frac{\hbar}{i}\frac{\partial}{\partial r} \tag{24.64}$$

If we substitute relations (24.63) and (24.64) into equations (24.54) and (24.55), and take equations (24.58), (24.59), and (24.60) into account, we obtain

$$(E - \mu c^2 + e\phi)F - \hbar c\left(\frac{d}{dr} + \frac{j + \frac{3}{2}}{r}\right)f = 0 \tag{24.65}$$

$$(E + \mu c^2 + e\phi)f + \hbar c\left(\frac{d}{dr} - \frac{j - \frac{1}{2}}{r}\right)F = 0 \tag{24.66}$$

and

$$(E - \mu c^2 + e\phi)G - \hbar c\left(\frac{d}{dr} - \frac{j - \frac{1}{2}}{r}\right)g = 0 \tag{24.67}$$

$$(E + \mu c^2 + e\phi)g + \hbar c\left(\frac{d}{dr} + \frac{j + \frac{3}{2}}{r}\right)G = 0 \tag{24.68}$$

So far, it has not been necessary to introduce the explicit form of the potential, but at this point we assume that the electron moves in the Coulomb field of a point nucleus of charge Ze:

$$\phi(r) = \frac{Ze}{r} \tag{24.69}$$

We also define the abbreviations

$$\lambda = j + \tfrac{1}{2}, \qquad E/\mu c^2 = \varepsilon, \qquad r/(\hbar/\mu c) = x, \qquad e^2/\hbar c = \alpha \tag{24.70}$$

The coupled radial equations then become

$$\left(\varepsilon - 1 + \frac{Z\alpha}{x}\right)F - \left(\frac{d}{dx} + \frac{\lambda + 1}{x}\right)f = 0 \tag{24.71}$$

$$\left(\varepsilon + 1 + \frac{Z\alpha}{x}\right)f + \left(\frac{d}{dx} - \frac{\lambda - 1}{x}\right)F = 0 \tag{24.72}$$

and

$$\left(\varepsilon - 1 + \frac{Z\alpha}{x}\right)G - \left(\frac{d}{dx} - \frac{\lambda - 1}{x}\right)g = 0 \tag{24.73}$$

$$\left(\varepsilon + 1 + \frac{Z\alpha}{x}\right)g + \left(\frac{d}{dx} + \frac{\lambda + 1}{x}\right)G = 0 \tag{24.74}$$

Obviously, these two sets of equations are obtained from one another by the transformation

$$F \to G, \quad f \to g, \quad \lambda \to -\lambda$$

and it suffices to consider equations (24.71) and (24.72).

The analysis of the radial equations proceeds as usual. Asymptotically, for $x \to \infty$ we find the behavior

$$F \sim f \sim e^{\pm \sqrt{1-\varepsilon^2}\, x}$$

For *bound states*, to which we shall devote our attention, we must require $|\varepsilon| \leqslant 1$ and choose the minus sign in the exponent. With the ansatz

$$F = e^{-\sqrt{1-\varepsilon^2}\, x}\, x^\gamma \sum_{v=0} a_v x^v \tag{24.75}$$

$$f = e^{-\sqrt{1-\varepsilon^2}\, x}\, x^\gamma \sum_{\rho=0} b_\rho x^\rho \tag{24.76}$$

we obtain by substitution

$$(\varepsilon - 1)a_{v-1} + Z\alpha a_v + \sqrt{1 - \varepsilon^2}\, b_{v-1} - (\lambda + 1 + \gamma + v)b_v = 0 \tag{24.77}$$

and

$$(\varepsilon + 1)b_{v-1} + Z\alpha b_v - \sqrt{1 - \varepsilon^2}\, a_{v-1} + (-\lambda + 1 + \gamma + v)a_v = 0 \tag{24.78}$$

For $v = 0$ and $a_{-1} = b_{-1} = 0$, we get

$$Z\alpha a_0 - (\lambda + 1 + \gamma)b_0 = 0, \quad (-\lambda + 1 + \gamma)a_0 + Z\alpha b_0 = 0 \tag{24.79}$$

or

$$(Z\alpha)^2 = \lambda^2 - (\gamma + 1)^2$$

Hence,

$$\gamma = -1 \pm \sqrt{\lambda^2 - (Z\alpha)^2} = -1 \pm \sqrt{(j + \tfrac{1}{2})^2 - (Z\alpha)^2}$$

The negative root for γ must be excluded because the corresponding wave function would be too singular at the origin to be admissible. Hence,

$$\gamma = -1 + \sqrt{(j + \tfrac{1}{2})^2 - (Z\alpha)^2} \tag{24.80}$$

Prövided that $Z < (1/\alpha) \simeq 137$, this will yield a real value for γ.[3]

The usual argument can now be made to show that both power series (24.75) and (24.76) must terminate at $x^{n'}$ (see Chapter 10).

Exercise 24.10. Carry out the study of the asymptotic behavior of (24.75) and (24.76), and show that the power series must terminate.

---

[3] For $j = \tfrac{1}{2}$, $-1 < \gamma < 0$ and the wave function is mildly singular at the origin, but not enough to disturb its quadratic integrability. See footnote 7 in Chapter 10.

From the recursion relations (24.77) and (24.78), we then obtain for $v = n' + 1$ (with $a_{n'+1} = b_{n'+1} = 0$),

$$b_{n'} = \sqrt{\frac{1 - \varepsilon}{1 + \varepsilon}} \, a_{n'} \tag{24.81}$$

From equations (24.77) and (24.78), we may simultaneously eliminate a_{v-1} and b_{v-1} to get

$$a_v[Z\alpha\sqrt{1 + \varepsilon} + (\lambda - 1 - \gamma - v)\sqrt{1 - \varepsilon}]$$
$$= b_v[Z\alpha\sqrt{1 - \varepsilon} + (\lambda + 1 + \gamma + v)\sqrt{1 + \varepsilon}] \tag{24.82}$$

Letting $v = n'$ and comparing with equations (24.81), we finally conclude that

$$\sqrt{\frac{1 - \varepsilon}{1 + \varepsilon}} = \frac{Z\alpha\sqrt{1 + \varepsilon} + (\lambda - 1 - \gamma - n')\sqrt{1 - \varepsilon}}{Z\alpha\sqrt{1 - \varepsilon} + (\lambda + 1 + \gamma + n')\sqrt{1 + \varepsilon}}$$

or

$$\sqrt{1 - \varepsilon^2}\,(1 + \gamma + n') = Z\alpha\varepsilon$$

or

$$\frac{E}{\mu c^2} = \left\{ 1 + \frac{(Z\alpha)^2}{[\sqrt{(j + \tfrac{1}{2})^2 - (Z\alpha)^2} + n']^2} \right\}^{-1/2} \tag{24.83}$$

This is the famous *fine structure formula* for the hydrogen atom. The quantum numbers j and n' assume the values

$$j = \tfrac{1}{2}, \tfrac{3}{2}, \tfrac{5}{2}, \ldots \, ; \quad n' = 0, 1, 2, \ldots$$

The principal quantum number n of the nonrelativistic theory of the hydrogen atom is related to n' and j by

$$n = j + \tfrac{1}{2} + n' \tag{24.84}$$

From (24.81) we have for $n' = 0$

$$\sqrt{1 - \varepsilon}\, a_0 - \sqrt{1 + \varepsilon}\, b_0 = 0$$

This relation between a_0 and b_0 is consistent with (24.79) only if $\lambda + 1 + \gamma > 0$, or $\lambda > -\sqrt{\lambda^2 - (Z\alpha)^2}$, hence $\lambda > 0$. The transformation $\lambda \to -\lambda$, which takes us from the states with $j = l + \tfrac{1}{2}$ to those with $j = l - \tfrac{1}{2}$, is therefore not permissible if $n' = 0$, and a solution of type (24.62) is thus not possible if n $= j + \tfrac{1}{2}$. Hence, for a given value of the principal quantum number n, there is only one state with $j = $ n $- \tfrac{1}{2}$, while there are two states of opposite parities for all $j < $ n $- \tfrac{1}{2}$. Since $\lambda = j + \tfrac{1}{2}$ appears squared in the energy formula, pairs of states with the same j but opposite parities (e.g., $2S_{1/2}$ and $2P_{1/2}$) remain degenerate in the approximation of the one-electron Dirac theory. Experiments have substantiated this formula and its radiative corrections (Lamb shift which removes the $2S_{1/2}$-$2P_{1/2}$ degeneracy) to very high accuracy, impressively vindicating the one-electron abbreviation of the full relativistic electron-positron theory.

Problems

1. Show that the vector operator

$$\mathcal{O} = \beta\mathbf{\Sigma} + (1 - \beta)\mathbf{\Sigma} \cdot \hat{\mathbf{p}}\hat{\mathbf{p}}$$

   satisfies the angular momentum commutation relations and that it commutes with the free Dirac particle Hamiltonian. Show that the eigenvalues of any component of $\mathcal{O}$ are ± 1.
   Apply the unitary transformation

$$e^{i(\theta/2)(-p_y\mathcal{O}_x + p_x\mathcal{O}_y)/\sqrt{p_x{}^2 + p_y{}^2}}$$

   to the spinors (23.85) and (23.86), and prove that the resulting spinors are eigenstates of H with sharp momentum and definite value of $\mathcal{O}_z$. Show that these states are the relativistic analogs of the nonrelativistic momentum eigenstates with "spin up" and "spin down."

2. Show that for a massless Dirac particle (neutrino) the matrix γ^5 is a constant of the motion, and exhibit its relation to the helicity operator.

3. Assume that the potential $\phi(r)$ in the Dirac Hamiltonian (24.51) is a square well of depth V_0 and radius a. Determine the continuity condition for ψ_e at $r = a$, and derive a transcendental equation for the minimum value of V_0 which just binds a particle of mass μ for a given value of a.

4. Expand the relativistic expression for the energy levels of a hydrogenic atom in powers of $Z\alpha$ to obtain the Bohr-Balmer formula and the first correction to the nonrelativistic energies. Compute the fine structure separation of the $2P$ doublet for hydrogen, and compare with the measured value.

5. Solve the Klein-Gordon equation for a spinless particle of mass μ and charge $-e$ in the presence of the Coulomb field of a point nucleus with charge Ze. Compare the fine structure of the energy levels with the corresponding results for the Dirac electron.

6. Consider a neutral spin $\frac{1}{2}$ Dirac particle with mass and with an intrinsic magnetic moment, and assume the Hamiltonian

$$H = c\mathbf{\alpha} \cdot \frac{\hbar}{i}\nabla + \beta\mu c^2 + \lambda B\beta\Sigma_z$$

   in the presence of a uniform constant magnetic field along the z-axis. Determine the important constants of the motion, and derive the energy eigenvalues. Show that orbital and spin motions are coupled in the relativistic theory but decoupled in a nonrelativistic limit. λ is the coupling constant.

7. If a Dirac electron is moving in a uniform constant magnetic field pointing along the z-axis, determine the energy eigenvalues and eigenspinors.

Index